Phytochemistry: An in-silico and in-vitro Update

Shashank Kumar • Chukwuebuka Egbuna
Editors

Phytochemistry: An in-silico and in-vitro Update

Advances in Phytochemical Research

Springer

Editors
Shashank Kumar
Department of Biochemistry and Microbial Sciences
Central University of Punjab
Bathinda, Punjab, India

Chukwuebuka Egbuna
Department of Biochemistry
Chukwuemeka Odumegwu Ojukwu University
Uli, Anambra, Nigeria

ISBN 978-981-13-6919-3 ISBN 978-981-13-6920-9 (eBook)
https://doi.org/10.1007/978-981-13-6920-9

This Springer imprint is published by the registered company Springer Nature Singapore Pte Ltd.
The registered company address is: 152 Beach Road, #21-01/04 Gateway East, Singapore 189721, Singapore

Acknowledgement

Dr. Shashank Kumar acknowledges the Department of Science and Technology, India, and the University Grants Commission, India, for providing financial support in the form of DST-SERB Grant [EEQ/2016/000350] and UGC-BSR Research Start-Up Grant [No. F.30–372/2017 (BSR)], respectively.

Contents

Part II Phytochemicals and in-vivo Updates

About the Editors

Shashank Kumar is working as an Assistant Professor at the Department of Biochemistry and Microbial Sciences, Central University of Punjab, Bathinda, India. Dr. Kumar obtained his B.Sc., M.Sc., and Ph.D. degrees in biochemistry from the Department of Biochemistry, University of Allahabad, Allahabad, India. He worked as a Postdoctoral Fellow at the Department of Biochemistry, King George's Medical University, Lucknow, India. He has about 50 published scientific papers/reviews/editorial articles in various national/international peer-reviewed journals of repute. He edited several books, such as "Concepts in Cell Signaling," etc. He has expertise in areas of free radical biology, cancer biology, characterization of plant products, xenobiotic metabolism, and microbiology. He is familiar with many biochemical techniques such as spectrophotometry, ELISA, electrophoresis, PCR, RT-PCR, flow cytometry, TLC, HPLC, LC-MS, cell culture, and microbiological techniques. He has presented his research findings in more than 25 national/international conferences and attended about 30 workshops at the majority of major universities and medical colleges throughout India. He is a life-time member of *Italo-Latin American Society of Ethnomedicine*, Indian Sciences Congress Association, India, and a member of Asian Council of Science Editors, Dubai, UAE, and Publication Integrity & Ethics, London. He has been awarded Junior/Senior and Research Associate Fellowships formulated and funded by various Indian agencies such as Indian Council of Medical Research, University Grants Commission, Council of Scientific, and Industrial Research India. He is working on identification and validation of novel natural inhibitors of cancer signaling pathways. His laboratory is involved in studying the efficacy of phytochemicals against drug resistance in cancer cells. He is also working on computer-based anticancer drug design.

Chukwuebuka Egbuna is a chartered chemist, a chemical analyst, and an academic researcher. He is a member of the Institute of Chartered Chemists of Nigeria (ICCON), the Nigerian Society of Biochemistry and Molecular Biology (NSBMB), the Society of Quality Assurance (SQA), USA, and the Royal Society of Chemistry (RSC), UK. He has been engaged in a number of roles at New Divine Favor

Pharmaceutical Industry Limited, Akuzor Nkpor, Anambra State, Nigeria, and Chukwuemeka Odumegwu Ojukwu University (COOU), Nigeria. He has collaboratively worked and published quite a number of research articles in the area of phytochemistry and its applications. Egbuna has edited a three-volume set book of phytochemistry with Apple Academic Press and working on other projects with Elsevier and Springer Nature. He is a reviewer and editorial board member of various journals. His primary research interests are in phytochemistry, food and medicinal chemistry, analytical chemistry, and nutrition and toxicology. He obtained his B.Sc. and M.Sc. degrees in biochemistry at Chukwuemeka Odumegwu Ojukwu University.

Part I
Phytochemicals and in-vitro Updates

Chapter 1
Phytochemicals as Sources of Drugs

Shahira M. Ezzat, Jaison Jeevanandam, Chukwuebuka Egbuna, Shashank Kumar, and Jonathan C. Ifemeje

1.1 Introduction

Plants have been used as medicines since ancient times, due to the presence of numerous phytocompounds that helps them to prevent and cure several diseases and disorders. They are utilized in different formulations such as herbal tea, extracts, decoctions, infusions, tincture or powder (Balick and Cox 1997; Thomas et al. 1999; Samuelsson 2004; Ujah 2019). Initially, humans started to utilize plants as food which was later segregated as medicinal plants with definite pharmacological action (Howes 2018). In the past, the methods for the application of a medicinal herb for certain ailment were mainly based on the history of the plant which was recorded in herbals. It has been reported from cave carvings and literatures that the medicinal plants are widely used in ancient health care systems, such as Ayurveda, traditional Chinese medications and several ancient medications from ancient civilizations

S. M. Ezzat
Department of Pharmacognosy, Faculty of Pharmacy, Cairo University, Cairo, Egypt

Department of Pharmacognosy, Faculty of Pharmacy, October University for Modern Sciences and Arts (MSA), 6th of October, Egypt

J. Jeevanandam
Department of Chemical Engineering, Curtin University Malaysia, Miri, Malaysia

C. Egbuna (✉)
Department of Biochemistry, Chukwuemeka Odumegwu Ojukwu University, Uli, Anambra, Nigeria
e-mail: egbuna.cg@coou.edu.ng

S. Kumar
Department of Biochemistry and Microbial Sciences, Central University of Punjab Bathinda, Punjab, India

J. C. Ifemeje
Department of Biochemistry, Faculty of Natural Sciences, Chukwuemeka Odumegwu Ojukwu University, Uli, Anambra, Nigeria

S. Kumar, C. Egbuna (eds.), *Phytochemistry: An in-silico and in-vitro Update*,
https://doi.org/10.1007/978-981-13-6920-9_1

including Mesopotamia, Persia, Sumeria, Egypt and Arab, as a source of medicinal drugs for the treatment of diseases. In Ayurvedic literatures such as Charaka Samhita and Sushruta Samhita of the first millennium BC, more than 700 medicinal plants are listed as a source of drugs (Shakya 2016) and about 7500 plants are estimated to be used in ancient Indian villages as local health traditions (Sarker et al. 2007). Likewise, traditional Chinese medical systems also utilized plants as a drug source for disease treatments, for more than a thousand years. It is noteworthy that China has 10% of the world's plant species which portrays their rich biodiversity. This diversity of plant species in China led to the publication of an ancient encyclopedia called Compendium of *Materia Medica* in 1593 which reports ~1000 species that are used in traditional Chinese medicinal practices (Yang et al. 2016a). Other olden civilizations, throughout the world, also have reported to have deep knowledge on using plants as medicinal source (Petrovska 2012).

Plants possess the ability to secrete secondary metabolites as a result of their metabolism, which is extracted as phytochemicals. This class of phytochemicals is required by plants for protection and maintenance but is not considered essential nutrients (Egbuna and Ifemeje 2015). These phytochemicals from medicinal plants are the source of a wide variety of modern natural drugs. Traditional medical systems possess enormous positive attributes to cure diseases; however, they utilize crude phytochemicals which may contain cytotoxic compounds (Thomas et al. 1999). Moreover, the developments in the extraction, purification and characterization of phytochemicals have opened doors for the production of several medicinal drugs which is better than traditional synthetic and conventional drugs. There are wide drug varieties of medicinal plants that are irreplaceable, even with the modern synthetic medicines. Furthermore, drugs from medicinal plants are highly bioactive, biocompatible, and bioavailable (Saraf 2010). The latest in-silico techniques as well as biomedical tools helps to bring out the maximum medicinal potential of plant extracts to make them as a significant drug source (Lagunin et al. 2014).

A plant is classified as a medicinal plant, when the whole plant, part of plants, their metabolites (crude extracts) or its purified and separated phytocompounds possess ability to cure diseases or reduce disease complications. There are numerous plants that are listed to contain medicinal properties with the help of traditional medical literatures and recent extensive researches in this field. The most important aspect of medicinal plant is their diverse availability for the large-scale production of drugs to match the growing population of patients with specific diseases (Naresh and Nagendraswamy 2016). It is noteworthy that various medicinal plants are obtained from wild forest and not from urban landscapes. Thus, medicinal gardens are gaining focus in recent times among botanists to make medically beneficial plants available for public usage in cities (Furlan et al. 2016). Pharmaceutical companies handle the need of large quantity of plants for scale-up production via systematic cultivation. High yielding medicinal plants are selected for a systematic cultivation process which serves as a source for natural drug products (Kumar et al. 2014). Organic farming was also recently introduced to avoid the interaction of fertilizers, pesticides or other chemicals while extracting the medicinal phytochemicals from the plants (Kala 2015). The phytochemicals with medicinal properties are extracted from the plant parts such as stem, root, bark, leaf, fruit, Heartwood and exudates. Literatures suggested that the quantity and type of phytochemicals that are extracted from the medicinal plants depends on the part from which they are obtained. Common phytochemicals such as essential oils, phenols, terpenoids, carotenoids, xanthophyll and flavonoids are present in plants which are extracted as a drug source for several diseases (Xiao 2015). Apart from common phytochemicals, unique species related phytocompounds namely andrographolides, amaranthine, allicin and various other exclusive drug components (Susantiningsih et al. 2012; Zhang et al. 2015). All these phytochemicals were proved to process excessive biomedical applications which are formulated as medicinal drugs for long lasting effects and enhanced disease treatments.

1.2 Drug Discovery Process

Drug discovery from natural products has brought about the isolation of valuable molecules. There are many examples of drugs isolated from plant sources, poppy seeds (seeds of *Papaver somniferum*), the source of morphine alkaloid which was isolated in the nineteenth century, avermectin the antiparasitics drug, quinine and artemisinin, which are used as antimalarials, lovastatin and its analogs that are used for lipid control, cyclosporine and rapamycins which are used as immunosuppressants in case of organ transplantation, paclitaxel and irinotecan the famous anticancer drugs (Harvey 2008). Nowadays, drug discovery is not only based on the isolation of active constituents, but also involves other techniques such as the high-performance liquid chromatography (HPLC) and standardization of herbal medicines using a marker compound (Newman et al. 2000; Butler 2004; Samuelsson 2004).

Drug discovery always begins with the collection and authentication of the selected plant (Fig. 1.1), and this should be done by a botanist, ethnobotanist, ethnopharmacologist, or plant ecologist. The plants of interest are either those species with known biological activity or used traditionally by man and thus their active compounds are required to be isolated. This approach depends on the attempt that the natural compounds purified from such herbs are probably more reliable than those gotten from plants having no history of human use (Katiyar et al. 2012). On the other hand, the collection may involve taxa that were not examined before and are randomly collected for a large screening program. In this respect, the intellectual property rights of the country from which the plants were collected must be protected (Baker et al. 1995). The second step involves the preparation of different plant extracts by the phytochemists and using the relevant pharmacological assays for the biological screening of these extracts. The third step is the purification and structure elucidation of the active compounds through bioassay-guided fractionation (Balunas and Kinghorn 2005). Recently, molecular biology has evolved as a fundamental technique in drug discovery through the usage of the proper screening assays directed towards physiologically relevant molecular targets.

However, according to Pan et al. (2013), drug discovery process from natural products can be categorized into three distinct phases, which are predrug stage, quasidrug stage, and full-drug stage:

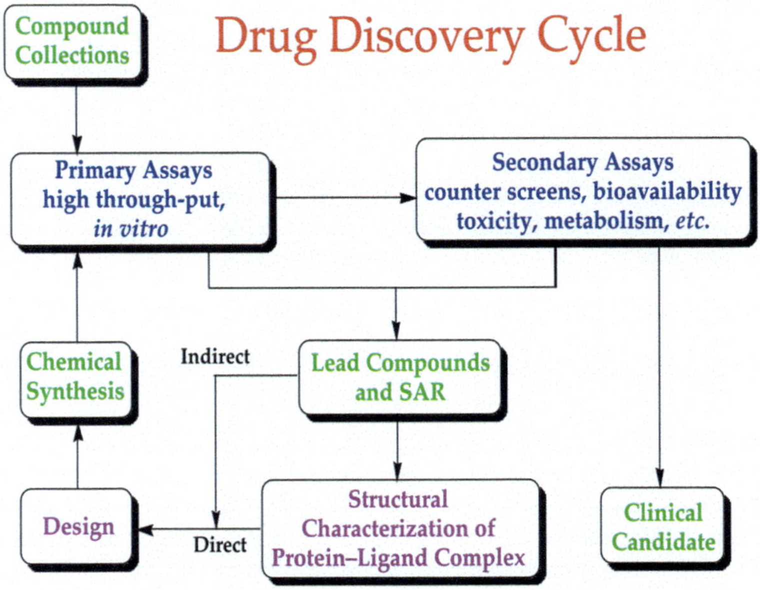

Fig. 1.1 Drug discovery cycle. (Source Boghog 2015. Licensed under Creative Commons Attribution-Share Alike 4.0 International)

1.2.1 Predrug Stage

Being the first stage, it involves the choice of the plants through one of three approaches, the edible plants, the traditionally used herbs, and scientific research (phytochemical analysis). The medical information about the therapeutic effects of herbs was mainly acquired by ancient people through many trials. Accordingly, Chinese herbal medicine and Indian herbal medicine, which were developed in ancient China, Japan, Korea, and India, are still influencing the modern healthcare system (Samy et al. 2008). The World Health Organization (WHO) stated that about 3.5–4 billion people in the whole world depend on herbal medicines for their health care, and about 85% of traditional medicine involves the use of plant extracts in what is named as the modern herbal medicine (Pan et al. 2013).

1.2.2 Quasidrug Stage

This stage includes extraction of plants, isolation of pure compounds, structure/composition elucidation, and evaluation of the biological activities in which isolates are to be used as lead molecules for further drug development (Sasidharan et al. 2011).

1.2.3 Full-Drug Stage

The high potential of natural compounds bioactivities, makes them the major source of components used for constructing hybrid molecules in the development of anticancer, antioxidant, and antimalarial drugs (Decker 2011). Therefore, the multinational pharmaceutical organizations commonly spend a yearly measure of US$ 110 billion trying to find new medications from natural products.

As earlier stated, natural products have long been the source of new chemical entities (NCEs). In the period between 1981 and 2002, approximately 28% of NCEs are purely natural products or are inspired by the molecules derived from natural sources including semi-synthetic analogs (Newman et al. 2003). Since 1994, half of the approved drugs have been based mainly on natural products. Another 20% of NCEs during this period are like natural product analogs, meaning that the compound was synthesized according to a structure of a natural product (Newman et al. 2003). During the years 2005–2007, 13 natural product related drugs were also approved (Harvey 2008). Natural products can serve as a precursor for synthetic new compounds, with different structures and frequently with numerous stereocenters that can be defying in synthesis (Clardy and Walsh 2004; Koehn and Carter 2005).

Many structural features found in natural products poses challenges in the new drug discovery, such as a high number of chiral centers, great steric complexity, large number of oxygen atoms, low number of aromatic atoms relative to the heavy atoms, great number of solvated hydrogen bond donors and acceptors, high molecular rigidity, and diversity of ring systems (Feher and Schmidt 2003; Piggott and Karuso 2004; Clardy and Walsh 2004; Koehn and Carter 2005). Furthermore, since the heightening of enthusiasm for combinatorial chemistry, numerous synthetic and medicinal scientists is investigating the making of natural compounds and natural compounds-like libraries (Ganesan 2004; Tan 2004).

Medicinal plant-derived drugs either serve as new drugs by themselves or serve as drug leads that could be optimized by medicinal and synthetic chemists (See Chap. 2 for more information). Drug leads are not necessarily the newest chemical structures isolated during natural product drug discovery, but may also be the known compounds with new biological activity (Balunas and Kinghorn 2005).

Thousands of new molecular targets have been identified since the sequencing of the human genome (Kramer and Cohen 2004). Finding new screening assays aimed to act on these targets, known compounds isolated from traditionally used medicinal plants may be of essence on newly validated molecular targets. This could be observed on three examples, cucurbitacin I, which has been found to be highly selective in inhibiting the JAK/STAT3 pathway in tumors with activated STAT3 (Blaskovich et al. 2003), h-lapachone, which selectively kills tumor cells and don't affect normal cells by direct checkpoint activation during the cell cycle (Li et al. 2003), and betulinic acid that has selective cytotoxicity on melanoma via p38 activation (Pisha et al. 1995; Tan et al. 2003; Cichewicz and Kouzi 2004).

1.3 Efficacy of Medicinal Plants

The medicines from plants are better in bioactivity and are found to possess curative properties against various diseases, compared to chemical and synthetic medicines. Thus, the efficacy of drugs from medicinal plants would be higher than conventional drugs. Although, these predominantly depends on the type of plant parts and their processing, phytochemicals as well as other specific drug component extraction and purification procedures involved in the drug formulation (Briskin 2000). Initially, pre-extraction preparations of plant parts including a selection of fresh or dried, grinded or powdered samples and process of drying are significant in yielding bioactive compounds which eventually affects the effectiveness of plant-based medicines (Borhan et al. 2013; Vongsak et al. 2013). Conventional methods, namely maceration, infusion, percolation, decoction, steam and hydro-distillation are used for the extraction of phytochemicals that are later converted into drugs. However, drawbacks such as longer extraction time and smaller phytochemical yield exists as challenges in these traditional extraction methods (Azwanida 2015). Major modifications in these conventional methods and coupling of two or more techniques, subsequently increases their efficiency and yield of phytochemicals from these methods. Recently, extraction procedures such as accelerated solvent extraction (Nastić et al. 2018), Soxhlet, solid phase, sonication-assisted, micro-assisted and supercritical fluid extractions are used for the enhanced phytochemical extraction with high yield (Wang and Weller 2006). Similarly, methods to isolate and purify individual phytochemicals such as chromatography, hyphenated techniques, crystallization, ion exchange and solvent extraction using partition coefficient were used to segregate biologically active phytocompounds. The planar, low-pressure column, high-speed countercurrent, high-performance liquid and crystallization are the recent modifications in the conventional isolation and purification methods to enhance the specific phytocompound separation process. The separation of specific phytochemicals with the medicinal property via purification process is highly beneficial to enhance their efficacy and reduce cytotoxicity (Sarker et al. 2005).

The medicinal drug entities extracted from phytocompounds possess medicinal properties against various microbial infections and diseases. The efficacy of a specific phytocompound is evaluated using in-vitro models or through in-silico methods which is further formulated and prescribed as medicine to be evaluated in in-vivo models, clinical trials and humans (Shobana and Vidhya 2016; Nivedha et al. 2017). It has been reported through several literatures that medicines formulated from plant-based compounds and chemicals possess antioxidant (Zhou et al. 2004), antibacterial (Khyade and Vaikos 2009), antifungal (da Silva et al. 2018), antiviral (Yang et al. 2016b), antialgal (Hussain et al. 2015), antimalarial (Chander et al. 2016), anti-diarrheal (Bellah et al. 2017), hypoglycemic (Loizzo et al. 2015), antitumor (Sahreen et al. 2015), anticancer (Clark and Lee 2016), anti-inflammatory (Van de Velde et al. 2016), anti-enteric (Kabir et al. 2017) and immuno-stimulant properties (Dügenci et al. 2003). The effectiveness of these properties in medicinal plants makes them to be a better drug candidate against diseases such as diabetes, neurodegenerative, cardiovascular, microbial infections and deficiencies

(Shankar et al. 2016). It also showed effective remedial potential against rare complications such as menstrual hemorrhage (Armand et al. 2018), menstrual disorders (Deoray and Page 2018), vaginal candidiasis (Sirilun et al. 2018), migraine headache (Nwidu et al. 2015) and Lafora (Wang et al. 2016). The dose and concentration-dependent efficacy of these plant extract-based medicines leads to their usage, despite the advent of numerous synthetic drug molecules (Elujoba et al. 2005). The latest computational methods, enhanced extraction, purification and characterization techniques in retrieving phytocompounds from plants and in formulating pharmaceutical drugs helps to improve the efficacy of medicines from plants (Atanasov et al. 2015).

The in-vitro analysis using bacteria, virus, algae and fungi helps to evaluate the antimicrobial properties of the phytocompounds. Disc diffusion assay, minimum inhibitory concentration (Discussed in subsequent chapters), calorimetric and spectroscopic analysis are used to analyze and optimize the antimicrobial activity of medicinal phytocompounds (Ahmad and Beg 2001). Similarly, egg, larvae and adult of mosquitos are reared in laboratory conditions to investigate the mosquitocidal properties of medicinal plant products (Chander et al. 2015). Normal and unhealthy cell lines of rabbit, mouse, monkey or human origin is used to establish the medicinal properties of phytocompounds in curing and reducing the complications of diseases (Dutra et al. 2016). The cell lines are utilized to evaluate the cytotoxicity of phytocompounds with remedial potential with (3-(4,5-Dimethylthiazol-2-yl)-2,5-diphenyl tetrazolium bromide) (MTT), water-soluble tetrazolium salts (WST-1), Alamar blue, neutral red uptake, lactate dehydrogenase, propidium iodide, protein or deoxyribonucleic acid (DNA) content measurement, 3H-thymidine incorporation and bromodeoxyuridine incorporation assay (Omidi et al. 2017). In-silico methods such as computer simulations and statistical tools are also used to investigate cytotoxicity and efficacy of plant-based medicines towards rare diseases in which cells cannot be raised in in-vitro conditions (Zengin et al. 2018). After obtaining the highly efficient phytocompound from in-vitro models and in-silico studies, it is subjected to in-vivo analysis to estimate their toxic reactions and medicinal potentials in animal models (Gambini et al. 2015). It is also proved in recent times that isolated phytocompounds shows high bioactivity with less cytotoxicity, compared to crude extracts (Sharma et al. 2016). Thus, natural drugs from medicinal plants are gradually capturing the pharmaceutical market of synthetic drugs due to their effective disease curing or microbial growth inhibiting ability. However challenges in large-scale production, stability and isolation of individual compounds still exists as drawbacks for commercialization of natural drugs from medicinal plants.

1.4 Safety of Drugs from Medicinal Plants

The reports from the World Health Organization (WHO) revealed that more than 80% of the world population uses plant-based medicines for healthcare and 25% of drugs in pharmaceutical market are derived from plants (Li et al. 2009). Recently, natural drugs from plants are tipped to be safer than the synthetic drugs. Despite these positive aspects, the lack of safety among natural phytomedicines is a significant drawback of concern. The common thoughts of safety and devoid is not true in all cases of phytomedicines as they lack quality control during production, disparities of active ingredients in distinct plant parts, imprecise identification of plant species and unclear nomenclature (Ekor 2014). The safety of extracted medicinal phytocompounds also depends on the geography, time and stages of plant harvest, soil, weather, nutrient levels and other conditions required for the growth of individual plants. The mixture of adulterants or contaminants in phyto-extracts and contamination via microbes as well as fungal toxins, heavy metals and pesticides also raises a question of safety towards phytomedicines (Brevoort 1998). Numerous literatures described that unsafe phytomedicine preparations, including lack of plant-drug interaction mechanism, adulteration and unsuitable formulation may lead to complications such as kidney disease, neurological, cardiovascular, dermatological effect, liver fibrosis, cellular and genotoxicity (Auerbach et al. 2012). Thus, cautious regulatory and safety monitoring procedure is advised as a prime importance to amplify the safety of phytomedicines (Saad et al. 2017).

In order to make sure the safety of natural phytomedicines is guaranteed, the international level standardization need be tightened for regulatory policies by healthcare professional and regulatory authorities. The route of administration, geographical origin, their compatibility and processing needs be evaluated to ensure the safety of phytomedicines by techniques mentioned in the previous section. Also, common binomial names for the medicinal plants needs be adopted to avoid challenges in identifying and collecting those plants. Therefore, a group of scientists from botany, phytochemistry, pharmacology and stakeholders are essential for the effective monitory of phytomedicinal safety. However, there exists weak regulation, high-profile safety concerns and lack of knowledge on the side-effects even in the developed countries. The side-effects can be caused due to over dosage, undeclared medicines, usage of different plant species, phytomedicines with synthetic drugs and contamination. Hence the need for regulation and safety measures to be implemented by each country which is quite difficult as a single plant species can be utilized either as medicine, functional diet or a diet supplement. It is advisable to form a global phytomedicine regulatory body in the future, which should draft safety rules for each continent to segregate medicinal plants, their formulation, dosage and ethical trials in humans.

The current regulatory policies have certain influence on the safety of the phytomedicines. In the UK, there are regulatory routes, including unlicensed phytomedicinal remedy until year 2011 that allowed unlicensed phytochemical to be available in the market. A 7-year transition period and simple licensing system was recently directed by the European Union to harmonize regulations on traditional phytomedicinal products. Also, the traditional phytomedicine registration scheme has been introduced in which a phytomedicine product is provided with a unique license depending on their efficacy, safety and quality. Similarly, Dietary Supplement Health and Education Act (DSHEA) in United States regulated natural drugs from plants that contain potential toxicity and side-effects towards humans (Ekor 2014).

Quality assessment of phytomedicines is highly essential in ensuring the safety of natural phytomedicines. The initial steps involved in quality assurance are authentication of plant species via genomic profiling, classical systematics, DNA Barcoding, utilization of good collection and agricultural practice, characterization of medicinal plant materials using micro and macroscopic techniques, identification and classification of medicinal plants via phytochemical profiling, better guidelines for documentation and authentication of medicinal plants and impurity profiling (Saharan 2011). Quality assurance of phytochemicals during extraction process can be achieved by Good Medical Practices (GMP) and Good Laboratory Practices (GLP) such as phytochemical profiling and standardization, harmonization via the analytical method in the entire supply chain, comprehensive quality measurement and transparency in the supply chain (Sharma 2015). These models of quality assurance are present to ensure safe plant formulations as medicines, functional foods, dietary supplements and nutraceuticals to correlate with international drug regulatory policies. However, the present models are only based on the efficacy and evidences of phytochemicals (Govindaraghavan and Sucher 2015). Thus, it is clear that strict quality control at each step of phytomedicine production with stringent regulatory measures will ensure the safety of natural drugs from plants, similar to synthetic drugs.

1.5 Current Uses of Natural Products as Drugs

Currently, several synthetic drugs and cosmetic consumers are gaining interest towards natural products due to the recent awareness about the toxic chemicals and reactions involved in those synthetic medicines. Among natural medicinal products from microbes, insects and animals, phytomedicines are accepted to possess high pharmaceutical significance due to their safety, ease production and wide availability. In recent years, plant based natural products are extensively employed in pharmaceutical, nutraceutical and dietary supplements. Nogueira et al. (2018) proposed molecular factories of plants for nutritional and industrial production of isoprenoids including tocopherols and carotenoids which is a major constituent of human diet. The plant-based natural products are widely used as medicines for the ailment of deadly diseases, including HIV and new diseases that are spreading in recent times (Vora et al. 2018). It is noteworthy that the natural phytoproducts are presently utilized in the treatment of neurodegenerative and cardiovascular diseases (Croft et al. 2018; Hussain et al. 2018) including Alzheimer's (Shal et al. 2018), diabetic neuropathy (Solanki et al. 2018) and cardiomyopathy (Uddandrao et al. 2018). The problem of multidrug-resistant and drug sensitive pathogens are tackled by natural plant-based drug products that are produced via modern safety measures and isolation methods (Shin et al. 2018; Vambe et al. 2018). The enhanced mosquitocidal property of natural phytomedicines helps to eradicate yellow fever (Nesterkina et al. 2018), Zika fever (Bajpai et al. 2018), tuberculosis (Gupta et al. 2018) and other protozoal diseases, namely leishmaniasis, schistosomiasis and trypanosomiasis (Simoben et al. 2018). Phytochemical based drug delivery systems are widely explored by researchers in the treatment of diseases such as cancer (Karthikeyan and Laxmanappa 2015), diabetes (Ríos et al. 2015), menstruation disorders (Steenkamp 2003) and sexually transmitting diseases (Hossan et al. 2010). The recent and modern techniques in the fabrication and formulation of natural drug products make phytomedicines as a potential tool to cure these diseases. Extensive research in ethnopharmacology field will lead plant based natural drugs to be utilizable in the future to cure rare diseases such as hyperprolactinemia, hypogonadism syndromes (Besser et al. 1976), uterine contractility (Ekstrom et al. 1992), menorrhagia (Kingman et al. 2004), progeria (Jeyam et al. 2011), Morgellons (Kontos et al. 2017), micropsia (Sayin 2016) and porphyria (Smith and Foster 2018).

The awareness about natural products among common people leads to the transformation of medicinal advertisements to rely on commercials that explains the natural origin of the product. Thus, numerous natural plant-based products will be available in the pharmaceutical market. The improvement in the safety measures of phytomedicinal production and mass cultivation of plants to retrieve medicine from them will enhance the usage of plant-based medicines in the future. Nanoformulations of active phytocompounds for the targeted delivery of these compounds in target location and nanoparticles synthesized via phytochemicals are the future of phyto-therapies and phytomedicines. Liposomes, micelles, nanoencapsulations, dendrimers and polymers are some of the nanoformulations that are available for the targeted delivery of phytochemicals (Jeevanandam et al. 2016a). Phytochemicals such as flavonoids, phenols, terpenoids, essential oils, xanthophyll and several novel phytocompounds are used for the synthesis of less toxic nanoparticles with enhanced bioactivity, bioavailability and biocompatibility (Jeevanandam et al. 2016b). Recently, virus nanoparticles and virus-like nanoparticles are also employed for the formulation of active medicinal phytochemical to carry and deliver them at the target site of infection or disease (Aljabali 2018). Hence, the future of pharmaceutical industry lies in the emergence of novel methods to isolate, characterize, purify and formulate compounds and chemicals with medicinal properties from plants to cure and reduce complications of numerous diseases.

1.6 Advances in Tools for the Screening of Medicinal Herbs

According to the most recent statistics for the time period from 1981 to 2014, about 42% of 1562 newly approved drugs are derived from natural medicinal plants (Newman and Cragg 2016). Meanwhile, in the period from 1981 to 2010, the proportion of natural bioactive compounds is more than one half of the approved 1073 new types of small molecule drugs (Atanasov et al. 2015). The complexity of the chemical compositions of natural compounds and great variability in their structures, coupled with nonspecific adsorptions, false positive results, the undetectable trace amounts of active components have been a challenging factor (Potterat and Hamburger 2013; Zhu et al. 2013; Wu et al. 2016a). These obstacles represent the actual challenges in the identification of bioactive components and prediction of their possible mechanisms of action. The good choice of in-vitro and in-vivo assays for screening of the plant bioactivity is a very important step in drug discovery from plants. The chosen bioassays should be characterized by simplicity together with good sensitivity and reproducibility. Numerous in-vitro models using purified proteins, cell-based target-oriented or phenotypic assays can be utilized for evaluation of the biological activities of natural products, in addition to the isolated tissues or organ models, and in-vivo preclinical animal models can be used (Atanasov et al. 2015).

In the past, the screening of plant-derived extracts and compounds utilized the forward pharmacological approach which starts with in-vivo animal tests, organ, tissue models, or bacterial preparations, followed by in-vitro testing to determine the mechanism of action. The forward pharmacology begins with determination of functional activity through phenotypic change detection in complex biological systems and then characterizes the molecular target of the active compounds, so it is named the phenotypic drug discovery. This traditional way of drug discovery was applied mainly before the development of the modern molecular biology techniques and the Human Genome Project (Takenaka 2001; Lee et al. 2012; Schenone et al. 2013; Zheng et al. 2013).

Now when the screening starts by testing of the plant-derived compounds ("libraries") against pre-characterized disease-relevant protein targets, in order to identify "hits", the biologically active compounds which are then studied using in-vivo animal models with the target of their validation this is named as a reverse pharmacology approach. The forward and reverse pharmacology approaches differ only in the stage when the assays are applied. The reverse pharmacology begins by identifying the promising pharmacological target to obtain the promising hit compounds which are then validated in-vivo and thus considered as a target-directed drug discovery (Takenaka 2001; Lee et al. 2012; Schenone et al. 2013; Zheng et al. 2013).

In-silico simulations are computational methods which can be used to suggest a protein ligand binding characteristic of a molecular structure that may be a known plant-derived compound. Compounds that give good results in the in-silico predictions can be promising candidates for further experimental work. Applying virtual screening for activity predictions have shown increasing rates of success (Hein et al. 2010). In-silico simulations can also act as valuable filter tools to predict new activities for an already known natural product as well as detection of ADME/T properties (Kaserer et al. 2014). These computational methods can be employed also to discover new binding sites on protein with already known structures. Pocket finders are used to identify solvent-accessible cavities in the protein surface that represent important ligand binding sites that can then be computationally analyzed (Rollinger et al. 2008).

The active compounds may exert their biological activity through regulating the body's signal transduction and maintaining normal metabolism. Or they may exert their action through the interactions with the disease-related drug targets, such as enzymes and receptors (Mulabagal and Calderon 2010; Wu et al. 2015b). Screening of the biologically active natural compounds targeting these enzymes and receptors, could not only provide new approaches for new drug discovery, but also help to know the mechanistic action of bioactive small natural molecules (Chen et al. 2016a, b).

In this respect, affinity ultrafiltration mass spectrometry (UF-LC/MS) is one of the potential techniques as it combines ultrafiltration with liquid chromatography-mass spectrometry (LC/MS), which achieve the affinity capture of active small molecules against biomolecule targets through high-throughput screening and rapid identification of biologically active compounds in the complex mixtures (Li et al. 2015; Tao et al. 2015).

In UF-LC/MS, the bio-affinity ultrafiltration process begins with the separation of the ligand– enzyme complexes from free components, and then those ligands released from the complexes will be further identified and quantified by high performance liquid chromatography mass spectrometry (HPLC–MS) analysis (Tang et al. 2015). Compared to the extremely complicated traditional procedures of screening for bioactive plant constituents, the UF-LC/MS method is not just very easy, but also greatly reduces the time required for screening, the consumptions of samples and expensive reagents (Chen and Guo 2017a, b; Li et al. 2009).

Affinity UF-LC/MS is a powerful method that can rapidly screen and identify small drug molecules bound to target proteins, and also be used for the screening of the leading compounds. Moreover, this technique can also identify the mechanism by which bioactive compounds exert their actions. For this reason, it is an important approach for drug discovery from the natural medicinal plants. There are still many challenges facing affinity UF-LC/MS. For example, it is mainly used for screening of small active molecules from medicinal plants at one or two protein targets. Moreover, reduction of the non-specific bindings of small drug molecules to ultra-filtration membrane so as to reduce the false positive results is an urgent issue to be solved in the future.

1.7 Problems and Way Forward

Despite the great prospects of natural products drug discovery, future attempts face many challenges. The process of drug discovery usually takes a long time, estimated from 10 years or more (Reichert 2003), which costs more than 800 million dollars (Dickson and Gagnon 2004). Although, much of this time and money are spent, especially on the purification and structure elucidation of the leads, however, one in 5000 lead compounds usually pass the clinical trials and be approved for use.

The first challenge in natural products drug discovery is the precise identification and nomenclature of the plant of interest, the step on which all the following steps are dependent. For this reason, a combination of methods are used for authentication of medicinal plants such as genetic, morphological and anatomical characterization in addition to the chemical characterization (Bucar et al. 2013). Accordingly, the collection of plant material and accurate documentation, botanical identification, as well as preparation of the herbarium vouchers are tasks that cannot be automated (David et al. 2015) and requires specialists who are not available anymore (Bucar et al. 2013).

The collections of medicinal plants from wild species have many difficulties, as the plant habitats may quickly vanish due to human over-consumption of these plants (David et al. 2015). In cases of imported plant material, many factors such as local wars, natural crisis, or changing laws for travelling between countries and the exportation of plant material may affect its accessibility.

When a plant becomes utilized as an herbal medicine or when one of its constituents proved its bioactivity, the plant becomes threatened by extensive collection and unstable harvesting systems (Cordell 2011). A typical example of this problem is the "Taxol supply crisis" (Kingston 2011). When Taxol proved clinically effective against ovarian cancer, the demand for Taxol suddenly increased. The process for its isolation from the barks of *Taxus brevifolia* L. (The western yew tree) starts with the collection of the bark, drying, preparation of plant extract, and purification of the compound. This process is time consuming. At this time, worries on the environmental effect of excessive bark gathering appeared (Kingston 2011). Although Taxol is meanwhile accessible via semi-synthesis, the problem of sustainable supply of Taxol still frequently occurs.

Limited availability of a bioactive, plant-derived natural product is also problematic, especially when it proves a very promising bioactivity and becomes a pharmaceutical lead. Natural products are usually isolated in small amounts which are deficient for the development of lead compounds. To improve the rate for natural compound isolation, new technologies should be incorporated. New applications of NMR and MS should be employed to facilitate compound isolation (Glish and Vachet 2003). Also, the use of high-throughput X-ray crystallography can be used in natural product discovery (Blundell et al. 2002). Coordinated effort with medicinal and synthesis experts is important to decide whether synthesis or semi-synthesis may be conceivable (Lombardino and Lowe 2004). Another strategy to enhance natural product compounds advancement may include the formation of natural product and natural-product-like libraries that join the characteristics of natural products with combinatorial chemistry (Koehn and Carter 2005).

Further complications appear because natural compounds usually have complex chemical structures with numerous chiral centers, which make the pathway for their synthesis or even derivatization very difficult (Henrich and Beutler 2013).

The incompatibility of natural products with high-throughput screening is another challenge in drug discovery (Koehn and Carter 2005). High-throughput screening of plant extracts followed by the identification of its biologically active compounds is highly difficult as it needs sample preparation and assay designs. In general, high-throughput screening may depend on cell-free or cell-based assays which should be reproducible, accurate, and reliable. The tested compounds should not decompose, precipitate, or interfere with assay reagents, but plant-derived compounds most probably fail to fulfil these requirements. Accordingly, sophisticated sample preparation or fractionation of the crude extracts prior to testing is highly demanded (Johnson et al. 2011; Maes et al. 2012).

1.8 Conclusion

Despite the challenges that faces scientists in natural product drug discovery, natural products and their derivatives have been sources for numerous clinically useful medicines. Therefore, natural products remain a potential lead compound and precursors for production of new medicines. The next chapter details the various phytochemicals and synthetic analogs, their mechanisms of action and structure-activity relationships. A table summarizing 150 natural plant-derived drugs, their uses, and sources were presented.

References

Ahmad I, Beg AZ. Antimicrobial and phytochemical studies on 45 Indian medicinal plants against multi-drug resistant human pathogens. J Ethnopharmacol. 2001;74(2):113–23.
Aljabali AA. Viral nanoparticles: a drug delivery platform. J Pharm Toxicol. 2018;1(1):1–2.
Armand Z, Abel A, Hanitra RR, Joseph RD, Mahandrimanana A. Ethnobotanical, antimicrobial and phytochemical screening of Euphorbia Thymoflia L.(Kinononono madiniky) to cure the menstrual hemorrhage of a woman Sakalava Bemazava in the Northern Region of Madagascar. Int J Plant Res. 2018;8(1):1–7.
Atanasov AG, Waltenberger B, Pferschy-Wenzig E, Linder T, Wawrosch C, Uhrin P, Temml V, Wang L, Schwaiger S, Heiss E, Rollinger JM, Schuster D, Breuss JM, Bochkov V, Mihovilovic MD, Kopp B, Bauer R, Dirsch VM, Stuppner H. Discovery and resupply of pharmacologically active plant-derived natural products: a review. Biotechnol Adv. 2015;33:1582–614.
Auerbach BJ, Reynolds SJ, Lamorde M, Merry C, Kukunda-Byobona C, Ocama P, Semeere AS, Ndyanabo A, Boaz I, Kiggundu V. Traditional herbal medicine use associated with liver fibrosis in rural Rakai, Uganda. PLoS One. 2012;7(11):e41737.
Azwanida N. A review on the extraction methods use in medicinal plants, principle, strength and limitation. Med Aromat Plants. 2015;4(3):3–8.
Bajpai VK, Chandra V, Kim N-H, Rai R, Kumar P, Kim K, Aeron A, Kang SC, Maheshwari D, Na M. Ghost probiotics with a combined regimen: a novel therapeutic approach against the Zika virus, an emerging world threat. Crit Rev Biotechnol. 2018;38(3):438–54.
Baker JT, Borris RP, Carte B, Cordell GA, Soejarto DD, Cragg GM, Gupta MP, Iwu MM, Madulid DR, Tyler VE. Natural products drug discovery and development: new perspectives on international collaboration. J Nat Prod. 1995;58(9):1325–57.
Balick MJ, Cox PA. Plants, people, and culture: the science of ethnobotany. New York: Scientific American Library; 1997. p. 228.
Balunas MJ, Kinghorn AD. Drug discovery from medicinal plants. Life Sci. 2005;78:431–41.
Bellah SF, Islam MN, Karim MR, Rahaman MM, Nasrin MS, Rahman MA, Reza AA. Evaluation of cytotoxic, analgesic, antidiarrheal and phytochemical properties of Hygrophila spinosa (T. Anders) whole plant. J Basic Clin Physiol Pharmacol. 2017;28(2):185–90.
Besser GM, Thorner MO. Bromocriptine in the treatment of the hyperprolactinaemia-hypogonadism syndromes. Postgrad Med J. 1976;52:64–70.
Blaskovich MA, Sun J, Cantor A, Turkson J, Jove R, Sebti SM. Discovery of JSI-124 (cucurbitacin I), a selective Janus kinase/signal transducer and activator of transcription 3 signaling pathway inhibitor with potent antitumor activity against human and murine cancer cells in mice. Cancer Res. 2003;63(6):1270–9.
Blundell TL, Jhoti H, Abell C. High-throughput crystallography for lead discovery in drug design. Nat Rev Drug Discov. 2002;1(1):45–54.

Boghog. Drug discovery cycle. 2015. Available: https://commons.wikimedia.org/wiki/File:Drug_discovery_cycle.svg. Accessed 15 Sept 2018.

Borhan M, Ahmad R, Rusop M, Abdullah S. Impact of nanopowders on exraction yield of *Centella asiatica*. Adv Mater Res Trans Tech Publ. 2013;667:246–50.

Brevoort P. Booming US botanical market: a new overview. HerbalGram. 1998;44:33–46.

Briskin DP. Medicinal plants and phytomedicines. Linking plant biochemistry and physiology to human health. Plant Physiol. 2000;124(2):507–14.

Bucar F, Wube A, Schmid M. Natural product isolation – how to get from biological material to pure compounds. Nat Prod Rep. 2013;30:525–45.

Butler MS. The role of natural product chemistry in drug discovery. J Nat Prod. 2004;67 (12):2141–53.

Chander MP, Pillai C, Sunish I, Vijayachari P. Antimalarial efficacy of nine medicinal plants traditionally used by the Karens of Andaman and Nicobar Islands, India. Bangladesh J Pharmacol. 2015;11:126–9.

Chander MP, Pillai C, Sunish I, Vijayachari P. Antimicrobial and antimalarial properties of medicinal plants used by the indigenous tribes of Andaman and Nicobar Islands, India. Microb Pathog. 2016;96:85–8.

Chen GL, Guo MQ. Rapid screening for α-glucosidase inhibitors from *Gymnema sylvestre* by affinity ultrafiltration-HPLC-MS. Front Pharmacol. 2017a;8:228–35.

Chen GL, Guo MQ. Screening for natural inhibitors of topoisomerases I from *Rhamnus davurica* by affinity ultrafiltration and high-performance liquid chromatography-mass spectrometry. Front Plant Sci. 2017b;8:1521–31.

Chen GL, Tian YQ, Guo MQ. Screening for inhibitors of topoisomerase I from *Lycoris radiata* by combining ultrafiltration with liquid chromatography/mass spectrometry. Rapid Commun Mass Spectrom. 2016a;30:95–9.

Chen GL, Tian YQ, Wu JL, Li N, Guo MQ. Antiproliferative activities of Amaryllidaceae alkaloids from *Lycoris radiata* targeting DNA topoisomerase I. Sci Rep. 2016b;6:38284–2383.

Cichewicz RH, Kouzi SA. Chemistry, biological activity, and chemotherapeutic potential of betulinic acid for the prevention and treatment of cancer and HIV infection. Med Res Rev. 2004;24(1):90–114.

Clardy J, Walsh C. Lessons from natural molecules. Nature. 2004;432(7019):829–37.

Clark R, Lee S-H. Anticancer properties of capsaicin against human cancer. Anticancer Res. 2016;36(3):837–43.

Cordell GA. Sustainable medicines and global health care. Planta Med. 2011;77:1129–38.

Croft KD, Yamashita Y, O'Donoghue H, Shirasaya D, Ward NC, Ashida H. Screening plant derived dietary phenolic compounds for bioactivity related to cardiovascular disease. Fitoterapia. 2018;126:22–8.

da Silva DT, Herrera R, Heinzmann BM, Calvo J, Labidi J. Nectandra grandiflora by-products obtained by alternative extraction methods as a source of phytochemicals with antioxidant and antifungal properties. Molecules. 2018;23(2):372.

David B, Wolfender JL, Dias DA. The pharmaceutical industry and natural products: historical status and new trends. Phytochem Rev. 2015;14(2):299–315.

Decker M. Hybrid molecules incorporating natural products: applications in cancer therapy, neurodegenerative disorders and beyond. Curr Med Chem. 2011;18(10):1464–75.

Deoray V, Page A. Treatment of menstrual disorders: a shift from synthetic drugs to drugs of natural origin. Int J Pharm Bio Sci. 2018;9(1):120–31.

Dickson M, Gagnon JP. Key factors in the rising cost of new drug discovery and development. Nat Rev Drug Discov. 2004;3(5):417–29.

Dügenci SK, Arda N, Candan A. Some medicinal plants as immunostimulant for fish. J Ethnopharmacol. 2003;88(1):99–106.

Dutra RC, Campos MM, Santos AR, Calixto JB. Medicinal plants in Brazil: pharmacological studies, drug discovery, challenges and perspectives. Pharmacol Res. 2016;112:4–29.

Egbuna C, Ifemeje JC. Biological functions and anti-nutritional effects of phytochemicals in living system. IOSR J Pharm Biol Sci. 2015;10(2):10–9.

Ekor M. The growing use of herbal medicines: issues relating to adverse reactions and challenges in monitoring safety. Front Pharmacol. 2014;4:1–10.
Ekstrom P, Akerlund M, Forsling M, Kindahl H, Laudanski T, Mrugacz G. Stimulation of vasopressin release in women with primary dysmenorrhoea and after oral contraceptive treatment – effect on uterine contractility. Br J Obstet Gynaecol. 1992;99(8):680–4.
Elujoba AA, Odeleye O, Ogunyemi C. Traditional medicine development for medical and dental primary health care delivery system in Africa. Afr J Trad Comp Altern Med. 2005;2(1):46–61.
Feher M, Schmidt JM. Property distributions: differences between drugs, natural products, and molecules from combinatorial chemistry. J Chem Inf Comput Sci. 2003;43(1):218–27.
Furlan V, Kujawska M, Hilgert NI, Pochettino ML. To what extent are medicinal plants shared between country home gardens and urban ones? A case study from Misiones, Argentina. Pharm Biol. 2016;54(9):1628–40.
Gambini J, Inglés M, Olaso G, Lopez-Grueso R, Bonet-Costa V, Gimeno-Mallench L, Mas-Bargues C, Abdelaziz K, Gomez-Cabrera M, Vina J. Properties of resveratrol: in vitro and in vivo studies about metabolism, bioavailability, and biological effects in animal models and humans. Oxidative Med Cell Longev. 2015;2015:837042.
Ganesan A. Natural products as a hunting ground for combinatorial chemistry. Curr Opin Biotechnol. 2004;15(6):584–90.
Glish GL, Vachet RW. The basics of mass spectrometry in the twenty-first century. Nat Rev Drug Discov. 2003;2(2):140–50.
Govindaraghavan S, Sucher NJ. Quality assessment of medicinal herbs and their extracts: criteria and prerequisites for consistent safety and efficacy of herbal medicines. Epilepsy Behav. 2015;52:363–71.
Gupta VK, Kaushik A, Chauhan DS, Ahirwar RK, Sharma S, Bisht D. Anti-mycobacterial activity of some medicinal plants used traditionally by tribes from Madhya Pradesh, India for treating tuberculosis related symptoms. J Ethnopharmacol. 2018;227:113–20.
Harvey AL. Natural products in drug discovery. Drug Discov Today. 2008;13:894–901.
Hein M, Zilian D, Sotriffer CA. Docking compared to 3D-pharmacophores: the scoring function challenge. Drug Discov Today Technol. 2010;7:e229–36.
Henrich CJ, Beutler JA. Matching the power of high throughput screening to the chemical diversity of natural products. Nat Prod Rep. 2013;30:1284–98.
Hossan S, Agarwala B, Sarwar S, Karim M, Jahan R, Rahmatullah M. Traditional use of medicinal plants in Bangladesh to treat urinary tract infections and sexually transmitted diseases. Ethnobot Res Appl. 2010;8:061–74.
Howes MJR. The evolution of anticancer drug discovery from plants. Lancet Oncol. 2018;19 (3):293–4.
Hussain H, John M, Al-Harrasi A, Shah A, Hassan Z, Abbas G, Rana UA, Green IR, Schulz B, Krohn K. Phytochemical investigation and antimicrobial activity of an endophytic fungus Phoma sp. J King Saud Univ-Sci. 2015;27(1):92–5.
Hussain G, Rasul A, Anwar H, Aziz N, Razzaq A, Wei W, Ali M, Li J, Li X. Role of plant derived alkaloids and their mechanism in neurodegenerative disorders. Int J Biol Sci. 2018;14(3):341.
Jeevanandam J, San Y, Chan YS, Danquah MK. Nano-formulations of drugs: recent developments, impact and challenges. Biochimie. 2016a;128:99–112.
Jeevanandam J, Chan YS, Danquah MK. Biosynthesis of metal and metal oxide nanoparticles. Chem Bio Eng Rev. 2016b;3(2):55–67.
Jeyam M, Sushma R, Sharanya M, Poornima V. Validating nutraceuticals to alleviate progeria using molecular docking studies. J Pharm Biomed Sci (JPBMS). 2011;13(13):1–7.
Johnson TA, Sohn J, Inman WD, Estee SA, Loveridge ST, Vervoort HC, et al. Natural product libraries to accelerate the high-throughput discovery of therapeutic leads. J Nat Prod. 2011;74:2545–55.
Kabir Y, Shekhar HU, Sidhu JS. Phytochemical compounds in functional properties of mangoes. In: Siddiq M, Brecht JK, Sidhu JS, editors. Handbook of mango fruit: production, postharvest science, processing technology and nutrition. Oxford: Wiley; 2017. p. 237–54.
Kala CP. Medicinal and aromatic plants: boon for enterprise development. J Appl Res Med Aromat Plants. 2015;2(4):134–9.

Karthikeyan S, Laxmanappa HS. Development of fourth generation ABC inhibitors from natural products: a novel approach to overcome cancer multidrug resistance. Anti-Cancer Agents Med Chem (Formerly Curr Med Chem-Anti-Cancer Agents). 2015;15(5):605–15.

Kaserer T, Temml V, Schuster D. Polypharmacology and adverse bioactivity profiles predict potential toxicity and drug-related ADRs. In: Wang J, Urban L, editors. Predictive ADMET. Hoboken: Wiley; 2014. p. 23–45.

Katiyar C, Gupta A, Kanjilal S, Katiyar S. Drug discovery from plant sources: an integrated approach. Ayu. 2012;33(1):10–9.

Khyade M, Vaikos N. Phytochemical and antibacterial properties of leaves of *Alstonia scholaris* R. Br Afr J Biotechnol. 2009;8(22):6434–6.

Kingman C, Kadir R, Lee C, Economides D. The use of levonorgestrel-releasing intrauterine system for treatment of menorrhagia in women with inherited bleeding disorders. BJOG Int J Obstet Gynaecol. 2004;111(12):1425–8.

Kingston DG. Modern natural products drug discovery and its relevance to biodiversity conservation. J Nat Prod. 2011;74:496–511.

Koehn FE, Carter GT. The evolving role of natural products in drug discovery. Nat Rev Drug Discov. 2005;4:206–20.

Kontos N, Beach SR, Smith FF, Greenberg DB. Psychosomatic conditions: somatic symptom. In: Massachusetts general hospital handbook of general hospital psychiatry e-book. Philadelphia: Elsevier Health Sciences; 2017. p. 161.

Kramer R, Cohen D. Functional genomics to new drug targets. Nat Rev Drug Discov. 2004;3 (11):965–72.

Kumar V, Babu S, Revale A, Meena R, Ranjan M, Desai B. Cultivation of medicinal plants in natural ecosystem in Gujarat (India): constraints and conservation need. J Plant Dev Sci. 2014;6 (3):425–35.

Lagunin AA, Goel RK, Gawande DY, Pahwa P, Gloriozova TA, Dmitriev AV, Ivanov SM, Rudik AV, Konova VI, Pogodin PV. Chemo-and bioinformatics resources for in silico drug discovery from medicinal plants beyond their traditional use: a critical review. Nat Prod Rep. 2014;31 (11):1585–611.

Lee JA, Uhlik MT, Moxham CM, Tomandl D, Sall DJ. Modern phenotypic drug discovery is a viable, neoclassic pharma strategy. J Med Chem. 2012;55:4527–38.

Li Y, Sun X, LaMont JT, Pardee AB, Li CJ. Selective killing of cancer cells by beta-lapachone: direct checkpoint activation as a strategy against cancer. Proc Natl Acad Sci U S A. 2003;100 (5):2674–8.

Li HL, Song FR, Xing JP, Tsao R, Liu ZQ, Liu SY. Screening and structural characterization of α-glucosidase inhibitors from hawthorn leaf flavonoids extract by ultrafiltration LC-DAD-MSn and SORI-CID FTICR MS. J Am Soc Mass Spectrom. 2009;20:1496–503.

Li SN, Tang Y, Liu CM, Zhang YC. Development of a method to screen and isolate potential α-glucosidase inhibitors from *Panax japonicus* C.A. Meyer by ultrafiltration, liquid chromatography, and counter-current chromatography. J Sep Sci. 2015;38:2014–23.

Loizzo MR, Pugliese A, Bonesi M, Tenuta MC, Menichini F, Xiao J, Tundis R. Edible flowers: a rich source of phytochemicals with antioxidant and hypoglycemic properties. J Agric Food Chem. 2015;64(12):2467–74.

Lombardino JG, Lowe JA III. The role of the medicinal chemist in drug discovery—then and now. Nat Rev Drug Discov. 2004;3(10):853–62.

Maes J, Verlooy L, Buenafe OE, deWitte PA, Esguerra CV, Crawford AD. Evaluation of 14 organic solvents and carriers for screening applications in zebrafish embryos and larvae. PLoS One. 2012;7:e43850.

Mulabagal V, Calderon AI. Development of an ultrafiltration-liquid chromatography/mass spectrometry (UF-LC/MS) based ligand-binding assay and an LC/MS based functional assay for *Mycobacterium tuberculosis* shikimate kinase. Anal Chem. 2010;82:3616–21.

Naresh Y, Nagendraswamy H. Classification of medicinal plants: an approach using modified LBP with symbolic representation. Neurocomputing. 2016;173:1789–97.

Nastić N, Borrás-Linares I, Lozano-Sánchez J, Švarc-Gajić J, Segura-Carretero A. Optimization of the extraction of phytochemicals from black mulberry (*Morus nigra* L.) leaves. J Ind Eng Chem. 2018;68:282–92.

Nesterkina M, Bernier UR, Tabanca N, Kravchenko I. Repellent activity of monoterpenoid esters with neurotransmitter amino acids against yellow fever mosquito, *Aedes aegypti*. Open Chem. 2018;16(1):95–8.

Newman DJ, Cragg GM. Natural products as sources of new drugs from 1981 to 2014. J Nat Prod. 2016;79:629–61.

Newman DJ, Cragg GM, Snader KM. The influence of natural products upon drug discovery. Nat Prod Rep. 2000;17(3):215–34.

Newman DJ, Cragg GM, Snader KM. Natural products as sources of new drugs over the period 1981–2002. J Nat Prod. 2003;66(7):1022–37.

Nivedha RP, Suryanarayanan V, Selvaraj C, Singh SK, Rajalakshmi M. Chemopreventive effect of saponin isolated from Gymnema sylevestre on prostate cancer through in silico and in vivo analysis. Med Chem Res. 2017;26(9):1915–25.

Nogueira M, Enfissi EM, Almeida J, Fraser PD. Creating plant molecular factories for industrial and nutritional isoprenoid production. Curr Opin Biotechnol. 2018;49:80–7.

Nwidu LL, Nwafor PA, Vilegas W. The aphrodisiac herb Carpolobia: a biopharmacological and phytochemical review. Pharmacogn Rev. 2015;9(18):132.

Omidi M, Fatehinya A, Farahani M, Akbari Z, Shahmoradi S, Yazdian F, Tahriri M, Moharamzadeh K, Tayebi L, Vashaee D. Chapter 7: Characterization of biomaterials. In: Tayebi L, Moharamzadeh K, editors. Biomaterials for oral and dental tissue engineering. Cambridge: Woodhead Publishing; 2017. p. 97–115.

Pan S, Zhou S, Gao S, Yu Z, Zhang S, Tang M, Sun J, Ma D, Han Y, Fong W, Ko K. New perspectives on how to discover drugs from herbal medicines: CAM's outstanding contribution to modern therapeutics. Evid Based Complement Alternat Med. 2013;2013:627375, 25 pages

Petrovska BB. Historical review of medicinal plants' usage. Pharmacogn Rev. 2012;6(11):1.

Piggott AM, Karuso P. Quality, not quantity: the role of natural products and chemical proteomics in modern drug discovery. Comb Chem High Throughput Screen. 2004;7(7):607–30.

Pisha E, Chai H, Lee IS, Chagwedera TE, Farnsworth NR, Cordell GA, Beecher CW, Fong HH, Kinghorn AD, Brown DM, Wani MC, Wall ME, Hieken TJ, Das Gupta TK, Pezzuto JM. Discovery of betulinic acid as a selective inhibitor of human melanoma that functions by induction of apoptosis. Nat Med. 1995;1(10):1046–51.

Potterat O, Hamburger M. Concepts and technologies for tracking bioactive compounds in natural product extracts: generation of libraries, and hyphenation of analytical processes with bioassays. Nat Prod Rep. 2013;30:546–64.

Reichert JM. Trends in development and approval times for new therapeutics in the United States. Nat Rev Drug Discov. 2003;2(9):695–702.

Ríos JL, Francini F, Schinella GR. Natural products for the treatment of type 2 diabetes mellitus. Planta Med. 2015;81(12/13):975–94.

Rollinger JM, Stuppner H, Langer T. Virtual screening for the discovery of bioactive natural products. Prog Drug Res. 2008;65:213–49.

Saad B, Zaid H, Shanak S, Kadan S. Anti-diabetes and anti-obesity medicinal plants and phytochemicals: safety, efficacy, and action mechanisms. Cham: Springer; 2017.

Saharan VA. Current status of regulations for herbal medicines in Europe, United States and India. J Nat Conscientia. 2011;2(3):406.

Sahreen S, Khan MR, Khan RA, Hadda TB. Evaluation of phytochemical content, antimicrobial, cytotoxic and antitumor activities of extract from Rumex hastatus D. Don roots. BMC Complement Altern Med. 2015;15(1):211.

Samuelsson G. Drugs of natural origin: a textbook of pharmacognosy. 5th ed. Stockholm: Swedish Pharmaceutical Press; 2004.

Samy RP, Pushparaj PN, Gopalakrishnakone P. A compilation of bioactive compounds from Ayurveda. Bioinformation. 2008;3(3):100–10.

Saraf S. Applications of novel drug delivery system for herbal formulations. Fitoterapia. 2010;81(7):680–9.

Sarker S, Nahar L. Chemistry for pharmacy students: general, organic and natural product chemistry. Somerset: Wiley; 2007.

Sarker SD, Latif Z, Gray AI. Natural products isolation: an overview. In: Sarker SD, Latif Z, Gray AI, editors. Natural products isolation. 2nd ed. Totowa: Humana Press; 2005

Sasidharan CY, Saravanan D, Sundram KM, Yoga LL. Extraction, isolation and characterization of bioactive compounds from plants' extracts. Afr J Tradit Complement Altern Med. 2011;8(1):1–10.

Sayin HU. Psychoactive plants used during religious rituals. In: Neuropathology of drug addictions and substance misuse. Amsterdam: Elsevier; 2016. p. 17–28.

Schenone M, Dancik V, Wagner BK, Clemons PA. Target identification and mechanism of action in chemical biology and drug discovery. Nat Chem Biol. 2013;9:232–40.

Shakya AK. Medicinal plants: future source of new drugs. Int J Herb Med. 2016;4(4):59–64.

Shal B, Ding W, Ali H, Kim YS, Khan S. Anti-neuroinflammatory potential of natural products in attenuation of Alzheimer's disease. Front Pharmacol. 2018;9:548.

Shankar RJ, Deol PK, Kaur IP. Editorial (thematic issue: perspectives and challenges in rational exploitation of phytochemicals in cure, control and management of diseases). Curr Pharm Des. 2016;22(27):4095–7.

Sharma S. Current status of herbal product: regulatory overview. J Pharm Bioallied Sci. 2015;7(4):293–6.

Sharma D, Pramanik A, Agrawal PK. Evaluation of bioactive secondary metabolites from endophytic fungus Pestalotiopsis neglecta BAB-5510 isolated from leaves of *Cupressus torulosa* D. Don. 3 Biotech. 2016;6(2):210.

Shin J, Prabhakaran VS, Kim KS. The multi-faceted potential of plant-derived metabolites as antimicrobial agents against multidrug-resistant pathogens. Microb Pathog. 2018;116:209–14.

Shobana S, Vidhya R. Evaluation of in vitro hemolytic activity of different parts of *Abutilon indicum* (Linn.). World J Pharm Pharm Sci. 2016;5(5):1182–96.

Simoben CV, Ntie-Kang F, Akone SH, Sippl W. Compounds from African medicinal plants with activities against selected parasitic diseases: schistosomiasis, trypanosomiasis and leishmaniasis. Nat Prod Bioprospect. 2018;8(3):151–69.

Sirilun S, Sivamaruthi BS, Kesika P, Pengkumsri N, Tuntisuwanno N, Chaiyasut K, Peerajan S, Chaiyasut C. Development and stability evaluation of vaginal suppository containing *glycyrrhiza glabra* l. for the treatment of *candida albicans* infection. Development. 2018;11(7):205–9.

Smith AG, Foster JR. The association between chemical-induced porphyria and hepatic cancer. Toxicol Res. 2018;7:647–63.

Solanki ND, Bhavsar SK, Pandya DT. Role of phytotherapy in diabetic neuropathy and neurodegeneration: from pathogenesis to treatment. J Phytopharmacol. 2018;7(2):152–16.

Steenkamp V. Traditional herbal remedies used by South African women for gynaecological complaints. J Ethnopharmacol. 2003;86(1):97–108.

Susantiningsih T, Ridwan R, Prijanti AR, Sadikin M, Freisleben HJ. Schizonticidal effect of a combination of *Amaranthus spinosus* L. and *Andrographis paniculata* Burm. f./Nees extracts in Plasmodium berghei-infected mice. Med J Indones. 2012;21(2):66–70.

Takenaka T. Classical vs reverse pharmacology in drug discovery. BJ Int. 2001;88(Suppl.2):7–10.. (discussion 49–50)

Tan DS. Current progress in natural product-like libraries for discovery screening. Comb Chem High Throughput Screen. 2004;7(7):631–43.

Tan Y, Yu R, Pezzuto JM. Betulinic acid-induced programmed cell death in human melanoma cells involves mitogen-activated protein kinase activation. Clin Cancer Res. 2003;9:2866–2875.

Tang Y, Liu CM, Ren JQ, Zhang YC, Li SN. Screening of α-glucosidase and xanthine oxidase inhibitors from *Garcinia mangostana* extract using ultrafiltration HPLC-ESI-MS. Lishishen Med Mate Med Res. 2015;26:2322–5.

Tao Y, Cai H, Li WD, Cai BC. Ultrafiltration coupled with high-performance liquid chromatography and quadrupole-time-of-flight mass spectrometry for screening lipase binders from different extracts of Dendrobium officinale. Anal Bioanal Chem. 2015;407:6081–93.

Thomas J, Joy P, Mathew S, Skaria BP. Development of medicinal and aromatic plants-outlook for the next millennium. Indian J Arecanut Species Med Plant. 1999;1(2):38–41.

Uddandrao VS, Brahmanaidu P, Nivedha P, Vadivukkarasi S, Saravanan G. Beneficial role of some natural products to attenuate the diabetic cardiomyopathy through Nrf2 pathway in cell culture and animal models. Cardiovasc Toxicol. 2018;18(3):199–205.

Ujah FO. Medicinal potentials of green tea. In: Egbuna C, Kumar S, Ifemeje JC, Kurhekar JV, editors. Phytochemistry. Volume 2: Pharmacognosy, nanomedicine, and contemporary issues. Toronto: Apple Academic Press; 2019. p. 257–70.

Vambe M, Aremu A, Chukwujekwu J, Finnie J, Van Staden J. Antibacterial screening, synergy studies and phenolic content of seven South African medicinal plants against drug-sensitive and-resistant microbial strains. S Afr J Bot. 2018;114:250–9.

Van de Velde F, Grace MH, Esposito D, Pirovani ME, Lila MA. Quantitative comparison of phytochemical profile, antioxidant, and anti-inflammatory properties of blackberry fruits adapted to Argentina. J Food Compos Anal. 2016;47:82–91.

Vongsak B, Sithisarn P, Mangmool S, Thongpraditchote S, Wongkrajang Y, Gritsanapan W. Maximizing total phenolics, total flavonoids contents and antioxidant activity of *Moringa oleifera* leaf extract by the appropriate extraction method. Ind Crop Prod. 2013;44:566–71.

Vora J, Patel S, Sinha S, Sharma S, Srivastava A, Chhabria M, Shrivastava N. Molecular docking, QSAR and ADMET based mining of natural compounds against prime targets of HIV. J Biomol Struct Dyn. 2018;7:1–16.

Wang L, Weller CL. Recent advances in extraction of nutraceuticals from plants. Trends Food Sci Technol. 2006;17(6):300–12.

Wang J, Liu Y, Li X-H, Zeng X-C, Li J, Zhou J, Xiao B, Hu K. Curcumin protects neuronal cells against status-epilepticus-induced hippocampal damage through induction of autophagy and inhibition of necroptosis. Can J Physiol Pharmacol. 2016;95(5):501–9.

Wu B, Song HP, Zhou X, et al. Screening of minor bioactive compounds from herbal medicines by in silico docking and the trace peak exposure methods. J Chromatogr A. 2016a;1436:91–9.

Wu SQ, Yang H, Li P. Application of the affinity ultrafiltration coupled with LC-MS technology in screening active components of traditional Chinese medicines. Acta Pharm Sin. 2016b;51:1060–7.

Xiao J. Phytochemicals in medicine and food. Phytochem Rev. 2015;14:317–20.

Yang L, Yang C, Li C, Zhao Q, Liu L, Fang X, Chen XY. Recent advances in biosynthesis of bioactive compounds in traditional Chinese medicinal plants. Sci Bull. 2016a;61(1):3–17.

Yang M, Lee G, Si J, Lee SJ, You HJ, Ko G. Curcumin shows antiviral properties against norovirus. Molecules. 2016b;21(10):1401.

Zengin G, Bulut G, Mollica A, Picot-Allain CMN, Mahomoodally MF. In vitro and in silico evaluation of Centaurea saligna (K. Koch) Wagenitz – an endemic folk medicinal plant. Comput Biol Chem. 2018;73:120–6.

Zhang X, Zhu Y, Duan W, Feng C, He X. Allicin induces apoptosis of the MGC-803 human gastric carcinoma cell line through the p38 mitogen-activated protein kinase/caspase-3 signaling pathway. Mol Med Rep. 2015;11(4):2755–60.

Zheng W, Thorne N, McKew JC. Phenotypic screens as a renewed approach for drug discovery. Drug Discov Today. 2013;18:1067–73.

Zhou K, Su L, Yu LL. Phytochemicals and antioxidant properties in wheat bran. J Agric Food Chem. 2004;52(20):6108–14.

Zhu HB, Liu S, Li X, Song FR, Liu ZQ, Liu S. Bioactivity fingerprint analysis of cyclooxygenase-2 ligands from radix Aconiti by ultrafiltration- UPLC-MSn. Anal Bioanal Chem. 2013;405:7437–45.

Chapter 2
Synthetic Analogs of Phytochemicals

Chukwuebuka Egbuna, Shahira M. Ezzat, Habibu Tijjani, and Vijay K. Srivastav

2.1 Introduction

The term '*synthetic analog*' or '*chemical analog*' refers to chemical compound that bear close resemblance to a naturally occurring chemical compound which differ slightly in chemical structure, property, and function. They are sometimes referred to as '*designer drugs*' because they are artificially modified from an existing drug. In the field of pharmacology, an analog can either be a functional analog or a structural analog. A functional analog is one which has similar properties (physical, chemical, biochemical, or pharmacological) as the natural compound, whereas a structural analog is a chemical compound with a slightly altered chemical structure compared to another. An example of pharmacological analogs are morphine, heroin and fentanyl (Fig. 2.1), which have the same mechanism of action as the naturally occurring morphine, but fentanyl is structurally quite different from the other two (Mutschler and Schäfer-Korting 2001).

C. Egbuna (✉)
Department of Biochemistry, Chukwuemeka Odumegwu Ojukwu University, Uli, Anambra, Nigeria
e-mail: egbuna.cg@coou.edu.ng

S. M. Ezzat
Department of Pharmacognosy, Faculty of Pharmacy, Cairo University, Cairo, Egypt

Department of Pharmacognosy, Faculty of Pharmacy, October University for Modern Sciences and Arts (MSA), 6th of October, Egypt

H. Tijjani
Natural Product Research Laboratory, Department of Biochemistry, Bauchi State University, Gadau, Bauchi State, Nigeria

V. K. Srivastav
Department of Pharmacy, Shri Govindram Seksaria Institute of Technology and Science, Indore, Madhya Pradesh, India

S. Kumar, C. Egbuna (eds.), *Phytochemistry: An in-silico and in-vitro Update*,
https://doi.org/10.1007/978-981-13-6920-9_2

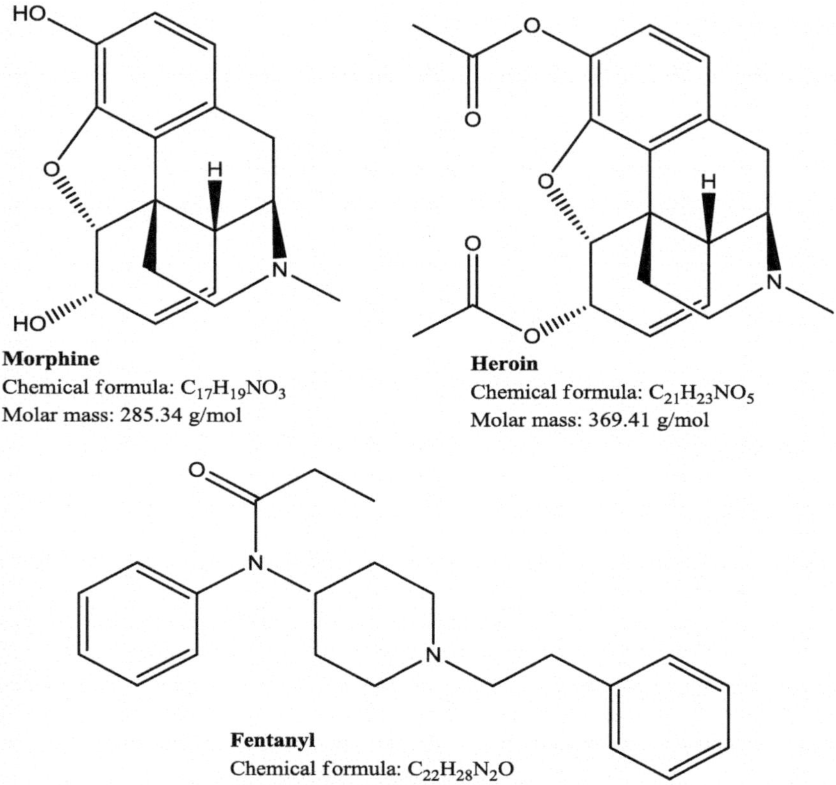

Morphine
Chemical formula: $C_{17}H_{19}NO_3$
Molar mass: 285.34 g/mol

Heroin
Chemical formula: $C_{21}H_{23}NO_5$
Molar mass: 369.41 g/mol

Fentanyl
Chemical formula: $C_{22}H_{28}N_2O$
Molar mass: 336.471 g/mol

Fig. 2.1 Functional analogs: morphine, heroin and fentanyl

Biosynthetically, plant natural products are categorized into numerous broad classes including alkaloids (compounds containing at least one basic nitrogen – typically present in a heterocyclic ring; more than 12,000 exists), polypeptides (compounds made up of numerous amino acid residues), terpenes or terpenoids (compounds containing a hydrocarbon skeleton formed from isoprene units), and polyketides (compounds made via sequential condensations of acetyl coenzyme A and resulting in an alternating pattern of carbonyl and methylene derived groups). Accordingly, plant secondary metabolites may serve as drug precursors or drug prototypes (Table 2.1). Recently, there is much developments in drug discovery from plants, including clinical trials for approving the use of such drugs and compounds as therapeutic remedies (Salim et al. 2008). In parts, this is due to the advent of routine High Throughput Screening (HTS) which has facilitated drug discovery pre-process.

Over the years, plant-derived natural compounds have been used as therapeutic agents, either by itself or through the derivation of more active analog which is an important process in drug discovery and development. The rich and complex

Table 2.1 Some important class of secondary metabolites

Class	Type	Examples
Alkaloids	Nitrogen-containing	Atropine, Berberine, Caffeine, Cocaine, Ephedrine, Galantamine, Morphine, Nicotine, Quinidine, Quinine, Reserpine, Vincamine, Vincristine, etc.
Non-protein amino acids (NPAAs)	Nitrogen-containing	Azatyrosine, Canavanine
Cyanogenic glycosides	Nitrogen-containing	Amygdalin, Dhurrin, Linamarin, Lotaustralin, Prunasin
Lectins, peptides and polypeptides	Nitrogen-containing	Concanavalin A
Terpenes	Without nitrogen. Contains isoprene units	Taxol, Azadirachtin, Artemisinin, Tetrahydro-cannabinol, Farnesol, Digitogenin, Carotene
Steroids and saponins	Without nitrogen	Cycloartenol
Flavonoids and tannins	Without nitrogen	Luteolin (a flavone), tannic acid
Phenylpropanoids, lignin, coumarins and lignans	Without nitrogen	Resveratrol

structural components of plants natural products (NPs) have proven invaluable since antiquity. The need for mass production of NPs to meet commercial demand has led scientists to consider total synthesis (a complete chemical synthesis of NPs, from simple, commercially available precursors without the aid of biological processes) and semi-synthesis of natural compounds (a partial chemical synthesis that involves the use of chemical compounds isolated from natural sources as the starting material for the production of other novel compounds which may have a distinct chemical or medicinal properties). Whichever, the goal is to arrive at a target molecule with low molecular weight, inexpensive starting materials for large scale production yet inexpensive and one which exhibits desirable curative potentials with fewer side effects.

Again, the resistance of NPs and the need for more potent drugs has inspired the synthesis of various functional analogs to a whole new level. The total synthesis of these natural products have overtime been a challenging goal for many chemists. Hence, the major breakthrough in medicinal chemistry is the ability of medicinal chemists to alter or modify either one or more atom, a functional group, or substructure to give structurally related compounds with more potent activities yet with lesser side effects than the original drug. Inspiringly, this has led to the search for an innovative and verifiable chemical methodologies based on insights from the study of the NPs biosynthetic pathways, resulting in the modifications of NPs and corresponding increase in the libraries of functional and structural analogs, which has helped to establish mechanisms of action and structure-activity relationships (SARs). One recent breakthrough is the semisynthesis of the anti-cancer agent paclitaxel from 10-deacetylbaccatin isolated from the needles of *Taxus baccata* (European yew).

2.2 Natural Products as Drug Precursors

Many small molecules of natural products can work as drug precursors, which can be converted by chemical modification or fermentation methods into the compound of interest. The synthesis of compounds with complex structures that have many chiral centers, would not be economically feasible, so to overcome problems such as the low yield of natural compounds and the high cost of chemical synthesis, the semisynthetic approach is the most appropriate alternative. In this section, we enumerated some plant secondary metabolites which are used as drug precursors, although they are not necessarily pharmacologically active in their original naturally occurring forms. The various plant derived drugs (approx. 150) were presented in Table 2.2.

A survey of plant-derived chemical compounds utilized as therapeutic agents in countries having WHO Traditional Medicine Centers indicated that, 80% of the 122 identified compounds are used for close ethnomedical purposes and are derived from only 94 plant species (Farnsworth et al. 1985). Notable examples are are khellin (**1**), isolated from *Ammi visnaga* (L) Lamk, from which chromolyn (**2**) was derived, which is used as bronchodilator in the form of sodium chromoglycate (Fabricant and Farnsworth 2001). Papaverine (**3**) from *Papaver somniferum* formed the basis for synthesis of verapamil (**4**), the antihypertensive drug (Fabricant and Farnsworth 2001).

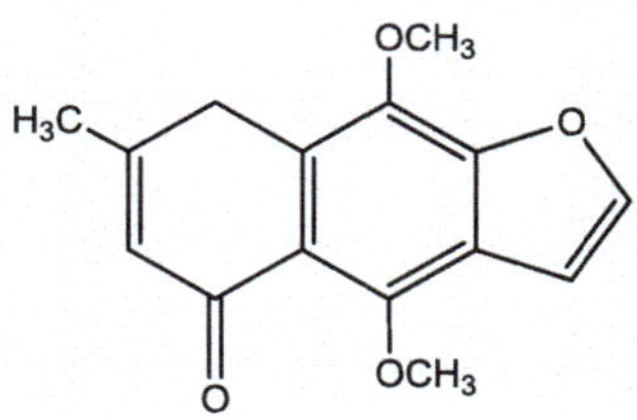

1

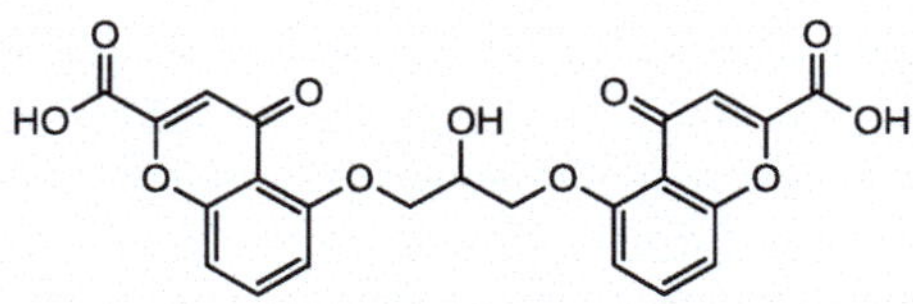

2

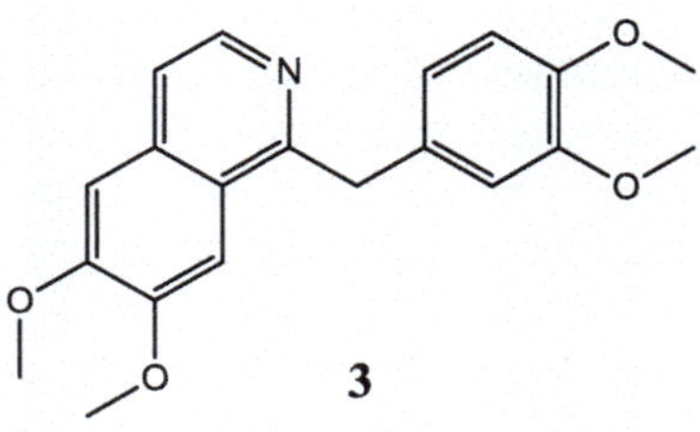

3

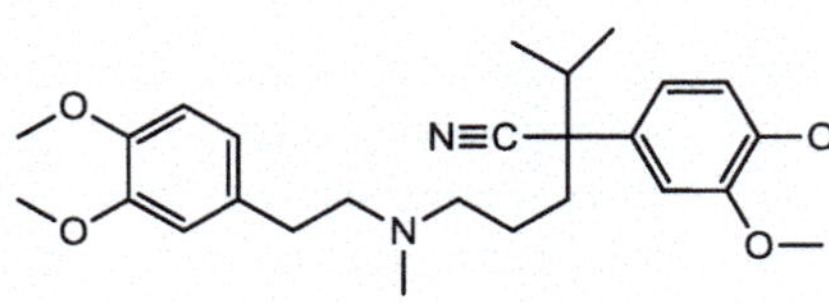

4

Papaver somniferum is also the source of painkillers such as morphine and codeine (Buss and Waigh 1995), in addition to the two alkaloids, thebaine and oripavine, which are isolated from the poppy capsule, and serve as intermediates for the biosynthesis of codeine and morphine and can be subjected to chemical synthesis that produces unnatural opiate drugs, such as nalbuphine, naloxone and buprenorphine (Fig. 2.2) (Fist et al. 2002; Machara et al. 2012; Rinner and Hudlicky

Table 2.2 Summary of plant natural compounds, medicinal uses, sources and analogs

S/ no.	Plant natural compounds (class)	Medicinal use	Plant sources	Analogs/derivatives	References
1.	Adoniside (cardenolides: cardiac glycoside)	Cardiotonic	*Adonis vernalis* L.	–	Maridass and John (2008)
2.	Aescin (mixture of saponin)	Anti-inflammatory, vasoconstrictor and vasoprotective effects. Used for short-term treatment of chronic venous insufficiency	*Aesculus hippocastanum* L.	–	Pittler and Ernst (2012)
3.	Aesculetin (coumarin)	Antidysentery	*Fraxinus rhynchophylla*	Scopoletin, isoscopoletin	Maridass and John (2008)
4.	Agrimophol	Antihelmintic	*Agrimonia eupatoria* L.	–	Maridass and John (2008)
5.	Ajmalicine (alkaloid)	Antihypertensive drug used in the treatment of high blood pressure	*Rauvolfia serpentina* L. Benth ex. Kurz	Yohimbine, rauwolscine, and other yohimban derivatives	Wink and Roberts (1998)
6.	Allyl isothiocyanate (organosulfur compound)	Rubefacient, speculated to possess some anticancer activities	*Brassica nigra* L. Koch	–	Taylor (2000) and Zhang (2010)
7.	Allantoin, first isolated in 1800 by the Italian physician Michele Francesco Buniva	Vulnerary	Several plants	–	Taylor (2000)
8.	Anabasine (pyridine and piperidine alkaloid)	Skeletal muscle relaxant	*Anabasis aphylla* L., Tree Tobacco (*Nicotiana glauca*)	Structural isomer of nicotine	Maridass and John (2008)
9.	Andrographolide	Bacillary dysentery	*Andrographis paniculata*	–	Taylor (2000)
10.	Anisodamine	Anticholinergic	*Anisodus tanguticus*	–	Maridass and John (2008)
11.	Anisodine	Anticholinergic	*Anisodus tanguticus*	–	Maridass and John (2008)

(continued)

Table 2.2 (continued)

S/ no.	Plant natural compounds (class)	Medicinal use	Plant sources	Analogs/derivatives	References
12.	Arecoline	Anthelmintic	*Areca catechu* L.	–	Taylor (2000)
13.	Artemisinin (terpenoids). Isolated 1972 by Tu Youyou, a Chinese scientist	Active against *Plasmodium falciparum* malaria	*Artemisia annua* (Chinese wormwood)	Artesunate, Artemether, Dihydroartemisinin, Artelinic acid, Artemotil, Arterolane	Dondorp et al. (2005)
14.	Asiaticoside	Vulnerary	*Centella asiatica* L.	–	Taylor (2000)
15.	Atropine. First isolated in 1833	Anticholinergic, used to treat certain types of nerve agent and pesticide poisonings	Nightshade plant family including deadly nightshade (*Atropa belladonna* L.), Jimson weed, and mandrake	Homatropine, atropine methonitrate, amprotropine, adiphenine, cyclopentolate, droperidol, ipratropium bromide, loperamide, methadone, pethidine	Ainsworth (2014)
16.	Benzyl benzoate	Scabicide	Several plants such as tuberose, hyacinth, Balsam of Peru and Tolu balsam	–	Taylor (2000)
17.	Berbamine	Treatment of chronic myeloid leukemia	*Berberis amurensis*	–	Nirmala et al. (2011)
18.	Berberine	Bacillary dysentery	*Berberis vulgaris* L.	–	Taylor (2000)
19.	Bergenin	Antitussive	*Ardisia japonica* Bl.	–	Taylor (2000)
20.	Beta-lapachone	Broad spectrum of antineoplastic activity against breast cancer, prostate cancer, lung cancer, pancreatic cancer and also in promyelocytic leukemic cells	*Tabebuia avellanedae*	–	Li et al. (2000)
21.	Betulinic acid (pentacyclic triterpenoid)	Anticancer, antibacterial, antimalarial, anti-HIV, anthelminthic, anti-inflammatory, and antioxidant properties	*Betula alba*	Bevirimat	Chowdhury et al. (2002)

22.	Borneol	Antipyretic, analgesic, anti-inflammatory	Several plants	–	Taylor (2000)
23.	Bromelain	Anti-inflammatory; proteolytic agent	*Ananas comosus* L.	–	Taylor (2000)
24.	Bruceatin (triterpene quassinoid)	Antineoplastic antibiotic, anti-tumor	*Brucea antidysenterica*	–	Nirmala et al. (2011)
25.	Caffeine (alkaloid), methylxanthine class. First isolated in 1821	Central nervous system (CNS) stimulant. A psychoactive drug	Found in over 60 plant species such as in coffee plants (*Coffea arabica* and *Coffea canephora*), Kola nut, *Camellia sinensis* L. etc.	Xanthines DMPX, 8-chlorotheophylline, theophylline, theobromine, dipropylcyclopentylxanthine; 8-Cyclopentyl-1,3-dimethylxanthine; 8-Phenyltheophylline	Nehlig et al. (1992)
26.	(+)-Catechin	Haemostatic	*Potentilla fragaroides* L.	–	Taylor (2000)
27.	Chymopapain	Proteolytic; mucolytic	*Carica papaya* L.	–	Taylor (2000)
28.	Cissampeline	Skeletal muscle relaxant	*Cissampelos pareira* L.	–	Taylor (2000)
29.	Camphor	Rubefacient	*Cinnamonum camphora* L	–	Taylor (2000)
30.	Camptothecin, CPT. Discovered in 1966 by M. E. Wall and M. C. Wani	Topoisomerase inhibitor	Bark and stem *of Camptotheca acuminate*	Topotecan, irinotecan	Samuelsson (2004)
31.	Capsaicin	An analgesic in topical ointments, nasal sprays (Sinol-M), and dermal patches to relieve pain, typically in concentrations between 0.025% and 0.1%. Agent used in making personal defense pepper spray	Chili peppers. Capsicum, members of the nightshade family, Solanaceae	Several capsaicinoids exists: capsaicin (69%), dihydrocapsaicin (22%), nordihydrocapsaicin (7%), homocapsaicin (1%), and homodihydrocapsaicin (1%)	Fattori et al. (2016)
32.	Castanospermine (indolizidine alkaloid)	Potent inhibitor of some glucosidase enzymes, antidiabetic, antiviral	*Castanospermum australe*	Celgosivir	Saul et al. (1985)

(continued)

Table 2.2 (continued)

S/ no.	Plant natural compounds (class)	Medicinal use	Plant sources	Analogs/derivatives	References
33.	Cocaine	Local anaesthetic	*Erythroxylum coca* Lam.	Alpha-Eucaine®, beta-Eucaine®, Othococaine, niravanin, amylocaine, procaine, procainamide and analogs (sulpiride, metoclopramide)	Salim et al. (2008)
34.	Codeine (alkaloid), first isolated in 1832 in France by Pierre Robiquet	Used to treat pain, as a cough medicine, and for diarrhea. Also sold in combination of other drugs such as paracetamol, diclofenac, phenacetin, caffeine	*Papaver somniferum* (opium poppy)	Nicocodeine, isocodeine and its derivatives	Prommer (2010)
35.	Colchicine	Antitumor agent; antigout; anti-inflammatory	*Colchicum autumnale* L.	–	Taylor (2000)
36.	Combretastatin (natural phenols)	Combretastatin family possess varying ability to cause vascular disruption in tumors	Bark of *Combretum caffrum* (South African Bush Willow)	Combretastatin A4 phosphate, ombrabulin	Pettit et al. (1987)
37.	Convallotoxin	Cardiotonic	*Convallaria majalis* L.	–	Taylor (2000)
38.	Curcuminoids (inear diarylheptanoid)	Antioxidant, choleretic	*Curcuma longa*	Curcumin, demethoxycurcumin, bisdemethoxycurcumin,	Jayaprakasha et al. (2006)
39.	Cynarin	Choleretic	*Cynara scolymus* L.	–	Taylor (2000)
40.	Cytisine (alkaloid)	Smoking cessation	*Cytisus laburnum* L.	Varenicline	Walker et al. (2014)
41.	Danthron	Laxative	*Cassia* spp.	–	Taylor (2000)
42.	Daphnoretin	Protein kinase C activator	*Wikstroemia indica*	–	Nirmala et al. (2011)

43.	Deserpidine	Antihypertensive; tranquilizer	*Rauvolfia canescens* L.	–	Taylor (2000)
44.	Demecolcine	Antitumor agent	*Colchicum autumnale* L.	–	Taylor (2000)
45.	Diosgenin (saponins)	Prodrug to progesterone. Used in early combined oral contraceptive pills	*Costus speciosus, Smilax menispermoidea*	–	GRIN (2008)
46.	Deslanoside	Cardiotonic	*Digitalis lanata* Ehrh.	–	Taylor (2000)
47.	Dicoumarol or dicumarol (derivative of coumarin). Isolated by Karl Link's in the 1920s	Used along with heparin, for the treatment of deep venous thrombosis	Found in decaying sweet-clover bcause of the conversion of coumarin to dicoumarol by Fungi (including *Penicillium, Aspergillus, Fusarium*, and *Muco*	A prototype to 4-hydroxycoumarin, warfarin, phenindione, ethyl biscoumacete, tesicam, piroxicam	Kresge et al. (2005)
48.	Digitalin	Cardiotonic	*Digitalis purpurea* L.	–	Taylor (2000)
49.	Digitoxin (cardiac glycoside). First description of the use of foxglove dates back to 1775	Cardiotonic. Anti-cancer but restricted by its narrow therapeutic index	*Digitalis purpurea* L.	Glycorandomization, digitoxigenin,	Sneader (2005) and Elbaz et al. (2012)
50.	Digoxin	Cardiotonic	*Digitalis lanata* Ehrh.	Acetyldigoxin	Taylor (2000)
51.	Ellipticine	Anti-tumor	*Ochrosia* spp.	–	Nirmala et al. (2011)
52.	Emetine	Anti-protozoal and to induce vomiting (emetic)	*Cephaelis ipecacuanha* (ipecac root)	Cephaeline, dehydroemetine	Taylor (2000)
53.	Emodin	Apoptosis of cancer cells by several pathways (lung, liver, ovarian and blood cancer)	Rhizome of rhubarb	–	Huang et al. (2009)
54.	Enoxolone (Pentacyclic triterpenoid)	Treatment of peptic, esophageal and oral ulceration and inflammation. In trial for its use for verbal fluency	Root of (*Glycyrrhiza glabra*) licorice plant	Carbenoxalone sodium, CBX	Connors (2012)
55.	Ephedrine (alkaloid), a sympathomimetic amine and substituted amphetamine. First isolated in 1885	Stimulant used to prevent low blood pressure during spinal anesthesia	*Ephedra sinica*	Amphetamine, methylphenidate, dexamphetamine	Bagchi and Preuss (2013)

(continued)

Table 2.2 (continued)

S/ no.	Plant natural compounds (class)	Medicinal use	Plant sources	Analogs/derivatives	References
56.	Etoposide	Antitumour agent	*Podophyllum peltatum* L.	–	Taylor (2000)
57.	Flavocoxid	Medical food therapy for osteoarthritis	*Acacia catechu*	–	Taylor (2000)
58.	Flavopiridol	Antocancer	Stems of *Amoorarohituka*	–	Nirmala et al. (2011)
59.	Galantamine	Treatment of patients with early-onset Alzheimer's	*Galanthus woronowii* Losinsk (snowdrop)	–	Taylor (2000)
60.	Genistein (isoflavone)	Antioxidant, anti-atherosclerosis, anticancer,	*Glycine max* Merr. (soybean)	Phenoxodiol	Si et al. (2007)
61.	Gitalin	Cardiotonic	*Digitalis purpurea* L.	–	Taylor (2000)
62.	Galanthamine	Treatment of cognitive decline in mild to moderate Alzheimer's disease and various other memory impairments	*Galanthus nivalis, Lycoris squamigera* Maxim	–	Birks (2006)
63.	Glaucaroubin	Amoebicide	*Simarouba glauc*	–	Taylor (2000)
64.	Glaucine	Antitussive	*Glaucium flavum* Crantz	–	Taylor (2000)
65.	Glaziovine	Antidepressant	*Ocotea glazovii* Mez	–	Taylor (2000)
66.	Glycyrrhizin	Sweetener	*Glycyrrhiza glabra* L.	–	Taylor (2000)
67.	Gossypol	Male contraceptive	*Gossypium* spp.	–	Taylor (2000)
68.	Guanidine, first isolated by Adolph Strecker via the degradation of guanine	Guanidinium hydroxide is the active ingredient in some non-lye hair relaxers	*Galega officinalis* L	Derivatives: arginine, triazabic clodecene, saxitoxin, and creatine. Metformin (dimethylbiguanide),	Güthner et al. (2005)
69.	Hemsleyadin	Bacillary dysentery	*Helmsleya amabilis* Diels	–	Taylor (2000)
70.	Hesperidin	Capillary fragility	*Citrus* spp.	–	Taylor (2000)

71.	Homoharringtonine	Treatment of chronic myeloid leukaemia	*Cephalotaxus harringtonii*	–	Sweetman (2012)
72.	Hydrastine	Hemostatic; astringent	*Hydrastis canadensis* L.	–	Taylor (2000)
73.	Hyoscamine	Anticholinergic	*Hyoscamus niger* L.	–	Taylor (2000)
74.	Huperzine A (sesquiterpene alkaloid)	Treatment of neurological conditions such as Alzheimer's disease	*Huperzia serrate*	–	Zangara (2003)
75.	Indirubin	Treatment of chronic myeloid leukemia	Indigo naturalis	Meisoindigo	Nirmala et al. (2011)
76.	Ingenol-3-angelate	Topical treatment of actinic keratosis, inducer of cell death and possesses prospects for reacting latent HIV	*Euphorbia peplus*	–	Jiang et al. (2015)
78.	Kainic acid	Ascaricide	*Digenea simplex*	–	Taylor (2000)
79.	Kawain	Tranquilizer	*Piper methysicum*	–	Taylor (2000)
80.	Khellin. Found in Egypt	For treating renal colic, kidney stones, coronary disease, bronchial asthma, vitiligo, and psoriasis	*Ammi visnaga*	Benzarone, benziodarone, amiodarone, cromolyn sodium, nedocromil	Shinde and Laddha (2014)
81.	Lanatosides A, B, C	Cardiotonic	*Digitalis lanata*	–	Taylor (2000)
82.	L-Dopa	Precursor for monoamine or catecholamine neurotransmitters dopamine, norepinephrine (noradrenaline), and epinephrine (adrenaline)	*Mucuna pruriens* (velvet bean), *Vicia faba* (broad bean)	Methyldopa, droxidopa	Ingle (2003)
83.	Lobeline (alkaloid)	Used as a smoking cessation aid. Respiratory stimulant	*Lobelia* spp.	Lobelanidine, Lobelanine (precursor of lobeline)	Stead et al. (2012)
84.	Monocrotaline	Antitumor agent	*Crotolaria sessiliflora* L.	–	Taylor (2000)

(continued)

Table 2.2 (continued)

S/ no.	Plant natural compounds (class)	Medicinal use	Plant sources	Analogs/derivatives	References
85.	Morphine (benzylisoquinoline Alkaloid) first isolated in 1803 and 1805 by Friedrich Sertürner	Used to treat both acute and chronic severe pain (analgesic)	*Papaver somniferum* (opium poppy) and other plants	More than 250 derivatives exists. Some are heroin, hydromorphone, oxymorphone, diprenorphine, nalorphine including the derivatives of thebaine	Courtwright (2009)
86.	Methyl salicylate	Rubefacient	*Gautheria procumbens* L.	–	Taylor (2000)
87.	Menthol	Rubefacient, antipruritic, used to relieve minor throat irritation, increase addictive potentials of tobacco	*Mentha* spp.	Has 8 possible structural isomers	Biswas et al. (2016)
88.	Neoandrographolide	Bacillary dysentery	*Andrographis paniculata* Nees	–	Taylor (2000)
89.	Nitisinone	Treatment of hereditary tyrosinemia type 1 (HT-1)	*Callistemon citrinus*	–	NORD (2014)
90.	Nicotine (alkaloid)	Potent parasympathomimetic stimulant and very addictive	*Nicotiana tabacum* L. and the nightshade family	–	Malenka et al. (2009)
91.	Noscapine	Antitussive	*Papaver somniferum* L.	–	Taylor (2000)
92.	Nordihydroguaiaretic acid	Antioxidant	*Larrea divaricata* Cav.	–	Taylor (2000)
93.	Ouabain	Cardiotonic	*Strophanthus gratus*	–	Taylor (2000)
94.	Paclitaxel/taxol (terpenoids). First isolated in 1971	Anticancer	*Taxus brevifolia* (the Pacific yew)	Docetaxel, 10-deacetylbaccatin, baccatin III	Fischer and Ganellin (2006)
95.	Pachycarpine	Oxytocic	*Sophora pachycarpa* Schrenk ex C.A. Meyer	–	Taylor (2000)
96.	Palmatine	Antipyretic, detoxicant	*Coptis japonica* Makino	–	Taylor (2000)
97.	Papain	Proteolytic; mucolytic	*Carica papaya* L.	–	Taylor (2000)

98.	Papaverine. Discovered in 1848 by Georg Merck	Antispasmodic, treatment of erectile dysfunction, acute mesenteric ischemia	*Papaver somniferum* L.	Eupaverin®, ethaverine, mebeverine, verapamil	William (1997)
99.	Phyllodulcin	Sweetener	*Hydrangea macrophylla*	–	Taylor (2000)
100.	Physostigmine (parasympathomimetic alkaloid). Discovered by Sir Robert Christison in in 1846	A reversible cholinesterase inhibitor. Used to treat glaucoma and delayed gastric emptying. Also an antidote for *Atropa belladonna* poisoning	*Physostigma venenosum* (Calabar bean)	Miotine, neostigmine, pyridostigmine	Roberts and Wink (1998)
101.	Phorbol	Promote apoptosis in a range of cancer cell	Croton oil from the seeds of *Croton tiglium* L.	–	Patries et al. (2009)
102.	Picrotoxin	Analeptic	*Anamirta cocculus* L.	–	Taylor (2000)
103.	Pilocarpine	Parasympathomimetic), and physostigmine (cholinesterase inhibitor)	*Pilocarpus jaborandi*	–	Salim et al. (2008)
104.	Pinitol	Expectorant	Several plants	–	Taylor (2000)
105.	Podophyllotoxin, PPT, also known as podofilox (a non-alkaloid toxin lignin). Isolated in 1880	Used to treat genital warts and *molluscum contagiosum.* Topical anti-viral agent	Rhizome of *Podophyllum peltatum* (American Mayapple)	Podophylotoxin benzylidene glucoside, teniposide, etoposide	Canel et al. (2000)
106.	Pomiferin	Anticancer	*Maclura pomifera* (Moraceae); *Dereeis Malaccensis*	–	Greenwell and Rahman (2015)
107.	Protopanaxadiol	Anticancer	Saponins of *Panax ginseng*	Protopanaxatriol, Panaxatriol	Nirmala et al. (2011)
108.	Protoveratrines A and B	Antihypertensive	*Veratrum album* L.	–	Taylor (2000)
109.	Pseudoephedrine	Sympathomimetic	*Ephedra sinica* Stapf.	–	Taylor (2000)
110.	Pseudoephedrine, nor-	Sympathomimetic	*Ephedra sinica* Stapf.	–	Taylor (2000)
111.	Quisqualic acid	Anthelmintic	*Quisqualis indica* L.	–	Taylor (2000)

(continued)

Table 2.2 (continued)

S/ no.	Plant natural compounds (class)	Medicinal use	Plant sources	Analogs/derivatives	References
112.	Quinine (aminoquinoline alkaloid). First isolated in 1820 by Pelletier and Caventou	Used to treat malaria and babesiosis. No longer available as a first-line treatment for malaria, recommended only when artemisinins are not available	Bark of *Cinchona* spp. tree	Chloroquine, primaquine, mefloquine, quinine bisulfate, quinine sulfate dehydrate, quinine hydrochloride, quinine dihydrochloride, quinine gluconate	Dorndorp et al. (2005) and Achan et al. (2009)
113.	Rescinnamine	Antihypertensive; tranquilizer	*Rauvolfia serpentina* L.	–	Taylor (2000)
114.	Reserpine	Antihypertensive; tranquilizer	*Rauvolfia serpentina* L.	–	Taylor (2000)
115.	Rhomitoxine	Antihypertensive	*Rhododendron molle*	–	Taylor (2000)
116.	Rorifone	Antitussive	*Rorippa indica* L.	–	Taylor (2000)
117.	Rotenone	Piscicide	*Lonchocarpus nicou*	–	Taylor (2000)
118.	Rotundine	Analgesic; sedative	*Stephania sinica* Diels	–	Taylor (2000)
119.	Rutin	Capillary fragility	*Citrus* spp.	–	Taylor (2000)
120.	Salicin	Analgesic	*Salix alba* L.	–	Taylor (2000)
121.	Salvinorin A	Used as an entheogen. Considered a dissociative	*Salvia divinorum* (Epling and Játiva)	Salvinorin B, C, D, E, F, 22-thiocyanato-salvinorin A, 2-ethoxymethyl salvinorin B	MacLean et al. (2013)
122.	Sanguinarine	Dental plaque inhibitor	*Sanguinaria canadensis* L.	–	Taylor (2000)
123.	Santonin	Ascaricide	*Artemisia maritima* L.	–	Taylor (2000)
124.	Scillarin A	Cardiotonic	*Urginea maritima* L.	–	Taylor (2000)
125.	Scopolamine	Sedative	*Datura metel* L.	–	Salim et al. (2008)
126.	Sennosides A and B	Laxative	*Cassia* spp.	–	Taylor (2000)
127.	Schischkinnin (indole alkaloid)	Antitumor	*Centaurea schischkinii*	–	Nirmala et al. (2011)

128.	Shikimic acid	Starting material for synthesis of Oseltamivir phosphate used for treatment and prophylaxis of influenza viruses A and B	*Illicium verum* Hook.f. (shikimi tree)	–	Taylor (2000)
129.	Silymarin	Antihepatotoxic	*Silybum marianum* L.	–	Taylor (2000)
130.	Sparteine	Oxytocic	*Cytisus scoparius* L.	–	Taylor (2000)
131.	Stevioside	Sweetener	*Stevia rebaudiana* Bertoni	–	Taylor (2000)
132.	Strychnine	CNS stimulant	*Strychnos nux-vomica* L.	–	Taylor (2000)
133.	Swainsonine (indolizidine alkaloid)	Potent inhibitor of Golgi alpha-mannosidase II, phytotoxin: cause of economic losses in livestock	Locoweed (also crazyweed and loco)	–	Stegelmeier et al. (1995)
134.	Teniposide	Antitumor agent	*Podophyllum peltatum* L.	–	Taylor (2000)
135.	Thebaine (paramorphine), an alkaloid	Not directly used for therapeutic purpose but rather a starting material for the production of morphine and derivatives, codeine and many other vital compounds	*Papaver bracteatum* (Iranian poppy)	Oxycodone, oxymorphone, nalbuphine, naloxone, naltrexone, buprenorphine, etorphine, and utorphanol	Novak et al. (2000)
136.	Tetrahydropalmatine THP, (isoquinoline alkaloid)	Analgesic, treatment of heart disease and liver damage	*Corydalis* spp. (Yan Hu Suo), *Stephania rotunda*	1-Stephdidine (SPD), tetrahydroberberine (THB), THPB-18	Hu and Jin (1999)
137.	Tetrandrine	Antihypertensive	*Stephania tetrandra* S. Moore	–	Taylor (2000)
138.	Theobromine	Diuretic, bronchodilator	*Theobroma cacao* L.	–	Taylor (2000)
139.	Theophylline	Diuretic; bronchodilator	*Camellia sinensis* L.	–	Taylor (2000)
140.	Thymol	Antifungal (topical)	*Thymus vulgaris* L.	–	Taylor (2000)

(continued)

Table 2.2 (continued)

S/ no.	Plant natural compounds (class)	Medicinal use	Plant sources	Analogs/derivatives	References
141.	Δ9-trans-tetrahydrocannabinol	THC is the principal psycho-active constituent of cannabis. Appetite stimulant for people with AIDS and antiemetic for people receiving chemotherapy	*Cannabis sativa* L.	1 out of 113 cannabinoids	Salim et al. (2008)
142.	Trichosanthin	Abortifacient	*Thymus vulgaris* L.	–	Taylor (2000)
143.	Tubocurarine	Skeletal muscle relaxant	*Chondodendron tomentosum* (a climbing vine)	–	Maridass and John (2008) and Egbuna and Ifemeje (2015)
144.	Vasicine (peganine)	Cerebral stimulant	*Adhatoda vasica Nees*	–	Maridass and John (2008)
145.	Valepotriates	Sedatifve	*Valeriana officinalis* L.	–	Maridass and John (2008)
146.	Vincamine	Cerebral stimulant	*Vinca minor* L.	–	Maridass and John (2008)
147.	Vincristine	Antitumor, antileukemic agent	*Catharanthus roseus* L.	–	Taylor (2000)
148.	Xanthotoxin	Leukoderma, vitiligo	*Ammi majus* L.	–	Taylor (2000)
149.	Yohimbine	Aphrodisiac	*Pausinystalia yohimbe*	–	Taylor (2000)
150.	Yuanhuacine	Abortifacient	*Daphne genkwa*	–	Taylor (2000)

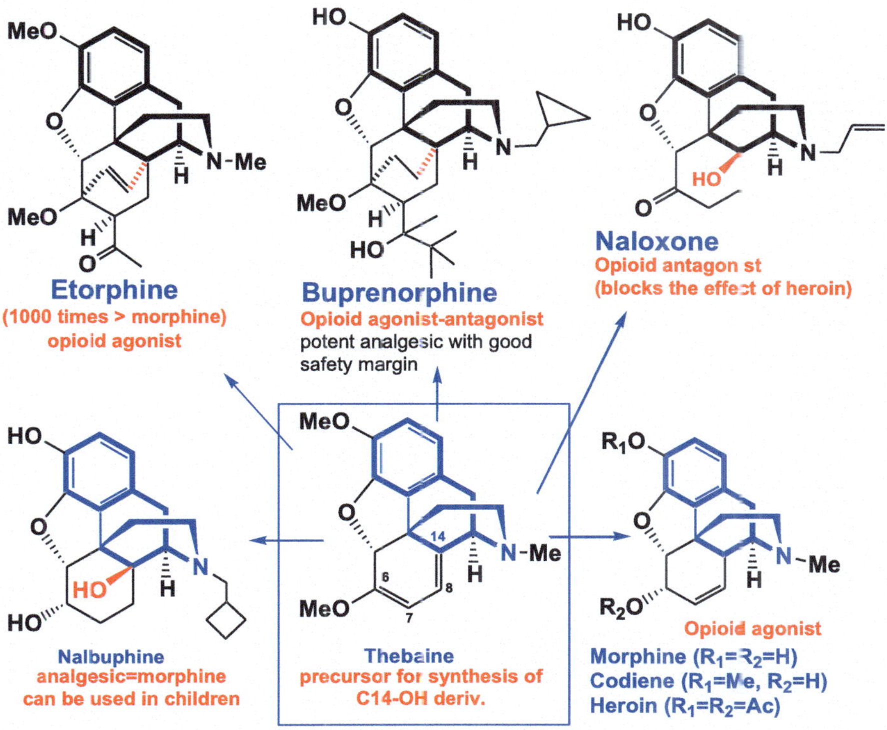

Fig. 2.2 Opiates synthesized from the baine alkaloid

2012). These derivatives possess substantial different pharmacological profiles: some are potent analgesics or non-addictive pain-killers, and others even serve as therapeutic agents for addiction and/or overdose to other opiates.

Caventou and Pelletier isolated the antimalarial alkaloid, quinine (**5**), from the barks of different Cinchona species in 1820 (Buss and Waigh 1995). Quinine became lead for the synthesis of antimalarial drugs such as chloroquine (**6**) and mefloquine (**7**) which replaced quinine in the mid-twentieth century, but due to the resistance developed for both drugs in many tropical regions, another plant was introduced to Traditional Chinese Medicine (TCM) for the treatment of fevers, this plant was *Artemisia annua* (Quinhaosu) that yielded a new natural product lead compound, known as artemisinin (**8**) that was mainly the antimalarial principle (Wongsrichanalai et al. 2002). Semisynthetic artemisinin analogs are now used for the treatment of malaria in many countries (O'Neill and Posner 2004). The totally synthetic analogue OZ277 (**9**) (Vennerstrom et al. 2004), and the dimeric analogue

(**10**) are the two most promising analogs. Single doses of these analogs were reported to cure malaria-infected mice, while artemisinin was much less effective (Posner et al. 2007). It is also relevant that artemisinin and related compounds also have significant anti-tumor activity in-vitro and have shown promise. As of the year 2018, human clinical trials and other clinical research had been conducted in lung, colorectal, breast, and cervical cancers, but the studies are still too small to generalize (Raffetin et al. 2018).

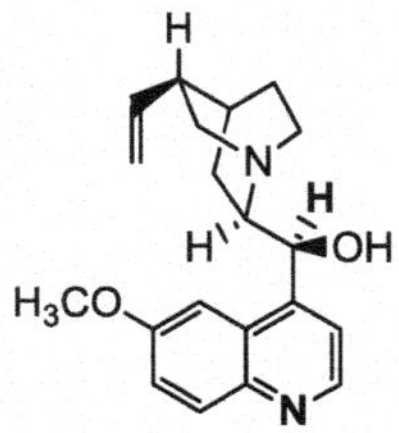

5

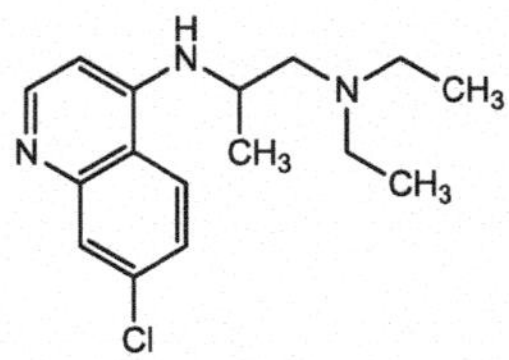

6

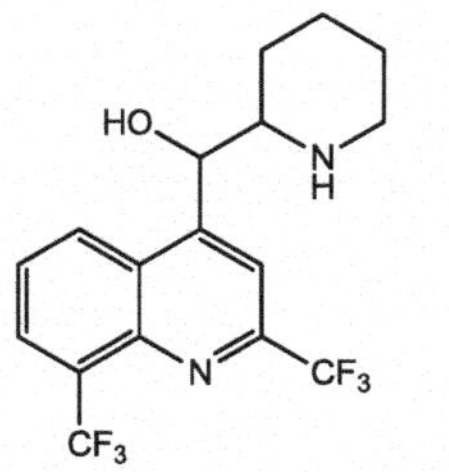

7

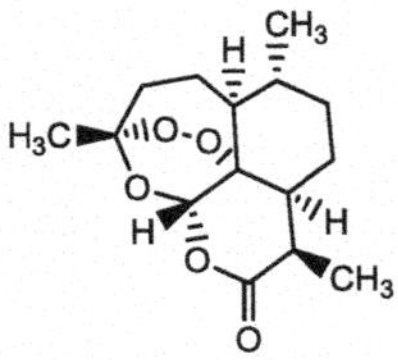

8

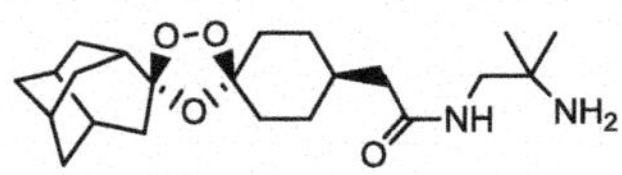

9

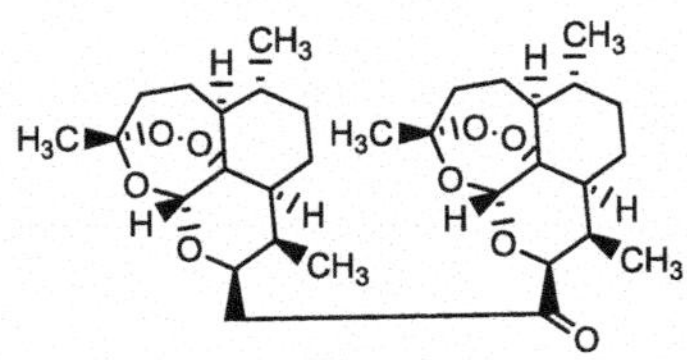

10

The anticancer drug paclitaxel (Taxol) (**11**) can be obtained by isolation from the barks of the slow-growing Pacific yew tree, *Taxus brevifolia* Nutt., but this method is neither economically feasible nor sufficient to produce the amounts that can meet the market demand (the bark yields only 0.014% w/w of pure taxol) (Kingston 2000). The total synthesis of paclitaxol has proven to be an ineffective method to produce sufficient quantities of paclitaxel. In addition, it was found too expensive to be applied in industry especially because it involves several steps for its synthesis as it has 11 stereocenters (Holton et al. 1995a, b; Nicolau and Guy 1995). Fortunately, 10-deacetylbaccatin III (**12**) which is isolated from *Taxus baccata* L. in relatively large amounts, can be converted chemically into paclitaxel by adopting semisynthetic method developed by Holton et al. (1995a, b). Bristol-Myers Squibb used this method for production of its needs of paclitaxel during 1993–2002 (Holton et al. 1995a, b; Kingston 2006). After that, paclitaxel was produced using a plant cell culture method.

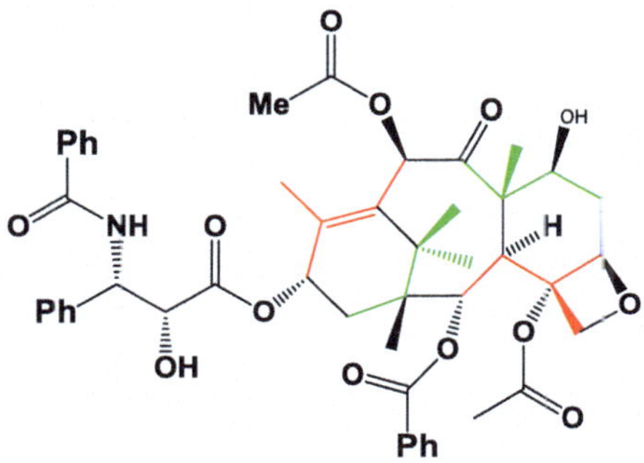

11

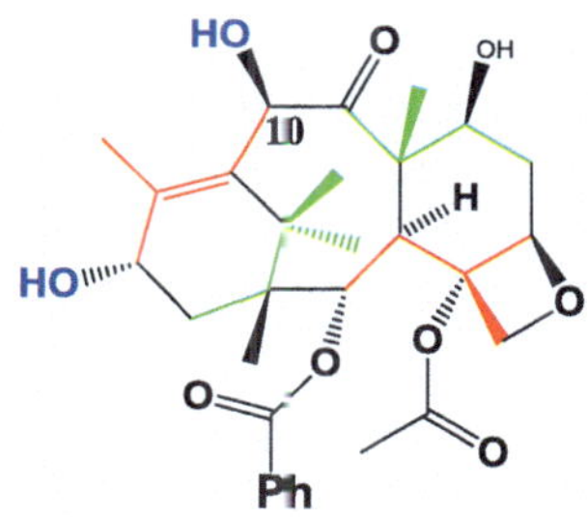

12

Diosgenin (**13**), the steroidal sapogenin that can be isolated from the tubers of *Dioscorea* species growing in Mexico and Central America, is an important starting material for the production of various steroid hormones such as progesterone (**14**).

Progesterone hormone (**14**), is used as a female oral contraceptive (Wall 1960). Natural progesterone could be isolated from sow ovaries but it has a very low yield (20 mg from 625 kg of ovaries), and also can be synthesized from cholesterol, but its yield was insufficient (Applezweig 1962). Progesterone itself is a key intermediate for the production of the important anti-inflammatory drug, cortisone (**15**). By microbial hydroxylation of progesterone at C-11, it is converted into 11-α-hydroxyprogesterone, which produces cortisone (**15**) by chemical reactions (Mancera et al. 1952; 1953).

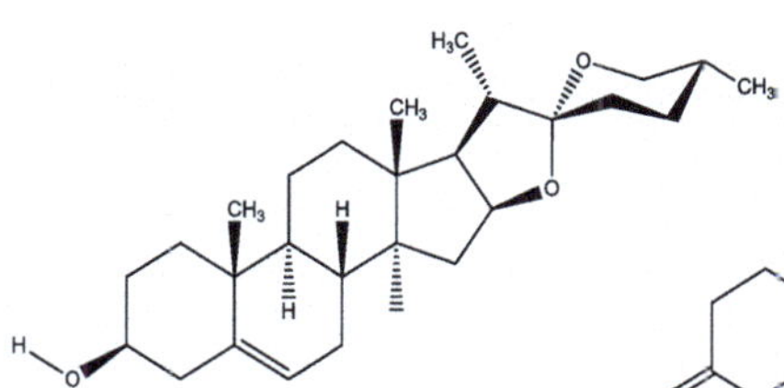

13

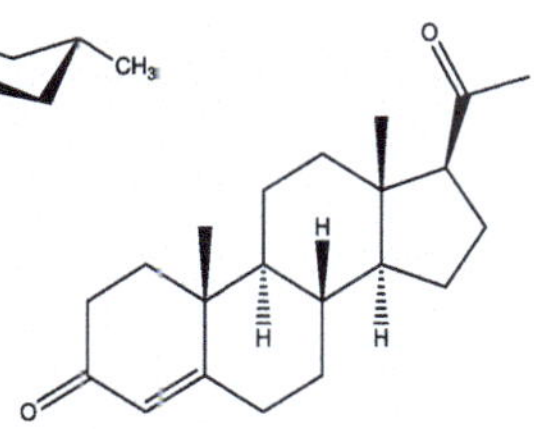

14

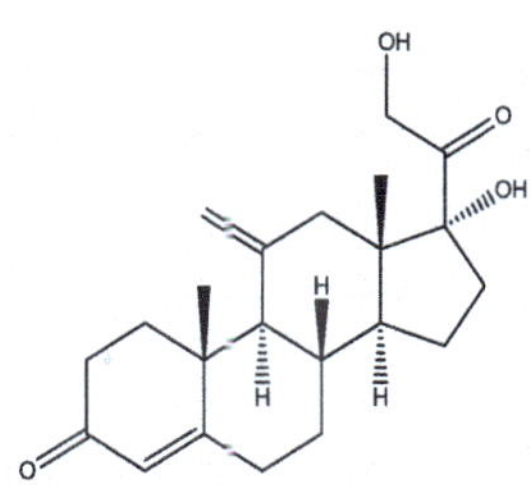

15

Oseltamivir phosphate (**16**) known as Tamiflu which is an orally active neuraminidase inhibitor developed for the prophylaxis and treatment of influenza viruses A and B (Ward et al. 2005; Graeme 2006). (−)-Shikimic acid (**17**) is the starting material for the semisynthesis of oseltamivir, which can be isolated in large quantities from the fruits of star anise (*Illicium verum* Hook.f.) (Yarnell 2005). Shikimic acid can also be gotten by the fermentation of genetically engineered *Escherichia coli* strains, which are deficient in the shikimate kinase gene (Kramer et al. 2003).

Currently, the drug manufacturers depend on both isolation and fermentation methods to obtain large quantities of shikimic acid (Yarnell 2005). Oseltamivir can be obtained by several alternative ways independent of shikimic acid (Abrecht et al. 2004; Yeung et al. 2006), but these alternatives are not economically feasible (Yarnell 2005).

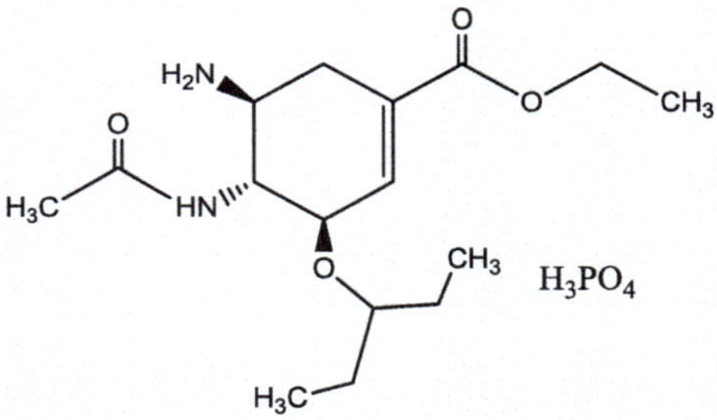

16

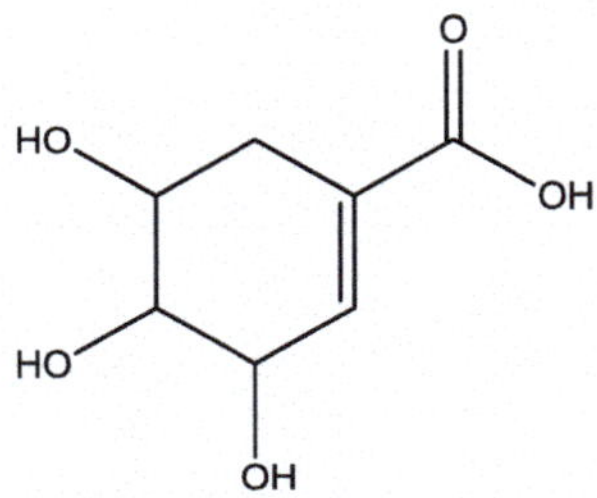

17

Angiotensin converting enzyme (ACE) inhibition has long been the cornerstone for therapy of cardiovascular disorders, chronic heart failure and hypertension, as well as prevention of nephropathy in diabetes mellitus (McMurray et al. 2005; Dagenais et al. 2006). L-Homophenylalanine (**18**) is the common building block of all significantly active ACE inhibitors such as enalapril (**19**), delapril (**20**), lisinopril, quinapril, ramipril, trandolapril, cilazapril and benzapril (Johnson et al. 1985; Liese et al. 2000). L-Homophenylalanine moiety present the central pharmacophore unit of all these compounds (Larissegger-Schnell et al. 2006). L-homophenylalanine is a crucial chiral intermediate for the synthesis of a variety of novel pharmaceuticals such as acetylcholinesterase inhibitor, β-lactam antibiotics, and neutral endopeptidase (NEP) inhibitor which overtime have contributed to human health preservation (Ahmad et al. 2009).

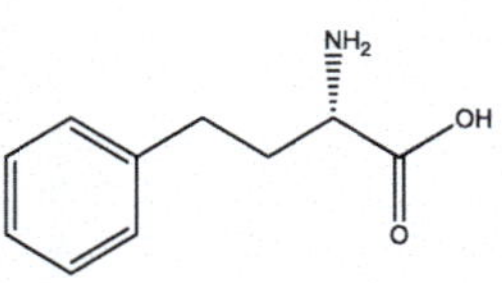

18

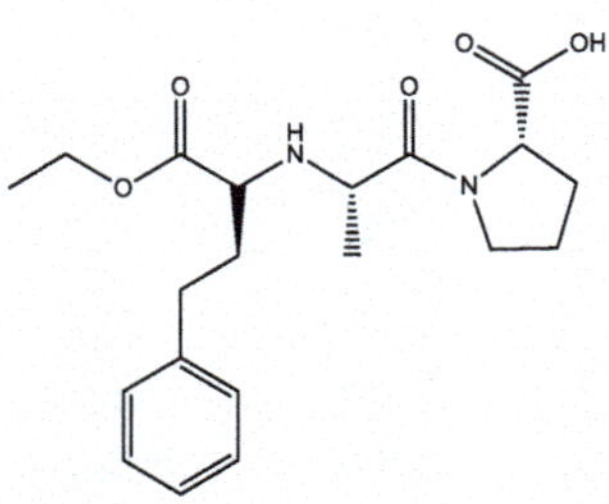

19

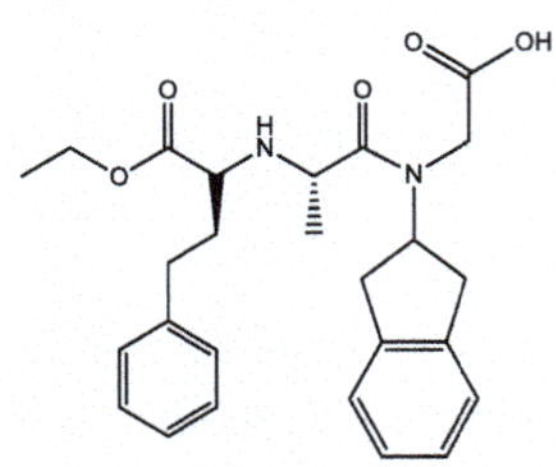

20

Ephedrine (**21**) is a major alkaloid gotten from *Ephedra sinica* (Ma Huang), a plant utilized in traditional Chinese medicine. It forms the basis for the synthesis of some important biomolecule such as the anti-asthma agents (β-agonist), salbutamol (**22**) and salmetrol (**23**) (Buss and Waigh 1995).

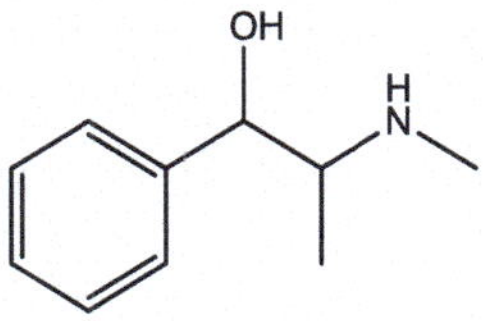

21

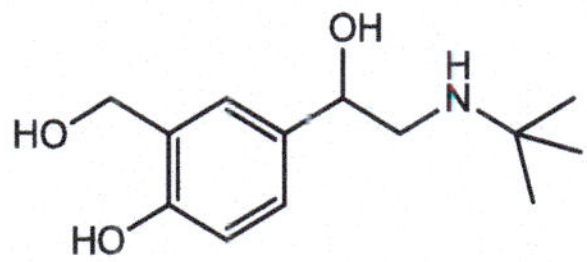

22

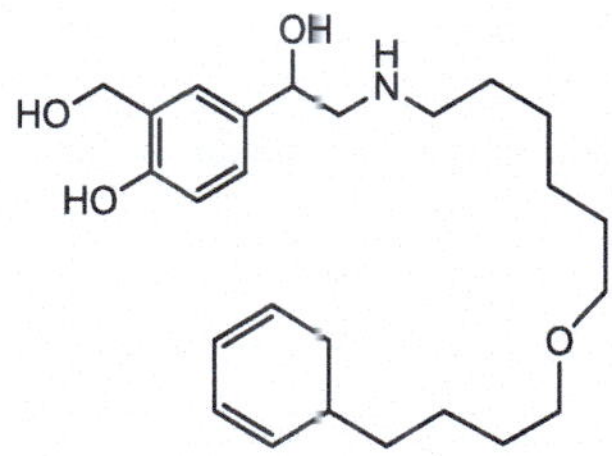

23

2.3 Drug Precursors from Marines

There are lots of drugs isolated from marine sources. An example is Ziconotide (**24**), a non-narcotic analgesic regarded as the first approved marine derived drug currently being marketed as Prialt® which is administered through the intrathecal injection route (Wallace 2006). Ziconotide was isolated from a constituent of the venom cone snail genus Conus used to stun their prey prior to capture. It is comprised of hundreds of peptides (Bulaj et al. 2003).

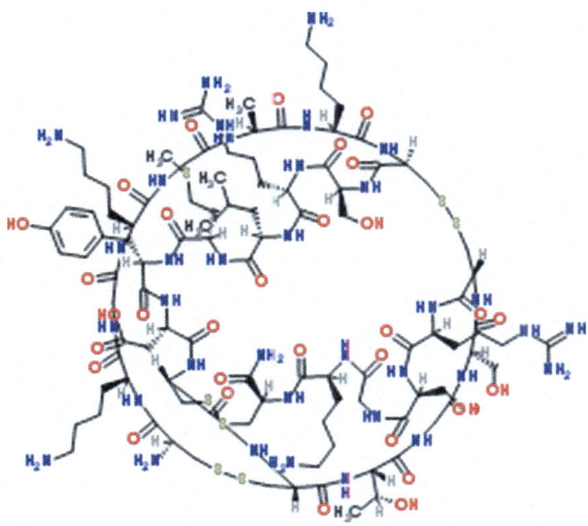

24

Ecteinascidin 743 (**25**) is a complex alkaloid isolated from the colonial tunicate *Ecteinascidia turbinate*. It received approval by the European Medicines Agency (EMEA) in September, 2007 for the treatment of soft tissue sarcomas (STS) after being granted Orphan Drug designation in Europe and the USA under the name

Yondelis®. Later on, it was approved in 2009 in the EU for treatment of relapsed ovarian cancer in conjunction with liposomal doxorubicin. Just like many other marine derived drugs, Ecteinascidin is faced with a low extraction yield. To make up for large yield, has led to the development of a semisynthetic route from the microbial product cyanosafracin B (Henriquez et al. 2005; Cuevas and Francesch 2009; Cuevas et al. 2012).

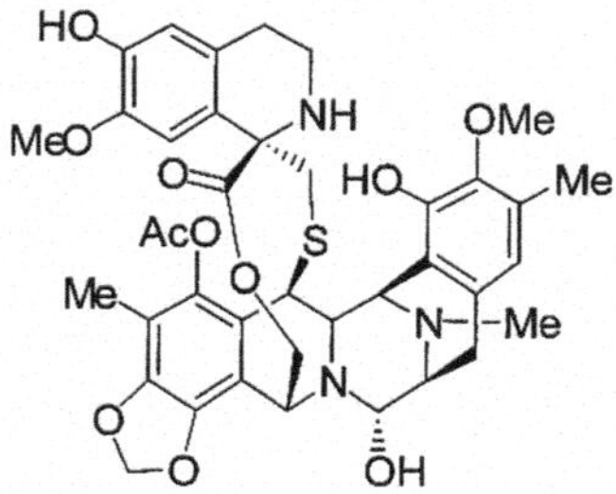

25

Halichondrin B (**26**) is a potent anticancer complex polyether isolated from several sponge sources in a very low yield (Yu et al. 2005a). The total synthesis of halichondrin B was impractical due to its complex structure, but fortunately the synthesis studies revealed that all or most of the potency of the parent compound halichondrin B is dependent on the right hand half of the structure. Accordingly, large-scale synthesis of the analogue, Eribulin (E7389) (**27**) provided adequate supplies (Yu et al. 2011), which was approved in November 2010 by the FDA for the treatment of refractory metastatic breast cancer.

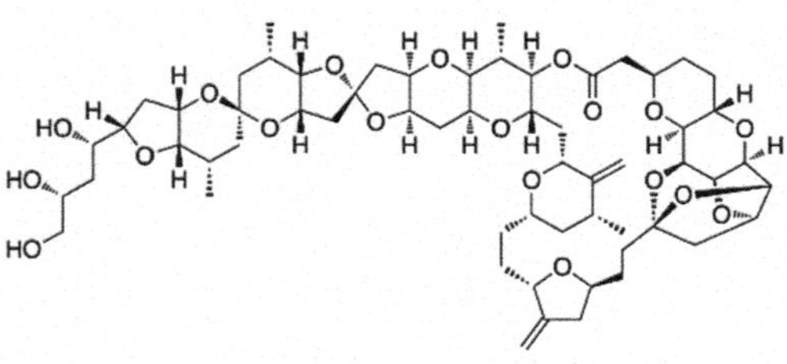

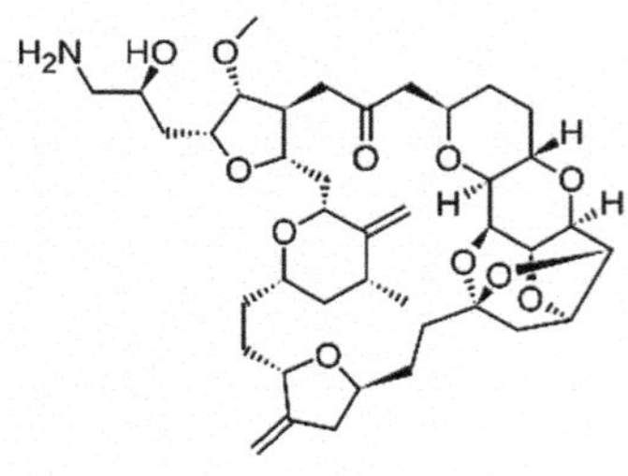

26 **27**

2.4 Drug Precursors from Microorganisms

Pyridomycin (**28**), a natural compound produced by *Dactylosporangium fulvum* has a strong activity against mycobacteria, as it inhibits the nicotinamide adenine dinucleotide (NADH)-dependent enoyl-ACP reductase (inhA) enzyme directly through the competitive inhibition of the NADH binding site, without activation by the katG enzyme (Riccardi and Pasca 2014). The inhibition of inhA enzyme causes an accumulation of long-chain fatty acids and cell death (Vilchèze and Jacobs 2007). By this mechanism, it is superior to Isoniazid, the first line drug for the treatment of tuberculosis (TB) that requires activation by *Mycobacterium tuberculosis* (MTB) catalase-peroxidase katG enzyme to form an INH-NAD complex which inhibits the nicotinamide adenine dinucleotide (NADH)-dependent enoyl-ACP reductase inhA enzyme (Riccardi and Pasca 2014).

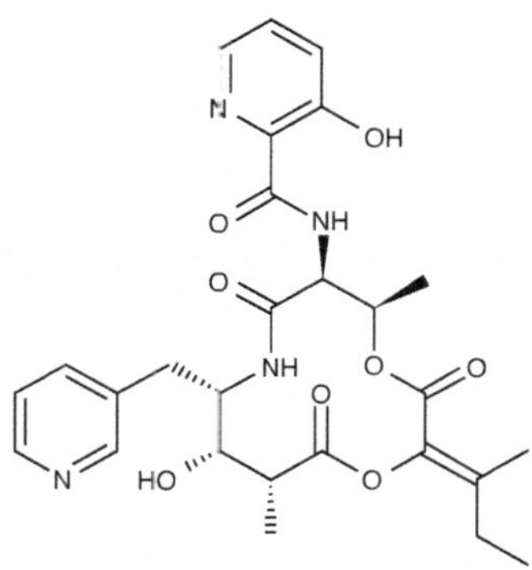

28

Rifapentine (**29**), a semisynthetic cyclopentyl rifamycin derivative, acts by binding the b-subunit of the RNA polymerase in MTB, a mechanism that is also utilized by rifampicin (Munsiff et al. 2006). It is more effective than rifampicin against MTB, both in-vitro and in-vivo with an MIC value in the range of 0.02–0.06 µg/ml (Chan et al. 2014). Both drugs (rifamycin and rifapentine) is known to exhibit cross resistance.

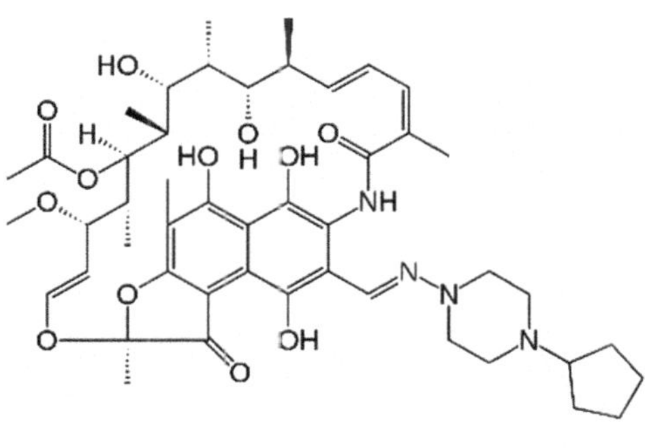

29

The quinolones are a broad class of cell-signaling compounds found naturally in many bacterial species such as *Pseudomonas* spp. (Pesci et al. 1999; Diggle et al. 2006), *Escherichia* spp. (Vial et al. 2008; Kwak et al. 2013). Nalidixic acid was the first discovered quinolone which was originally derived from quinolone by-product of quinine distillation (Bisacchi 2015). There are up to four generations of quinolone antibiotics, including the fluoroquinolones and diarylquinolones which are of particular interest to TB drug development.

Griselimycin (**30**) is a cyclic peptide antibiotic discovered in the 1960s. It is produced by two strains of Streptomyces. The natural peptide has poor solubility, but more stable analogs of grislimycin, particularly cyclohexylgriselimycin (**31**), are active against intracellular *M. tuberculosis* and drug-resistant strains (Kling et al. 2015).

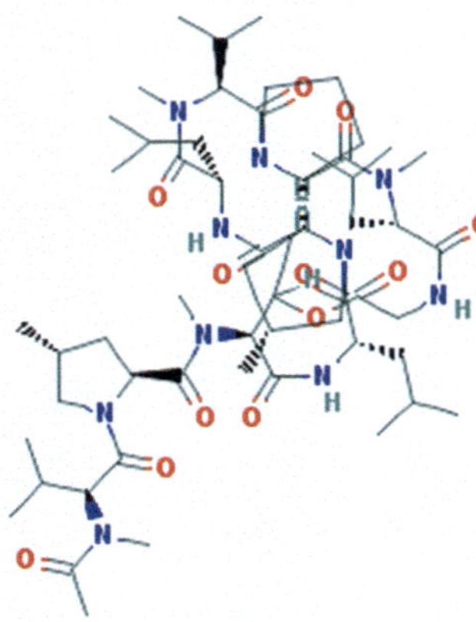

30

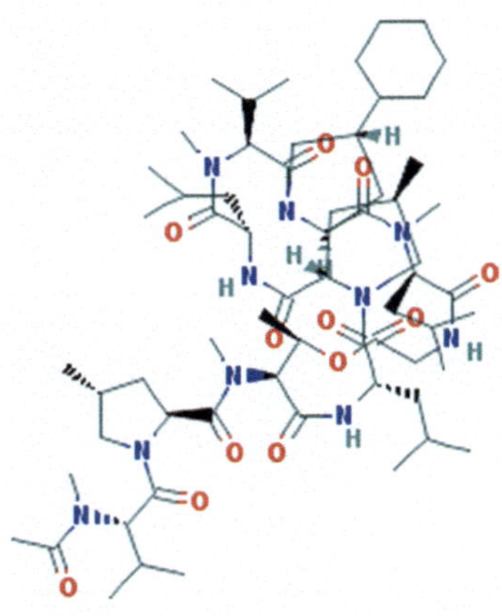

31

Sansanmycins, an important antibiotic that acts by the inhibition of cell wall biosynthesis, are members of the uridylpeptide family (Xie et al. 2007; Li et al. 2011). The modifications of sansanmycin through semisynthetic and biosynthetic means has led to the development of derivatives with greater activity against virulent *M. tuberculosis*, including MDR and XDR strains (Shi et al. 2016).

2.5 Plant Secondary Metabolites as Drug Prototypes

Plants secondary metabolites contributed 23% of the identified drug prototypes in 1996 from minerals, plants, animals, microbes, and chemical sources (Sneader 1996). Applying advanced organic chemistry, many analogs could be prepared from these drug prototypes which in many cases are safer, more potent drugs and sometimes the developed drugs are new compounds with novel pharmacological properties. In the following section, we discussed some the most famous plant-derived drug prototypes.

Podophyllotoxin (**32**) and camptothecin (**33**) are antineoplastic compounds isolated from plants, but unfortunately they are too toxic or not water soluble enough for clinical application (Lee and Xiao 2005; Rahier et al. 2005).

Accordingly, there was a demand for the synthesis of analogs with wider therapeutic margins such as topotecan (Hycamtin) (**34**) and etoposide (Vepesid) (**35**) which have been developed in consequence (Lee and Xiao 2005; Rahier et al. 2005).

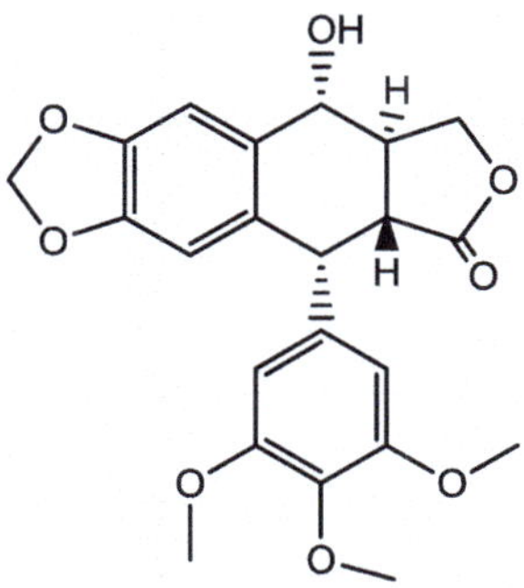

32

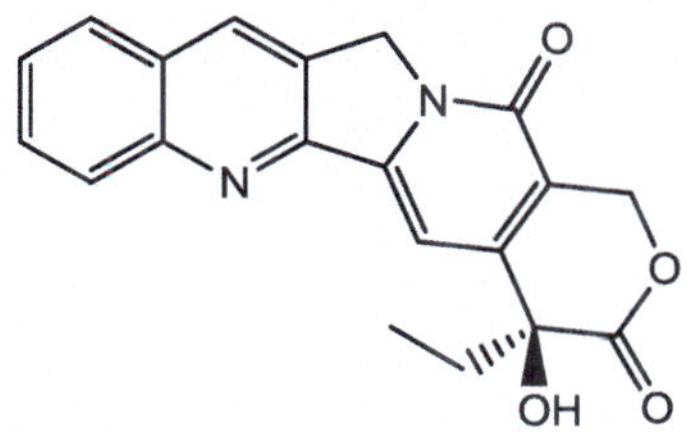

33

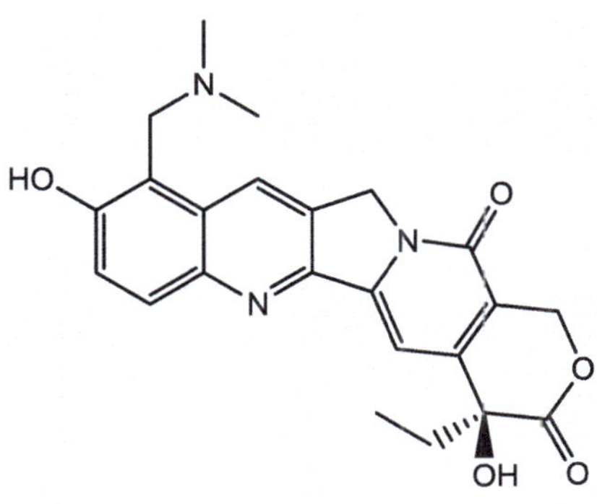

34

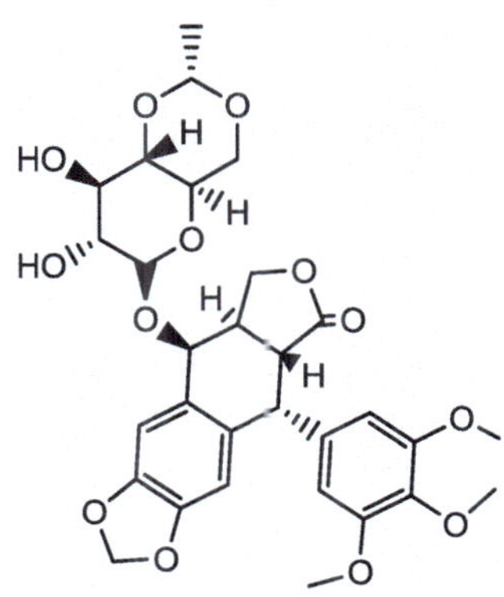

35

Paclitaxel (**11**) and camptothecin (**33**) have unique mechanism of action as anticancer drugs, thus there is always increasing interest in the clinical development of more derivatives of the two compounds (Cragg and Newman 2004; Butler 2005; Rahier et al. 2005; Kingston 2006). Docetaxel (**36**) (taxol analog and old under the brand name Taxotere) is mainly used just like paclitaxel as a regimen for treatment of metastatic and early-stage breast cancer. Docetaxel has a tert-butoxycarbonyl group instead of the benzoyl group on the taxol side chain and a hydroxyl group instead of the acetoyl group on the taxol B ring, so it becomes more water soluble and has shorter infusion time. Topotecan (**34**) and irinotecan (**37**) were synthesized based on the parent molecule camptothecin and they constituted over 30% of the total global sales of cytotoxic drugs (Oberlies and Kroll 2004). Irinotecan (**37**) has a major role in the management of metastatic colorectal cancer, while topotecan (**34**), has antitumor activity comparable with paclitaxel in patients with recurrent ovarian cancer and is an established treatment in second-line or salvage settings (Oberlies and Kroll 2004).

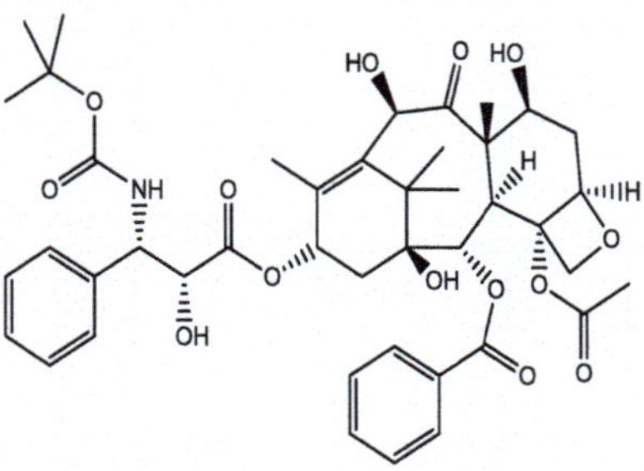

36

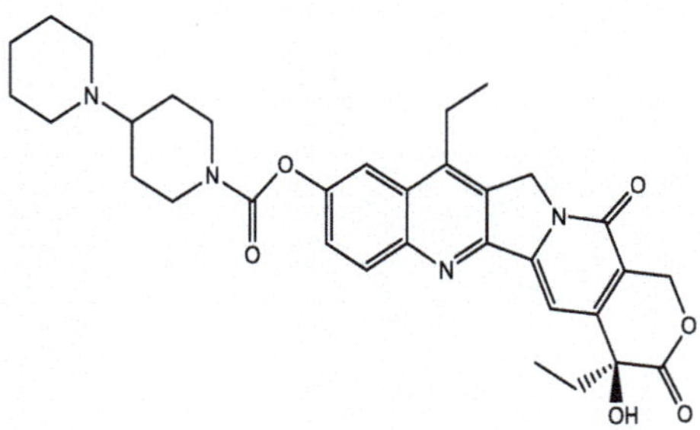

37

Guanidine (**38**) the main constituents of *Galega officinalis* L. is a natural product with significant antihyperglycemic activity. Clinical trials have shown that it is too toxic for human use. Many derivatives of guanidine have been synthesized, among which the famous antidiabetic drug metformin (dimethylbiguanide) (**39**) proved its effectiveness for treatment of type II diabetes over the years.

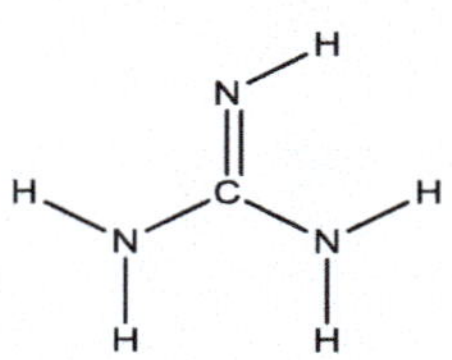

38

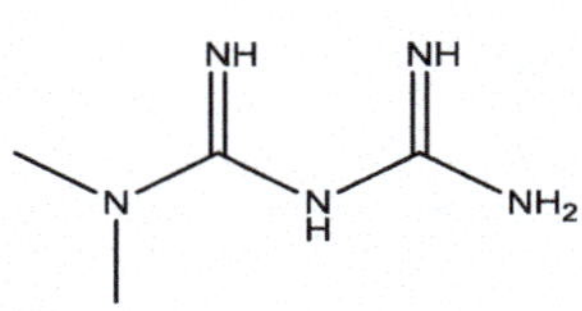

39

Atropine (**40**), the competitive antagonist of muscarinic acetylcholine receptors (antimuscarinic agent) is produced due the racemization of the tropane alkaloid (−)-hyoscyamine, during the extraction process from *Atropa belladonna*. Atropine has mydriatic action thus sometimes is used in ophthalmology, in addition of being an antispasmodic drug (Sneader 1996). It is also used as a pre-anesthetic medication to reduce bronchial and salivary secretions, and to block the bradycardia associated with the administration of anesthetic drugs (Sneader 1996). A large number of synthetic atropine analogs were prepared leading to the introduction of novel drugs with different biological activities and therapeutic applications than atropine. Among these analogs are droperidol (**41**) which has antipsychotic effect, methadone (**42**), a morphine substitute for treatment of addiction, pethidine (**43**), used as analgesic drug, ipratropium bromide (**44**), bronchodilator drug for the treatment of asthma, and the antidiarrheal drug loperamide (**45**) (Sneader 1996).

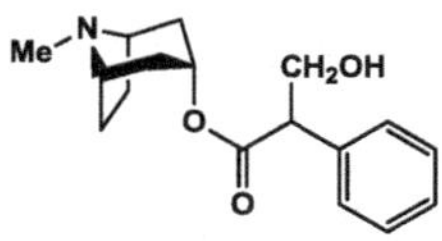

40

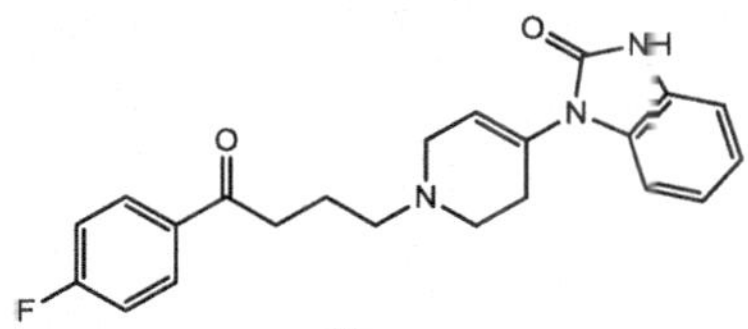

41

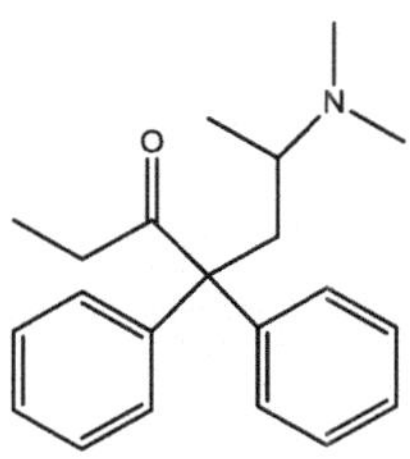

42

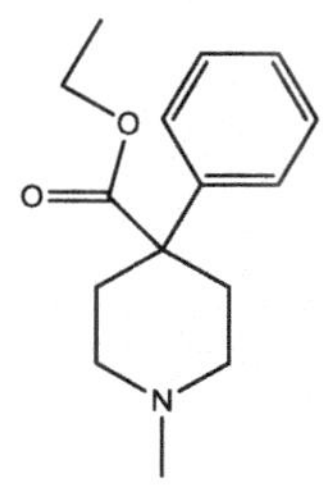

43

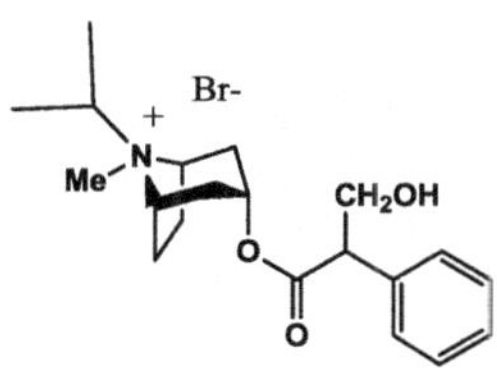

44

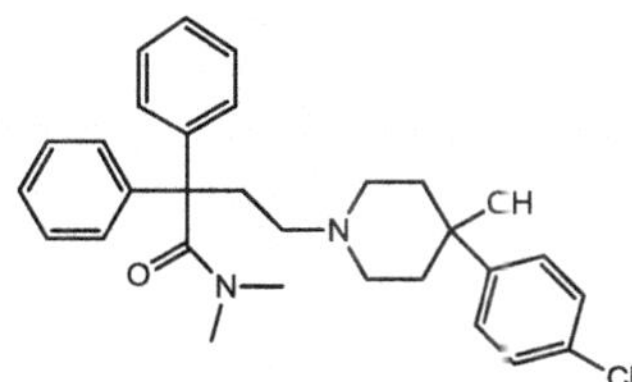

45

2.6 Conclusion

Since ancient time, medicinal plants have a great importance in human life as it helps fulfill man primary health care needs. The diversity in the plant kingdom and the presence of huge number of secondary metabolites with diverse pharmacological activity makes it the most focused area of research amongst scientists for new drug discovery. The present chapter described various important drug developments from plant origin. Still there is a lot of scope of development from plant origin using advanced techniques and novel drug discovery tools. Tracer techniques, high- throughput screening, and combinatorial chemistry can be very helpful in new drug development from plant origin. The case studies specifically describes how natural product can be modified to a compound with increased yield and efficacy.

References

Abrecht S, Harrington P, Iding H, Karpf M, Trussadi R, Wirz B, Zutter U. The synthetic development of the anti-influenza neuraminidase inhibitor oseltamivir phosphate (Tamiflu®): a challenge for synthesis & process research. Chimia. 2004;58:621.

Achan J, Tibenderana JK, Kyabayinze D, et al. Effectiveness of quinine versus artemether-lumefantrine for treating uncomplicated falciparum malaria in Ugandan children: randomised trial. Br Med J. 2009;338:b2763.

Ahmad AL, Oh PC, Abd Shukor SR. Sustainable biocatalytic synthesis of L-homophenylalanine as pharmaceutical drug precursor. Biotechnol Adv. 2009;27:286–96.

Ainsworth S. Neonatal formulary: drug use in pregnancy and the first year of life. New York: Wiley; 2014. p. 94. ISBN:9781118819593

Applezweig N. Steroid drugs. New York: McGraw-Hill; 1962.

Bagchi D Preuss HG. Obesity epidemiology, pathophysiology, and prevention. 2nd ed. Boca Raton: CRC Press: 2013. p. 692. ISBN 9781439854266.

Birks J. Cholinesterase inhibitors for Alzheimer's disease. Cochrane Database Syst Rev. 2006;1: CD005593. https://doi.org/10.1002/14651858.CD005593.

Bisacchi GS. Origins of the quinolone class of antibacterials: an expanded "Discovery Story". J Med Chem. 2015;58:4874–82.

Biswas L, Harrison E, Gong Y, Avusula R, Lee J, Zhang M, Rousselle T, Lage J, Liu X. Enhancing effect of menthol on nicotine self-administration in rats. Psychopharmacology. 2016;233:3417–27.

Bulaj G, Buczek O, Goodsell I, Jiminez EC, Kranski J, Nielsen JS, Garrett JE, Olivera BM. Efficient oxidative folding of conotoxins and the radiation of venomous cone snails. Proc Natl Acad Sci U S A. 2003;100:14562–8.

Buss AD, Waigh RD. Natural products as leads for new pharmaceuticals. In: Wolff ME, editor. Burger's medicinal chemistry and drug discovery, Principles and practice, vol. 1. New York: Wiley; 1995. p. 983–1033.

Butler MS. Natural products to drugs: natural product derived compounds in clinical trials. Nat Prod Rep. 2005;22:162.

Canel C, Moraes RM, Dayan FE, Ferreira D. Molecules of interest: podophyllotoxin. Phytochemistry. 2000;54(2):115–20. https://doi.org/10.1016/s0031-9422(00)00094-7.

Chan JG, Bai X, Traini D. An update on the use of rifapentine for tuberculosis therapy. Expert Opin Drug Deliv. 2014;11:421–31.

Chowdhury AR, Mandal S, Mittra B, Sharma S, Mukhopadhyay S, Majumder HK. Betulinic acid, a potent inhibitor of eukaryotic topoisomerase I: identification of the inhibitory step, the major functional group responsible and development of more potent derivatives. Med Sci Monit. 2002;8(7):BR254–65. PMID 12118187

Connors BW. Tales of a dirty drug: carbenoxolone, gap junctions, and seizures. Epilepsy Curr. 2012;12(2):66–8. https://doi.org/10.5698/1535-7511-12.2.66. PMC 3316363 Freely accessible. PMID 22473546

Courtwright DT. Forces of habit drugs and the making of the modern world. 1st ed. Cambridge, MA: Harvard University Press; 2009. p. 36–7. ISBN:978-0-674-02990-3

Cragg GM, Newman DJ. A tale of two tumor targets: topoisomerase I and tubulin. The Wall and Wani contribution to cancer chemotherapy. J Nat Prod. 2004;67:232.

Cuevas C, Francesch A. Development of Yondelis® (trabectedin, et-743). A semisynthetic process solves the supply problem. Nat Prod Rep. 2009;26:322–37.

Cuevas C, Francesch A, Galmarini CM, Aviles P, Munt S. Ecteinascidin-743 (Yondelis(r)), Aplidine(r), and Irvalec(r). In: Cragg GM, Kingston DGI, Newman DJ, editors. Anticancer agents from natural products. 2nd ed. Boca Raton: Taylor and Francis; 2012. p. 291–316.

Dagenais GR, Pogue J, Fox K, Simoons ML, Yusuf S. Angiotensin-converting-enzyme inhibitors in stable vascular disease without left ventricular systolic dysfunction or heart failure: a combined analysis of three trials. Lancet. 2006;368:581–8.

Diggle SP, Cornelis P, Williams P, Camara M. 4-quinolone signalling in *Pseudomonas aeruginosa*: old molecules, new perspectives. Int J Med Microbiol. 2006;296:83–91.

Dondorp A, Nosten F, Stepniewska K, Day N, White N. Artesunate versus quinine for treatment of severe falciparum malaria: a randomised trial. Lancet. 2005;366(9487):717–25. https://doi.org/10.1016/S0140-6736(05)67176-0. PMID 16125588

Dorndorp A, Nosten F, Stepniewska K, et al. Artesunate verus quinine for treatment of severe falciparum malaria: a randomised trial. Lancet. 2005;366(9487):717–25.

Egbuna C, Ifemeje JC. Biological functions and anti-nutritional effects of phytochemicals in living system. IOSR J Pharm Biol Sci. 2015;10(2):10–9.

Elbaz HA, Stueckle TA, Tse W, Rojanasakul Y, Dinu CZ. Digitoxin and its analogs as novel cancer therapeutics. Exp Hematol Oncol. 2012;1(1):4. https://doi.org/10.1186/2162-3619-1-4. PMC 3506989 Freely accessible. PMID 23210930

Fabricant DS, Farnsworth NR. The value of plants used in traditional medicine for drug discovery, environ. Health Perspect. 2001;109(supplement):69–75.

Farnsworth NR, Akerele RO, Bingel AS, Soejarto DD, Guo Z. Medicinal plants in therapy. Bull World Health Organ. 1985;63:965–81.

Fattori V, Hohmann MS, Rossaneis AC, Pinho-Ribeiro FA, Verri WA. Capsaicin: current understanding of its mechanisms and therapy of pain and other pre-clinical and clinical uses. Molecules. 2016;21(7):844. https://doi.org/10.3390/molecules21070844. PMID 27367653

Fischer J, Ganellin CR. Analogue-based drug discovery. Weinheim: Wiley; 2006. p. 512. ISBN:9783527607495

Fist AJ, Byrne CJ, Gerlach WL. Harvesting poppy capsules of stably reproducing papaver somniferum to produce straw containing thebaine and oripavine in at least 50% by weight of alkaloid mixture formed, chemically extracting thebaine and/or oripavine. 2002. Available: http://www.google.com/patents/US6376221. Accessed 24 June 2018.

Graeme L. Antiviral drugs for influenza: Tamiflu® past, present and future. Future Virol. 2006;1:577.

Greenwell M, Rahman P. Medicinal plants: their use in anticancer treatment. Int J Pharm Sci Res. 2015;6:4103–12. https://doi.org/10.13040/IJPSR.0975-8232.6(10).4103-12.

GRIN (Germplasm Resources Information Network). Agricultural Research Service (ARS), United States Department of Agriculture (USDA). 2008. Accessed 16 May 2018.

Guthner T, Mertschenk B, Schulz B. Guanidine and derivatives, Ullmann's encyclopedia of industrial chemistry. Weinheim: Wiley-VCH; 2005. https://doi.org/10.1002/14356007.a12_545.pub2.

Henriquez R, Faircloth G, Cuevas C (2005) Ecteinascidin 743 (et-743); Yondelis™, Aplidin and Kahalalide F, Cragg GM, Kingston DGI, Newman DJ Anticancer agents from natural products, Taylor and Francis: Boca Raton 215–240.

Holton RA, Biediger RJ, Boatman PD. Semisynthesis of taxol and taxotere. In: Suffness M, editor. Taxol, science and applications. Boca Raton: CRC Press; 1995a. p. 97.

Holton RA, Somoza C, Kim H-B, Liang F, Biediger RJ, Boatman PD, Shindo M, Smith CC, Kim S, Nadizadeh H, Suzuki Y, Tao C, Vu P, Tang S, Zhang P, Murthi KK, Gentile LN, Liu JH. The total synthesis of paclitaxel starting with amphor. In: Georg GI, Chen TT, Ojima I, Vyas DM, editors. ACS symposium series 583 (taxane anticancer agents, basic science and current status). Washington, DC: American Chemical Society; 1995b. p. 288.

Hu J, Jin G. Effect of tetrahydropalmatine analogs on Fos expression induced by formalin-pain. Acta Pharmacologica Sinica. 1999;20(3):193–200.

Huang Z, Chen G, Shi P. Effects of emodin on the gene expression profiling of human breast carcinoma cells. Cancer Detect Prev. 2009;32(4):286–91.

Ingle PK. L-DOPA bearing plants. Nat Prod Radiance. 2003;2(3):126–33.

Jayaprakasha GK, Rao LJ, Sakariah KK. Antioxidant activities of curcumin, demethoxycurcumin and bisdemethoxycurcumin. Food Chem. 2006;98(4):720–4. https://doi.org/10.1016/j.foodchem.2005.06.03.

Jiang G, Mendes EA, Kaiser P, Wong DP, Tang Y, Cai I, et al. Synergistic reactivation of latent HIV expression by Ingenol-3-angelate, PEP005, targeted NF-kB signaling in combination with JQ1 induced p-TEFb activation. PLoS Pathog. 2015;11(7):e1005066. https://doi.org/10.1371/journal.ppat.1005066.

Johnson AL, Price WA, Wong PC, Vavala RF, Stump JM. Synthesis and pharmacology of the potent angiotensin-converting enzyme-inhibitor N-[1(S)-(ethoxycarbonyl)-3- phenylprophyl]-(S)-alanyl-(S)-pyroglutamic acid. J Med Chem. 1985;28:1596–602.

Kingston DGI. Recent advances in the chemistry of taxol. J Nat Prod. 2000;63:726.

Kingston DGI. Taxol and its analogs. In: Cragg GM, Kingston DGI, Newman DJ, editors. Anticancer agents from natural products. Boca Raton: Taylor and Francis; 2006. p. 89.

Kling A, Lukat P, Almeida DV, Bauer A, Fontaine E, Sordello S, et al. Targeting DnaN for tuberculosis therapy using novel griselimycins. Science. 2015;348:1106–12.

Kramer M, Bongaerts J, Bovenberg R, Kremer S, Muller U, Orf S, Wubbolts M, Raeven L. Metabolic engineering for microbial production of shikimic acid. Metab Eng. 2003;5:277.

Kresge N, Simoni RD, Hill RL. Sweet clover disease and warfarin review. 2005. Jbc.org. Accessed 26 May 2018.

Kwak YG, Jacoby GA, Hooper DC. Induction of plasmid-carried qnrS1 in *Escherichia coli* by naturally occurring quinolones and quorum-sensing signal molecules. Antimicrob Agents Chemother. 2013;57:4031–4.

Larissegger-Schnell B, Glueck SM, Kroutil W, Faber K. Enantio-complementary deracemization of (+/−)-2-hydroxy-4-phenylbutanoic acid and (+/−)-3-phenyllactic acid using lipase-catalyzed kinetic resolution combined with biocatalytic racemization. Tetrahedron. 2006;62:2912–6.

Lee K-H, Xiao Z. Podophyllotoxins and analogs. In: Cragg GM, Kingston DGI, Newman DJ, editors. Anticancer agents from natural products. Boca Raton: Taylor and Francis; 2005. p. 71.

Li Y, Li CJ, Yu D, Pardee AB. Potent induction of apoptosis by βlapachone in human multiple myeloma cell lines and patient cells. Mol Med. 2000;6:1008–15.

Li Y-B, Xie Y-Y, Du N-N, Lu Y, Xu H-Z, Wang B, et al. Synthesis and in vitro antitubercular evaluation of novel sansanmycin derivatives. Bioorg Med Chem Lett. 2011;21:6804–7.

Liese A, Seelbach K, Wandrey C. Industrial biotransformations. Weinheim: Wiley-VCH Verlag GmbH; 2000.

Machara A, Werner L, Endoma-Arias MA, Cox DP, Hudlicky T. Improved synthesis of buprenorphine from thebaine and/or oripavine via palladium- catalyzed N-demethylation/acylation and/or concomitant O-demethylation. Adv Synth Catal. 2012;354:613–26.

MacLean K, Johnson M, Reissig C, Prisinzano T, Griffiths R. Dose-related effects of salvinorin A in humans: dissociative, hallucinogenic, and memory effects. Psychopharmacology. 2013;226(2):381–92. https://doi.org/10.1007/s00213-012-2912-9.

Malenka RC, Nestler EJ, Hyman SE. Chapter 9: Autonomic nervous system. In: Sydor A, Brown RY, editors. Molecular neuropharmacology: a foundation for clinical neuroscience. 2nd ed. New York: McGraw-Hill Medical; 2009. p. 234. ISBN:9780071481274.

Mancera O, Zaffaroni A, Rubin BA, Sondheimer F, Rosenkranz G, Djerassi C. STEROIDS. XXXVII. A ten step conversion of progesterone to cortisone. J Am Chem Soc. 1952;74:3711.

Mancera O, Ringold HJ, Djerassi C, Rosenkranz G, Sondheimer F. Steroids. XLIII. A ten step conversion of progesterone to cortisone. The differential reduction of pregnane-3,20-diones with sodium borohydride. J Am Chem Soc. 1953;75:1286.

Maridass M, John A. Origins of plant derived medicines. Ethnobot Leafl. 2008;12:373–87.

McMurray J, Cohen-Solal A, Dietz R, Eichhorn E, Erhardt L, Hobbs FDR, et al. Practical recommendations for the use of ACE inhibitors, beta-blockers, aldosterone antagonists and angiotensin receptor blockers in heart failure: putting guidelines into practice. Eur J Heart Fail. 2005;7:710–21.

Munsiff SS, Kambili C, Ahuja SD. Rifapentine for the treatment of pulmonary tuberculosis. Clin Infect Dis. 2006;43:1468–75.

Mutschler E, Schäfer-Korting M. Arzneimittelwirkungen (in German). 8th ed. Stuttgart: Wissenschaftliche Verlagsgesellschaft; 2001. p. 214f, 286. ISBN:3-8047-1763-2

Nehlig A, Daval JL, Debry G. Caffeine and the central nervous system: mechanisms of action, biochemical, metabolic and psychostimulant effects. Brain Res Brain Res Rev. 1992;17 (2):139–70. https://doi.org/10.1016/0165-0173(92)90012-B. PMID 1356551

Nicolau KC, Guy RK. The total synthesis of paclitaxel by assembly of the ring system. In: Georg GI, Chen TT, Ojima I, Vyas DM, editors. ACS symposium series 583 (taxane anticancer agents, basic science and current status). Washington, DC: American Chemical Society; 1995. p. 302.

Nirmala MJ, Samundeeswari A, Sankar PD. Natural plant resources in anti-cancer therapy-a review. Res Plant Biol. 2011;1(3):01–14.

NORD (National Organization for Rare Disorders). Physician's guide to tyrosinemia type 1 archived 2014-02-11 at the Wayback Machine. 2014

Novak B, Hudlicky T, Reed J, Mulzer J, Trauner D. Morphine synthesis and biosynthesis-an update. Curr Org Chem. 2000;4(3):343–62. https://doi.org/10.2174/1385272003376292.

O'Neill PM, Posner GH. A medicinal chemistry perspective on artemisinin and related endoperoxides. J Med Chem. 2004;47:2945–64.

Oberlies NH, Kroll DJ. Camptothecin and taxol: historic achievements in natural products research. J Nat Prod. 2004;67:129.

Patries M, Herst JE, Davis PN, Michael VB, David SR. Haematologica. 2009;94:928–34. https://doi.org/10.3324/haematol.2008.003996.

Pesci EC, Milbank JB, Pearson JP, McKnight S, Kende AS, Greenberg EP, et al. Quinolone signaling in the cell-to-cell communication system of *Pseudomonas aeruginosa*. Proc Natl Acad Sci. 1999;96:11229–34.

Pettit GR, Singh SB, Niven ML, Hamel E, Schmidt JM. Isolation, structure, and synthesis of combretastatins A-1 and B-1, potent new inhibitors of microtubule assembly, derived from Combretum caffrum. J Nat Prod. 1987;50(1):119–31. https://doi.org/10.1021/np50049a016. PMID 3598594

Pittler MH, Ernst E. Horse chestnut seed extract for chronic venous insufficiency. Cochrane Database Syst Rev. 2012;11:CD003230. https://doi.org/10.1002/14651858.CD003230.pub4. ISSN:1469-493X. PMID 23152216

Posner GH, Paik IH, Chang W, Borstnik K, Sinishtaq S, Rosenthal AS, Shapiro TA. Malaria-infected mice are cured by a single dose of novel artemisinin derivatives. J Med Chem. 2007;50:2516–9.

Prommer E. Role of codeine in palliative care. J Opioid Manag. 2010;7(5):401–6. PMID 22165039

Raffetin A, Bruneel F, Roussel C, Thellier M, Buffet P, Caumes E, Jauréguiberry S. Use of artesunate in non-malarial indications. Med Mal Infect. 2018; https://doi.org/10.1016/j.medmal.2018.01.004. PMID 29422423

Rahier NJ, Thomas CJ, Hecht SM. Camptothecin and its analogs. In: Cragg GM, Kingston DGI, Newman DJ, editors. Anticancer agents from natural products. Boca Raton: Taylor and Francis; 2005. p. 5.

Riccardi G, Pasca MR. Trends in discovery of new drugs for tuberculosis therapy. J Antibiot. 2014;67:655–9.

Rinner U, Hudlicky T. Synthesis of morphine alkaloids and derivatives. Top Curr Chem. 2012;309:33–66.

Roberts MF, Wink M. Alkaloids: biochemistry, ecology, and medicinal applications. New York: Plenum Press; 1998. p. 38. ISBN:978-1-4419-3263-1

Salim AA, Chin YW, Kinghorn AD. Drug discovery from plants. In: Ramawat KG, Merillon JM, editors. Bioactive molecules and medicinal plants. Berlin: Springer; 2008. https://doi.org/10.1007/978-3-540-74603-4_1

Samuelsson G. Drugs of natural origin: a textbook of pharmacognosy. 5th ed. Stokkholm: Swedish Pharmaceutical Press; 2004. ISBN:91-974318-4-2

Saul R, Ghidoni JJ, Molyneux RJ, Elbein AD. Castanospermine inhibits alpha-glucosidase activities and alters glycogen distribution in animals. PNAS. 1985;82(1):93–7. https://doi.org/10.1073/pnas.82.1.93.

Shi Y, Jiang Z, Lei X, Zhang N, Cai Q, Li Q, et al. Improving the N-terminal diversity of sansanmycin through mutasynthesis. Microb Cell Factories. 2016;15:77.

Shinde PB, Laddha KS. Development of new isolation technique and validated HPLC method development for khellin. Indian J Nat Prod Resour. 2014;5(1):40–3.

Si H, Liu D, Si H, Liu D. Phytochemical genistein in the regulation of vascular function: new insights. Curr Med Chem. 2007;14(24):2581–9. https://doi.org/10.2174/092986707782023325. PMID 17979711

Sneader W. Drug prototypes and their exploitation. Chichester: Wiley; 1996.

Sneader W. Drug discovery: a history. Chichester: Wiley; 2005. p. 107. ISBN:978-0-471-89980-8

Stead LF, Hughes JR, Hughes. Lobeline for smoking cessation. Cochrane Database Syst Rev. 2012;2:CD000124.

Stegelmeier BL, Molyneux RJ, Elbein AD, James LF. The lesions of locoweed (*Astragalus mollissimus*), swainsonine, and castanospermine in rats. Vet Pathol. 1995;32(3):289–98. https://doi.org/10.1177/030098589503200311.

Sweetman S, editor. Omacetaxine mepesuccinate. Martindale: the complete drug reference. Medicines complete. London: Pharmaceutical Press; 2012.

Taylor L. Plant based drugs and medicines. 2000. Available: http://www.rain-tree.com/plantdrugs.htm#.Wy3iq9JKhdg. Accessed 23 June 2018.

Vennerstrom JL, Arbe-Barnes S, Brun R, Charman SA, Chiu FCK, Chollet J, et al. Identification of an antimalarial synthetic trioxolane drug development candidate. Nature. 2004;430:900–4.

Vial L, Lepine F, Milot S, Groleau MC, Dekimpe V, Woods DE, et al. Burkholderia pseudomallei B. thailandensis, and B. ambifaria produce 4- hydroxy-2-alkylquinoline analogues with a methyl group at the 3 position that is required for quorum-sensing regulation. J Bacteriol. 2008;190:5339–52.

Vilchèze C, Jacobs WR Jr. The mechanism of isoniazid killing: clarity through the scope of genetics. Annu Rev Microbiol. 2007;61:35–50.

Walker N, Howe C, Glover M, McRobbie H, Barnes J, Nosa V, Parag V, Bassett B, Bullen C. Cytisine versus nicotine for smoking cessation. N Engl J Med. 2014;371(25):2353–62. https://doi.org/10.1056/nejmoa1407764. PMID 2551770

Wall ME. Am Perfumer Aromatics. 1960;75:63.

Wallace MS. Ziconotide: a new nonopioid intrathecal analgesic for the treatment of chronic pain. Expert Rev Neurother. 2006;6:1423–8.

Ward P, Small I, Smith J, Suter P, Dutkowski R. Oseltamivir (Tamiflu®) and its potential for use in the event of an influenza pandemic. J Antimicrob Chemother. 2005;55:5.

William HB. Justus von Liebig: the chemical gatekeeper. Cambridge: Cambridge University Press; 1997. p. 120.

Wink M, Roberts MW. Alkaloids: biochemistry, ecology, and medicinal applications. New York: Plenum Press; 1998. ISBN:0-306-45465-3

Wongsrichanalai C, Pickard AL, Wernsdorfer WH, Meshnick SR. Epidemiology of drug-resistant malaria. Lancet Infect Dis. 2002;2:209–18.

Xie Y, Chen R, Si S, Sun C, Xu H. A new nucleosidyl-peptide antibiotic, sansanmycin. J Antibiot. 2007;60:158–61.

Yarnell A. Complexity of Tamiflu manufacturing may hamper on demand production. Chem Eng News. 2005;83:22.

Yeung YY, Hong S, Corey EJ. A short enantioselective pathway for the synthesis of the anti-influenza neuramidase inhibitor oseltamivir from 1,3-butadiene and acrylic acid. J Am Chem Soc. 2006;128:6310.

Yu MJ, Kishi Y, Littlefield BA. Discovery of e7389, a fully synthetic macrocyclic ketone analog of halichondrin b. In: Cragg GM, Kingston DGI, Newman DJ, editors. Anticancer agents from natural products. Boca Raton: Taylor and Francis; 2005a. p. 241–65.

Yu MJ, Zheng W, Seletsky BM, Littlefield BA, Kishi Y. In: Macor JA, editor. Case history: discovery of eribulin (halaven™), a halichondrin b analogue that prolongs overall survival in patients with metastatic breast cancer, Annual reports in medicinal chemistry, vol. 46. Amsterdam: Academic; 2005b. p. 227–41.

Zangara A. The psychopharmacology of huperzine A: an alkaloid with cognitive enhancing and neuroprotective properties of interest in the treatment of Alzheimer's disease. Pharmacol Biochem Behav. 2003;75(3):675–86. https://doi.org/10.1016/S0091-3057(03)00111-4. PMID 12895686

Zhang Y. Allyl isothiocyanate as a cancer chemopreventive phytochemical. Mol Nutr Food Res. 2010;54(1):127–35.

Chapter 3
Phytochemicals as Anticancer Drugs: Targeting the Microtubular Network of Cancer Cells

Rajni Khan, Aroni Chatterjee, Hiya Ghosh, Sonia Kapoor, and Shashikant Ray

3.1 Introduction

Physicians and local practitioners throughout the ages have relied upon herbal medications for the treatment of different type of diseases. Be it among the cultured civilizations or tribal societies, the use of plant derived medications has always remained a major priority. Ayurveda, the antiquated Vedic literature of India, defines itself as the explorative science of maintaining good and prosperous health by utilizing natural methodologies. It speaks of the accumulated traditional and experimental insights necessary to cure diseases (Jain et al. 2010). These herbal drug formulations have been found to be effective in the treatment of a wide range of medical conditions including cancer (Jain et al. 2010). Current drug development programmes dependent on ayurvedic ideas is progressively gaining worldwide acknowledgment in the modern healthcare framework. Recent endeavors in pharmacology has successfully supported the use of plant derived drugs as they are nontoxic to normal human cells and are also better endured (Mahima et al. 2012). Metabolites derived from plants, for example, flavonoids, alkaloids, phenolics,

R. Khan
Department of Physiology, University of Calcutta, Kolkata, West Bengal, India

A. Chatterjee
Department of Virology, ICMR Virus Unit, Kolkata, West Bengal, India

H. Ghosh
Department of Endocrinology and Metabolism, IPGMER and SSKM Hospital, Kolkata, West Bengal, India

S. Kapoor
Amity Institute of Molecular Medicine and Stem Cell Research, Amity University, Noida, Uttar Pradesh, India

S. Ray (✉)
Department of Biotechnology, School of Life Sciences, Mahatma Gandhi Central University, Motihari, Bihar, India
e-mail: shashikantray@mgcub.ac.in

S. Kumar, C. Egbuna (eds.), *Phytochemistry: An in-silico and in-vitro Update*,
https://doi.org/10.1007/978-981-13-6920-9_3

glycosides, tannins, gums and oils play out several pharmacological functions in human body. A few of these phytochemicals or their changed forms have demonstrated critical antitumor potential (Akindele et al. 2015). Vincristine, paclitaxel, vinblastine, elliptinium, camptothecin, colchicinamide, gossypol, curcumol, lycobetain, ipomeanol, etc. are some very important phytomolecules whose efficacy has been tested in multiple disease scenarios (AlSinani et al. 2016). Microtubules are dynamic cytoskeletal proteins of filamentous nature that are of essential therapeutic focus in malignant chemotherapy (Nogales 2001). Microtubule restricting drugs have been a significant piece of the pharmacopeia of cancer for quite a long time, and until the appearance of target based treatment, microtubules were the main options as a remedial target in cancer. Insensitivity to mechanisms of chemoresistance, improved tumor specificity, and decreased neurotoxicity, are the three primary goals in the current exploration for new and advanced microtubule restricting molecules (Jordan et al. 1998). Microtubules perform some essential jobs in cell trafficking, proliferation, signaling, and programmed migration of eukaryotic cells. In mammalian cells, microtubules assume remarkable roles amid interphase and cell division. Microtubules which establish the mitotic spindle are impeccably sensitive to therapeutic inhibitors (Giannakakou et al. 2000). This clarifies why agents that affect microtubule function have turned out to be profoundly useful in tumor patients. The *vinca* alkaloids, distinguished more than 50 years ago and the taxanes, first isolated right around 40 years back are presently used in the treatment of solid tumors like ovarian tumor, Non-small cell lung carcinoma (NSCLC), Kaposi's sarcoma and also in case of hematological malignancies (Jordan and Wilson 2004; Mollinedo and Gajate 2003).

3.2 Microtubules

α and β tubulin subunits together self-assemble to form protofilaments. Ten to 15 of these protofilaments join together to form a 24 nm wide hollow cylinder called microtubule. Microtubules are the most important functional components of the cytoskeleton. They are found in every eukaryotic cell, and they are associated with mitosis, cell motility, intracellular transport, and support of cell shape (Jordan and Wilson 2004). Microtubules are dynamic structures i.e. they are in a constant state of growth and shortening, termed polymerization and depolymerization respectively. They are very essential and critical for the proper functioning and maintenance of eukaryotic cells. Hence, understanding their structure, dynamics and how their functional features are regulated by cells is vital.

3.2.1 Microtubule Dynamics

Microtubules are to a great extent dynamic structure made out of αβ tubulin molecules that are consistently integrated or shed into the cytoplasm as the microtubules significantly polymerize and de-polymerize at the same time. Nucleation and elongation are the two main events that conjure the polymerization of the tubulin heterodimers (Mollinedo and Gajate 2003). First of all, a polymerization core is generated which is thereafter elongated by reversible addition of tubulin dimers. At steady state, the microtubule polymer length is kept constant by the equal rate of polymerization and depolymerization (Zhou and Giannakakou 2005). This continuous characteristic episode of rapid growth and shrinkage is called as dynamic instability. Dynamic instability can be portrayed as a consolidated capacity of four factors: the rate of microtubule growth, the rate of shortening, the frequency of transition from a growing or paused (neither developing nor shortening) state to a shortening state (an occasion referred to as "dynamic catastrophe"), and the frequency of transition from a shortening state to a growing or paused state (termed "rescue"). It is generally accepted that hydrolysis of GTP sustains the dynamic instability of microtubules (Mitchison and Kirschner 1984a). Structurally comparative subunits of 55 kDa each, the α-and β-monomers both have a GTP binding locale that can harbor one molecule of GTP. Hydrolysis of GTP powers the "dynamic" conduct of microtubules and has crucial implications for a plenty of functions performed by microtubules. The GTP bound β-tubulin subunit present within the free tubulin molecule attaches with the microtubule polymer (Pasquier et al. 2006). The GTP is thereafter hydrolyzed to GDP which still remains bound to the tubulin. When the rate of tubulin subunit integration is very high, even before the GTP of the last included subunit is hydrolyzed, another GTP-bound subunit is added to the microtubule polymer. While when the rate of subunit addition is moderate, adequate time gap enables the GTP to get hydrolyzed, thereby uncovering the GTP cap. The conformational changes that arises in the subunits due to the hydrolysis of the bound GTP molecule, causes a significant reduction in their affinity to bind with the neighboring molecules in the polymer. This causes separation of the subunits and the microtubular filament is constrained into a bent shape. This uniformly puts a strain on the straight protofilament and makes it turn outward symmetrically from the microtubule (Correia and Lobert 2001). Multiple studies have demonstrated that the depolymerization rate of microtubules is around 100 times slower from an end containing GTP-tubulin in contrast to the one containing GDP-tubulin. GTP-tubulin gradually polymerizes into linear protofilaments, which instigates the gradual development of the microtubule, while GDP-tubulin results in depolymerization and bent protofilaments. The firm tube-shaped structure of the microtubules is maintained by the presence of a protective GTP cap at the tip (Mitchison and Kirschner 1984a) (Fig. 3.1).

3.2.2 Treadmilling

Microtubules are always in a state of dynamic growth, developing from one end while degrading from the other yet keeping the overall polymer length constant. Tubulin dimers are continuously added to the plus end while they are constitutively released from the minus end. This dynamic conduct is defined as the phenomenon of treadmilling (Mitchison and Kirschner 1984a). The rapid microtubular treadmilling

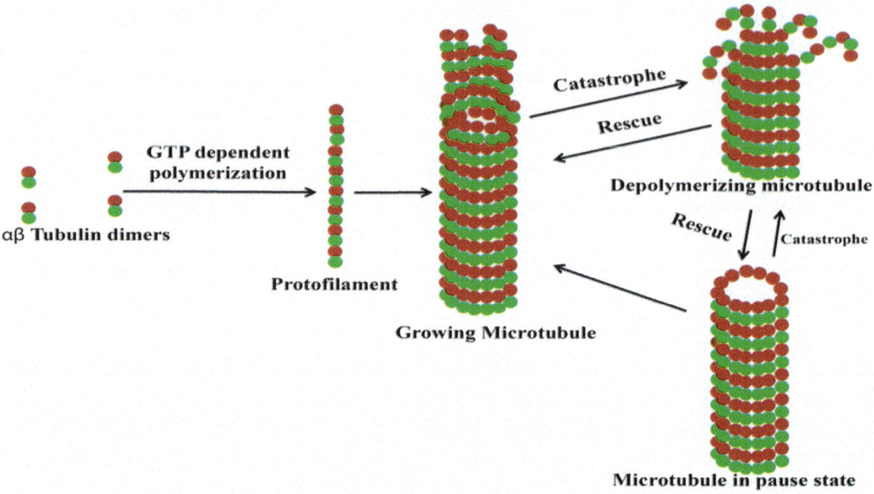

Fig. 3.1 Regulation of tubulin assembly and disassembly

has been observed to happen amid metaphase and anaphase, where it plays an extremely important role in the relay of intracellular signals from the kinetochore to the poles (Mitchison and Kirschner 1984b). The phenomenon of treadmilling and dynamic instability are not mutually exclusive and frequently happen in concert (Mitchison and Kirschner 1984b). There is an unpredictable arrangement of cellular mechanisms that decide the dynamics of microtubules in-vivo, including the level of expression of various tubulin isotypes, the activity of MAPs, and posttranslational changes associated with each component (Margolis and Wilson 1998). There are essentially six isotypes of α-tubulin and seven isotypes of β-tubulin present in humans, and the level of expression of each isotype fluctuates among different tissues and cells along with their posttranslational changes including polyglutamylation, phosphorylation, acetylation, etc. The suitability of the tubulin-binding drugs varies in case of different malignancies according to their affinities for the various tubulin subtypes expressed (Cassimeris et al. 1987).

3.2.3 Functions of Microtubules

3.2.3.1 Microtubules Are the Main "Building Blocks" That Form the Cytoskeleton

Which is the cell's framework within which all components of the cell are held in position or allowed to move within certain constraints (Ke et al. 2000)?

3.2.3.2 Movement of Materials and Structures

Within cells e.g. they help to form the mitotic spindle apparatus via which chromosomes move to two poles during cell division by mitosis. This occurs during the "prophase" of mitosis (Ye et al. 1998).

3.2.3.3 Locomotion in Protists and Bacteria via Cilia and Flagella Composed of Microtubules

To form cilia or flagella, microtubules arrange themselves in a "9 + 2" array. Each of the two central microtubules consists of a single microtubule with 13 protofilaments arranged to form the wall of a circular tube. Each of the outer nine consists of a pair of microtubules that share a common wall (Brinkley 1985).

3.3 Microtubules as Anticancer Drug Targets

Tubulin Binding Agents (TBAs) are a group of novel microtubule targeting anticancer drugs (Heidemann 2006). The increasing resistance of cancer cells to conventional anticancer drugs has rendered their efficacy vulnerable (Kops et al. 2005). After the complete elucidation of the three dimensional structure of the αβ-tubulin dimer, it became much easier to categorically design microtubule targeting modified drugs and analyze their binding and therapeutic potential (Weisenberg et al. 1968; Gould and Borisy 1977). A serious issue that is to be considered in this regard is which one(s) of the tubulin species, their isotypes, or their post-translationally altered structures, ought to be particularly focused while targeting cancer cells (Walker et al. 1998). This chapter depicts a clearer picture of our current knowledge regarding microtubular dynamics as well as their mechanism of functioning in neoplastic cells and also provides an update on the current status of the microtubule-targeting class of phytochemicals with anticancer properties. The major focus of the new class of tubulin targeting drugs has been on the βIII isotype. It is found to be overexpressed in most cancer cells and works as a survival factor related with the dynamic instability of microtubules. Other targets include the γ-tubulin, the most important microtubule nucleating protein; and spastin which is the microtubule separating enzyme, associated with motility and proliferation of cancer cells.

3.4 Microtubule Targeted Drugs

3.4.1 Two Classes: Stabilizers and Destabilizers

Microtubule targeting anti-mitotic drugs can be categorized under two heads; Stabilizers and destabilizers (Valiron et al. 2001). Stabilizers at high concentrations stimulate the polymerization of microtubules. Paclitaxel, docetaxel and epothilones are some of the examples which hold special interest in cancer therapeutics. Destabilizers on the other hand are drugs that at high concentrations hinder with the process of polymerization of microtubules and include few well known compounds like the Vinca alkaloids (vincristine, vinblastine, vinorelbine, vinflunine and vindesine), colchicine, halichondrins, estramustine, cryptophycins, and combretastatins that are under clinical examination or currently used for treating neoplastic conditions (Cassimeris 1993). The grouping of drugs as 'stabilizers' or 'destabilizers' of microtubule is based on their capacity to enhance or reduce microtubule polymerization at high levels (Al-Taweel et al. 2015). The most imperative activity of the two classes of drugs is the suppression of spindle microtubule element functions which completely arrests mitosis at the metaphase-anaphase transition stage and activates apoptotic cell death (Aziz et al. 2014).

3.5 Phytochemicals as Anti-cancer Drugs

A lot of studies carried out throughout the world have reported the presence of anticancer properties in many chemical compounds extracted from plant sources. These phytochemicals have been promoted as anticancer drugs giving a distinctive emphasis on those which interfere with the microtubule stabilization process and thereby inhibiting apoptosis in cancer cells (Badmus et al. 2015). The first clinically formulated and used phytochemicals were isolated from the plant *Catharanthus roseus* and named as Vinblastine and Vincristine. They were used in the treatment of breast, lung and blood cancers along with other anticancer drugs in combination. Paclitaxel isolated from the bark of the tree *Taxus brevifolia*, is one of the most widely used and effective phytochemical compound against ovarian, cervical, breast and pancreatic cancers (Nobili et al. 2009).

3.5.1 Apoptotic Pathways Induced by the Phytochemicals

Polyphenols are natural compounds with universal availability. They are present in rich quantities in plant extracts and organic food products like grains, fruits and vegetables. Flavonoids (catechins, flavones, isoflavones, and anthocyanins) and carotenoids are active anti-oxidants and have been found to cause apoptosis in cancer cells (Hatfield et al. 2014). Some worldwide investigations have

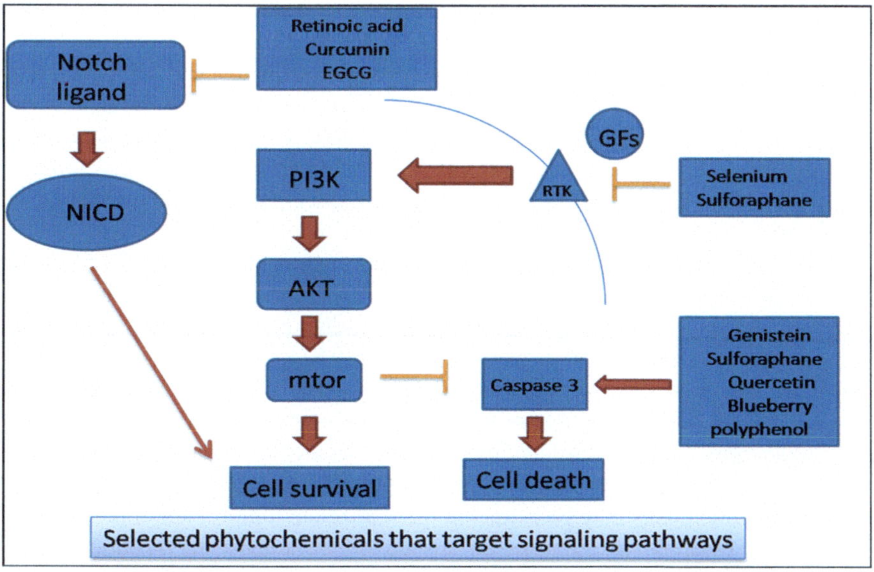

Fig. 3.2 Selected phytochemicals along with the signaling pathways targeted by them

recommended that phenolic compounds or their concentrates can actively impact the survival of cancer stem cells too (Hatfield et al. 2014). Polyphenols have the ability to instigate programmed cell death via caspase-3 in malignant cells by means of some crucial signaling networks or their apparent cross-talk (Fig. 3.2).

Akt/mTOR signaling is known to be crucial for the survival and metastasis of cancer cells. Akt inhibition causes a much-preferred reduction in cancer cell motility and induction of apoptosis. Selenium works naturally as selenoproteins that scavenge reactive oxygen species. Selenium inclusions were found to trigger apoptosis in leukemic cells and it was shown that the apoptosis was controlled through Akt/mTOR signaling (Li et al. 2010). The correct fundamental mechanisms supplementing this selenium-intervened cancer cell apoptosis anticipates future investigations. Likewise, sulforaphane found in cruciferous vegetables, has been shown to decrease the feasibility of pancreatic and breast malignancy cases by influencing Hedgehog or Wnt/β-catenin signaling pathways (Graham et al. 2000). A few investigations have likewise shown that sulforaphane can downregulate Akt signaling in different carcinomas (Bishayee and Block 2015).

3.6 Prospect of Clinical Development of New Microtubule Targeted Phytochemicals

Due to minimal cost, unwavering quality, and limited side effects the world is gradually progressing towards natural alternative therapeutics. Due to the development of multidrug resistance in diseases, scientists have escalated their endeavors for the search and advancement of phytopharmaceuticals against serious health disorders like cancer, diabetes, tuberculosis etc. Although earth is covered with many flora but only few floral species and their constitutive phytochemicals have been screened for these purpose. However, only a limited number out of them have proved out to be useful at the clinical level (Amin et al. 2009). All the anticancer phytochemicals that have been depicted in this chapter needs additional clinical investigations for their viability and toxicological documentation. They should be transformed to drugs with adequate bioavailability.

India is one of the biggest exporter and producer of medicinal plants and many plant-based health care items (Vetrivel et al. 2009). A considerable scientific effort has been put forward to establish the efficacy of Indian plants and their products in the fight against some dreadful diseases. They should be further investigated for their anticancer potential. The immense therapeutic qualities of these medicinal plants have made them highly sought after throughout the world and thereby has rendered their biodiversity extremely vulnerable. Owing to the expanding demand and deforestation, over abuse of these plants proceeds around the world. Due to this, a portion of the therapeutic plants may go extinct in the immediate future. Thus at this hour, it is extremely essential to monitor the sustainability of these therapeutic plants. Suitable seed protection, germplasm conservation, and cryopreservation are promising techniques for this purpose.

Plant derived compounds like roscovitine, flavopiridol, betulinic acid, silvestrol, and combretastatin are as of now in clinical or preclinical phases of drug development attributable to their anti-cancer impacts (Diederich and Cerella 2016). Alvocidib (Flavopiridol) is a prospective anti-cancer candidate undergoing clinical trials, which acts by blocking cell division and induction of apoptosis. DJ-927, TPI-287 and 9-amino camptothecin, are analogues of camptothecin undergoing rigorous clinical trials. Tafluposide 105 and NK 611 (Both analogues of epipodophyllotoxin) are in phase 1 clinical trials (Diederich and Cerella 2016). BMS-188797, DHA-paclitaxel and some other analogues of paclitaxel are also in various stages of clinical experimentation. Other important candidates include the combretastatin analogues, CA4PO4, AVE- 8064, AVE-8063 which are currently in different stages of clinical trials (Shanmugam et al. 2016; Wani et al. 1971).

3.7 Conclusion

With the development of new cutting age technologies and modern crystallographic techniques, huge advancement has been made in the development of more sophisticated anti-mitotic drugs with greater efficacies at both cellular and organism level. There still remains a lot to learn especially about their mechanism of activity alone or in combination with other photochemical. Microtubules appear to be a troublesome target for crystallization; hence the crystal structure of the microtubule binding drugs while bound to the microtubules remains hard to be elucidated.

As an outcome, it turns out to be extremely difficult to reach firm inferences from these solved structures regarding their binding efficiency and the exact binding conditions of the drugs inside tubulin and the consequences of these binding on the larger microtubule structure itself. Different drugs affect the microtubular structure in different ways. Taxol remains bound at the inside (lumenal) edge of α-tubulin but doesn't influence the GTP hydrolysis or exchange scale of the dimer, and in this manner bolsters microtubule stability and development (Schiff et al. 1979). Vinca drugs bind to the interdimer interface, adjusting the collaboration among neighboring heterodimers, and in this manner makes a kink in the natural vector to such an extent that it never again lines up with the straight axis of the microtubular hub (Schiff et al. 1979). Colchicine binds to the free heterodimer, at the intradimer interface, adjusting the adaptation of the heterodimer, opening a wedge between the two monomers, and altogether changing the association of the vector with protofilaments (Desai and Mitchison 1997). It binds at a site neighboring the non-hydrolyzable GTP and permits an increase in GTP hydrolysis and exchange scale, hence diminishing polymerization and increasing depolymerization of the microtubules (Erickson and O'Brien 1992). Since just the straight GTP-bound tubulin conformation can be incorporated into the cross-sectional lattice of the microtubule, any drug that supports a straight vector, will function as a polymerizer and help microtubule development, while any drug that interferes with this vector will promote depolymerization and limit microtubule development (Sa and Das 2008). Indeed, even minute changes in the microtubule structure can cast huge effects on their biological capacity and dynamics. At the point when seen at an atomic level, this viewpoint will ideally prompt novel insights into the comprehension of the activities of antimitotic drugs on the structure of microtubules, which will, in turn, rationalize the judicious plan of developing new and more compelling chemotherapeutic agents.

Acknowledgements Sonia Kapoor acknowledges the grant by DST-SERB, and Department of Biotechnology, Govt. of India for funding.

References

Akindele AJ, Wani ZA, Sharma S, Mahajan G, Satti NK, Adeyemi OO, Mondhe DM, Saxena AK. In vitro and in vivo anticancer activity of root extracts of Sansevieria liberica Gerome and Labroy (Agavaceae). Evid Based Complement Alternat Med. 2015;2015:560404. https://doi.org/10.1155/2015/560404.

AlSinani SS, Eltayeb EA, Coomber BL, Adham SA. Solamargine triggers cellular necrosis selectively in different types of human melanoma cancer cells through extrinsic lysosomal mitochondrial death pathway. Cancer Cell Int. 2016;16:11. https://doi.org/10.1186/s12935-016-0287-4.

Al-Taweel AM, Perveen S, Fawzy GA, Ibrahim TA, Khan A, Mehmood R. Cytotoxicity assessment of six different extracts of Abelia triflora leaves on A-549 human lung adenocarcinoma cells. Asian Pac J Cancer Prev. 2015;16:4641–46445. https://doi.org/10.7314/APJCP.2015.16.11.4641.

Amin A, Gali-Muhtasib H, Ocker M, Schneider-Stock R. Overview of major classes of plant-derived anticancer drugs. Int J Biomed Sci. 2009;5:1–11.
Aziz MY, Omar AR, Subramani T, Yeap SK, Ho WY, Ismail NH, Ahmad S, Alitheen NB. Damnacanthal is a potent inducer of apoptosis with anticancer activity by stimulating p53 and p21 genes in MCF-7 breast cancer cells. Oncol Lett. 2014;7:1479–84.
Badmus JA, Ekpo OE, Hussein AA, Meyer M, Hiss DC. Antiproliferative and apoptosis induction potential of the methanolic leaf extract of *Holarrhena floribunda* (G. Don). Evid Based Complement Alternat Med. 2015;2015:756482. https://doi.org/10.1155/2015/756482.
Bishayee A, Block K. A broad-spectrum integrative design for cancer prevention and therapy: the challenge ahead. Semin Cancer Biol. 2015;35:S1–4.
Brinkley BR. Microtubule organizing centers. Annu Rev Cell Biol. 1985;1:145–72. https://doi.org/10.1146/annurev.cb.01.110185.001045.
Cassimeris L. Regulation of microtubule dynamic instability. Cell Motil Cytoskeleton. 1993;26:275–81. https://doi.org/10.1002/cm.970260402.
Cassimeris LU, Walker RA, Pryer NK, Salmon ED. Dynamic instability of microtubules. BioEssays. 1987;7:149–54. https://doi.org/10.1002/bies.950070403.
Correia JJ, Lobert S. Physiochemical aspects of tubulin-interacting antimitotic drugs. Curr Pharm Des. 2001;7:1213–28.
Desai A, Mitchison TJ. Microtubule polymerization dynamics. Annu Rev Cell Dev Biol. 1997;13:83–117. https://doi.org/10.1146/annurev.cellbio.13.1.83.
Diederich M, Cerella C. Non-canonical programmed cell death mechanisms triggered by natural compounds. Semin Cancer Biol. 2016;41:4–34. https://doi.org/10.1016/j.semcancer.2016.06.001.
Erickson HP, O'Brien ET. Microtubule dynamic instability and GTP hydrolysis. Annu Rev Biophys Biomol Struct. 1992;21:145–66. https://doi.org/10.1146/annurev.bb.21.060192.001045.
Giannakakou P, Sackett D, Fojo T. Tubulin/microtubules: still a promising target for new chemotherapeutic agents. J Natl Cancer Inst. 2000;92:182–3.
Gould RR, Borisy GG. The pericentriolar material in Chinese hamster ovary cells nucleates microtubule formation. J Cell Biol. 1977;73:601–15.
Graham JG, Quinn ML, Fabricant DS, Farnsworth NR. Plants used against cancer – an extension of the work of Jonathan Hartwell. J Ethno Pharmacol. 2000;73:347–77.
Hatfield DL, Tsuji PA, Carlson BA, Gladyshev VN. Selenium and selenocysteine: roles in cancer, health, and development. Trends Biochem Sci. 2014;39:112–20.
Heidemann S. Microtubules, leukemia, and cough syrup. Blood. 2006;107:2216–7. https://doi.org/10.1182/blood-2005-12-5142.
Jain R, Kosta S, Tiwari A. Ayurveda and cancer. Pharmacogn Res. 2010;2:393–4. https://doi.org/10.4103/0974-8490.75463.
Jordan MA, Wilson L. Microtubules as a target for anticancer drugs. Nat Rev Cancer. 2004;4:253–65. https://doi.org/10.1038/nrc1317.
Jordan A, Hadfield JA, Lawrence NJ, McGown AT. Tubulin as a target for anticancer drugs: agents which interact with the mitotic spindle. Med Res Rev. 1998;18:259–96.
Ke Y, Ye K, Grossniklaus HE, Archer DR, Joshi HC, Kapp JA. Noscapine inhibits tumor growth with little toxicity to normal tissues or inhibition of immune responses. Cancer Immunol Immunother. 2000;49:217–25.
Kops GJ, Weaver BA, Cleveland DW. On the road to cancer: aneuploidy and the mitotic checkpoint. Nat Rev Cancer. 2005;5:773–85. https://doi.org/10.1038/nrc1714.
Li Y, Zhang T, Korkaya H, Liu S, Lee HF, Newman B, Yu Y, Clouthier SG, Schwartz SJ, Wicha MS, Sun D. Sulforaphane, a dietary component of broccoli/broccoli sprouts, inhibits breast cancer stem cells. Clin Cancer Res. 2010;16:2580–90. https://doi.org/10.1158/1078-0432.CCR-09-2937.

Mahima RA, Deb R, Latheef SK, Abdul Samad H, Tiwari R, Verma AK, Kumar A, Dhama K. Immunomodulatory and therapeutic potentials of herbal, traditional/indigenous and ethnoveterinary medicine. Pak J Biol Sci. 2012;15:754–74.

Margolis RL, Wilson L. Microtubule treadmilling: what goes around comes around. BioEssays. 1998;20:830–6. https://doi.org/10.1002/(SICI)1521-1878(199810)20:10<830::AID-BIES8>3.0.CO;2-N.

Mitchison T, Kirschner M. Microtubule assembly nucleated by isolated centrosomes. Nature. 1984a;312:232–7.

Mitchison T, Kirschner M. Dynamic instability of microtubule growth. Nature. 1984b;312:237–42.

Mollinedo F, Gajate C. Microtubules, microtubule-interfering agents and apoptosis. Apoptosis. 2003;8:413–50.

Nobili S, Lippi D, Witort E, Donnini M, Bausi L, Mini E, Capaccioli S. Natural compounds for cancer treatment and prevention. Pharmacol Res. 2009;59:365–78. https://doi.org/10.1016/j.phrs.2009.01.017.

Nogales E. Structural insight into microtubule function. Annu Rev Biophys Biomol Struct. 2001;30:397–420.

Pasquier E, Honore S, Braguer D. Microtubule-targeting agents in angiogenesis: where do we stand? Drug Resist Updat. 2006;9:74–86.

Sa G, Das T. Anti cancer effects of curcumin: cycle of life and death. Cell Div. 2008;3:14. https://doi.org/10.1186/1747-1028-3-14.

Schiff PB, Fant J, Horwitz SB. Promotion of microtubule assembly in vitro by taxol. Nature. 1979;277:665–7.

Shanmugam MK, Lee JH, Chai EZ, Kanchi MM, Kar S, Arfuso F, Dharmarajan A, Kumar AP, Ramar PS, Looi CY, Mustafa MR, Tergaonkar V, Bishayee A, Ahn KS, Sethi G. Cancer prevention and therapy through the modulation of transcription factors by bioactive natural compounds. Semin Cancer Biol. 2016;41:35–47. https://doi.org/10.1016/j.semcancer.2016.03.005.

Valiron O, Caudron N, Job D. Microtubule dynamics. Cell Mol Life Sci. 2001;58:2069–84. https://doi.org/10.1007/PL00000837.

Vetrivel U, Subramanian N, Pilla K. InPACdb – Indian plant anticancer compounds database. Bioinformation. 2009;4:71–4.

Walker RA, O'Brien ET, Pryer NK, Soboeiro MF, Voter WA, Erickson HP, Salmon ED. Dynamic instability of individual microtubules analyzed by video light microscopy: rate constants and transition frequencies. J Cell Biol. 1998;107:1437–48.

Wani MC, Taylor HL, Wall ME, Coggon P, McPhail AT. Plant antitumor agents. VI. The isolation andstructure of taxol, a novel antileukemic and antitumor agent from Taxusbrevifolia. J Am Chem Soc. 1971;93:2325–7.

Weisenberg RC, Borisy GG, Taylor EW. The colchicine-binding protein of mammalian brain and its relation to microtubules. Biochemistry. 1968;7:4466–79.

Ye K, Ke Y, Keshava N, Shanks J, Kapp JA, Tekmal RR, Petros J, Joshi HC. Opium alkaloid noscapine is an antitumor agent that arrests metaphase and induces apoptosis in dividing cells. Proc Natl Acad Sci U S A. 1998;95:1601–6.

Zhou J, Giannakakou P. Targeting microtubules for cancer chemotherapy. Curr Med Chem Anticancer Agents. 2005;5:65–71.

Chapter 4
Reprogramming of Tumor Associated Immune Cells by Phytochemicals: In-vitro Approaches for Cancer Treatment

Pradip Kumar Jaiswara, Vishal Kumar Gupta, Shiv Govind Rawat, Pratishtha Sonker, and Ajay Kumar

4.1 Introduction

Immune system is a complex network of specified cells, tissues and organs which prevents the development of microbial infections and transformed cells in an organism through complex processes. Moreover, myriads of biomolecules belonging to different classes of proteins, lipids, and carbohydrates regulate the immune system in an orchestrated manner (Wolowczuk et al. 2008). Therefore, any alterations in this system results into the development of several pathological conditions such as cancer, autoimmune disorder, allergy and immunodeficiency diseases (Caspi 2008; Her and Kavanaugh 2016; Wong et al. 2017). Henceforth, immune system maintains the homeostasis of the body through eliminating infectious, toxic and allergic agents without harming the host tissues (Chaplin 2010).

Further, immunosuppression has also been identified as a major culprit behind the progression of cancer as immune cells, particularly, macrophages, NK cells and Tc cells are found to be suppressed in the tumor-bearing host (Whiteside 2008). The immunosuppressive state is developed due to the up-regulated expression of immunosuppressive molecules by tumor cells as well as surrounding tumor-associated cells. Studies also indicate the direct or indirect involvement of compromised tumor-associated immune cells in tumor progression. Therefore, reactivation of suppressed antitumor immune system has emerged as a novel oncotherapeutic strategy.

Phytochemicals are bioactive compounds, synthesized by various plants, have several therapeutic values, including immunomodulatory, anticancerous, antioxidant, and antimicrobial activities (Sithranga Boopathy and Kandasamay 2010; Shukla and Mehta 2014; Adeonipekun et al. 2016). Moreover, several in-vivo and in-vitro studies have also indicated the immunomodulatory action of various phytochemicals in boosting the immune system of cancer patients either by altering the

P. K. Jaiswara · V. K. Gupta · S. G. Rawat · P. Sonker · A. Kumar (✉)
Department of Zoology, Banaras Hindu University, Varanasi, India
e-mail: ajayzoo@bhu.ac.in

S. Kumar, C. Egbuna (eds.), *Phytochemistry: An in-silico and in-vitro Update*,
https://doi.org/10.1007/978-981-13-6920-9_4

number or functions of immune cells of tumor microenvironment. In this book chapter, we have discussed about the alterations in the behavior of tumor-associated immune cells by phytochemicals in-vitro.

4.2 Behavior of Immune Cells in Tumor-Bearing Host

Tumor microenvironment plays an indispensable role in tumor development and its progression. It is composed of various cellular components such as tumor cells, immune cells, stromal cells, blood vessels, and non-cellular components, including cytokines, chemokines, and metabolites. Tumor-derived factors include lactate, inflammatory mediators, chemokines and cytokines which help in the recruitment and suppression of immune cells at the site of the tumor (Yuan et al. 2016) (Fig. 4.1). Tumor cells reprogram the cellular components of immune system by altering their number and function. In addition, tumor cells also suppress the expression of major histocompatibility complex and co-stimulatory molecules which help them in evading the immune surveillance, an immune mechanism helps in the identification and clearance of transformed cells (Driessens et al. 2009; Seager et al. 2017).

Macrophages are major constituents of mononuclear phagocyte system which play a crucial role in providing the innate immunity and activating the adaptive immunity (Richards et al. 2013). They also play a vital role in the regulation of various physiological processes, including angiogenesis and removal of apoptotic cells (Nielsen and Schmid 2017).

The macrophages are sensitive to the surrounding environment and have polarizing potential in response to various intra- and extracellular factors such as IFN-γ, IL-4, IL-13 and lipopolysaccharide (LPS). In presence of IFN-γ and LPS, macrophages polarize into M1 macrophages whereas in presence of IL-4 and IL-13, they polarize into M2 macrophages. M1 macrophages secrete various pro-inflammatory cytokines namely, IFN-γ, TNF-α and IL-12 and found to be involved in the antigen presentation, activation of Tc and NK cells (Nielsen and Schmid 2017; Roszer 2015). However, M2 macrophages also called as alternatively activated-M2 macrophages, help in angiogenesis, tissue repairing and remodeling through VEGF or EGF (Nielsen and Schmid 2017). The macrophages associated to cancer are basically M2-like macrophages and commonly called as tumor-associated macrophages (TAM), which induces tumor progression via producing the tumor-promoting factors such as IL-10, TGF-β, and VEGF (Yang and Zhang 2017). They promote cancer development either via augmenting proliferation, angiogenesis, and invasion and metastasis or by suppressing the antineoplastic activity of Tc and NK cells (Yang and Zhang 2017; Dahlberg et al. 2015). Therefore, macrophage polarization plays a crucial role in the regulation of cancer progression.

The natural killer cell is another important cellular component of the innate immune system, having a unique potential for the identification and elimination of neoplastic and virally-infected cells (Langers et al. 2012). NK cells are large,

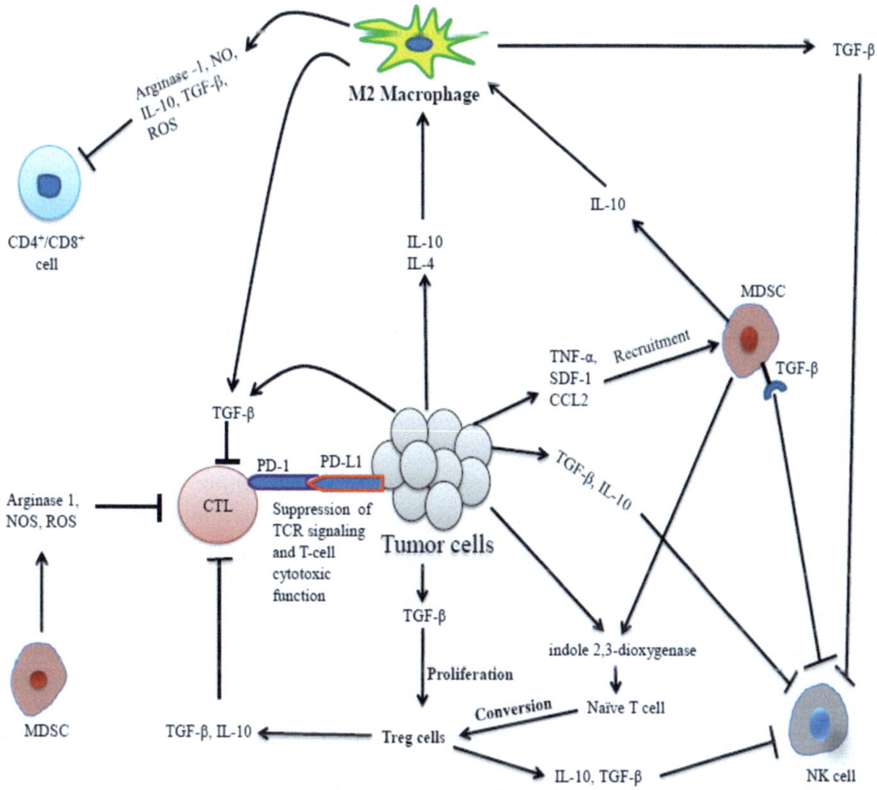

Fig. 4.1 Factors responsible for the development of immunosuppressive state in tumor-bearing host: Tumor cell-associated molecules or derived factors lead to suppression of innate and adaptive immune cells of the tumor-bearing host by modulating their infiltration, proliferation, differentiation, and function

granular lymphocytes, usually identified as $CD3^-CD56^+$ cells in humans or $CD3^-NKp46^+$ or $CD3^-NK1.1^+$ cells in mice. Further, human NK cells are classified on the basis of their immunophenotype and function into two subpopulations: $CD56^{dim}$ and $CD56^{bright}$. $CD56^{dim}CD16^+$ NK cells have high cytotoxic activity than CD56 ^{bright} $CD16^+$ NK cells due to the presence of a large number of perforin and granzyme containing granules. In addition, $CD56^{dim}CD16^+$ NK cells also perform antibody-dependent cell cytotoxicity (ADCC) to tumor cells. Further, approximately 90% of NK cells of peripheral blood belong to $CD56^{dim}CD16^+$ subpopulation whereas $CD56^{bright}$ $CD16^+$ subpopulation constitutes only 10% of total NK cells which are mainly involved in the secretion of large amounts of cytokines, including IFN-γ. NK cell recognizes abnormal cells such as tumor cells and virus-infected cells on the basis of deficient MHC-I and inducible ligand secreted by target cells (Langers et al. 2012; Waldhauer and Steinle 2008; Wu and Lanier 2003). However, immunosuppressive factors of tumor microenvironment suppress proliferation, maturation and functional activities of NK cells. TGF-β, primarily secreted by tumor cells, tumor-associated macrophages and Treg cells, is reported as one of the crucial immunosuppressive factors responsible for the suppression of the antineoplastic action of NK cells in tumor-bearing host. Further, a study indicates that melanoma cells show resistivity to NK cell-mediated killing when low number of NK cells (represents infiltrated NK cells at the tumor site) co-cultured with melanoma cells via elevating the expression of HLA-I molecules and decreasing the expression of NKG2D ligands on the surface of resistant melanoma cells (Balsamo et al. 2012). Therefore, tumor microenvironment is one of the key factors for the modulated NK cell-mediated antitumor immunity.

Several evidences suggest the crucial role of tumor microenvironment in the modulation of cell-mediated immunity. Immunosuppressive tumor microenvironment constitutes several cellular and non-cellular components, including Tregs, MDSCs, TAMs, tumor-associated neutrophils (TANs), indole 2,3-dioxygenase (IDO), tumor-associated PD-L1, IL-10, TGF-β, VEGF, and galectins. The cellular and non-cellular components of TME alter cell-mediated immunity either directly or indirectly through inhibition of infiltration, survival, proliferation, differentiation and function of cytotoxic T cells along with switching the function of Th cells towards tumor-promoting type 2 immune response (Chew et al. 2012; Devaud et al. 2013; Gajewski et al. 2013).

Recent reports suggest a vital role of regulatory T (Treg) cells in cancer progression. Treg cells usually inhibit the activity of other T cells and thus play a crucial role in the prevention of autoimmune diseases by providing immunological tolerance (Corthy 2009). This immunosuppressive nature of Treg cells promotes unhindered progression of cancer cells by providing immunosuppressed microenvironment to tumor cells through inhibiting the activities of Th and Tc cells.

4.3 Role of Phytochemicals in Immunomodulation

Phytochemicals are bioactive compounds synthesized by plants as secondary metabolites. Plants synthesize phytochemicals in response to infection, climatic or nutritional changes which help them in providing the protection during the unfavorable conditions. Phytochemicals belong to different classes of compounds, namely terpenes, terpenoids, alkaloids, flavonoids, glycosides, and coumarins. Phytochemicals not only provide protection to plants but also have several therapeutic values which include memory enhancing, cholesterol lowering, anti-diabetic and anticancerous activities. They are being used against various cancers due to their antiproliferative, antiangiogenic and antimetastatic properties. Further, several in-vitro and in-vivo studies indicate the immunomodulatory action of various phytochemicals. Immunmodulators are the compounds which can stimulate or suppress the immune response by modulating the constituents of the immune system. Thus, immunomodulators control immune response either acting as immunoadjuvants, immunostimulants or immunosuppresants (Cundell 2014). Below we have discussed the immunomodulatory role of few phytochemicals.

Curcumin is one of the most popular phytochemicals extracted from the rhizome of *Curcuma longa*. It acts as anti-cancer, anti-oxidant, anti-angiogenic, anti-proliferative and pro-apoptotic agent. It also shows immunomodulatory action as it modulates proliferation, differentiation, and function of various immune cells. Curcumin has been reported in the suppression of activated macrophages by lowering the secretion of pro-inflammatory cytokines such as TNF-α and IL-1β via inhibition of NF-κB pathway in a dose-dependent manner (Chen et al. 2008). Curcumin has also been observed in the inhibition of nitric oxide synthase (iNOS) in LPS and IFN-γ activated macrophages (Brouet and Ohshima 1995). Further, curcumin is reported in the modulation of tumoricidal activity of macrophages. An investigation by Bhaumik et al. also indicated that curcumin-treatment lowers the tumor cytotoxic ability of macrophages via suppressing the production of nitric oxide (Bhaumik et al. 2000). Therefore, the anti-tumor activity of curcumin is due to its direct tumor cell killing ability but not through immunostimulation. Further, a recent study by Zhou et al. suggests the vital role of curcumin in macrophage polarization via inhibiting TLR-4 signaling pathway. In this study, researchers found a dose-dependent inhibition of LPS/IFN-γ-induced M1 macrophage polarization along with the declined expression of pro-inflammatory cytokines, namely TNF-α, IL-6, and IL-12B (p40) at transcriptional as well as translational level via inhibiting the TLR4/MAPK/NF-κB signaling pathway (Zhou et al. 2015). Furthermore, curcumin has also been implicated in the polarization of macrophages towards M2 phenotype via increasing the expression of M2 markers such as macrophage mannose receptor (MMR), arginase-1 (Arg-1) and peroxisome proliferator-activated receptor-γ (PPAR-γ) and production of IL-4 and IL-13 (Gao et al. 2015). Furthermore, curcumin is also implicated in the suppression of T cell activation. A study by Kliem et al. demonstrated the suppression of T cell activation via inhibiting nuclear factor of activated T cells (NFAT) and NFAT-regulated cytokines expression through blocking Ca^{2+} mobilization (Kliem et al. 2012). In this study, they found that lower doses of curcumin (7–28 μM) suppressed the cell surface expression of CD69 (T cell activation marker) as well as T cell cytokines, namely IL-2 and IFN-γ in ionomycin and PMA-4 activated Jurkat T cells and peripheral blood T cells in a dose-dependent manner. On the other hand, apoptosis of Jurkat and peripheral blood T cells were observed at higher doses of curcumin (>30 μM). Further, they confirmed that curcumin inhibits cytokine production by activated T cells through inhibition of Ca^{2+} mobilization and NFAT. Therefore, they explored a novel mechanism behind the immunosuppressive action of curcumin. In addition, a report indicates curcumin-dependent suppression of Treg cell ($CD4^{+}CD25^{+}$ cells) activity occurs due to the decreased expression of CTLA-4 and Foxp3 (Zhao et al. 2012). Further, a recent study by Fiala suggests the role of curcumin in augmenting the cancer cell killing ability of NK cells. Although, inhibition of IFN-γ production by NK cells has been observed in curcumin-treated NK cells (Fiala 2015). Curcumin has also been reported in augmenting the tumoricidal activity of NK cells against breast cancer via increasing the population of $CD56^{dim}CD16^{+}$ NK cells as well as enhancing the production of IFN-γ by them through the up-regulated expression of pstat4 and pstat5 (Lee and Cho 2018). However, a study by Bill et al. suggests the inhibition of NK cells function by curcumin (Bill et al. 2009). In this study, they observed a declined production of IFN-γ by NK cells when healthy donor NK cells co-treated with curcumin and IL-12. Further, they also found decreased granzyme B release and IFN-γ secretion by curcumin-treated healthy donor NK cells when exposed to A375 melanoma cells. In a similar kind of study, Halder and colleagues showed inhibited production of IFN-γ by curcumin-treated NK cells co-cultured with pancreatic cancer cells (Halder et al. 2015).

Resveratrol is another phytochemical found in various dietary products and plants including grape wine, red wine, and peanuts. It also plays a diverse role in the regulation of various pathological conditions through its antimicrobial, chemopreventive, anticancer/proapoptotic, anti-inflammatory and antioxidant properties. Several reports have suggested the role of resveratrol in the modulation of immune cell function and activation (Leiro et al. 2002; Schwager et al. 2017; Zou et al. 2013; Espinoza et al. 2016; Falchetti et al. 2001; Li et al. 2014). Leiro et al. have demonstrated the modulated anti-oxidant and phagocytic activities of rat macrophage in a dose-dependent manner. They had reported that resveratrol inhibited the production of reactive oxygen intermediates (ROI) and reactive nitrogen intermediates (RNI) by macrophages at the concentration of 1–10 μM and 10–100 μM, respectively (Leiro et al. 2002). Recently, Schawager et al. have shown the anti-inflammatory action of resveratrol through inhibiting the production of pro-inflammatory cytokines by murine and human macrophages. In this study, authors have demonstrated that resveratrol reduces the expression of IL-1β, IL-6, IL-12(p70) and TNF-α in LPS-activated murine macrophage cell line at transcriptional as well as translational level. In addition, resveratrol also diminished PGE_2, IL-1β, IL-6 and CXCL10/IP-10 in LPS-activated human monocytic leukemia cell line at gene and protein level (Schwager et al. 2017). Furthermore, resveratrol modulates T-cell mediated immune response. Nagendran et al. have shown the inhibition of proliferation and activation of T cells via lowering the level of TNF-α and IFN-γ (Nagendran et al. 2013). Zou et al. have shown in-vitro and in-vivo inhibition of T-cell activation by resveratrol through inhibiting the expression of Th1 and Th2 cytokines, namely IL-2, IFN- γ, IL-4, and IL-5 (Zou et al. 2013). Furthermore, Espinoza et al. have showed that in-vitro treatment of PBMCs with resveratrol in presence of TCR stimulators led to increased number of Treg cells ($CD3^+CD4^+$ $Foxp3^+$ cells representing classical or conventional Treg cells) in culture medium but they did not observe any alteration in the number of Treg cells by resveratrol treatment in absence of TCR stimulation (Espinoza et al. 2016). Apart from macrophages and T cells, resveratrol is also involved in the modulation of NK cell activity. A study by Falchetti et al. indicates the role of resveratrol in the augmentation of antitumor activity of NK cells. In this study, isolated peripheral blood mononuclear cells (PBMCs) were treated with various concentrations of resveratrol for 18 h and further tumor cell killing ability of NK cell of PBMCs were estimated against myelogenous leukemia (K562 cells). Interestingly, they observed an increased tumor cell killing ability of NK cells-treated with the lower dose of resveratrol concentrations (0.075–1.25 μg/ml). However, the higher concentration of resveratrol exhibits suppressed cytolytic activity of NK cells (5 and 20 μg/ml) (Falchetti et al. 2001). Similar kind of findings has also been reported by Li et al. as they found enhanced expression of NKG2D and IFN-γ at gene as well as at protein level in NK cells-treated with lower concentrations (1.56–3.13 μM) of resveratrol along with increased tumor cell killing ability. However, resveratrol suppressed the viability and increased the apoptosis of NK cells at higher concentrations (Li et al. 2014). Therefore, both of these two studies suggested the biphasic action of resveratrol on the tumor cell killing ability of NK cells. In addition, Lu and Chen also reported augmented tumor cell killing ability of NK cells which are pre-incubated with lower doses of resveratrol (Lu and Chen 2010). They also observed increased expression of perforin and phosphorylated ERK-1/2 and JNK by resveratrol in a dose-dependent manner.

Flavonoids are low-molecular-weight phenolic compounds, found in different parts of the plant. Currently, more than 5000 secondary metabolites are identified as flavonoids such as quercetin, genistein, catechin, gallocatechin, epigallocatechin and epigallocatechin gallate. Further, their diverse biological roles have been identified in plants as well as in animals. They protect plants from biotic and abiotic stresses. Further, they also act as detoxifying agents as well as show anti-microbial activities (Panche et al. 2016). Furthermore, flavonoids also show antioxidant, anticancer, hepatoprotective and anti-inflammatory activities in animals (Kumar and Pandey 2013). In addition, they play a significant role in the immune regulation. Several in-vitro studies have suggested the role of flavonoids in the inhibition of pro-inflammatory cytokines such as TNF-α, IL-2, IFN-γ, IL-12 and IL-6 and affects the activities of phagocytic cells and NK cells (Kim et al. 2004; Nair et al. 2006; Watson et al. 2005; López- Posadas et al. 2008; Verbeek et al. 2004; Lee et al. 2009). Furthermore, several studies also suggest the role of flavonoids in the modulation of cellular immunity. In-vitro reports indicate the inhibition and promotion of Th1 and Th2-type immune response, respectively in the presence of flavonoids (Murr et al. 2005). Moreover, they have also been involved in the lymphocyte proliferation and inhibition of apoptosis through their antioxidant activities (Carrero et al. 1998). Studies have also shown the increased cytotoxic activity and proliferation of NK cell in the presence of flavonoids. Han et al. have reported that total flavonoids (FAC) isolated from *Astragalus complanatus* increases the cytotoxicity and proliferation of NK92 cells in-vitro. In this study, when they incubated NK92 cells with FAC for 72 h then they found an augmented proliferation of NK92 cells and their tumor cell killing ability against K562 cells in a dose-dependent manner. Further, they observed the maximum proliferation of NK92 cells at 50 and 100 μg/ml concentrations of FAC. In addition, they also showed increased expression of NK cell surface markers such as CD56, CD25, and CD69 and activating receptors such as NKG2D, NKP46, and NKP44 after treatment of NK92 cells with 50 and 100 μg/ml concentrations of FAC (Han et al. 2015). Recently, Hou et al. have also reported that total flavonoids from *Hippophae rhamnoides* (TFH) enhance the cytotoxic ability of NK cell against tumor cells via increasing the expression of NK cell activating receptors, namely NKp44 and NKp46, and perforin and granzymes B (Hou et al. 2017). Further, several flavonoids such as quercetin, and myricetin have also been reported in enhancing the activity of NK cells (Bae et al. 2010; Burkard et al. 2017). Therefore, these studies suggest a wide opportunity to harness the therapeutic potential of flavonoids against different pathological conditions.

4.3.1 *Modulation of Tumor-Associated Immune Cells Behavior by Phytochemicals*

Recently, several studies have been done to establish the role of phytochemicals in the modulation of immune responses for designing the novel therapeutic strategies for the treatment of various diseases. Moreover, immunomodulatory role of phytochemicals have also been observed in tumor-bearing hosts. Reports indicate that phytochemicals modulate infiltration, proliferation, survival, differentiation as well as the function of tumor-associated immune cells and thus they play a crucial role in tumor regression via restoring the antitumor immunity of the host.

4.3.1.1 Repolarization of Tumor-Associated Macrophages by Phytochemicals

There are several studies which indicate the modulation of macrophage function and their polarization by phytochemicals but only few reports indicate the role of phytochemicals in the reactivation of tumor-associated macrophages. In a recent report, Sun et al. have shown that resveratrol inhibits the polarization of human monocyte-derived macrophage into M2 (Sun et al. 2017). In this, study, they observed that lung cancer conditioned medium-induced M2 polarized macrophages were changed to M1 polarized macrophages after the treatment with resveratrol as they detected the increased level of IL-12 and TNF-α while a decreased IL-10 level. Further, they also observed decreased expression of M2 macrophage markers, namely MRC1, CCL24, chil3, and Retnla. In addition, they revealed that macrophage-induced proliferation of lung cancer cells was inhibited by resveratrol probably due to repolarization of M2 macrophages into M1 macrophages.

4.3.1.2 Modulation of Tumor-Associated T Cell by Phytochemicals

Cancerous cells escape from immunosurveillance by promoting immunosuppression through altering the population of T and NK cell and their functions. Reports also indicate suppressed effector T and NK cell responses whereas increased number and function of Treg cells in the vicinity of tumor due to elevated production of various inflammatory cytokines and chemokines by tumor cells, other immune and stromal cells (Smith and Kang 2013).

There are limited studies which indicate the T cell modulatory actions of phytochemicals in-vitro. An investigation by Bhattacharyya et al. indicates the immune-potentiating role of curcumin for T cells. In this study, investigators firstly isolated the T cells from the peripheral blood of healthy mice and incubated them with culture supernatant of curcumin-treated and untreated tumor cells and observed an increased tumor cell killing ability of T cells incubated with culture supernatant of curcumin-treated tumor cells (Bhattacharyya et al. 2010). In another investigation, the same group has shown the anti-apoptotic nature of curcumin for T cells incubated with cell-free Ehrlich's ascites carcinoma (EAC) and renal cell carcinoma (RCC) supernatant. They observed that when curcumin-treated human or mouse peripheral T cells incubated with culture supernatant of EAC and RCC, both $CD4^+$ and $CD8^+$ T cells show resistivity towards the tumor-induced apoptosis (Bhattacharyya et al. 2007). Moreover, they showed Bcl2-dependent inhibition of tumor-induced T cell apoptosis by curcumin (Bhattacharyya et al. 2007). Further, a recent study by Zou et al. suggested that curcumin participates in the conversion of $FOXP3^+$ regulatory T cells into Th1 cells (Zou et al. 2018). In this study, they treated peripheral Treg cells isolated from the lung cancer patients with curcumin for 6 days in-vitro and analyzed Th1 and Treg populations by flow cytometry and observed high percentage of Th1 cells. Therefore, they suggested the role of suppressed FOXP3 activities in the conversion of Treg cells into the Th1 cells by curcumin (Zou et al. 2018).

Another phytochemical, resveratrol has also shown in-vivo and in-vitro immunomodulatory actions on T cells. A study by Yang et al. suggests modulation of number and function of T cells by resveratrol (Yang et al. 2008). In this study, authors showed that ex-vivo treatment of splenocytes isolated from EG-7 bearing C57BL/6 mice with different concentrations of resveratrol decreased the number of Treg cells in concentration-dependent manner (Yang et al. 2008). Moreover, they also observed up-regulated expression of IFN-γ in $CD8^+$ T cells by resveratrol as they found an increased number of IFN-γ-expressing $CD8^+$ T cells in cells treated with resveratrol which were isolated from the lymph node of EG-7 bearing C57BL/6 mice (Yang et al. 2008).

4.3.1.3 Modulation of NK Cell Activity of Tumor-Bearing Host by Phytochemicals

Natural killer cell serves as one of the key components of the first line of defense system and eliminates the virus infected and transformed cells. NK cell preferentially identify and kill MHC-I deficient self-altered cells (Wu and Lanier 2003). NK cell activities depend on specific receptors present on its surface. NK cell expresses stimulatory and inhibitory receptors on its cell surface. When NK cell recognizes self MHC-I molecule, inhibitory receptors of NK cells don't allow the activation of NK cell and therefore, provide the signal for self-tolerance and limits the risk of autoimmune diseases (Mandal and Viswanathan 2015). NK cell recognized altered cells undergo lysis through two predominant pathways: (i) perforin and granzyme-mediated apoptosis and (ii) death receptor-mediated apoptosis (Mandal and Viswanathan 2015; Zamia et al. 1998). The MHC-I deficient transformed cell releases stress markers which are recognized by the activating receptors of NK cell, namely NKG2D, NKp30, NKp44, NKp46, and DC226. In addition, NK cell also modulates the activity of other immune cell such as dendritic and T cells via releasing cytokines and chemokines, including IFN-γ, TNF-α, CCL5, and IL-6 (Guillerey et al. 2016; Sungur and Murphy 2014). Further, several studies indicate a direct association between the altered NK cell function and cancer progression. In the tumor microenvironment, tumor-associated macrophages, myeloid-derived suppressor cells, Treg cells are either directly or indirectly involved in the suppression of antitumor activity of NK cells. The reactivation of NK cells either via applying synthetic or natural components is one of the emerging immunotherapeutic strategies for the treatment of various cancers. Further, due to least side effects and cost-effectiveness, the potential of phytochemicals to restore the antineoplastic property of NK cells are currently being studied for designing effective and long lasting immunotherapeutic approaches against cancer. There are very limited phytochemicals whose effects on the tumoricidal activity of NK cells have been observed. Further, studies indicate that curcumin has the ability to restore the suppressed tumor cytotoxic activity of natural killer cell which is generated due to the immunosuppressive tumor microenvironment. A study by Bhaumik et al. suggests that curcumin enhances the tumoricidal activity of NK cell (Bhaumik et al. 2000). In this study, investigators analyzed the tumor-killing ability of NK cells isolated from the spleen of normal, immune (AK-5 tumor rejected animals) and curcumin-treated animals on 5, 10, 15 and 20 days after AK-5 tumor transplantation. Further, they found an elevated cytotoxic activity of NK cell isolated from curcumin-administrated tumor bearing animals against AK-5 and YAC-1 as compared to immune NK cells (Bhaumik et al. 2000).

4.4 Conclusion

Phytochemicals not only exert cytotoxic activity against tumor cells directly but also limit tumor growth indirectly via modulating infiltration, proliferation, survival, apoptosis, differentiation, and function of various immune cells in the vicinity of tumor. Therefore, phytochemicals could emerge as promising immunotherapeutic agents against cancer.

Acknowledgement We thankfully acknowledge fellowship support to Pradip Kumar Jaiswara [Award No. 1002/(SC)(CSIR-UGC NET DEC. 2016] and Shiv Govind Rawat [Award No. 09/013 (0772/2018-EMR-I)] from CSIR, New Delhi and Pratishtha Sonker [Award No. F117.1/201516/ RGNF201517SCUTT4822/(SAIII/Website)] from UGC, New Delhi. Funding from University Grants Commission and Department of Science & Technology, New Delhi, India, in the form of UGC-Start-Up Research (F. No. 30-370/2017 (BSR)) and Early Career Research Award (ECR/2016/001117) is highly acknowledged. Financial support from ISLS and UGC-UPE, Banaras Hindu University is also acknowledged. We also acknowledge UGC-CAS and DST-FIST program to the Department of Zoology, Banaras Hindu University, India.

References

Adeonipekun PA, Adeniyi TA, David ED. Antimicrobial properties and melissopalynology, proximate and elemental analyses of honey samples from three different ecozones in Nigeria. Notulae Sci Biol. 2016;8(3):326–33. https://doi.org/10.15835/nsb839844.

Bae JH, Kim JY, Kim MJ, Chang SH, Park YS, Son CH, Park SJ, Chung JS, Lee EY, Kim SH, Kang CD. Quercetin enhances susceptibility to NK cell-mediated lysis of tumor cells through induction of NKG2D ligands and suppression of HSP70. J Immunother. 2010;33:391–401. https://doi.org/10.1097/CJI.0b013e3181d32f22.

Balsamo M, Vermi W, Parodi M, Pietra G, Manzini C, Queirolo P, Lonardi S, Augugliaro R, Moretta A, Facchetti F, Moretta L, Mingari MC, Vitale M. Melanoma cells become resistant to NK-cell-mediated killing when exposed to NK-cell numbers compatible with NK-cell infiltration in the tumor. Eur J Immunol. 2012;42(7):1833–42. https://doi.org/10.1002/eji.201142179.

Bhattacharyya S, Mandal D, Saha B, Sen GS, Das T, Sa G. Curcumin prevents tumor-induced T cell apoptosis through Stat-5a-mediated Bcl-2 induction. J Biol Chem. 2007;282(22):15954–64. https://doi.org/10.1074/jbc.M608189200.

Bhattacharyya S, Md Sakib Hossain D, Mohanty S, Sankar Sen G, Chattopadhyay S, Banerjee S, Chakraborty J, Das K, Sarkar D, Das T, Sa G. Curcumin reverses T cell-mediated adaptive immune dysfunctions in tumor-bearing hosts. Cell Mol Immunol. 2010;7(4):306–15. https://doi.org/10.1038/cmi.2010.11.

Bhaumik S, Jyothi MD, Khar A. Differential modulation of nitric oxide production by curcumin in host macrophages and NK cells. FEBS Lett. 2000;483(1):78–82. https://doi.org/10.1016/S0014-5793(00)02089-5.

Bill MA, Bakan C, Benson DM Jr, Fuchs J, Young G, Lesinski GB. Curcumin induces proapoptotic effects against human melanoma cells and modulates the cellular response to immunotherapeutic cytokines. Mol Cancer Ther. 2009;8(9):2726–35. https://doi.org/10.1158/1535-7163.MCT-09-0377.

Brouet I, Ohshima H. Curcumin, an anti-tumour promoter and anti-inflammatory agent, inhibits induction of nitric oxide synthase in activated macrophages. Biochem Biophys Res Commun. 1995;206(2):533–40. https://doi.org/10.1006/bbrc.1995.1076.

Burkard M, Leischner C, Lauer UM, Busch C, Venturelli S, Frank J. Dietary flavonoids and modulation of natural killer cells: implications in malignant and viral diseases. J Nutr Biochem. 2017;46:1–12. https://doi.org/10.1016/j.jnutbio.2017.01.006.

Carrero P, Ortega H, Martinez-Botas J, Gomez-Coronado D, Lasuncion MA. Flavonoid-induced ability of minimally modified low-density lipoproteins to support lymphocyte proliferation. Biochem Pharmacol. 1998;55(7):1125–9. https://doi.org/10.1016/S0006-2952(97)00635-7.

Caspi RR. Immunotherapy of autoimmunity and cancer: the penalty for success. Nat Rev Immunol. 2008;8(12):970–6. https://doi.org/10.1038/nri2438.

Chaplin DD. Overview of the immune response. J Allergy Clin Immunol. 2010;125(2 Suppl 2):S3–23. https://doi.org/10.1016/j.jaci.2009.12.980.

Chen D, Nie M, Fan MW, Bian Z. Anti-inflammatory activity of curcumin in macrophages stimulated by lipopolysaccharides from Porphyromonas gingivalis. Pharmacology. 2008;82(4):264–9. https://doi.org/10.1159/000161127.

Chew V, Toh HC, Abastado JP. Immune microenvironment in tumor progression: characteristics and challenges for therapy. J Oncol. 2012;2012:608406. https://doi.org/10.1155/2012/608406.

Corthay A. How do regulatory T cells work? Scand J Immunol. 2009;70(4):326–36. https://doi.org/10.1111/j.1365-3083.2009.02308.x.

Cundell RD. Herbal phytochemicals as immunomodulators. Curr Immunol Rev. 2014;10(2):64–81. https://doi.org/10.2174/1573395510666140915213156.

Dahlberg CI, Sarhan D, Chrobok M, Duru AD, Alici E. Natural Killer Cell-Based Therapies Targeting Cancer: Possible Strategies to Gain and Sustain Anti-Tumor Activity. Front. Immunol. 2015;6:605. https://doi.org/10.3389/fimmu.2015.00605.

Devaud C, John LB, Westwood JA, Darcy PK, Kershaw MH. Immune modulation of the tumor microenvironment for enhancing cancer immunotherapy. Oncoimmunology. 2013;2(8):e25961. https://doi.org/10.4161/onci.25961.

Driessens G, Kline J, Gajewski TF. Costimulatory and coinhibitory receptors in anti-tumor immunity. Immunol Rev. 2009;229(1):126–44. https://doi.org/10.1111/j.1600-065X.2009.00771.x.

Espinoza JL, Nguyen VH, Ichimura H, Pham TT, Nguyen CH, Pham TV, Elbadry MI, Yoshioka K, Tanaka J, Trung LQ, Takami A, Nakao S. A functional polymorphism in the NKG2D gene modulates NK-cell cytotoxicity and is associated with susceptibility to Human Papilloma Virus-related cancers. Sci Rep. 2016;6:39231. https://doi.org/10.1038/srep39231.

Falchetti R, Fuggetta MP, Lanzilli G, Tricarico M, Ravagnan G. Effects of resveratrol on human immune cell function. Life Sci. 2001;70(1):81–96. https://doi.org/10.1016/S0024-3205(01)01367-4.

Fiala M. Curcumin and omega-3 fatty acids enhance NK cell-induced apoptosis of pancreatic cancer cells but curcumin inhibits interferon-gamma production: benefits of omega-3 with curcumin against cancer. Molecules. 2015;20(2):3020–6. https://doi.org/10.3390/molecules20023020.

Gajewski TF, Schreiber H, Fu YX. Innate and adaptive immune cells in the tumor microenvironment. Nat Immunol. 2013;14(10):1014–22. https://doi.org/10.1038/ni.2703.

Gao S, Zhou J, Liu N, Wang L, Gao Q, Wu Y, Zhao Q, Liu P, Wang S, Liu Y, Guo N, Shen Y, Yuan Z. Curcumin induces M2 macrophage polarization by secretion IL-4 and/or IL-13. J Mol Cell Cardiol. 2015;85:131–9. https://doi.org/10.1016/j.yjmcc.2015.04.025.

Giurisato E, Xu Q, Lonardi S, Telfer B, Russo I, Pearson A, Finegan KG, Wang W, Wang J, Gray NS, Vermi W, Xia Z, Tournier C. Myeloid ERK5 deficiency suppresses tumor growth by blocking protumor macrophage polarization via STAT3 inhibition. Proc Natl Acad Sci U S A. 2018;115(12):E2801–10. https://doi.org/10.1073/pnas.1707929115.

Guillerey C, Huntington ND, Smyth MJ. Targeting natural killer cells in cancer immunotherapy. Nat Immunol. 2016;17(9):1025–36. https://doi.org/10.1038/ni.3518.

Halder RC, Almasi A, Sagong B, Leung J, Jewett A, Fiala M. Curcuminoids and omega-3 fatty acids with anti-oxidants potentiate cytotoxicity of natural killer cells against pancreatic ductal adenocarcinoma cells and inhibit interferon gamma production. Front Physiol. 2015;6:129. https://doi.org/10.3389/fphys.2015.00129.

Han R, Wu WQ, Wu XP, Liu CY. Effect of total flavonoids from the seeds of Astragali complanati on natural killer cell function. J Ethnopharmacol. 2015;173:157–65. https://doi.org/10.1016/j.jep.2015.07.017.

Her M, Kavanaugh A. Alterations in immune function with biologic therapies for autoimmune disease. J Allergy Clin Immunol. 2016;137(1):19–27. https://doi.org/10.1016/j.jaci.2015.10.023.

Hou D, Wang D, Ma X, Chen W, Guo S, Guan H. Effects of total flavonoids of sea buckthorn (*Hippophae rhamnoides* L.) on cytotoxicity of NK92-MI cells. Int J Immunopathol Pharmacol. 2017;30(4):353–61. https://doi.org/10.1177/0394632017736673.

Kim HP, Son KH, Chang HW, Kang SS. Anti-inflammatory plant flavonoids and cellular action mechanisms. J Pharmacol Sci. 2004; 96(3):229–45.

Kliem C, Merling A, Giaisi M, Kohler R, Krammer PH, Li-Weber M. Curcumin suppresses T cell activation by blocking Ca2+ mobilization and nuclear factor of activated T cells (NFAT) activation. J Biol Chem. 2012;287(13):10200–9. https://doi.org/10.1074/jbc.M111.318733.

Kumar S, Pandey AK. Chemistry and biological activities of flavonoids: an overview. Sci World J. 2013;2013:162750. https://doi.org/10.1155/2013/162750.

Langers I, Renoux VM, Thiry M, Delvenne P, Jacobs N. Natural killer cells: role in local tumor growth and metastasis. Biologics: Targets Ther. 2012;6:73–82. https://doi.org/10.2147/BTT.S23976.

Lee HH, Cho H. Improved anti-cancer effect of curcumin on breast cancer cells by increasing the activity of natural killer cells. J Microbiol Biotechnol. 2018;28(6):874–82. https://doi.org/10.4014/jmb.1801.01074.

Lee JK, Kim SY, Kim YS, Lee WH, Hwang DH, Lee JY. Suppression of the TRIF-dependent signaling pathway of Toll-like receptors by luteolin. Biochem Pharmacol. 2009;77(8):1391–400. https://doi.org/10.1016/j.bcp.2009.01.009.

Leiro J, Alvarez E, Garcia D, Orallo F. Resveratrol modulates rat macrophage functions. Int Immunopharmacol. 2002;2(6):767–74. https://doi.org/10.1016/S1567-5769(02)00014-0.

Li Q, Huyan T, Ye LJ, Li J, Shi JL, Huang QS. Concentration-dependent biphasic effects of resveratrol on human natural killer cells in vitro. J Agric Food Chem. 2014;62(45):10928–35. https://doi.org/10.1021/jf502950u.

Lopez-Posadas R, Ballester I, Abadia-Molina AC, Suarez MD, Zarzuelo A, Martinez-Augustin O, Sanchez de Medina F. Effect of flavonoids on rat splenocytes, a structure-activity relationship study. Biochem Pharmacol. 2008;76(4):495–506. https://doi.org/10.1016/j.bcp.2008.06.001.

Lu CC, Chen JK. Resveratrol enhances perforin expression and NK cell cytotoxicity through NKG2D-dependent pathways. J Cell Physiol. 2010;223(2):343–51. https://doi.org/10.1002/jcp.22043.

Mandal A, Viswanathan C. Natural killer cells: in health and disease. Hematol Oncol Stem Cell Ther. 2015;8(2):47–55. https://doi.org/10.1016/j.hemonc.2014.11.006.

Murr C, Schroecksnadel K, Winkler C, Ledochowski M, Fuchs D. Antioxidants may increase the probability of developing allergic diseases and asthma. Med Hypotheses. 2005;64(5):973–7. https://doi.org/10.1016/j.mehy.2004.11.011.

Nagendran J, Bleackley C, Ross DB, West LJ, Dyck JR. Resveratrol attenuates stimulated t-cells activation and proliferation: a novel therapy against cellular rejection in cardiac transplantation. J Heart Lung Transplant. 2013;32(4S):S248–9. https://doi.org/10.1016/j.healun.2013.01.642.

Nair MP, Mahajan S, Reynolds JL, Aalinkeel R, Nair H, Schwartz SA, Kandaswami C. The flavonoid quercetin inhibits proinflammatory cytokine (tumor necrosis factor alpha) gene expression in normal peripheral blood mononuclear cells via modulation of the NF-kappa beta system. Clin Vaccine Immunol. 2006;13(3):319–28. https://doi.org/10.1128/CVI.13.3.319-328.2006.

Nielsen SR, Schmid MC. Macrophages as key drivers of cancer progression and metastasis. Mediat Inflamm. 2017;2017:9624760. https://doi.org/10.1155/2017/9624760.

Panche AN, Diwan AD, Chandra SR. Flavonoids: an overview. J Nutr Sci. 2016;5:e47. https://doi.org/10.1017/jns.2016.41.

Richards DM, Hettinger J, Feuerer M. Monocytes and macrophages in cancer: development and functions. Cancer Microenviron. 2013;6(2):179–91. https://doi.org/10.1007/s12307-012-0123-x.

Roszer T. Understanding the mysterious M2 macrophage through activation markers and effector mechanisms. Mediat Inflamm. 2015;2015:816460. https://doi.org/10.1155/2015/816460.

Schwager J, Richard N, Widmer F, Raederstorff D. Resveratrol distinctively modulates the inflammatory profiles of immune and endothelial cells. BMC Complement Altern Med. 2017;17(1):309. https://doi.org/10.1186/s12906-017-1823-z.

Seager RJ, Hajal C, Spill F, Kamm RD, Zaman MH. Dynamic interplay between tumour, stroma and immune system can drive or prevent tumour progression. Converg Sci Phys Oncol. 2017;3:034002. https://doi.org/10.1088/2057-1739/aa7e86.

Shukla S, Mehta A. Comparative phytochemical analysis and in vivo immunomodulatory activity of various extracts of *Stevia rebaudiana* leaves in experimental animal model. Front Life Sci. 2014;8(1):55–63. https://doi.org/10.1080/21553769.2014.961615.

Sithranga Boopathy N, Kathiresan K. Anticancer drugs from marine flora: an overview. J Oncol. 2010;2010:214186. https://doi.org/10.1155/2010/214186.

Smith HA, Kang Y. The metastasis-promoting roles of tumor-associated immune cells. J Mol Med (Berl). 2013;91(4):411–29. https://doi.org/10.1007/s00109-013-1021-5.

Sun L, Chen B, Jiang R, Li J, Wang B. Resveratrol inhibits lung cancer growth by suppressing M2-like polarization of tumor associated macrophages. Cell Immunol. 2017;311:86–93. https://doi.org/10.1016/j.cellimm.2016.11.002.

Sungur CM, Murphy WJ. Positive and negative regulation by NK cells in cancer. Crit Rev Oncog. 2014;19(1–2):57–66. https://doi.org/10.1615/CritRevOncog.2014010805.

Verbeek R, Plomp AC, van Tol EA, van Noort JM. The flavones luteolin and apigenin inhibit in vitro antigen-specific proliferation and interferon-gamma production by murine and human autoimmune T cells. Biochem Pharmacol. 2004;68(4):621–9. https://doi.org/10.1016/j.bcp.2004.05.012.

Waldhauer I, Steinle A. NK cells and cancer immunosurveillance. Oncogene. 2008;27(45):5932–43. https://doi.org/10.1038/onc.2008.267.

Watson JL, Vicario M, Wang A, Moreto M, McKay DM. Immune cell activation and subsequent epithelial dysfunction by Staphylococcus enterotoxin B is attenuated by the green tea polyphenol (-)-epigallocatechin gallate. Cell Immunol. 2005;237(1):7–16. https://doi.org/10.1016/j.cellimm.2005.08.030.

Whiteside TL. The tumor microenvironment and its role in promoting tumor growth. Oncogene. 2008;27(45):5904–12. https://doi.org/10.1038/onc.2008.271.

Wolowczuk I, Verwaerde C, Viltart O, Delanoye A, Delacre M, Pot B, Grangette C. Feeding our immune system: impact on metabolism. Clin Dev Immunol. 2008;2008:639803. https://doi.org/10.1155/2008/639803.

Wong GK, Heather JM, Barmettler S, Cobbold M. Immune dysregulation in immunodeficiency disorders: the role of T-cell receptor sequencing. J Autoimmun. 2017;80:1–9. https://doi.org/10.1016/j.jaut.2017.04.002.

Wu J, Lanier LL. Natural killer cells and cancer. Adv Cancer Res. 2003;90:127–56. https://doi.org/10.1016/S0065-230X(03)90004-2.

Yang L, Zhang Y. Tumor-associated macrophages: from basic research to clinical application. J Hematol Oncol. 2017;10(1):58. https://doi.org/10.1186/s13045-017-0430-2.

Yang Y, Paik JH, Cho D, Cho JA, Kim CW. Resveratrol induces the suppression of tumor-derived CD4+CD25+ regulatory T cells. Int Immunopharmacol. 2008;8(4):542–7. https://doi.org/10.1016/j.intimp.2007.12.006.

Yuan Y, Jiang YC, Sun CK, Chen QM. Role of the tumor microenvironment in tumor progression and the clinical applications (review). Oncol Rep. 2016;35:2499–515. https://doi.org/10.3892/or.2016.4660.

Zamai L, Ahmad M, Bennett IM, Azzoni L, Alnemri ES, Perussia B. Natural killer (NK) cell-mediated cytotoxicity: differential use of TRAIL and Fas ligand by immature and mature primary human NK cells. J Exp Med. 1998;188(12):2375–80. https://doi.org/10.1084/jem.188.12.2375.

Zhao GJ, Lu ZQ, Tang LM, Wu ZS, Wang DW, Zheng JY, Qiu QM. Curcumin inhibits suppressive capacity of naturally occurring CD4+CD25+ regulatory T cells in mice in vitro. Int Immunopharmacol. 2012;14(1):99–106. https://doi.org/10.1016/j.intimp.2012.06.016.

Zhou Y, Zhang T, Wang X, Wei X, Chen Y, Guo L, Zhang J, Wang C. Curcumin modulates macrophage polarization through the inhibition of the toll-like receptor 4 expression and its signaling pathways. Cell Physiol Biochem. 2015;36(2):631–41. https://doi.org/10.1159/000430126.

Zou T, Yang Y, Xia F, Huang A, Gao X, Fang D, Xiong S, Zhang J. Resveratrol inhibits CD4+ T cell activation by enhancing the expression and activity of Sirt1. PLoS One. 2013;8(9):e75139. https://doi.org/10.1371/journal.pone.0075139.

Zou JY, Su CH, Luo HH, Lei YY, Zeng B, Zhu HS, Chen ZG. Curcumin converts Foxp3+ regulatory T cells to T helper 1 cells in patients with lung cancer. J Cell Biochem. 2018;119(2):1420–8. https://doi.org/10.1002/jcb.26302.

Chapter 5
Recent Advances in Phytochemicals and Their Synergistic Role in Multiple Myeloma

Sharmistha Singh and Astha Dwivedi

5.1 Introduction

Medicinal science supports the vital role of phytochemicals in several diseases. Phytochemicals are natural compound inbuilt with anti –inflammatory and anticancer properties. Phytochemical were widely used since history in treatment of many deadliest diseases such as cancers. Phytochemicals have been approved by its self-determined potency of antioxidants beneficial for health. MM is classified in two categories symptomatic and asymptomatic. Asymptomatic is characterized as smoldering multiple myeloma (SMM) indicates the monoclonal protein level in blood found to be approx. 3 g dl which consider as more than 10% of plasma malignant cells in the bone marrow without showing any symptoms, it represent the overall risk of progression to MM of 10% per year for the first 5 year. SMM becomes more advanced with related to Monoclonal gammopathy of undetermined significance (MGUS) is a condition in which myeloma protein, or an abnormal protein present in the blood. SMM is still untreatable and it just kept in routine observation and therapeutic measures (Landgren et al. 2009) waited to get converted in symptomatic multiple myeloma. MGUS is an only marker found in blood for diagnosis of multiple myeloma and is categorized in the presence of CRAB criteria.

MM is advanced symptomatic malignant disease of the plasma cells in blood (terminally differentiated B-Lymphocytes) commonly it's an incurable malignancy, despite high dose chemotherapy, autologous stem cell transplant and novel agents. Advanced MM are symptomatic with uncontrolled growth of monoclonal plasma cells in the bone marrow, and more significantly leading towards nonfunctional intact immunoglobulin or immunoglobulin chain. According to WHO classification, advanced MM have been associated with many factors with monoclonal gammopathy, solitary plasmocytoma of bone and systemic light chain amyloidosis (Swerdlow et al. 2008). MM can also been diagnosed by using CRAB Criteria

S. Singh (✉) · A. Dwivedi
Department of Biochemistry, Faculty of Science, University of Allahabad, Allahabad, India

S. Kumar, C. Egbuna (eds.), *Phytochemistry: An in-silico and in-vitro Update*,
https://doi.org/10.1007/978-981-13-6920-9_5

(Hyper calcemia, Renal failure, Anemia, Bone lesions) for end organ damage, as follows Calcium >11.5 mg/dl, Renal insufficiency creatinine>2 mg/dl, Hemoglobin <10 mg/dl, Bone lesions bone marrow has <10% clonal plasma. MM accounts for around 1% of all cancers worldwide and 10–15% of all hematological neoplasma. MM occurence median age at onset is 71 years for men and 72 years for women. The risk of progression of MM is much higher in older age groups. However onset before the age of 45 is rare approx 3–5% of cases. The average risk of progression to MM is around 1% per annum worldwide (Landgren et al. 2009; Kyle et al. 2006).

5.2 Diagnostic Markers, Cytogenetic Abnormalities and Mutation Causes Multiple Myeloma

For diagnosis and prognosis of all patients with MGUS, Smoldering or active multiple myeloma, Solitary plasmacytoma and AL amyloidosis in which progression of asymptomatic SMM to symptomatic MM consider as plasma cell proliferation disorder, hence bone marrow aspiration is recommended if bone marrow show plasma cell <10% and binucleate and multinucleate plasma cells are seen considered as advanced multiple myeloma. Measurement of free light chains (FLCs) aids in the diagnosis and monitoring of Multiple myeloma. When there is increased polyclonal immunoglobulin production and/or renal impairment, concentration of both kappa and lambda light chain can increase by 30 to 40-fold but the kappa and lambda ratio remains unchanged. In contrast, tumor produces monoclonal excess of only one type of light chain, often with bone marrow suppression of alternate light chains resulting in abnormal kappa/lambda ratio. The combination of serum protein electrophoresis (SPE) and immunofixation (IFE) are highly sensitive but fail to identify a few patients with Light Chain Multiple Myeloma and all patients of Non Secretory Multiple Myeloma Combination of free light chain assay with SPE and IFE improves the sensitivity to 99%. Immunofixation electrophoresis (IFE) identifies the 'M' spike as IgG, Lamba. Lactate dehydrogenase (LDH) is a non- specific enzyme found in most organ. LDH measurement are used in the diagnosis and treatment of liver disease, cardiac disease, tumor of Lungs /kidney and Hematologic disorders.

Cytogenetic abnormalities are found to mutate the normal cells into malignant cells, with respect to addition of chromosome abnormalities, copy number abnormalities, genetic mutation, and gene expression signature. Method used by fluorescence in-situ Hybridization (FISH). Many of these alterations are associated with the changes to cell growth, apoptosis, metabolism and the epigenetic of MM cells. Studies reported that almost all MM patients display cytogenetic abnormalities and cytogenetics has become one of the most important prognostic factors (17p (p53) deletion and (4:14). Mutated genes participate two third in occurrence of MM patients namely *(ACTG1, RB1, CYLD, PRDM1, TRAF3, BRAF, FAM46C, DIS3, TP53, NRAS, and KRAS)*, with frequent intra clonal heterogenecity that plays a critical role in disease outcome and drug resistance (Seidl et al. 2003; Debes-Marun et al. 2003). Sustained molecular signal pathways such as activation of NF-kB can promote MM cell proliferation, mediate the secretion of IL-6 and the expression of adhesion molecules, up regulate anti-apoptotic proteins, inhibit death receptor pathway and promote angiogenesis, contributing to the proliferation of malignant myeloma cells and their resistance to apoptosis. Furthermore drug targeting NF-KB can prevent NF-kB activation, promote apoptosis and improve prognosis, providing a broad prospect for the treatment of MM (Li et al. 2014).

5.3 Role of Phytochemicals as Chemo Preventive and Its Synergistic Effect in Multiple Myeloma

The benefits of phytochemicals in MM are understood in literature as well as in clinical trials. Phytochemicals have been reported to suppress uncontrolled multiplication of myeloma cells, and also down regulate transcription factors, activate chemokines, cell surface adhesion molecules, induces apoptotic pathway in cancer cell lines. Phytochemicals has antioxidant and anti-inflammatory activity had found to be beneficial in suppression of abnormal plasma cell growth.

MM is incurable but highly treatable by several drug are based on different categories: Immunomodulating drugs are extensively used in treatment of MM they include Thalidomide, Lenalidomide, Pomalidomide. Protease inhibitors drugs include Bortezomib, Carfilzomib, Ixazomib. Histone Deacetylase (HDAC) inhibitors drugs includes Panobinostat and Monoclonal antibodies Darzalex and Elotuzumab.

Combination of Bortezomib and Thalidomide (corticosteroids) are highly used in the initial treatment of advanced multiple myeloma, these drug combination are targeted towards the abnormal proteins formed by myeloma plasma cells. These targeted therapy are not 100% chemotherapy because these drugs are not killing the abnormal protein instead of that they are slowing the growth rate of abnormal protein. Many studies have reported the synergistic effect of phytochemicals and drug in treatment of Multiple myeloma. Targeted therapy may have side effects of drugs on patients suffering from MM. Amelioration of phytochemicals with specific drugs may help MM patients in suppressing the side effects of drugs and upgrading the immune system (Table 5.1).

Phytochemical used as a treatment for cancers, widely and historically (Cragg and Newman 2005). Specific plants and there phytochemicals were investigated and processed for their anti-inflammatory and anti-cancerous properties (Anupama et al. 2014). The need for new chemotherapeutics agents with increased efficacy and decreased toxicity has led to the development of novel drugs leads for the treatment of cancer and reduces side effect of chemotherapy, evidence thus exist for use of phytochemicals as novel chemotherapeutics agents as several studies in cancer signaling pathways and other mechanism explains the role and therapeutic of phytochemicals in living system. In-vitro studies resulted several success in targeted therapy along with phytochemical molecules suppressing cancer signaling pathway

Table 5.1 Phytochemicals show synergistic effect with drugs used in treatment of multiple myeloma

Combined therapy		Synergistic effect in clinical trails	Type of cancer
Phytochemicals	Drugs		
Ursolic acid	Thalidamide, Bortezomib, Cisplatin	↓metastasis, ↓angiogenesis, ↓Inflammation, ↑ apoptosis	Multiple myeloma, leukemia, cervical cancer
Curcumin	Bortezomib	↓cell proliferation	Multiple myeloma
		↑cytotoxicity	
		↓inflammation	
Betulinic acid	Thalidomide, Doxorubicin, Taxol, Cisplatin	↑apoptosis	Lymphoma, acute lymphoma leukemia, multiple myeloma, ovarian cancer
		↓cell proliferation	
Capsaicin	Bortezomib, Doxorubicin	↓cancer signaling pathway,	Multiple myeloma, leukemia
		↑apoptosis	
Cannabis	Bortezomib,	↑apoptosis	Multiple myeloma, lymphoma
	Carfilozomib	↓cell proliferation	
Apigenin	Bortezomib,	↑apoptosis,	Multiple myeloma, acute lymphoma leukemia, ovarian cancer
	Epotoside, Doxorubicin	↓anti apoptotic protein	
		↓inflammation	
Resveratrol	Bortezomib, Temozolomide	↓inflammatory cells,	Multiple myeloma, leukemia, brain cancer
		↑immune system	
Genistein	Bortezomib, Thalidomide, Carfilzomib	↓topoisomerase II,	Multiple myeloma, lymphoma
		↓activate cancer signaling pathway, inflammation, ↑Apoptosis	

and finally inducing apoptosis (Harrison et al. 2014). Phytochemicals involved in suppressing the molecular signaling pathway and triggering the apoptosis in cancer cells in Multiple myeloma are discussed in the subsequent sections.

5.3.1 Ursolic Acid (UA)

Ursolic acid is one of such compound that participates in the prevention of cancer, found in waxy peels of fruits has been demonstrated to induce apoptosis in myeloma cells (Manu and Kuttan 2008). Structure of UA consist C-30 isoprenoid in a pentacyclic triterpenoid. The molecular mechanism of UA can establish a potent protective effect a novel foundation to treat hematological cancer. UA targets both intra and extra cellular molecules responsible for progression of apoptosis, autophagy, angiogenesis and inflammation. Further study demonstrate that UA suppress the cancerous cells such as HepG2, which inhibit the focal adhesion signaling pathway along with other cancer biomarkers such as *ICAM-1, VCAM-1, E –selectin, FAK, paxillin and PTEN.*

In another study UA reported as potent anti-cancer phytochemicals reduces tumor growth in-vivo and down regulate malignant protein expression such as *VEGF, ICAM-1, MM-9* in-vitro studies. UA activates several genes which help to induce apoptosis and autophagy in cancer cells and suppress anti-apoptotic proteins. Moreover in study of cancer cell line found that UA reduces cell proliferation, invasion and metastasis growth, by directly suppressing NF-$_{K}$B and STAT- 3 signaling pathways. UA shows synergistic therapeutic effect with anti-cancer drugs used in chemotherapy and radiotherapy; hence it can be an effective approach to cure cancer such as multiple myeloma. Recent study is on continuation to postulates UA as potent inhibitor of myeloma cells and inducing apoptosis. UA found to be an potent anti-cancerous phytochemical has great efficacy and safety profile, but due to its limited water solubility and low bioavailability, it has not yet been successfully employed in clinical trials (Manu and Kuttan 2008).

5.3.2 *Curcumin*

Curcumin is extensively used as an herbs and spice widely since historically, curcumin is composed of diferuloymethane present in rhizome part of *Curcuma longa* plant, curcumin compound are loaded with immense beneficial properties such as anti-cancerous, they are extracted naturally from the rhizome part, this dried yellow powder have potent safety efficacy towards cancer like disease. *Curcuma longa* also known as turmeric is a tropical plant native to southern or south eastern tropical Asia. It is a perennial herb belonging to the ginger family (Aggarwal et al. 2005). In-vitro study found that curcumin inhibit malignant cell proliferation, along with down regulate the transcription factor *NF-$_{K}$B, AP-1*, and early growth response gene-1. This phytochemical also proven to be effective in suppressing the inflammatory expression of lipoxygenase (LOX), cyclooxygenase-2 (COX-2), NO synthase, matrix metalloproteinase such as urokinase plasminogen activator, tumour necrosis factor (TNF-α), cachexin or cachectic and cell adhesion molecules (CAM).

UA also capable to inhibit the deregulated expression of c-Jun-N- terminal kinase (JNK) pathway, protein tyrosine kinases, serine/Theorine kinases and several other protein pathways. Nuclear factor kappa –light –chain enhancer of activated B cells (NF-κB), NF-κB signaling pathway is shown to up-expressed in multiple myeloma.

Bortezomib 26S protease inhibitor an anti-cancer drug achieved via blocking NF-κB signaling pathway (Adams 2004). Synergistic effect of phytochemicals have been extensively studied and reported, such recent study shows curcumin and bortezimib synergistic effect on multiple myeloma in clinical trials in-vivo, the daily dose of curcumin could significantly down regulate the expression of NF-κB in MM resulting in decreased cell proliferation and increased cellular apoptosis. Numerous reports suggest that curcumin act as chemopreventive and chemotherapeutics effects by reducing the side effects of chemotherapy. The mechanism by which curcumin slows down the disease activity in multiple myeloma is still unclear. Many studies revealed the curcumin effect on malignant cells by down regulation of

Interleukin-6 (a growth factor for myeloma cells and also an inflammatory cytokine) and suppression of receptor activator NF-κB signaling. Curcumin also effect cellular immunity. Curcumin show antioxidant and anti-inflammatory properties and found to suppress tumor initiation, promotion and metastasis.

5.3.3 Betulinic Acid

Betulinic acid is a natural product that show to exhibit an anti-tumor effect without cytotoxicity and used for treatment of human cancer (Haywood et al. 2015). It is a naturally occurring pentacyclictriterpenoid in many plant species, can be obtained in quantity from the bark or core of some plant species. Betulinic acid is an experimental antineoplastic agent that induces apoptosis in myeloma cells in-vitro and in-vivo, as well as in neuro ectodermal tumor cell lines in-vitro. Wick et al. (1999) *3-beta –hydroxy-lup-20(29)m-ene-28-oic acid* (betulinic acid), an C-28 a carboxylic acid derivative of the ubiquitin triterpenenbetulin, is a member of lupine type triterpenes. However, oxidized derivatives betulinic acid possesses a number of intriguing pharmacological effects. One characteristic features of betulinic acid *cytotoxicity* is its ability to trigger the mitochondrial pathway of apoptosis independent of their *p53* status in cancer cells. Betulinic acid also induces proteosome dependent degradation of Spl, Sp3 and Sp4 proteins. Sp proteins are activated in many tumor/cancer cells, and are associated with uncontrolled cell proliferation, angiogenesis and anti- apoptotic pathway (Haywood et al. 2015). It is hypothesized that the anticancer activity of betulinic acid may be due to down regulation of Sp proteins. Betulinic acid derivatives were designed synthesized and evaluated their anti- cancerous activities against MGC-803, PC3, A375, Bcap -37 and A431 cell lines and the study suggest that the C-28 amino substituted BetA derivatives possess stronger anti cell proliferative ability. Betulinic acid is a natural compound along with high in-vitro cytotoxicity towards myeloma cells. However major drawback is its poor solubility in water, through which the compound hamper an effective in-vivo cancer study (Wick et al. 1999). Betulinic acid has been found to be anti –cancerous phytochemical However, betulinic acid function as cytotoxic phytochemical in several types of cancer cells.

5.3.4 Capsaicin

Capsaicin is a bioactive phytochemical abundant in red and chili pepper. Capsaicin exhibited to down regulate the expression of several genes involved in cancer cell survival, growth arrest, angiogenesis and metastasis. Capsaicin targets several cell signaling pathway to down regulate progressive oncogenes and up regulate tumour suppressor genes such as p53, p38 in several types of cancer models. In multiple myeloma capsaicin helps in down regulate *STAT 3* target genes (BCL2, surviving) and increases *HIFI* (Hypoxia inducible factor 1) alpha degradation and reduces *VEGF* (vascular *endothelial* growth factor). Several studies revealed the potential effects of capsaicin as chemopreventive and chemotherapeutic (Amantini et al. 2009). Capsaisin found to induce apoptosis in-vivo study in different types of blood cancers such as leukemia, plasma cell dysphrasia while leaving the normal cell unharmed. Capsaisin appears to induce apoptosis in over 40 distinct cancer cell lines (Bley et al. 2012). The p53 tumour suppression gene is a well-known as anti-metastasis gene i.e. frequent mutated in many carcinomas. Capsaicin was found to induce p53 phosphorylation at the Ser 15 residue (Ito et al. 2004) and enhanced acetylation through down regulation of sirtuin which is responsible for activation of apoptosis. Capsacin represent synergistic effect with anti-cancer drugs, reduce the risk of toxicity and improve the cellular immunity. Hence capsacin display strong anticancer activity through targeting cancers activated signaling pathway and cancer associated genes in different tumor stages including initiation, promotion, progression and metastasis.

5.3.5 *Cannabis*

Cannabis widely known as *marijuana,* is genus of flowering plant in the family cannabaceae. Cannabis plants manufacture a group of chemicals called cannabinoids, which causes mental and physical hallucination when consumed. Cannabinoids were secreted by glandular trichomes that occur most abundantly on the floral calyxes and bracts of female plants (Mahlberg and Eun 2001). As a drug it usually comes in the form of dried flower buds (marijuana) resin (hashish).

Cannabis is a potent phytochemical also raises the quality of phytochemical literatures and research related to the management of cancer therapy in Multiple myeloma. Cannabidiol (CBD), cannabinol, cannabichromene, cannabigerol, tetrahydrocannabivirin, and delta-8-THC are just some of the additional cannabinoids that have been identified.

Cannabis plants which contain the psychotropic chemical tetrahydrocannibal (THC) and weakly psychotropic chemical cannabinol (CBD), the potential benefits of these plants and these chemicals in management of cancers (Cragg and Newman 2005). In medical science cannabis cannabis and its constituent cannabinoids are often used in the treatment of patients suffering from cancers such as acute lymphoma leukemia, multiple myeloma. Cannabis is used to reduce nausea and vomiting during chemotherapy (Watt et al. 1962). A number of small studies of smoked marijuana found that it can be helpful in treating nausea and vomiting from cancer chemotherapy and it is also helpful in treatment of pain caused by cancer. Recently, researchers reported that tetra hydro cannabinol (THC) and other cannabinoids (CBD) slow down progression of cancer cell and may cause death in certain types of cancer cells growing in lab dishes. Some animal studies also revealed certain cannabinoids as an potent inhibitors of cancer cells, reduces spread of cancer cells. There have been some early clinical ongoing practices of cannabinoids in treatment of myelomas cells and more studies are planned. Cannabinoids can be safe in the treatment of cancer as they exhibit potent potential for controlling and curing the deadliest disease. Relying on marijuana alone as treatment while avoiding or delaying conventional medical care for cancer may have serious health consequences.

5.3.6 Apigenin

Apigenin is a potent flavone usually present in vegetables such as parsley, celery, chamomile and Egyptian plant *Moringa peregrine* (El-Alfy et al. 2011). This phytochemical demonstrate cytotoxic activities against cancers comparable to that of doxorubicin an anti-cancer drug. Apigenin proved to be chemo preventive and also mediates synergistic effects with anticancer drugs, induces apoptosis as well as autophagy but persuade resistance against chemotherapy in human cancer cells (Chung et al. 2007). Apigenin also plays vital role in triggering apoptosis in myeloma cell by targeting apoptotic protein Bax/Bim, along with anti-cancer drugs widely used in chemotherapy is an effective strategy in treatment of cancers. Apigenin increased the Bax/Bim ratio in favor of cell apoptosis in hematological cancer cells and also proven chemo preventive in acute leukemia treatment.

Apigenin reported as very effective and relevant tested flavonoids which helps to induce apoptosis in cancer growing cells. Furthermore in-vitro studies reported Apigenin as an deregulator of expression of topoisomerase, an inhibitors anti- cancer therapy that are involved in many aspects of chromosomal mutations including replication and transcription and hence resulting in cancer such as multiple myeloma (Wang et al. 1999). Beside several study are in favor of Apigenin act as a chemo preventive phytochemical, induces apoptosis in lymphocytic and monocytic leukemias, mediated by induction of protein kinase C (PKC) (Vargo et al. 2006). Apigenin is an potent inhibitor of platelet function through several mechanisms including blockade of TXA receptors (TPs). Some studies suggest that the TXA_2 downstream metabolite of these cyclooxygenase along with its thromboxane receptor (TA) contribute to mediating this progression and hence TP activation stimulate tumor cell proliferation. Apigenin potency tends to inhibit TXA receptor or thromboxane receptor (TP). Apigenin found to be most potent beneficial phytochemical among all in suppressing carcinogen toxicity. The mode of mechanism of apigenin are in several pathways such as inhibiting the disturbed cell cycle, diminishing oxidative stress, improving the efficacy of detoxification enzymes, inducing apoptosis, and stimulating the immune system (Yang et al. 2001). However, based on the above highlighted findings apigenin have been proven as a potential anti-cancer phytochemical in cancer treatment, further more investigation and tools should be applied for better results regarding with this phytochemical.

5.3.7 Resveratrol

Resveratrol is a naturally occurring polyphenol that provides a number of anti-aging health benefits including improved metabolism, cardioprotective and cancer prevention. Resveratrol (*trans*-3,5,4′-trihydroxystilbene) is a phytoalexin found in many plant species, including those often consumed by humans such as grapes, peanuts, and berries; it is produced in plants in response to mechanical injury, fungal infection, and UV radiation (Langcake and Pryce 1976). Resveratrol is postulated to function as a potential signaling pathway modulator and also affect multiple signal transduction pathway associated with malignant cell proliferation or carcinogenesis. The link between inflammation and cancer is well understood, these inflammatory processes contribute in the development and progression of cancer. A key mediator NF-$_K$B an inflammation –induced cellular transformation transcription factor, seems to be suppressed by the effect of Resveratrol attributing it in inactivated form Resveratrol also block Interleukin-1B, survival (Bcl-2, Bcl-XL, XIAP, c-1AP and survivin), and HO-1 induced NF-$_K$B dependent signaling related proliferation (cyclin D1), Tumour necrosis factor, angiogenesis (vascular endothelial growth factor) and metastasis. Resveratrol also activate transcription factor 3 (ATF3) plays as an anti-tumorigenic role, hence regarded as an novel molecular target in carcinoma cells. Several studies including laboratory, have implicated the activation of p53 –dependent pathways mediated by the increased expression of pro apoptotic protein such as (Bax, Bak, Bim, PUMA, NOXA etc.) observed anti proliferative effect of resveratrol. This phytochemical also reported to induce apoptosis via activation of both intrinsic (mitochondrial mediated) and extrinsic (death receptor mediated) pathways.

Resveratrol appeared to be anti-inflammatory proven as a potent inhibitors of cyclooxygenase (COX) and hydro-peroxidas functions. Resveratrol reported in reducing hypoxia inducible factor-1α and matrix metallopeptidase-9 (MMP-9) found in tumor progression, from angiogenesis, and also inhibits dextran sulfate sodium (DSS) induced colitis through down regulation of p38 (mitogen activated protein kinase), prostaglandin E synthase 1, iNOS and COX 2. Resveratrol also participate in signaling pathways by inhibiting Wnt signaling and beta-catenin localization in myeloma derived cells. Another study found that resveratrol at a concentration of 10 μM or more induces apoptosis in normal cells as well as cancer cells which demonstrated a potential cytotoxic effect on normal cells (Langcake and Pryce 1976).

5.3.8 Genistein

It is soy derived isoflavone and had shown protective effect against malignant cells. Geinstein found to inhibit cell proliferation, cause cell cycle arrest at G2/M phase and induce apoptosis in cancer cells (Choi et al. 2007). Geistein act as antioxidant and anthelmintic, genistein has been found to have anti-angiogenic effects (blocking

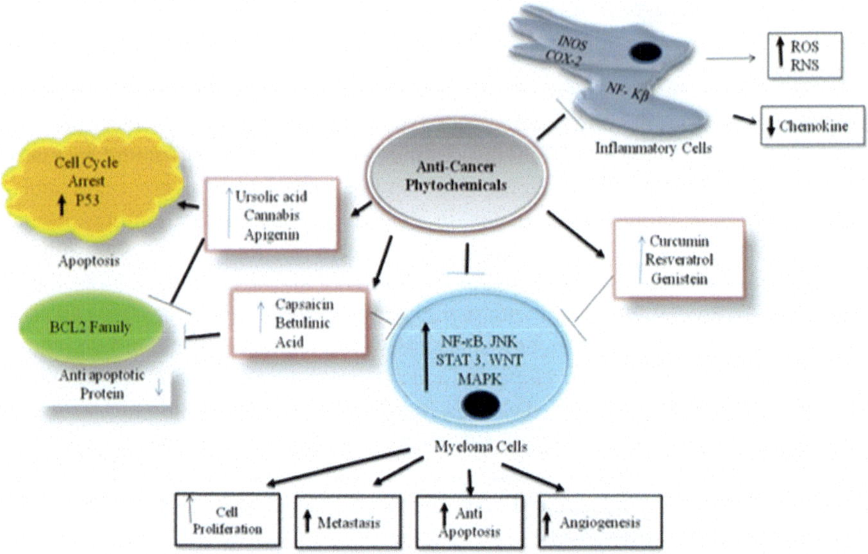

Fig. 5.1 Anti-cancerous phytochemicals triggers apoptotic pathways and deactivate inflammation and molecular signaling pathway in progressive myeloma cells

formation of new blood vessels), and also block the uncontrolled cell growth associated with cancer, most likely by inhibiting the enzymes that regulate cell division and cell survival (growth factors). Geinstein belongs *Leguminosae* family, with the largest quantity identifies *Fructus Sophorae* and Sub prostrate Sophora. Geinstein is able to induce programmed cell death, enhances the anti-cancer efficacy and inhibit angiogenesis, and thus is a promising cancer chemopreventive agent. Geinstein also show synergistic effects with anti-cancer cells used in treatment of multiple myeloma, by suppressing the NF-$_K$B activated receptors and improving cellular immunity.

Among all flavoinoids genistein exhibit the most potent inhibitory effect on VEGF expression is a signal protein produced by cells that stimulate the formation of blood vessels. Soy isoflavones interfere with production of IL-6, which affects immune homeostasis and inflammatory reactions (Dijsselbloem et al. 2004). In-vitro and in-vivo studies show that genistein has been found to be useful in treatment leukemia. Moreover, the in-vitro and in-vivo studies demonstrate that the phytochemical have inhibitory effect on various human cancers, and can enhance the anti-tumor activity of chemotherapeutics agents by regulating cellular signal transduction pathways resulting in the induction of apoptosis cell death. Study conducted on genistein represent inhibition of uncontrolled cell growth, induced apoptosis pathway, inhibition of NF-$_K$B, also activate miR-29b expression. Several emerging experimental evidences shows the potency of genistein could induce cancer cell death by regulating multiple signaling pathways including Akt, NF-$_K$B, MAPK, Wnt, androgen receptor (AR), p53 and Notch signaling all of which found to be malfunctioned in cancer cells. Genistein also mediate to suppress androgen receptor signaling, which contributes to carcinogenesis and cancer progression through regulating transcription of androgen responsive genes. Genistein activity was chiefly functioned as a tyrosine kinase inhibitor by inhibiting DNA topoisomerase II. Aberrant IL-6 expression has been associated with various pathologic tumor progression, inflammatory conditions and chemo-resistance. Collectively, by modulating inflammatory signaling components including NF-$_{K}$B, VEGF and IL-6, genistein may be used as chemopreventive agent in future clinical trials (Fig. 5.1).

5.4 Conclusion

This chapter clearly demonstrate the role and properties of anti-cancerous phytochemicals and can mediate potential therapeutic effects in multiple myeloma (a rare blood cancer), possibly through suppression of the NF-$_{K}$B, STAT-3, JNK, MAPK, WNT activation signaling pathways in myeloma cells and activation of apoptotic tumor suppression genes. Recent studies also demonstrate the synergistic effect of phytochemicals with anti- cancerous drugs used in treatment of multiple myeloma and hence to modulate the efficacy of drugs used against cancer. Furthermore research may use these phytochemicals in the designing of better pathways for suppression of myeloma cells and enhancing the strength of human immune system which is affected more in cancer patients.

References

Adams J. The proteasome: a suitable anti neoplastic target. Nat Rev Cancer. 2004;4:349–60.

Aggarwal BB, Kumar A, Aggarwal MS, et al. Curcumin derived from turmeric *Curcuma longa*: a spice for all seasons. In: Phytopharmaceuticals in cancer chemoprevention. Boca Raton: CRC Press; 2005. p. 349–87.

Amantini C, Ballarini P, Caprodossi S, et al. Triggering of transient receptor potential type 1 (TRPV1) by capsaicin induces FAS/CD95-mediated apoptosis of urothelial cancer cells in an ATM-dependent manner. Carcinogenesis. 2009;30:1320–9.

Anupama N, Madhumitha G, Rajesh KS. Role of dried fruits of *Carissa carandas* as anti-inflammatory agents and the analysis of phytochemical constituents by GC-MS. BioMed Res Int. 2014;2014:6.

Bley KB, Boorman G, Mohammad B, et al. Comprehensive review of the carcinogenic and anticarcinogenic potential of capsaicin. Toxicol Pathol. 2012;40:847–73.

Choi EJ, Kim T, Lee MS. Pro-apoptotic effect and cytotoxicity of genistein and genistin in human ovarian cancer sk-ov-3 cells. Life Sci. 2007;80(15):1403–8.

Chung CS, Jiang Y, Cheng D, et al. Impact of adenomatous polyposis coli (APC) tumor suppressor gene in human colon cancer cell lines on cell cycle arrest by apigenin. Mol Carcinog. 2007;46 (9):773–82.

Cragg GM, Newman DJ. Plants as a source of anticancer agents. J Ethnopharmacol. 2005;100:72–9.

Debes-Marun CS, Dewald GW, Bryant S, et al. Chromosome abnormalities clustering and its implications for pathogenesis and prognosis in myeloma. Leukemia. 2003;17:427–36.

Dijsselbloem N, Vanden BW, De Naeyer A, et al. Soy isoflavonephyto-pharmaceuticals in interleukin-6 affections. Multi-purpose nutraceuticals at the crossroad of hormone replacement, anti-cancer and anti-inflammatory therapy. Biochem Pharmacol. 2004;68(6):1171–85.

El-Alfy TS, Ezzat SM, Hegazy AK, et al. Isolation of biologically active constituents from Moringaperegrina (Forssk.) Fiori. (family: Moringaceae) growing in Egypt. Pharmacogn Mag. 2011;7(26):109–15.

Harrison AM, Thalji NM, Greenberg AJ, et al. Rituximab for non-Hodgkin's lymphoma: a story of rapid success in translation. Clin Transl Sci. 2014;1:82–6.

Haywood, Good P, Khan S, et al. Corticosteroids for the management of cancer-related pain in adults. Cochrane Database Syst Rev. 2015;4:CD010756.

Ito K, Nakazato T, Yamato K, et al. Induction of apoptosis in leukemic cells by homovanillic acid derivative, capsaicin, through oxidative stress: implication of phosphorylation of p53 at Ser-15 residue by reactive oxygen species. Cancer Res. 2004;64:1071–8.

Kyle RA, Therneau TM, Rajkumar SV, et al. Prevalence of monoclonal gammopathy of undetermined significance. N Engl J Med. 2006;354:1362–9.

Landgren O, Kyle RA, Pfeifer RM, et al. Monoclonal gammopathy of undetermined significance (MGUS) consistently precedes multiple myeloma: a prospective study. Blood. 2009;113:5412–7.

Langcake P, Pryce RJ. Production of resveratrol by vitis-vinifera and other members of vitaceae as a response to infection or injury. Physiol Plant Pathol. 1976;9:77–86.

Li Z, Yang Z, Peng X, et al. Nuclear factor-κB is involved in the protocadherin-10-mediated pro-apoptotic effect in multiple myeloma. Mol Med Rep. 2014;10:832–8.

Mahlberg Paul G, Soo Kim E. THC (tetrahyrdocannabinol) accumulation in glands of Cannabis (Cannabaceae). Hemp Rep. 2001;3:17.

Manu KA, Kuttan G. Ursolic acid induces apoptosis by activating p53 and caspase-3 gene expressions and suppressing NF-B mediated activation of bcl-2 in B16F-10melanoma cells. Int Immunopharmacol. 2008;8:974–81.

Seidl S, Kaufmann H, Drach J. New insights into the pathophysiology of multiple myeloma. Lancet Oncol. 2003;4:557–64.

Swerdlow SH, Campo E, Harris NL, et al. WHO classification of tumours of haematopoietic and lymphoid tissues. Lyon: IARC Press; 2008.

Vargo MA, Voss OH, Poustka F. Apigenin-induced-apoptosis is mediated by the activation of PKCdelta and caspases in leukemia cells. Biochem Pharmacol. 2006;72:681–92.

Wang IK, Lin-Shiau SY, Lin JK. Induction of apoptosis by apigenin and related flavonoids through cytochrome c release and activation of caspase-9 and caspase-3 in leukaemia HL-60 cells. Eur J Cancer. 1999;35:1517–25.

Watt JM, Breyer BW, Gerdina M. The medicinal and poisonous plants of Southern and Eastern Africa. 2nd ed. Edinburgh: Oxford University Press; 1962.. Pub. E & S Livingstone

Wick W, Grimmel C, Wagenknecht B. Betulinic acid-induced apoptosis in glioma cells: a sequential requirement for new protein synthesis, formation of reactive oxygen species, and caspase processing. J Pharmacol Exp Ther. 1999;289:1306–12.

Yang CS, Landau JM, Huang MT, et al. Inhibition of carcinogenesis by dietary polyphenolic compounds. Annu Rev Nutr. 2001;21:381–406.

Chapter 6
Drug Resistance in *Plasmodium* sp. and Novel Antimalarial Natural Products-Emerging Trends

Aswathy Narayanan, Kirthana M. V. Sindhe, and Laxmi Shanker Rai

6.1 Global Impact of Malaria

Malaria remains a major cause of death and morbidity worldwide. It has been ranked fourth among major infectious diseases in causing deaths after pneumococcal acute respiratory infections, HIV/AIDS and tuberculosis. Malaria continues to kill over a million people mainly in tropical countries and accounts for approximately 2.6% of the total disease burden of the world. *Plasmodium falciparum* accounts for the majority of malarial mortality, though the less virulent *P. vivax* and *P. ovale* also contribute significantly to morbidity. According to the WHO reported in 2016, 91 countries reported a total of 216 million cases of malaria, an increase of 5 million cases as compared to 2015. Although malaria incidence has fallen globally since 2010, the rate of decline has stalled and even reversed in some regions since 2014. Mortality rates have followed a similar pattern. The rapid increase in resistance of *plasmodium* to current antimalarials has made treatment ineffective and has caused major issues in the eradication of malaria. In addition, mosquitoes resistant to pesticides, restriction in the use of chemical sprays and lack of effective vaccines pose grave threats to global health. With global warming there is a sudden and dramatic resurgence of malaria in many countries. Moreover, no new synthetic antimalarial drug has been discovered over the past four decades. This warrants the urgent need for the development of new antimalarials.

A. Narayanan
Molecular Biology and Genetics Unit, Jawaharlal Nehru Centre for Advanced Scientific Research, Bangalore, India

K. M. V. Sindhe
Department of Medicine, University of California, San Francisco, CA, USA

L. S. Rai (✉)
Department of Mycology, Institut Pasteur, Paris, France
e-mail: laxmi.rai@pasteur.fr

S. Kumar, C. Egbuna (eds.), *Phytochemistry: An in-silico and in-vitro Update*,
https://doi.org/10.1007/978-981-13-6920-9_6

6.2 Life Cycle of the Malarial Parasite and Clinical Features of Malaria

The malarial parasite has a complicated and fascinating lifecycle which promises different aspects to investigate in detail. The parasite completes its lifecycle in two different hosts – humans and female Anopheles mosquitoes. Life cycle of the parasite can be divided into three different stages:

6.2.1 Infection of a Human Host

When a plasmodium-infected female Anopheles mosquito bites human, it inoculates sporozoites into the human skin. Sporozoites enter liver cells and grow into tissue schizonts which rupture and release merozoites to the blood stream thus completing the pre-erythrocytic stage. At this stage, *P. vivax* and *P. ovale* can remain dormant as hypnozoites in the liver and relapses may occur within a week, or even years later.

6.2.2 Asexual Stage

Following the pre-erythrocytic stage, asexual stage begins in erythrocytes in which merozoites develop into ring and trophozoites that later convert to schizonts. Finally, the mature schizonts burst and release merozoites in blood stream. At this stage, patients show clinical symptoms leading to illness and complications.

6.2.3 Sexual Stage

Merozoites may also progress into sexual erythrocytic stages and develop into immature gametocytes, which are the precursors of male and female gametes. These are ingested by an Anopheles mosquito through a blood meal and undergo fertilization in the mosquito midgut forming oocyst. Oocyst subsequently undergoes multiple rounds of asexual replication resulting in the production of sporozoites. Rupture of the mature oocyst releases the sporozoites into the body cavity of the mosquito which migrates to the salivary glands, ready to infect another human host, thus completing the life cycle (Scherf et al. 2008).

Malaria is an acute febrile illness characterized by fever, headache, chills, muscular aches, weakness, vomiting, cough, diarrhea and abdominal pain. If patients are not treated at this stage, malaria develops into a severe illness which includes symptoms such as seizures, coma, metabolic acidosis, pulmonary edema, pneumonia and renal failure. Any of these complications can develop rapidly and progress to death within hours or days. All the clinical manifestations of malaria are associated with the asexual erythrocytic or blood stage parasites. In an individual lacking immunity, symptoms appear in 7 days or longer after the infectious mosquito bite. Clinical relapses may occur weeks to months after the first infection, if proper treatment regime is not followed by the patient. The patient experiences relapse at regular intervals of 48–72 h which is named as "short term relapses". Some patients experience "long term relapses" after a gap of 20–60 days or more. In malaria caused by *P. vivax* and *P. ovale*, relapse occur due to reactivation of the dormant parasites called hypnozoites in the liver. In *P. falciparum* and *P. malariae* infections, recrudescence occurs because of the surviving parasites in red blood cells.

6.3 Diagnosis of Malaria

The clinical diagnosis of malaria depends on the symptoms of the patient which can vary from abdominal pain to anorexia. The detection is cumbersome considering the wide range of symptoms that also overlap with many other diseases. The laboratory diagnosis attains significance here as it can reduce the time for diagnosis thus preventing the transmission/spread of the disease. The peripheral blood from the patient is used to make thin and thick smears (thick for detecting the pathogen and thin for species confirmation), stained (usually with Giemsa) and viewed under the microscope to detect the parasite. A more sensitive microscopic technique is Quantitative Buffy Coat (QBC) test, in which parasite DNA is stained with fluorescent dyes, (e.g. acridine orange) detected by epi-fluorescent microscopy. Rapid Detection Kits which detect plasmodium-specific antigens are also gaining popularity as they reduce the time for diagnosis. The patient blood flows through a membrane containing malaria antigen-specific antibodies. The kits can be species-specific or can detect a pan-plasmodium protein like lactate dehydrogenase. Serological tests like Immunofluorescence antibody testing (IFA) are also in use, owing to their specificity and sensitivity. Molecular techniques like polymerase chain reaction (PCR) are being standardized to detect the pathogen at the DNA level, aiming to reduce the diagnosis time even further. Other molecular techniques at their infancy for malaria detection are real-time PCR, microarray and flow cytometry (Tangpukdee et al. 2009).

6.4 Natural Antimalarial Drugs Derived from Plants

Historically, plants have been a prominent source of antimalarial drugs like quinine and artemisinin (Saxena et al. 2003). Compounds from *Cinchona* species are well known for their antimalarial properties and the constituent alkaloid, quinine remained the major antimalarial drug for many years but later this natural product was replaced by a series of synthetic drugs including eight aminoquinolines, four

Table 6.1 Major anti-malarial natural drugs from plants

Plant name	Plant family	Plants part	Drug name	References
Cinchona	Rubiaceae	Bark of cinchona tree	Quinine	Jiménez-Díaz et al. (2009)
Artemisia annua	Asteraceae	Leaves, stems and inflorescence	Artemisinin	Mueller et al. (2004)
Cryptolepis sanguinolenta	Apocynaceae	Root extract	Cryptolepine	Grellier et al. (1996)
Curcuma longa	Zingiberaceae	Rhizome	Curcumin	Wells (2011)
Nauclea pobeguinii	Rubiaceae	Bark	Strictosamide	Mesia et al. (2010)
Artabotrys uncinatus	Annonaceae	Root extract	Yingzhaosu A	Wells (2011)
Varnonia amygdalina	Asteraceae	Extract from leaves and root bark	Varnodalin	Challand and Willcox (2009)
Dichroa febrifuga	Hydrangeaceae	Root extract	Febrifugine	Jang et al. (1946)
Azadirachta indica	Meliaceae	Leaves	Gedunin	MacKinnon et al. (1997)
Argemone mexicana	Papaveraceae	Leaves	Protopine	Willcox et al. (2007)
			Allocryptopine	
			Berberine	

aminoquinolines and folic acid synthesis inhibitors. Plant products continue to make an immense contribution to malaria chemotherapy, either directly as antimalarial agents, or as important lead compounds for the discovery of more potent antimalarial drugs (Kaur et al. 2009; Chinsembu 2015). The Chinese traditional treatment of malaria includes the use of *Artemisia annua* and its active compound artemisinin. Artemisinin has more chemotherapeutic index than chloroquine and it is also used in choloquinine resistant strains of malaria. Another plant used as an antimalarial drug is a Chinese medicinal plant, *Dichroea febrifuga*. The active drug isolated from this plant is febrifugine. Febrifungine has been used against *P. vivax* and *P. ovale* but its harmful effect on liver makes it unacceptable as antimalarial drug. The important useful antimalarials derived from plants are given in Table 6.1.

6.5 Established Antimalarial Drug Classes and Their Sources

Apart from the decoctions and the isolated compounds, the plants themselves can be used as phytopharmaceuticals. Plant components can also provide templates for the development of structurally simpler analogues that serve as effective antimalarials. A vast majority of prescription drugs used today contain compounds that are directly or indirectly derived from plants (Oksman-Caldentey and Inzé 2004).

Antimalarials derived from plants are classified into Alkaloids, Terpenes, Quassinoids, Flavonoids, Limonoids, Chalcones, Peptides, Xanthones, quinones, coumarines and Lignans.

6.5.1 Alkaloids

Alkaloids are one of the most important classes of natural antimalarials since ancient time. The outstanding example in this class is quinine from *Cinchona succirubra* (Rubiaceae) used more than three centuries to treat malaria and its mechanism of action is hypothesized to inhibit parasite heme detoxification. Indole alkaloids are derived from the stem and root bark of *Monodora angolensis* and *Isolona cauliflora* plants. They are also isolated from the leaves of *Glassocalyx brevipes* and *Strychnos usambarensis*. Fruits of the plant *Picralikma nitida* are also used to isolate indole alkaloids to treat malaria.

Naphthyl isoquinolines have been used as antimalarials. Leaves of *Ancistrocladus robertsoniorum* and *Ancistrocladus tanzaniensis* are main source of naphthyl isoquinolines. Root of Vepris *uguenensis* and *Toddalia asiatica* are also used to extract furoquinolines. Other alkaloid classes isolated from different plants that have been used as antimalarial medicines are bisbenzyl isoquinolines, protoberberines, aporphines and manzamines.

6.5.2 Terpenoids

Terpenoids also show promising antimalarial activity and are classified as Sesquiterpenes, Triterpenes and Diterpenes. Leaves of African plants *Vernonia amygdalina* and *Vernonia brachycalyx* are important source of sesquiterpenes. The fruits of *Reneilmia cincinnata*are also used to isolate this class of antimalarials. Diterpenes are isolated from the seeds of *Aframomum zambesiacum.*

6.5.3 Quassinoids

Quassinoids are a group of degraded triterpenes found in the family Simaroubaceae. These bioactive phytochemicals belong to the triterpene chemical family. The main active compounds in quassinoids are ailanthinone, glaucarubinone, and holacanthone. Extracts of *Ailanthus altissima*, member of simaroubaceae have been used against *P. falciparum* in-vitro. A study suggests that root bark of *Simaba orinocensis* is also an important source of quassinoids (Muhammad et al. 2004).

6.5.4 Flavonoids

Several bioactive flavonoids have been derived from medicinal plants growing in Africa. It is believed that flavonoids exhibit antimalarial activity by inhibiting the influx of L-glutamine and myoinositol into infected erythrocytes. Stem bark of *Erythrina abyssinica* and root bark of *Friesodielsia obovata* are the sources of antimalarials flavonoids. Stem bark of *Morus mesozygia* and *Erythrina sacleuxii* are also major sources of antimalarial flavonoids.

6.5.5 Limonoids

Limonoids are produced by Meliaceae species. A well-known representative of this family is *Azadirachta indica* (Neem tree), a widely used antiplasmodial plant in Asia and is known to contain nimbolide, the active antimalarial principle (Rochanakij et al. 1985).

6.5.6 Chalcones

Chalcones are active in-vitro against *P. falciparum* and are shown to exhibit antimalarial action by protease inhibition. Few known Chalcones are Licochalcone A isolated from *Glycyrrhiza inflat*a (Chen et al. 1994), (+)-Nyasol isolated from *Asparagus africanus* (Oketch-Rabah et al. 1997) and pinostrobin from *Cajanus cajan* (Ducker-Eshun et al. 2004).

6.5.7 Antimalarial Peptides

Antimalarial peptides (AMPs) isolated from various sources have been proven to possess antimalarial activity. Fungal cyclic tetrapeptide (apicidin) isolated from *Fusarium pallidoroseum*, is orally active against *P. berghei* at nanomolar concentrations. It inhibits protozoal histone deacetylase (HAD) thereby interfering with transcriptional control and cell proliferation (Singh et al. 1996).

Dermaseptin S4 (ALWMTLLKKVLKAAAKAALNAVLVGANA) and Dermaseptin S3 (ALWKNMLKGIGKLAGKAALGAVKKLVGAES) are antimicrobial peptides isolated from frog skin which are active against the parasite (Ghosh et al. 1997). Two cyclodepsipeptides, beauvericin and beauvericin Awere isolated from the insect pathogenic fungus *Paecilomyces tenuipes BCC 1614*, exhibiting moderate antiplasmodial activities against K1 strain (Nilanonta et al. 2000). Jasplakinolide (Jonckers et al. 2002), a cyclic peptide isolated from the marine sponge Jaspis sp. markedly decreased parasitemia of *P. falciparum* by virtue of an apical protrusion that appears to interfere with the erythrocyte invasion by the merozoites (Mizuno et al. 2002).

6.5.8 Xanthones

Xanthones are phenolic secondary metabolites derived from Calophyllum and Garcinia species of the Clusiaceae family, found to be active against the parasites.

6.5.9 Quinones

Quinones have been found to possess antiplasmodial activity (Carvalho et al. 1988). Among naphthoquinones isolated from higher plants, Plumbagin possess the strongest antimalarial activity against *P. falciparum* (Likhitwitayawuid et al. 1993).

6.5.10 Coumarins

Coumarins such as 20-epicycloisobrachy coumarinone epoxide, I clausarin, dentatin, O-methylexostemin phenylcoumarins etc. (Kaur et al. 2009) have shown good antimalarial activity.

6.5.11 Synthetic Compounds

Quinine which was used from ancient times was later replaced by synthetic compounds like Pamaquin, Chloroquin, Piperquin etc. Some of them like pamaquin were terminated because to their high toxicity. Another major drug artemisinin, a sesquiterpenoid obtained from plant used in Chinese traditional medicine. Though a potent drug, it had a short pharmacological span as it gets metabolized fast. So, chemical derivatives of Artemesinin, like artesunate are developed to overcome these limitations. It is also derivatized to produce dihydroartemisinin and Artemether. Dimers, trimers and tetramers of artemisinin are also produced to increase the drug potency. A different approach is to have conjugate hybrids of different drugs (for example, dihydroartemisinin and quinine) (Pinheiro Luiz et al. 2018).

6.6 Bioprospecting and Antimalarial Drugs

Bioprospecting refers to identification and commercialization of biologically active compounds from the nature. It aims at systematic search for such compounds, utilization of traditional knowledge to generate income, commercialization of biodiversity and proper management of resources.

6.6.1 Phytomedicines

The classical pharmacological approach involves the transfer of a potential drug from the laboratory to the hospitals (bench-to-bedside). This process is time consuming and laborious. The traditional medicine has witnessed the use of plant decoctions to treat different illnesses including malaria. Reverse pharmacological approach (Bedside-to-bench) cuts down the time used by the classical approach by making use of the traditional medicine. The antimalarial decoctions are now extensively studied to identify and estimate their activity.

6.6.2 Semisynthetic Phtyomedicines

Semisynthetic phtyomedicines with improved qualities are also being developed. The compounds found in these plants may be too complex to manufacture or too expensive to extract. The strategy to use the plants themselves as medicines in such cases paves way to the development of phytomedicines. The mode of production should be cost-effective and efficient. It is equally important to ensure that the phytomedicines are safe as there are numerous unknown compounds also being administered. The major challenges in the use of phytomedicines are:

(a) Lack of complete botanical information about the plants.
(b) Unpredictability in the effects of abiotic and biotic factors on the plant and in turn, its antimalaria activity.
(c) Variations in the concentration of the active ingredient based on the plant part or the batch used.

Thorough evaluation and standardization are crucial before introducing a phytomedicine to clinics.

6.6.3 Anti-malarial Agents from Endophytic Fungi

Endophytic fungi, present mainly in phyla Ascomycota and Basidiomycota, are associated with different plants. Endophytic fungi are being identified as the prominent sources of structurally unique bioactive natural products. The secondary metabolites isolated from endophytic fungi, possessing antimalarial potential may compose the base for the synthesis of novel drugs which might be utilized to withstand malaria and its resistance. Epoxycytochalasin H a Secondary metabolites isolated from *Diaporthe miriciae* has high antimalarial activity against the chloroquine-resistant strain of *P. falciparum*, with an IC50 approximatly 3.5-fold lower than that with chloroquine (Ferreira et al. 2017).

Fusarium sp. endophytic in marine algae and *Nigrospora* sp. endophytic in trees have also show anti-plasmodium activity (Kaushik et al. 2014).

Endophytic fungi such as *Paecilomyces lilacinus*, *Penicillium janthinellum* and *Paecilomyces* sp. isolated from the bark of *Symphonia globulifera* show antiplasmodial potency (Ateba et al. 2018).

6.6.4 Antimalarial Drug from Endophytic Bacteria

Endophytic bacteria, such as *Streptomyces*, have the potential to act as a source for novel bioactive molecules with medicinal properties. Three endophytic streptomyces isolates named SUK8, SUK10 and SUK27 which were obtained from the Malaysian plants *Scindapsus hederaceus*, *Shorea ovalis* and *Zingiber spectabile* respectively have antimalarial activity (Baba et al. 2015).

6.7 Potential Drug Targets and Semisynthetic Compounds

Apicoplasts are vestigial non-photosynthetic plastids, a characteristic feature of the Apicomplexan parasites, including *P. falciparum*. These membrane- bound organelles are absent in the hosts thus any drug targeting molecules within the Apicoplast will be highly specific in action. Enzymes involved in several metabolic pathways (Fatty acid, isoprenoid, haem biosynthesis, Fe-S clusters) located in Apicoplasts have been considered as drug targets (Fig. 6.1). Fatty acid biosynthesis involves multiple enzymes like Fatty Acid Synthases (FAS), beta-keto acyl ACP synthase I/II (Fab B/F), beta-keto acyl ACP reductase (FabG) and beta- hydoxyacyl ACP dehydratase (FabZ). These enymes when inhibited can affect different stages of the life cycle of the malaria parasite. Some of them are observed to be stage-specific. For instance, FAS II is crucial for the liver stage development (Carlton et al. 2002). Another enzyme, Pyruvate dehydrogenase is essential for the transition from liver stage to the blood stage (Pei et al. 2010).

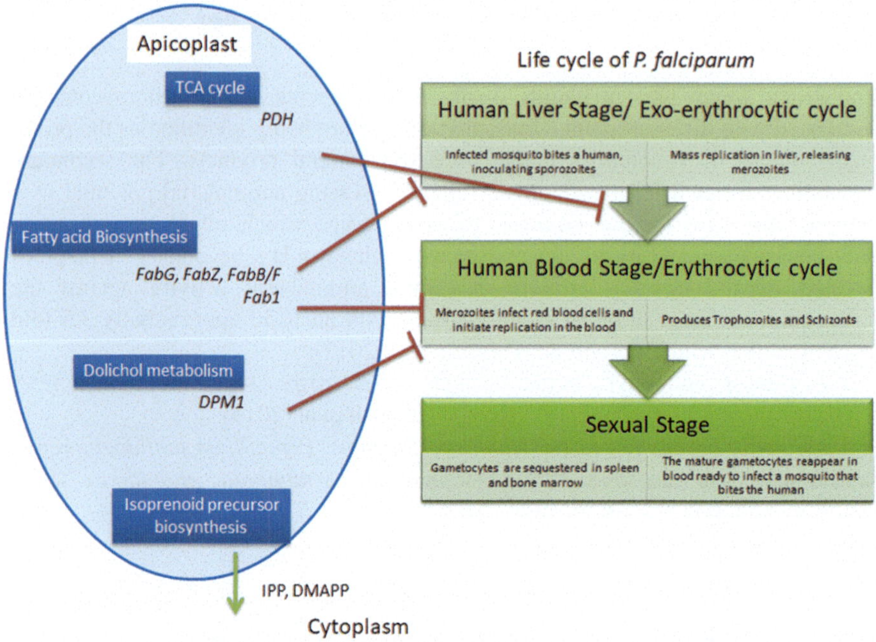

Fig. 6.1 Different pathways operating in the apicoplast and potential drug targets. Inhibition of these targets affect specific stages of the life-cycle of *P. falciparum* (shown in red)

Isoprenoid biosynthetic pathway is also subjected to thorough analysis to identify potential drug targets. There are different steps which can be blocked in the parasite to prevent its survival and multiplication. Dolichol gets phosphorylated to form dolichol phosphate which is important in the synthesis of glycoproteins. Glycoproteins are required for the differentiation of the parasite in the erythrocytes (Kimura et al. 1996). It also acts as a donor of mannose in the form of Dolichol-P-Mannose, synthesized by dolichol phosphate mannose synthase (DPM). The precursors isopentenyl pyrophosphate (IPP) and dimethylallyl pyrophosphate (DMAPP) produced in the apicoplast are exported into the cytoplasm where they are consumed by the isoprenoid and prenylated protein synthesis pathways. They can also be synthesized by the nonmevalonate pathway- the enzymes of which are also under consideration as drug targets (Cassera et al. 2004).

Rapid multiplication of the pathogen in the erythrocytes demands vast supplies of phospholipids for the synthesis of new cell membranes. The precursors are transported from host cells and are converted into phospholipids by the pathogen's enzymatic pathways. The PMT gene in *P. falciparum* encodes the phosphoethanolamine methyltransferase that specifically methylates phosphoethanolamine to phosphocholine (p-Cho). The gene is essential for the organism's survival and multiplication (Via et al. 2005; Holz 1977; Pessi and Ben Mamoun 2006). Compounds that target the transport of either raw materials or conversion of them to the required products- alone or in combination- will make attractive drug candidates for the treatment of malaria.

Prenylation is a post-translational modification that requires protein farnesyl transferases (PFTs). Many PFT inhibitors (PFTIs) are under study to evaluate their antiparasite activity (Olepu et al. 2008).

6.8 Challenges and Future Perspectives

Highly adaptability and increase in drug resistance behaviour of *P. falciparum* continues to be a major problem in fight against malaria. The parasite resistance to anti-malarial agents has presented a major barrier to successful disease management in endemic areas and has probably contributed to the resurgence of infection and the increase in malaria-related deaths in recent years. Efforts to develop new antimalarial drugs have increased substantially in the last few years, both as a result of awareness of the global importance of fighting against malaria and the dedicated public –private partnership strategy to discover, develop and deliver new drugs. However, multidrug resistance of *P. falciparum* has led to an urgent need for the development and implementation of new drugs for the treatment of the malaria parasite. Artemisinin and its derivatives are becoming important and used preferably in combination with a second antimalarial agent to increase the efficacy and combat the resistance of the parasite. However, cost production and pharmacological issues associated with artemisinin derivatives and potential partner drugs are hindering the implementation of combination therapies. Moreover, the development of novel antimalarials is an expensive and slow process and the implementation of new drug policies need resources which are not easily available in the third world.

Resistance should be the prime determinant of the use of a drug. It is critical that decisions regarding which drug regimen to change to, and how to implement the changes, are made in a way that maximizes the benefit to the patient while minimizing the risk of the development of drug resistance.

Mass production of the existing natural compounds should go parallel with the search for new drug candidates. Traditional medicines are used by ~80% of Africans as a first response to ailment. Many of the traditional medicines (for instance, the local plants in the Amazon region) have demonstrable anti-plasmodial activities. Protecting the effective use of a drug must rank as a number-one priority for research and control programs. It is suggested that rigorous evaluation of traditional medicines involving controlled clinical trials in parallel with agronomical development for more reproducible levels of active compounds could improve the availability of drugs at an acceptable cost and a source of income in malaria endemic countries (Ginsburg and Deharo 2011). Ventures like African Network for Drugs and Diagnostics Innovation (ANDI) aim at bridging the gap between the research and marketing of antimalarial drugs obtained from such plant resources. Endophytes that widely colonize healthy tissues of plants have been shown to synthesize a great variety of secondary metabolites that might possess antiplasmodial benefits.

The development of functional assays to investigate essential molecular mechanisms during malaria's liver stage is critical for the validation of new targets to drive antimalarial drug discovery. One underexplored avenue is the manipulation of essential host targets to kill or block *Plasmodium* parasites.

6.9 Conclusion

Malaria is still a major threat, difficult to diagnose and treat, owing to its general symptoms and emerging multidrug resistance. Judicious use of the existing drugs and identification of new drugs can help in the combat against malaria. Improvement of the diagnostic methods can increase their sensitivity, making the early detection possible. On a final note, it is essential that any policy decisions to implement more expensive antimalarials are funded by substantially increased inputs from donor nations. An integrated approach of bioprospecting, efficient manufacture, policy making, appropriate distribution and careful use of the drugs can help in the long-standing combat against malaria.

Acknowledgments A.N acknowledges the Jawaharlal Nehru Centre for Advanced Scientific Research (JNCASR), Bangalore. K.S acknowledges Department of Medicine, University of California. L.S.R acknowledges Institut Pasteur.

References

Ateba JET, Toghueo RMK, Awantu AF, Mba'ning BM, Gohlke S, Sahal D, Rodrigues-Filho E, Tsamo E, Boyom FF, Sewald N, Lenta BN. Antiplasmodial properties and cytotoxicity of endophytic fungi from *Symphonia globulifera* (Clusiaceae). J Fungi. 2018;4(2):E70.

Baba MS, Zin NM, Hassan ZAA, Latip J, Pethick F, Hunter IS, et al. In vivo antimalarial activity of the endophytic actinobacteria, Streptomyces SUK 10. J Microbiol. 2015;53(12):847–55.

Carlton JM, Angiuoli SV, Suh BB, Kooij TW, Pertea M, Silva JC, Ermolaeva MD, Allen JE, Selengut JD, Koo HL, Peterson JD, Pop M, Kosack DS, Shumway MF, Bidwell SL, Shallom SJ, van Aken SE, Riedmuller SB, Feldblyum TV, Cho JK, et al. Genome sequence and comparative analysis of the model rodent malaria parasite *Plasmodium yoelii*. Nature. 2002;419(6906):512–9.

Carvalho LH, Rocha EMM, Raslan DS, Oliveira AB, Krettli AU. In vitro activity of natural and synthetic naphthoquinones against erythrocytic stages of *Plasmodium falciparum*. Braz J Med Biol Res. 1988;21:485–7.

Cassera MB, Gozzo FC, D'Alexandri FL, et al. The methylerythritol phosphate pathway is functionally active in all intraerythrocytic stages of *Plasmodium falciparum*. J Biol Chem. 2004;279(50):51749–59.

Challand S, Willcox M. A clinical trial of the traditional medicine *Vernonia amygdalina* in the treatment of uncomplicated malaria. J Altern Complement Med. 2009;15(11):1231–7.

Chen M, Theander TG, Christensen BS, Hviid L, Zhai L, Kharazmi A. Licochalcone A, a new antimalarial agent, inhibits in vitro growth of the human malaria parasite *Plasmodium falciparum* and protects mice from *P. yoelii* infection. Antimicrob Agents Chemother. 1994;38:1470–5.

Chinsembu KC. Plants as antimalarial agents in Sub-Saharan Africa. Acta Trop. 2015;152:32–48.

Ducker-Eshun G, Jaroszewski JW, Asomaning WA, Oppong-Boachie F, Brøgger Christensen S. Antiplasmodial constituents of *Cajanus cajan*. Phytother Res. 2004;18:128–30.

Ferreira M, Cantrell C, Wedge D, Gonçalves V, Jacob M, Khan S, Rosa CA, Rosa L. Antimycobacterial and antimalarial activities of endophytic fungi associated with the ancient and narrowly endemic neotropical plant Vellozia gigantea from Brazil. Mem Inst Oswaldo Cruz. 2017;112:692–7.

Ghosh JK, Shaool D, Guillaud P, Ciceron L, Mazier D, Kustanovich I, et al. Selective cytotoxicity of dermaseptin S3 toward intraerythrocytic *Plasmodium falciparum* and the underlying molecular basis. J Biol Chem. 1997;272:31609–16.

Ginsburg H, Deharo E. A call for using natural compounds in the development of new antimalarial treatments – an introduction. Malar J. 2011;10(Suppl 1):S1.

Grellier P, Ramiaramanana L, Millerioux V, Deharo E, Schrével J, Frappier F, Trigalo F, Bodo B, Pousset J-L. Antimalarial activity of cryptolepine and isocryptolepine, alkaloids isolated from *Cryptolepis sanguinolenta*. Phytother Res. 1996;10:317–21.

Holz GG Jr. Lipids and the malarial parasite. Bull WHO. 1977;55(2–3):237–48.

Jang CS, Fu FY, Wang CY, Huang KC, Lu G, Chou TC. Ch'ang Shan, a Chinese antimalarial drug. Science. 1946;103(2663):59.

Jiménez-Díaz MB, Mulet T, Viera S, Gómez V, Garuti H, Ibáñez J, Alvarez-Doval A, Shultz LD, Martínez A, Gargallo-Viola D, Angulo-Barturen I. Improved murine model of malaria using *Plasmodium falciparum* competent strains and non-myelodepleted NOD-scid IL2Rgammanull mice engrafted with human erythrocytes. Antimicrob Agents Chemother. 2009;53(10):4533–6.

Jonckers THM, van Miert S, Cimanga K, Bailly C, Colson P, De PauwGillet MC, van den Heuvel H, Claeys M, Lemière F, Esmans EL, Rozenski J, Quirijnen L, Maes L, Dommisse R, Lemière GL, Vlietinck A, Pieter L. Synthesis, cytotoxicity, and antiplasmodial and antitrypanosomal activity of new neocryptolepine derivatives. J Med Chem. 2002;45:3497–508.

Kaur K, Jain M, Kaur T, Jain R. Antimalarials from nature. Bioorg Med Chem. 2009;17:3229–56.

Kaushik N, Murali TS, Sahal D, Suryanarayanan T. A search for antiplasmodial metabolites among fungal endophytes of terrestrial and marine plants of southern India. Acta Parasitol. 2014;59:745–57.

Kimura EA, Couto AS, Peres VJ, Casal OL, Katzin AM. N-linked glycoproteins are related to schizogony of the intraerythrocytic stage in *Plasmodium falciparum*. J Biol Chem. 1996;271(24):14452–61.

Likhitwitayawuid K, Angerhofer CK, Cordell GA, Pezzuto JM. Cytotoxic and antimalarial bisbenzylisoquinoline alkaloids from *Stephania erects*. J Nat Prod. 1993;56:30–8.

MacKinnon S, Durst T, Arnason JT. Antimalarial activity of tropical Meliaceae extracts and Gedunin derivatives. J Nat Prod. 1997;60(4):336–41.

Mesia M, Cimanga RK, Dhooge L, Cos P, Apers S, Totte J, Tona GL, Pieters L, Vlietinck A, Maes L. Antimalarial activity and toxicity evaluation of a quantified *Nauclea pobeguinii* extract. J Ethnopharmacol. 2010;131(1):10–6.

Mizuno Y, Makioka A, Kawazu S, Kano S, Kawai S, Akaki M, Aikawa M, Ohtomo H. Effect of jasplakinolide on the growth, invasion, and actin cytoskeleton of *Plasmodium falciparum*. Parasitol Res. 2002;88(9):844–8.

Mueller MS, Runyambo N, Wagner I, Borrmann S, Dietz K, Heide L. Randomized controlled trial of a traditional preparation of *Artemisia annua* L. (Annual Wormwood) in the treatment of malaria. Trans R Soc Trop Med Hyg. 2004;98(5):318–21.

Muhammad I, Bedir E, Khan SI, Tekwani BL, Khan AI, Takamatsu S, Pelletier J, Walker L. A new antimalarial quassinoid from *Simaba orinocensis*. J Nat Prod. 2004;67(5):772–7.

Nilanonta C, Isaka M, Kittakoop P, Palittapongarnpim P, Kamchonwongpaisan S, Pittayakhajonwut D, Tanticharoen M, Thebtaranonth Y. Antimycobacterial and antiplasmodial cyclodepsipeptides from the insect pathogenic fungus *Paecilomyces tenuipes* BCC 1614. Planta Med. 2000;66:756–8.

Oketch-Rabah HA, Dossaji SF, Christensen SB, Frydenvang K, Lemmich E, Cornett C, Olsen CE, Chen M, Kharazmi A, Theander T. Antiprotozoal compounds from *Asparagus africanus*. J Nat Prod. 1997;60(10):1017–22.

Oksman-Caldentey KM, Inzé D. Plant cell factories in the post-genomic era: new ways to produce designer secondary metabolites. Trends Plant Sci. 2004;9:433–40.

Olepu S, Suryadevara PK, Rivas K, et al. 2-Oxo-tetrahydro-1,8-naphthyridines as selective inhibitors of malarial protein farnesyltransferase and as anti-malarials. Bioorg Med Chem Lett. 2008;18(2):494–7.

Pei Y, Tarun AS, Vaughan AM, Herman RW, Soliman JM, Erickson-Wayman A, Kappe SH. Plasmodium pyruvate dehydrogenase activity is only essential for the parasite's progression from liver infection to blood infection. Mol Microbiol. 2010;75(4):957–71.

Pessi G, Ben Mamoun C. Pathways for phosphatidylcholine biosynthesis: targets and strategies for antimalarial drugs. Future Med Future Lipidol. 2006;1(2):173–80.

Pinheiro Luiz CS, Feitosa LM, Da Silviera FF, Boechat N. Current antimalarial therapies and advances in the development of semi-synthetic artemisinin derivatives. Ann Braz Acad Sci. 2018;90(1 Suppl. 2):1251–71.

Rochanakij S, Thebtaranonth Y, Yenjai C, Yuthavong Y. Nimbolide, a constituent of Azadirachta indica, inhibits *Plasmodium falciparum* in culture. Southeast Asian J Trop Med Public Health. 1985;16(1):66–72.

Saxena S, Pant N, Jain DC, Bhakuni RS. Antimalarial agents from plant sources. Curr Sci. 2003;85:1314–29.

Scherf A, Lopez-Rubio JJ, Riviere L. Antigenic variation in *Plasmodium falciparum*. Annu Rev Microbiol. 2008;62:445–70.

Singh SB, Zink DL, Polishook JD, Dombrowski AW, Darkin-Rattray SJ, Schmatz DM, Goetz MA. Apicidins: novel cyclic tetrapeptides as coccidiostats and antimalarial agents from *Fusarium pallidoroseum*. Tetrahedron Lett. 1996;37(45):8077–80.

Tangpukdee N, Duangdee C, Wilairatana P, Krudsood S. Malaria diagnosis: a brief review. Korean J Parasitol. 2009;47(2):93–102.

Via HJ, Ben Mamoun C, Sherman IW. Plasmodium lipids: metabolism and function in molecular approaches to malaria. Washington, DC: ASM Press; 2005. p. 327–52.

Wells TNC. Natural products as starting points for future antimalarial therapies: going back to our roots? Malar J. 2011;10(Suppl 1):S3.

Willcox ML, Graz B, Falquet J, Sidibé O, Forster M, Diallo D. *Argemone mexicana* decoction for the treatment of uncomplicated falciparum malaria. Trans R Soc Trop Med Hyg. 2007;101(12):1190–8.

Chapter 7
Bacterial Cell Division Machinery: An Insight for Development of New Antibacterial Agent

Rajni Khan and Shashikant Ray

7.1 Introduction

Antibiotic resistance against bacterial diseases demands the discovery of new inhibitors to cope up with this problem. In the recent years, it has evolved into a confounding problem for developing nations, involved in fighting a plethora of infectious diseases. Plant extracts have been used to treat infectious diseases since ancient times. In the current era, natural products from plants, semi synthetic compound, small cationic peptides and peptide-based inhibitors are being used to develop new antibacterial agents for a targeted therapy. The targeting of the conserved and essential protein tubulin has been successfully used to hinder cell division in cancer, therefore a constitutively expressed homologue of tubulin in bacteria; FtsZ, may prove to be a novel target for hampering bacterial cell division will help to design novel inhibitors against bacterial cells. Binary fission is the most common mode of bacterial cell division. It leads to the splitting of a single bacterium into two identical daughter cells, which are the replica of their parent cell. Cytokinesis is the process in which combination of several coordinated events occurs. These events include chromosome duplication, cell elongation, and chromosome segregation. Bacterial cell division, which is called septation, happens after the completion of DNA replication and nucleoid segregation (Wu and Errington 2004). The Z-ring forms at the midpoint of a pre-divisional bacterial cell, which finally constricts giving rise to two daughter cells (Fig. 7.1). The septum is formed through the inward growth of cytoplasmic membrane and cell wall material from opposing directions at the central plane of the cell.

R. Khan
Department of Physiology, University of Calcutta, Kolkata, West Bengal, India

S. Ray (✉)
Department of Biotechnology, School of Life Sciences, Mahatma Gandhi Central University, Motihari, Bihar, India
e-mail: shashikantray@mgcub.ac.in

S. Kumar, C. Egbuna (eds.), *Phytochemistry: An in-silico and in-vitro Update*,
https://doi.org/10.1007/978-981-13-6920-9_7

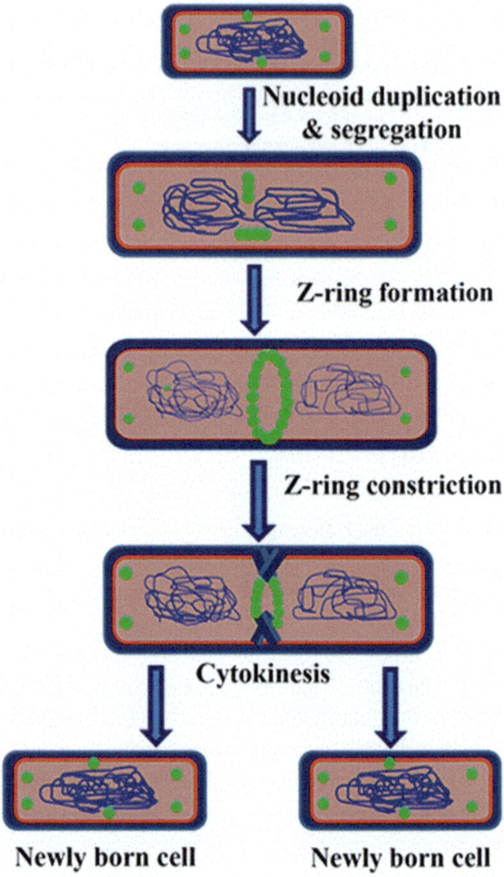

Fig. 7.1 Schematic representation of different steps of cytokinesis during binary fission

Cytokinesis of a bacterial cell starts when septum formation occurs. A divisional region is formed by a contractile Z-ring during the septation process. Bacterial cytokinesis depends on the formation of a complex divisome, which is generated through the chronological assembly of miscellaneous groups of proteins on the Z ring, which acts as the foundation for the future division septum (Buddelmeijer and Beckwith 2002; Errington et al. 2003; Goehring and Beckwith 2005; Harry et al. 2006).

Nevertheless, many facts regarding the nature of the divisome such as the molecular function, the regulatory mechanisms that control the cell cycle, the positioning of Z-ring at the mid cell position, the divisome complex assembly at the center and the appropriate time for successful cell division are poorly understood and require further in-depth investigations. In this chapter we have discussed the role of FtsZ proteins in bacterial cell division. Further we have discussed the important positive and negative regulators of FtsZ assembly and different physiological factors which govern the formation of the divisome for successful cell division. We have also discussed about the important FtsZ targeting synthetic and natural inhibitor, their binding site on FtsZ and their mode of action.

7.2 FtsZ

FtsZ, the most prominent protein of the bacterial cell cycle was first identified from screening of temperature sensitive *Escherichia coli* mutants (Boyd et al. 1968). These mutants showed filamentous morphology and were unable to proliferate under non-permissive temperature. Therefore, these mutants were designated to belong to the *fts* (filamentous temperature sensitive) family of genes (van den Ent and Lowe 2000). In the newly generated cells, FtsZ is found to be present in the monomeric state and migrates to the mid-cell position during the initiation of the division process when the cell elongates and the process of DNA segregation begins. Slowly, lateral protofilaments are formed through the sequential polymerization of these monomers, and subsequently, these pro-filaments bundle together to form a ring-like structure at the mid-cell position known as the Z-ring. The Z-ring serves as the fundamental skeleton for the localization of the cell division associated proteins and assists the configuration of the divisome complex (Hale and de Boer 1999; Ma et al. 1996). FtsZ in the Z-ring undergoes continuous exchange with that of the monomeric FtsZ from the cytoplasm (Stricker et al. 2002).

7.2.1 Structure of FtsZ

Crystal structure of FtsZ was first solved by Lowe et al. (Lowe and Amos 1998). It has two main globular domains which are connected by a helix H7 (Fig. 7.2). The N-terminal contains the GTP binding motifs and previous reports have shown that the N terminal domain is indispensable for the localization of FtsZ to the membrane (Ma et al. 1996). The GTP binding domain consists of six parallel β-sheets, which are similar to Rossmann folds (protein structural motifs that bind to nucleotides). The bound nucleotide is also in contact with six other loops called T-loops. These loops exhibit sequence homology with eukaryotic cytoskeleton protein tubulin in a structure based alignment and also bind to nucleotide in the same way as in tubulin, hence the name T-loops (Nogales et al. 1998). When GTP binds with FtsZ the T7 loop protrudes out of the FtsZ monomer and binds with another GTP bound FtsZ monomer and plays a role in nucleotide hydrolysis. The asparagines residue at 238 position in the T7 loop was found to be essential for GTP hydrolysis and substitution mutation by alanine made FtsZ incapable of polymerization (Dai et al. 1994). The C terminal of FtsZ is a parallel β-sheet surrounded by two helices and contains mostly acidic residues (Lowe 1998; Lowe and Amos 1998). The C-terminal part is not essential for polymerization of FtsZ but is required for the interaction of FtsZ with its other accessory proteins like FtsA, SepF, ZipA and EzrA (Ma et al. 1996). In the case of C-terminal truncated FtsZ, it was seen that the protein localized and formed polymers at the mid cell but it failed to form stable and normal rings (Ma and Margolin 1999).

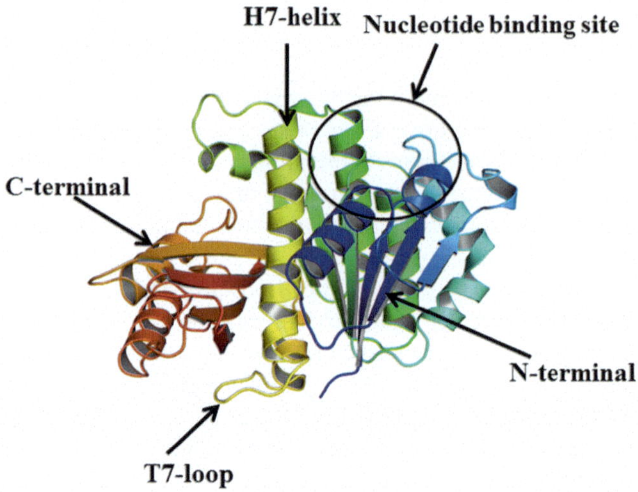

Fig. 7.2 Crystal structure of FtsZ. FtsZ (PDB ID: 2VXY) was taken from protein data bank and then drawn in PyMol (DeLano 2002)

7.3 Structural and Functional Similarities Between Ftsz and Tubulin

Although FtsZ shows lower level of sequence similarities with tubulin, both exhibit a similar 3-dimensional (Clement et al. 2009) structure (Fig. 7.3). The similarities between FtsZ and tubulin were identified on the basis of structural and functional data (Scheffers and Driessen 2001). FtsZ and tubulin both have similar GTP binding motif GGGTGS/TG (Mukherjee et al. 1993; RayChaudhuri and Park 1992) and ability to hydrolyze GTP, forming protofilaments (Mukherjee and Lutkenhaus 1994) and microtubules, respectively. FtsZ from thermophilic archaeon and α-β tubulin from bovine brain shows structural homology (Lowe 1998; Nogales et al. 1998). Both have T3 loop which is found to be involved in γ-phosphate hydrolysis of GTP. When FtsZ and tubulin bind with GTP the T7-loop region forms contact with anther GTP bound FtsZ or tubulin monomer (Nogales 1999) and causes a change in their structural conformation.

7.4 FtsZ Polymerization Dynamics

FtsZ is a GTP binding protein with GTPase activity. The GTP molecule binds at N-terminal domain of FtsZ. During polymerization the two FtsZ monomers associate such that the GTP lies at their interface, causing the monomers to assemble in a head to tail manner forming protofilaments (Lowe 1998; Lowe and Amos 1998). As mentioned earlier the N-terminal of FtsZ contains six parallel β-sheets surrounded by

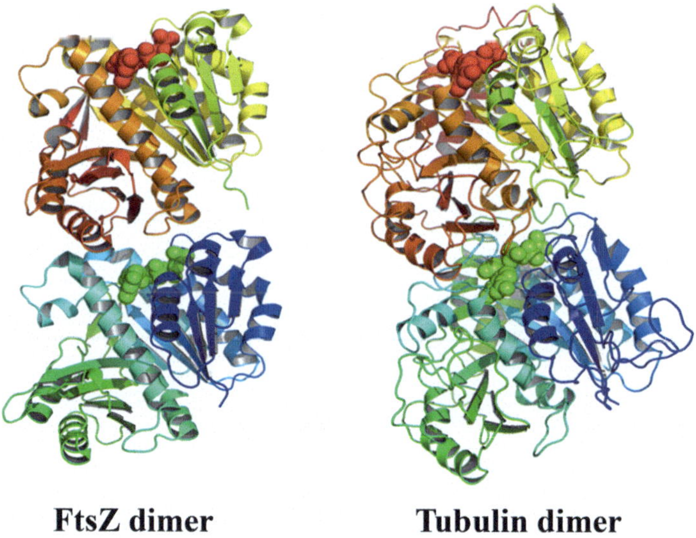

Fig. 7.3 FtsZ, homologue of eukaryotic cytoskeleton protein tubulin. The crystal structure of FtsZ (left, 1W5B) and tubulin (right, 1TUB) have been taken from protein data bank and rendered using PyMol (DeLano 2002)

two and three helices on each side (Lowe 1998; Lowe and Amos 1998). These β-sheets bind to GTP that constitutes the GTP binding domain. The possibility of FtsZ having a GTPase activity was pointed out by the presence of the sequence, GGGTGTG (Lowe 1998; Lowe and Amos 1998) which resembles a sequence in tubulin (GGGTGSG) responsible for nucleotide binding. FtsZ assembly dynamics involves three major steps:

(a) Assembly of FtsZ monomers
(b) Steady state turnover
(c) Disassembly

The second step that is the turnover state is powered by GTP hydrolysis. The hydrolysis of each GTP molecule requires many steps like GTP binding, GTP hydrolysis and phosphate and GDP release. The final product during the turnover is determined by the slowest step in the reaction. If the hydrolysis rate is fast and release is slow then GDP and inorganic phosphate will build up in the polymer, but if the product release is fast then GTP bound state will increase in the polymer. The process of GTP hydrolysis reaction is made complicated by the individual subunits because they have to undergo the process of assembly and disassembly. The assembly state of the polymer affects the hydrolysis reaction (release of GDP and Pi) as GTP is sandwiched between two subunits of FtsZ monomers.

FtsZ, the key molecule in bacterial cytokinesis, forms dynamic polymers in-vitro. FtsZ polymerizes in the presence of GTP in a head to tail manner forming long protofilaments (Kumar et al. 2010). These protofilaments assemble into bundles and

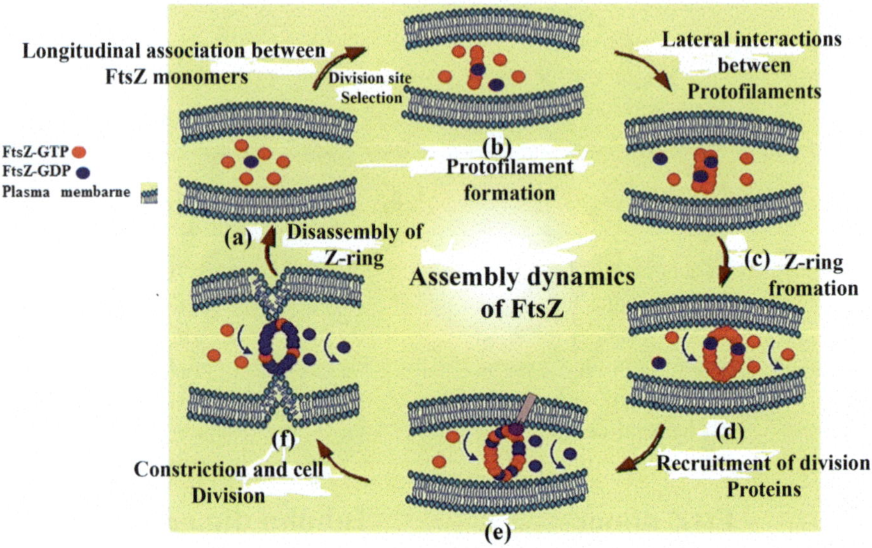

Fig. 7.4 FtsZ assembly dynamics during binary fission. (**a**) FtsZ monomers bound with FtsZ after the nucleoid segregation (**b**) Protofilament formation and its stabilization by its modulatory proteins (**c**) Formation of Z-ring (**d**) Recruitment of different modulatory proteins at the site of Z-ring to form complete septa (**e**) Constriction of Z-ring for cell division (**f**) Disassembly of Z-ring into FtsZ monomers

sheets by lateral interactions. Once the GTP bound FtsZ binds to the other monomer the GTP at the interface is hydrolyzed and Pi released (Fig. 7.4). In the absence of GTP, the protofilaments depolymerize or disassemble. The polymerization and bundling of FtsZ polymers could be stimulated by different agents including Mg^{2+}, Ca^{2+}, monosodium glutamate, dextran DEAE (Beuria et al. 2003; Lowe 1998; Lowe and Amos 1999; Santra et al. 2004; Santra et al. 2005). Furthermore, glutamate strongly stabilized the polymers against dilution-induced disassembly and decreased the GTPase activity of FtsZ (Beuria et al. 2003). Calcium induced FtsZ polymerization and induced bundling of FtsZ polymers.

7.5 Factors Affecting Z Ring Positioning and Its Polymerization Assembly

FtsZ serves as a key protein for bacterial cytokinesis and acts as a cytoskeletal element that provides the energy for constriction of the cytoplasmic membrane (Erickson et al. 2010). Till now, there are several proteins which were discovered in different bacterial species which help in either the formation of Z-ring or tethering of Z-ring to plasma membrane for formation of stable septa, and known as positive regulators of FtsZ assembly. However, there are a group of proteins known as negative regulators of FtsZ assembly that ensure that Z-ring assembly does not take place at an inappropriate time and position. The intricate interaction between these proteins is responsible for spatial and temporal control of cytokinesis. The balanced concentration of positive and negative regulatory factors dictates the assembly and disassembly of FtsZ.

7.5.1 *Stabilizing Proteins (Positive Regulators)*

In bacterial cytokinesis, the first event is the FtsZ assembly which is a prerequisite for localization of other cell division proteins. It is been reported that the Z ring itself is not sufficiently stable and requires assistance by other proteins like FtsA, ZipA, SepF and ZapA for the formation of a stable and complete septa (Cabeen and Jacobs-Wagner 2010). These are the proteins which either help in anchoring FtsZ to the inner membrane or help in stabilizing the protofilaments during cell divisions. These proteins are generally recruited either in the initial phases or late phases of the division process and assist in successful cytokinesis.

7.5.1.1 Function of Positive Regulators of FtsZ Assembly

7.5.1.1.1 FtsA

FtsA is a highly conserved membrane anchoring protein which helps in tethering FtsZ to the inner cytoplasmic membrane (Pichoff and Lutkenhaus 2005). It is a 47 kDa protein, member of actin/hexokinase/HSP70 superfamily and possesses ATPase activity (Bork et al. 1992). The absence of FtsA in cell leads to defective septum formation and the morphology of the cells becomes filamentous (Beall and Lutkenhaus 1992; Pichoff and Lutkenhaus 2007). Yeast two-hybrid studies have shown that FtsA interact directly with FtsZ (Din et al. 1998). Since FtsA is one of the first proteins to be localized to the division septum, it is instrumental in the proper recruitment of the other downstream proteins to the cell division machinery. In addition to its membrane anchoring function, FtsA also stabilizes the FtsZ protofilaments and leads to an enhancement in their assembly (Wang et al. 1997). In cells, the ratio of FtsZ to FtsA is important for proper cell division. In *E. coli* this ratio is 1:100 owing to the presence of regulator ZipA whereas in the case of *B. subtilis* it is 1:5 (Dai and Lutkenhaus 1992; Feucht et al. 2001). The purified form of FtsA is phosphorylated in *E. coli*, but not in *B. subtilis* (Feucht et al. 2001).

A recent study has shown that FtsA acts synergistically with FtsQLB complex to help FtsZ to form complete divisome.

7.5.1.1.2 ZipA

ZipA is another early cell division protein which stabilizes FtsZ assembly and helps in anchoring FtsZ to the membrane in *E. coli* cells. ZipA was predicted to bind to the C-terminal tail region of FtsZ and influence the assembly of FtsZ through hydrophobic interactions (Kuchibhatla et al. 2011; Pichoff and Lutkenhaus 2002). It is localized to Z ring in a FtsZ dependent and FtsA independent manner (Hale and de Boer 1999). Z- ring can form in the absence of either ZipA or FtsA, but not when both are absent in the cell. A heat sensitive mutation *ftsZ84* can be rescued by overexpression of ZipA. ZipA has been reported to cause bundling of FtsZ similar to divalent cations (Mukherjee and Lutkenhaus 1999).

7.5.1.1.3 ZapA

ZapA is also an early cell division protein which is conserved in most bacteria (Gueiros-Filho and Losick 2002). It is an 85 amino acid cytoplasmic protein with 50% of the structure being coiled coil. In the presence of FtsA and ZipA, ZapA has no significant role in the assembly of FtsZ. ZapA directly interacts with FtsZ and enhances the bundling of FtsZ protofilaments. It has been reported that the function of ZapA is redundant and deletion mutants of *zapA* fail to show elongated phenotype both in *E.coli* and *B. subtilis* cells (Gueiros-Filho and Losick 2002).

7.5.1.1.4 ZapB

ZapB is a non-essential division protein which has been reported to enhance FtsZ assembly (Ebersbach et al. 2008). The recruitment of ZapB to the cell division machinery was independent of the membrane anchoring protein FtsA and ZipA, but it required FtsZ for its localization.

7.5.1.1.5 ZapC

ZapC is an essential component of divisiome complex of *E. coli* (Durand-Heredia et al. 2011). Its deletion or overexpression leads to the abnormal septum formation (Durand-Heredia et al. 2011; Hale et al. 2011). *E.coli* FtsZ directly binds to ZapC in-vitro and promotes its assembly. It has been reported that unlike the other regulators of FtsZ assembly ZapC does not bind to the C-terminal tail of FtsZ (Bhattacharya et al. 2015). In a recent study it has been reported that the two negative regultors ClpXP and ClpAP regulate the level of ZapC in *E. coli* by the process of proteolysis (Buczek et al. 2016).

7.5.1.1.6 ZapD

ZapD is an important protein which enhances Z-ring formation by increasing the lateral interaction of FtsZ protofilaments (Durand-Heredia et al. 2012). Recent studies explored the interaction between *E. coli* FtsZ and ZapD interaction (Huang et al. 2016). They have found that ZapD forms functional dimer which interacts with the C-terminal domain of FtsZ (Roach et al. 2016). By using structural and site directed mutation analysis it was found that charged interaction was mostly involved between FtsZ and ZapD (Roach et al. 2016).

7.5.1.1.7 SepF

SepF is encoded by *ylmF* gene which is conserved in Gram-positive bacteria and acts as both early and late cell division protein (Ishikawa et al. 2006). In a recent study a homologue of SepF is reported in Mycobacterium species and alteration of SepF caused filamentous morphology in *Mycobacterium smegmatis* which depicts the divisional defects (Gola et al. 2015). SepF complements the function of FtsA in *B. subtilis* cells. If *ylmF* is overexpressed in the cells with a mutant *ftsA* it is seen that the cells show a normal phenotype. On the other hand the cells with mutated *ylmF* and *ftsA* show complete inhibition of cytokinesis. The overlapping functional role of FtsA and SepF in Z-ring formation indicates that SepF also participates in enhancing Z-ring formation in early stages (Ishikawa et al. 2006). From recent studies it has been found that SepF helps in septa formation in late stages of cell division. SepF binds to the C-terminal tail of the FtsZ and helps in assembly of Z-ring (Ishikawa et al. 2006; Krol et al. 2012). SepF increases the FtsZ assembly by decreasing the hydrolysis of GTP; it also increases the bundling of FtsZ by increasing the lateral interactions between the FtsZ protofilaments (Singh et al. 2008).

7.5.1.1.8 WhmD

WhmD is an essential protein which plays important role in cell division in *Mycobacterium smegmatis* species (Gomez and Bishai 2000). It is an interacting protein of FtsZ. WhmD directly interacts with *M. smegmatis* FtsZ in-vitro and promotes the assembly of purified FtsZ (Bhattacharya et al. 2017). Further the role of WhmD in septation has been proven in-vivo (Gomez and Bishai 2000).

7.5.2 Role of Negative Regulators in FtsZ Assembly

Similar to positive regulators, negative regulators also play a crucial role in cell division. The assembly and formation of Z-ring is inhibited by negative regulators which are present in the bacterial cells. These proteins assist in the disassembly of the Z-ring during the constriction process and also prevent the formation of Z-ring at inappropriate places throughout the cell, thereby ensuring proper division. Negative regulators antagonize the action of positive regulators to maintain the subunit turnover in the Z ring (Adams and Errington 2009).

The most prominent among the negative regulators are the Min proteins and the nucleoid occlusion systems. The Min proteins comprising of MinC, MinD and MinE (DivIVA in *Bacillus subtilis*) help to regulate the formation of the Z-ring at the mid-cell position. MinCD is present as a concentration gradient in the cell, the highest concentration being at the poles and the lowest concentration being at the mid-cell position (Hu et al. 1999; Huang et al. 1996; Pichoff and Lutkenhaus 2001). Therefore, the Z-ring is unable to form at the poles and is assembled at the center of the cell resulting in successful cell division (Taviti and Beuria 2017). MinC directly interacts with FtsZ, however, it requires MinD for its inhibitory activity (Shiomi and Margolin 2007). MinE is essential for oscillation of the MinCD from one pole to the other (Raskin and de Boer 1999). Further, over-expression of MinC and MinD was found to be lethal for the cells and the cell became filamentous, thereby affecting the process of cytokinesis (Ramirez-Arcos et al. 2004) and it binds with H10 helix of FtsZ (Taviti and Beuria 2017).

Nucleoid occlusion also inhibits the assembly of FtsZ by preventing the formation of Z ring. Nucleoid occlusion system suggests that, the conformational state and the activity of DNA guides the cell about the time and place of division with the spatial discontinuity generated by separation of two nucleoids forming a signal for the place of division. Compact nucleoid present at the center of the cell inhibits the formation of the FtsZ ring in its vicinity thus inhibiting cell division. It has been suggested that presence of transcription, translation and translocation of membrane proteins in the vicinity of the nucleoids results in the inhibition of assembly of FtsZ there (Woldringh 2002; Zaritsky and Woldringh 2003). Once the DNA replication and segregation has occurred, free space is generated at the mid cell position where the negative effect of nucleoid sufficiently decreases and hence division can occur at that position. This model also correlates the termination of the replication event with the beginning of peptidoglycan synthesis in *E.coli*, thus suggesting that nucleoid activity also provides signals for the time of division (Woldringh et al. 1991). This coupling of nucleoid separation with cell division is maintained even in the absence of *min* locus indicating that Min proteins are not involved in it (Dassain and Bouche 1994).

7.5.2.1 EzrA

EzrA was found to be conserved across low GC Gram-positive bacteria (Haeusser et al. 2004) and is a transmembrane protein with N-terminus in the plasma membrane and C-terminus in the cytoplasm (Levin et al. 1999). It is mainly distributed all over the plasma membrane. It is known to cause aberrant FtsZ assembly at cell poles (Land et al. 2014). The inhibition of cell division caused by overexpression of MinCD is suppressed by *ezr* null mutation (Levin et al. 2001) EzrA is a membrane-anchoring protein which is attached with a single N-terminal transmembrane helix and has four coiled coil motifs. In-vitro, EzrA was found to inhibit FtsZ polymerization and destabilize the preformed FtsZ protofilaments. EzrA interacts with the C-terminal tail of FtsZ and inhibits the assembly and bundling of FtsZ protofilaments (Singh et al. 2007).

7.5.2.2 SulA

SulA is a soluble protein present in *E. coli*. SulA is activated during DNA damage repair mechanism initiated by SOS response. In SOS response *lexA* repressor binds with SOS box and represses the synthesis of *sos* gene. RecA protein binds to a single stranded damaged DNA and induces lexA repressor for auto proteolysis (Fernandez De Henestrosa et al. 2000). SulA protein is activated by *lexA* gene. When the concentration of SulA increases in the cell it inhibits Z-ring formation. SulA has been reported to bind with the T7 loop of FtsZ and block protofilament interface resulting in inhibition of polymerization. The binding of SulA and FtsZ are concentration dependent. In-vitro, 1:4 ratio of SulA to FtsZ can inhibit polymer assembly without affecting the GTP hydrolysis activity of FtsZ (Trusca et al. 1998). At a 1:1 ratio of SulA to FtsZ, both assembly and GTP hydrolysis are inhibited (Trusca et al. 1998). It has been reported that binding of SulA to FtsZ inhibits its binding with another molecule of FtsZ. When SulA binds with one side of FtsZ it prevents the binding of FtsZ to another FtsZ. SulA can prevent assembly at relatively low concentrations. Consequently, it can be speculated that even when slightly induced SulA can rapidly block cytokinesis under in-vivo conditions.

7.5.2.3 UgtP

UgtP is an accessory protein of FtsZ which regulates the size of bacteria according to its nutrient status. It is widely conserved among the bacterial species (Weart et al. 2007). During nutrient enriched condition, UgtP is distributed throughout the cytoplasm with a higher concentration found near the Z-ring. It has been proposed that during high nutrient content UDP-glucose strengthens the interaction between FtsZ and UgtP at the Z-ring, resulting in the maturation of Z-ring and increase in cell size. However during poor nutrient condition UgtP is unable to interact with FtsZ and hence cell size was found to be small (Changes in the oligomerization potential of the division inhibitor UgtP co-ordinate Bacillus subtilis cell size with nutrient availability 2012). In-vitro, it was found to be interacting with itself or FtsZ and inhibited the assembly of FtsZ and prevented the lateral interaction between FtsZ protofilaments (Changes in the oligomerization potential of the division inhibitor UgtP co-ordinate Bacillus subtilis cell size with nutrient availability 2012).

7.5.2.4 MciZ

MciZ, a peptide having 40 amino acid residues, has been shown to express during bacterial sporulation (Araujo-Bazan et al. 2016; Handler et al. 2008). It was found to inhibit the Z-ring formation of mother sporangium cells during sporulation and also inhibited the assembly of FtsZ in-vitro in a GTP-dependent manner. Further, a substitution mutation near the GTP-binding pocket of FtsZ has been found to confer resistance to the inhibitory effect of MciZ raising the possibility that GTP might occlude the binding of FtsZ to MciZ (Handler et al. 2008; Ray et al. 2013).

7.5.2.5 ClpX

ClpX is a protein of the ClpXP protease which was reported to interact with FtsZ and inhibit its polymerization. ClpXP has ATP dependent protease activity (Camberg et al. 2009). It preferentially binds with FtsZ polymers and degrades them to maintain normal cell division. Further, the N-terminal residue of ClpX was found to interactwith C-terminal tail of FtsZ. However unlike other regulators of FtsZ, ClpX did not have any effect on the GTPase activity of FtsZ (Buske and Levin 2012). Moreover the overexpression of ClpXP resulted in filamentation of bacterial cells.

The crystal structure and sequences of all FtsZ homologues have revealed that FtsZ mainly consists of five distinct domains: the N-terminal domain, the highly conserved globular core, an unstructured C-terminal linker, a short, conserved "C-terminal tail" (CTT), and a highly variable set of residues, also known as C-terminal variable region (CTV) present in the sequential order from N- terminal to the C-terminal (Buske and Levin 2012). The N-terminal domain is unstructured and poorly conserved (Fig. 7.5). The globular core of the FtsZ protein consists of GTP binding site and T7 synergy loop which are required for the GTPase activity. The residues essential for the longitudinal subunit bonds at the dimer interface and lateral interactions between protofilaments, are present in the core region.

The C-terminal linker region comprises 50–300 residues (varies in different bacterial species). CTT consists of 11 residues, also the functional analogue of the eukaryotic protein tubulin's C- terminal tail. The CTT is the critical region for interaction of FtsZ with other cell division proteins like FtsA, ZipA, SepF, EzrA, MinC, ClpX etc. Deletion or mutation of CTT leads to the disruption of protein-protein interaction resulting in inhibition of FtsZ assembly to form the Z-ring during bacterial cell division. Asp-373, Ile-374, Phe-377, and Leu-378 are four conserved residues of CTT region required for the interaction of FtsZ with FtsA, MinC, and ZipA (Buske and Levin 2012). CTV is the highly variable domain, spanning 6–10 residues, depending on which FtsZ can be positive, negative or neutral in charge. In absence of other modulatory proteins, CTV is sufficient to aid the lateral interactions between FtsZ protofilaments. Mutations in CTV region has shown aberrant effects on the assembling capacity of FtsZ in *B. subtilis* and *E. coli*, which specifies the importance of this highly variable region in FtsZ assembly and cell division (Buske and Levin 2012). Till now the binding site of few regulatory proteins on FtsZ is well

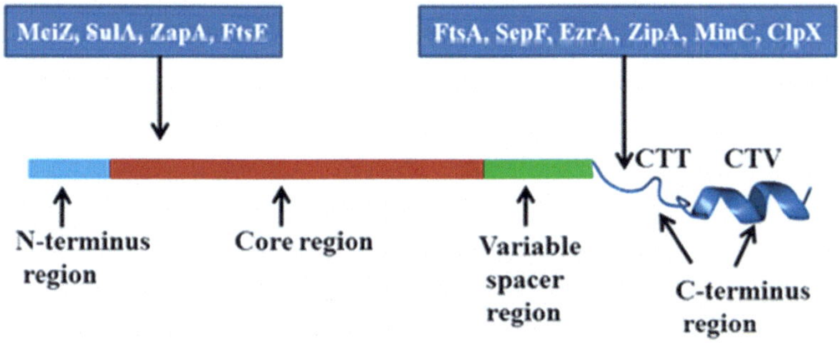

Fig. 7.5 Structural domains of FtsZ and binding site of its regulators. The schematic represents the arrangement of different domains of FtsZ and the binding site of some of the positive and negative regulators

characterized. Exploring the binding of different regulatory proteins will provide an idea to design novel inhibitors.

Apart from positive and negative regulators of FtsZ assembly there are several small molecule inhibitors that are known to hamper bacterial cell division by targeting FtsZ assembly (Jindal et al. 2013; Kapoor and Panda 2009; Lock and Harry 2008). These small molecule inhibitors inhibit bacterial cytokinesis either by enhancing or decreasing FtsZ polymerization (Jindal et al. 2013; Kapoor and Panda 2009; Lock and Harry 2008). These small molecule inhibitors have been identified naturally or chemically synthesized specifically to target FtsZ assembly. Tubulin has three well characterized binding pockets for small molecules and it appears that FtsZ also harbors certain sites which are preferable binding sites of many known small molecule inhibitors (Fig. 7.6).

Since FtsZ has GTPase activity important for forming protofilaments, targeting the GTP site could be used for the development of new FtsZ targeted inhibitors. As discussed above, successful formation of formation of the FtsZ ring requires support from their modulatory proteins. Any disruption of the interaction of FtsZ with its interacting proteins causes improper division and abnormal septa formation. In past few years by using structure activity relationship, many group identified new chemical moieties which targeted the GTP site of FtsZ or perturbed the interaction between FtsZ and its modulatory proteins to inhibit the bacterial cytokinesis machinery. A list of important FtsZ targeted synthetic and phytochemical inhibitors and their mode of action is given in Table 7.1.

7.6 Future Prospectus and Conclusion

The massive failure of the clinically approved antibiotics in controlling the spread of pathogenesis caused by bacteria necessitates the development of new antibacterials with a novel mode of action, and not just merely variants of older drugs. The failure is markedly due to the increase in the antibiotic resistant strains of pathogenic

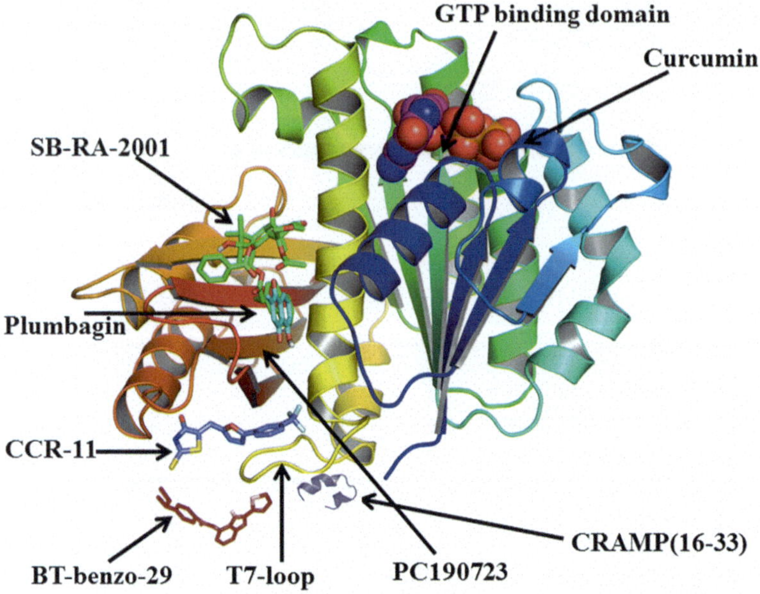

Fig. 7.6 Predicated putative binding site of different FtsZ targeted inhibitors on FtsZ. Illustration of the probable binding site of FtsZ inhibitors on the crystal structure of *B. subtilis*. The crystal structure of FtsZ (PDB ID: 2VXY) was taken from protein data bank and then drawn in PyMol (DeLano 2002)

bacteria. Generally, the antibiotics used to treat bacterial infections target either of these four processes, i.e. protein synthesis, cell wall synthesis, nucleic acid synthesis or folate synthesis (Lock and Harry 2008). But, development of clinically approved drugs that target bacterial cell division is still underway.

Cell division is the most vital event that occurs in any organism for continuous survival of its kind. Therefore, developing drugs that can target the proteins involved in bacterial cell division could be the solution to restrict bacterial viability. The most vital protein that initiates and facilitates the proper bacterial cell division is FtsZ. FtsZ is a conserved bacterial cell division protein present in almost all bacteria. It is a structural homologue of eukaryotic cytoskeleton protein tubulin which is already efficacious and used in cancer chemotherapy and, one of the most important components of bacterial divisome. The bacterial cytokinesis is initiated by the assembly of the FtsZ protein into a contractile ring known as the Z ring which marks the start of further processes of cell division. Hence, FtsZ protein is a very important target of the antibacterial drugs to impede the bacterial cell division and viability (Buske and Levin 2012; Camberg et al. 2009; Jindal et al. 2013; Kapoor and Panda 2009; Lock and Harry 2008) (Fig. 7.7).

Although the proteins FtsZ and tubulin have structural similarity, the sequence similarity is only 10–20%. Hence, the specific inhibitors developed against FtsZ are likely to perturb the function of the eukaryotic protein, tubulin. FtsZ and its

Table 7.1 List of FtsZ targeted inhibitors and their mode of action

Inhibitor	Mode of action	Targeted organism	Minimum inhibitory concentration (MIC)	Sourceof Inhibitor	References
Viriditoxin	Inhibits FtsZ polymerization assembly	*Streptococcus*, *Enterococcus* and *Staphylococcus*	MICs varies between 2–32 μg/mL	Microbial fermentation broths and plant extracts	Wang et al. (2003)
8-bromoguanosine 5′-triphosphate (Br-GTP)	Inhibits FtsZ polymerization and GTPase activity in-vitro	Not reported	Not reported	Chemically synthesized GTP analogue	Lappchen et al. (2005)
PC190723	Promotes FtsZ polymerization and suppresses GTPase activity of FtsZ	*B. subtilis 168*, MDR Methicillin-resistant *S. aureus*	2.8 μM for *B. subtilis* 168 and *S. aureus*	Taxol derivative	Haydon et al. (2008)
Amikacin	Perturbs Z-ring formation	*E. coli*	MICs varies between 0.15–256 μg/ml	Aminoglycoside	Possoz et al. (2007)
SRI-3072	Inhibits MtbFtsZ polymerization	*M. tuberculosis*	0.15 μg/mL for *M. tuberculosis*	Chemically synthesized	White et al. (2002)
CCR-11	Inhibits FtsZ polymerization and GTPase activity in-vitro	*B. subtilis* 168 cells	3 μM for *B. subtilis* 168	Rhodanine derivative	Singh et al (2012)
Totarol	Inhibits FtsZ polymerization	*B. subtilis* 168 cells	2 μM for *B. subtilis* 168	Diterpenoid phenol	Jaiswal et al. (2007)
Sanguinarine	Inhibits FtsZ polymerization and suppresses GTPase activity of FtsZ	*S. aureus, B. subtilis* 168, *E. coli* K12 cells	3.12 μg/ml for S. aureus	Roots of the herbaceous plant *Sanguinaria canadensis*	Beuria et al. (2005)

(continued)

Table 7.1 (continued)

Inhibitor	Mode of action	Targeted organism	Minimum inhibitory concentration (MIC)	Sourceof Inhibitor	References
Berberine	Inhibits FtsZ polymerization and suppresses GTPase activity of FtsZ	*E. coli*, MDR *M. tuberculosis,* Methicillin-resistant *S. aureus*	1.5 mM for *E. coli* K12 cells	Plant alkaloid	Boberek et al. (2010)
Curcumin	Inhibits FtsZ polymerization and increases GTPase activity of FtsZ	*S. aureus, B. subtilis* 168, *E. coli* K12 cells	100 μM for *B. subtilis* 168	Plant product	Rai et al. (2008)
OTBA	Promotes FtsZ polymerization and suppresses GTPase activity of FtsZ	*B. subtilis* 168	100 μM for *B. subtilis* 168	3-{5-[4-oxo-2-thioxo-3-(3-trifluoromethylphenyl)-thiazolidin-5-ylidenemethyl]-furan-2-yl}-benzoic acid	Beuria et al. (2009)
Taxanes	Inhibits FtsZ polymerization	*M. tuberculosis* starins H37Rv and IMCJ946. K2	MICs varies between 1.25–2.5 μg/ml	Chemically synthesized	Huang et al. (2006)
Chrysophaentins	Inhibits FtsZ polymerization and suppresses GTPase activity of FtsZ	MSRA *E. faecium*, and vancomycin resistant *E. faecium* and *S. aureus* cells	MICs varies between 1.5–4 μg/ml	Isolated from marine alga *Chrysophaem taylori*	Keffer et al. (2013)
SB-RA-2001	Promotes FtsZ polymerization and increases GTPase activity of FtsZ	*B. subtilis* 168, *M. smegmatis* mc^2 cells	38 μM for *B. subtilis* 168 and 60 μM for *M. smegmatis* mc^2	Taxane derivative	Singh et al. (2014)
Plumbagin	Inhibits FtsZ polymerization and suppresses GTPase activity of FtsZ	*B. subtilis* 168, *M. smegmatis* mc^2 cells	29 μM for *B. subtilis* 168 and 31 μM for *M. smegmatis* mc^2	5-hydroxy-2-methyl-1, 4-naphthoquinone (secondary plant metabolite)	Bhattacharya et al. (2013)

BT-benzo-29	Inhibit FtsZ polymerization and suppresses GTPase activity of FtsZ	*B. subtilis* 168, *M. smegmatis* mc^2 cells	8 µM for *B. subtilis* 168	N-(4-sec- butylphenyl)-2-(thiophen-2-yl)-1Hbenzo [d]imidazole-4-carboxamide	Ray et al. (2015)
CRAMP(16–33)	Inhibit FtsZ polymerization and suppresses GTPase activity of FtsZ	*B. subtilis* 168, *E.coli k12* cells	20 µM for *B. subtilis* 168 and 50 µM for *E. coli K12 cells*	Cathelin-related antimicrobial derived peptide	Ray et al. (2014)
Doxorubicin	Inhibit FtsZ polymerization and suppresses GTPase activity of FtsZ	*E. coli* BL21, *E. coli* MG1655, *B. subtilis* and *S. aureus* cells	20 µM for *E. coli* BL21, 40 µM for *E. coli* MG1655, 10 µM for *B. subtilis* and 5 µM for *S. aureus* cells	Anthracycline molecule	Panda et al. (2015)

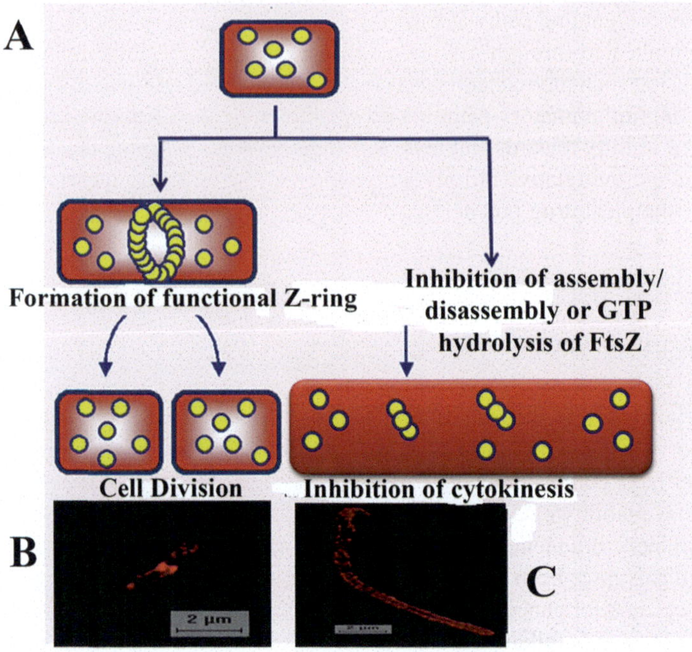

Fig. 7.7 Perturbation of functional Z-ring formation hampers bacterial cytokinesis. (**a**) The circle represents FtsZ monomer. FtsZ targeted inhibitors inhibit complete septa formation by targeting FtsZ assembly leads to the filamentous morphology of bacterial cell. (**b**) FtsZ was visualized (red) using antibody in *B. subtilis* 168 cells (**c**) The *B. subtilis* 168 cells achieved filamentous morphology in the presence of FtsZ targated inhibitor CRAMP (16–33)

eukaryotic analogue have many common properties, but their different structures, biochemistry, regulation and interacting members enable us to find compounds that inhibit FtsZ and not tubulin. For example, the alkoxy carbonyl aminopyridine compound SRI-3072 is potent against susceptible and drug-resistant strains of *M. tuberculosis*. It inhibits FtsZ but not tubulin polymerization (Lock and Harry 2008; White et al. 2002).

Viriditoxin, a naturally isolated compound selectively inhibits FtsZ polymerization and GTPase activity in-vitro and in-vivo with no toxicity to eukaryotic cells, making it a leading compound for drug discovery (White et al. 2002).

Further, the assembly of divisome and regulation of division process requires the sequential interaction of other proteins such as FtsA, ZipA, ZapA, MinC and other stabilizing and destabilizing proteins, with FtsZ. An inhibition of this protein-protein interaction could be a potential target of the antibacterials. Significant progress has already been made in the discovery of small molecule inhibitors that prevent the protein-protein interaction. Moreover, the external presence of these cell division proteins on the bacterial membrane makes them easily accessible to antibacterials.

The genome of more than 850 bacteria has already been sequenced (www.ncbi.nlm.nih.gov/genomes/lproks.cgi) and many more are expected to be sequenced in near future. As FtsZ is conserved in most of the bacteria it reveals that it is essential protein. The inhibitors, which target FtsZ or FtsZ interacting proteins, may act as broad-spectrum antibiotics. These broad-spectrum antibiotics are very effective in treating infections caused by several bacteria, so we can use one antibiotic which will act against all bacteria. This demands targeting a highly conserved protein like FtsZ and its interacting proteins which will lead to the development of new antibacterial drugs.

Acknowledgements S.R. acknowledges Indian Institute of Technology Bombay, India and Mahatma Gandhi Central University, Motihari, Bihar. S. R also sincerely thanks to Ankit Rai for critical reading of the manuscript.

References

Adams DW, Errington J. Bacterial cell division: assembly, maintenance and disassembly of the Z ring. Nat Rev Microbiol. 2009;7:642–53. https://doi.org/10.1038/nrmicro2198.

Araujo-Bazan L, Ruiz-Avila LB, Andreu D, Huecas S, Andreu JM. Cytological profile of antibacterial FtsZ inhibitors and synthetic peptide MciZ. Front Microbiol. 2016;7:1558. https://doi.org/10.3389/fmicb.2016.01558.

Beall B, Lutkenhaus J. Impaired cell division and sporulation of a Bacillus subtilis strain with the ftsA gene deleted. J Bacteriol. 1992;174:2398–403.

Beuria TK, Krishnakumar SS, Sahar S, Singh N, Gupta K, Meshram M, Panda D. Glutamate-induced assembly of bacterial cell division protein FtsZ. J Biol Chem. 2003;278:3735–41. https://doi.org/10.1074/jbc.M205760200.

Beuria TK, Santra MK, Panda D. Sanguinarine blocks cytokinesis in bacteria by inhibiting FtsZ assembly and bundling. Biochemistry. 2005;44:16584–93. https://doi.org/10.1021/bi050767+.

Beuria TK, Singh P, Surolia A, Panda D. Promoting assembly and bundling of FtsZ as a strategy to inhibit bacterial cell division: a new approach for developing novel antibacterial drugs. Biochem J. 2009;423:61–9. https://doi.org/10.1042/BJ20090817.

Bhattacharya A, Jindal B, Singh P, Datta A, Panda D. Plumbagin inhibits cytokinesis in Bacillus subtilis by inhibiting FtsZ assembly – a mechanistic study of its antibacterial activity. FEBS J. 2013;280:4585–99. https://doi.org/10.1111/febs.12429.

Bhattacharya A, Ray S, Singh D, Dhaked HP, Panda D. ZapC promotes assembly and stability of FtsZ filaments by binding at a different site on FtsZ than ZipA. Int J Biol Macromol. 2015;81:435–42. https://doi.org/10.1016/j.ijbiomac.2015.08.030.

Bhattacharya D, Kumar A, Panda D. WhmD promotes the assembly of Mycobacterium smegmatis FtsZ: a possible role of WhmD in bacterial cell division. Int J Biol Macromol. 2017;95:582–91. https://doi.org/10.1016/j.ijbiomac.2016.11.056.

Boberek JM, Stach J, Good L. Genetic evidence for inhibition of bacterial division protein FtsZ by berberine. PLoS One. 2010;5:e13745. https://doi.org/10.1371/journal.pone.0013745.

Bork P, Sander C, Valencia A. An ATPase domain common to prokaryotic cell cycle proteins, sugar kinases, actin, and hsp70 heat shock proteins. Proc Natl Acad Sci U S A. 1992;89:7290–4.

Boyd D, Nixon R, Gillespie S, Gillespie D. Screening of *Escherichia coli* temperature-sensitive mutants by pretreatment with glucose starvation. J Bacteriol. 1968;95:1040–50.

Buczek MS, Cardenas Arevalo AL, Janakiraman A. ClpXP and ClpAP control the *Escherichia coli* division protein ZapC by proteolysis. Microbiology. 2016;162:909–20. https://doi.org/10.1099/mic.0.000278.

Buddelmeijer N, Beckwith J. Assembly of cell division proteins at the *E. coli* cell center. Curr Opin Microbiol. 2002;5:553–7.
Buske PJ, Levin PA. Extreme C terminus of bacterial cytoskeletal protein FtsZ plays fundamental role in assembly independent of modulatory proteins. J Biol Chem. 2012;287:10945–57. https://doi.org/10.1074/jbc.M111.330324.
Cabeen MT, Jacobs-Wagner C. The bacterial cytoskeleton. Annu Rev Genet. 2010;44:365–92. https://doi.org/10.1146/annurev-genet-102108-134845.
Camberg JL, Hoskins JR, Wickner S. ClpXP protease degrades the cytoskeletal protein, FtsZ, and modulates FtsZ polymer dynamics. Proc Natl Acad Sci U S A. 2009;106:10614–9. https://doi.org/10.1073/pnas.0904886106.
Changes in the oligomerization potential of the division inhibitor UgtP co-ordinate Bacillus subtilis cell size with nutrient availability. Mol Microbiol. 2012;86:594–610. https://doi.org/10.1111/mmi.12007.
Clement MJ, et al. The stathmin-derived I19L peptide interacts with FtsZ and alters its bundling. Biochemistry. 2009;48:9734–44. https://doi.org/10.1021/bi900556a.
Dai K, Lutkenhaus J. The proper ratio of FtsZ to FtsA is required for cell division to occur in *Escherichia coli*. J Bacteriol. 1992;174:6145–51.
Dai K, Mukherjee A, Xu Y, Lutkenhaus J. Mutations in ftsZ that confer resistance to SulA affect the interaction of FtsZ with GTP. J Bacteriol. 1994;176:130–6.
Dassain M, Bouche JP. The min locus, which confers topological specificity to cell division, is not involved in its coupling with nucleoid separation. J Bacteriol. 1994;176:6143–5.
DeLano WL. The PyMOL molecular genetics graphics system. San Carlos: DeLano Scientific LLC; 2002.
Din N, Quardokus EM, Sackett MJ, Brun YV. Dominant C-terminal deletions of FtsZ that affect its ability to localize in caulobacter and its interaction with FtsA. Mol Microbiol. 1998;27:1051–63.
Durand-Heredia JM, Yu HH, De Carlo S, Lesser CF, Janakiraman A. Identification and characterization of ZapC, a stabilizer of the FtsZ ring in *Escherichia coli*. J Bacteriol. 2011;193:1405–13. https://doi.org/10.1128/JB.01258-10.
Durand-Heredia J, Rivkin E, Fan G, Morales J, Janakiraman A. Identification of ZapD as a cell division factor that promotes the assembly of FtsZ in *Escherichia coli*. J Bacteriol. 2012;194:3189–98. https://doi.org/10.1128/JB.00176-12.
Ebersbach G, Galli E, Moller-Jensen J, Lowe J, Gerdes K. Novel coiled-coil cell division factor ZapB stimulates Z ring assembly and cell division. Mol Microbiol. 2008;68:720–35. https://doi.org/10.1111/j.1365-2958.2008.06190.x.
Erickson HP, Anderson DE, Osawa M. FtsZ in bacterial cytokinesis: cytoskeleton and force generator all in one. Microbiol Mol Biol Rev. 2010;74:504–28. https://doi.org/10.1128/MMBR.00021-10.
Errington J, Daniel RA, Scheffers DJ. Cytokinesis in bacteria. Microbiol Mol Biol Rev. 2003;67:52–65, table of contents
Fernandez De Henestrosa AR, Ogi T, Aoyagi S, Chafin D, Hayes JJ, Ohmori H, Woodgate R. Identification of additional genes belonging to the LexA regulon in *Escherichia coli*. Mol Microbiol. 2000;35:1560–72.
Feucht A, Lucet I, Yudkin MD, Errington J. Cytological and biochemical characterization of the FtsA cell division protein of Bacillus subtilis. Mol Microbiol. 2001;40:115–25.
Goehring NW, Beckwith J. Diverse paths to midcell: assembly of the bacterial cell division machinery. Curr Biol. 2005;15:R514–26. https://doi.org/10.1016/j.cub.2005.06.038.
Gola S, Munder T, Casonato S, Manganelli R, Vicente M. The essential role of SepF in mycobacterial division. Mol Microbiol. 2015;97:560–76. https://doi.org/10.1111/mmi.13050.
Gomez JE, Bishai WR. whmD is an essential mycobacterial gene required for proper septation and cell division. Proc Natl Acad Sci USA. 2000;97:8554–9. https://doi.org/10.1073/pnas.140225297.

Gueiros-Filho FJ, Losick R. A widely conserved bacterial cell division protein that promotes assembly of the tubulin-like protein FtsZ. Genes Dev. 2002;16:2544–56. https://doi.org/10.1101/gad.1014102.

Haeusser DP, Schwartz RL, Smith AM, Oates ME, Levin PA. EzrA prevents aberrant cell division by modulating assembly of the cytoskeletal protein FtsZ. Mol Microbiol. 2004;52:801–14. https://doi.org/10.1111/j.1365-2958.2004.04016.x.

Hale CA, de Boer PA. Recruitment of ZipA to the septal ring of *Escherichia coli* is dependent on FtsZ and independent of FtsA. J Bacteriol. 1999;181:167–76.

Hale CA, Shiomi D, Liu B, Bernhardt TG, Margolin W, Niki H, de Boer PA. Identification of *Escherichia coli* ZapC (YcbW) as a component of the division apparatus that binds and bundles FtsZ polymers. J Bacteriol. 2011;193:1393–404. https://doi.org/10.1128/JB.01245-10.

Handler AA, Lim JE, Losick R. Peptide inhibitor of cytokinesis during sporulation in Bacillus subtilis. Mol Microbiol. 2008;68:588–99. https://doi.org/10.1111/j.1365-2958.2008.06173.x.

Harry E, Monahan L, Thompson L. Bacterial cell division: the mechanism and its precison. Int Rev Cytol. 2006;253:27–94. https://doi.org/10.1016/S0074-7696(06)53002-5.

Haydon DJ, et al. An inhibitor of FtsZ with potent and selective anti-staphylococcal activity. Science. 2008;321:1673–5. https://doi.org/10.1126/science.1159961.

Hu Z, Mukherjee A, Pichoff S, Lutkenhaus J. The MinC component of the division site selection system in *Escherichia coli* interacts with FtsZ to prevent polymerization. Proc Natl Acad Sci U S A. 1999;96:14819–24.

Huang J, Cao C, Lutkenhaus J. Interaction between FtsZ and inhibitors of cell division. J Bacteriol. 1996;178:5080–5.

Huang Q, et al. Targeting FtsZ for antituberculosis drug discovery: noncytotoxic taxanes as novel antituberculosis agents. J Med Chem. 2006;49:463–6. https://doi.org/10.1021/jm050920y.

Huang KH, Mychack A, Tchorzewski L, Janakiraman A. Characterization of the FtsZ C-terminal variable (CTV) region in Z-ring assembly and interaction with the Z-ring stabilizer ZapD in *E. coli* cytokinesis. PLoS One. 2016;11:e0153337. https://doi.org/10.1371/journal.pone.0153337.

Ishikawa S, Kawai Y, Hiramatsu K, Kuwano M, Ogasawara N. A new FtsZ-interacting protein, YlmF, complements the activity of FtsA during progression of cell division in Bacillus subtilis. Mol Microbiol. 2006;60:1364–80. https://doi.org/10.1111/j.1365-2958.2006.05184.x.

Jaiswal R, Beuria TK, Mohan R, Mahajan SK, Panda D. Totarol inhibits bacterial cytokinesis by perturbing the assembly dynamics of FtsZ. Biochemistry. 2007;46:4211–20. https://doi.org/10.1021/bi602573e.

Jindal B, Bhattacharya A, Panda D. Inhibitors of bacterial cell partitioning. In: Antibiotics- targets, mechanisms and resistance. Weinheim: Wiley-VCH; 2013. p. 151–82.

Kapoor S, Panda D. Targeting FtsZ for antibacterial therapy: a promising avenue. Expert Opin Ther Targets. 2009;13:1037–51. https://doi.org/10.1517/14728220903173257.

Keffer JL, Huecas S, Hammill JT, Wipf P, Andreu JM, Bewley CA. Chrysophaentins are competitive inhibitors of FtsZ and inhibit Z-ring formation in live bacteria. Bioorg Med Chem. 2013;21:5673–8. https://doi.org/10.1016/j.bmc.2013.07.033.

Krol E, van Kessel SP, van Bezouwen LS, Kumar N, Boekema EJ, Scheffers DJ. Bacillus subtilis SepF binds to the C-terminus of FtsZ. PLoS One. 2012;7:e43293. https://doi.org/10.1371/journal.pone.0043293.

Kuchibhatla A, Bhattacharya A, Panda D. ZipA binds to FtsZ with high affinity and enhances the stability of FtsZ protofilaments. PLoS One. 2011;6:e28262. https://doi.org/10.1371/journal.pone.0028262.

Kumar K, Awasthi D, Berger WT, Tonge PJ, Slayden RA, Ojima I. Discovery of anti-TB agents that target the cell-division protein FtsZ. Future Med Chem. 2010;2:1305–23. https://doi.org/10.4155/fmc.10.220.

Land AD, Luo Q, Levin PA. Functional domain analysis of the cell division inhibitor EzrA. PLoS One. 2014;9:e102616. https://doi.org/10.1371/journal.pone.0102616.

Lappchen T, Hartog AF, Pinas VA, Koomen GJ, den Blaauwen T. GTP analogue inhibits polymerization and GTPase activity of the bacterial protein FtsZ without affecting its eukaryotic homologue tubulin. Biochemistry. 2005;44:7879–84. https://doi.org/10.1021/bi047297o.

Levin PA, Kurtser IG, Grossman AD. Identification and characterization of a negative regulator of FtsZ ring formation in Bacillus subtilis. Proc Natl Acad Sci U S A. 1999;96:9642–7.

Levin PA, Schwartz RL, Grossman AD. Polymer stability plays an important role in the positional regulation of FtsZ. J Bacteriol. 2001;183:5449–52.

Lock RL, Harry EJ. Cell-division inhibitors: new insights for future antibiotics. Nat Rev Drug Discov. 2008;7:324–38. https://doi.org/10.1038/nrd2510.

Lowe J. Crystal structure determination of FtsZ from methanococcus jannaschii. J Struct Biol. 1998;124:235–43. https://doi.org/10.1006/jsbi.1998.4041.

Lowe J, Amos LA. Crystal structure of the bacterial cell-division protein FtsZ. Nature. 1998;391:203–6. https://doi.org/10.1038/34472.

Lowe J, Amos LA. Tubulin-like protofilaments in Ca2+-induced FtsZ sheets. EMBO J. 1999;18:2364–71. https://doi.org/10.1093/emboj/18.9.2364.

Ma X, Margolin W. Genetic and functional analyses of the conserved C-terminal core domain of *Escherichia coli* FtsZ. J Bacteriol. 1999;181:7531–44.

Ma X, Ehrhardt DW, Margolin W. Colocalization of cell division proteins FtsZ and FtsA to cytoskeletal structures in living *Escherichia coli* cells by using green fluorescent protein. Proc Natl Acad Sci U S A. 1996;93:12998–3003.

Mukherjee A, Lutkenhaus J. Guanine nucleotide-dependent assembly of FtsZ into filaments. J Bacteriol. 1994;176:2754–8.

Mukherjee A, Lutkenhaus J. Analysis of FtsZ assembly by light scattering and determination of the role of divalent metal cations. J Bacteriol. 1999;181:823–32.

Mukherjee A, Dai K, Lutkenhaus J. *Escherichia coli* cell division protein FtsZ is a guanine nucleotide binding protein. Proc Natl Acad Sci U S A. 1993;90:1053–7.

Nogales E. A structural view of microtubule dynamics. Cell Mol Life Sci. 1999;56:133–42.

Nogales E, Downing KH, Amos LA, Lowe J. Tubulin and FtsZ form a distinct family of GTPases. Nat Struct Biol. 1998;5:451–8.

Panda P, Taviti AC, Satpati S, Kar MM, Dixit A, Beuria TK. Doxorubicin inhibits *E. coli* division by interacting at a novel site in FtsZ. Biochem J. 2015;471:335–46. https://doi.org/10.1042/BJ20150467.

Pichoff S, Lutkenhaus J. *Escherichia coli* division inhibitor MinCD blocks septation by preventing Z-ring formation. J Bacteriol. 2001;183:6630–5. https://doi.org/10.1128/JB.183.22.6630-6635.2001.

Pichoff S, Lutkenhaus J. Unique and overlapping roles for ZipA and FtsA in septal ring assembly in *Escherichia coli*. EMBO J. 2002;21:685–93.

Pichoff S, Lutkenhaus J. Tethering the Z ring to the membrane through a conserved membrane targeting sequence in FtsA. Mol Microbiol. 2005;55:1722–34. https://doi.org/10.1111/j.1365-2958.2005.04522.x.

Pichoff S, Lutkenhaus J. Identification of a region of FtsA required for interaction with FtsZ. Mol Microbiol. 2007;64:1129–38. https://doi.org/10.1111/j.1365-2958.2007.05735.x.

Possoz C, Newmark J, Sorto N, Sherratt DJ, Tolmasky ME. Sublethal concentrations of the aminoglycoside amikacin interfere with cell division without affecting chromosome dynamics. Antimicrob Agents Chemother. 2007;51:252–6. https://doi.org/10.1128/AAC.00892-06.

Rai D, Singh JK, Roy N, Panda D. Curcumin inhibits FtsZ assembly: an attractive mechanism for its antibacterial activity. Biochem J. 2008;410:147–55. https://doi.org/10.1042/BJ20070891.

Ramirez-Arcos S, et al. Conserved glycines in the C terminus of MinC proteins are implicated in their functionality as cell division inhibitors. J Bacteriol. 2004;186:2841–55.

Raskin DM, de Boer PA. Rapid pole-to-pole oscillation of a protein required for directing division to the middle of *Escherichia coli*. Proc Natl Acad Sci U S A. 1999;96:4971–6.

Ray S, Kumar A, Panda D. GTP regulates the interaction between MciZ and FtsZ: a possible role of MciZ in bacterial cell division. Biochemistry. 2013;52:392–401. https://doi.org/10.1021/bi301237m.

Ray S, Dhaked HP, Panda D. Antimicrobial peptide CRAMP (16-33) stalls bacterial cytokinesis by inhibiting FtsZ assembly. Biochemistry. 2014;53:6426–9. https://doi.org/10.1021/bi501115p.

Ray S, Jindal B, Kunal K, Surolia A, Panda D. BT-benzo-29 inhibits bacterial cell proliferation by perturbing FtsZ assembly. FEBS J. 2015;282:4015–33. https://doi.org/10.1111/febs.13403.

RayChaudhuri D, Park JT. *Escherichia coli* cell-division gene ftsZ encodes a novel GTP-binding protein. Nature. 1992;359:251–4. https://doi.org/10.1038/359251a0.

Roach EJ, Wroblewski C, Seidel L, Berezuk AM, Brewer D, Kimber MS, Khursigara CM. Structure and mutational analyses of *Escherichia coli* ZapD reveal charged residues involved in FtsZ filament bundling. J Bacteriol. 2016;198:1683–93. https://doi.org/10.1128/JB.00969-15.

Santra MK, Beuria TK, Banerjee A, Panda D. Ruthenium red-induced bundling of bacterial cell division protein, FtsZ. J Biol Chem. 2004;279:25959–65. https://doi.org/10.1074/jbc.M312473200.

Santra MK, Dasgupta D, Panda D. Deuterium oxide promotes assembly and bundling of FtsZ protofilaments. Proteins. 2005;61:1101–10. https://doi.org/10.1002/prot.20671.

Scheffers D, Driessen AJ. The polymerization mechanism of the bacterial cell division protein FtsZ. FEBS Lett. 2001;506:6–10.

Shiomi D, Margolin W. The C-terminal domain of MinC inhibits assembly of the Z ring in *Escherichia coli*. J Bacteriol. 2007;189:236–43. https://doi.org/10.1128/JB.00666-06.

Singh JK, Makde RD, Kumar V, Panda D. A membrane protein, EzrA, regulates assembly dynamics of FtsZ by interacting with the C-terminal tail of FtsZ. Biochemistry. 2007;46:11013–22. https://doi.org/10.1021/bi700710j.

Singh JK, Makde RD, Kumar V, Panda D. SepF increases the assembly and bundling of FtsZ polymers and stabilizes FtsZ protofilaments by binding along its length. J Biol Chem. 2008;283:31116–24. https://doi.org/10.1074/jbc.M805910200.

Singh P, Jindal B, Surolia A, Panda D. A rhodanine derivative CCR-11 inhibits bacterial proliferation by inhibiting the assembly and GTPase activity of FtsZ. Biochemistry. 2012;51:5434–42. https://doi.org/10.1021/bi201813u.

Singh D, Bhattacharya A, Rai A, Dhaked HP, Awasthi D, Ojima I, Panda D. SB-RA-2001 inhibits bacterial proliferation by targeting FtsZ assembly. Biochemistry. 2014;53:2979–92. https://doi.org/10.1021/bi401356y.

Stricker J, Maddox P, Salmon ED, Erickson HP. Rapid assembly dynamics of the Escherichia coli FtsZ-ring demonstrated by fluorescence recovery after photobleaching. Proc Natl Acad Sci U S A. 2002;99:3171–5. https://doi.org/10.1073/pnas.052595099.

Taviti AC, Beuria TK. MinD directly interacts with FtsZ at H10 helix suggests a model for robust activation of MinC to destabilize FtsZ polymers. Biochem J. 2017;474(18):3189–205. https://doi.org/10.1042/BCJ20170357.

Trusca D, Scott S, Thompson C, Bramhill D. Bacterial SOS checkpoint protein SulA inhibits polymerization of purified FtsZ cell division protein. J Bacteriol. 1998;180:3946–53.

van den Ent F, Lowe J. Crystal structure of the cell division protein FtsA from Thermotoga maritima. EMBO J. 2000;19:5300–7. https://doi.org/10.1093/emboj/19.20.5300.

Wang X, Huang J, Mukherjee A, Cao C, Lutkenhaus J. Analysis of the interaction of FtsZ with itself, GTP, and FtsA. J Bacteriol. 1997;179:5551–9.

Wang J, et al. Discovery of a small molecule that inhibits cell division by blocking FtsZ, a novel therapeutic target of antibiotics. J Biol Chem. 2003;278:44424–8. https://doi.org/10.1074/jbc.M307625200.

Weart RB, Lee AH, Chien AC, Haeusser DP, Hill NS, Levin PA. A metabolic sensor governing cell size in bacteria. Cell. 2007;130:335–47. https://doi.org/10.1016/j.cell.2007.05.043.

White EL, Suling WJ, Ross LJ, Seitz LE, Reynolds RC. 2-Alkoxycarbonylaminopyridines: inhibitors of *Mycobacterium tuberculosis* FtsZ. J Antimicrob Chemother. 2002;50:111–4.
Woldringh CL. The role of co-transcriptional translation and protein translocation (transertion) in bacterial chromosome segregation. Mol Microbiol. 2002;45:17–29.
Woldringh CL, Mulder E, Huls PG, Vischer N. Toporegulation of bacterial division according to the nucleoid occlusion model. Res Microbiol. 1991;142:309–20.
Wu LJ, Errington J. Coordination of cell division and chromosome segregation by a nucleoid occlusion protein in Bacillus subtilis. Cell. 2004;117:915–25. https://doi.org/10.1016/j.cell.2004.06.002.
Zaritsky A, Woldringh CL. Localizing cell division in spherical *Escherichia coli* by nucleoid occlusion. FEMS Microbiol Lett. 2003;226:209–14.

Chapter 8
Immunomodulatory Potential of Phytochemicals: Recent Updates

Nidhi Sharma, Herschel S. Dhekne, and Sabyasachi Senapati

8.1 Introduction

Immune system acts as a well-defined protective shield against noxious external or internal intervening agents. Its structural and functional complexities have evolved over several million years from single cell prokaryotes to vertebrates, such as in mammals. Complex immune system in higher mammals including human has broadly two functional modules, i.e., (i) cellular immune system, and (ii) humoral immune system. Several cellular signaling molecules and immune system specific molecules induce specific immunomodulation to maintain homeostasis. Such balanced conditions provide optimal niche required for overall health. Pathogenesis of different forms of tissue inflammation (through external antigen or auto-antigen) and several forms of cancers occur due to defects in immune homeostasis. Therefore, this system has been targeted to induce specific immune response through therapeutic agents to heal any imbalance. Synthetic molecules such as organic or inorganic compounds and induced biomolecules such as biologics are generally used as therapeutics to induce desired immune response. Modern as well as ancient folk or traditional medicines often include plant extracts to treat diseases. Several recent evidences suggested role of these plant extracts in immune system modulation.

Phytochemicals are the naturally occurring secondary metabolites present in abundance in plants. They do not have any nutritional importance but they are essential for the growth and maintenance of plants. They provide protection to plants against herbivores, insects, pathogens, micro-organisms etc. (Devappa et al. 2015). Phytochemicals play significant role in fertilization, nitrogen fixation and pollination as well as impart characteristic texture, color, flavor and fragrance to plants

N. Sharma · S. Senapati (✉)
Department of Human Genetics and Molecular Medicine, Central University of Punjab, Bathinda, Punjab, India

H. S. Dhekne
Department of Biochemistry, Stanford University, Stanford, CA, USA

S. Kumar, C. Egbuna (eds.), *Phytochemistry: An in-silico and in-vitro Update*,
https://doi.org/10.1007/978-981-13-6920-9_8

(Molyneux et al. 2007). They are extensively present in beverages derived from plants, nuts, fruits, herbs, vegetables, legumes, etc. With the evolution human has learnt the ways to harvest and manipulate these phytochemicals for his own benefits (Barbieri et al. 2017).

Success of any therapeutics is correlated with the evidence on its precise cellular target at molecular level. Lack of such detailed knowledge on mode of action of phytochemicals followed by failure of many clinical trials with therapeutic dietary additives cast a shadow of doubt on the efficacy of phytochemicals. Knowing defined molecular targets for phytochemicals is a great challenge for pharmacists, biochemists and cell biologists.

8.2 Phytochemicals in Human Health

Phytochemicals the non-nutrient plant products obtained from the fruits, grains, vegetables, and other parts of plants have shown tremendous health benefits. Phytochemicals (carotenoids and flavonoids) present in large amount in fruits and vegetables have shown remarkable antioxidant properties and are beneficial in treating many chronic diseases such as cancer, diabetes etc. (Liu 2003). Brief accounts of such evidences are discussed below.

8.2.1 Cardiovascular Diseases (CVD)

CVDs are the leading cause of death worldwide. It has been evaluated that healthy eating and physical exercise can prevent the 90% of CVDs. According to a systematic review, intake of flavonoids and flavonols is inversely associated with the risk of CVDs. For 10 mg/d intake of flavonol reduces the risk of CVDs by 5%. According to this study intake of all flavonoids i.e. anthocyanin, flavonols, flavones, proanthocyanidin, flavon-3-ols can significantly reduce the risk of CVDs (Leitzmann 2016). According to a meta-analysis conducted on 452,564 individuals among them 7233 individuals were of coronary heart diseases (CHD), lower intake of flavonoids is associated with the high risk of CHD (Leitzmann 2016). Additionally phytochemicals play vital role in reducing the blood pressure, platelet aggregation and modulate the synthesis and absorption of cholesterol. As inflammation is a critical factor in CVDs thus anti-inflammatory property of phytochemicals prove to be beneficial in this context (Liu 2003).

8.2.2 Diabetes

Nearly 400 million people are suffering from diabetes and among them 90% are of type-2 diabetes. A healthy diet, no use of tobacco, normal weight, and physical exercise are prerequisite for the prevention and cure of diabetes. According to a prospective meta-analysis where 18,146 individuals were diabetic among 284,806 study participants, intake of 500 mg/day of flavonoids was found associated with 5% reduction in diabetes risk. Beside this intake of β-carotene is also inversely associated with the risk of diabetes (Leitzmann 2016).

8.2.3 Cancer

Nutrition is one of the external factors which contribute to about one third of different types of cancer. According to a large prospective study which followed up for 20 years has revealed that lycopene, α and β-carotene and total carotenoids can significantly lower the risk of cancer by 18–28%. It has been noted that there is an inverse association between the intake of carotenoids and the risk of breast cancer. Women with high carotenoid content in plasma are at low risk of breast cancer (Leitzmann 2016). Similarly a prospective study on 9959 subjects was carried out in Finland, on the association of flavonoids intake and the risk of cancer. The outcome of the study has revealed that the high intake of flavonoid reduces the risk of all type of cancer and after a follow up for 24 years it has been found out that risk of lung cancer is declared to 50% in individuals having intake of flavonoids (Liu 2003).

8.3 Source and Bioavailability of Phytochemicals

As it has been discussed earlier that phytochemicals are the plant products derived from the various parts (fruit, flower, stem, root, leaf, bark) of the plant. Few examples of various sources of phytochemicals are discussed below.

Flavonoids obtained from the bark, stem, roots, flowers, grains, vegetables, tea, sepals and wine. Anthocyanin (cyaniding, peonidin) obtained from black current, black and blue berries. Flavonones (naringenin, hesperidin) obtained from orange juice, lemon juice and grape juice while flavones (apigenin, luteolin) obtained from parsley, celery and capsicum. Isoflavonoids such as Catechin present in abundance in berries, cocoa, tea, apple while flavonols like quercetin, kaempferol, myrcetin is present in apple, onion, tea, wine, curly kele, leek and reservetol is widely obtained from the berries, grapes laves of eucalyptus etc. (Egert and Rimbach 2011). Lycopene a carotenoid obtained from the red papaya, watermelon, tomatoes and red carrots while mushrooms are the rich source of polysaccharides, proteins and vitamin-D (Iqbal et al. 2017). Beside this dry fruits are the good source of flavonoids and carotenoids (Chang et al. 2016).

Terpenes like mono and sesquiterpene are widely obtained from essential oil while others are present in resins, waxes, rubber etc. A diverse group of cucurbitacins belong to triterpene are present in vegetables like squash, melon, cucumber, eggplant, gourds and pumpkin (Las Heras et al. 2003). Tocotrienols of vitamin-E family obtained from the cereal grains e.g. rye, barley, oats and edible oils (Vasanthi et al. 2012). Alkaloids namely caffeine which is present in our daily food items and beverages like coffee, tea, cocoa. Morphine which is used as a stimulant and codeine having sedative and cough suppressant properties are obtained from poppy plant while *Nicotiana tabacum* is a rich source of tobacco (Matsuura and Fett-Neto 2017; Barbieri et al. 2017).

According to Food and Drug Administration (FDA, USA) bioavailability is the "rate and extent to which the therapeutic moiety is absorbed and becomes available to the site of drug action".

Absolute bioavailability: It is the exact amount of substance of the ingested material which reaches the circulation.

Relative bioavailability: It is described as the bioavailability of a substance from one source as compared to another source (Holst and Williamson 2008).

Bioavailability of phytochemicals is low as compare to the macromolecules because they are taken as xenobiotic by the body. After absorption phase I and mainly phase II biotransformation is responsible for their low bioavailability. In the colon the gut microbiota acts upon the phytochemicals and produces a large amount of beneficial bioactive compounds (Holst and Williamson 2008). Bioavailability of phenols depends upon the source and processing of the food, chemical composition of the food, its molecular and physiochemical properties etc. Additionally conditions of the gastrointestinal (GI) tract like pH, amount of secretion, age of the individual and gut microbiota are the vital factors that influence the absorption as well the digestion of phytochemicals. Beside this interaction between the components of food particles and their interaction with the secretion of GI tract affect the bioavailability of phytochemicals (Epriliati and Ginjom 2012). Interaction of tannin proteins is depending upon the charge on protein, size, protein confirmation and their hydrophilic nature. Protein with high molecular weight or loose confirmation or hydrophobic in nature are more prone to tannin's precipitation. Bioaccesibility, is the amount that actually absorbed by the lumen of the gut is prerequisite for the bioavailability of phytochemicals. Low bioaccesibility affect the bioavailability of particular compound significantly (Holst and Williamson 2008).

8.4 Application of Nanotechnology Based Approach

Nanotechnology is a science deals with the study of matter with size 100 nm or less than that (Wang et al. 2014). Nanomolecules are used for the delivery of various vaccines, enzymes, phytochemicals and nutrients. They are used for the treatment of various diseases like cancer as they are specific to their target, increases the bioavailability, stability and solubility of the compound (Wang et al. 2014; Aqil et al. 2013). Commonly used biocompatible and biodegradable nanomolecules are micelles, nanoemulsions, liposomes, nanoparticles etc. (Wang et al. 2014).

8.4.1 Liposomes

Liposomes are the spherical shaped structures having a diameter of 20–1200 nm besides this their size may vary from 25 nm to 2.5 μM (Xie et al. 2016; Aqil et al. 2013). They can be mono or multilayered structures having an internal aqueous core and outer bilayer made up of phospholipids (Xie et al. 2016) (Fig. 21.4.1) (Fig. 8.1).

The advantage of using liposome is that it can carry both hydrophobic (embedded in the outer layer) and hydrophilic (enclosed in the aqueous core) compounds (Aqil et al. 2013; Wang et al. 2014; Chuan et al. 2015; Xie et al. 2016). Cholesterol is used to enhance the physical characteristics of liposomes. Once the phytochemicals loaded into the liposomes it enhances their stability, circulation and aqueous circulation and reduces their side effects (Chuan et al. 2015).

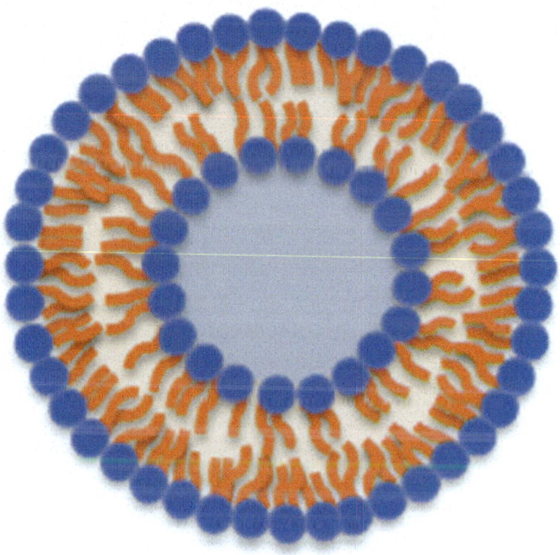

Fig. 8.1 Structure of bilayer liposome (Reproduced from Wang et al. 2014)

Drug delivery via liposomes can be given orally, subcutaneously, intravenously and topically (Chuan et al. 2015). The compound enclosed in the liposomes is protected from all the external modulations like inhibitors and enzymes. Liposomes deliver the drug inside the cell by the process of endocytosis. Liposomes are coated with inert polymer which provide them stability and help them to evade the reticuloendothelial system as well as the van der waal's, hydrophobic and electrostatic forces (Aqil et al. 2013).

8.4.2 Nanoemulsion

Emulsion is a mixture of two immiscible liquids like emulsion of oil in water. An emulsifier having amphipathic in nature is used to spread the two immiscible liquids and to stabilize the emulsion structure (Fig. 8.2).

Beside this nanoemulsion need surfactants and high energy to maintain the diameter i.e. <100 nm and to reduce the surface tension. Oil in water type nanoemulsion is used to carry hydrophobic phytochemicals such as genistein, resveratrol, curcumin and quercetin to cancer cells (Wang et al. 2014).

The advantage of nanoemulsion is to increase the stability, improve the aqueous solubility and absorption of hydrophobic phytochemicals (Chuan et al. 2015).

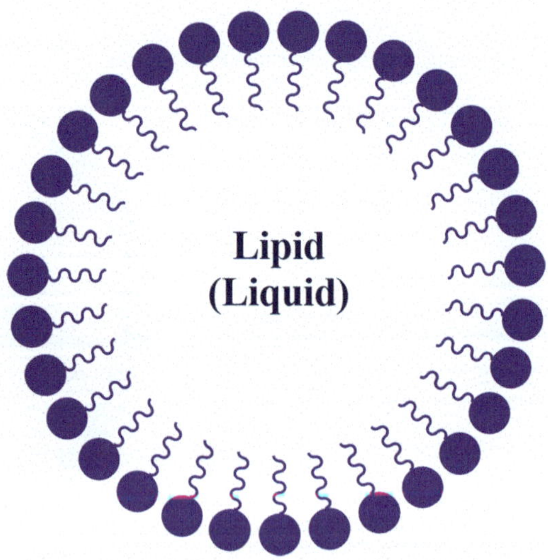

Fig. 8.2 Structure of emulsion. Core is hydrophobic while outer shell is hydrophilic (Reproduced from Chuan et al. 2015)

Fig. 8.3 Structure of micelle. Core is hydrophobic made up of lipids and while hydrophilic heads are facing outward (Reproduced from Wang et al. 2014)

8.4.3 *Micelles*

Micelles are amphipathic molecules that arrange themselves in a spherical structure when present in aqueous solution. Size of micelle is ranges between 20 and 80 nm in diameter (Wang et al. 2014). Their structure arrangement consists of a hydrophobic core of a lipid and monolayer of phospholipids (Fig. 8.3).

Like nanoemulsion surfactant and co-surfactant are used for the synthesis of micelles. Like liposomes and nanoemulsion, micelle also enhances the bioavailability, specificity, blood circulation and aqueous solubility of hydrophobic compound and reduces the side-effects and increases the permeability across the physiological barrier (Wang et al. 2014; Chuan et al. 2015). Release of drug compound from micelle depends upon various factors like its partition co-efficient, stability of micelle, rate of diffusion of drug, concentration, molecular weight and phytochemical characteristics of drug (Aqil et al. 2013).

8.4.4 *Nanoparticles*

These are the particles of size 10–1000 μm. They are of two types on the basis of material they are made up of i.e. biodegradable high molecular polymer nanopolymer [poly(lactic-co-glycolic acid) i.e. PLGA] and natural polymer nanoparticle which include chitosan, albumin and gelatin cellulose nanoparticles (Xie et al. 2016). PLGA nanoparticles form lactic and glycolic acid after hydrolysis.

It increases the drug efficiency and has been approved by FDA (Wang et al. 2014; Chuan et al. 2015). Various phytochemicals such as quercetin, resveratrol and curcumin are delivered through PLGA nanoparticles (Chuan et al. 2015). Use of nanoparticles for drug delivery improves the oral availability of poorly soluble drugs and enhances the drug uptake by tissues after parenteral administration. Due to small in size they can leave the vascular system ad reach to the site of inflammation. Beside this they have potential to increase the delivery of particular drugs across the biological membrane (Aqil et al. 2013).

8.4.5 Solid Lipid Nanoparticles (SLNs) and Nanostructured Lipid Carriers (NLCs)

They are lipid molecules consisting of water, surfactant, lipids or co-surfactant. They are used as an alternative carrier to the traditional carrier system like nanoemulsion, liposome etc. (Wang et al. 2014; Chuan et al. 2015). Structurally they have a hydrophilic shell made up of phospholipid and surfactant while the hydrophobic core is made up of waxes, triglycerides and fatty acid (Fig. 8.4).

NLC is a new generation lipid monocarriers that can overcome the SLN's limitations such as increases the release potential, less loading capacity and discharge of drug during storage (Chuan et al. 2015).

Thus nanotechnology has shown a great promise as a carrier of phytochemicals as it has a potential to increase the stability and solubility of phytochemicals, enhances the circulation and bioavailability, increases their permeation, absorption and retention in the targeted tissue. Beside this there are some challenges in using the nanotechnology. Firstly, synthesis of functional nanoparticles is an expensive task,

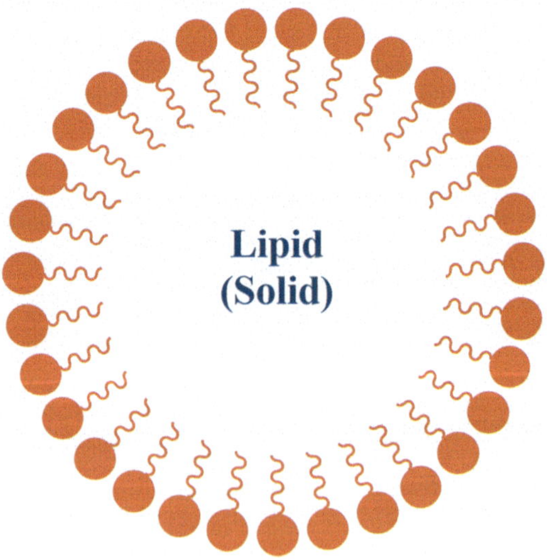

Fig. 8.4 Structure of SLNs. Core is hydrophobic while outer shell is hydrophilic (Reproduced from Wang et al. 2014)

secondly there are many barriers in the delivery of compound by nanoparticles such as blood-brain barrier, extracellular matrix, pressure gradient of interstitial fluid etc. Another relevant issue is the toxicity associated with them as nanoparticles are comprise of proteins, peptides, fragments of antibodies and nucleic acids which may pose as an antigen and may enhance immunotoxicity. Additionally, an optimized formulation, on the basis of physiochemical property of particular phytochemical, is required to enhance their efficacy, specificity and reducing the cost and other side effect.

8.5 Inflammation and Regulation of the Immune System

There have been a handful of spectacular achievements and advances for some well studied compounds with anti-cancer properties such as Quercetin and Myricetin (flavonoids found at high concentration in red wine) and Equol (Soybean isoflavonedaidzein derivate) which inhibit mitogen-activated protein kinase (MAPK) pathway and curcumin (from turmeric) which inhibits proteasomal activity. A group of these phytochemicals induce competitive inhibition of enzyme activity while others bind allosteric pockets to inhibit co-factor interaction. For example, Quercetin is thought to block the allosteric pockets of MEK1 without affecting ATP binding while Curcumin occupies the ATP binding pocket of DYRK2 in the 26S proteasome (Surh 2003; Jana et al. 2004; Banerjee et al. 2018).

In this section we will be discussing mechanisms underpinning phytochemical bioactivity in their roles in inflammation and immune cell regulation.

8.6 Polyphenols and Cell Signaling

The polyphenols have diverse structures, and their function relies on specific interaction with target proteins, which implies that different polyphenols have different target proteins. More than 4000 such compounds have been identified and subclasses of polyphenols (particularly flavonoids) are shown to have anti-inflammatory properties. These compounds directly modulate activity of intracellular signaling and transcription factor proteins such as NF-κB in immune cells and thereby affect the expression and secretion of cytokines. Resveratrol (3,5,4′ -trihydroxyl-*trans*-stilbene), a phytoalexin is one such compound found in many plants especially in grape skins that can potentially modulate signaling pathways by affecting NF-κB activity, a key transcription factor that regulates expression of myriad of inflammatory genes (Banerjee et al. 2002).

Catechins are polyphenols found in green tea extracts, which have potent anti-inflammatory and anti-thrombogenic properties. As mentioned above, MAPK pathways are central targets to immune cell modulation. All catechins show competitive inhibition of transforming growth factor beta activated kinase or TAK-1, a MAP3K, at its kinase domain. TAK-1 regulates cell survival through nuclear factor kB (NF-κB) dependent pathway and also modulates cytokine expression. Epigallocatechin-3-gallate (EGCG) and Epi-catechin-3 gallate (EGC) but not Epicatechin (EC) inhibit pro-inflammatory cytokines IL6, IL8 and reduce COX-2 expression (Fechtner et al. 2017). Some of these catechins have been shown to directly suppress IL1b induced iNOS signaling in chondrocyte cells and reduced osteoclast differentiation in animal models of rheumatoid arthritis, a chronic inflammatory disorder. This example show the importance of studying cytokine secretion in non-immune cells such as chondrocytes which are primarily involved in inflammatory cytokine signaling in arthritis patients (Saqib et al. 2018; Oka et al. 2012).

Cells sense amino acids by the mTOR and AKT signaling pathways that regulate the localization and activity of lysosomes and thereby modulate autophagy and cell survival. Several immunomodulatory and anti-proliferative drugs target PAM (PI3K-AKT-mTOR) pathway such as Gallic acid (*Phalericamacrocarpa*), Diosgeneic (*Dioscorea*spp), 3,3 dindolyl methane (cruciferous vegetables), epigallocatechin gallate (green tea) and the famous example of Curcumin (Turmeric). Curcumin has been shown to down regulate the downstream substrates of mTOR such as RAPTOR and RICTOR as well decrease substrate phosphorylation and mTOR auto-phosphorylation (Mozaffarian and Wu 2018; Lou et al. 2010).

One of the important mechanisms of anti-inflammatory activity of flavonoids comes from their interaction with the Toll like receptors (TLRs) and the cytosolic aryl hydrocarbon receptor (AhR), a nuclear receptor. Flavonoids bind to certain TLRs on the cell surface, which are internalized and transported to the nucleus via AhR interaction and affect anti-inflammatory changes (Hoensch and Weigmann 2018). In cell culture models, all catechins along with Gallic acid potently generate free radicals and hydrogen peroxide (H_2O_2) which block topoisomerase-2 function. EGC is found to have less potency compared to EGCG in poisoning topo-2 activity. This inhibition is attributed to the galloyl group of EGCG. These differences in potencies and targets reflects the importance of subtle structural differences and underscores the need for more molecular docking studies of these compounds. Many of these compounds produce hermetic or biphasic effects in cell culture experiments and hence need to be better studied in in-vivo and their specific protein interactions need to be well characterized (Suzuki et al. 2001; López-Lázaro et al. 2011; Chirumbolo 2012).

Could polyphenols function in gastrointestinal diseases? Unfortunately, the gut uptake for many polyphenolic compounds is weak and is limited by the ability of these compounds to cross the junctional barrier of the intestinal epithelium as well to be taken up by epithelial cells. While this can be a disadvantage, the upside of this limitation allows for high concentrations to build up in the lumen of the gut to affect and sustain anti-inflammatory properties on some cell types specifically in this region. A clinical trial using Apigenin (from chamomile extract) as a nutritional supplement showed a reduced remission of chronic IBD compared to the placebo group. This is consistent with its successful usage as an anti-inflammatory molecule in rodent models of colitis (Langhorst et al. 2013; Aldini et al. 2014). Importantly, gut microbiota could be directly affected by this build-up of flavonoids in the lumen.

A dysbalance in the gut microbiota has been implicated in various inflammatory disorders including inflammatory bowel's disease and in some cases can indicate a pre-inflammatory status of the patient. It will be very interesting to understand how these individuals respond to flavonoid treatments and whether that can lead to tease out the anti-inflammatory microbiome (Man 2018).

8.7 Immune Cell Modulation

Inflammation is mainly modulated by macrophages, which on activation come in two flavors, M1 and M2. M1 are induced by IFN-G, CMCSF, LPS and are Th1 responsive pro-inflammatory type; M2 are induced by IL4, IL13, IL10, TGF-B and are Th2 responsive anti-inflammatory type. One of the best known synthetic repressor of inflammatory macrophages is Dexamethasone, a synthetic glucocorticoid. It binds glucocorticoid receptor in the cytosol inducing a structural change that triggers translocation to the nucleus and activates plethora of anti-inflammatory genes (Saqib et al. 2018).

Polyphenols such as Reservatrol, Curcumin, Gernaiin, Quercetin, Naringeninetc have been shown to reverse the M1/M2 switch in experimental conditions and induce M2 macrophage activity primarily by downregulating iNOS and COX2 expression in some cell types and affecting AKT phosphorylation in others (Yang et al. 2017). All have a direct or indirect effect on NF-kB p65 subunit nuclear translocation. For example, Naringenin directly modulates the translocation of the NF-kB p65 subunit to the nucleus thereby affecting expression of its target genes (Raza et al. 2013). Lupeol, a pentacyclictripenoid from fruit plants inhibits latent membrane protein -1 (LMP1) induced NF-kB activation and IFN-G production that indirectly stops macrophage differentiation to M1 type (Zhu et al. 2016). Mallibatol A (MA), an oligo stilbene from *Hope malibato* leaves can reverse specifically microglial activation by inhibiting the nuclear receptor PPAR-γ activity, an effect with implications in ameliorating brain injuries such as neurodegeneration, infection borne encephalitis and stroke (Saqib et al. 2018; González-Reyes et al. 2017).

BIO or 6-bromoindirubin-3′-oxime, a synthetic derivative of 6-bromoindirubin (from Mediterranean Mollusk) is a glycogen synthase kinase-3 inhibitor and plays an important role in not only stem cell regeneration but also in myocardial infarction and neuroinflammation by polarizing macrophages. BIO can convert the high iNOS expressing M1 cells to high arginase-1 expressing M2 cells. Arginase-1 is a target for IL4 pathway that can effectively block inflammation in cultured cells (Kim et al 2016; Song et al. 2013). In fact, in an Alzheimer's disease model where amyloid plaques were induced in the hippocampal region of the mouse brain, increased number of arginase-1 expressing microglia was correlated with a decrease in the amyloid plaques in the brain (Cherry et al. 2015). Moreover, differentiated microglia with upregulated arginase-1 also provided enhanced neuroprotective cytokines such as glia derived neurotropic factor (GDNF), a lack of which may cause dopaminergic neuron death.

Sterol metabolism, trafficking and storage can have drastic effects on eurkaryotc cells. Many diseases are associated with cholesterol trafficking and storage such as Neiman-Pick disease, Tangier's disese which are also compounded with neurodegeneration. High accumulation of cholesterol can disrupt inflammasome signaling in macrophages and lymphocytes (Westerterp et al. 2017). Caspase-1 activation cleaves the cytokine precursors of interleukins priming them for secretion in inflammatory response. Quercetin, which can regulate PPAR-g signaling and is an M1 inhibitor can also directly stabilizes ATP-binding cassette transporter (ABCA1), a cholesterol exporter whose downregulation expression has been correlated to atherogenesis, a cardiovascular disorder whose pathology is aggravated by the inflammatory response to the atherosclerotic vascular system as well as neurodegenerative Tangier's disease (Chang et al. 2012).

Leukotrienes are metabolites of the arachidonic pathway produced by C5-lipooxgenase. Leukotriene A4 is a key compound in the biosynthesis whose amino acid modifications give rise to other leukotrienes such as B2, B4, C4 etc. These are powerful pro-inflammatory molecules which not only induce inflammation, but also cause vasoconstriction, secretion of lysosomal enzymes, leukocyte and neurotrophil migration, and along with thromboxanes and prostaglandins have strong myotropic and anaphylactic effects (Larussa et al. 2017; Haeggstrom 2018). The most well-known antagonists of leukotriene induced inflammation are corticosteroids that block lipoxygenase activity. Phytochemicals that directly interact with leukotriene synthesis by binding Leukotriene A4 hydrolase have a great potential in acute and chronic inflammatory disorders (Goldyne et al. 1984; Sirois et al. 1984). In the past decade, several studies have identified inhibitors of Leukotriene A4 hydrolase using various biophysical techniques such as metabolomics biased X-Ray crystallography (Davies et al. 2009), and biochemical methods where phytochemicals were tagged with cyanogen bromide and by reverse binding assay, targets pulled down to detect proteins bound tightly to chemicals by mass spectrometry.

Many flavonoids have been identified that interact with various steps in the lipoxygenase pathway. Cocoa extract with high flavonoid content was shown to reduce vasoconstriction and induce bronchodilation. Also, Epicatechins were identified that directly inhibited the enzyme activity of leukotriene A4 hydrolase (Sies et al. 2005) thereby reducing the formation of other pro-inflammatory and anaphylactic leukotrienes. But inhibitions can also occur at upstream targets such as arachidonic acid metabolism by inhibiting the phospholipase-A2 (PLA_2), lipoxygenase and cyclooxygenase. Quercetin, for instance, can inhibit $PLA_{2,}$ in the migrating neutrophils (Kim et al. 2004). Interestingly, Quercetin and myricetin, but Naringen, could inhibit snake venom $PLA_{2.}$ There has been a lot of effort also in finding inhibitors of cyclooxygenase pathway. This could be a double-edged sword in certain conditions as has been shown by Aspirin which stops inflammation by inhibiting COX2, but by doing so increases the substrates for lipoxygenase and thereby increasing inflammation during anaphylaxis.

8.8 Multi-target Specificity

As is clear from this chapter, phytochemicals are known to bind multiple proteins. Prima facie, this looks like a disadvantage due to the lack of specificity, low sensitivity, requirement for high dosage and possible side effects. For example, Epigallocatechin gallate (EPCG) can bind multiple protein targets such as IGF1R (immune cell growth and homeostasis), GRP78 (ER stress), HSP90 (protein folding and secretion), FYN (target to non-steroidal anti-inflammatory drugs) and ZAP70 kinase (trigger signaling pathways for T-cell maturation and function). Each of these has been implicated through myriad pathways in promoting inflammation. But the upside to using such phytochemicals is the synergistic and additive effects they have on preventing inflammatory pathology. Usually, phytochemicals are known to have low binding affinity towards single protein targets. But fine tuning the drug structure can help refine the relative affinity and avidity to their various targets and thereby help phytochemical drug potencies.

8.9 Anti-cancerous Activity

Cancer belongs to the category of fatal diseases that are a bane to human life in the present world. Cancer therapies as well as various chemotherapeutic drugs given to treat cancer have come along with their side effects such as non-specificity, toxicity, limited bioavailability etc. (Iqbal et al. 2017). These drugs lead to the dysfunction of many prominent organelles such as mitochondria. To minimize such side-effects phytochemicals have been explored as they often show less or not toxicity and thus are better options for cancer treatment. There are many plant derived products such as phenols, terpenes, alkaloids, minerals, vitamins, glycosides, oils, primary and secondary metabolites, saponins etc. that are effective against cancer. These plant products execute their anti cancer activity either by inhibiting enzyme, proteins, and signaling pathways that are activating cancer cells or by promoting apoptosis (Bax, Bak, Bid), DNA repair proteins (p21, p27, p53) and anti-oxidant enzymes (GSH, GST, GPxn) (Iqbal et al. 2017) (Fig. 21.9.1).

8.10 Phytochemicals with Anti-cancer Activity

Several novel phytochemicals having anti-cancer properties have been identified and few of them along with their anti-cancer approach have been discussed here (Iqbal et al. 2017).

8.10.1 *Alkaloids*

Alkaloids are the nitrogen containing compounds of plants origin having vital role in human physiology. Some of the important alkaloids carrying anti-cancerous properties are discussed below (Rolin 2012).

8.10.1.1 Vinca Alkaloids (VA)

These phytochemicals, belong to the apocynacea family, are extracted from *Catharanthus roseus* and are used for the treatment of various type of cancer like breast, lung, liver, testes etc. There are two natural (Vincristine, Vinblastin) and three semisynthetic (Vindesine, Vinorelbine and vinflunine) Vas that are in use for cancer treatment (Moudi et al. 2013; Safarzadeh et al. 2014; Iqbal et al. 2017; Emanuela et al. 2018). Among all of this Vinblastin (VBL), Vincristine (VCR) binds to the tubulin heterodimer and disturbs the functioning of microtubule which causes the cell cycle arrest at metaphase (Moudi et al. 2013; Xie et al. 2016; Iqbal et al. 2017; Emanuela et al. 2018). Both vinblastine and vincristine are approved as an anti-cancer drug by FDA in 1961 and 1963 respectively and since then has been widely used in treating various type of cancer (Xie et al. 2016; Emanuela et al. 2018). Vinorelbine (VRL), a second generation was found to be more effective and less neurotoxic as compare to vinblastine and vincristine. Its injectable form is also available and has been extensively used for clinical purposes. Vinflunine (VFN), a fluorinated and most advanced form of Vas which is least neurotoxic and has excellent anti-neoplastic activity. Indeed, its evaluation through clinical trials is still going on beside this it has been authorized in Europe for the treatment of metastatic carcinoma of urothelial tract (Moudi et al. 2013; Emanuela et al. 2018). These drugs are not recommended to patients who are planning pregnancy as well as to pregnant and lactating women. Patients who are on this medication should refrain from any other vaccination (Moudi et al. 2013). Presently, various semisynthetic derivatives of VA are available in the market that can be used alone or with other phytochemicals to combat cancer (Iqbal et al. 2017; Emanuela et al. 2018).

8.10.1.2 Camptothecin (CPT)

Camptothecin is a strong chemotherapeutic agent isolated from the bark of *Camptothica acuminate* of Nyssaceae family. It carries out its anti-tumor activity by inhibiting topoisomerase-1 (Top-1), a DNA-relaxing enzyme (Takimoto et al. 1998; Iqbal et al. 2017). CPT stabilizes the cleavable complex of DNA and TOP-1 which prevent its resealing and leads to the accumulation of cleavable complex. As the replication fork interact with the cleavable complex it arrest the replication process by introducing double stranded breaks (Dancy and Eisenhauer 1996; Pizzolato and Saltz 2003; Tomicic and Kaina 2013; Martino et al. 2017; Hevener et al. 2018).

8.10.1.3 Ellipticine

Ellipticine, a member of apocynacea family is a strong anti-tumor agent extracted from the stem, leaf, root and bark of *Ochrosia elliptica* (first in 1959), *Bleekeria vitensis, Ochrosia borbonica, Excavatia coccinea* (Nirmala et al. 2011; Siu 2011; Kuo et al. 2005; Isah 2016; Iqbal et al. 2017). It has been used to cure breast cancer, colon cancer, myeloma, leukemia, melanoma etc. (Isah 2016; Iqbal et al. 2017). The primary mode of action of ellipticine includes (a) DNA intercalation, which induces anti-proliferative affect, (b) inhibition of topoisomerase-II activity (Siu 2011; Isah 2016; Iqbal et al. 2017), (c) inhibition of kinase activity of CDK2, (d) inducing ER stress, and (e) covalent alkylation of macromolecules (Kuo et al. 2005). In-vitro studies on cancer cell lines have shown positive results of ellipticine and its derivatives (Isah 2016). After evaluation few derivatives of ellipticine (elliptinium acetate, retelliptine dihydrochloride, and 2-methyl-9-hy-droxyellipticinium) have been withdrawn from clinical trials due to its toxic effects (Wang and Hu 2012). Many of its derivatives have passed phase I and II clinical trials while elliptinium one of the derivative of ellipticine is under clinical trials in France to evaluate its promising anti-tumor property against breast cancer (Iqbal et al. 2017).

8.10.1.4 Colchicine

A secondary metabolite obtained from *Gloriosa superba* and *Colchicine autumnale* (Colchicaceae). Besides cancer it has been used in the treatment of various type of diseases such as gout, arthritis etc. It acts by binding to tubulin, disturb the microtubule dynamics, and block the cell cycle and causes apoptosis (Nirmala et al. 2011; Lin et al. 2016; Iqbal et al. 2017). Due to its toxic effects colchicine is not recommended for the cancer treatment therefore its semisynthetic derivatives (deacetylcolchicine, colchicoside, colchicinamide, thiocolchicocide) have been synthesized which are less toxic and are used against cancer various type of cancer (breast, leukemia, melanoma etc.) (Nirmala et al. 2011; Iqbal et al. 2017)

8.10.1.5 Berberine

An alkaloid carrying strong anti-cancer properties with reference to the clinical trials and its potency (Iqbal et al. 2017). It is isolated from the rhizome and root of different plant species such as *Berberis vulgaris, Rhizome coptidis, Berberis aquifolium, Hydratis conadensis, Tinospora cordifolia* etc. it has been used to treat various type of cancer such as breast cancer, liver cancer, colorectal cancer, prostate cancer, lung cancer etc. (Nirmala et al. 2011; Iqbal et al. 2017; McCubrey et al. 2017). Berberine carries out its anti-tumor activity by inducing apoptosis and arresting cell cycle (G2/M phase) in case of liver, colorectal and breast cancer. Besides this it arrests the activity of anti-apoptotic protein (Bcl2, c-IAP1) as well as turn on the apoptosis inducing proteins (p53, p21, caspase 3 and 9) (Iqbal et al. 2017). Both in-vitro and in-vivo studies have shown that berberine can overcome the doxorubicin resistance of breast cancer cells in a dose dependent manner (Pan et al. 2017).

8.10.2 Phenols

Phenols are the naturally occurring chemical compounds produced by the plants having hydroxyl group and provide protection to plants in stress conditions (Bhattacharya et al. 2010). Besides this they have been used extensively for cancer treatment. Phenols are further divided into six categories (**a**) flavones (e.g. luteolin, apigenin), (**b**) flavan-3-ols (e.g., catechin, theaflavin, and gallic esters of catechin), (**c**) flavanones (e.g., naringenin, hesperidin), (**d**) flavonols (e.g., quercetin, kaempferol, myricetin), (**e**) anthocyanidins (e.g., pelargonidin, uppress), and (**d**) isoflavones (e.g., genistein, daidzein) (Le Marchand 2002; Romagnolo and Selmin 2012).

8.10.2.1 Flavonoids as an Anti-cancer Agent

Both in-vitro and in-vivo studies have documented the anti-cancerous properties of flavonoids. They act by hampering the processes like inflammation, invasion, proliferation, metastasis, and angiogenesis of cancerous cells or by disrupting the activation of tumor activator protein-1 (AP-1) and pro-inflammatory nuclear factor kB (NF-kB) during apoptosis and cell cycle arrest. Being associated with the malignancy and cancer advancement phosphatidylinositol 3-kinase (PI3K)/Akt, mitogen-activated protein kinase (MAPK), b-catenin pathways and protein kinase C (PKC) are the main targets of flavonoids (Romagnolo and Selmin 2012).

Different modes of action of Flavonoids are outlined below:

8.10.2.1.1 Interference in the Metabolism of Carcinogens

Several carcinogens require metabolizing enzymes such as cytochrome P450 (phase I enzyme) to get transformed into more reactive form which makes them more competent to bind with cDNA. The flavonols (kaempferol, galangin, and quercetin) and flavone (apigenin) act by inhibiting the enzyme cytochrome P450 (CYP1A family) which play vital role in the activation of various human carcinogens such as heterocyclic amines and polycyclic hydrocarbons. Flavonols (quercetin) and flavanones (naringenin) inhibits the enzymatic activity of CYP3A4 (P450 enzyme) which metabolizes various carcinogens in the liver. Recent in-vivo and in-vitro studies have revealed that tea catechins enhances the antioxidant and detoxifying activity of various enzymes such as quinonereductase, glutathione *S*-reductase, glutathione peroxidase, glutathione reductase, and catalase (Le Marchand 2002; Chahar et al. 2011; Wang et al. 2011).

8.10.2.1.2 Inhibiting Protein Kinases (Pks)

Flavonoids can exert their anti-cancer effect by inhibiting the activity of various protein kinases (serine, threonine, and tyrosine) that play important role in cell growth, cell proliferation and signal transduction (Batra and Sharma 2013).

8.10.2.1.3 Inhibiting Pro-oxidant Enzymes

Reactive oxygen species (ROS) formation is the hallmark of cancer progression which participates in the activation of proto-oncogene (c-JUN, c-Myc, and c-FUS) and interfere the process of apoptosis. Flavonoids acts as an anti-oxidant compounds by inhibiting the pro-oxidant enzymes i.e. xanthine oxidase (XO), uppress enases (COXs), and Lipoxygenases (LOX) whose catalytic activity comprises of ROS production. Beside this flavonoids disrupt the activity of ornithine decarboxylase which eventually decreases the biosynthesis of polyamines as well DNA and protein synthesis (Batra and Sharma 2013).

8.10.2.1.4 Inhibiting Angiogenesis

Flavonoids reported to be the inhibitors of angiogenesis by which they disrupt the transfer of oxygen and nutrient to the rapidly dividing cancerous cells which ultimately leads to death. Isoflavonones (biochanin A, daidzein, genistin, and genistein) have shown positive effect by arresting cell cycle, inhibiting angiogenesis and inducing apoptosis on murine as well as human bladder cancer lines (Chahar et al. 2011; Batra and Sharma 2013).

8.10.2.1.5 Promoting Cell Cycle Arrest and Apoptosis

Flavonoids may hamper the signaling pathway of NF-kB, which is activated by various processes (endotoxins, UV light, carcinogens, X-rays, cytokines etc.) that further stimulate the NF-kB targeted genes responsible for invasion, inhibiting apoptosis, metastasis, inflammation etc. These NF kB targeted genes inhibit the Bclx, BCl-2, matrix metaloproteinases (MMP-2 and 9), vascular endothelial growth factor (VEGF), cyclin D1. Flavonoids can inhibit the activation of AP-1 i.e. Activator protein-1 which further involve in the regulation of various processes like angiogenesis, differentiation, metastasis, invasion, proliferation and survival

(Chahar et al. 2011; Batra and Sharma 2013). Beside this they act by arresting the cell cycle at different phases (G1, S, S-G2 and G2-S) by down-regulating the cyclins and cyclin-dependent kinases i.e. via activation of p21, p27 and p53 which ultimately leads to cell death (Wang et al. 2011).

Some common flavonoids with their anti-cancerous activities are discussed below:

8.10.2.2 Cyanidin Glycoside

The glycosides are obtained from plums, apples, red berries, blackberries, grapes, etc. and act by inhibiting cell division, removing free radicals and as an anti-oxidant through COX-2 and iNOS gene expression. It interrupts the COX-2 enzyme synthesis in case of colon cancer and promotes apoptosis in prostate cancer while in case of lung and bladder cancer it stops the expression of MMP-9 (Iqbal et al. 2017).

8.10.2.3 Saffron

Saffron is also known as *Crocus sativus L.* and a potent anti-cancer agent used against breast, liver, lung, skin cancer. It carries out its activity through different means such as inhibiting the enzymatic activity of COX-2, iNOX, promoting apoptosis, disrupting the expression of MMP-2 and 9, reducing the serum level of TNF-α, IL-β, CDK-2, cyclin-A and D or by regulating caspase-3, 8, 9 as well as of BCL-2/Bax ratio (Iqbal et al. 2017).

8.10.2.4 Gingerol

It is extracted from the fresh rhizome of *Zingiber officinale* having positive effect in ovarian, colon, breast, and pancreatic cancer. Its mechanism of action includes inhibiting the phosphorylation and nuclear translocation of IκBα and NF-Kb respectively which ultimately decreases the expression of TNF-α and iNOS (Iqbal et al. 2017).

8.10.2.5 Resveratrol

A polyphenol present in grapes, blueberries, peanut, mulberries etc. and play significant role in various type of cancer treatment such as liver, lung, breast, prostate, colorectal, pancreatic etc. Its mechanism of action includes increased expression of proteins p53, Bax and reduced expression of Bcl-2, COX, AP-1, NF-κB, cytokines etc. (Iqbal et al. 2017).

8.10.2.6 Curcumin

Another important phenolic compound isolated from *Curcuma longa* having an inhibitory effect on the growth of human glioblastoma cells by regulating various cellular and nuclear factors or by increasing the expression of various genes and their products such as p16, p21, p53, Bax, caspase-3, 7, 9 etc. Besides this they also act by decreasing the expression of p65, NF-κB, cyclin-D, Bcl-2 etc. (Siu D 2011; Iqbal et al. 2017).

8.10.2.7 Chalcone

A flavonoid present in fruits and vegetables and acts as an anti-cancerous agent by increasing the expression of pro-apoptotic proteins (Bax, Bak, Bid) and reducing the expression of anti-apoptotic proteins (Bcl-2). It has been in use to treat liver, lung, colon, breast, prostate cancer (Iqbal et al. 2017).

Besides this there are number of flavonoids like genistein, oroxylin, quercetin, apigenin, kaempferol, chrysin etc. having significant anti-cancerous properties and have been used for the treatment of various type of cancer (Le Marchand 2002; George et al. 2017; Iqbal et al. 2017).

8.10.3 Terpenes

Terpenoids represent the largest class of naturally occurring plant products with promosing anti-cancerous properties. Both in-vitro and in-vivo studies have documented positive results on ovarian, breast, leukemia and prostate cancer. They are classified on the basis of number of isoprene units thay are made up of such as Hemiterpene (C_5), Monoterpene (C_{10}), Sesquiterpene (C_{15}), Diterpene (C_{20}), Seterpene (C_{20}), triterpene (C_{30}), tetraterpene (C_{40}), polyterpene $(C_5)_n$, $n > 8$.

8.10.3.1 Role of Various Terpenoids in Cancer Treatment

8.10.3.1.1 Monoterpenoids

These terpenoids found in floral scents, essential oil of mint, cherry, citrus fruits etc. Limonene, present in the oil of caraway seeds, peel oil of lemon, orange, and citrus is an established chemotherapeutic agents against cancerous cells. Carvone of caraway seed oil has shown positive effects in case of chemically induced lung and stomach cancer (Wang et al. 2005). In-vivo studies have shown that among the two active forms of limonene (D and L form) D-form is known to be effective against liver, colon, pancreas, skin, and stomach cancer. Its anti-tumor activities are attributed to its anti-oxidant, angiogenic and pro-apoptotic properties (Yang and Ping Dou 2010; Huang et al. 2012). In case of hepatic cell cancer it uppress the activity of HMC-CoA reductase (enzyme responsible for cholesterol biosynthesis) (Thoppil and Bishayee 2011; Huang et al. 2012).

8.10.3.1.2 Sesquiterpenes

Sesquiterpenes are one of the members of terpenoid family with promising anti-tumor properties. Few examples of sesquiterpenes with their efficacy against cancerous cells are discussed below.

8.10.3.1.2.1 β-Elemene

An estrogenic sesquiterpene isolated from *Curcuma wenyujin* (chinese herb) is effective against various type of cancer. Its anti-cancerous activity includes inducing apoptosis via targeting p21, p53 and caspases (3, 7, 9 and 10), inhibiting angiogenesis and the expression of VEGF, reducing the expression of CD34 and phosphorylation of Akt, increased expression of E3 ubiquitin ligases i.e. c-Cbl and Cbl-b in gastric tumor (Iqbal et al. 2017).

8.10.3.1.2.2 Artemisinin

It belongs to the estrogenic sesquiterpene extracted from a chinese herb named *Artemisia annua L.* It inhibits the cell proliferation in leukemia, breast cancer, ovarian cancer, colon cancer, lung and gastric cancer etc. It inhibits cell proliferation in breast cancer cells by reducing the expression of Erα without affecting Erβ expression (Kiyama 2017). It also execute its anti-cancerous action by arresting cell cycle at G1 phase, promoting apoptosis via activating p38 and caspases, and by impairing the process of angiogenesis, invasion and metastasis. Its efficacy as anti-tumor agent is same in case of parent and MDR (Multidrug resistance) cancerous cells. Artemisinin's anti-cancerous properties have been reported in various xenograft animal models (Huang et al. 2012).

8.10.3.1.2.3 Sesquiterpene Lactones

They comprises of various groups known for their biologically important compound. Their anti-tumor property is used to treat breast cancer, prostate cancer, liver cancer, esophagus and lung cancer (Iqbal et al. 2017). Dehydrocostuslactone (DHE), a sesquiterpene lactone, shows its anti-cancerous activity against various hepatic cancerous cell line of human origin such as PRF/PCL/5 and HepG2. DHE inhibits cell proliferation in liver cancer cell lines by inducing apoptosis via increasing pro-apoptotic and caspases proteins and decreasing anti-apoptotic proteins as well as by inducing endoplasmic reticulum stress (Thoppil and Bishayee 2011).

8.10.3.1.3 Diterpenoids

They belong to a large group of terpene family execute their anti-cancer mechanism by promoting apoptosis, microtubule stabilization and mitotic arrest. Few important diterpenes with their anti-tumor properties are discussed here.

8.10.3.1.3.1 Taxanes

Paclitaxel (taxol) and docetaxel, (second generation taxanes) are highly efficient anti-cancer agents (Yang and Ping Dou 2010). Taxol, a mitotic inhibitor was first extracted in 1962 (Xie et al. 2016; Bernabeu et al. 2017) from the bark and leaves of *Taxusbaccata, T. bravifolia, T. Canadensis and Corylusavellana* and in use to treat various type of cancer (lung, ovarian, breast) (Iqbal et al. 2017; Bernabeu et al. 2017; Xie et al. 2016). Taxol binds with β-tubulin that stabilizes the microtubule and arrest the cell cycle at mitotic pahse. Docetaxel is a semisynthetic derivative extracted from *T. baccata* is more potent than taxol as it affects the centrosome of mitotic spindle. It also promote stabilization of microtubule and arrest their disassembly which causes the remarkable reduction in the number of free tubulins, check mitotic cell division and finally prevent proliferation of cancer cells (Weaver 2014; Xie et al. 2016; Bernabeu et al. 2017; Iqbal et al. 2017). It has been used widely in prostate, lung, gastric and breast cancer treatment (Nabholtz and Gligorov 2005; Xie et al. 2016; Iqbal et al. 2017; Bernabeu et al. 2017). In 1992 FDA has approved this drug for the treatment of breast, neck, prostate, head and gastric cancer while in 2005 taxol has been approved by FDA for the treatment of metastatic cancer (Safarzadeh et al. 2014).

8.10.3.1.3.2 Triptolide

A diterpene extracted from the roots of *Tripterygium wilfordii Hook F* (a chinese herb) (Yang and Ping Dou 2010; Huang et al. 2012). At low concentration it inhibits the cell growth while at high concentration it induces apoptosis (Yang and Ping Dou 2010). It inhibits the various transcription factors NF-κB, NF-AT, HSF-1 and p53. Triptolide prevents the TNF-α induced transactivation of NF-κB which further inhibits the activation of c-IAP1 and c-IAP2 (inhibitor of apoptosis protein). In-vivo studies have confirmed the anti-tumor property of triptolide on xenograft animal models. Besides this, it also inhibits the RNA polymerases I and II and ultimately prevent the *de novo* synthesis of mRNA. Thus triptolide show its anti-cancerous activities by hitting through various targets (Yang and Ping Dou 2010; Huang et al. 2012).

8.10.3.1.3.3 Oridonin

Oridonin, a diterpenoid with many beneficial effects is obtained from the *Rabdosiuru besons* (chinese herb). It has been reported to be used against osteoma, skin cancer, liver cancer and colorectal cancer. Besides this it prevents the growth of non-hodgkin's lymphoma, primary T-cell leukemia, chronic lymphocytic leukemia and acute lymphoblastic leukemia (Huang et al. 2012). Oridonin induces apoptosis in cancer cells by inhibiting NF-κB signaling pathway. It also acts in a dose dependent manner thus up-regulate and down-regulate the expression of p53, Bax, and Bcl2 respectively. Moreover studies on colorectal cancer model has illustrated that oridonin reduces the AP-1 which further downregulate the p38 and NF-κB pathway (Huang et al. 2012). It has been investigated that oridonin induces autophagy besides apoptosis in cancer cells which has been reported in MCF-7 cells. However the reason behind the association of these two events is still not clear and more studies are required in this field (Yang and Ping Dou 2010).

8.10.3.1.4 Triterpenoids

A member of terpene family with strong anti-cancerous properties. Bothe in-vitro and in-vivo studies have shown potent anti-tumor properties of triterpenes against brest cancer, leukemia, and pancreatic cancer. Few triterpenes with their chacterstick anti-tumor properties are discussed here.

8.10.3.1.4.1 Celastrol

A triterpene isolated from the bark of *Triterygium wilfordii Hook F* (*Lei gong teng*) and from the *Triterygium regeli*, a substitute of *Lei gong teng* (Yang and Ping Dou 2010; Iqbal et al. 2017). Celstrol promote apoptosis via caspase-3 and inhibits heat shock protein (Hsp) by preventing its interaction with cdc37 in case of prostate, ovary and lung cancer. Besides this it has been reported that it inhibits topoisomerase, VEGFR, IκBα kinase, proteasome, NF-κB etc. (Yang and Ping Dou 2010; Huang et al. 2012). In-vivo study using nude mice with prostate cancer has demonstrated that celastrol treatment has significantly (65–93%) prevented the tumor growth. Thus celastrol acts by targeting NF-κB, Hsp, and ubiquitin proteasome pathways to cause cell death. Moreover, it inhibits the expression of TNF-induced genes which are associated with invasion, angiogenesis, progression and anti-apoptosis (Yang and Ping Dou 2010).

8.10.3.1.4.2 Ursolic Acid

It is the main component of herbal plants such as Oscimum (Basil) leaves, fruits, flowers, leaves and berries of *Eugenia jumbolana, Rosemarinus officinalis, Callus vulgaris* and *Eriobotrya japonica* (Yang and Ping Dou 2010; Iqbal et al. 2017). Ursolic acid has strong anti-cancerous properties as well as significant therapeutic applications. Its pro-apoptotic action on colorectal cancer (HCT116 cell line) involve reduction in the level of Bcl-2, NF-κB and MMP (Matrix metaloprotease), a pro-metastatic protein (Iqbal et al. 2017). It has been investigated that the basil leaf extract prevents the migration, morphogenesis, proliferation, anchorage independent growth, and activate COX-2 protein in breast cancer cells (Yang and Ping Dou 2010).

8.10.3.1.4.3 Cucurbitacins

They are tetracyclic triterpenes obtained from the Cucucrbitaces and Cruciferae family of plants. They prevent the activation of pro-inflammatory agents i.e. NF-κB and iNOS (Patlolla and Rao 2012). Cucurbitacin B is one of the most abundant forms of cucurbitacins and reduces the cell survival rate in Hep-G2 (liver cancer cell line), level of Bcl-2 protein (pro-apoptotic protein) (Thoppil and Bishayee 2011). It also promote cell killing by inhibiting the phosphorylation of STAT-3 which ultimately disturb the JAK2/STAT3 pathway (Huang et al. 2012; Thoppil and Bishayee 2011).

8.10.3.1.5 Tetraterpenoids

Carotenoids are the most common representative of tetraterpenoids that impart bright colors to the plants and animals. It has been reported that carotenoids play important role in reducing the risk of various type of cancer. Pre-clinical studies have unfolded the therapeutic role of various carotenoids such as α and β carotenoids, lutein, lycopene, zeaxanthin, fucoxanthin, canthaxanthin etc. in cancer treatment (Huang et al. 2012).

8.10.3.1.5.1 Lycopene

It is a bright red color pigment found in carrots, red papaya, tomatoes etc. and possesses remarkable anti-neoplastic properties (Xie et al. 2016; Iqbal et al. 2017). It down regulates the expression of Bcl-2 and Erk proteins by disrupting the signaling pathway of PI3K/Akt in case of stomach and pancreatic cancer. Besides this it increases the expression of anti-oxidant enzymes (GSH, GST, GPxn) in case of prostate, breast, colon and endothelial cancer (Iqbal et al. 2017). Their anti-oxidant activity includes (i) ROS scavenging which further inhibits DNA damage and lipid peroxidation, (ii) activation of anti-oxidant enzymes, (iii) inhibition of angiogenesis. According to the recent studies lycopene regulate the expression of Hsp and cell cycle proteins (Huang et al. 2012). Besides this it suppresses the expression of Bcl-2 and arrest the cell cycle at G2/M phase in MCF-7 cell lines (Yang and Ping Dou 2010). Similarly another study on Hep3B (liver cancer cell line) has shown that lycopene exerts its inhibitory effect via arresting DNA damage and cell cycle at G0/G1 phase. Additionally, it prevents invasion and migration in SK-Hep-1 (liver adenocarcinoma cells) (Thoppil and Bishayee 2011).

8.10.3.1.6 Other Tetraterpenoids

Many carotenoids of this category exhibit anti neoplastic properties but the deep understanding about these carotenoids is comparatively limited. B-carotene promotes apoptosis in breast cancer cells by increasing the expression of peroxisome proliferator-activated receptor gamma (PPARγ). Additionally it increases the ROS production which leads to mitochondrial dysfunctioning and promote the release of cytochrome-c which ultimately contributes to the apoptosis (Yang and Ping Duo 2010; Huang et al. 2012). In-vitro studies have shown that α-carotene inhibits the chemically prompted malignant tumor and growth of cancerous cells. Similarly, in-vivo studies have revealed that α-carotene is more effective than β-carotene against liver cancer and chemically prompted skin and lung cancer (Ziegler et al. 1996). Few carotenoids are reported to modulate drug resistance while violaxanthin, lutein, antheraxanthin, fucoxanthin etc. showed moderate effect on the reversal of multidrug resistance (MDR) in the cancerous cells (Yang and Ping Duo 2010; Huang et al. 2012).

8.11 Conclusion and Future Perspective

Phytochemicals have been extensively used in medicine since ancient time. Several traditional and folk medicines were based on unexplained role of these phytochemicals. Scientific investigations in last several decades have uncovered the molecular roles of these phytochemicals. Recent in-vitro and in-vivo studies have highlighted their role in immune system modulation. Role of phytochemicals in treating infectious, acute as well as chronic diseases were studied extensively. However, their effectiveness, biosafety and bioavailability have not been fully investigated. Application of systems biology approach to model the multidimentional targets of these chemicals would be a much effective approach. This would further pave the path for application of phytochemicals in personalized medicine to minimize the adverse/side effects and maximize the effectiveness.

References

Aldini R, Micucci M, Cevenini M, et al. Anti-inflammatory effect of phytosterols in experimental murine colitis model: prevention, induction, remission study. PLoS One. 2014;9(9):e108112.

Aqil F, Munagala R, Jeyabalan J, et al. Bioavailability of phytochemicals and its enhancement by drug delivery systems. Cancer Lett. 2013;334(1):133–41.

Banerjee S, Bueso-Ramos C, Aggarwal BB. Suppression of 7,12-dimethylbenz(a)anthracene-induced mammary carcinogenesis in rats by resveratrol: role of nuclear factor-κB, cyclooxygenase 2, and matrix metalloprotease. Cancer Res. 2002;62:4945–54.

Banerjee S, Ji C, Mayfield JE, Goel A, et al. Ancient drug curcumin impedes 26S proteasome activity by direct inhibition of dual-specificity tyrosine-regulated kinase 2. Proc Natl Acad Sci U S A. 2018;115(32):8155–60.

Barbieri R, Coppo E, Marchese A, et al. Phytochemicals for human disease: an update on plant-derived compounds antibacterial activity. Microbiol Res. 2017;196:44–68.

Batra P, Sharma AK. Anti-cancer potential of flavonoids: recent trends and future perspectives. Biotech. 2013;3(6):439–59.

Bernabeu E, Cagel M, Lagomarsino E, et al. Paclitaxel: what has been done and the challenges remain ahead. Int J Pharm. 2017;526(1–2):474–95.
Bhattacharya A, Sood P, Citovsky V. The roles of plant phenolics in defence and communication during Agrobacterium and Rhizobium infection. Mol Plant Pathol. 2010;11(5):705–19.
Chahar MK, Sharma N, Dobhal MP. Flavonoids: a versatile source of anticancer drugs. Pharmacogn Rev. 2011;5(9):1.
Chang YC, Lee TS, Chiang AN. Quercetin enhances ABCA1 expression and cholesterol efflux through a p38-dependent pathway in macrophages. J Lipid Res. 2012;53(9):1840–50.
Chang SK, Alasalvar C, Shahidi F. Review of dried fruits: phytochemicals, antioxidant efficacies, and health benefits. J Funct Foods. 2016;21:113–32.
Cherry JD, Olschowka JA, O'Banion MK. Arginase 1+ microglia reduce Aβ plaque deposition during IL-1β-dependent neuroinflammation. J Neuroinflammation. 2015;12:203.
Chirumbolo S. Plant phytochemicals as new potential drugs for immune disorders and cancer therapy: really a promising path? J Sci Food Agric. 2012;92(8):1573–7.
Chuan LI, Jia Z, Yu-Jiao ZU, et al. Biocompatible and biodegradable nanoparticles for enhancement of anti-cancer activities of phytochemicals. Chin J Nat Med. 2015;13(9):641.
Dancey J, Eisenhauer EA. Current perspectives on camptothecins in cancer treatment. Br J Cancer. 1996;74:327–38.
Davies DR, Mamat B, Magnusson OT, et al. Discovery of leukotriene A4 hydrolase inhibitors using metabolomics biased fragment crystallography. J Med Chem. 2009;52(15):4694–715.
Devappa RK, Rakshit SK, Dekker RF. Forest biorefinery: potential of poplar phytochemicals as value-added co-products. Biotechnol Adv. 2015;33(6):681–716.
Egert S, Rimbach G. Which sources of flavonoids: complex diets or dietary supplements? Adv Nutr. 2011;2(1):8–14.
Emanuela M, Giuseppe C, Sonia C, et al. Vinca alkaloids and analogues as anti-cancer agents: looking back, peering ahead. Bioorg Med Chem Lett. 2018;28:2816–26.
Epriliati I, Ginjom IR. Bioavailability of phytochemicals. In: Phytochemicals – a global perspective of their role in nutrition and health. Rijeka: InTech; 2012.
Fechtner S, Singh A, Chourasia M. Molecular insights into the differences in anti-inflammatory activities of green tea catechins on IL-1β signaling in rheumatoid arthritis synovial fibroblasts. Toxicol Appl Pharmacol. 2017;329:112–20.
George VC, Dellaire G, Rupasinghe HV. Plant flavonoids in cancer chemoprevention: role in genome stability. J Nutr Biochem. 2017;45:1–14.
Goldyne ME, Burrish GF, Poubelle P. Arachidonic acid metabolism among human mononuclear leukocytes. Lipoxygenase-related pathways. J Biol Chem. 1984;259(14):8815–9.
González-Reyes RE, Nava-Mesa MO, Vargas-Sánchez K. Involvement of astrocytes in Alzheimer's disease from a neuroinflammatory and oxidative stress perspective. Front Mol Neurosci. 2017;10:427.
Haeggström JZ. Leukotriene biosynthetic enzymes as therapeutic targets. J Clin Invest. 2018;128 (7):2680–90.
Hevener KE, Verstak TA, Lutat KE. Recent developments in topoisomerase targeted cancer chemotherapy. Acta Pharmaceutica Sinica B. 2018;8(6):844–61.
Hoensch HP, Weigmann B. Regulation of the intestinal immune system by flavonoids and its utility in chronic inflammatory bowel disease. World J Gastroenterol. 2018;24(8):877–81.
Holst B, Williamson G. Nutrients and phytochemicals: from bioavailability to bioefficacy beyond antioxidants. Curr Opin Biotechnol. 2008;19(2):73–82.
Huang M, Lu JJ, Huang MQ. Terpenoids: natural products for cancer therapy. Expert Opin Investig Drugs. 2012;21(12):1801–18.
Iqbal J, Abbasi BA, Mahmood T. Plant-derived anticancer agents: a green anticancer approach. Asian Pac J Trop Biomed. 2017;7:1129–50.
Isah T. Anticancer alkaloids from trees: development into drugs. Pharmacogn Rev. 2016;10(20):90.
Jana NR, Dikshit P, Goswami A, et al. Inhibition of proteasomal function by curcumin induces apoptosis through mitochondrial pathway. J Biol Chem. 2004;279(12):11680–5.

Kim HP, Son KH, Chang HW, et al. Anti-inflammatory plant flavonoids and cellular action mechanisms. J Pharmacol Sci. 2004;96(3):229–45.
Kim YS, Jeong HY, Kim AR, et al. Natural product derivative BIO promotes recovery after myocardial infarction via unique modulation of the cardiac microenvironment. Sci Rep. 2016;6:30726.
Kiyama R. Estrogenic terpenes and terpenoids: pathways, functions and applications. Eur J Pharmacol. 2017;815:405–15.
Kuo PL, Hsu YL, Chang CH, et al. The mechanism of ellipticine-induced apoptosis and cell cycle arrest in human breast MCF-7 cancer cells. Cancer Lett. 2005;223(2):293–301.
Langhorst J, Varnhagen I, Schneider SB, et al. Randomised clinical trial: a herbal preparation of myrrh, chamomile and coffee charcoal compared with mesalazine in maintaining remission in ulcerative colitis – a double-blind, double-dummy study. Aliment Pharmacol Ther. 2013;38:490–500.
Larussa T, Imeneo M, Luzza F. Potential role of nutraceutical compounds in inflammatory bowel disease. World J Gastroenterol. 2017;23(14):2483–92.
Las Heras B, Rodriguez B, Bosca L, Villar AM. Terpenoids: sources, structure elucidation and therapeutic potential in inflammation. Curr Top Med Chem. 2003;3(2):171–85.
Le Marchand L. Cancer preventive effects of flavonoids – a review. Biomed Pharmacother. 2002;56 (6):296–301.
Leitzmann C. Characteristics and health benefits of phytochemicals. Complement Med Res. 2016;23(2):69–74.
Lin ZY, Kuo CH, Wu DC, et al. Anticancer effects of clinically acceptable colchicine concentrations on human gastric cancer cell lines. Kaohsiung J Med Sci. 2016;32(2):68–73.
Liu RH. Health benefits of fruit and vegetables are from additive and synergistic combinations of phytochemicals. Am J Clin Nutr. 2003;78(3):517S–20S.
López-Lázaro M, Calderón-Montaño JM, Burgos-Morón E. Green tea constituents (-)-epigallocatechin-3-gallate (EGCG) and gallic acid induce topoisomerase I- and topoisomerase II-DNA complexes in cells mediated by pyrogallol-induced hydrogen peroxide. Mutagenesis. 2011;26(4):489–98.
Lou JR, Zhang XX, Zheng J, et al. Transient metals enhance cytotoxicity of curcumin: potential involvement of the NF-kappaB and mTOR signaling pathways. Anticancer Res. 2010;30 (9):3249–55.
Man SM. Inflammasomes in the gastrointestinal tract: infection, cancer and gut microbiota homeostasis. Nat Rev Gastroenterol Hepatol. 2018;15(12):721–37. https://doi.org/10.1038/s41575-018-0054-1.
Martino E, Della Volpe S, Terribile E, et al. The long story of camptothecin: from traditional medicine to drugs. Bioorg Med Chem Lett. 2017;27(4):701–7.
Matsuura HN, Fett-Neto AG. Plant alkaloids: main features, toxicity, and mechanisms of action. In: Plant toxins. Dordrecht: Springer; 2017. p. 243–61.
McCubrey JA, Lertpiriyapong K, Steelman LS, et al. Effects of resveratrol, curcumin, berberine and other nutraceuticals on aging, cancer development, cancer stem cells and microRNAs. Aging (Albany NY). 2017;9(6):1477.
Molyneux RJ, Lee ST, Gardner DR, et al. Phytochemicals: the good, the bad and the ugly? Phytochemistry. 2007;68(22–24):2973–85.
Moudi M, Go R, Yien CYS, et al. Vinca alkaloids. Int J Prev Med. 2013;4(11):1231.
Mozaffarian D, Wu JHY. Flavonoids, dairy foods, and cardiovascular and metabolic health: a review of emerging biologic pathways. Circ Res. 2018;122(2):369–84.
Nabholtz JM, Gligorov J. The role of taxanes in the treatment of breast cancer. Expert Opin Pharmacother. 2005;6(7):1073–94.
Nirmala MJ, Samundeeswari A, Sankar PD. Natural plant resources in anti-cancer therapy-a review. Res Plant Biol. 2011;1(3):1–14.
Oka Y, Iwai S, Amano H, et al. Tea polyphenols inhibit rat osteoclast formation and differentiation. J Pharmacol Sci. 2012;118(1):55–64.

Pan Y, Zhang F, Zhao Y. Berberine enhances chemosensitivity and induces apoptosis through dose-orchestrated AMPK signaling in breast cancer. J Cancer. 2017;8(9):1679.

Patlolla JMR, Rao CV. Triterpenoids for cancer prevention and treatment: current status and future prospects. Curr Pharm Biotechnol. 2012;13(1):147–55.

Pizzolato JF, Saltz LB. The camptothecins. Lancet. 2003;361(9376):2235–42.

Raza SS, Khan MM, Ahmad A, et al. Neuroprotective effect of naringenin is mediated through suppression of NF-κB signaling pathway in experimental stroke. Neuroscience. 2013;230:157–71.

Rolin D. Metabolomics coming of age with its technological diversity, vol. 67. Oxford: Academic; 2012.

Romagnolo DF, Selmin OI. Flavonoids and cancer prevention: a review of the evidence. J Nutr Gerontol Geriatr. 2012;31(3):206–38.

Safarzadeh E, Shotorbani SS, Baradaran B. Herbal medicine as inducers of apoptosis in cancer treatment. Adv Pharm Bull. 2014;4(Suppl 1):421.

Saqib U, Sarkar S, Suk K, et al. Phytochemicals as modulators of M1-M2 macrophages in inflammation. Oncotarget. 2018;9(25):17937–50.

Sies H, Schewe T, Heiss C, et al. Cocoa polyphenols and inflammatory mediators. Am J Clin Nutr. 2005;81(1):304S–12S.

Sirois P, Saura C, Salari H, et al. Comparative effects of etodolac, indomethacin, and benoxaprofen on icosanoid biosynthesis. Inflammation. 1984;8(4):353–6.

Siu D. Natural products and their role in cancer therapy. Med Oncol. 2011;28(3):888–900.

Song Y, Dou H, Gong W, et al. Bis-N-norgliovictin, a small-molecule compound from marine fungus, inhibits LPS-induced inflammation in macrophages and improves survival in sepsis. Eur J Pharmacol. 2013;705:49–60.

Surh Y-J. Cancer chemoprevention with dietary phytochemicals. Nat Rev Cancer. 2003;10:768–80.

Suzuki K, Yahara S, Hashimoto F, et al. Inhibitory activities of (-)-epigallocatechin-3-O-gallate against topoisomerases I and II. Biol Pharm Bull. 2001;24(9):1088–90.

Takimoto CH, Wright J, Arbuck SG. Clinical applications of the camptothecins. Biochim Biophys Acta (BBA)-Gene Struct Expr. 1998;1400(1–3):107–19.

Thoppil RJ, Bishayee A. Terpenoids as potential chemopreventive and therapeutic agents in liver cancer. World J Hepatol. 2011;3(9):228.

Tomicic MT, Kaina B. Topoisomerase degradation, DSB repair, p53 and IAPs in cancer cell resistance to camptothecin-like topoisomerase I inhibitors. Biochim Biophys Acta (BBA)-Rev Cancer. 2013;1835(1):11–27.

Vasanthi HR, ShriShriMal N, Das DK. Phytochemicals from plants to combat cardiovascular disease. Curr Med Chem. 2012;19(14):2242–51.

Wang W, Hu Y. Small molecule agents targeting the p53-MDM2 pathway for cancer therapy. Med Res Rev. 2012;32(6):1159–96.

Wang G, Tang W, Bidigare RR. Terpenoids as therapeutic drugs and pharmaceutical agents. In: Natural products. Totowa: Humana Press; 2005. p. 197–227.

Wang S, Meckling KA, Marcone MF, et al. Can phytochemical antioxidant rich foods act as anti-cancer agents? Food Res Int. 2011;44(9):2545–54.

Wang S, Su R, Nie S, et al. Application of nanotechnology in improving bioavailability and bioactivity of diet-derived phytochemicals. J Nutr Biochem. 2014;25(4):363–76.

Weaver BA. How Taxol/paclitaxel kills cancer cells. Mol Biol Cell. 2014;25(18):2677–81.

Westerterp M, Gautier EL, Ganda A, et al. Cholesterol accumulation in dendritic cells links the inflammasome to acquired immunity. Cell Metab. 2017;25(6):1294–304.

Xie J, Yang Z, Zhou C, et al. Nanotechnology for the delivery of phytochemicals in cancer therapy. Biotechnol Adv. 2016;34(4):343–53.

Yang H, Ping Dou Q. Targeting apoptosis pathway with natural terpenoids: implications for treatment of breast and prostate cancer. Curr Drug Targets. 2010;11(6):733–44.

Yang X, Xu S, Qian Y, et al. Resveratrol regulates microglia M1/M2 polarization via PGC-1α in conditions of neuroinflammatory injury. Brain Behav Immunol. 2017;64:162–72.

Zhu Y, Li X, Chen J, et al. The pentacyclic triterpene Lupeol switches M1 macrophages to M2 and ameliorates experimental inflammatory bowel disease. Int Immunopharmacol. 2016;30:74–84.
Ziegler RG, Colavito EA, Hartge P, et al. Importance of α-carotene, β-carotene, and other phytochemicals in the etiology of lung cancer. J Natl Cancer Inst. 1996;88(9):612–5.

Chapter 9
Health Benefits and Pharmacological Molecular Properties of Isoflavandiol (Equol): In-silico and in-vitro Updates

Pushpendra Singh, Prem P. Kushwaha, and Shashank Kumar

9.1 Introduction

EQ (7-hydroxy-3-(40-hydroxyphenyl)-chroman) is an orally bioavailable, nonsteroidal estrogen natural compound formed by metabolism of the isoflavonoid daidzein with the help of human intestinal microflora. A comparison study of various isoflavones with EQ concluded that EQ has chiral C at the C'3 position in furan ring that led to generate the enantiomeric forms, S-EQ and R-EQ. Further, it has disclosed that human intestinal bacteria synthesize only S-EQ from daidzein (Magee 2011; Tanaka et al. 2009). Moreover, it has potential chemo protective via modulating estrogen receptor (ER) activities. S-EQ isoform binds to the beta isoform of ER and activates apoptosis. Moreover, EQ accentuates the expression of various ER-responsive genes in a tissue-specific manner. Many investigations have been initiated to reveal the significance and the mechanism of soy isoflavones on their anticancer effects (Liu et al. 2017; Itsumi et al. 2016; Sareddy and Vadlamudi 2015). Soy isoflavones have Chemo-preventive activity mediated via various mechanisms which includes interruption in growth signaling, angiogenesis inhibition and induction of apoptosis (Hedelin et al. 2016; Mahmoud et al. 2014). Further, EQ reduces the cell proliferation in different cancer including breast, hepatocellular, ovarian, and prostate cancer (Liang et al. 2014; Liu et al. 2017; Taghizadeh et al. 2015; Wang and Ren 2014; Zheng et al. 2012).

P. Singh
Tumor Biology Laboratory, National Institute of Pathology, New Delhi, India

P. P. Kushwaha · S. Kumar (✉)
Department of Biochemistry and Microbial Sciences, Central University of Punjab, Bathinda, Punjab, India

S. Kumar, C. Egbuna (eds.), *Phytochemistry: An in-silico and in-vitro Update*,
https://doi.org/10.1007/978-981-13-6920-9_9

9.2 Structure and Bioavailability of EQ

Chemically EQ is 7-hydroxy-3-(49-hydroxyphenyl)-chroman with 242.27 Daltons molecular weight and $C^{15}H^{14}O^{3}$ molecular composition. Like other flavonoids, it has three basic heterocyclic rings namely A, B, and C. Furthermore, two of these rings (A and B) have two hydroxyl rings at each one. The central furan ring (Ring C) has single unreactive oxygen (comparatively inert). Although the structure shows two hydroxyl groups and one oxygen molecule, physiochemically, it shows insolubility and nonpolar property. During in-vitro experiments, the physiochemical property should be taken care of. Ring A and B give two phenolic ring structure to EQ but despite that, the structure exhibits poor UV absorption (Kumar and Pandey 2013). The carbon (of C Ring) at which ring B is attached is a chiral carbon and produce two enantiomers (S-(+) and R-(+) EQ) of the compound. EQ is different from its precursors (isoflavone, daidzein, genistein, and glycitein) having natural diastereoisomer (S-(−) EQ) produced by intestinal bacteria. The racemic mixture of EQ (A unit) and separate enantiomer at an equal concentration (A unit) might have different extent of biological properties (Setchell and Clerici 2010) (Fig. 9.1).

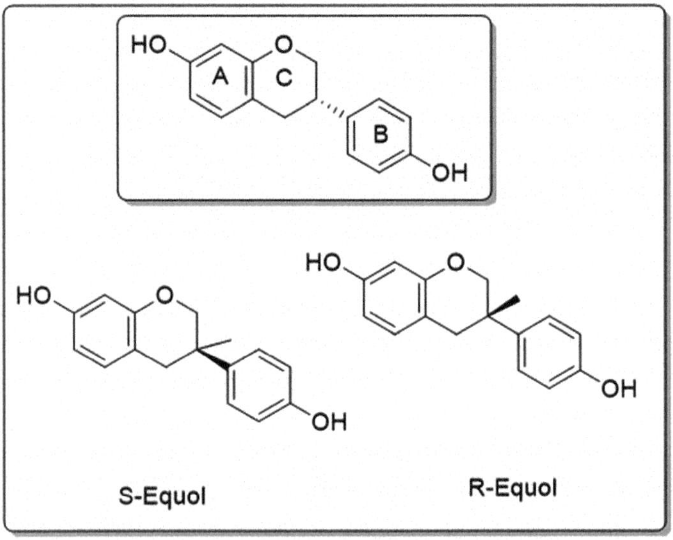

Fig. 9.1 Structure of EQ (7-hydroxy-3-(49-hydroxyphenyl)-chroman). The carbon (of C Ring) at which ring B is attached is a chiral carbon and produce two enantiomers (S-(+) and R-(+) equol) of the compound. EQ is different from its precursors (isoflavone, daidzein, genistein and glycitein) having natural diastereoisomer (S-(−) EQ) produced by intestinal bacteria

9.3 EQ and Prostate Cancer

Prostate cancer is fourth most widespread cancer and has high prevalence and metastasis rate. It ranked as the second most often diagnosed cancer between men in most of the western and Asian countries (Sharma et al. 2016). Moreover, environmental and hormonal factors are independent risk factors which have a key role in prostate cancer development. A previous study identified that natural product lycopene and isoflavone as defensive factors against prostate cancer (Miyanaga and Akaza 2015). Furthermore, soy isoflavones reduced the threat of prostate cancer via different defense activities such as antioxidant, DNA repair, impediment of metastasis and angiogenesis, enhancing sensitivity of radio and chemotherapeutic agents (Mahmoud et al. 2014). Moreover, isoflavones strengthens the effect of radiation therapy, and modulate the epigenetic regulation of various oncogenes and tumor suppressor gene (Ahmad et al. 2013). Further, meta-analysis study clarifies that phytoestrogens lowers the possibility of prostate cancer (Zhang et al. 2017). EQ also is known as a Japanese meal. Consequently, it has shown that EQ is the most active aglycone, converted from daidzein by human intestinal flora. Moreover researchers showing interest to increase the concentration of EQ by inducing the EQ producing bacterium in human (Sugiyama et al. 2013). EQ may act as anti-oxidant effect facilitated through various phosphatase (Zheng et al. 2012). Forkhead box O3 is a member of most important transcription factors that have tumor suppressor activity in prostate cancer. Furthermore, studies showed that EQ targeted the FOXO3a that led to reduces cell survival, cell cycle progression, and apoptosis. Moreover, treatment with EQ reserved the PC3 xenograft tumor growth of in BALB/c nude mice (Lu et al. 2016). EQ may appear to chemo-preventive for prostate cancer mediated via Skp2-mediated androgen receptor activity (Itsumi et al. 2016).

9.4 EQ and Breast Cancer

Breast cancer is second in the line of most widespread cancers in the world and most prevalent cancer in women. Epidemiological studies showed that EQ has anti-estrogenic properties that led to breast cancer preventives activity-mediated by up-regulation of eIF4GI translation protein (de la Parra et al. 2015). Furthermore, it has shown that EQ widely consumed by menopausal women, that reduces the breast cancer initiation and progression (Jiang et al. 2013). Moreover, EQ is a radiosensitizing agent that enhances cell death following irradiation, and therefore dropping the number of surviving cells (Taghizadeh et al. 2015). Consequently, it has reported that EQ intake reduces the breast cancer risk by influencing EQ producing gut bacteria in human cells (Tseng et al. 2013; Virk-Baker et al. 2014). Further, anticancer effect of EQ reported in breast and liver cancer by up-regulation of Bax and down-regulation of Bcl-2 (Liang et al. 2014). The EQ could inhibit the proliferation via down-regulating the expression of Bcl-2, VEGF, p-ERK1/2 and p-p38 protein (Wang and Ren 2014). EQ also work as chemopreventive agents when used with anti-inflammatory compounds in postmenopausal women (Obiorah et al. 2014). Further, it has confirmed that EQ may be used with the combination of tamoxifen's to enhance the apoptosis activity (Charalambous et al. 2013). Moreover, S-equol inhibits the invasion of breast cancer cells (MDA-MB-231) by down-regulating MMP-2 expression (Magee et al. 2014). Furthermore, EQ tends to alter transcription by demethylation and acetylation of histone proteins in cell lines of breast cancer (Dagdemir et al. 2013). Moreover, EQ stimulates breast cancer with bone micro-metastasis in mice that increase apoptosis and decrease metastasis activity (Yang et al. 2015a, b). The various study concluded that EQ intake lowered the possibility of breast cancer, osteoporosis, heart related diseases and menopausal symptoms by effecting endocrine-disrupting properties (Patisaul 2017). Equol (soy isoflavone daidzein's bacterial metabolite) is produced by 30–50% of humans and augmented induction of apoptosis (Ono et al. 2017). Furthermore, most impressive, development of multidrug resistance and toxicity are the foremost problem in the cancer treatment. P-glycoprotein, MRP1, MRP2 and BCRP are an independent risk factor for drug resistance. Reverri (2015) have shown that EQ diminished the activity of P-glycoprotein, MRP1, MRP2 and BCRP (Rigalli et al. 2017).

9.5 EQ and Cardiovascular Disease (CVD)

CVD is among the foremost causes of premature death in the world and food and drug administration (FDA) recommended cholesterol-lowering properties of soy protein for controlling CVD. Furthermore, study intimates that soy nuts have positive outcomes for attenuating endothelial dysfunction which may helpful for protection of CVD, osteoporosis, and cancers (Hazim et al. 2016; Magee 2011; Reverri et al. 2015; Richter et al. 2017; Vafeiadou et al. 2006). Moreover, cardioprotective effect of EQ lowers the activation of PMN and PMN-platelet interaction. Isoflavones activate the endothelial cells for NO production that leads to boost Nrf2-Keap1 mediated antioxidant activity (Mann et al. 2009). Epidemiological evidence implies that EQ intake activates CVD preventive eNOS, EDHF, and redox-sensitive gene expression (Bonacasa et al. 2011; Gil-Izquierdo et al. 2012). Oxidative stress and Endothelial dysfunction are implicated various metabolic syndromes such as diabetes and obesity. Epidemiological mark revealed that intake of soy food diminishes the chances of cardiovascular mediated metabolic risk (Reverri et al. 2015). Furthermore, Double-blind randomized controlled trial research presented that EQ intake reduced the chance of cardiovascular disease (Hazim et al. 2016).

9.6 EQ and Menopause

The menopause is the end stage of menstruation in every women life and often involves troublesome symptoms, including vasomotor symptoms, libido, insomnia, fatigue, and joint pain. Estrogen depletion is a significant determinant of bone loss in women at the postmenopausal stage. Isoflavones have the ability to reduce the bone loss in women at postmenopausal stages (Pawlowski et al. 2015). In-vitro analysis have shown that EQ have anti-oxidative and vasorelaxant activity which can be beneficial against hypertension (Liu et al. 2016). EQ production was considerably related with reduced levels of uric acid, arterial stiffness and an increased ratio of arachidonic acid to eicosapentaenoic acid which may recover menopause-related health problem in women (Davinelli et al. 2017; Yoshikata et al. 2017). Previous reports suggest that the O-desmethylangolensin non-producer phenotype is linked to obesity in menopausal women (Miller et al. 2017). EQ producing bacteria inside human gut increasing the endogenous formation of this bioactive compound that triggering of intestinal health-related beneficial effects (Guadamuro et al. 2017; Vazquez et al. 2017).

9.7 EQ and Glucose Metabolism/Diabetes

Diabetes mellitus (DM) is the complicated metabolic disorder defined by hypoglycaemic and hyperglycemia and affecting the world population. T1DM (Type 1 diabetes mellitus) develops due to the destruction of pancreatic β cells and T2DM (type 2 diabetes mellitus) is characterized by insulin resistance. Even though T2DM is much more prevalent, both types of diabetes may involve in various cardiovascular complications (microvascular and macrovascular complications). Various phytochemicals show anti-oxidative, anti-inflammatory and anti-osteoporotic activities pertinent to prevention of diseases, such as cancer, cardiovascular disease, and osteoporosis. It is documented that EQ isoflavones act as influential antioxidants led to reducing oestrogenic effects in diabetic women (Vedavanam et al. 1999). Moreover, study shown that soy protein (EQ) increase glycaemic control in adults types 2 diabetes. Thus, EQ may play the substantial role in the management of T2DM programme (Gobert et al. 2010). Moreover, it was suggested that diabetes prevention activity of EQ mediated by activation of AMPK activation, GLUT4 translocation, and glucose uptake in muscle cells. Consequently, it has shown that EQ overcoming insulin resistance development. EQ also plays an antihyperglycemic role like other resveratrol and aspalathin (Cheong et al. 2014; Minakawa et al. 2011; Son et al. 2013). Furthermore, EQ producers showed lower leptin than nonproducers (inverse correlation) in the diabetic females (Sakane et al. 2014). S equol boosts human b-cell function mediated via activating cAMP-PKA signaling showed that potential for T2DM prevention (Horiuchi et al. 2017). Further, results demonstrate that EQ augments adipocyte differentiation through PPARγ activation (Cho et al. 2010). It is showed that EQ increases antioxidant system through PTEN (Kunigal et al. 2007). Further, data recommended that EQ with genistein and daidzein repressed prostate cancer cell migration and invasion. EQ reduces the metastatic activity by inhibition of secretion of matrix metalloproteinase-2, matrix metalloproteinase-9, and urokinase-type plasminogen activator (Zheng et al. 2012). Further, it has shown that may EQ contribute to a abridged level of OX-LDL-stimulated apoptosis that associated to reduced intracellular ROS (reactive oxygen species) (Kamiyama et al. 2009). Moreover, EQ affects the antioxidant system in both the short-term and long-term by activation of stimulation of catalase and SOD (Choi 2009; Pazourekova et al. 2016). EQ and O-desmethylangolensin (another metabolite of daidzein) reduce the bone metabolism that led to increasing EQ producing intestinal microflora in humans (Landete et al. 2016; Uehara 2013). Soya includes a large number of compounds such as stilbenes, isoflavones, and lignans that play a protecting role against many diseases (cancer, cardiovascular disease, and metabolic syndrome (Landete et al. 2016). Compounds such as, genistein, daidzein, and equol have a more effective free radical scavengers activity that extends the antioxidant activity (Kladna et al. 2016). EQ repressed the lipopolysaccharide (LPS) induced TLR4, MAPK, NF-κB-mediated transcription in murine microglia cells that show neuroprotective role (Subedi et al. 2017).

9.8 EQ- Protein Interactions: In-silico Molecular Mechanism

EQ-protein interactions are not enough explored to date in published literature list. In this context, our aim to investigate the EQ-protein interactions and ADME/T attributes of the EQ by using Computer-Aided Drug Design (CADD).

9.8.1 Molecular Docking

9.8.1.1 Preparation of EQ Molecules

Protocol for molecular docking taken from our earlier published literature (Singh and Bast 2014a, b, 2015a, b; Singh et al. 2016). The structure of EQ had drawn by ChemBioDraw Ultra 12 and prepared using Ligprep module and then performed optimized potential for liquid simulations (OPLS_2005) force field of the Maestro 9.6 (Schrödinger, LLC, New York, NY) (Jorgensen et al. 1996; Jorgensen and Tirado-Rives 1988; Shivakumar et al. 2010). The unspecified chirality of EQ was applied and low energy conformation generated. The last stage of EQ preparation used in geometry optimization through MacroModel and further used for docking process.

9.8.1.2 Preparation of Protein Molecules

We studied molecular docking of IGF1R (PDB; 1K3A), IR (PDB; 3ETA), EGFR (PDB; 1M17), VEGFR1 (PDB; 3HNG), VEGFR2 (PDB; 2OH4), STAT3 (PDB; 1BG1), PI3K (PDB; 3S2A), PDK1 (PDB; 3RWQ), Akt (PDB; 3MV5), CXCR2 (PDB; 4JL7), CXCR4 (PDB; 3OE6), CCR5 (PDB; 4MBS), mTOR (PDB; 4DRI), TGFβ RI (PDB; 2X7O), TGFβ RII (PDB; 1KS6), TAK1 (PDB; 4L3P), AR (PDB; 2PIV), ER (PDB; 3ERT), γ-secretase (PDB; 4Y6K), PARP1 (PDB; 1UK1), Smoothened 7TM (PDB; 7QIM), and Wnt (PDB; 1JDH) against EQ. GLIDE (Maestro 9.6) was used to dock EQ with the X-ray crystal structure of a selected proteins. The X-ray structure of the chosen protein retrieved from the PDB and protein preparing application corrected the arrangement of the protein by addition of hydrogen atoms, assigning bond orders, and charge fixing (Singh and Bast 2014a, b, 2015a, b; Singh et al. 2016).

9.8.1.3 Receptor Grid Generation

After the protein preparation receptor grid was generated to assign active site in protein (Singh and Bast 2014a, b, 2015a, b; Singh et al. 2016).

9.8.2 *Grid-Based Ligand Docking with Energetics (GLIDE) Molecular Docking*

Molecular docking was conducted using GLIDE software from Schrödinger suite (Halgren et al. 2004; Friesner et al. 2004, 2006) (Singh and Bast 2014a, b, 2015a, b; Singh et al. 2016). We study interactions profile of EQ with 20 overexpressed proteins (The protein involved in different disease) employing molecular docking approach. In-silico data for EQ-protein interactions generated and presented in Table 9.1. Docking study of EQ was performed to know their interaction with the different protein and efficiency of docking process defined by measuring RMSD (root mean square deviation). Molecular docking process distinguishes the docking free energy value (Gscore) against different protein molecules and more than −8 Kcal/mol values were taken into consideration. EQ showed hydrophobic, and hydrogen bond interactions with flap and dyad approach that are important in computer-aided drug design (CADD).

Gscore from docking EQ with VEGFR2 (PDB; 2OH4) is −8.27 Kcal/mol. 2OH4-EQ interactions showed that amino acids Leu838, Val846, Ala864, Val897, Leu1033, Phe919, Phe1045 and Cys1043 form hydrophobic interactions. Furthermore, Glu915 and Asp1044 are negatively charged, and Lys866 and Lys918 are positively charged amino acid which involved in interactions. Moreover, Cys917 involved in H-bond backbone and Glu883 involved in H-bond side chain at the

Table 9.1 Lowest binding energy for the EQ-protein interactions as detected by GLIDE molecular docking

Receptors	GScore	Lipo EvdW	HBond	Electro
VEGFR2	−8.27	−4.66	−0.75	−0.7
IGFR1	−5.5	−3.57	−1.33	−0.62
IR	−6.65	−5	−0.91	−0.24
EGFR	−5.58	−3.28	−1	−0.94
STAT3	−3.91	−1.21	−1.78	−0.84
PI3K	−8.11	−4.34	−1.45	−0.91
PDK1	−4.15	−3.11	−1.53	−0.82
Akt	−3.74	−3.13	−1.51	−0.72
TGFβR1	−6.86	−4.71	−0.99	−0.57
TGFβR2	−4.36	−2.16	−1	−0.56
TAK1	−7.47	−3.15	−1.5	−0.65
CCR2	−3.37	−1.6	−1.42	−0.49
CCR4	−4.7	−2.3	−1.38	−0.65
CCR5	−5.22	−2.91	0	−0.33
AR	−8.22	−4.65	1.78	−0.85
ER	−8.86	−4.88	−1.48	−0.92
GAMA Scretease	−4.49	−3.61	−0.32	−0.19
PARP	−4.62	−3.22	−0.75	−0.27
Smoothened 7TM	−5.8	−3.17	−0.78	−0.86
Wnt	−2.99	−2.04	−1.05	−0.6

IGF1R (PDB; 1KA), IR (PDB; 3ETA), EGFR (PDB; 1M17), VEGFR1 (PDB; 3HNG), VEGFR2 (PDB; 2OH4) STAT3 (DB; 1BG1), PI3K (PDB; 3S2A), PDK1 (PDB; 3RWQ), Akt (PDB; 3MV5), TGFβ RI (PDB; 2X7O), GFβ RII (PDB; 1KS6), TAK1 (PDB; 4L3P), CXCR2 (PDB; 4JL7), CXCR4 (PDB; 3OE6), CCR5 (PDB; 4MBS), AR (PDB; 2PIV), ER (PDB; 3ERT), γ-secretase (PDB; 4Y6K), PARP1 (PDB; 1UK1), Smoothened 7TM (PDB; 7QIM), and Wnt (PDB; 1JDH)
Ligand; EQ
GScore; Glide Extra Precision Docking Scores (kcal/mol)
Lipophilic E Vdw; Chemscore lipophilic pair term and fraction of the total protein-ligand vdw energy
HBond hydrogen-bonding term, *Electro* electrostatic rewards

interactions sites (Fig. 9.2). Gscore from molecular docking EQ with PI3K (PDB; 3S2A) is −8.11 Kcal/mol. Moreover, the structure of PI3K complexed with EQ described the importance of amino acids Met804, Pro810, Trp812, Glu880, Ile831, Ile881, Val882, Ala885, Ile879, Met 953, Phe961, and Asp964 in hydrophobic interactions. Amino acid Val882 and Ala885 committed for backbone hydrogen bonding. Moreover, amino acid Trp812 and Tyr867 were involved in π-π interactions of active site (Fig. 9.3).

Molecular docking results of AR (PDB; 2PIV) against EQ revealed that the EQ has good Gscore −8.22 Kcal/mol. Furthermore, interactions regarding protein-ligand of 2PIV with the EQ divulge that Leu704, Asn705, Leu 707, Val746, Met749, Phe764, Met780, Leu873, Phe876, and Thr877 involved in the many interactions such as hydrophobic interactions, hydrogen bond, and pi-pi stacking interactions. Moreover, docking results of ER (PDB; 3ERT) against EQ disclosed

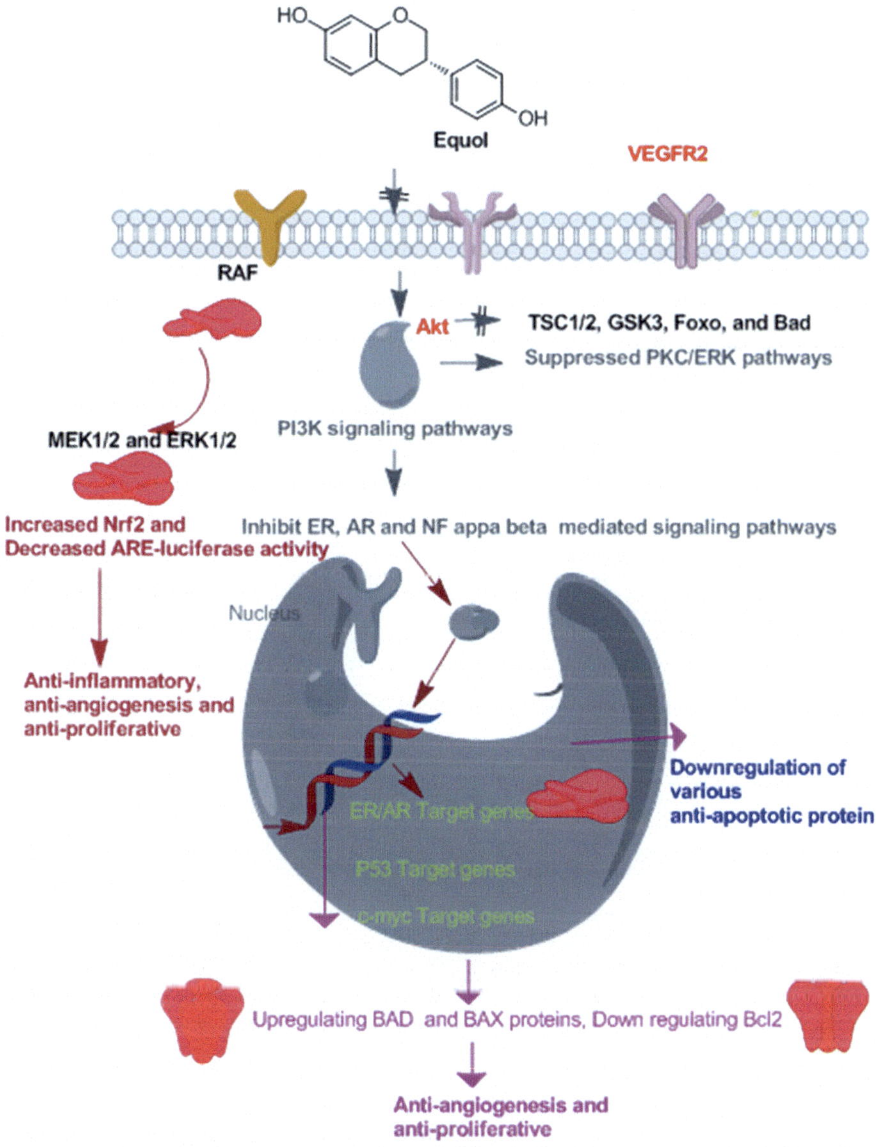

Fig. 9.2 Mechanism of action of EQ. EQ reduces the cell proliferation through the VEGFR2, PI3K and AR/ER pathways. EQ may bind to VEGFR2 and reduces the activity of downstream PI3K activity. In another way, it may also block the activity of AR/ER

that the EQ has Gscore −8.86 Kcal/mol. Amino acids Met343, Leu346, Leu347, Leu349, Ala350, Asp351, Trp383, Leu384, Leu387, and Leu525, are responsible for hydrophobic interactions. Moreover, amino acid Glu353, and Hid 524, involved in H-bond side chain in addition to Gly521 backbone H-bond (Fig. 9.4).

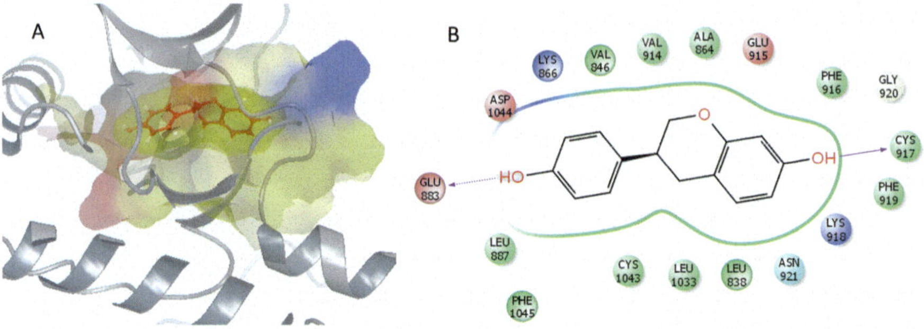

Fig. 9.3 (**a**) Ribbon presentation of VEGFR2 (PDB; 2OH4) molecule with EQ. (**b**) Protein-ligand interactions profile of 2OH4 with EQ. 2OH4-EQ interactions showed that amino acids Leu838, Val846, Ala864, Val897, Phe919, Leu1033, Cys1043, and Phe1045 form hydrophobic interactions during interactions. Furthermore, Glu915 and Asp1044 are negatively charged, Lys866 and Lys918 are positively charged amino acid which involved in interactions. Cys917 involved in H-bond back bone and Glu883 involved in H-bond side chain at the interactions sites

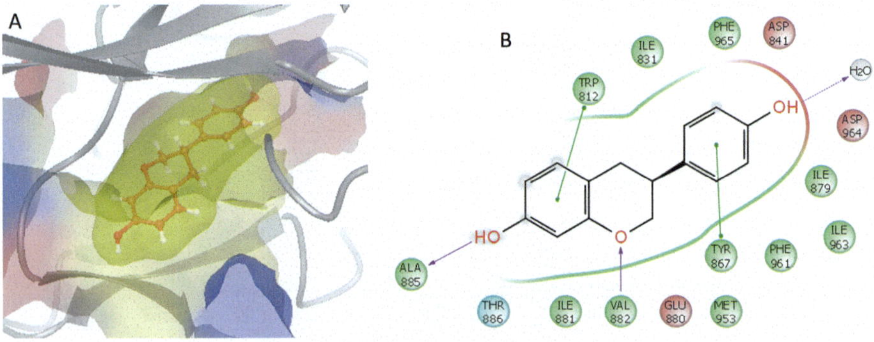

Fig. 9.4 (**a**) Ribbon presentation of PI3K (PDB; 3S2A) molecule with EQ. (**b**) Protein-ligand interactions profile of 3S2A with EQ. Moreover, the structure of PI3K complexed with EQ described the importance of amino acids Met804, Pro810, Trp812, Glu880, Ile831, Ile881, Val882, Ala885, Ile879, Met 953, Phe961, and Asp964 in hydrophobic interactions. Amino acid Val882 and Ala885 committed for backbone hydrogen bonding. Moreover, amino acid Trp812 and Tyr867 were occupied in π-π interactions at an active site

Soy isoflavone induced arrest in growth and apoptosis in prostate cancer cell occurs via antioxidant defense, and estrogen antagonism (Mahmoud et al. 2014). Soy isoflavones diminish the biological effect of the AR (androgen receptor) in prostate cancer (Itsumi et al. 2016). Moreover, EQ inhibits gastric cancer (MGC-803) cells proliferation via Akt dephosphorylation at Thr450 (Yang et al. 2015a, b). Further, EQ regulated the c-Myc function via upregulation of eIF4GI (translation initiation factor 4GI) (de la Parra et al. 2015). EQ inhibited the proliferation of colon cancer cell by antioxidant activities (Cai et al. 2017). Moreover, EQ provokes mitochondria-dependent apoptosis in gastric cancer MGC-803 cells via the continued activation of the ERK1/2 (Yang et al. 2016). EQ enhances AR (androgen

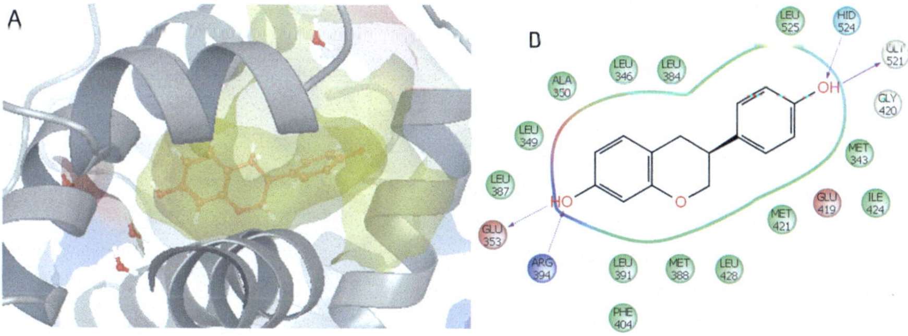

Fig. 9.5 (**a**) Ribbon presentation of ER (PDB; 3ERT) molecule with EQ. (**b**) Protein-ligand interactions profile of 3ERT with EQ. 3ERT-EQ interactions showed that amino acids Met343, Leu346, Leu347, Leu349, Ala350, Asp351, Trp383, Leu384, Leu387, and Leu525, included in the hydrophobic interactions. Moreover, amino acid Glu353, and Hid 524, involved in H-bond side chain in addition to Gly521 back bone H-bond

receptor) degradation mediated by Skp2 by in prostate cancer (Itsumi et al. 2016). Isoflavones decrease the ERK1/2 expression in prostate cancer via VEGFR signaling which may be driven to the anticancer activity of soy (Clubbs and Bomser 2007). Further, EQ could hinder the proliferation of the MCF-7, and its inhibitory effect may be due to inducing apoptosis, arresting the cell cycle, down-regulating the expression of VEGFR and p-ERK1/2 (Wang and Ren 2014). Furthermore, 6-methoxy EQ represses VEGF and FGF2 driven endothelial cells proliferation via MEK1/2 and ERK1/2 (Bellou et al. 2012). ER inhibitors or PI3K/Akt pretreatment of cells increased Nrf2 and reduced ARE-luciferase activity. Thus, beneficial health effects of S-(−) EQ might present a novel strategy against cardiovascular diseases (Zhang et al. 2013). Further, it discussed that EQ stimulates cell growth via ER and Akt/PI3K mediated cell growth (Fig. 9.5) (Liu et al. 2014).

9.8.3 ADME/T (Absorption, Distribution, Metabolism, Excretion, and Toxicity)

ADME properties of EQ were predicted by applying QikProp application of Maestro 9.6. (Jorgensen and Duffy 2002; Lu et al. 2004). Moreover, this application also calculates properties for Jorgensen's 'Rule of 3' and Lipinski's 'Rule of 5' to filter out undesirable compounds (Singh and Bast 2014a, b, 2015a, b; Singh et al. 2016).

ADME attributes of the EQ observed by applying QikProp module in Schrodinger as shown in Table 9.2. Moreover, various ADME/T properties of EQ were in the range of Lipinski's rule of five, establishing as the best compound for cancer prevention. In ADME studies, negative values for the QPlog HERG (blockage of K^+ channels) and QPlogBB (brain blood partition coefficient) of EQ showed best anti-cancer compound. Moreover, EQ has the best limit for Caco-2 cell permeability (QPPCaco),

Table 9.2 ADME/T properties of EQ

Molecule	Mw	Dipole	Donor HB	Accpt HB	Volume
CID91469 (Equol)	241.27	2.98	2.9	474.85	798.66

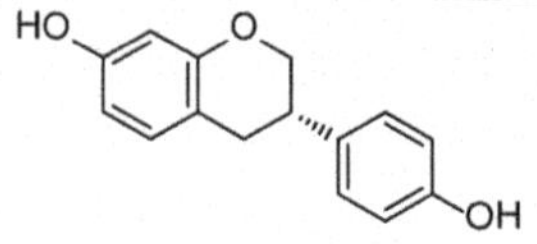

Equol, CID91469

Molecule	QPlog HERG (acceptable range: above -5.0)	QPP Caco (nm/sec) <25-poor >500-great	QPlog BB (-3-1.2)	QPP MDCK (nm/sec) <25-poor >500-great	QPlog Khsa (Acceptable range: -1.5 to 1.5).	Percent Human Oral Absorption >80 % -High <25 % -Poor
CID91469 (Equol)	-4.863	910.085	-0.505	446.809	0.171	96.074

Molecular weight
Volume; Predicted no. of hydrogen bonds which can be accepted by the solute from water molecules in aqueous solution
Hydrogen bond donors (<5)
Hydrogen bond acceptors (<10)
Predicted IC50 values for blockage of HERG K+ channels; (acceptable range: above −5.0)
QPP Caco-Estimated apparent Caco-2 cell permeability in nm/sec. Caco-2 cells, a model for the study of gut blood barrier; (nm/sec) <25-poor >500- great
Q P log BB- Predicted brain/blood partition coefficient;
QPP MDCK -estimated apparent MDCK cell permeability in nm/sec. MDCK cells are regarded as a good mimic for the blood-brain barrier; (nm/sec) <25-poor >500-great
Q P log KP- Predicted skin permeability; Q P log K_{hsa}- Estimation of binding to human serum albumin; (acceptable range: −1.5 to 1.5)
Percentage of human oral absorption; (<25% is poor and >80% is high)

MDCK cell permeability (QPPMDCK), and percentage of human oral absorption value 910, 446.80 and 96 respectively which convince the Lipinski's rule of five. Further, estimation of chemical carcinogenicity through ensemble learning methods (CarcinoPred-EL) was done by using three approaches RF, SVM and XGBoost (http://ccsipb.lnu.edu.cn/toxicity/CarcinoPred-EL). These result showed that EQ is non-carcinogen compound. The toxicity of EQ predicted by using PreADMET online server and predicted value summarized in Tables 9.3 and 9.4 (Vimal et al. 2017). Ames test predicts the mutagenicity of test compound with the help of different strains of *S. typhimurium* including TA98, TA100, and TA1535. The positive value of result showed that EQ has non-mutagenic properties of various strains.

Table 9.3 In-silico bioactivity of compounds (http://www.molinspiration.com/cgi-bin/properties)

Target activity	Score
GPCR ligand	−0.02
Ion channel modulator	0.02
Kinase inhibitor	−0.05
Nuclear receptor ligand	0.4
Protease inhibitor	−0.29
Enzyme inhibitor	0.24

Table 9.4 Toxicity and drug-likeness of EQ

Toxicity ID	Value	Drug-likeness ID	Value
Algae at	0.0395569	CMC like rule	Qualified
Ames test	Mutagen	Lead like rule	Violated
Carcino mouse	Positive	Lead like rule violations	1
Carcino rat	Negative	MDDR like rule violation fields	No rotatable bonds
Daphnia at	0.105265	MDDR like rule violations	1
HERG inhibition	Medium risk	Rule of five	Suitable
Medaka at	0.016674		
Minnow at	0.0127925		
TA100 10RLI	Positive		
TA100 NA	Negative		
TA1535 10RLI	Negative		
TA1535 NA	Negative		

Algae at (level of toxic accumulation in algae), Ames test (mutagenicity of a compound), Carcino Mouse (carcinogenicity bioassay in mouse), Carcino Rat (carcinogenicity bioassay in rat), Daphnia at (Acute daphnia (fish) toxicity), HERG inhibition (potassium channel inhibition), Medaka (*Oryzias latipes*) Japanese Small aquarium fishes used for toxicity testing, Minnow (*Pimephales promelas*) a small fish model in the area of aquatic toxicology

9.8.4 *In-silico Bioactivity of Compounds (Molinspiration)*

Molinspiration is an online service for analysis of bioactivity score for the six protein molecules viz., kinase inhibitors GPCR ligands, enzymes, ion channel modulators, and nuclear receptors. Therefore, bioactivity of EQ was demonstrated using Molinspiration (Jarrahpour et al. 2012).

9.8.5 *In-silico Toxicity Prediction PreADMET (https://preadmet.bmdrc.kr/toxicity)*

A pre ADME/T toxicity prediction is a tool for drug design. Therefore, the toxicity of EQ was calculated using PreADMET (Kartasasmita et al. 2014; Muster et al. 2008). Further, estimation of chemical carcinogenicity utilizing ensemble learning methods

(CarcinoPred-EL) was done by using three approaches RF, SVM, and XGBoost. Moreover, drug-likeness and toxicity were studied through PreADMET (https://preadmet.bmdrc.kr/toxicity).

9.8.6 *Ligand-Based Virtual Screening (LBVS)*

LBVS is among the most accepted techniques for the identification of best target compound. Here we employed this technique for the detection of EQ similar compounds which have analogous pharmacological profiles. USR-VS executed LBVS to cataloging unknown active molecules for an interested target. There are two forms USR (Ballester 2011) and USRCAT (Schreyer and Blundell 2012) available for perspective virtual screening (Ballester and Richards 2007; Li et al. 2016).

Prospective applications of USR and USRCAT are used in various previous studies to search inhibitor for cancer-promoting phosphatases, p53-MDM2, and arylamine N-acetyltransferases inhibitors (Ballester et al. 2009; Hoeger et al. 2014; Patil et al. 2014). Further antibacterial hit identification is also done by using this technique (Ballester et al. 2012). Here, EQ utilized a template and hundred hits taking into consideration for ligands selection. Further, we found ultrafast shape recognition (USR) score (0.95367342), USRCAT (0.60377538) and Tanimoto (0.18691589), respectively to identify the new best compounds and molecular docking are performed. Table 9.1 explained that VEGFR2, PI3K, and ER have the best energy (compare to all selected proteins) score against EQ. Thus, docking study of these compounds (selected from USR and USRCAT) was performed to know their interaction with the ER. Free energy value (Gscore) against VEGFR2, PI3K, and ER are calculated and more than −8 Kcal/mol taken into consideration. Docking results including several interactions energy are provided in Table 9.5.

9.9 Conclusion

EQ is an orally bioavailable, nonsteroidal estrogen natural compound having chiral C at C'3 of the furan ring. EQ health benefits related to positive outcome in various diseases such as osteoporosis, prostate cancer, breast cancer, menopausal symptoms glucose metabolism/diabetes and the cardiovascular risk biomarkers low-density lipoprotein cholesterol. Chemo-preventive properties of soy isoflavones have shown via various mechanisms such as the interruption of growth signaling, angiogenesis inhibition and induction of apoptosis. Further, EQ reduces the cell proliferation in different cancer including breast, hepatocellular, ovarian, and prostate cancer.

Table 9.5 Lowest binding energy for the Ligand-based virtual screening (EQ EQ similar compounds selected from USR and USRCAT)-protein interactions as detected by GLIDE molecular docking

PDB	Lig.	ZINC ID	GScore	Glide emodel	HBond	LipophilicEvdW	Electro
		95088258	−8.69	−32.622	−0.9	−3.051	−1.177
		20216878	−8.591	−49.701	−1.297	−2.891	−1.279
	USR	87199384	−8.215	−41.463	−0.9	−3.231	−0.856
		82351094	−8.051	−39.646	−0.013	−3.877	−0.662
2OH4		406936	−8.003	−37.81	−0.9	−2.953	−0.863
		8703291	−10.695	−52.402	−2.402	−3.693	−0.862
		72661838	−9.846	−59.552	−1.833	−3.874	−0.898
	USRCAT	41661642	−9.18	−48.728	−0.963	−4.309	−0.526
		16887367	−9.416	−49.109	−1.519	−3.275	−0.957
		65337757	−8.959	−52.824	−1.004	−4.56	−0.517
		48964142	−9.694	−51.628	−1.443	−3.038	−0.222
		71773700	−9.602	−47.099	−1.88	−3.561	−0.803
	USR	65337747	−9.568	−46.737	−1.8	−3.536	−0.652
		40537834	−8.993	−49.63	−1.33	−3.716	−0.583
3S2A		6747792	−9.261	−47.916	−1.265	−3.509	−0.826
		5535734	−10.417	−46.089	−1.33	−3.356	−0.657
		86895709	−9.016	−37.728	−1.314	3.086	−0.467
	USRCAT	87101621	−8.742	−42.262	−1.33	−2.449	−0.514
		44713373	−8.52	−40.133	−1.296	−2.682	−0.204
		86196851	−8.363	−30.964	−1.027	−3.501	−0.426
		95088258	−8.438	−27.609	−0.419	−3.062	−0.256
		26420321	−8.136	−37.637	−0.732	−2.695	−0.321
3ERT	USR	95091324	−7.946	−24.749	−0.461	−2.959	−0.269
		20216881	−7.381	−34.193	−0.697	−2.527	−0.8
		87199384	−7.292	−32.949	−0.583	−2.974	−0.303

Molecular docking was done preliminarily to examine the target of EQ. Analysis of docking poses noted that EQ has better energy scores against VEGFR2, PI3K, AR, and ER. Protein-EQ interactions outline highlighted that hydrophobic interaction is the leading force at the active site, in addition to hydrogen bonding and π-π stacking interactions. Outcomes of this review could be accommodating to identify protein-EQ interactions that are needed to improve the cell growth inhibition as well as ADME properties. ADME of EQ was inside the criteria of Lipinski's rule of five, establishing as the best compound for cancer prevention. In ADME studies, negative values for the QPlog HERG and QPlogBB of EQ showed best anti-cancer compound. Moreover, EQ has the best limit for Caco-2 cell permeability, MDCK cell permeability, and percentage human oral absorption value which convince the Lipinski's rule of five. Additional investigational studies are essential for the validation of our study.

Acknowledgments We would like to thank Central University of Punjab, Bathinda, Punjab, (India) and Director in-charge, National Institute of Pathology, New Delhi (India) for supporting this study with infrastructural requirements. This study was also supported by a Centenary-Post Doctoral Research Fellowship Grant-in-Aid from Indian Council of Medical Research (ICMR), Government of India awarded to PS. PPK acknowledges financial support from UGC-CSIR in the form of Senior Research Fellowship.

Conflict of Interest The authors declare that no financial or commercial conflict of interest.

References

Ahmad A, Biersack B, Li Y, Bao B, Kong D, Ali S, Banerjee S, Sarkar FH. Perspectives on the role of isoflavones in prostate cancer. AAPS J. 2013;15(4):991–1000.

Ballester PJ. Ultrafast shape recognition: method and applications. Future Med Chem. 2011;3 (1):65–78.

Ballester PJ, Richards WG. Ultrafast shape recognition for similarity search in molecular databases. In: Proceedings of the Royal Society of London. Series A. Mathematical, physical and engineering sciences, vol. 2081. London: The Royal Society; 2007. p. 1307–21.

Ballester PJ, Westwood I, Laurieri N, Sim E, Richards WG. Prospective virtual screening with ultrafast shape recognition: the identification of novel inhibitors of arylamine N-acetyltransferases. J Royal Soc Interface. 2009;7(43):335–42.

Ballester PJ, Mangold M, Howard NI, Robinson RLM, Abell C, Blumberger J, Mitchell JB. Hierarchical virtual screening for the discovery of new molecular scaffolds in antibacterial hit identification. J Royal Soc Interface. 2012;9(77):3196–207.

Bellou S, Karali E, Bagli E, Al-Maharik N, Morbidelli L, Ziche M, Adlercreutz H, Murphy C, Fotsis T. The isoflavone metabolite 6-methoxyequol inhibits angiogenesis and suppresses tumor growth. Mol Cancer. 2012;11(35):1–11.

Bonacasa B, Siow RC, Mann GE. Impact of dietary soy isoflavones in pregnancy on fetal programming of endothelial function in offspring. Microcirculation. 2011;18(4):270–85.

Cai YF, Zhang HM, Niu WY, Zou YQ, Ma DF. Effects of equol on colon cancer cell proliferation. Beijing Da Xue Xue Bao. 2017;49(3):383–7.

Charalambous C, Pitta CA, Constantinou AI. Equol enhances tamoxifen's anti-tumor activity by induction of caspase-mediated apoptosis in MCF-7 breast cancer cells. BMC Cancer. 2013;13:238.

Cheong SH, Furuhashi K, Ito K, Nagaoka M, Yonezawa T, Miura Y, Yagasaki K. Antihyperglycemic effect of equol, a daidzein derivative, in cultured L6 myocytes and ob/ob mice. Mol Nutr Food Res. 2014;58(2):267–77.

Cho KW, Lee OH, Banz WJ, Moustaid-Moussa N, Shay NF, Kim YC. Daidzein and the daidzein metabolite, equol, enhance adipocyte differentiation and PPARgamma transcriptional activity. J Nutr Biochem. 2010;21(9):841–7.

Choi EJ. Evaluation of equol function on anti- or prooxidant status in vivo. J Food Sci. 2009;74(2): H65–71.

Clubbs EA, Bomser JA. Glycitein activates extracellular signal-regulated kinase via vascular endothelial growth factor receptor signaling in nontumorigenic (RWPE-1) prostate epithelial cells. J Nutr Biochem. 2007;18(8):525–32.

Dagdemir A, Durif J, Ngollo M, Bignon YJ, Bernard-Gallon D. Histone lysine trimethylation or acetylation can be modulated by phytoestrogen, estrogen or anti-HDAC in breast cancer cell lines. Epigenomics. 2013;5(1):51–63.

Davinelli S, Scapagnini G, Marzatico F, Nobile V, Ferrara N, Corbi G. Influence of equol and resveratrol supplementation on health related quality of life in menopausal women: a randomized, placebo-controlled study. Maturitas. 2017;96:77–83.

de la Parra C, Borrero-Garcia LD, Cruz-Collazo A, Schneider RJ, Dharmawardhane S. Equol, an isoflavone metabolite, regulates cancer cell viability and protein synthesis initiation via c-Myc and eIF4G. J Biol Chem. 2015;290(10):6047–57.

Friesner RA, Banks JL, Murphy RB, Halgren TA, Klicic JJ, Mainz DT, Repasky MP, Knoll EH, Shelley M, Perry JK. Glide: a new approach for rapid, accurate docking and scoring. 1. Method and assessment of docking accuracy. J Med Chem. 2004;47(7):1739–49.

Friesner RA, Murphy RB, Repasky MP, Frye LL, Greenwood JR, Halgren TA, Sanschagrin PC, Mainz DT. Extra precision glide: docking and scoring incorporating a model of hydrophobic enclosure for protein-ligand complexes. J Med Chem. 2006;49(21):6177–96.

Gil-Izquierdo A, Penalvo JL, Gil JI, Medina S, Horcajada MN, Lafay S, Silberberg M, Llorach R, Zafrilla P, Garcia-Mora P, Ferreres F. Soy isoflavones and cardiovascular disease epidemiological, clinical and -omics perspectives. Curr Pharm Biotechnol. 2012;13(5):624–31.

Gobert CP, Pipe EA, Capes SE, Darlington GA, Lampe JW, Duncan AM. Soya protein does not affect glycaemic control in adults with type 2 diabetes. Br J Nutr. 2010;103(3):412–21.

Guadamuro L, Dohrmann AB, Tebbe CC, Mayo B, Delgado S. Bacterial communities and metabolic activity of faecal cultures from equol producer and non-producer menopausal women under treatment with soy isoflavones. BMC Microbiol. 2017;17(1):93.

Halgren TA, Murphy RB, Friesner RA, Beard HS, Frye LL, Pollard WT, Banks JL. Glide: a new approach for rapid, accurate docking and scoring. 2. Enrichment factors in database screening. J Med Chem. 2004;47(7):1750–9.

Hazim S, Curtis PJ, Schar MY, Ostertag LM, Kay CD, Minihane AM, Cassidy A. Acute benefits of the microbial-derived isoflavone metabolite equol on arterial stiffness in men prospectively recruited according to equol producer phenotype: a double-blind randomized controlled trial. Am J Clin Nutr. 2016;103(3):694–702.

Hedelin M, Lof M, Sandin S, Adami HO, Weiderpass E. Prospective study of dietary phytoestrogen intake and the risk of colorectal cancer. Nutr Cancer. 2016;68(3):388–95.

Hoeger B, Diether M, Ballester PJ, Köhn M. Biochemical evaluation of virtual screening methods reveals a cell-active inhibitor of the cancer-promoting phosphatases of regenerating liver. Eur J Med Chem. 2014;88:89–100.

Horiuchi H, Usami A, Shirai R, Harada N, Ikushiro S, Sakaki T, Nakano Y, Inui H, Yamaji R. S-Equol activates cAMP signaling at the plasma membrane of INS-1 pancreatic beta-cells and protects against streptozotocin-induced hyperglycemia by increasing beta-cell function in male mice. J Nutr. 2017;147(9):1631–9.

Itsumi M, Shiota M, Takeuchi A, Kashiwagi E, Inokuchi J, Tatsugami K, Kajioka S, Uchiumi T, Naito S, Eto M. Equol inhibits prostate cancer growth through degradation of androgen receptor by S-phase kinase-associated protein 2. Cancer Sci. 2016;107(7):1022–8.

Jarrahpour A, Fathi J, Mimouni M, Hadda TB, Sheikh J, Chohan Z, Parvez A. Petra, Osiris and Molinspiration (POM) together as a successful support in drug design: antibacterial activity and biopharmaceutical characterization of some azo Schiff bases. Med Chem Res. 2012;21 (8):1984–90.

Jiang Y, Gong P, Madak-Erdogan Z, Martin T, Jeyakumar M, Carlson K, Khan I, Smillie TJ, Chittiboyina AG, Rotte SC, Helferich WG, Katzenellenbogen JA, Katzenellenbogen BS. Mechanisms enforcing the estrogen receptor beta selectivity of botanical estrogens. FASEB J. 2013;27(11):4406–18.

Jorgensen WL, Duffy EM. Prediction of drug solubility from structure. Adv Drug Deliv Rev. 2002;54(3):355–66.

Jorgensen WL, Tirado-Rives J. The OPLS [optimized potentials for liquid simulations] potential functions for proteins, energy minimizations for crystals of cyclic peptides and crambin. J Am Chem Soc. 1988;110(6):1657–66.

Jorgensen WL, Maxwell DS, Tirado-Rives J. Development and testing of the OPLS all-atom force field on conformational energetics and properties of organic liquids. J Am Chem Soc. 1996;118 (45):11225–36.

Kamiyama M, Kishimoto Y, Tani M, Utsunomiya K, Kondo K. Effects of equol on oxidized low-density lipoprotein-induced apoptosis in endothelial cells. J Atheroscler Thromb. 2009;16 (3):239–49.

Kartasasmita R, Musfiroh I, Muhtadi A, Ibrahim S. Binding affinity of asiatic acid derivatives design against inducible nitric oxide synthase and ADMET prediction. J Appl Pharm Sci. 2014;4(02):75–80.

Kladna A, Berczynski P, Kruk I, Piechowska T, Aboul-Enein HY. Studies on the antioxidant properties of some phytoestrogens. Luminescence. 2016;31(6):1201–6.

Kumar S, Pandey AK. Chemistry and biological activities of flavonoids: an overview. Sci World J. 2013;2013:1–16.

Kunigal S, Lakka SS, Gondi CS, Estes N, Rao JS. RNAi-mediated downregulation of urokinase plasminogen activator receptor and matrix metalloprotease-9 in human breast cancer cells results in decreased tumor invasion, angiogenesis and growth. Int J Cancer. 2007;121 (10):2307–16.

Landete JM, Arques J, Medina M, Gaya P, de Las Rivas B, Munoz R. Bioactivation of phytoestrogens: intestinal bacteria and health. Crit Rev Food Sci Nutr. 2016;56(11):1826–43.

Li H, Leung K-S, Wong M-H, Ballester PJ. USR-VS: a web server for large-scale prospective virtual screening using ultrafast shape recognition techniques. Nucleic Acids Res. 2016;44 (W1):W436–41.

Liang XL, Li M, Li J, Wang XL. Equol induces apoptosis in human hepatocellular carcinoma SMMC-7721 cells through the intrinsic pathway and the endoplasmic reticulum stress pathway. Anti-Cancer Drugs. 2014;25(6):633–40.

Liu H, Hu C, Wu X, Li Z. Equol elicits estrogenic activities via PI3K/akt pathway in the estrogen receptor-positive MCF-7 cells. Mol Cell Toxicol. 2014;10(3):285–91.

Liu ZM, Ho SC, Chen YM, Xie YJ, Huang ZG, Ling WH. Research protocol: effect of natural S-equol on blood pressure and vascular function – a six-month randomized controlled trial among equol non-producers of postmenopausal women with prehypertension or untreated stage 1 hypertension. BMC Complement Altern Med. 2016;16:89.

Liu J, Viswanadhapalli S, Garcia L, Zhou M, Nair BC, Kost E, Tekmal R, Li R, Rao MK, Curiel T, Vadlamudi RK, Sareddy GR. Therapeutic utility of natural estrogen receptor beta agonists on ovarian cancer. Oncotarget. 2017;8(30):50002–14.

Lu JJ, Crimin K, Goodwin JT, Crivori P, Orrenius C, Xing L, Tandler PJ, Vidmar TJ, Amore BM, Wilson AG. Influence of molecular flexibility and polar surface area metrics on oral bioavailability in the rat. J Med Chem. 2004;47(24):6104–7.

Lu Z, Zhou R, Kong Y, Wang J, Xia W, Guo J, Liu J, Sun H, Liu K, Yang J, Mi M, Xu H. S-equol, a secondary metabolite of natural anticancer isoflavone daidzein, inhibits prostate cancer growth in vitro and in vivo, though activating the Akt/FOXO3a pathway. Curr Cancer Drug Targets. 2016;16(5):455–65.

Magee PJ. Is equol production beneficial to health? Proc Nutr Soc. 2011;70(1):10–8.

Magee PJ, Allsopp P, Samaletdin A, Rowland IR. Daidzein, R-(+) equol and S-(−) equol inhibit the invasion of MDA-MB-231 breast cancer cells potentially via the down-regulation of matrix metalloproteinase-2. Eur J Nutr. 2014;53(1):345–50.

Mahmoud AM, Yang W, Bosland MC. Soy isoflavones and prostate cancer: a review of molecular mechanisms. J Steroid Biochem Mol Biol. 2014;140:116–32.

Mann GE, Bonacasa B, Ishii T, Siow RC. Targeting the redox sensitive Nrf2-Keap1 defense pathway in cardiovascular disease: protection afforded by dietary isoflavones. Curr Opin Pharmacol. 2009;9(2):139–45.

Miller LM, Lampe JW, Newton KM, Gundersen G, Fuller S, Reed SD, Frankenfeld CL. Being overweight or obese is associated with harboring a gut microbial community not capable of metabolizing the soy isoflavone daidzein to O-desmethylangolensin in peri- and postmenopausal women. Maturitas. 2017;99:37–42.

Minakawa M, Kawano A, Miura Y, Yagasaki K. Hypoglycemic effect of resveratrol in type 2 diabetic model db/db mice and its actions in cultured L6 myotubes and RIN 5F pancreatic β-cells. J Clin Biochem Nutr. 2011;48(3):237–44.
Miyanaga N, Akaza H. Prostate cancer prevention. Gan to kagaku ryoho Cancer Chemother. 2015;42(5):538–43.
Muster W, Breidenbach A, Fischer H, Kirchner S, Müller L, Pähler A. Computational toxicology in drug development. Drug Discov Today. 2008;13(7):303–10.
Obiorah IE, Fan P, Jordan VC. Breast cancer cell apoptosis with phytoestrogens is dependent on an estrogen-deprived state. Cancer Prev Res. 2014;7(9):939–49.
Ono M, Ejima K, Higuchi T, Takeshima M, Wakimoto R, Nakano S. Equol enhances apoptosis-inducing activity of genistein by increasing Bax/Bcl-xL expression ratio in MCF-7 human breast cancer cells. Nutr Cancer. 2017;69(8):1300–7.
Patil SP, Ballester PJ, Kerezsi CR. Prospective virtual screening for novel p53–MDM2 inhibitors using ultrafast shape recognition. J Comput Aided Mol Des. 2014;28(2):89–97.
Patisaul HB. Endocrine disruption by dietary phyto-oestrogens: impact on dimorphic sexual systems and behaviours. Proc Nutr Soc. 2017;76(2):130–44.
Pawlowski JW, Martin BR, McCabe GP, McCabe L, Jackson GS, Peacock M, Barnes S, Weaver CM. Impact of equol-producing capacity and soy-isoflavone profiles of supplements on bone calcium retention in postmenopausal women: a randomized crossover trial. Am J Clin Nutr. 2015;102(3):695–703.
Pazourekova S, Lucova M, Nosal R, Drabikova K, Harmatha J, Smidrkal J, Jancinova V. Equol effectively inhibits toxic activity of human neutrophils without influencing their viability. Pharmacology. 2016;97(3–4):138–45.
Reverri EJ, LaSalle CD, Franke AA, Steinberg FM. Soy provides modest benefits on endothelial function without affecting inflammatory biomarkers in adults at cardiometabolic risk. Mol Nutr Food Res. 2015;59(2):323–33.
Richter CK, Skulas-Ray AC, Fleming JA, Link CJ, Mukherjea R, Krul ES, Kris-Etherton PM. Effects of isoflavone-containing soya protein on ex vivo cholesterol efflux, vascular function and blood markers of CVD risk in adults with moderately elevated blood pressure: a dose-response randomised controlled trial. Br J Nutr. 2017;117(10):1403–13.
Rigalli JP, Scholz PN, Tocchetti GN, Ruiz ML, Weiss J. The phytoestrogens daidzein and equol inhibit the drug transporter BCRP/ABCG2 in breast cancer cells: potential chemosensitizing effect. Eur J Nutr. 2017;58(1):39–150.
Sakane N, Kotani K, Tsuzaki K, Takahashi K, Usui T, Uchiyama S, Fujiwara S. Equol producers can have low leptin levels among prediabetic and diabetic females. Ann Endocrinol. 2014;75 (1):25–8.
Sareddy GR, Vadlamudi RK. Cancer therapy using natural ligands that target estrogen receptor beta. Chin J Nat Med. 2015;13(11):801–7.
Schreyer AM, Blundell T. USRCAT: real-time ultrafast shape recognition with pharmacophoric constraints. J Cheminform. 2012;4(1):27.
Setchell KD, Clerici C. Equol: history, chemistry, and formation. J Nutr. 2010;140(7):1355s–62s.
Sharma M, Lawson J, Karunanayake C, Dosman J, Punam P. Prostate cancer, farming and other risk factors: a mini review. J Pros Canc. 2016;1(109):2.
Shivakumar D, Williams J, Wu Y, Damm W, Shelley J, Sherman W. Prediction of absolute solvation free energies using molecular dynamics free energy perturbation and the OPLS force field. J Chem Theory Comput. 2010;6(5):1509–19.
Singh P, Bast F. In silico molecular docking study of natural compounds on wild and mutated epidermal growth factor receptor. Med Chem Res. 2014a;23(12):5074–85.
Singh P, Bast F. Multitargeted molecular docking study of plant-derived natural products on phosphoinositide-3 kinase pathway components. Med Chem Res. 2014b;23(4):1690–700.
Singh P, Bast F. Screening and biological evaluation of myricetin as a multiple target inhibitor insulin, epidermal growth factor, and androgen receptor; in silico and in vitro. Investig New Drugs. 2015a;33(3):575–93.

Singh P, Bast F. Screening of multi-targeted natural compounds for receptor tyrosine kinases inhibitors and biological evaluation on cancer cell lines, in silico and in vitro. Med Oncol. 2015b;32(9):1–18.

Singh P, Singh RS, Rani A, Bast F. Homology modeling of chemokine CCR7, molecular docking, and in vitro studies evidenced plausible immunotherapeutic anticancer natural compounds. Med Chem Res. 2016;25(10):2410–24.

Son MJ, Minakawa M, Miura Y, Yagasaki K. Aspalathin improves hyperglycemia and glucose intolerance in obese diabetic ob/ob mice. Eur J Nutr. 2013;52(6):1607–19.

Subedi L, Ji E, Shin D, Jin J, Yeo JH, Kim SY. Equol, a dietary daidzein gut metabolite attenuates microglial activation and potentiates neuroprotection in vitro. Nutrients. 2017;9(3):207.

Sugiyama Y, Masumori N, Fukuta F, Yoneta A, Hida T, Yamashita T, Minatoya M, Nagata Y, Mori M, Tsuji H, Akaza H, Tsukamoto T. Influence of isoflavone intake and equol-producing intestinal flora on prostate cancer risk. Asian Pac J Cancer Prev. 2013;14(1):1–4.

Taghizadeh B, Ghavami L, Nikoofar A, Goliaei B. Equol as a potent radiosensitizer in estrogen receptor-positive and -negative human breast cancer cell lines. Breast Cancer. 2015;22 (4):382–90.

Tanaka M, Fujimoto K, Chihara Y, Torimoto K, Yoneda T, Tanaka N, Hirayama A, Miyanaga N, Akaza H, Hirao Y. Isoflavone supplements stimulated the production of serum equol and decreased the serum dihydrotestosterone levels in healthy male volunteers. Prostate Cancer Prostatic Dis. 2009;12(3):247.

Tseng M, Byrne C, Kurzer MS, Fang CY. Equol-producing status, isoflavone intake, and breast density in a sample of U.S. Chinese women. Cancer Epidemiol Biomarkers Prev: Publ Am Assoc Cancer Res. 2013;22(11):1975–83.

Uehara M. Isoflavone metabolism and bone-sparing effects of daidzein-metabolites. J Clin Biochem Nutr. 2013;52(3):193–201.

Vafeiadou K, Hall WL, Williams CM. Does genotype and equol-production status affect response to isoflavones? Data from a pan-European study on the effects of isoflavones on cardiovascular risk markers in post-menopausal women. Proc Nutr Soc. 2006;65(1):106–15.

Vazquez L, Guadamuro L, Giganto F, Mayo B, Florez AB. Development and use of a real-time quantitative PCR method for detecting and quantifying equol-producing bacteria in human faecal samples and slurry cultures. Front Microbiol. 2017;8:1155.

Vedavanam K, Srijayanta S, O'Reilly J, Raman A, Wiseman H. Antioxidant action and potential antidiabetic properties of an isoflavonoid-containing soyabean phytochemical extract (SPE). Phytother Res. 1999;13(7):601–8.

Vimal A, Pal D, Tripathi T, Kumar A. Eucalyptol, sabinene and cinnamaldehyde: potent inhibitors of salmonella target protein L-asparaginase. 3 Biotech. 2017;7(4):258.

Virk-Baker MK, Barnes S, Krontiras H, Nagy TR. S-(−) equol producing status not associated with breast cancer risk among low isoflavone-consuming US postmenopausal women undergoing a physician-recommended breast biopsy. Nutr Res. 2014;34(2):116–25.

Wang M, Ren G. Study on the inhibiting effects of equol on MCF-7 cells proliferation and its molecular mechanisms. J Hyg Res. 2014;43(1):11–5.

Yang X, Belosay A, Hartman JA, Song H, Zhang Y, Wang W, Doerge DR, Helferich WG. Dietary soy isoflavones increase metastasis to lungs in an experimental model of breast cancer with bone micro-tumors. Clin Exp Metastasis. 2015a;32(4):323–33.

Yang ZP, Zhao Y, Huang F, Chen J, Yao YH, Li J, Wu XN. Equol inhibits proliferation of human gastric carcinoma cells via modulating Akt pathway. World J Gastroenterol. 2015b;21 (36):10385–99.

Yang Z, Zhao Y, Yao Y, Li J, Wang W, Wu X. Equol induces mitochondria-dependent apoptosis in human gastric cancer cells via the sustained activation of ERK1/2 pathway. Mol Cell. 2016;39 (10):742–9.

Yoshikata R, Myint KZ, Ohta H. Relationship between equol producer status and metabolic parameters in 743 Japanese women: equol producer status is associated with antiatherosclerotic

conditions in women around menopause and early postmenopause. Menopause. 2017;24 (2):216–24.
Zhang T, Liang X, Shi L, Wang L, Chen J, Kang C, Zhu J, Mi M. Estrogen receptor and PI3K/Akt signaling pathway involvement in S-(−) equol-induced activation of Nrf2/ARE in endothelial cells. PLoS One. 2013;8(11):e79075.
Zhang Q, Feng H, Qluwakemi B, Wang J, Yao S, Cheng G, Xu H, Qiu H, Zhu L, Yuan M. Phytoestrogens and risk of prostate cancer: an updated meta-analysis of epidemiologic studies. Int J Food Sci Nutr. 2017;68(1):28–42.
Zheng W, Zhang Y, Ma D, Shi Y, Liu C, Wang P. (+/−) Equol inhibits invasion in prostate cancer DU145 cells possibly via down-regulation of matrix metalloproteinase-9, matrix metalloproteinase-2 and urokinase-type plasminogen activator by antioxidant activity. J Clin Biochem Nutr. 2012;51(1):61–7.

Chapter 10
Anticancer and Neuroprotective Activity of Chrysin: Recent Advancement

Pushpendra Singh, Ravi S. Singh, Prem P. Kushwaha, and Shashank Kumar

10.1 Introduction

Chrysin (5, 7-Dihydroxyflavone) is a flavone (flavonoid) found in the *Passiflora caerulea*, *Passiflora incarnata* (blue passion flower), and in *Oroxylum indicum*. In addition, it also found in honey, honeycomb, chamomile, and in the mushroom (Pleurotus ostreatus). Chrysin is used for body building, and has reported that it boosts the testosterone levels (Ciftci et al. 2012; Kumar and Pandey 2013). Several studies have shown, chrysin possesses diverse biological properties such as antioxidant, anti-inflammatory, anti-apoptotic and anti-cancer (Kasala et al. 2015; Mani and Natesan 2018; Mantawy et al. 2017). More recent evidence from Lim et al. shows chrysin attenuated the human ovarian cancer by the activation of PI3K and MAPK pathways (Lim et al. 2018). Doxorubicin (DOX) is the commonly used for treatment of various human neoplasms. Chrysin supplements neutralises the DOX-induced oxidative stress by decreasing lipid peroxidation, and up-regulating the antioxidant enzymes. Furthermore chrysin also attenuated DOX-induced apoptosis by decreasing expression of p53, Bax, Puma, and increasing expression of Bcl-2 (Mantawy et al. 2017). Elegant study of Mantawy et al. also showed Chrysin attenuated the DOX-induced cardiomyopathy by p53 dependent apoptotic, MAPK and NF-kB pathways while augmenting the VEGF/AKT pathway (Mantawy et al. 2017) (Fig. 10.1). Further, In china chrysin is the most valuable Chinese medicine has been used to induce apoptosis in leukemic cells (Owen et al. 2017). More

P. Singh
Tumor Biology Laboratory, National Institute of Pathology, New Delhi, India

R. S. Singh
Department of Biochemistry, Microbiology and Immunology, University of Saskatchewan, Saskatoon, SK, Canada

P. P. Kushwaha · S. Kumar (✉)
Department of Biochemistry and Microbial Sciences, Central University of Punjab, Bathinda, Punjab, India

S. Kumar, C. Egbuna (eds.), *Phytochemistry: An in-silico and in-vitro Update*,
https://doi.org/10.1007/978-981-13-6920-9_10

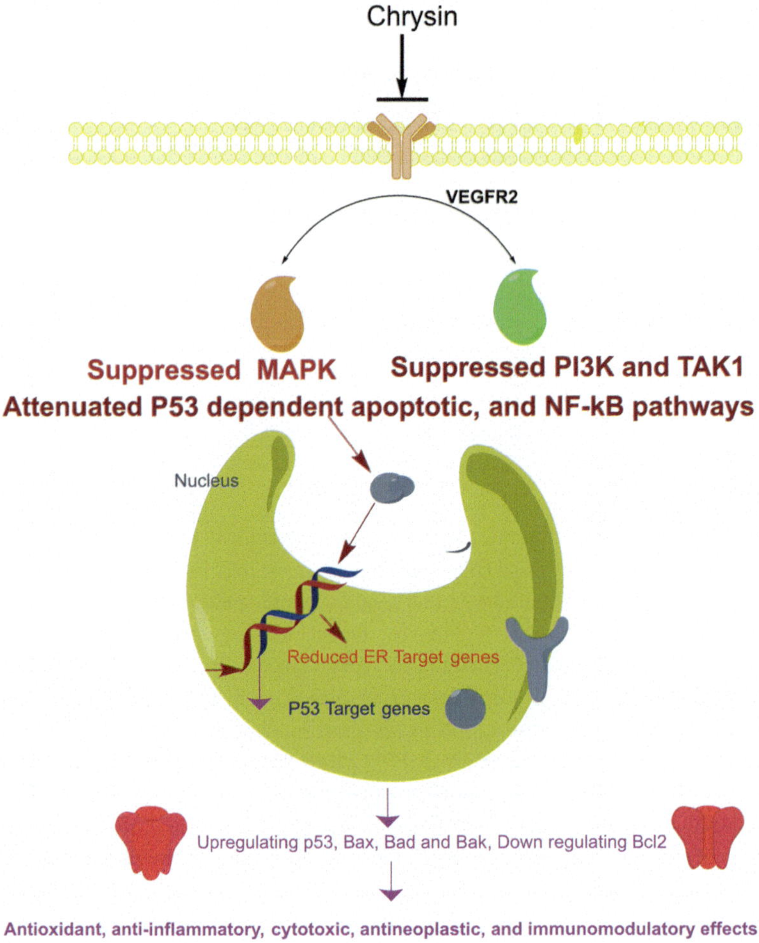

Fig. 10.1 Anticancer activity of Chrysin through the destruction of MAPK, PI3K and nuclear factor-κB signaling

recently evidence suggest that chrysin target Wnt/β-catenin, Notch, Hedgehog and EGFR pathways to eliminates leukemic stem cells (LSCs) and cancer stem cells, (Manupati et al. 2017; Owen et al. 2017). In addition, study has confirmed that Chrysin treatment induces the expression of proapoptotic proteins, including p53, Bax, Bad and Bak, while it decreases the level of antiapoptotic protein Bcl-2 in hepatocellular carcinoma (Zhang et al. 2016). However, the effectiveness of chrysin in cancer treatment depends on its bioavailability and solubility. Poor intestinal absorption of Chrysin, limit its effectiveness. Chrysin has a number of pharmacological activities, such as:

1. Antiasthmatic activity through the destruction of inducible nitric oxide synthase (iNOS) and nuclear factor-κB, (Wadibhasme et al. 2011), inhibition of histone

deacetylase (Szkudelski 2001), DNA topoisomerases (Russo et al. 2012), Cardioprotective activity via advancing post-ischemic functional recovery (Nabavi et al. 2015)
2. Anti-inflammatory activity via blocking histamine release and pro-inflammatory cytokine expression (Rehman et al. 2013)
3. Prevention of osteoporosis by activation of ER/MAPK (Zeng et al. 2013)
4. Anticancer activity by endorsing the cell death induced by TNF, TRAIL and increasing TRAIL-induced degradation of caspases-3 and caspases-8 (Anandhi et al. 2014; Li et al. 2015).
5. Suppressive effect on vascular endothelial growth factor-induced angiogenesis (Lin et al. 2006, 2010)
6. Antihypercholesterolemic activity (Balta et al. 2015)
7. Antidiabetogenic (Ramírez-Espinosa et al. 2017)
8. Antihypertensive (Balasuriya and Rupasinghe 2012; Villar et al. 2002)
9. Blocking metastatic progression in breast cancer cells (Lirdprapamongkol et al. 2013)
10. Neuroprotective (Gresa-Arribas et al. 2010; Nabavi et al. 2015; Zhang et al. 2015) (Fig. 10.2).

Biologically active chrysin has 15-carbon basic polyphenolic skeleton. Chrysin occupies 2 benzene rings (A, B) that are connected with third, oxygen-containing (C) ring. Moreover, Chrysin possesses a C–C double bond (C2–C3), and carbonyl group (C4). Based on these buildings blocks, Chrysin classified as flavones. Chrysin also holds hydroxyl groups at C-5 and C-7 (Fig. 10.3). Further, structure-activity relationship study, shown that biological activity of Chrysin (anti-inflammatory and anti-oxidant activity) is due to B and C-ring. The anti-oxidant activity of chrysin involved 3′, 4′ hydroxylation, a double bond between C-2 and C-3, and existence of a carbonyl group on C-4 (Mattison et al. 2017; Nabavi et al. 2015; Samarghandian et al. 2017; Wang et al. 2017a, b). Chrysin targets various pathways as shown in Table 10.1.

Xu et al. (2017) found that chrysin has antitumor activity against hepatocellular carcinoma (HCC) via inhibiting tumor glycolysis and apoptosis. Further, Chrysin has anti-metastatic activities on metastatic triple-negative breast cancer (TNBC) cells, suggests that Chrysin might be a potential therapeutic candidate to treat advanced or metastatic breast cancer (Grudzien 2010).

10.2 Combination Therapy and Nanoparticle Study of Chrysin

Chrysin has antioxidant, and anticarcinogenic activity (Chassagne et al. 2018). Moreover, it also controls normal testicular morphology and enhances the quantity of expanding ovarian follicles (Campos et al. 2018). Chrysin treatment reverses depressive-like behaviours in hypothyroid female mice by modulating the levels of

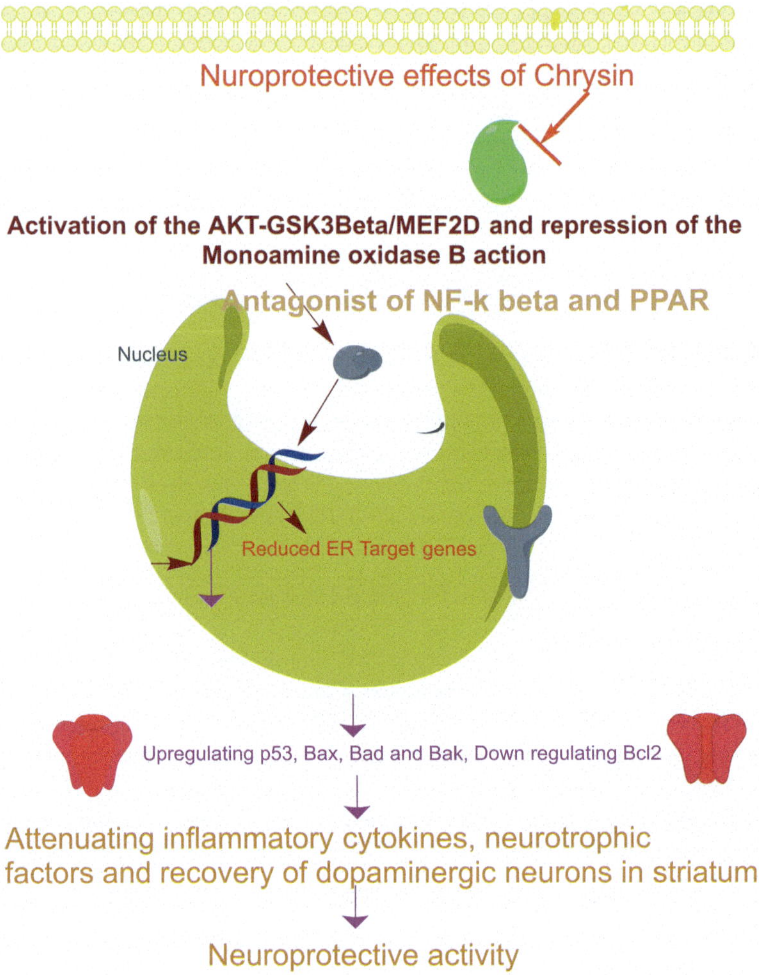

Fig. 10.2 Nuroprotective activity of Chrysin through the destruction of Akt, PPAR and nuclear factor-κB signaling

5-HT and dopamine (Bortolotto et al. 2018). Further, the combination of natural products along with currently approved cancer therapies added an additional advantage and they considered as promising targets for treatment of cancer (Rajagopal et al. 2018). Nano-encapsulation of curcumin and chrysin enhanced delivery of these compounds to cancer cells and hence it can be assumed that PLGA-PEG nanoparticles improve anticancer effects of curcumin-chrysin by enhancing bioavailability and the solubility of these drugs (Bagheri et al. 2018).

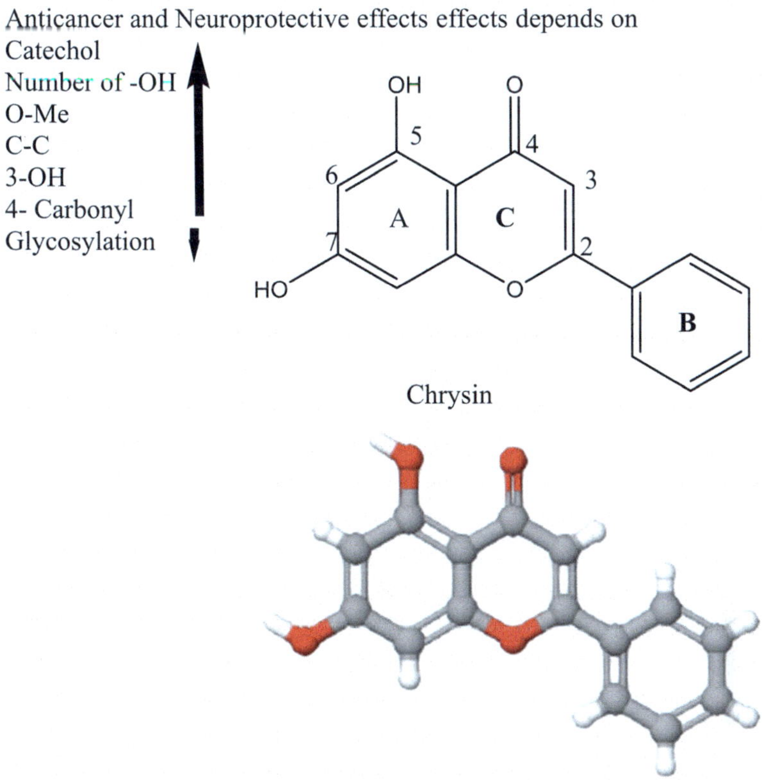

Fig. 10.3 Structure-activity relationship study of Chrysin. Biologically active chrysin has 15-carbon basic polyphenolic skeleton. Chrysin occupies 2 benzene rings (**a** and **b**) that are connected with third, oxygen-containing (**c**) ring. Moreover, Chrysin possesses a C–C double bond (C2–C3), and carbonyl group (C4). Based on these buildings blocks, Chrysin classified as flavones

10.3 Neurodegenerative Diseases

Neurodegenerative diseases are most common among older people. World Health Organization (WHO) estimated the portion of older people will increase from 11% to 22% during next 40 years throughout the world (Forman et al. 2004). Many studies have demonstrated that phytochemicals possess neuroprotective effects and also mitigate neurodegeneration (Nabavi et al. 2015). Recently it has demonstrated that chrysin has neuroprotective activity in parkinsonian mouse model, via attenuating inflammatory cytokines, neurotrophic factors and recovery of dopaminergic neurons in striatum (Goes et al. 2018) (Table 10.2). In addition, chrysin possess a potent anti-amyloidogenic and neurotrophic effects (Nabavi et al. 2015). Further, Chrysin exerts beneficial effect on PD, via multiple pathways including activation of the AKT-GSK3β/MEF2D and repression of the MAO-B action (Guo et al. 2016).

Table 10.1 Effect of chrysin in different cancer

Cancer	Targets	References
Ovarian cancer	PI3K/MAPK	Kumar and Pandey (2013)
Ovarian cancer	Antioxidant and anticancer	Patel et al. (2016)
Cervical cancer	Antioxidant and anticancer	Laishram et al. (2015)
Cervical cancer and ovarian cancer	Antioxidant and anticancer agents	Mistry et al. (2015)
Malignant glioma cell	Antiproliferative and apoptotic activity	Jia et al. (2017)
Colon adenocarcinoma model	Cell cycle arrest in G2/M	Leon et al. (2015)
Male Sprague-Dawley rats	Akt/MAPK	Mantawy et al. (2017)
Lung adenocarcinoma	Antiproliferative activity	Fonseca et al. (2017)
Leukemic stem cells	Overcome chemoresistance	Owen et al. (2017)
Gastric cancer	Alters microRNAs expression	Mohammadian et al. (2017)
Gastric cancer	Blocked AP-1	Xia et al. (2015a, b)
Gastric cancer	Suppressing early growth response-1	Xia et al. (2015a, b)
Anticancer assay	Reduce NRF2 expression	Stepanic et al. (2015)
Breast cancer	Down-regulation of cyclin D1 and hTERT	Maasomi et al. (2017)
Breast cancer	Cell cycle to be arrested in the G0/G1 phase	Wang et al. (2017a, b)
Breast cancer	EGFR inhibitor	Manupati et al. (2017)
Breast cancer	Antiproliferative effect	Samarghandian et al. (2016)
Breast cancer	Inhibitors of DNA methyltransferases	Kanwal et al. (2016)
Breast cancer	Induced apoptosis	Xuan et al. (2016)
Breast cancer	Induce apoptosis	Yang et al. (2013)
Breast and liver cancer	Induced apoptosis	Czyzewska et al. (2016)
Chr-loaded nanofibers	Reduced overexpression of IL-6, IL-1β,	Deldar et al. (2018)
Glioblastoma cells	Inhibiting export of arsenic	Gulden et al. (2017)
Breast cancer stem cell	Epidermal growth factor receptor	Manupati et al. (2017)
Hepatocellular carcinoma	Hexokinase-2	Xu et al. (2017)
CLL B-lymphocytes	Inhibiting mitochondrial complex II and V	Salimi et al. (2017)
Prostate cancer	Inducing ROS and ER stress	Ryu et al. (2017)
Mouse 4T1 tumor model	Metalloproteinase gene expression	Mohammadi et al. (2017)
Tongue squamous cell carcinoma		Czyzewska et al. (2016)
Hepatocellular carcinoma	Suppressing PI3K/Akt/Nrf2 and ERK/Nrf2	Gao et al. (2013)
Skin cancer	Inhibiting cyclin-dependent kinases	Liu et al. (2013)
Neuronal stem cells	Inhibit indoleamine 3,5-dioxygenase-1	Chen et al. (2012)
Breast cancer	Histone deacetylase 8 inhibitor	Sun et al. (2012)
In-silico	Inhibitors of carbonyl reductase 1	Arai et al. (2015)

Table 10.2 Neurodegenerative diseases and chrysin

	Targets	Model	References
Parkinson's disease	Inflammatory cytokines, neurotrophic factors and recovery of dopaminergic neurons in striatum	Mouse model	Goes et al. (2018)
Parkinson's disease	Increased nuclear factor-erythroid 2-related factor 2 protein expression and transcriptional activity	Zebrafish and mice	Zhang et al. (2015)
Parkinson's disease	Anti-oxidant therapies	Primary rat	Mercer et al. (2005)
Parkinson's disease	Chrysin rescued the dopaminergic neurons loss and alleviated the decrease in dopamine level	In-vivo model	Guo et al. (2016)
Multiple sclerosis	Decrease possible side effects and to develop new medicines for MS	Systematic review	Sanadgol et al. (2017)
Huntington's disease	Improving the antioxidant status	Male wistar rats	Thangarajan et al. (2016)
Neuroprotective	Potent anti-amyloidogenic and neurotrophic effects	Review	Nabavi et al. (2015)
Neurodegeneration,	Estrogen receptor antagonists	MCF-7	Resende et al. (2013)
Chronic cerebral hypoperfusion	Decreased lipid peroxide, reduced the increased activities of superoxide dismutase	Rat 2VO	He et al. (2012)
Neuroinflammation	Decrease CCAAT/enhancer binding proteins (C/EBPs) β and δ transcription factor	Microglial cells	Gresa-Arribas et al. (2010)
Neuroprotective	Anti-oxidative and anti-inflammatory	Review	Zeinali et al. (2017)
Alzheimer's disease	Anticholinesterase activity	In-silico	Ali et al. (2017)
Alzheimer's disease	Lipid peroxidation and acetylcholine esterase were increased significantly	Rat	Vedagiri and Thangarajan (2016)

It's have been also shown that chrysin has been used as antagonist of NF-kB and of peroxisome proliferator-activated receptor gamma (PPAR-γ) (Yao et al. 2017). Also, chrysin prevents apoptosis by upregulating the Bcl-2 and downregulating the pro-apoptotic (Bax, Bad) (Sanadgol et al. 2017). The control of neuroinflammation is a potential target to be considered in the treatment of neurodegenerative disease.

It is therefore important to find anti-inflammatory drugs and study new targets that inhibit neuroinflammation. Chrysin pre-treatment protects neurons via inhibiting nitric oxide and tumor necrosis factor-α productions without affecting the cyclooxygenase-2 gene in lipopolysaccharide of *E. coli* and interferon-γ-treated mouse primary microglial cells. These effects were associated to a decrease in C/EBPδ protein level, mRNA expression, and DNA-binding activity, with no effect on C/EBPβ and p65 nuclear protein levels or DNA-binding activity, pointing out C/EBPδ as a possible mediator of Chrysin effects. Consequently, C/EBPδ is a possible target to act against neuroinflammation in neurodegenerative processes

(Resende et al. 2013) (Table 10.2). Chrysin significantly inhibited the release of nitric oxide (NO) and proinflammatory cytokines such as tumor necrosis factor-α (TNF-α) and interleukin-1β (IL-1β). The expressions of inducible NO synthase (iNOS) and cyclooxygenase-2 (COX-2) were also significantly inhibited by chrysin (He et al. 2012). The previous research has shown that pretreatment with chrysin appeared in (i) improved nuclear factor-erythroid 2-related factor 2 (ii) inflection of cellular redox status with the upregulated expression of antioxidant enzymes and (iii) reduced levels of malondialdehyde. A combination therapy involving protocatechuic acid and chrysin exhibits its enhanced neuroprotective outcomes through antioxidant cytoprotection and anti-inflammation (Gresa-Arribas et al. 2010). Chrysin prevented adhesion of monocytes to bEnd.3 cells after the treatment of to LPS. Nuclear factor-κ B (NF-κB), p38 mitogen-activated kinase (MAPK), and c-Jun N-terminal kinase, which are all stimulated by LPS, were significantly repressed by chrysin. The anti-inflammatory effects of chrysin recommend a plausible therapeutic application in neurodegenerative diseases, such as multiple sclerosis, septic encephalopathy, and allergic encephalomyelitis (Zeinali et al. 2017).

10.4 Future Perspective of Chrysin

Molecular docking was performed to study the ligand-protein interaction of most activated protein which are involved in cancer and neurodegenerative. Molecular docking protocol adapted from our earlier published literature (Singh and Bast 2014a, b, 2015a, b; Singh et al. 2016). The structure of Chrysin optimized using Ligprep module of the Maestro11.2 and optimized potential for liquid simulations (OPLS_2005) force field applied (Jorgensen et al. 1996; Jorgensen and Tirado-Rives 1988; Shivakumar et al. 2010).

We studied molecular docking of IR (PDB; 3ETA), IGF1R (PDB; 1K3A), EGFR (PDB; 1M17), VEGFR1 (PDB; 3HNG), TGFβ RI (PDB; 2X7O), TGFβ RII (PDB; 1KS6), TAK1 (PDB; 4L3P), CXCR2 (PDB; 4JL7), CXCR4 (PDB; 3OE6), CCR5 (PDB; 4MBS), AR (PDB; 2PIV), ER (PDB; 3ERT), γ-secretase (PDB; 4Y6K), VEGFR2 (PDB; 2OH4) STAT3 (PDB; 1BG1), PI3K (PDB; 3S2A), PDK1 (PDB; 3RWQ), Akt (PDB; 3MV5), mTOR (PDB; 4DRI), PARP1 (PDB; 1UK1), Smoothened 7TM (PDB; 7QIM), and Wnt (PDB; 1JDH) against chrysin. Chrysin docked with the X-ray crystal structure of selected protein by employing GLIDE Maestro 9.6. The X-ray structure of the chosen protein retrieved from the PDB. Maestro 9.6 protein preparation wizard application executed many corrections of raw PDB file (Singh and Bast 2014a, b, 2015a, b; Singh et al. 2016).

10.4.1 Receptor Grid Generation

For generating receptor-grid file of different PDB, Van der Waal radii of receptor atoms and partial atomic charge set as the default setting of Maestro and molecular docking performed (Singh and Bast 2014a, b, 2015a, b; Singh et al. 2016).

10.4.2 Glide (Grid-Based Ligand Docking with Energetics) Molecular Docking

Molecular docking was conducted using GLIDE software from Schrödinger suite (Friesner et al. 2004, 2006; Singh and Bast 2014a, b, 2015a, b; Singh et al. 2016).

10.4.3 Absorption, Distribution, Metabolism, Excretion, and Toxicity (ADME/T)

ADME properties of Chrysin were predicted using QikProp application of Maestro 11.2 (Jorgensen and Duffy 2002; Lu et al. 2004; Singh and Bast 2014a, b, 2015a, b; Singh et al. 2016).

10.5 Results and Discussion

10.5.1 Interaction of Chrysin with VEGFR2

Figure 10.4a, b depict the interaction of chrysin with VEGFR2. Lys866 was found to interact with VEGFR2 via face-to-face π-π stacking interaction. Additionally Val912 formed an H-bond. Several hydrophobic residues in the binding pocked such as Leu887, Ile913, Val914, Val914, Leu1033, Cys1043, and Phe1045 were stabilized the interaction with the Chrysin.

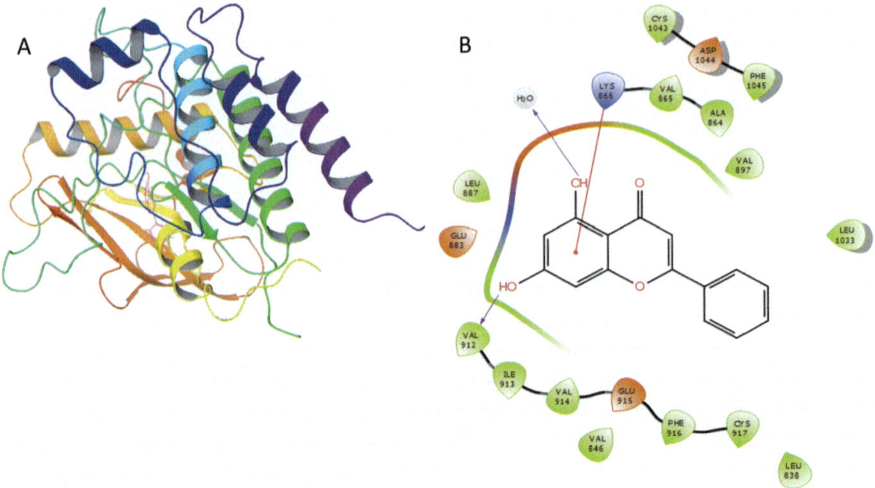

Fig. 10.4 (**a**) Ribbon presentation of VEGFR2 (PDB; 2OH4) protein molecule with Chrysin (**b**) Protein-ligand interactions profile of VEGFR2 with Chrysin

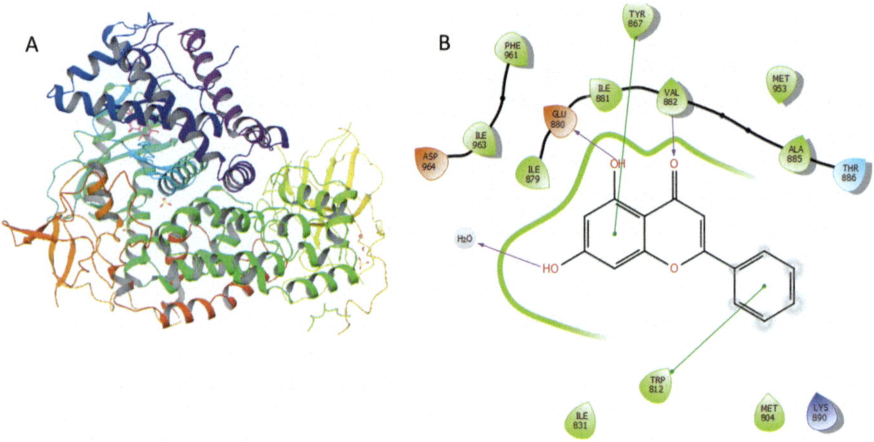

Fig. 10.5 (**a**) Ribbon presentation of PI3K protein molecule with Chrysin (**b**) Protein-ligand interactions profile of PI3K with Chrysin

10.5.2 Interaction of Chrysin with PI3K

The binding pose of Chrysin with PI3K is shown in Fig. 10.5a, b. Trp812 and Tyr867 were found to interact with PI3K via face-to-face π-π stacking interaction. Additionally, Glu880 and Val882 formed an H-bond with PI3K. Several hydrophobic residues in the binding pocket such as Met804, Ile831, Tyr867, Ala885, Ile879, Met 953, Phe961 and Ile963 were stabilized the interaction with the Chrysin.

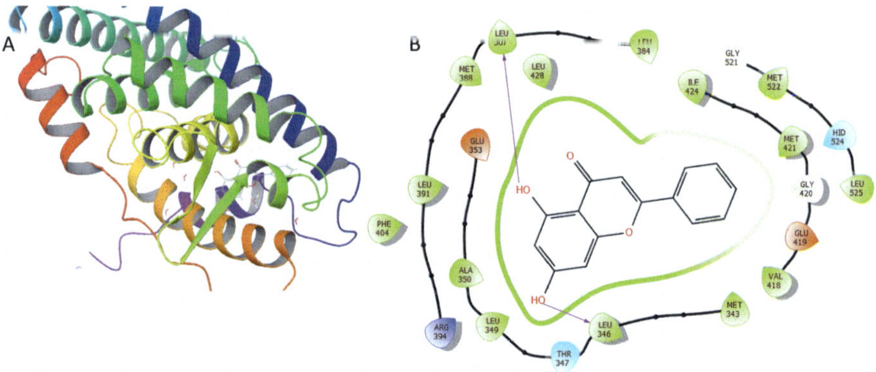

Fig. 10.6 (**a**) Ribbon presentation of ER protein molecule with Chrysin (**b**) Protein-ligand interactions profile of ER with Chrysin

10.5.3 Interaction of Chrysin with TAK1

Molecular docking result of TAK1 against chrysin has better Gscore −8.44 Kcal/mol. Moreover, protein-ligand interactions showed that amino acids Val42, Ala 46, Phe47, Ala61, Leu81, Val90, Met104, Glu105, Tyr106, Ala107, Ala107, Cys174, and Phe176 involved in hydrophobic interaction and hydrogen bonding.

10.5.4 Interaction of Chrysin with ER

Gscore for chrysin when docked with the ER, is −8.05 Kcal/mol (Fig. 10.6a, b). Furthermore, protein-ligand interactions of ER with Chrysin stated that amino acid Leu701, Leu704, Leu701, Asp705, Leu707, Trp741, Phe764, Thr877, Leu873, Met880, AND Met887, form hydrophobic interactions and hydrogen bonding.

Chrysin diminished the enhanced activity of HIF-1α, VEGFR2 in high glucose-exposed retinal endothelial cells in db/db mouse eyes (Kang et al. 2016). It was confirmed that chrysin-induced antimetastatic activity by inhibiting the Akt signaling pathway in metastatic triple-negative breast cancer (Yang et al. 2014). Chrysin may be useful to overcome anticancer drug resistance by down-regulating Nuclear factor-E2-related factor 2 (Nrf2), PI3K/Akt pathways (Gao et al. 2013; Woo et al. 2004). Chrysin induced apoptosis by the activation of PI3K and MAPK in ovarian cancer cells (Woo et al. 2004). Chrysin reduces the cell proliferation through the TGF-β1/Smad pathway in chronic liver diseases (Balta et al. 2015). Furthermore, chrysin treatment suppressed TGF-β, fibronectin and collagen-IV protein expressions in renal tissues (Ahad et al. 2014). Chrysin progress the mitochondrial-mediated apoptosis via ER stress in prostate cancer (Ryu et al. 2017).

Mw	Dipole	Donor HB	Accpt HB	Volume	FOSA	FISA	PISA	WPSA
254	5.65	1	3	799	0	149.122	328.997	0

CID5281607 (Chrysin)

QPlogPC16	QPlogPoct	QPlogPw	QPlogPo/w	QPlogS	CIQPlogS
9.19	12.899	8.1	2.367	-3.607	-4.123

QPlog HERG (acceptable range: above -5.0)	QPP Caco (nm/sec) <25-poor >500-great	QPlog BB (-3-1.2)	QPP MDCK (nm/sec) <25-poor >500-great	QPlog Khsa (Acceptable range: -1.5 to 1.5).	Percent Human Oral Absorption >80 % -High <25 % -Poor
-5.26	384.037	-0.876	175.836	0.138	87.174

Fig. 10.7 Evaluation of drug-like properties of the lead molecules by Qikprop

10.5.5 ADME Properties of Chrysin

Pharmacokinetic characteristics were predicted applying QikProp and are described in the Figs. 10.7 and 10.8. The results confirm that chrysin was predicted to have good solubility, metabolism and cell permeation. The predicted binding to human serum albumin (HAS), blocking of HERG K+ channels, Caco2 permeability values were within range of 95% of the known drugs.

10.5.6 Ligand-Based Virtual Screening (LBVS)

LBVS is a method for the detection of the similar compound. There are two forms USR (Ballester 2011) and USRCAT (Schreyer and Blundell 2012) available for virtual outlook screening of analogous compounds (Ballester 2011; Ballester and Richards 2007; Li et al. 2016). There are various studies where USR and USRCAT are practiced to identify comparable compounds to cancer markers and antibacterial (Ballester et al. 2010, 2012; Hoeger et al. 2014; Patil et al. 2014). Here, Chrysin applied a template and hundred hits leading into consideration for ligands assortment. Moreover, we observed ultrafast shape recognition (USR) score (0.95367342), USRCAT (0.60377538) and Tanimoto (0.18691589), sequentially to identify the new reliable compounds and then molecular docking was performed. Table 10.3 explained that ER showed remarkable Lipo EvdW, HBond and Electro values against Chrysin. Therefore, docking investigation of these compounds (selected from USR and USRCAT) was presented to know their interaction with the ER (Table 10.4).

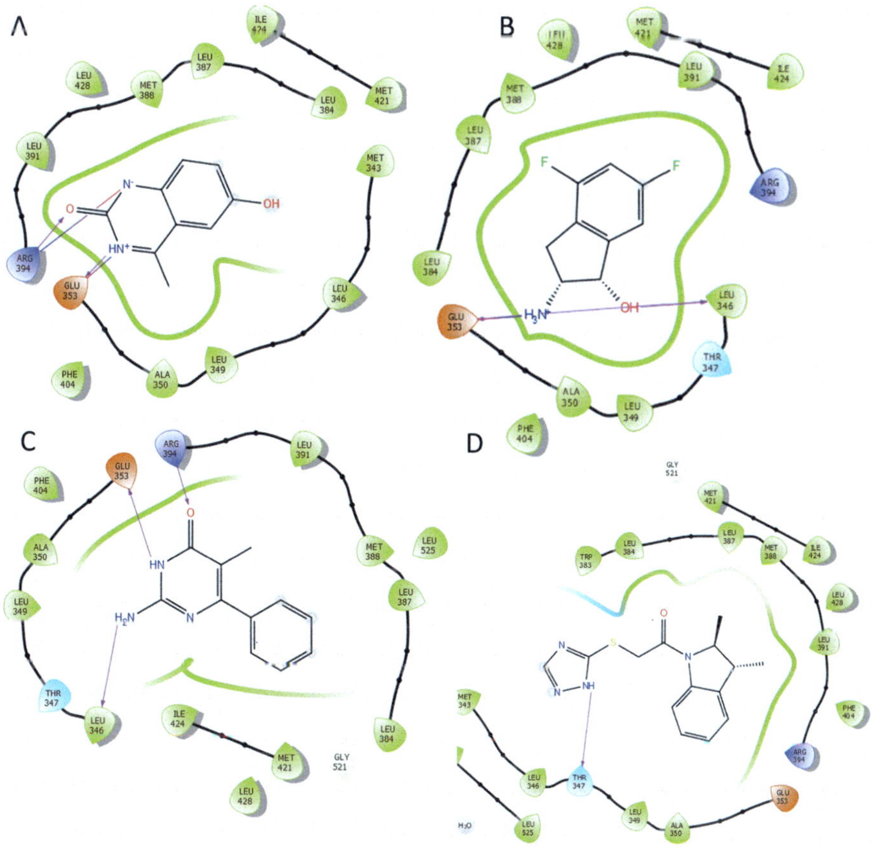

Fig. 10.8 (**a**) Protein-ligand interactions profile of ER with 06493707 (**b**) Protein-ligand interactions profile of ER with 08699413 (**c**) Protein-ligand interactions profile of ER with 17327441 (**d**) Protein-ligand interactions profile of ER with 38936403

10.6 Conclusion

Chrysin is a secondary metabolite of polyphenol, has been using as traditional herbal medicine. Chrysin has various bioactive effects including anti-viral, anti-inflammatory, cardioprotective, anti-diabetic, anti-cancer, anti-aging etc. Molecular docking result of VEGFR2, PI3K, TAK1 and ER against chrysin has better Gscore (Kcal/mol). Further ADME properties confirm that chrysin have good solubility, binding to human serum albumin (HAS), blocking of HERG K^+ channels, Caco2 permeability values were within range of 95% of the known drugs.

Table 10.3 Binding affinity score of chrysin with different receptors

Receptors	GScore	Lipo EvdW	HBond	Electro
IGFR1	−5.47	−3.72	−1.18	−0.43
IR	−6.69	−5.35	−1.12	−0.34
EGFR	−7.57	−4.6	−1.51	−0.36
VEGFR1	−6.68	−5.24	−0.48	−0.13
VEGFR2	−8.88	−4.97	−0.91	−0.72
STAT3	−4.27	−1.42	−2.19	−0.8
PI3K	−8.14	−3.77	−1.63	−0.58
PDK1	−5.29	−3.43	−2.1	−0.56
Akt	−1.83	−2.17	−1.66	−0.61
TGFβR1	−7.01	−4.6	−0.64	−1.05
TGFβR2	−4.95	−3.43	−0.7	−0.34
TAK1	−8.44	−4.75	−1.74	−0.57
CXCR2	−4.02	−1.53	−1.63	−0.82
CXCR4	−5.7	−2.5	−2.09	−0.72
CCR5	−6.71	−2.95	−1.18	−0.35
AR	−6.68	−4.5	−1.01	−0.5
ER	−8.05	−5.35	−1.18	−0.18
GAMA scretease	−5.64	−3.37	−1.03	−0.15
PARP	−6.66	−3.36	−1.77	−0.98
Smoothened 7TM	−5.56	−3.49	−1.43	−1.16
Wnt	−3.05	−2.14	−1.18	−0.34

Table 10.4 Glide score of different ligands against ER

Ligand	ZINC ID	Glide gscore	Glide emodel	XP HBond	XP electro
	6493707	−9.371	−31.765	−0.949	−0.414
	8699413	−8.99	−30.362	−1.476	−0.521
	4203552	−8.429	−27.09	−1.764	−1.682
	42063044	−8.438	−37.974	−1.013	−0.703
USR	86722103	−7.972	−31.996	−0.7	−0.234
	90225630	−7.849	−39.519	−1.165	−0.601
	86721951	−7.698	−30.102	−0.7	−0.245
	38537133	−7.682	−39.769	−0.732	−0.738
	59072810	−7.627	−21.471	−4.673	−0.5
	86722106	−7.601	−20.598	−2.416	−0.5
	17327441	−8.983	−44.908	−1.089	−0.459
	38936403	−9.163	−54.611	−0.77	−0.581
	8700025	−8.847	−48.728	−0.601	−0.403
	59408899	−8.784	−39.599	−0.9	−0.506
USRCAT	8691825	−8.674	−45.311	−0.645	−0.696
	18217770	−8.446	−45.642	−0.612	−0.486
	38936402	−8.936	−56.565	−1.06	−0.634
	25492688	−8.492	−46.857	−0.594	−0.528
	37206048	−8.608	−4.259	−46.95	0
	139628	−8.069	−4.001	−37.637	−1.3

Acknowledgments We would like to thank Central University of Punjab, Bathinda, Punjab, (India) and Director in charge, National Institute of Pathology, New Delhi (India) for supporting this study with infrastructural requirements. This study was also supported by a Centenary-Post Doctoral Research Fellowship Grant-in-Aid from Indian Council of Medical Research (ICMR), Government of India awarded to PS. PPK acknowledges financial support from UGC-CSIR in the form of Senior Research Fellowship.

References

Ahad A, Ganai AA, Mujeeb M, Siddiqui WA. Chrysin, an anti-inflammatory molecule, abrogates renal dysfunction in type 2 diabetic rats. Toxicol Appl Pharmacol. 2014;279(1):1–7.

Ali M, Muhammad S, Shah MR, Khan A, Rashid U, Farooq U, Ullah F, Sadiq A, Ayaz M, Ali M, Ahmad M. Neurologically potent molecules from *Crataegus oxyacantha*; isolation, anticholinesterase inhibition, and molecular docking. Front Pharmacol. 2017;8:327.

Anandhi R, Thomas PA, Geraldine P. Evaluation of the anti-atherogenic potential of chrysin in Wistar rats. Mol Cell Biochem. 2014;385(1–2):103–13.

Arai Y, Endo S, Miyagi N, Abe N, Miura T, Nishinaka T, Terada T, Oyama M, Goda H, El-Kabbani O, Hara A. Structure–activity relationship of flavonoids as potent inhibitors of carbonyl reductase 1 (CBR1). Fitoterapia. 2015;101:51–6.

Bagheri R, Sanaat Z, Zarghami N. Synergistic effect of free and nano-encapsulated chrysin-curcumin on inhibition of hTERT gene expression in SW480 colorectal cancer cell line. Drug Res. 2018;68(06):335–43.

Balasuriya N, Rupasinghe HV. Antihypertensive properties of flavonoid-rich apple peel extract. Food Chem. 2012;135(4):2320–5.

Ballester PJ. Ultrafast shape recognition: method and applications. Future Med Chem. 2011;3(1):65–78.

Ballester PJ, Richards WG. Ultrafast shape recognition for similarity search in molecular databases. In: Proceedings of the Royal Society of London. Series A. Mathematical, physical and engineering sciences, vol. 463, 2081. London: The Royal Society; 2007. p. 1307–21.

Ballester PJ, Westwood I, Laurieri N, Sim E, Richards WG. Prospective virtual screening with Ultrafast Shape Recognition: the identification of novel inhibitors of arylamine N-acetyltransferases. J Royal Soc Interface. 2010;7(43):335–42.

Ballester PJ, Mangold M, Howard NI, Robinson RL, Abell C, Blumberger J, Mitchell JB. Hierarchical virtual screening for the discovery of new molecular scaffolds in antibacterial hit identification. J R Soc Interface. 2012;9(77):3196–207.

Balta C, Herman H, Boldura OM, Gasca I, Rosu M, Ardelean A, Hermenean A. Chrysin attenuates liver fibrosis and hepatic stellate cell activation through TGF-β/Smad signaling pathway. Chem Biol Interact. 2015;240:94–101.

Bortolotto VC, Pinheiro FC, Araujo SM, Poetini MR, Bertolazi BS, de Paula MT, Meichtry LB, de Almeida FP, de Freitas Couto S, Jesse CR, Prigol M. Chrysin reverses the depressive-like behavior induced by hypothyroidism in female mice by regulating hippocampal serotonin and dopamine. Eur J Pharmacol. 2018;822:78–84.

Campos MS, Ribeiro NC, de Lima RF, Santos MB, Vilamaior PS, Regasini LO, Biancardi MF, Taboga SR, Santos FC. Anabolic effects of chrysin on the ventral male prostate and female prostate of adult gerbils (*Meriones unguiculatus*). Reprod Fertil Dev. 2018;30:1180–91.

Chassagne F, Haddad M, Amiel A, Phakeovilay C, Manithip C, Bourdy G, Deharo E, Marti G. A metabolomic approach to identify anti-hepatocarcinogenic compounds from plants used traditionally in the treatment of liver diseases. Fitoterapia. 2018;127:226–36.

Chen SS, Corteling R, Stevanato L, Sinden J. Polyphenols inhibit indoleamine 3, 5-dioxygenase-1 enzymatic activity—a role of immunomodulation in chemoprevention. Discov Med. 2012;14 (78):327–33.

Ciftci O, Ozdemir I, Aydin M, Beytur A. Beneficial effects of chrysin on the reproductive system of adult male rats. Andrologia. 2012;44(3):181–6.

Czyżewska U, Siemionow K, Zaręba I, Miltyk W. Proapoptotic activity of propolis and their components on human tongue squamous cell carcinoma cell line (CAL-27). PLoS One. 2016;11(6):e0157091.

Deldar Y, Pilehvar-Soltanahmadi Y, Dadashpour M, Montazer Saheb S, Rahmati-Yamchi M, Zarghami N. An in vitro examination of the antioxidant, cytoprotective and anti-inflammatory properties of chrysin-loaded nanofibrous mats for potential wound healing applications. Artif Cells Nanomed Biotechnol. 2018;46(4):706–16.

Fonseca SF, Padilha NB, Thurow S, Roehrs JA, Savegnago L, de Souza MN, Fronza MG, Collares T, Buss J, Seixas FK, Alves D. Ultrasound-promoted copper-catalyzed synthesis of bis-arylselanyl chrysin derivatives with boosted antioxidant and anticancer activities. Ultrason Sonochem. 2017;39:827–36.

Forman MS, Trojanowski JQ, Lee VM. Neurodegenerative diseases: a decade of discoveries paves the way for therapeutic breakthroughs. Nat Med. 2004;10(10):1055.

Friesner RA, Banks JL, Murphy RB, Halgren TA, Klicic JJ, Mainz DT, Repasky MP, Knoll EH, Shelley M, Perry JK, Shaw DE. Glide: a new approach for rapid, accurate docking and scoring. 1. Method and assessment of docking accuracy. J Med Chem. 2004;47(7):1739–49.

Friesner RA, Murphy RB, Repasky MP, Frye LL, Greenwood JR, Halgren TA, Sanschagrin PC, Mainz DT. Extra precision glide: docking and scoring incorporating a model of hydrophobic enclosure for protein– ligand complexes. J Med Chem. 2006;49(21):6177–96.

Gao AM, Ke ZP, Shi F, Sun GC, Chen H. Chrysin enhances sensitivity of BEL-7402/ADM cells to doxorubicin by suppressing PI3K/Akt/Nrf2 and ERK/Nrf2 pathway. Chem Biol Interact. 2013;206(1):100–8.

Goes AT, Jesse CR, Antunes MS, Ladd FV, Ladd AA, Luchese C, Paroul N, Boeira SP. Protective role of chrysin on 6-hydroxydopamine-induced neurodegeneration a mouse model of Parkinson's disease: involvement of neuroinflammation and neurotrophins. Chem Biol Interact. 2018;279:111–20.

Gresa-Arribas N, Serratosa J, Saura J, Solà C. Inhibition of CCAAT/enhancer binding protein δ expression by chrysin in microglial cells results in anti-inflammatory and neuroprotective effects. J Neurochem. 2010;115(2):526–36.

Grudzien P. Notch signaling is important in the survival, proliferation, and self-renewal of the putative breast cancer stem cell population. Chicago: Loyola University Chicago; 2010.

Gülden M, Appel D, Syska M, Uecker S, Wages F, Seibert H. Chrysin and silibinin sensitize human glioblastoma cells for arsenic trioxide. Food Chem Toxicol. 2017;105:486–97.

Guo B, Zheng C, Cai W, Cheng J, Wang H, Li H, Sun Y, Cui W, Wang Y, Han Y, Lee SM. Multifunction of chrysin in Parkinson's model: anti-neuronal apoptosis, neuroprotection via activation of MEF2D, and inhibition of monoamine oxidase-B. J Agric Food Chem. 2016;64 (26):5324–33.

He XL, Wang YH, Bi MG, Du GH. Chrysin improves cognitive deficits and brain damage induced by chronic cerebral hypoperfusion in rats. Eur J Pharmacol. 2012;680(1–3):41–8.

Hoeger B, Diether M, Ballester PJ, Köhn M. Biochemical evaluation of virtual screening methods reveals a cell-active inhibitor of the cancer-promoting phosphatases of regenerating liver. Eur J Med Chem. 2014;88:89–100.

Jia WZ, Zhao JC, Sun XL, Yao ZG, Wu HL, Xi ZQ. Additive anticancer effects of chrysin and low dose cisplatin in human malignant glioma cell (U87) proliferation and evaluation of the mechanistic pathway. J BUON. 2017;20(5):1327–36.

Jorgensen WL, Duffy EM. Prediction of drug solubility from structure. Adv Drug Deliv Rev. 2002;54(3):355–66.

Jorgensen WL, Tirado-Rives J. The OPLS [optimized potentials for liquid simulations] potential functions for proteins, energy minimizations for crystals of cyclic peptides and crambin. J Am Chem Soc. 1988;110(6):1657–66.

Jorgensen WL, Maxwell DS, Tirado-Rives J. Development and testing of the OPLS all-atom force field on conformational energetics and properties of organic liquids. J Am Chem Soc. 1996;118 (45):11225–36.

Kang MK, Park SH, Kim YH, Lee EJ, Antika LD, Kim DY, Choi YJ, Kang YH. Dietary compound chrysin inhibits retinal neovascularization with abnormal capillaries in db/db mice. Nutrients. 2016;8(12):782.

Kanwal R, Datt M, Liu X, Gupta S. Dietary flavones as dual inhibitors of DNA methyltransferases and histone methyltransferases. PLoS One. 2016;11(9):e0162956.

Kasala ER, Bodduluru LN, Madana RM, Gogoi R, Barua CC. Chemopreventive and therapeutic potential of chrysin in cancer: mechanistic perspectives. Toxicol Lett. 2015;233(2):214–25.

Kumar S, Pandey AK. Chemistry and biological activities of flavonoids: an overview. Sci World J. 2013;2013:1–16.

Laishram S, Moirangthem DS, Borah JC, Pal BC, Suman P, Gupta SK, Kalita MC, Talukdar NC. Chrysin rich Scutellaria discolor Colebr. induces cervical cancer cell death via the induction of cell cycle arrest and caspase-dependent apoptosis. Life Sci. 2015;143:105–13.

León IE, Cadavid-Vargas JF, Tiscornia I, Porro V, Castelli S, Katkar P, Desideri A, Bollati-Fogolin M, Etcheverry SB. Oxidovanadium (IV) complexes with chrysin and silibinin: anticancer activity and mechanisms of action in a human colon adenocarcinoma model. J Biol Inorg Chem. 2015;20(7):1175–91.

Li X, Wang JN, Huang JM, Xiong XK, Chen MF, Ong CN, Shen HM, Yang XF. Chrysin promotes tumor necrosis factor (TNF)-related apoptosis-inducing ligand (TRAIL) induced apoptosis in human cancer cell lines. Toxicol In Vitro. 2015;25(3):630–5.

Li H, Leung KS, Wong MH, Ballester PJ. USR-VS: a web server for large-scale prospective virtual screening using ultrafast shape recognition techniques. Nucleic Acids Res. 2016;44(W1): W436–41.

Lim W, Ryu S, Bazer FW, Kim SM, Song G. Chrysin attenuates progression of ovarian cancer cells by regulating signaling cascades and mitochondrial dysfunction. J Cell Physiol. 2018;233 (4):3129–40.

Lin CM, Chang H, Li SY, Wu IH, Chiu JH. Chrysin inhibits lipopolysaccharide-induced angiogenesis via down-regulation of VEGF/VEGFR-2 (KDR) and IL-6/IL-6R pathways. Planta Med. 2006;72(08):708–14.

Lin CM, Shyu KG, Wang BW, Chang H, Chen YH, Chiu JH. Chrysin suppresses IL-6-induced angiogenesis via down-regulation of JAK1/STAT3 and VEGF: an in vitro and in vivo approach. J Agric Food Chem. 2010;58(11):7082–7.

Lirdprapamongkol K, Sakurai H, Abdelhamed S, Yokoyama S, Maruyama T, Athikomkulchai S, Viriyaroj A, Awale S, Yagita H, Ruchirawat S, Svasti J. A flavonoid chrysin suppresses hypoxic survival and metastatic growth of mouse breast cancer cells. Oncol Rep. 2013;30(5):2357–64.

Liu H, Liu K, Huang Z, Park CM, Thimmegowda NR, Jang JH, Ryoo IJ, He L, Kim SO, Oi N, Lee KW. A chrysin derivative suppresses skin cancer growth by inhibiting cyclin-dependent kinases. J Biol Chem. 2013;288:25924–37.

Lu JJ, Crimin K, Goodwin JT, Crivori P, Orrenius C, Xing L, Tandler PJ, Vidmar TJ, Amore BM, Wilson AG, Stouten PF. Influence of molecular flexibility and polar surface area metrics on oral bioavailability in the rat. J Med Chem. 2004;47(24):6104–7.

Maasomi ZJ, Soltanahmadi YP, Dadashpour M, Alipour S, Abolhasani S, Zarghami N. Synergistic anticancer effects of silibinin and chrysin in T47D breast cancer cells. Asian Pac J Cancer Prev Asian Pac J Cancer Prev. 2017;18(5):1283.

Mani R, Natesan V. Chrysin: sources, beneficial pharmacological activities, and molecular mechanism of action. Phytochemistry. 2018;145:187–96.

Mantawy EM, Esmat A, El-Bakly WM, ElDin RA, El-Demerdash E. Mechanistic clues to the protective effect of chrysin against doxorubicin-induced cardiomyopathy: plausible roles of p53, MAPK and AKT pathways. Sci Rep. 2017;7(1):1–13.

Manupati K, Dhoke NR, Debnath T, Yeeravalli R, Guguloth K, Saeidpour S, De UC, Debnath S, Das A. Inhibiting epidermal growth factor receptor signalling potentiates mesenchymal–epithelial transition of breast cancer stem cells and their responsiveness to anticancer drugs. FEBS J. 2017;284(12):1830–54.

Mattison CP, Rai R, Settlage RE, Hinchliffe DJ, Madison C, Bland JM, Brashear S, Graham CJ, Tarver MR, Florane C, Bechtel PJ. RNA-Seq analysis of developing pecan (*Carya illinoinensis*) embryos reveals parallel expression patterns among allergen and lipid metabolism genes. J Agric Food Chem. 2017;65(7):1443–55.

Mercer LD, Kelly BL, Horne MK, Beart PM. Dietary polyphenols protect dopamine neurons from oxidative insults and apoptosis: investigations in primary rat mesencephalic cultures. Biochem Pharmacol. 2005;69(2):339–45.

Mistry BM, Patel RV, Keum YS, Kim DH. Chrysin–benzothiazole conjugates as antioxidant and anticancer agents. Bioorg Med Chem Lett. 2015;25(23):5561–5.

Mohammadi Z, Zak MS, Seidi K, Barati M, Akbarzadeh A, Zarghami N. The effect of chrysin loaded PLGA-PEG on metalloproteinase gene expression in mouse 4t1 tumor model. Drug Res. 2017;67(04):211–6.

Mohammadian F, Pilehvar-Soltanahmadi Y, Alipour S, Dadashpour M, Zarghami N. Chrysin alters microRNAs expression levels in gastric cancer cells: possible molecular mechanism. Drug Res. 2017;67(09):509–14.

Nabavi SF, Braidy N, Habtemariam S, Orhan IE, Daglia M, Manayi A, Gortzi O, Nabavi SM. Neuroprotective effects of chrysin: from chemistry to medicine. Neurochem Int. 2015;90:224–31.

Owen HC, Appiah S, Hasan N, Ghali L, Elayat G, Bell C. Phytochemical modulation of apoptosis and autophagy: strategies to overcome chemoresistance in leukemic stem cells in the bone marrow microenvironment. Int Rev Neurobiol. 2017;135:249–78. Academic

Patel RV, Mistry B, Syed R, Rathi AK, Lee YJ, Sung JS, Shinf HS, Keum YS. Chrysin-piperazine conjugates as antioxidant and anticancer agents. Eur J Pharm Sci. 2016;88:166–77.

Patil SP, Ballester PJ, Kerezsi CR. Prospective virtual screening for novel p53–MDM2 inhibitors using ultrafast shape recognition. J Comput Aided Mol Des. 2014;28(2):89–97.

Rajagopal C, Lankadasari MB, Aranjani JM, Harikumar KB. Targeting oncogenic transcription factors by polyphenols: a novel approach for cancer therapy. Pharmacol Res. 2018;130:273–91.

Ramírez-Espinosa JJ, Saldaña-Ríos J, García-Jiménez S, Villalobos-Molina R, Ávila-Villarreal G, Rodríguez-Ocampo AN, Bernal-Fernández G, Estrada-Soto S. Chrysin induces antidiabetic, antidyslipidemic and anti-inflammatory effects in athymic nude diabetic mice. Molecules. 2017;23(1):1–8.

Rehman MU, Tahir M, Khan AQ, Khan R, Lateef A, Qamar W, Ali F, Sultana S. Chrysin suppresses renal carcinogenesis via amelioration of hyperproliferation, oxidative stress and inflammation: plausible role of NF-κB. Toxicol Lett. 2013;216(2–3):146–58.

Resende FA, de Oliveira AP, de Camargo MS, Vilegas W, Varanda EA. Evaluation of estrogenic potential of flavonoids using a recombinant yeast strain and MCF7/BUS cell proliferation assay. PLoS One. 2013;8(10):e74881.

Russo P, Del Bufalo A, Cesario A. Flavonoids acting on DNA topoisomerases: recent advances and future perspectives in cancer therapy. Curr Med Chem. 2012;19(31):5287–93.

Ryu S, Lim W, Bazer FW, Song G. Chrysin induces death of prostate cancer cells by inducing ROS and ER stress. J Cell Physiol. 2017;232(12):3786–97.

Salimi A, Roudkenar MH, Seydi E, Sadeghi L, Mohseni A, Pirahmadi N, Pourahmad J. Chrysin as an anti-cancer agent exerts selective toxicity by directly inhibiting mitochondrial complex II and V in CLL B-lymphocytes. Cancer Investig. 2017;35(3):174–86.

Samarghandian S, Azimi-Nezhad M, Borji A, Hasanzadeh M, Jabbari F, Farkhondeh T, Samini M. Inhibitory and cytotoxic activities of chrysin on human breast adenocarcinoma cells by induction of apoptosis. Pharmacogn Mag. 2016;12(Suppl 4):S436.

Samarghandian S, Farkhondeh T, Azimi-Nezhad M. Protective effects of chrysin against drugs and toxic agents. Dose Response. 2017;15(2):1–10.

Sanadgol N, Shahraki Zahedani S, Sharifzadeh M, Khalseh R, Reza Barbari G, Abdollahi M. Recent updates in imperative natural compounds for healthy brain and nerve function: a systematic review of implications for multiple sclerosis. Curr Drug Targets. 2017;18 (13):1499–517.

Schreyer AM, Blundell T. USRCAT: real-time ultrafast shape recognition with pharmacophoric constraints. J Cheminform. 2012;4(1):1–12.

Shivakumar D, Williams J, Wu Y, Damm W, Shelley J, Sherman W. Prediction of absolute solvation free energies using molecular dynamics free energy perturbation and the OPLS force field. J Chem Theory Comput. 2010;6(5):1509–19.

Singh P, Bast F. In silico molecular docking study of natural compounds on wild and mutated epidermal growth factor receptor. Med Chem Res. 2014a;23(12):5074–85.

Singh P, Bast F. Multitargeted molecular docking study of plant-derived natural products on phosphoinositide-3 kinase pathway components. Med Chem Res. 2014b;23(4):1690–700.

Singh P, Bast F. High-throughput virtual screening, identification and in vitro biological evaluation of novel inhibitors of signal transducer and activator of transcription 3. Med Chem Res. 2015a;24(6):2694–708.

Singh P, Bast F. Screening of multi-targeted natural compounds for receptor tyrosine kinases inhibitors and biological evaluation on cancer cell lines, in silico and in vitro. Med Oncol. 2015b;32(9):233.

Singh P, Singh RS, Rani A, Bast F. Homology modeling of chemokine CCR7, molecular docking, and in vitro studies evidenced plausible immunotherapeutic anticancer natural compounds. Med Chem Res. 2016;25(10):2410–24.

Stepanic V, Gasparovic AC, Troselj KG, Amic D, Zarkovic N. Selected attributes of polyphenols in targeting oxidative stress in cancer. Curr Top Med Chem. 2015;15(5):496–509.

Sun LP, Chen AL, Hung HC, Chien YH, Huang JS, Huang CY, Chen YW, Chen CN. Chrysin: a histone deacetylase 8 inhibitor with anticancer activity and a suitable candidate for the standardization of Chinese propolis. J Agric Food Chem. 2012;60(47):11748–58.

Szkudelski T. The mechanism of alloxan and streptozotocin action in B cells of the rat pancreas. Physiol Res. 2001;50(6):537–46.

Thangarajan S, Ramachandran S, Krishnamurthy P. Chrysin exerts neuroprotective effects against 3-Nitropropionic acid induced behavioral despair—Mitochondrial dysfunction and striatal apoptosis via upregulating Bcl-2 gene and downregulating Bax—Bad genes in male wistar rats. Biomed Pharmacother. 2016;84:514–25.

Vedagiri A, Thangarajan S. Mitigating effect of chrysin loaded solid lipid nanoparticles against amyloid β25–35 induced oxidative stress in rat hippocampal region: an efficient formulation approach for Alzheimer's disease. Neuropeptides. 2016;58:111–25.

Villar IC, Jimenez R, Galisteo M, Garcia-Saura MF, Zarzuelo A, Duarte J. Effects of chronic chrysin treatment in spontaneously hypertensive rats. Planta Med. 2002;68(9):847–50.

Wadibhasme PG, Ghaisas MM, Thakurdesai PA. Anti-asthmatic potential of chrysin on ovalbumin-induced bronchoalveolar hyperresponsiveness in rats. Pharm Biol. 2011;49(5):508–15.

Wang TY, Li Q, Bi KS. Bioactive flavonoids in medicinal plants: structure, activity and biological fate. Asian Pac J Pharmacol. 2017a;13(1):12–23.

Wang Z, Deng X, Xiong S, Xiong R, Liu J, Zou L, Lei X, Cao X, Xie Z, Chen Y, Liu Y. Design, synthesis and biological evaluation of chrysin benzimidazole derivatives as potential anticancer agents. Nat Prod Res. 2017b;1–10.

Woo KJ, Jeong YJ, Park JW, Kwon TK. Chrysin-induced apoptosis is mediated through caspase activation and Akt inactivation in U937 leukemia cells. Biochem Biophys Res Commun. 2004;325(4):1215–22.

Xia Y, Lian S, Khoi PN, Yoon HJ, Han JY, Chay KO, Kim KK, Jung YD. Chrysin inhibits cell invasion by inhibition of Recepteur d'origine Nantais via suppressing early growth response-1 and NF-κB transcription factor activities in gastric cancer cells. Int J Oncol. 2015a;46 (4):1835–43.

Xia Y, Lian S, Khoi PN, Yoon HJ, Joo YE, Chay KO, Kim KK, Do Jung Y. Chrysin inhibits tumor promoter-induced MMP-9 expression by blocking AP-1 via suppression of ERK and JNK pathways in gastric cancer cells. PLoS One. 2015b;10(4):e0124007.

Xu D, Jin J, Yu H, Zhao Z, Ma D, Zhang C, Jiang H. Chrysin inhibited tumor glycolysis and induced apoptosis in hepatocellular carcinoma by targeting hexokinase-2. J Exp Clin Cancer Res. 2017;36(1):44.

Xuan HZ, Zhang JH, Wang YH, Fu CL, Zhang W. Anti-tumor activity evaluation of novel chrysin–organotin compound in MCF-7 cells. Bioorg Med Chem Lett. 2016;26(2):570–4.

Yang F, Jin H, Pi J, Jiang JH, Liu L, Bai HH, Yang PH, Cai JY. Anti-tumor activity evaluation of novel chrysin–organogermanium (IV) complex in MCF-7 cells. Bioorg Med Chem Lett. 2013;23(20):5544–51.

Yang B, Huang J, Xiang T, Yin X, Luo X, Huang J, Luo F, Li H, Li H, Ren G. Chrysin inhibits metastatic potential of human triple-negative breast cancer cells by modulating matrix metalloproteinase-10, epithelial to mesenchymal transition, and PI3K/Akt signaling pathway. J Appl Toxicol. 2014;34(1):105–12.

Yao Y, Ni Y, Zhang J, Wang H, Shao S. The role of Notch signaling in gastric carcinoma: molecular pathogenesis and novel therapeutic targets. Oncotarget. 2017;8(32):53839.

Zeinali M, Rezaee SA, Hosseinzadeh H. An overview on immunoregulatory and anti-inflammatory properties of chrysin and flavonoids substances. Biomed Pharmacother. 2017;92:998–1009.

Zeng W, Yan Y, Zhang F, Zhang C, Liang W. Chrysin promotes osteogenic differentiation via ERK/MAPK activation. Protein Cell. 2013;4(7):539–47.

Zhang Z, Li G, Szeto SS, Chong CM, Quan Q, Huang C, Cui W, Guo B, Wang Y, Han Y, Siu KM. Examining the neuroprotective effects of protocatechuic acid and chrysin on in vitro and in vivo models of Parkinson disease. Free Radic Biol Med. 2015;84:331–43.

Zhang Q, Ma S, Liu B, Liu J, Zhu R, Li M. Chrysin induces cell apoptosis via activation of the p53/Bcl-2/caspase-9 pathway in hepatocellular carcinoma cells. Exp Ther Med. 2016;12(1):469–74.

Chapter 11
Screening of Natural Antidiabetic Agents

Chukwuebuka Egbuna, Santwana Palai, Israel Ehizuelen Ebhohimen, Andrew G. Mtewa, Jonathan C. Ifemeje, Genevieve D. Tupas, and Toskë L. Kryeziu

11.1 Introduction

Diabetes is a disorder characterized by abnormal glycemic control. It is the most prevalent metabolic disorder in the world with about 5% of the world's population suffering one form of the disease. According to the International Diabetes Federation (2017), 1 in 2 adults with diabetes remains undiagnosed and the increasing trend in the number of adults with diabetes (20–79 years) spans 151 million patients in 2000, 194 million in 2003, 246 million in 2006, and 285 million in 2009, 366 million in

C. Egbuna (✉)
Department of Biochemistry, Chukwuemeka Odumegwu Ojukwu University, Uli, Anambra, Nigeria
e-mail: egbuna.cg@coou.edu.ng

S. Palai
Department of Veterinary Pharmacology and Toxicology, Orissa University of Agriculture and Technology, Bhubaneswar, Odisha, India

I. E. Ebhohimen
Department of Chemical Sciences (Biochemistry), Samuel Adegboyega University, Ogwa, Edo State, Nigeria

A. G. Mtewa
Department of Chemistry, Institute of Technology, Malawi University of Science and Technology, Thyolo, Malawi

J. C. Ifemeje
Department of Biochemistry, Faculty of Natural Sciences, Chukwuemeka Odumegwu Ojukwu University, Uli, Anambra, Nigeria

G. D. Tupas
College of Medicine, Department of Pharmacology, Davao Medical School Foundation, Inc., Davao City, Philippines

T. L. Kryeziu
Department of Pharmacy, Faculty of Medicine, University of Prishtina "Hasan Prishtina", Prishtina, Kosovo

S. Kumar, C. Egbuna (eds.), *Phytochemistry: An in-silico and in-vitro Update*,
https://doi.org/10.1007/978-981-13-6920-9_11

2011, 382 million in 2013, 415 million in 2015 to 425 million in 2017, with a projection of 629 million sufferers by the year 2045. In 2017, the global total healthcare expenditure by people with diabetes (20–79 years) increased by 54 billion US dollars (from US$673 in 2015 to US$727 in 2017) (International Diabetes Federation 2017).

Currently, there are four types of Diabetes: (i) Type 1 Diabetes mellitus (T1DM) formerly known as Insulin Dependent Diabetes mellitus (IDDM) – which is primarily characterized by β-cell destruction resulting to insulin deficiency; (ii) Type 2 Diabetes mellitus (T2DM) formerly known as Non-insulin Dependent Diabetes mellitus (NIDDM) – which is noted for its characteristic insulin resistance and a defect in insulin secretion; (iii) Type 3 DM is known as Gestational Diabetes mellitus (GDM) – a pregnancy induced-diabetes due to placental hormones which promotes insulin resistance (Tripathi and Verma 2014), and (iv) Type 4 DM-Diabetes due to specific causes such as pancreatic diseases, monogenic diabetes syndrome, drug or chemically-induced diabetes or those resulting from organ transplantation (American Diabetes Association 2015; WHO 2017). The hormone insulin plays a central role in the uptake of glucose into cells from blood circulation. A defect in the secretion of the hormone inhibit glucose uptake. This accounts for the high blood glucose concentration in the blood which is not available for use in metabolic processes. This defines T1DM. The onset is usually at a tender age (<20 years) and a management strategy is the external supply of insulin. The onset is multifactorial and genetic predisposition is generally required. In the other category, insulin may be present but glucose uptake is still inhibited. This class is called T2DM and the onset is usually later in life though some exceptions to onset have been reported. Approximately 90% of all cases of diabetes in developed and developing countries are T2DM (Shori 2015; WHO 2017). Criteria for diagnosis are based on glycosylated hemoglobin (HbA1C) and or plasma glucose levels. Diagnosis of Diabetes is considered in any of the following conditions: HbA1C level $\geq$6.5%, fasting plasma glucose (FPG) $\geq$126 mg/dL, 2-h post-prandial glucose $\geq$200 mg/dL after a 75-g oral glucose tolerance test (OGTT) or a random plasma glucose $\geq$200 mg/dL (Chaudhury et al. 2017). The above conditions should be accompanied by one or more clinical manifestations of diabetes such as polydipsia, polyuria, fatigue, slow healing of wounds and blurred vision. Hyperosmolar coma and protein glycation are established complications of uncontrolled diabetes mellitus. In the former, induced diuresis by metabolic stress in an undiagnosed patient results to loss of water, glucose and electrolytes. Such condition causes further increase in insulin resistance and hyperglycaemia resulting to extreme dehydration and coma. High concentration of blood glucose favours glycation of proteins as well as protein assembly and binding to other matrix molecules which account for the development of diabetic complications like cardiovascular problems and cataract. The coordinated treatment of diabetes and its comorbidities has proven to be too complex to be managed by a single anti-hyperglycermic agent due to its multifactorial nature (Defronzo 2009), hence the need for more potent antidiabetic drugs.

Drugs used for the management of DM aim at stabilizing or controlling blood glucose concentration within normal range. Antidiabetics have metamorphosed from drugs obtained by testing, through in-vitro and in-vivo testing to current high throughput screening in-silico techniques. Older antidiabetics focused on insulin supply or stimulation of synthesis to manage Type I and the three major defects associated with Type 2DM such as deranged activity of pancreatic cells, increased hepatic glucose output and faulty insulin response. These strategies are greatly limited by the side effects (hypoglycaemia, weight gain) associated with chronic use. To overcome these challenges, new therapies based on the interactions between lipids, carbohydrates and proteins (enzymes, ion channels, transporters, and receptors) need to be developed. Analysis of membrane microdomains in the normal and diseased subject as well as genetic studies has unveiled a great deal of possibilities in anti-diabetic drug design and development. Identifying molecular targets significantly increase efficacy and reduce the toxicity of drugs.

11.1.1 Potentials of Medicinal Plants as Source of Antidiabetic Agents

The plant kingdom harbors a wide variety of natural substances that have significant hypoglycemic properties with few or no documented side effects (Shori 2015). Medicinal plants are known to play a crucial role in the treatment of diabetes, particularly in developing countries where most people have limited resources and poor access to modern medicine. Natural antidiabetic agents work by modulating the expression, synthesis or degradation of insulin. The induction of insulin release is the main mechanism for some agents. Others could work by increasing the number of islet cells as well as producing antioxidants. Many other mechanisms such as inhibition of some enzymes involved in the release of insulin in diabetic patients have been utilized to develop methods for the screening of medicinal plants.

11.1.2 Screening Targets for Antidiabetic Agents

According to Brayer et al. (1995), during the digestion of starch in humans, there are several stages involved. Initially, salivary α-amylase catalyzes the breakdown of polymeric starch into shorter oligomers. Digestion further continues upon reaching the gut where the partially digested starch is then extensively hydrolyzed into smaller oligosaccharides by the α-amylase synthesized in the pancreas and excreted into the lumen. The resultant mixture of oligosaccharides, including maltose, maltotriose, and a number of α-(l–6) and ~(1–4) oligoglucans, then passes through the mucous layer to the brush border membrane. At this point, a number of α-glucosidases acts to further degrade the oligosaccharides to glucose. This glucose is then absorbed and enters the bloodstream by means of a specific transport system. Postprandial hyperglycemia has been proposed as an independent risk factor for cardiovascular disease (Cavarape et al. 2001). Therefore, control of postprandial hyperglycemia is

Table 11.1 Drug classes for diabetes mellitus management

Drug class	Mechanism of action
α-glucosidase inhibitors	Inhibit the action of α-glucosidase thus inhibiting intestinal dextrin absorption, e.g. acarbose
Amylin analogues	Regulate post-prandial glucose concentration by inhibiting glucagon, delaying gastric emptying and signalling satiety, e.g. pramlintide
Dipeptidyl peptidase-4 (DPP-4) inhibitors	Slow down the inactivation of and degradation of GLP-1, e.g. alogliptan
Incretin mimetics	Stimulate glucose dependent insulin release, e.g. albiglutide
Insulin	Stimulate glucose uptake
Meglitinides	Stimulate insulin secretion in the pancreas, e.g. nateglinide
Non-sulfonylureas	Inhibit glucose synthesis in the liver, stimulate tissue glucose uptake and increase insulin-receptor binding, e.g. metformin
Sodium-glucose co-transporter (SGLT) -2 inhibitors	Inhibit glucose reabsorption in renal tubules e.g. canagliflozin
Sulfonylureas	Stimulate insulin secretion by β-cells, e.g. chlorpropamide

suggested to be important in the treatment of diabetes and prevention of cardiovascular complications. The digestion process of starch before it gets to the blood stream are utilized as a factor in developing methods for anti-diabetic screening. This is because a potential inhibitor of any of the enzymes at each stage will prove valuable in delaying or reducing the breakdown of starch into glucose for the body to absorb.

Anti-diabetic molecular agents act on various targets, usually receptors close to the most affected areas. The most common molecular targets are G protein-coupled receptors (GPCR), dipeptidly peptidase (DPP), free fatty acid receptors (FFAR), protein tyrosine phosphatase 1B (PTP1B), aldose reductase, glycogen phosphorylase (GP), glucagon receptor (GCGr), fructose-1,6-bisphosphatase (FBPase), α-glucosidase, phosphoenolpyruvate carboxykinase (PEPCK) (Kerru et al. 2018), peroxisome proliferator activated receptor-γ (PPARγ) agonists (Varmaan et al. 2016) glucose transporter (GLUT) and sodium glucose co-transporter (SGLT) (Asano et al. 2004). The GLUT4 isoform of the glucose transporter group is of note-worth significance for being capable of maintaining the homeostasis of the metabolism of glucose due to its involvement in the uptake of glucose into skeletal muscles in the presence of insulin. Molecular groups that contain GLUT4 inducing agents, can potentially improve sensitivity to insulin, thus, possess great promise in the control of diabetes (Asano et al. 2004; Huang and Czech 2007).

The new classes of drugs administered alone or in combination are effective and can generate a specific therapeutic effect. This approach will effectively deliver the proposed personalized treatment of diabetes mellitus. A summary of the main classes of anti-diabetic drugs and their mechanism of actions are presented in Table 11.1. Like other disease conditions, increased understanding of molecular targets for diabetes mellitus is key to developing anti-diabetics. The continuous advancement in understanding of the biochemical processes that define the disorder at the molecular level will continue to increase the number of leads available for high throughput screening. This will result in the development of cost effective and safe drugs.

Table 11.2 Anti-diabetic targets in fat metabolism

Targets	Metabolic relevance
Adipocytes	I. Adipokines have anti-diabetic property. For example; leptin enhances insulin sensitivity
	II. High concentration of intracellular lipids inhibit insulin sensitivity
	III. TNF-α inhibit insulin action
	IV. Chronic high concentration of non-esterified fatty acids reduce insulin secretion
Peroxisome proliferator-activated receptor	I. Affect transcription and expression of numerous targets in adipocytes, cells and tissues
	II. Central regulator of lipid homeostasis and molecular target for hypertriglyceridaemia drugs
	III. Oppose lipid induced steric resistance
Enzymes of fat metabolism	I. Involve in the maintenance of average calorific value thus suppressing obesity
Stearoyl-CoA desaturase (SCD) 1	I. Rate limiting enzyme in lipogenesis
	II. Deficiency results in increased lipid oxidation
Diaclyglycerol acyl transfer-ase 1	I. Catalyse a major step in triacylglycerol synthesisDirect effect on insulin sensitivity
11β hydroxysteroid dehydrogenase	I. Converts inactive cortisone to active cortisol in the liver. High cortisol concentration induces insulin resistance
β_3-adrenergic receptors	I. Specifically enhance insulin action in adipose tissue

11.1.3 Fat Metabolism: A New Target in Diabetic Therapy

The role of fat metabolism in diabetes has gained attention in recent times. Visceral adipose tissue is now classified as an endocrine organ since a number of substances that might play a role in the development of insulin resistance are generated from it. Relevant factors that can modify the interaction of visceral fats and insulin action in-vivo are good targets (Table 11.2) currently being explored.

11.2 In-silico Screening Approach for Bioactive Compounds

The starting point is the identification of a suitable target associated with a disease process. A suitable target could be DNA, RNA or proteins. The next step is to validate the target. This helps to identify if the target exhibit a sufficient level of confidence and pharmacological relevance relative to the disease condition under-study. Target validation provides information that determine parameters that define drug leads. Leads are chemical agents that can modulate target activity as desired. Leads can be sourced from natural products, structure aided molecular design, modification of products of naturally occurring compounds, large databases of

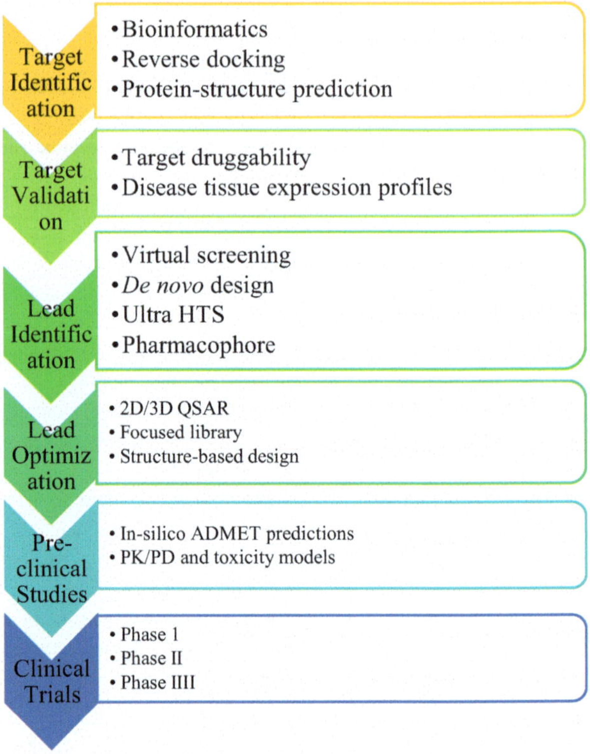

Fig. 11.1 In-silico drug discovery process

chemical compounds with biological activity and based on the understanding of the disease process. The compounds are subjected to high- throughput screening (HTS) to select only compounds showing a dose dependent target modulation. The lead at this stage is then optimized by synthesizing and testing derivatives of lead to develop structure activity relationships (SARs). Quantitative structure activity relationships (QSARs) techniques are currently available. These are cheaper analytical tools for the evaluation of physicochemical properties that are used to refine the leads. The refined lead can be further subjected to the remaining processes in the scheme (Fig. 11.1).

11.2.1 In-silico Tools and Importance in Anti-diabetic Drug Discovery

Available tools have quickly developed from 2D-structure activity prediction to current 3D-packages that effectively explain molecular interactions between a receptor and lead. A number of tools are available today (https://www.click2drug.org). Current concern is the development of effective multiparameter selection criteria that will cater for the multidisciplinary facets of this approach. The deployment of advanced biomolecule characterization tools and the possibility of matching data in-silico with pharmacophores has greatly improved diabetes management. The plethora of in-silico tools available in the twenty-first century has made drug discovery, drug repurposing, structure directed drug design a continuum.

11.2.1.1 Ligand-based Tools

Ligand-based tools are the common tools in in-silico drug development due to the significant role of ligands in interacting with proteins. Ligand-based tools are modeling tools currently being developed which include the evaluation of both pharmacophores and drug-like properties for virtual molecular screening from databases and bioactivity modeling (Basu et al. 2016; Zatsepin et al. 2016). There are various optimization algorithms under the ligand-based approach which can be used individually and sometimes in combination with each other. Among these algorithms are the support vector machines (SVMs) (Heikamp and Bajorath 2013), genetic algorithms (GAs) (Hao et al. 2012), monte carlo simulations (MCs) (Schuller and Schneider 2008), the k-nearest neighbor (kNN) (Rayan et al. 2010), the neural networks (NNs) (Lusci et al. 2013) and the simulated annealing (SA) (Schuller and Schneider 2008). One of the most common algorithm combinations is between the Monte carlo simulations and simulated anneaing (MSSA) (Shahaf et al. 2016). In the in-silico approach, a database that has more molecules is more convenient to model. Sometimes, depending on the nature of complexity, sets of known actives and inactive ligands are required to be embedded in the software to set baselines for the test ligand (Garciα-Sosa et al. 2012). The large numbers of ligands involved subsequently require a correspondingly large number of descriptors that limit efficiency in indexing (Zeidan et al. 2017).

11.2.1.2 Indexing Tools

To overcome indexing limitations and other challenges arising from the huge database of molecules with varying molecular descriptors, more algorithms are still being developed to complement the above or work independently. One of these algorithms is the iterative stochastic elimination (ISE) which checks potential problematic structural and ligand outputs (Rayan et al. 2004; Rayan et al. 2013). ISE uses algorithms that are capable of detecting protein side-chain conformations (Glick et al. 2002) and proton locations (Glick and Goldblum 2000), predicting molecule likeness to drugs (Rayan et al. 2013) and chemical indexing for their hERG liabilities (Rayan et al. 2010) among other capabilities in a more efficient way. In some research work, ISE was reported to be very effective in identifying potential anti diabetic drugs from natural products despite their nature of molecular complexity (Zeidan et al. 2017).

11.2.2 Validation of in-silico Methods

In-silico methods are tricky in nature as they depend on the baseline data fed to the softwares. This is evident from the differences observable in results from one through to another of the same function. Validation of in-silico software is critically important (Varmaan et al. 2016). As such, in-silico tools should never be used until well documented to have been validated using in-vivo obtained data. Sometimes, other researchers validate their software using another pre-validated tools available. The same applies to reporting of in-silico data. Each time results are to be reported, they should be accompanied by corresponding test conditions and reference standards used to enable repeatability of tests.

11.3 In-vitro Methods for Screening of Antidiabetic Agents

11.3.1 Pancreatic α-Amylase Enzyme

Pancreatic α-amylase arbitrarily cleaves the α(1–4) glycosidic bonds of amylose to produce dextrin, maltose, or maltotriose. Pancreatic α-amylase differ in action with α-glucosidase because it breaks down complex starches to smaller entities in the lumen of the small intestine such as oligosaccharides, with further breakdown of disaccharides and trisaccharides to glucose and additional monosaccharides is catalyzed by the membrane-bound intestinal α-glucosidases in the intestine.

11.3.1.1 Assay for Pancreatic α-Amylase Inhibitory Activity

Assay for pancreatic α-amylase inhibitory activity starts with pretreatment of enzyme with herbal extracts mixtures. The contents of the pre-treatment mixture is 2.5 ml of 0.1 M sodium phosphate buffer (pH 7.4) containing 100 U of pancreatic alphα-amylase and 50–250 μg ml^{-1} extracts. Following the incubation of the mixture at 37 °C for 30 min. The addition of 1.0 ml DNSA reagent and boiling for 10 min on the water bath terminates the reaction. The final volume of reaction mixture in each tube is made to 10 ml with distilled water and absorbance is measured at 540 nm on UV-Vis spectrophotometer. The reaction mixture in the control tube is prepared without any plant extract. The % Inhibition is calculated according to the formula given in Eq. 11.1.

$$\%Inhibition = \frac{\text{A540 Control} - \text{A540 sample}}{\text{A540 Control}} \; x\ 100 \qquad (11.1)$$

11.3.2 α-Glucosidase

The enzyme α-glucosidase is typically located on the brush border of the intestine. It breaks down starch and disaccharides to glucose by acting on the α(1→4) glycosidic bonds. Compounds such as acarbose, an α-glucosidase inhibitor, competitively and reversibly inhibits α-glucosidase in the intestines. This is typically a measure to avoid the breakdown of carbohydrate. Natural compounds in this category are used as anti-diabetic agents in combination with additional compounds for treatments.

11.3.2.1 Assay for α-Glucosidase Inhibitory Activity

The inhibitory activity is determined by incubating a 1 ml solution of starch substrate (2% w/v maltose or sucrose) with 0.2 M Tris buffer of pH 8.0 and numerous concentrations of herbal extract at 37 °C for 5 min. The reaction is initialized by the addition of 1 ml of α-glucosidase enzyme (1 U/ml) to it followed by incubation for 40 min at 35 °C. Then the reaction is terminated by the addition of 2 ml of 6N HCl. The intensity of the colour is read at 540 nm (Eq. 11.2).

$$\%Inhibition = \frac{Absorbance\ of\ control - Absorbance\ of\ sample}{Absorbance\ of\ control}\ x\ 100 \quad (11.2)$$

Where, Absorbance of control is the absorbance of the control reaction (covering every reagent excluding the test model). A raw enzyme solution is used as the control (Manikandan et al. 2016).

11.3.3 Sucrase

The invertase, is a digestive enzyme secreted in the intestine. It is localized at the mucosal brush border of the intestine and catalyzes the hydrolysis of sucrose into fructose and glucose. The capability of plant natural compounds to lower the postprandial blood glucose and insulin levels in diabetic patients is considered by its ability to prevent the actions of sucrase (Dsouza and Lakshmidev 2015). Colorimetrically, the glucose released into the adjacent solution by the action of sucrase on sucrose is measured.

11.3.3.1 Assay of Sucrase Inhibitory Activity

The enzyme solution (10 μl) and variable concentrations of the aqueous extract (100–500 μg) is incubated together for 10 min at 37 °C, and the volume is made up to 200 μL with maleate buffer (pH 6.0). The enzyme reaction is initiated by adding 100 μl sucrose solution (60 mM) and is allowed to incubate for 30 min. The reaction is brought to a stop by adding 200 μL of 3,5- dinitrosalysilic acid reagent and treating the mixture in a boiling water bath for several minutes. The absorbance of the solution is measured at 540 nm (Honda and Hara 1993). The percentage inhibitory activities are calculated using the formula (Eq. 11.2).

11.3.4 Dipeptidyl Peptidase-4 (DPP4)

DPP4 otherwise known as adenosine deaminase complexing protein 2 or CD26 (cluster of differentiation 26) is a serine protease that, in humans, is encoded by the DPP4 gene (Kameoka et al. 1993). DPP4 is a legalized drug target for type 2 diabetes since it plays a key role in glucose metabolism where it is accountable for the degradation of incretins, a group of metabolic hormones that stimulate a reduction in blood glucose levels. Incretins are frequently released after the ingesting of a meal and enhance the secretion of insulin released from pancreatic β-cells of the islets of Langerhans by a blood glucose-dependent mechanism. Prominent illustrations of incretins are the intestinal peptides glucagon-like peptide-1 (GLP-1) and gastric inhibitory peptide (also known as: glucose-dependent insulinotropic polypeptide or GIP). Both GLP-1 and GIP are swiftly neutralized by the enzyme dipeptidyl peptidase-4 (DPP4). DPP4 inhibitors can induce insulin secretion, lower blood glucose and HbA1C levels, lessen the apoptosis and enhance the proliferation of β-cells. Five DPP4 inhibitors: linagliptin, saxaglptin, valdagliptin, alogliptin and to finish sitagliptin, have been released on the market (Deacon 2011).

11.3.4.1 Assay of DPP4-Inhibitory Activity

The test well contain 10 mL compound solution (diluted to the concentration of 10 mmol/L) and 50 mL DPP4 solution (2 mU/mL in 50 mmol/L Tris–HCl buffer, pH 8.0) in a 96-well plate. The enzyme reaction is initiated by the addition of 40 mL Gly-Pro-pnitroanilide solution (0.26 mmol/L in HEPES buffer pH 7.05). The change in absorbance at 405 nm is monitored for 60 min at 37 °C. Compound solution is replaced by the distilled water in the negative control well. The percentage of inhibition is calculated by the formula given in Eq. 11.2.

11.3.5 Glycation Inhibition

According to Sattar et al. (2012), the incubation of proteins with glucose leads to non-enzymatic glycation and creation of amadori products recognized as an initial glycation product. Non-enzymatic glycation of proteins or Maillard reaction is aggravated in diabetes mellitus due to hyperglycemia and leads to numerous difficulties such as impaired vision, hepatic and cardiovascular diseases, nerve damage and kidney failure (Chen et al. 2011; Hussain et al. 2011). Lately, considerable attention has been drawn on natural and artificial glycation inhibitors to postpone the onset or advance of diabetic complications (Sattar et al. 2012).

11.3.5.1 Preparation for Glycation Inhibition Assay

To study the outcome of plant extract on protein glycation process, 15 combinations of 0.075 M phosphate buffer saline (PBS) with glucose (G_1: 5.5, G_2: 25, G_3: 50 mM), plant extract (I_1:5, I_2:10, I_3:20, I_4: 40 g/L as inhibitor) and protein (bovine albumin) is incubated at 37 °C for 5 weeks. Samples are evaluated after 3rd and 5th weeks of incubation. Glucose concentration is measured and samples dialyzed to eliminate free glucose. Free glucose is the major hindrance in the approximation of glycation level. During post-dialysis, glucose is once more estimated to determine final glucose levels. Total proteins in the samples after dialysis were then determined by biuret (Sattar et al. 2012).

11.3.5.2 Assay for Non-enzymatic and Enzymatic Glycation (Collective)

One milliliter of dialyzed sample (total protein = 10 mg/mL) is used for enzymatic and non-enzymatic glycation. Three test tubes were organized for each of reduced and non-reduced samples. A 0.1 mL of $NaBH_4$ is added in reduced samples and 0.1 mL of 0.01 N NaOH is then added to non-reduced samples. Every test tube is left for 30 min at 37 °C. After half an hour, 1 droplet of 1N HCl is added to each test tube, followed by the addition of 0.5 mL oxalic acid. Tubes are capped and autoclaved for half an hour at 124 °C (115 lb/in.^{2} pressure). Tubes are cooled to room temperature and then placed in ice. In each individual tube, 0.5 mL of 40% trichloroacetic acid (chilled) is added. Subsequently, samples are centrifuged for 15 min at 15,000 rpm. Supernatant (1.5 mL) is added with 0.5 mL recently prepared thiobarbituric acid. Subsequently, the samples are incubated at 37 °C in water bath for 15 min and absorbance determined at 443 nm (Sattar et al. 2012).

11.3.5.3 Assay for Enzymatic Glycation

For the purpose of determining enzymatic glycation, (EG), 0.1 mL of 0.01N NaOH comprising of 400 molar excess of $NaBH_4$ is used. After the reduction, the glycation level is determined by the matching procedure as revealed above. Non-enzymatic glycation (NEG) is calculated as follows: NEG = (NEG +EG) – EG (Sattar et al. 2012).

11.3.6 Glucose Diffusion Inhibition

The potential of plant extracts to hinder glucose diffusion into the external solution can be examined at set time intervals according to Gallagher et al. (2003).

11.3.6.1 Assay for Glucose Diffusion Inhibition

In a dialysis tube (6 cm × 15 mm), 15 mL of a solution of glucose and NaCl (0.15 M) is introduced and the appearance of glucose in the external solution is measured. The sealed tube is positioned in a centrifuge tube comprising 45 mL of 0.15 M NaCl. The tube is positioned on an orbital shaker at room temperature. Glucose concentrations is measured by glucose oxidase equipment technique. Incremental areas under the glucose curves (AUC) were calculated by trapezoidal rule (Gallagher et al. 2003; Sattar et al. 2012).

11.3.7 Angiotensin I-Converting Enzyme (ACE) Inhibition

Angiotensin-converting enzyme is a vital constituent of the renin–angiotensin system (RAS), which controls blood pressure by modifying the capacity of the fluids in the organism. It changes the hormone angiotensin I to the dynamic vasoconstrictor angiotensin II. Angiotensin II has been shown to upsurge hepatic glucose production and reduce insulin sensitivity. Patients who utilize either an ACE inhibitor (ACEI) or angiotensin receptor blocker (ARB) might involve a reduced incidence of new-onset type 2 diabetes (Gillespie et al. 2005). ACE inhibitors and angiotensin II-receptor blockers are recommended to evade circulatory disease and nephropathy in patients with type 2 diabetes. As there are prospects to this, the co-administration of dipeptidyl peptidase 4 (DPP4) inhibitors and angiotensin-converting enzyme inhibitors have been reported to be linked with adverse hemodynamic effects.

11.3.7.1 Assay for ACE Inhibition

ACE inhibition assayed is determined by a method modified by Kwon et al. (2006), using hippuryl-histidyl-leucine (HHL) as a substrate and the ACE-I enzyme (e.g. from rabbit lung). A volume of 50 μL of sample extracts is incubated with 100 μL of 1 M NaCl-borate buffer (pH 8.3) containing 2 mU of ACE-I solution at 37 °C for 10 min. After the pre-incubation, 100 μL of a 5 mU substrate (HHL) solution is added to the reaction mixture. Test solutions is incubated at 37 °C for 1 h. The reaction is terminated with the addition of 150 μL of 0.5 N HCl to the reaction.

11.3.8 Protein Tyrosine Phosphatases

11.3.8.1 Protein Tyrosine Phosphatases

Protein tyrosine phosphatases (PTP) are a group of enzymes that cleaves phosphate groups from phosphorylated tyrosine residues on proteins. Lately, the protein tyrosine phosphatase PTP-1B has been revealed to be a negative regulator of the insulin signaling pathway, telling that inhibitors of this enzyme might be advantageous in the handling of type 2 diabetes. Mice missing PTP-1B are resilient to both diabetes and overweightness (Kennedy 1999).

11.3.8.2 Assay for PTP 1B Inhibition

According to Saidu et al. (2017), the components were thawed on an ice bath with the exclusion of BIOMOL RED™ that is kept at room temperature. The substrate ('IR5' Insulin receptor b, residues 1142–1153, Py1146) is reconstituted to a concentration of 1.5 mM by assay buffer and distilled H_2O. The assay buffer, 100 mM MES, pH 6.0 comprising 300 mM NaCl, 2 mM EDTA, 2 mM DTT and 0.1% NP-40 is diluted with equivalent volume of distilled H_2O and preserved on ice. The PTP 1B (human recombinant) is prepared in ×1 cold assay buffer. Stock of 10 mM of suramin (a known inhibitor) is prepared in assay buffer. The assay mixture is prepared in 96 well plate which contains 10 μg per 100 μL assay mixture of the sample. The plate reader is read at 620 nm, and all the assay protocol is completed in harmony with manufacturer's instructions.

11.3.9 GLUT4 Translocation

Glucose transporter type 4 (GLUT-4), a membrane protein is an insulin-regulated glucose transporter found largely in adipose tissues and striated muscle (skeletal and cardiac) that continuously cycles between intracellular stores and the plasma membrane (PM). The insulin-stimulated glucose uptake mainly results from the translocation of GLUT4 from an intracellular storage pool to the plasma membrane in some ways through exocytosis and fusion with the PM. In type 2 diabetes mellitus, fat cells and muscle fibers seem to be oblivious to the hormone, a flaw that has been linked with decreased GLUT4 translocation (Wang et al. 1998). One supportive indication that agents proficient in enhancing the actions of GLUT4 will help ameliorate type 2 diabetes through the cumulative uptake of glucose, is the reorganization of RAC1, a GTPase also triggered by insulin. Rac1 works by stimulating the reform of the cortical Actin cytoskeleton (JeBailey et al. 2007), which permits the GLUT4 vesicles to be implanted into the plasma membrane (Sylow et al. 2014). A RAC1 Knockout mouse has condensed glucose acceptance in muscle tissue.

11.3.9.1 Assay for Determination of Surface GLUT4myc

Surface myc tagged GLUT4 is measured in whole, non permeabilized cells as formerly defined (Zaid et al. 2009) using anti-myc antibody followed by secondary antibody conjugated to horseradish peroxidase. Cells grown in 24-well plates for 1 day followed by addition of the herbal extracts for 20 h. Serum starved for 3 h is treated with 1 μM insulin for 20 min, washed twice with ice-cold phosphate buffer saline (PBS), fixed for 10 min with 3% paraformaldehyde, incubated with glycine 0.1 M, blocked 10 min with 3% (v/v) goat serum and reacted with polyclonal anti-myc antibody (1:200) for 1 h at 4 °C, washed 10 times with PBS and reacted with horseradish peroxidase-bound goat anti-rabbit secondary antibody (1:1000) for 1 h at 4 °C, and washed 15 times with PBS. Cells are then incubated with 1 mL ophenylenediamine dihydrochloride reagent and kept to develop for 20–30 min in the linear range in the dark at room temperature. The reaction is terminated with the addition of 1 ml/well of 3 N HCl. Supernatants are collected and absorbance then measured at 492 nm. Background absorbance obtained in the absence of antimyc antibody is subtracted from all values.

11.3.10 Glucose Induced Insulin Secretion Assay (GSIS)

GSIS assay is used to evaluate the potentials of an extract to modulate glucose-stimulated insulin secretion. These cells release insulin in response to physiological concentrations of glucose in a dose-dependent manner.

11.3.10.1 Insulin Secretion Assays

According to Ansarullah et al. (2011), the isolated islets are cultured at 37 °C in a moistened atmosphere of 5% CO_2 in air in RPMI-1640 medium comprising 10% FBS and antibiotics. Islets are seeded at a concentration of 100 or 50 islets per well in 24-well plates (Falcon, NJ) and permitted to attach overnight prior to acute examinations. Shafts are washed multiple times with Krebs-Ringer bicarbonate buffer (KRB; 115 mM NaCl, 4.7 mM KCl; 1.3 mM $CaCl_2$, 1.2 mM KH_2PO_4, 1.2 mM $MgSO_4$, 24 mM $NaHCO_3$, 10 mM HEPES, 1 g/L BSA, 1.1 mM glucose; pH 7.4) and pre-incubated for 1 h at 37 °C. If otherwise specified, wells are then incubated for 1 h in 1 mL KRB with 4.5 mM or 16.7 mM glucose and OI extract/FRF (10, 50, 100 or 250 μg/mL). Aliquots are detached from each well, centrifuged (2000 rpm for 5 min, at 4 °C), and analyzed for insulin with mouse insulin ELISA equipment and protein concentration is then determined.

11.3.11 Cell Lines as Model for in-vitro Antidiabetic Screening

The use of cell lines in antidiabetic screening offers an alternative to the practice of using research laboratory animal to study physiological and pathophysiological processes in diverse cell types associated with diabetes. There are different classes of cell lines available for such research. They can be grouped as:

1. Insulin-secreting cell lines: The most widely used insulin-secreting cell lines are rat insulinoma cell line (RIN), insulinoma cell line (INS-1)
2. Transformed β-cell lines and insulin-secreting β-cell lines derived from transgenic mice: Transgenic C57BL/6 mouse insulinoma cell line (MIN), Hamster pancreatic β-cells (HIT), Beta-hyperplastic islet-derived cells (βHC), Beta-tumour cells (βTC).
3. Insulin-secreting cells of non-islet cell origin.

11.3.12 Glucose Uptake Assay

According to Harbilas et al. (2009), glucose uptake assay is used to determine if a plant extracts had the potentials of an insulin-like or insulin-sensitizing effects on glucose acceptance, C2C12 myotubes and 3T3-L1 adipocytes were treated with the maximum non-hazardous concentration of herbal extracts and rate of glucose uptake is determined by calculating the integration over time of 3H-labelled deoxyglucose, a nonmetabolizable equivalent of glucose. According to Both the C2C12 and 3T3-L1 cell lines express GLUT-1 and GLUT-4 glucose transporters when distinguished and exhibit insulin-regulated glucose uptake.

11.3.13 Glucose Uptake Assay

As described by Harbilas et al. (2009), cells grown-up in 12-well plates to 80% confluence is treated with either vehicle (0.1% DMSO) alone, otherwise with extracts solubilized in vehicle, or else with a positive control solubilized in vehicle, for 1 h or 18 h. Metformin concentrations at 400 mmol/L can be used as positive controls for the C2C12 cell line, fenugreek seed ethanolic extracts at 75 mg/L can be used for the 3T3-L1 cell lines. Two hours before the 1 h treatment, differentiation medium is replaced with serum-free medium. The initial 15 h of the 18-h long treatment is achieved on a comprehensive differentiation medium. The medium is then replaced with serum-free medium containing fresh treatment for the remaining 3 h of the treatment. After an extended 18-h treatment, cells are rinsed twice with KPB (Krebs phosphate buffer, in mmol/L: 20 Hepes, 4.05 Na_2HPO_4, 0.95 NaH_2PO_4, pH 7.4, 136 NaCl, 5 Glucose, 4.7 KCl, 1 $CaCl_2$, and 1 $MgSO_4$) warmed in the presence of 1–100 nmol/L insulin at KPB for 30 min at 37 °C or without the presence of insulins. Afterwards, the cells are positioned on ice with rapid rinse three times continuously in cold-glucose KPB. The procedure continues with cells being lysed in 500 mL of 0.1 mol/L NaOH for 30 min and lastly scraped. Lysates, together with a trivial quantity of water (1 mL), are added to 4 mL of the Ready-Gel 586,601 liquid scintillation cocktail, and in conclusion measured for the incorporated radioactivity into a scintillation counter.

11.3.14 Intracellular Calcium Levels

As reported by Ansarullah et al. (2011), the intracellular calcium concentration is measured using furα-2 AM. Furα-2 AM crosses cell membranes in addition as soon as inside the cell, the acetoxymethyl groups detached by cellular esterase which generate furα-2, the fluorescent calcium indicator. Islets were incubated in calcium free Hanks' Balanced Salt Solution (HBSS) with 5 μM furα-2AM at 37 °C for 30 min in shaking water bath. After washing (2×) with calcium free HBSS, islets were suspended in complete HBSS and treated with numerous concentrations of FRF for 60 min in a shaking water bath. The fluorescence measurement is determined by a spectrophotometer which records a length of 500 nm for dual excitation of wavelength from 340 to 380 nm. Calcium ions are expressed as nmole/50 islet equivalents.

11.4 Cell Viability Studies

A viability assay is an assay to determine the capability of organs, cells or tissues to maintain or recover viability. Treating cells with the cytotoxic compound can result in a diversity of cell fates counting necrosis. The screening of antidiabetic agents frequently goes hand in hand with cytotoxicity examinations. This is because an effective antidiabetic agent should not be poisonous to cells of the body or else such agents will be considered not appropriate for use. Cytotoxicity can be monitored using MTT (3-(4,5-Dimethylthiazol-2-yl)-2,5-d iphenyl tetrazolium bromide) cytotoxicity assay and LDH (lactate dehydrogenase) leakage assay among other proliferation assays.

11.4.1 MTT Assay

The MTT assay is a colorimetric assay for evaluating cell metabolic activity. Cells with a low metabolism reduce small amount of MTT from a yellow tetrazole to reduced purple formazan. In contrast, swiftly dividing cells display high rates of MTT reduction.

MTT assay is based on the protocol described by Mosmann (1983). The assay is optimized for the cell lines used in the experiments. MTT is applied to assess cell viability as described by Kadan et al. (2013). Cells (2×10^4/well) were plated in 100 μL of medium/well in 96-well plates and were permitted to attach to the plate for 24 h. Plant extracts were added at growing concentrations (0–2 mg/ml) for 20 h. The cell medium was then substituted with 100 μL fresh medium/well containing 0.5 mg/ml MTT and cultivated for extra 4 h darkened in the cell's incubator. The supernatant is removed and 100 μL isopropanol/HCl (1 mM HCl in 100% isopropanol) were added per well. The absorbance at 570 nm is measured with microtiter plate reader. Two wells per plate lacking cells served as blank. The effect of the plants extracts on cell viability is expressed using Eq. 11.3.

$$\%Viability = \frac{A620\ nm\ of\ plant\ extract\ treated\ sample}{A620\ nm\ of\ nontreated\ sample}\ x\ 100 \qquad (11.3)$$

11.4.2 LDH Leakage Assay

LDH is a stable enzyme common in all cells which can be readily detected when cell membranes are no longer intact.

11.4.2.1 Assay for LDH

The permeability of cellular membranes following the exposures is determined by calculating the quantity of released LDH enzyme from (L6-GLUT4myc) cells. The proportion of LDH activity in the supernatant of treated cells to the LDH activity released in the control cells is stated to as cell membrane rupture. L6-GLUT4myc cells (2×10^4/well) were plated in 100 μL of medium/well in 96-well plates and were allowed to attach to the plate for 24 h. Cells were then pickled with cumulative concentrations of the plant extracts (0–2 mg/ml). The extracellular LDH activity is measured in the medium after 24 h. Therefore, 50 μL from each well is then transferred to a new-fangled 96 well plate; the enzyme response is carried out according to the manufacture equipment. The effect of the plants extracts on cell viability is expressed using Eq. 11.4.

$$\%Viability = \frac{A492\ nm\ of\ plant\ extract\ treated\ sample}{A492\ nm\ of\ nontreated\ sample}\ x\ 100 \qquad (11.4)$$

11.5 Animal Models for Screening of Antidiabetic Agents

The use of animal models to assess the pathophysiology state of a disease is considered a veritable tool for drug development. Animals such as mouse, rat, guinea pig and rabbit can acts as relevant model for the study of diabetics. These can help in elucidating the changes occurring during the natural development of diabetes, monitoring and treatment. Thus, these models are used for testing, developing and/or screening of antidiabetic agents. Furthermore, works are done to get techniques and develop chemicals that can alter the functioning of pancreas for changes in the body and thereby inducing diabetes mellitus. Diabetic models are developed by chemical, surgical and genetic interventions in these rodents. Chemical method of induction of diabetes comprises of the use of chemicals like streptozotocin, alloxan, dithizone, monosodium glutamates, ferric nitrilotriacetate, ditizona, antiinsulin serum etc. out of which alloxan and streptozotocin are the most commonly used ones (Mendez and Ramos 1994). The other methods are through surgery or by the use of genetically developed diabetic rat. Although, it is quite difficult to select the right model to screen a new compound for its efficacy as antidiabetic agent (Thatte 2009; Etuk 2010).

11.5.1 Chemically Induced Model

The chemicals used for the induction of diabetes specifically target β-cells; temporarily inhibit production and secretion of insulin and thus reducing the efficacy of insulin.

11.5.1.1 Streptozotocin

Streptozotocin is mostly used for inducing diabetes in rats (Junod et al. 1967). Streptozotocin is derived from *Streptomyces achromogens* and is effective against a large number of bacteria apart from being used as a chemical to induce diabetes (Vavra et al. 1959; Wu and Yan 2015). Streptozotocin is a nitrosourea antibiotic used clinically as an antineoplastic compound in the therapeutics of pancreatic β-cell carcinoma. Streptozotocin causes alkylation and then fragmentation of DNA. It changes and finally destroys β-cells, causing precipititation of insulin–dependent diabetes. Streptozotocin expresses its effect by damaging β-cells of pancreas, causing hypoinsulinemia, i.e., insulin deficiency and also shows polydipsia, polyuria and hyperglycemia (Wu and Huan 2008).

Streptozotocin has to pass into the pancreatic cell which occurs with help of GLUT2 (Glucose transporter 2), a facilitated glucose transporter of low affinity. Hence, GLUT2 free insulin producing cell become resistant to streptozotocin. Streptozotocin can only damage those cells of liver and kidney which express GLUT2 (Lenzen 2008). The more glucose sensitive β-cells allow more entry of

Streptozotocin via GLUT2 transporter finally resulting in Streptozotocin toxicity (Eleazu et al. 2013). Streptozotocin is similar in structure to N-acetyl glucosamine having a glucose moiety. N-acetyl glucosamine causes deleterious effects while the glucose moiety is non-toxic in nature (Lenzen 2008; Goyal et al. 2016).

Streptozotocin is the commonly used chemical for inducing experimental diabetes in rodents. It can be used alone or in combination of other drugs or with dietary interventions for induction of both types of diabetes. Diabetes induction is done in rodents by single parenteral dose of Streptozotocin injection by three approaches as follows:

1. Streptozotocin injection after nicotinamide administration (type 1 diabetes)
2. Increased fat diet with low amount of Streptozotocin (type 2 diabetes)
3. Streptozotocin injection during neonatal period (Wu and Yan 2015).

The following observations was made by some scientists:

1. Streptozotocin at 60 mg/kg dose was administered intravenously in Wistar rats. This caused degeneration of β-cells and swelling of pancreas and thus inducing diabetes within 2–4 days. There was marked histopathological changes in the β-cells of pancreas causing diabetes (Motyl and McCabe 2009).
2. Each rat was intraperitoneally injected Streptozotocin at 60 mg/kg body weight in 0.1 ml saline under anasesthesia. For confirming diabetes induction, samples of fasting blood was taken from dorsal vein of tail of rat and tested by glucose oxidase method before and after Streptozotocin administration. Those rats that showed fasting blood sugar higher than 250 mg/dl were separated as diabetic (Selim and Selim 2013).
3. Wistar rats of both sexes weighing 170–200 g were made diabetic by injecting Streptozotocin 40–60 mg/kg body weight dissolved in a citrate-saline solution, pH 4.5 in the tail vein. The female rats kept in another group were injected intravenously 3-O-methyl glucose at 1.65 mg immediately prior to Streptozotocin at 60 mg/kg to prevent the latter's diabetogenic effect (Ganda et al. 1976). To measure the fasting blood glucose, blood from tail vein of nonfasted ether anesthetized – control as well as diabetic animals were drawn (Malhotra et al. 1981).
4. Streptozotocin at 45 mg/kg body weight in citrate buffer (0.1 M) with pH 4.5 was administered intraperitoneally in the male Wistar rats to precipitate diabetes while control rats were given citrate buffer alone. After 3 days, fasting blood was taken from the rats' eyes through retro-orbital plexus to determine plasma glucose levels. The animals with the increased plasma glucose levels than 200 mg/dl are confirmed to be diabetic (Gondi and Prasada 2015).
5. Streptozotocin at 20 mg/kg body weight in a citrate buffer, pH 4.2, intraperitoneally was administered in mice for 5 days consecutively (Pechhold et al. 2001).
6. Streptozotocin at 40 or 60 μg/g body weight in 0.1 M citrate buffer were given intraperitoneally for 5 days daily in 10 weeks Male BALB/c mice to induce diabetes. The mice kept as control were injected with buffer only. After 7 days blood glucose was examined in blood from the lateral saphenous vein with a glucometer. The mice which had glucose concentration greater than 300 mg/dl were considered diabetic (Motyl and McCabe 2009).

7. Streptozotocin at 25 mg/kg body weight in CASC (citric acid-sodium citrate) buffer was injected intraperitoneally in high fat and sugar diet fed rats for 4 weeks to induce diabetes. Again, Streptozotocin at 40 mg/kg was repeated to each rat 4 weeks later. The glucose in blood was measured weekly after 8, 12 and 16 corresponding weeks (Yang et al. 2013).

11.5.1.2 Alloxan

Dunn, Sheehan and McLetchi were the first scientists in 1943 to demonstrate that alloxan can be used to develop a diabetic model (Dunn et al. 1943). Chemically, alloxan is 2,4,5,6-tetraoxypyrimidine; 5,6-dioxyuracil. The name comes from the word "allantoin" which is actually the byproduct of uric acid generally excreted by foetus in the allantois. Alloxan induces experimental diabetes by destruction of β-cells in the islets of pancreases selectively which produce insulin. The plasma insulin concentration decreases. Ultrastructural changes occur in β cell resulting in necrosis of the cell leading to its death.

Dose rate of alloxan:

The dose depends on strain of animal and route of administration of the drug (Szkudelski 2001; Wang et al. 2011). The following doses has been of use:

1. Rat: 40–200 mg/kg i.v or i.p.
2. Mice: 50–200 mg/kg i.v or i.p.
3. Rabbit: 100–150 mg/kg i.v.
4. Dog: 50–75 mg/kg i.v.

Alloxan causes response in three phases as follows:

1. Phase I – early hyperglycemia: about 1–4 h; sudden decrease or block of insulin.
2. Phase II – hyperglycemia: up to 48 h; convulsions.
3. Phase III – chronic diabetics.

Alloxan is similar to glucose in structural shape. Alloxan monohydrate is a β-cell destructor existing in aqueous medium. Glucose being a hydrophilic molecule is not able to cross the lipid layer of plasma membrane to go to the cytosol by itself. So, it takes help of GLUT2 (glucose transporter 2), a transport protein transported by facilitated diffusion (Ighodaro et al. 2017). The diabetes induced by administration of alloxan is insulin dependent diabetes. Alloxan shows antidiabetic effect by free radical formation. It rapidly enters into the β-cell and forms free radicals which β-cells cannot defend and are seriously affected (Nerup et al. 1994). There exists a redox cycle where alloxan is changed to dialuric acid and vice-versa for the generation of superoxide radical and again changes to H_2O_2 and then to hydroxyl radicals casing breaking of DNA of β-cells (Szkudelski 2001). However, the liver is not susceptible to ROS (reactive oxygen species) damage as its cells have better defense mechanisms (Malaisse et al. 1982; Mathews and Leiter 1999). β-cells are also damaged by oxidation of –SH groups of glucokinase (im Walde et al. 2002) and intracellular calcium homeostasis disturbances (Kim et al. 1994). Alloxan acts as a diabetogenic agent in a very narrow range. Slight over dosing causes toxicity (Szkudelski 2001; King 2012).

11.5.1.3 Gold Thioglucose

Gold thioglucose is a derivative of glucose sugar which shows hyperphagia and induces severe obesity which leads to Type-2 diabetes. In normal mouse, gold thioglucose can cause the development of obesity and thus induces diabetes.

Gold thioglucose, a neurotoxic glucose analogue rapidly destroys leptin receptor-positive hypothalamic neurons within 24 h, including those that regulate energy expenditure and food intake (satiety) leading to disturbed feed intake. C57BLKs, DBA/2 and BDF1 mice treated with gold thioglucose rapidly gain weight and there is significant increase in non-fasting blood sugar in 8–12 weeks. These animals showed impaired and reduced insulin secretion after glucose intake in pancreatic islets (Karasawa et al. 2011). Gold thioglucose-induces obesity-related type 2-diabetes in the apoE-KO mouse model after peripherally injecting a single dose of it which gives rise to insulin resistance leading to obesity.

11.5.1.4 Atypical Antipsychotics

During the 1950s, some connection between antipsychotic drugs and diabetes was suspected. In 1960s, it was accepted that there exists some connection between antipsychotic drugs and diabetes and there was emergence of a new term "Phenothiazine Diabetes".

The atypical antipsychotics influences metabolism of glucose irrespective of presence of psychiatric illness. This connection was proven by anecdotal case reports, drug safety studies, pharmacoepidemiological studies and prospective studies (Holt and Peveler 2006). It is still a very big challenge to understand the role of antipsychotics in hyperglycemia.

Antipsychotic medications affect energy intake and/or expenditure, insulin resistance and/or β-cell function. Besides the influence of the pharmacological profile of the drug and underlying psychiatric disease, the response of any individual to antipsychotic treatment is also determined by the environmental factors (food intake, smoking, etc.) and genetic factors for weight maintenance, insulin resistance, functioning of pancreas etc.

The intake of atypical antipsychotic drugs leads to different weight gains (Newcomer 2005). Thus, there arises difference in risk of hyperglycemia. This risk increases with increase in adiposity. Clinical reports say that this weight gain may not be a factor in cases of new-onset diabetes that occur during treatment. It is not very clear if the antipsychotics like clozapine has any effect on glucose metabolism, β-cells ability to secrete insulin (Bergman and Ader 2005).

Antipsychotic medications associated diabetes has two groups of patients (Scheen and De Hert 2004):

11.5.1.4.1 Rapid Onset After Start of Antipsychotic Medications

Here, the patients are younger, less overweight and more females are affected. Ketoacidotic coma occurs through defective insulin secretion (Bergman and Ader 2005).

After the withdrawal of the antipsychotic medication, the disorder becomes reversible. So, a functional defect instead of β-cell destruction is confirmed. This group needs further investigations on physiology and medications.

11.5.1.4.2 Classical Onset After Start of Antipsychotic Medications

The predisposing factors like gain in weight, increased abdominal fat etc. are important factors. The progress of the diabetes is started by increase in insulin resistance (Newcomer 2005). However, the progression to diabetes is more rapid than obese cases without psychiatric disease. The secreted insulin causing lesser compensation of β-cells is whether backed by central or peripheral mechanism is still unclear (Scheen and De Hert 2007).

11.5.1.5 Dithizone

Dithizone is an organosulphur chelating agent. It chelates lead, zinc, mercury. It has chemical name of 8-(p- toluene- sulfonylamino)– quinoline (8- TSQ). It can be employed to know the purity of pancreatic islet preparation of human for medication of type 1 diabetes mellitus. It is helpful to induce diabetes in species like mice, cat, rabbit and golden hamster. These animals when dithizonised show unchanged levels of copper and magnesium and higher levels of zinc, and potassium (Tripathi and Verma 2014). Dithizone can pass through membranes and form complex zinc inside liposomes. This releases protons that can enhance diabetogenicity. When dithizone like drugs are added to lipid vescicles of insulin storage granules of pancreatic β-cells having zinc ion within them, at pH 6, acidification and release of proton occurs. Then, the solubilisation of insulin is initiated which increases the osmotic stress further. Thereafter granules rupture and diabetes sets in (Epand et al. 1985).

11.5.2 Surgically Induced Model

Several techniques are used surgically for induction of diabetes. This could be done by duodenal-jejunal bypass and pancreatectomy (removing the pancreases completely). The process of pancreatomy is used in species like rats, pigs, dogs, primates etc. to find out effects of antidiabetic drugs (Choi et al. 2004; Rees and Alcolado 2005; Masiello 2006).

11.5.2.1 Duodenal-Jejunal Bypass

Duodenal-jejunal bypass model reverses type-2 diabetes mellitus in Goto-Kakizaki rats. This decrease in diabetes can be known by decrease in resistance to insulin in skeletal muscle studied by insulin signal and glucose disposal measurement in skeletal muscle (Gavin et al. 2011). It showed that proximal small intestine bypass does not increase glucose distribution in skeletal muscle. Nowadays, bariatric surgery has emerged as a method to decrease and reverse progress and effects of diabetes along with its normal use for weight loss. It caused improvement in hyperglycemia after surgery in obese patients. This positive effect can be seen in both human and animal models. The weight loss helps in having improved control of glucose in obese diabetics and restores euglycemia after surgery. In both non-obese diabetics (DePaula et al. 2009) and animal models, antidiabetic effects are seen without weight loss.

The speedier recovery and long term remission suggests surgery specific and weight loss independent mechanism. This leads to lots of positive consequences like better quality of patient life and less expenditure. So, researchers are becoming more interested in these mechanisms underlying remission of diabetes type 2. Knowing the pathogenesis and effect of surgeries on type 2 diabetes helps in development of new procedures, devices and drugs. The simulating drugs of these mechanisms will help develop new powerful medicines for treatment of diabetes without surgery (Allen et al. 2013).

11.5.2.2 Non-obese Partial Pancreatectomy

Pancreatectomy is the removal of the pancreas completely done to induce diabetes. This pancreatectomy is used in rats, pigs, dogs and primates (Choi et al. 2004; Rees and Alcolado 2005; Masiello 2006). Nearly all experiments in diabetes are performed in rodents. The dog was used as complete pancreatectomy model by Oskar Minkowski for first time but now its use is obsolete (Ozturk et al. 1999).

Partial pancreatectomy (70–90% dissection) of pancreas is done in rats, dogs, pigs, rabbit (McNeil 1999; Sasaki et al. 2000). Hyperglycemia of moderate level is seen but insulin level and body weights are not decreased. It is precipitating a mild form of diabetes. Partial pancreatectomy is done, but large (more than 80%) resection is required for milder hyperglycemia. Any additional resection results in hypoinsulinemia (Masiello 2006).

The animal model was made partially diabetic by putting a clamp between the body and tail of the pancreas. The specific portion of pancreas is made diabetic with total loss of insulin secreting β-cells. The rest of pancreases remain normal. For occulting the circulation, a vascular clamp is put in between body and tail of the pancreas. Alloxan at 200 mg/kg I/V was administered followed by dextrose at 0.5 g/kg I/V after 4 min. After 2 min the clamp was removed. The survivors were sacrificed in 4–12 weeks of surgery and animal were found fully not diabetic as they had normal β-cells in head and body potion of the pancreas. The disadvantages of this technique are:

1. Technical expertise is needed,
2. Surgical theatre with appropriate requirements is needed,
3. As it is a major surgery with risk of infection, post-operative antibiotics and analgesics are to be given,
4. Pancreatic enzyme supplement is to be given to prevent malabsorption and loss of pancreatic counter response to hypoglycemia.

11.5.3 Genetically Induced Diabetic Animal Model

The type-2 diabetes in spontaneous diabetic animals is achieved from the animals with one or numerous inherited mutations which is spread from one generation to another. A notable example is the db/db mice. This can be achieved by the selection from non-diabetic out bred animals through repetitive breeding over numerous generation (e.g., BB rat, Tsumara Suzuki Obese Diabetes mouse etc.). Diabetes is expressed in these rodents as single or multigene defects in KK mouse, db/db mouse, or Zucker fatty rat. The individual gene defect might be due to dominant genes (e.g., KK/A mice), recessive genes (e.g., db/db mouse) or multiple genes (Rodrigues 2016). The majority of human population suffers from Type- 2 diabetes which is a combination of ecological and numerous gene defects though certain subtype of diabetes with acknowledged causes e.g., maturity onset diabetes of youth due to defects in glucokinase gene and this single gene defects might cause type-2 diabetes. Thus, polygenic origin animals represent the human condition more accurately when compared to monogenic gene defect animals (Ktorza et al. 1997; McIntosh and Pederson 1999).

11.5.3.1 Zucker Fatty Rat

The Zucker fatty (ZF) rat is the model that results from a simple autosomal recessive (fa) gene on chromosome no. 5. It shows hyperphagia and early and rapid onset of obesity at 4 week of age and augmented progress of subcutaneous fat depots. It is linked with minor hyperglycaemia, hyperlipidaemia, hyperinsulinaemia, mild glucose intolerance, insulin resistance and moderate hypertension (Shafrir 2003). The hyperphagia has been attributed to hypothalamic defect in leptin receptor signaling. The variations in Zucker rats comprise diminished growth hormone and prolactin levels also (Shafrir 2003; Srinivasan and Ramarao 2007).

11.5.3.2 Zucker Diabetic Fatty Rat

This is a sub – strain of Zucker fatty rat. It is a selective inbred for hyperglycaemia and is vastly beneficial for the examination of mechanism of type 2 diabetes. The male Zucker diabetic fatty rat develops diabetes due to failure to recompense for insulin resistance. It is a lesser amount of overweight nonetheless supplementary insulin resistant. The male rats are further prone to the progress of diabetes which is generally at 7–10 week after birth. This model is currently obtainable with Charles River laboratories, Indianapolis, USA. A much missense mutation appears in Zucker fatty rat, it is about leptin receptors that cause augmented risk of overweightness manifestation without diabetes. The animal with mutation is associated with disruption of normal islet architecture, β-cell degranulation, and increased β-cell death. All animals develop overweightness, insulin resistance and non-insulin reliant on Diabetes Mellitus between 7 and 10 weeks of age in this strain (Kahn 2000). The male Zucker diabetic fatty rats spontaneously becomes type-2 diabetic but females develop diabetes when fed with diabetogenic high-fat diet. The study of hepatic molecular profiles of non-diabetic obese Zucker diabetic fatty rats shows that the hepatic sex alterations might contribute to the sex-based evolution of diabetes in rats (Gustavsson et al. 2011).

11.5.3.3 Goto-Kakizaki Rat

The Goto-Kakizaki rat was developed from Wistar rats with abnormal glucose tolerance which is repeated over several generations and shows profoundly defective insulin secretion leading to basal hyperglycemia. It is generally used for the assessment of type-2 diabetes. It has the characteristics of non-obesity, hypoinsulinaemia, normolipidaemia, hyperglycaemia, lessened glucose tolerance and an early onset of diabetic difficulties (Goto and Kakizaki 1981). In the adult Goto-Kakizaki rats, total pancreatic β-cell mass is reduced by 60% along with identical degree of decline in pancreatic insulin supplies. The defective beta cell mass and function in the Goto-Kakizaki rat model results from insufficiency of pancreatic growth factors essential for the growth and progress of fetal pancreatic cells during gestation. It may also be attributed to secondary loss of β-cell differentiation due to chronic exposure to hyperglycaemia signifying that type 2 diabetes is dependent on genetic factors, involving transgenerational epigenetic responses (Miralles and Portha 2001). The defects in β-cells, impaired insulin sensitivity in the liver, skeletal muscle and adipose tissues has also been reported. Impaired insulin secretion, hepatic glucose overproduction, i.e., hepatic insulin resistance are first events in diabetic Goto-Kakizaki rats. This mostly contributes to the development of hyperglycaemia rather than the peripheral muscle and adipose tissue insulin resistance. Goto-Kakizaki rat is one of the finest categorized animal models used for studying the relation of variations in β-cell mass, type-2 diabetes occurrence and complications particularly diabetic nephropathy. The islets became irregular. The immunoreactions of β-cells against anti-insulin are weakened. The insulin deficiency in Goto-Kakizaki rats is not triggered by simple dysfunction and/or degeneration of β-cells nonetheless rather by additional complex events inside cells (Momose et al. 2006).

11.5.3.4 LEW.1WR1 Rats

The LEW.1WR1 rat is a recent model used for studying autoimmune diabetes and arthritis, particularly collagen-induced arthritis. It is free of spontaneous thyroiditis. The LEW.1WR1 rats have an hereditary tendency to both disorders that can be improved by varying some factors. Spontaneous diabetes happens within age of 59 days in LEW.1WR1 rats. The disease shows symptoms of ketonuria, glycosuria, hyperglycemia and polyuria. The islets of acutely diabetic rats lack β-cells, whereas α- and δ-cell populations are spared (Mordes et al. 2005).

11.5.3.5 New Zealand Obese (NZO) Mouse

The New Zealand Obese mouse is an animal model of diabetes and overweightness. The animal is attained by selective inbreeding with the parents with agouti coat color. It exhibits a polygenic group of symptoms like obesity, hyperphagia, mild hyperglycaemia, impaired glucose tolerance, hyperinsulinaemia and insulin resistance. During initial 2 months of life, the body mass increases quickly due to hyperphagia. In New Zealand Obese mice, hyperleptinaemia and leptin resistance which might account for hyperphagia is seen. With advance in age of animals, hyperglycaemia and impaired glucose tolerance increase continuously. Up to 90% of beta cells displayed an obvious hypertrophy and hyperplasia. The early primary defect for causing diabetes is a reduced glycogen synthase activity in liver. This mouse is shown to have hepatic insulin resistance from an early age. It increased gluconeogenesis and hepatic glucose production linked to the abnormality in the regulation of gluconeogenic enzyme known as fructose-1,6-bisphosphatase. These mice have high incidence of autoimmune illness. This strain of mice has become a valuable model for studying the relationship between overweightness, autoimmunity, and diabetes (Knouff and Auwerx 2004).

11.5.3.6 NONcNZO10 Mouse

The NONcNZO10 mouse model is a very important model with the characteristics of type 2 diabetes. It is developed by joining five genomic intervals holding NZO/H1Lt diabetogenic quantifiable trait loci on top of the non-obese non-diabetic hereditary background at Jackson laboratory, Maine (Haskell et al. 2002). The NZO males acting as parent show the undesirable phenotypes of hyperphagia, morbid overweightness, deprived fertility, and a variable incidence of hyperglycaemia but the established NONcNZO10 males aren't hyperphagic, develop additional modest levels of overweightness and reproduce normally. They weigh significantly fewer than the NZO males and additional than NON-males. All NONcNZO10 males develop long-lasting hyperglycaemia by 12–20 week due to the condensed rate of mass gain compared to NZO. The pancreatic islets show the variations as parallel to diabetic NZO males. Liver steatosis and upsurge in serum triglycerides are realized. Females are unaffected by diabetes. Unlike db/db and ob/ob mice which are monogenic and exhibit a morbid obesity, these mice are polygenic, not hyperphagic. They have a normal leptin/leptin receptor axis but do not exhibit hypercorticism, show no obvious thermoregulatory defects as in humans with obesity/diabetes syndrome (Haskell et al. 2002). This model has gained popularity recently for the investigation of diabetes and obesity and in drug testing for the treatment of type 2 diabetes.

11.5.3.7 ob/ob Mouse

The ob/ob mouse (obese mouse likewise recognized as Lepob) is characterized by a hereditary monogenic autosomal recessive mutation on chromosome no. 6 in C57BL/6J, originating from the Bar Harbor, Jackson laboratory (Shafrir 2003). The mutation in ob/ob mice is recognized in leptin genetic factor, which encodes for leptin. Mutant homozygous mice exhibit a swift increase in body mass and might range three times of the normal weight. There is a very early detection of impaired thermogenesis at 10 days. There is also marked hyperphagia and decreased energy expenditure resulting in increased carcass lipid content with obvious obesity by approximately 4 weeks. It is characterized by diabetic like symptoms of hyperglycaemia, mild impaired glucose tolerance, severe hyperinsulinaemia, sub fertility and impaired wound healing. Hyperglycaemia is only transient on the C57BL/6J background. However, mice become severely diabetic with regression of islets and have early death, when ob gene is expressed on C57BL/KS background. Insulin resistance is associated with hepatic glucose overproduction, decreased activity of glycolytic and glycogen synthesis enzymes and increased lipogenesis in the liver, increased activity of gluconeogenic enzymes (McNeil 1999)

11.5.3.8 C57BL/6J Mice

Type-2 diabetes can be induced by merely administering high fat feed to non-obese, non-diabetic C57BL/6J mouse strain. It is considered by amplified overweightness, hyperinsulinemia, insulin resistance and glucose intolerance. They can develop resistance to diabetes and obesity if given a fat rich diet. When given high-fat diets, C57BL/6J (B6) mice develop severe obesity and diabetes but A/J mice get obese and diabetes-resistant. The harshness of diabetes is directly proportionate to overweightness and diabetes and is entirely revocable by plummeting the consumption of dietary fat (Parekh et al. 1998). When these mice are administered orally active inhibitor of dipeptidyl peptidase-IV, it normalized glucose tolerance with increased insulin secretion (Winzell and Ahrén 2004).

11.5.3.9 db/db Mice

The db/db mouse also identified as leprdb is initially derived after an autosomal recessive mutation on chromosome 4 in mice of C57BL/KsJ strain. The mutation in this diabetic animal is due to db inheritable factor that translates for the leptin receptors. These over-secretor mice become overweight, hyperglycaemic, hyperinsulinemic and insulin resilient within 30 days of age. They develop diseases such as hyperglycaemia and hypoinsulinaemia around the age of 3–4 months (Nuss and Wagman 2000). These db/db mice have been roughly used for the investigation of type-2 diabetes and for screening of agents like insulin mimetic and insulin sensitizers (Knouff and Auwerx 2004).

11.5.3.10 KK Mouse: KK (Kuo Kondo) Mouse

The Kuo Kondo mouse is likewise termed as Japanese KK mouse. This is a model for obesity and type 2 diabetes. These rodents are hyperphagic and hyperinsulinaemic. They show insulin resistance and show moderate obesity by 2 months which goes to the maximum at 4–5 months of age. Insulin resistance leads to onset of obesity. There is increase in pancreatic insulin content due to increase in number and size of pancreatic islets. Histologically on contrary, degranulation of beta cells and hypertrophy of islets are found (Vogel and Vogel 1997; Reddi and Camerini-Davalos 1988).

11.5.3.11 KK Obese Mouse

This mouse carries both lethal yellow obese (Ay) and diabetic genetic factor dissimilar to KK mouse that carries only the diabetic DNA segment. The mouse homozygous for the yellow spontaneous transmutation passes away before implantation or soon after that. KK/Ay heterozygous shows severe obesity, hyperglycaemia, hyperinsulinaemia and glucose intolerance by 8 weeks of age. The strain KK/Ay maintained at Upjohn colony (KK/Upj-Ay) are hyperphagic, hyperinsulinaemic, insulin resistant serve as a good model for obesity and type 2 diabetes and for screening various classes of antidiabetic agents (Umezawa et al. 2003).

11.6 Conclusion

Various in-silico and in-vitro methods for screening of natural antidiabetic agents have been considered. Considering the type of drug to be evaluated, the appropriate model is to be selected. Some of the important animal models described in this chapter share similar characteristic features of diabetes. Results from using them may not be the same as diabetes in human, but are an important tool for investigating genetic, endocrine, metabolic, morphologic changes and other underlying mechanisms.

References

Allen RE, Hughes TD, Ng JL, Ortiz RD, Ghantous MA, Bouhali O, Frogue P, Arredouani A. Mechanisms behind the immediate effects of Roux-en-Y gastric bypass surgery on type 2 diabetes. Theor Biol Med Model. 2013;10(45):1–19.

American Diabetes Association. Diabetes Care. 2015; 38(Supplement 1): S8-S16. https://doi.org/10.2337/dc15-S005. Available: http://care.diabetesjournals.org/content/38/Supplement_1/S8 Retrieved: 9th August, 2018.

Ansarullah BB, Dwivedi M, Laddha NC, Begum R, Hardikar AA, Ramachandran A. Antioxidant rich flavonoids from Oreocnide integrifolia enhance glucose uptake and insulin secretion and protects pancreatic β-cells from streptozotocin insult. BMC Complement Altern Med. 2011;11:126. https://doi.org/10.1186/1472-6882-11-126.

Asano T, Ogihara T, Katagiri H, Sakoda H, Ono H, Fujishiro M, Anai M, Kurihara H, Uchijima Y. Glucose transporter and Na+ / glucose cotransporter as molecular targets of anti-diabetic drugs. Curr Med Chem. 2004;11(20):2717–24.

Basu A, Sohn YS, Alyan M, Nechushtai R, Domb AJ, Goldblum A. Discovering novel and diverse iron-chelators in silico. J Chem Inf Model. 2016;56:2476–85.

Bergman RN, Ader M. Atypical antipsychotics and glucose homeostasis. J Clin Psychiatry. 2005;66:504–14.

Brayer GD, Luo Y, Withers SG. The structure of human pancreatic α-amylase at 1.8 A resolution and comparisons with related enzymes. Protein Sci. 1995;4:1730–42.

Cavarape A, Feletto F, Mercuri F, Quagliaro L, Damante G, Ceriello A. High-fructose diet decreases catalase mRNA levels in rat tissues. J Endocrinol Investig. 2001;24(11):838–45.

Chaudhury A, Duvoor C, Dendi VSR, Kraleti S, Chada A, Ravilla R, et al. Clinical review of antidiabetic drugs: implications for type 2 diabetes mellitus management. Front Endocrinol. 2017;8(6):1–12. https://doi.org/10.3389/fendo.2017.00006.

Chen L, Magliano DJ, Zimmet PZ. The worldwide epidemiology of type 2 diabetes mellitus-present and future perspectives. Nat Rev Endocrinol. 2011;8(4):228–36.

Choi SB, Park CH, Choi MK, Jun DW, Park S. Improvement of insulin resistance and insulin secretion by water extracts of Cordiceps militaris, phellinus linteus and paecilomyce tenuipes in 90% pancreatectomized rats. J Biotech Biochem. 2004;68:2257–64.

Deacon CF. Dipeptidyl peptidase-4 inhibitors in the treatment of type 2 diabetes: a comparative review. Diabetes Obes Metab. 2011;13:7–18.

Defronzo RA. Banting lecture. From the triumvirate to the ominous octet: a new paradigm for the treatment of type 2 diabetes mellitus. Diabetes. 2009;58:773–95.

DePaula AL, Macedo AL, Mota BR, Schraibman V. Laparoscopic ileal interposition associated to a diverted sleeve gastrectomy is an effective operation for the treatment of type 2 diabetes mellitus patients with BMI 21–29. Surg Endosc. 2009;23:1313–20.

Dsouza D, Lakshmidev N. Models to study in vitro antidiabetic activity of plants: a review. Int J Pharm Bio Sci. 2015;6(3):732–41.

Dunn JS, Sheehan HL, McLetchie NGB. Necrosis of islets of Langerhans produced experimentally. Lancet. 1943;241(6242):484–7. https://doi.org/10.1016/S0140-6736(00)42072-6.

Eleazu CO, Eleazu KC, Chukwuma S, Essien UN. Review of the mechanism of cell death resulting from streptozotocin challenge in experimental animals, its practical use and potential risk to humans. J Diabetes Metab Disord. 2013;12:60c67.

Epand RM, Stafford AR, Tyers M, Nieboer E. Mechanism of action of diabetogenic zinc chelating agents. Model Syst Stud Mol Pharmacol. 1985;27(3):366–74.

Etuk EU. Animals models for studying diabetes mellitus. Agric Biol J N Am. 2010;1:130–4.

Gallagher AM, Flatt PR, Duffy G, Abdelwahab YHA. The effects of traditional antidiabetic plants on in vitro glucose diffusion. Nutr Res. 2003;23(3):413–24.

Ganda OP, Rossini AA, Likc AA. Studies on streptozotocin diabetes. Diabetes. 1976;25:596–603.

Garcia-Sosa AT, Maran U, Hetenyi C. Molecular property filters describing pharmacokinetics and drug binding. Curr Med Chem. 2012;19:1646–62.

Gavin TP, Sloan RC, Lukosius EZ, et al. Duodenal-jejunal bypass surgery does not increase skeletal muscle insulin signal transduction or glucose disposal in Goto-Kakazaki type 2 diabetic rats. Obes Surg. 2011;21(2):231–7. https://doi.org/10.1007/s11695-010-0304-y.

Gillespie EL, White CM, Kardas M, Lindberg M, Coleman CI. The impact of ACE inhibitors or angiotensin II type 1 receptor blockers on the development of new-onset type 2 diabetes. Diabetes Care. 2005;28(9):2261–5.

Glick M, Goldblum A. A novel energy-based stochastic method for positioning polar protons in protein structures from X-rays. Proteins. 2000;38:273–87.

Glick M, Rayan A, Goldblum A. A stochastic algorithm for global optimization and for best populations: a test case of side chains in proteins. Proc Natl Acad Sci. 2002;99:703–8.

Gondi M, Prasada RU. Ethanol extract of mango (*Mangifera indica* L.) peel inhibits α-amylase and α-glucosidase activities, and ameliorates diabetes related biochemical parameters in streptozotocin (STZ)-induced diabetic rats. J Food Sci Technol Mysore. 2015;52(12):7883–93.

Goto Y, Kakizaki M. The spontaneous diabetes rat: a model of non insulin dependent diabetes mellitus. Proc Jpn Acad. 1981;57:381–4.

Goyal SN, Reddy NM, Patil KR, Nakhate KT, Ojha S, Patil CR, Agrawal YO. Challenges and issues with streptozotocin-induced diabetes – a clinically relevant animal model to understand the diabetes pathogenesis and evaluate therapeutics. Chem Biol Interact. 2016;244:49–63.

Gustavsson C, Soga T, Wahlström E, Vesterlund M, Azimi A, et al. Sex-dependent hepatic transcripts and metabolites in the development of glucose intolerance and insulin resistance in zucker diabetic fatty rats. J Mol Endocrinol. 2011;47(2):129–43.

Hao M, Zhang S, Qiu J. Toward the prediction of FBPase inhibitory activity using chemoinformatic methods. Int J Mol Sci. 2012;13:7015–37.

Harbilas D, Martineau LC, Harris CS, Adeyiwolα-Spoor DCA, et al. Evaluation of the antidiabetic potential of selected medicinal plant extracts from the Canadian boreal forest used to treat symptoms of diabetes: part II. Can J Physiol Pharmacol. 2009;87:479–92.

Haskell BD, Flurkey K, Duffy TM, Sargent EE, Leiter EH. The diabetes- prone NZO/H1Lt strain. I. Immunophenotypic comparison to the related NZB/B1NJ and NZW/LacJ strains. Lab Investig. 2002;82:833–42.

Heikamp K, Bajorath J. Comparison of confirmed inactive and randomly selected compounds as negative training examples in support vector machine-based virtual screening. J Chem Inf Model. 2013;53:1595–601.

Holt AIG, Peveler RC. Association between antipsychotic drugs and diabetes. Diabetes Obes Metab. 2006;8:125–35.

Honda M, Hara Y. Inhibition of rat small intestinal sucrase and α-glucosidase activities by tea polyphenols. Biosci Biotechnol Biochem. 1993;57:123–4.

Huang S, Czech MP. The GLUT4 glucose transporter. Cell Metab. 2007;5(4):237–52. https://doi.org/10.1016/j.cmet.2007.03.006.

Hussain F, Arif M, Sheikh MA. Prevalence of diabetic retinopathy in Faisalabad, Pakistan: a population based study. Turk J Med Sci. 2011;41(4):735–42.

Ighodaro OM, Adeosun AM, Akinloye OE. Alloxan-induced diabetes, a common model for evaluating the glycemic-control potential of therapeutic compounds and plants extracts in experimental studies. Medicina. 2017;53:365–74.

im Walde SS, Dohle C, Schott-Ohly P, Gleichmann H. Molecular target structures in alloxan-induced diabetes in mice. Life Sci. 2002;71:1681–94.

International Diabetes Federation. IDF diabetes atlas – 8th edition. Powerpoint presentation slide 3–4. 2017. Available: http://diabetesatlas.org/component/attachments/?task=download&id=271. Retrieved: 9th Aug 2018.
JeBailey L, Wanono O, Niu W, Roessler J, Rudich A, Klip A. Ceramide- and oxidant-induced insulin resistance involve loss of insulin-dependent Rac-activation and actin remodeling in muscle cells. Diabetes. 2007;56(2):394–403.
Junod A, Lambert AE, Orci L, Pictet R, Gonet AE, et al. Studies of the diabetogenic action of streptozotocin. Proc Soc Exp Biol Med. 1967;126(1):201–5.
Kahn SE. The importance of the beta-cell in the pathogenesis of type 2 diabetes mellitus. Am J Med. 2000;108(6a Suppl):2S–8S.
Kameoka J, Tanaka T, Nojima Y, Schlossman SF, Morimoto C. Direct association of adenosine deaminase with a T cell activation antigen, CD26. Science. 1993;261(5120):466–9.
Karasawa H, Takaishi K, Kumagae Y. Obesity-induced diabetes in mouse strains treated with gold thioglucose: a novel animal model for studying β-cell dysfunction. Obesity (Silver Spring, Md). 2011;19(3):514–21.
Kennedy BP. Role of protein tyrosine phosphatase-1B in diabetes and obesity. Biomed Pharmacother. 1999;53(10):466–70.
Kerru N, Singh-Pillay A, Awolade P, Singh P. Current anti-diabetic agents and their molecular targets: a review. Eur J Med Chem. 2018;152:436–88.
Kim HR, Rho HW, Park BH, Park JW, Kim JS, Kim UH, et al. Role of Ca2+ in alloxan-induced pancreatic beta-cell damage. Biochim Biophys Acta. 1994;1227:87–91.
King AJ. The use of animal models in diabetes research. Br J Pharmacol. 2012;166:877–94.
Knouff C, Auwerx J. Peroxisome proliferator-activated receptor –Î³ calls for activation in moderation: lessons from genetics and pharmacology. Endocr Rev. 2004;25(6):899–918.
Ktorza A, Bernard C, Parent V, Penicaud L, Froguel P, et al. Are animal models of diabetes relevant to the study of the genetics of noninsulin-dependent diabetes in humans? Diabetes Metab. 1997;23(Suppl 2):38–46.
Kwon YI, Vattem DV, Shetty K. Evaluation of clonal herbs of Lamiaceae species for management of diabetes and hypertension. Asia Pac J Clin Nutr. 2006;15:107–18.
Lenzen S. The mechanisms of alloxan-and streptozotocin-induced diabetes. Diabetologia. 2008;51:216e226.
Lusci A, Pollastri G, Baldi P. Deep architectures and deep learning in chemoinformatics: the prediction of aqueous solubility for drug-like molecules. J Chem Inf Model. 2013;53:1563–75.
Malaisse WJ, Malaisse-Lagae F, Sener A, Pipeleers DG. Determinants of the selective toxicity of alloxan to the pancreatic β cell. Proc Natl Acad Sci U S A. 1982;79:927–30.
Malhotra A, Penpargkul S, Fein FS, Sonnenblick EH, Scheuer J. The effect of streptozotocin induced diabetes in rats on cardiac contractile proteins. Circ Res. 1981;49(6):1243–50.
Manikandan R, Anand AV, Kumar S, Pushpa. Phytochemical and in vitro antidiabetic activity of *Psidium Guajava* leaves. Pharm J. 2016;8(4):392–4.
Masiello P. Animal models of type11 diabetes with reduced pancreatic β-cell mass. Int J Biochem Cell Biol. 2006;38:873–93.
Mathews CE, Leiter EH. Constitutive differences in antioxidant defense status distinguish alloxan-resistant and alloxan-susceptible mice. Free Radic Biol Med. 1999;27:449–55.
McIntosh CHS, Pederson RA. Non-insulin dependent animal models of diabetes mellitus. In: McNeil JH, editor. Experimental models of diabetes. Boca Raton: CRC Press LLC; 1999. p. 337–98.
McNeil JH. Experimental models of diabetes. Boca Raton: CRC Press LLC; 1999.
Mendez JD, Ramos HG. Animal models in diabetes research. Arch Med Res. 1994;25(4):367–75.
Miralles F, Portha B. Early development of beta cells is impaired in the GK rat model of type 2 diabetes. Diabetes. 2001;50:S84–8.
Momose K, Nunomiya S, Nakata M, Yada T, Kikuchi M, et al. Immunohistochemical and electron-microscopic observation of β-cells in pancreatic islets of spontaneously diabetic Goto-Kakizaki rats. Med Mol Morphol. 2006;39(3):146–53.

Mordes JP, Guberski DL, Leif JH, Woda BA, Flanagan JF, et al. LEW.1WR1 rats develop autoimmune diabetes spontaneously and in response to environmental perturbation. Diabetes. 2005;54(9):2727–33.

Mosmann T. Rapid colorimetric assay for cellular growth and survival: application to proliferation and cytotoxicity assays. J Immunol Methods. 1983;65:55–63.

Motyl K, McCabe LR. Streptozotocin, type I diabetes severity and bone. Biol Proced Online. 2009;11:296–315.

Nerup J, Mandrup-Poulsen T, Helqvist S, Andersen HU, Pociot F, Reimers JI, et al. On the pathogenesis of IDDM. Diabetologia. 1994;37(Suppl. 2):S82–9.

Newcomer JW. Second-generation (atypical) antipsychotics and metabolic effects: a comprehensive literature review. CNS Drugs. 2005;19(Suppl 1):1–93.

Nuss JM, Wagman AS. Recent advances in therapeutic approaches to type2 diabetes. Annu Rep Med Chem. 2000;35:211–20.

Ozturk Y, Altan VM, Yildizoğlu-Ari N. Effects of experimental diabetes and insulin on smooth muscle functions. Pharmacol Rev. 1999;48:69–112.

Parekh PI, Petro AE, Tiller JM, Feinglos MN, Surwit RS. Reversal of diet-induced obesity and diabetes in C57BL/6J mice. Metabolism. 1998;47(9):1089–96.

Pechhold K, Patterson NB, Blum C, Fleischacker CL, Boehm BO, Harlan DM. Low dose streptozotocin-induced diabetes in rat insulin promoter-mCD80-transgenic mice is T cell autoantigen-specific and CD28 dependent. J Immunol. 2001;166:2531–9.

Rayan A, Noy E, Chema D, Levitzki A, Goldblum A. Stochastic algorithm for kinase homology model construction. Curr Med Chem. 2004;11:675–92.

Rayan A, Falah M, Raiyn J, Mawassi H, Raiyn N. Assessing drugs for their cardio-toxicity. Lett Drug Des Discov. 2010;7:409–14.

Rayan A, Falah M, Raiyn J, Da'adoosh B, Kadan S, Zaid H, Goldblum A. Indexing molecules for their hERG liability. Eur J Med Chem. 2013;65:304–14.

Reddi AS, Camerini-Davalos RA. Hereditary diabetes in the KK mouse: an overview. Adv Exp Med Biol. 1988;246:7–15.

Rees DA, Alcolado JC. Animal models of diabetes mellitus. Diabet Med. 2005;22:359–70.

Rodrigues R. A comprehensive review: the use of animal models in diabetes research. J Anal Pharm Res. 2016;3(5):1–5.

Saidu Y, Muhammad SA, Abbas AY, Onu A, Tsado IM, Muhammad L. In vitro screening for protein tyrosine phosphatase 1B and dipeptidyl peptidase IV inhibitors from selected Nigerian medicinal plants. J Intercult Ethnopharmacol. 2017;6(2):154–7. https://doi.org/10.5455/jice.20161219011346.

Sasaki S, et al. Intra peritoneally implanted artificial pancrease with trans karyotic beta-cells on micro carrier beads in a diffusion chamber improve hyperglycemia after 90%pancreatectomy in rats. In Vivo. 2000;14:535–41.

Sattar NA, Hussain F, Iqbal T, Sheikh MA. Determination of in vitro antidiabetic effects of *Zingiber officinale* Roscoe. Braz J Pharm Sci. 2012;48(4):601–7.

Scheen AJ, De Hert M. Risque de diabète sucré sous antipsychotiques atypiques. Med Hyg (Geneve). 2004;62:1591–6.

Scheen AJ, De Hert M. Abnormal glucose metabolism in patients treated with antipsychotics. Diabetes Metab. 2007;33:169–75.

Schuller A, Schneider G. Identification of hits and lead structure candidates with limited resources by adaptive optimization. J Chem Inf Model. 2008;48:1473–91.

Selim SA, Selim AO. Effect of streptozotocin-induced diabetes mellitus on the. Egypt J Histol. 2013;36:103–13.

Shafrir E. Diabetes in animals: contribution to the understanding of diabetes by study of its etiopathology in animal models. In: Porte D, Sherwin RS, Baron A, editors. Diabetes mellitus. New York: McGraw-Hill; 2003. p. 231–55.

Shahaf N, Pappalardo M, Basile L, Guccione S, Rayan A. How to choose the suitable template for homology modelling of GPCRs: 5-HT7 receptor as a test case. Mol Inform. 2016;35:414–23.

Shori AB. Screening of antidiabetic and antioxidant activities of medicinal plants. J Integr Med. 2015;13(5):297–305.

Srinivasan K, Ramarao P. Animal models in type 2 diabetes research an overview. Indian J Med Res. 2007;125:451–72.

Sylow L, Kleinert M, Pehmøller C, Prats C, Chiu TT, Klip A, Richter EA, Jensen TE. Akt and Rac1 signaling are jointly required for insulin-stimulated glucose uptake in skeletal muscle and downregulated in insulin resistance. Cell Signal. 2014;26(2):323–31.

Szkudelski T. The mechanism of alloxan and streptozotocin action in β cells of the rat pancreas. Physiol Res. 2001;50:537–46.

Thatte U. Still in search of herbal medicine. Indian J Pharm. 2009;41:1–3.

Tripathi V, Verma J. Different models used to induce diabetes a comprehensive review. Int J Pharm Pharm Sci. 2014;6:29–32.

Umezawa K, Kawakami M, Watanabe T. Molecular design and biological activities of protein tyrosine phosphatase inhibitors. Pharmacol Ther. 2003;99:15–24.

Varmaan V, Rani Y, Avupati V. In silico based virtual screening for bioactive PPAR-y agonists. Indo Am J Pharm Sci. 2016;3(10):1237–75.

Vavra JJ, Deboer C, Dietz A, Hanka LJ, Sokolski WT. Streptozotocin, a new antibacterial antibiotic. Antibiot Annu. 1959;7:230–5.

Vogel HG, Vogel WH. Drug discovery and evaluation; pharmacological assays. Heidelberg/Berlin: Springer; 1997.

Wang Q, Khayata Z, Kishib K, Ebinab Y, Klipa A. GLUT4 translocation by insulin in intact muscle cells: detection by a fast and quantitative assay. FEBS Lett. 1998;427(2):193–7.

Wang TJ, Larson MG, Vasan RS, Cheng S, Rhee EP, McCabe E, et al. Metabolite profiles and the risk of developing diabetes. Nat Med. 2011;17(4):448–53.

WHO. Global Health Estimates (GHE), 2000–2015 estimates. Geneva: World Health Organization; 2017. Available: http://www.who.int/mediacentre/factsheets/fs312/en/. Accessed on 26 July 2018.

Winzell MS, Ahrén B. The high-fat diet-fed mouse: a model for studying mechanisms and treatment of impaired glucose tolerance and type 2 diabetes. Diabetes. 2004;53(Suppl 3):S215–9.

Wu KK, Huan Y. Streptozotocin-induced diabetic models in mice and rats. Curr Protoc Pharmacol. 2008;5:47.. 41–45.47. 14

Wu J, Yan L. Streptozotocin-induced type 1 diabetes in rodents as a model for studying.pdf>. Diabetes, Metab Syndr Obes Targets Ther Diabetes, Metab Syndr Obes: Targets Ther. 2015;8:181–8.

Yang H, Fan S, Song D, Wang Z, Ma S, Li S, Li X, Xu M, Xu M, Wang X. Long-term streptozotocin-induced diabetes in rats leads to severe damage of brain blood vessels and neurons via enhanced oxidative stress. Mol Med Rep. 2013;7(2):431–40.

Zaid H, Talior-Volodarsky I, Antonescu C, Liu Z, Klip A. GAPDH binds GLUT4 reciprocally to hexokinase-II and regulates glucose transport activity. Biochem J. 2009;419(2):475–84.

Zatsepin M, Mattes A, Rupp S, Finkelmeier D, Basu A, Burger-Kentischer A, Goldblum A. Computational discovery and experimental confirmation of TLR9 receptor antagonist leads. J Chem Inf Model. 2016;56:1835–46.

Zeidan M, Rayan M, Zeidan N, Falah M, Rayan A. Indexing natural products for their potential anti-diabetic activity: filtering and mapping discriminative physicochemical properties. Molecules. 2017;22(9):1563.

Chapter 12
In-vitro Models to Assess Antioxidant Potential

Ramesh Kumar, Ashutosh Gupta, Risha Ganguly, and Abhay K. Pandey

12.1 Introduction

The events of World War II directly led to the genesis of free radical biochemistry and the deadly effects of ionizing radiations were attributed to the formation of reactive oxygen species (ROS). Since then ROS and reactive nitrogen species (RNS) have earned bad reputation (Gilbert 1981). ROS also include free radicals and non-radical derivatives of oxygen (Namiki 1990). During respiration in biological systems, a substantial fraction of oxygen is partly reduced. Such partially reduced oxygen molecules and their derivatives are known as ROS, which are extremely reactive pro-oxidants and lethal. The reactivity of ROS can cause functional impairment in biological systems, eliciting a number of degenerative complications like mutagenesis, circulatory disturbances, carcinogenesis and aging (Halliwell and Gutteridge 1985). Inside the cell, free radicals are routinely produced in the metabolism within the peroxisomes, mitochondria through xanthine oxidase, phagocytosis, inflammation processes, physical exercise, arachidonate pathways and ischemia. Many external factors stimulate the production of free radicals viz., smoking, pollutants, drugs, radiation, pesticides, industrial solvents and ozone. It is an irony that essential elements for survival such as oxygen have harmful effects on the human body through ROS generation by prooxidative enzyme systems, lipid peroxidation, glycoxidation etc. which creates an imbalance leading to oxidative stress (Lobo et al. 2010; Kumar and Pandey 2015). However, living systems have natural enzymatic and non-enzymatic antioxidant defenses which counteract the harmful effects of free radicals and other oxidants (El-Bahr 2013; Kumar and Pandey 2014).

An antioxidant in biological systems has been defined as any compound that when present at small level compared to that of an oxidizable substrate, considerably retards or averts oxidation of the substrate (Halliwell 1990). But in food systems, antioxidants have been categorized as compounds, which are able to prevent or delay

R. Kumar · A. Gupta · R. Ganguly · A. K. Pandey (✉)
Department of Biochemistry, University of Allahabad, Allahabad, India

S. Kumar, C. Egbuna (eds.), *Phytochemistry: An in-silico and in-vitro Update*,
https://doi.org/10.1007/978-981-13-6920-9_12

the oxidation of easily oxidizable substances viz., lipids in small amounts (Chipault 1962). Antioxidants may interfere at any of the three main stages of the free radical mediated oxidative processes i.e., initiation, propagation, and termination (Cui et al. 2004; Kumar et al. 2013a; Sharma et al. 2018). In foods, it is necessary to measure the effectiveness of natural antioxidants for preservation and defense against oxidative destruction to evade harmful changes and loss of nutritional and commercial values. Thus, it is desirable to have a rapid method for qualitative and quantitative assessment of the potential antioxidant capacity in food products. Such a method would be a useful tool to make a choice among diverse species, varieties, degree of maturation, and culture conditions, so as to obtain high content of natural antioxidants in food products (Leonardi et al. 2000). Techniques that measure total antioxidant activities of foods can be used to assess the relationship between important dietary factors and disease prevention, thus reducing human suffering by helping to pick vital dietary sources of antioxidants. Various in-vitro methods used to study the antioxidant property of food products, plant extracts, commercial antioxidants and pure compounds are described below.

12.2 Peroxyl Radical Scavenging Capacity Assay

One of the most important radicals generated due to oxidative stress is the peroxyl radical ($ROO^{\cdot}$). It is associated with the oxygenation of unsaturated fatty acids which is a reactive target in the biological system. Peroxyl radicals are naturally produced by the electrophilic addition of molecular oxygen and the activated singlet oxygen by abstraction of an allylic hydrogen atom from the 1,4-pentadien system of the major part of unsaturated fatty acid carbon chains and is further stabilized by a double-bond conjugation. $ROO^{\cdot}$is involved in further decomposition and free radical stabilization till production of fatty acid rancidity marker compounds. Hence, these events explain the loss of membrane function triggered by the damage to the lipid bilayer and eventually undesirably affecting the biological function. Generation of free radicals and reaction monitoring are the most vital factors for the evaluation of antioxidant activity against peroxyl radicals. Free radicals can be generated either by the action of enzymes like lipoxygenase, or by the thermal decomposition of an azo-bis compound (Fig. 12.1a) (Scalzo et al. 2012). A sensitive assay has been developed to monitor peroxyl radical scavenging capacity of water- and lipid-soluble antioxidant compounds, food stuffs and plant extracts. The assay is based on the degree of inhibition of oxidation of the dye dichlorofluorescin diacetate (DCFH-DA) by antioxidants that have the competence to scavenge peroxyl radicals, produced by thermal degradation of 2,2′-azobis(amidinopropane) (ABAP). Thermal degradation of ABAP produces peroxyl radicals ($ROO^{\bullet}$) which oxidize non-fluorescent dichlorofluorescein (DCFH) to fluorescent dichlorofluorescein (DCF) (Fig. 12.1b). The degree of retardation of DCFH oxidation, by antioxidants that scavenge peroxyl radicals is used as the basis for calculation of antioxidant activity (Adom and Liu 2005).

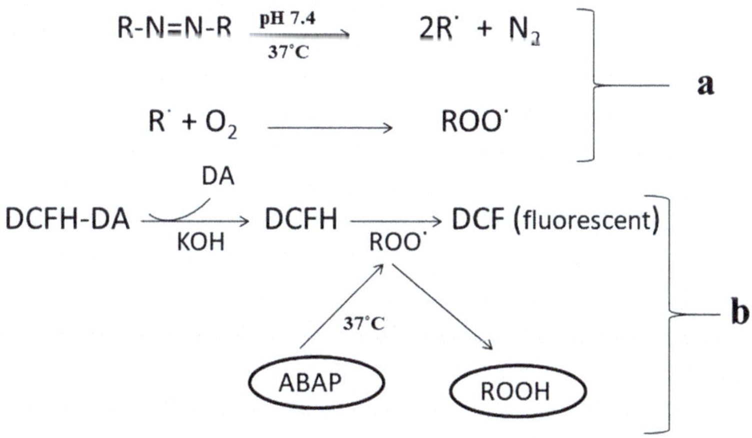

Fig. 12.1 Mechanism of peroxyl radical production and scavenging action. (**a**) Thermal decomposition of an azo-bis compound (R–N=N–R) by oxygen at physiological pH to yield peroxyl radical (ROO˙). (**b**) Reaction mechanism of peroxyl radical scavenging capacity assay. *Abbreviations*: *DCFH-DA* dichlorofluorescindiacetate, *ABAP* 2,2′-azobis(amidinopropane), *DCFH* nonfluorescentdichlorofluorescein, *DCF* fluorescent dichlorofluorescein

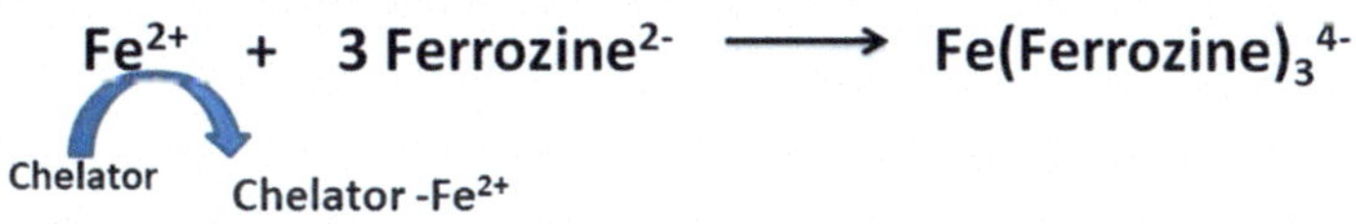

Fig. 12.2 Reaction mechanism of metal ion chelating assay

12.3 Metal Ion Chelating Assay

The metal ion chelating assay is based on the binding affinity of Ferrozine with Fe^{2+} ions. Ferrozine is a sulfonated derivative of the parent compound 3-(2-pyridyl)-5,6-bis-(4-phenylsulfonic acid)-1,2,4-triazine and forms a pink colored water soluble iron (II) complex (Carter 1971) (Fig. 12.2). Thus, the chelating activity of the plant extracts/samples and standard antioxidant compounds (BHA, BHT and EDTA) for ferrous ions is measured based on this mechanism. Redox chemistry of iron has been implicated in both the occurrence and the rate of lipid peroxidation (LPO). Reaction of Fe_3^+ with lipid hydroperoxides produces radicals that begin a chain reaction by reacting with other molecules producing MDA, a typical peroxidation marker. LPO also inflicts damage to unsaturated fatty acids results in decreased membrane fluidity and consequently leading to several pathological alterations. The transition metal ion, Fe_2^+ has the capacity to move single electrons by which it can permit the production and propagation of many radical reactions, even beginning with comparatively non-reactive radicals. To circumvent ROS generation related with redox active metal catalysis, chelation of the metal ions is the major approach. Iron can trigger LPO by the Fenton reaction and also speeds up peroxidation by breaking down lipid hydroperoxides into peroxyl and alkoxyl radicals that can abstract hydrogen and further propagate the peroxidation chain reaction. Ferrozine can quantitatively form complexes with Fe_2^+. However, formation of pink-red colored complexes decreases in the presence of compounds possessing chelating ability which can be measured spectrophotometrically at 562 nm. Therefore, monitoring the rate of reduction of color intensity helps to measure the chelating activity of the co-existing chelator present in the reaction mixture (Dinis et al. 1994; Kumar and Pandey 2012).

12.4 Superoxide Anion Radical Scavenging Capacity Assay

Superoxide anion radical ($O_2^{\bullet -}$) isproduced by the reduction of molecular oxygen. It is also formed in aerobic cells due to leakage of electrons from the electron transport chain. Superoxide radicals ($O_2^{\bullet -}$) are also produced by activated phagocytes such as monocytes, eosinophils, neutrophils and macrophages and play an important role in the phagocytosis of bacteria. In living organisms, $O_2^{\bullet -}$ is removed by the enzymes called superoxide dismutases (SOD). Thus, SOD keeps in check the excess production of superoxide radicals which may lead to oxidative damage (Halliwell and Gutteridge 1985). The in-vitro superoxide radical scavenging activity is based on the chemical anion radical generating system which includes phenazine methosulfate (PMS), nicotinamide adenine dinucleotide (NADH) and nitroblue tetrazolium (NBT). Superoxide radicals are produced within PMS-NADH systems by the oxidation of NADH and are examined by the reduction of NBT. Briefly, the generation of superoxide anion is measured in a reaction mixture containing NADH, NBT and PMS in phosphate buffered saline (PBS) at pH 7.4. The reduction of NBT is detected by measuring the change in absorbance at 560 nm (Liu et al. 1997; Shareef et al. 2014).

12.5 Hydroxyl Radical ($OH^{\bullet}$) Scavenging Assay

In cell metabolism the hydroxyl radicals ($^{\bullet}OH$) are the extremely reactive creations of ROS generated by successive 1-electron reduction of molecular oxygen (O_2). Hydroxyl radicals are mainly accountable for the cytotoxic effects in aerobic organisms. Several studies have reported protection from the harmful effects of hydroxyl radical by several medicinal plants, thus endorsing their antioxidant properties. The hydroxyl radical is one of the strongest known oxidants. It is capable of nonspecifically oxidizing all classes of biological macromolecules including lipids, carbohydrates, proteins, and nucleic acids at almost diffusion-limited rates. Therefore, disturbance in $^{\bullet}OH$ homeostasis may inflict oxidative injury that leads to many diseases like atherosclerosis, diabetes, cancer, arthritis etc. (Ozyurek et al. 2008). $OH^{\bullet}$ radicals are mainly generated in the body via the Haber-Weiss reaction where the superoxide anion radical ($O_2^{\bullet -}$) reduces cellular Fe(III) to Fe(II). Thus, it initiates the Fenton reaction between Fe^{2+} and $H_2O_2^-$ (Fig. 12.3).

$$O_2^{\cdot -} + H_2O_2 \longrightarrow O_2 + OH^{\cdot} + OH^- \quad \textbf{(Haber-Weiss Reaction)}$$

$$\left.\begin{array}{l} O_2^{\cdot -} + Fe^{3+} \longrightarrow Fe^{2+} + O_2 \\ Fe^{2+} + H_2O_2 \longrightarrow Fe^{3+} + OH^{\bullet} + OH \end{array}\right\} \textbf{(Fenton Reaction)}$$

Fig. 12.3 Hydroxyl radical generation by Haber-Weiss reaction and Fenton reaction

The hydroxyl radical scavenging assay is based on the inhibition of p-nitrosodimethylaniline (pNDA) bleaching. Hydroxyl radicals are generated through Fenton reaction in which iron (II) is oxidized by hydrogen peroxide to iron (III) and are capable of bleaching pNDA specifically. The scavenging of these radicals is measured by the extent of inhibition of bleaching in the presence of antioxidants (Kunchandy and Rao 1990).

12.6 DPPH Radical Scavenging Assay

DPPH (2,2-diphenyl-1-picrylhydrazyl) is an organic compound composed of stable free radical molecules. DPPH assay has been mostly used as a reliable, quick, and reproducible method for screening in-vitro antioxidant activity of pure compounds, food products as well as natural plant extracts. The violet coloured DPPH is converted to a yellow product in the presence of antioxidants. Antioxidants react with DPPH, a nitrogen-centered free radical, which is a stable, and convert it to α, α-diphenyl-*β*-picryl hydrazine. The degree of discoloration indicates the scavenging potentials of the antioxidant extracts. The decrease in absorbance by the DPPH radical with increasing concentration of the extracts in dose dependent manner results in the rapid discolouration of the purple DPPH, suggesting that compounds present in food material or plant extracts have radical scavenging antioxidant activity owing to its proton donating ability. For assessment of the antioxidant prowess of compounds or extracts, they are permitted to react with DPPH in methanolic medium. A deep violet color of the DPPH radical in solution due to a strong absorption band centered at about 520 nm and it turn into colorless or pale yellow after neutralization (Fig. 12.4). The percentage inhibition (% scavenging activities) at diverse concentrations of the test compounds or extracts are calculated using the formula (Eq. 12.1).

$$(\%\text{Inhibition}) = \left[(A_c - A_s)\right] \times 100 \qquad (12.1)$$

where A_c and As are the absorbance values of the control and the sample, respectively. Three replicates are made for each sample and results are expressed as mean ± SD.

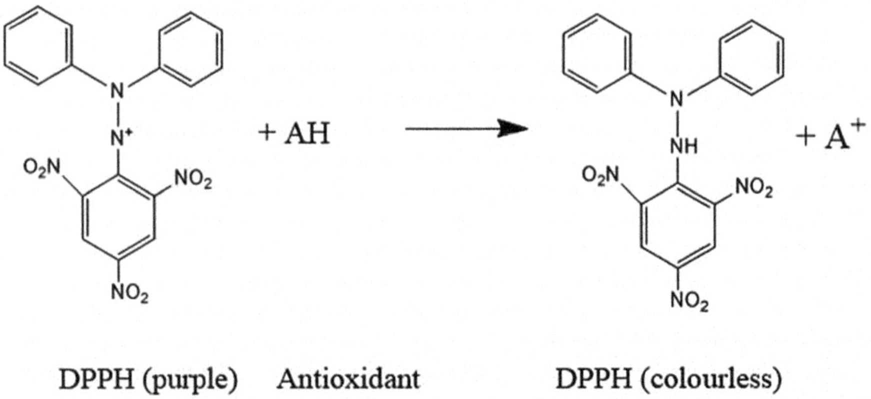

Fig. 12.4 Reaction mechanism of DPPH radical scavenging assay

The reduction of DPPH molecules could be correlated with the number of available hydroxyl groups. Chemical constituents present in food products or plant extract fractions exhibit significantly higher inhibition percentage (stronger hydrogen –donating ability) which can be positively correlated with total phenolic content (Sharma et al. 2016).

12.7 ABTS Radical Scavenging Assay

ABTS [2,2′-azinobis-(3-ethylbenzothiazoline-6-sulfonic acid)] is a chemical compound commonly used as a free radical in in-vitro antioxidant assays. Generation of the ABTS radical cation forms the basis of ABTS radical scavenging assay, which measures the total antioxidant activity of sample compounds. Initially, this method was based on the stimulation of metmyoglobin with H_2O_2 in the presence of ABTS to generate the radical cation, in the presence/absence of antioxidants, in which case the faster reacting antioxidants could also add to the reduction of the ferryl myoglobin radical. Therefore, the results in this method could be ambiguous. The decolorization technique is a relatively more targeted and absolute form of the assay as the radical is directly generated in a stable form prior to reaction with the antioxidants. In this system, the reaction between ABTS and potassium persulfate results in the direct formation of the blue/green ABTS chromophore which has absorption maxima at 734 nm (Fig. 12.5). Addition of antioxidants to the pre-formed radical cations reduces it to ABTS, which varies with the concentration and activity of the antioxidant as well as the reaction duration. Therefore, the magnitude of decolorization as % inhibition of the ABTS radical cation is measured with respect to concentration and time and is computed in comparison with the reactivity of a standard under the same conditions. The method is applicable for the study of both water-soluble and lipid-soluble antioxidants, pure compounds, food and plant extracts (Re et al. 1999).

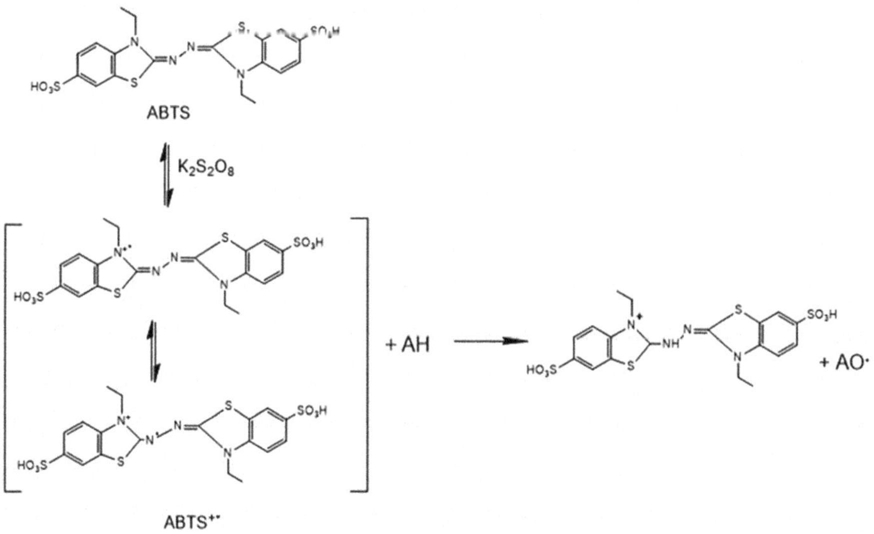

Fig. 12.5 Oxidation of ABTS molecule by potassium persulfate ($K_2S_2O_8$) to generate radical cation $ABTS^{+\bullet}$ (absorption maxima 734 nm) and its reaction with an antioxidant compound (AH)

Fig. 12.6 Fe^{2+} TPTZ complex

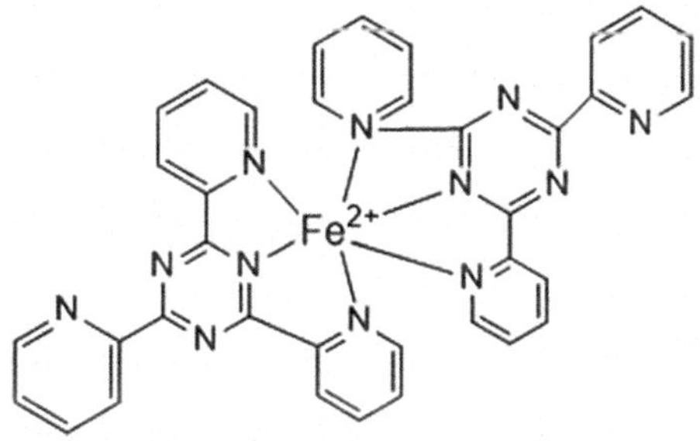

12.8 FRAP Assay

The ferric reducing antioxidant power (FRAP) assay is another in-vitro antioxidant capacity assay. At low pH, when a ferric-tripyridyltriazine (FeIII-TPTZ) complex is reduced to the ferrous (FeII) form, a dark blue color develops which absorbs spectrophotometrically at 593 nm (Fig. 12.6). This reaction is nonspecific, and any half-reaction with a less-positive redox potential than the Fe-III/Fe-II-TPTZ half-reaction, under similar reaction conditions, will impel the FeIII-TPTZ reduction. The complex is reduced in the presence of an antioxidant which acts as a reductant and blue color is developed under favorable conditions. Ferrozine is commonly used with ascorbic acid, which acts as an antioxidant and a reductant to measure iron (Benzie and Strain 1996; Sharma et al. 2014, 2017a). FRAP values can be calculated by relating the change in absorbance of the reaction mixture with those obtained from increasing concentrations of Fe^{3+} using calibration curve and expressed as mM of Fe^{2+} equivalents per kg (solid food) or per litre (liquid) of sample.

12.9 Nitric Oxide (NO) Radical Scavenging Assay

Nitric oxide (NO) and RNS are free radicals that originate from the interplay of NO with oxygen or ROS. NO exhibits reactivity with certain proteins and other free radicals such as superoxide due to its unpaired electron. The production of NO at lower levels plays crucial role in physiological processes. However, the sustained NO production can cause damaging effect in tissues and plays role in the pathological manifestation of disease states like inflammation and endotoxin shock (Gomes et al. 2006). Moreover, NO and $O_2^{\bullet -}$, produced by cells of the immune system during the oxidative burst, may react with each other, generating significant amounts of the highly reactive peroxynitrite anion ($ONOO^-$), which is a strong oxidative molecule. Therefore, toxicity of NO greatly increases when it reacts with the superoxide radical, and hence scavenging of NO by antioxidants is crucial for biological system homeostasis (Carr et al. 2000; Boora et al. 2014). The NO radical scavenging assay is based on the reaction of NO produced from aqueous sodium nitroprusside solution with oxygen to generate nitrite ions, at physiological pH. The nitrite ions are further quantified by Griess Illosvoy reaction (Hazra et al. 2008). Nitrite is detected and measured by formation of a red pink colour upon reaction with Griess reagent, which typically comprises of naphthylethylene diamine dihydrochloride, sulphanilamide and phosphoric acid.

12.10 Hydrogen Peroxide (H_2O_2) Scavenging Capacity Assay

Among ROS, H_2O_2 is a relatively stable, non-radical oxidant, having the ability to diffuse across biological membranes. It is produced by reduction of molecular oxygen via two electrons or by dismutation with or without the enzyme SOD (Fig. 12.7).

The biological system generates the superoxide anion radical $O_2^{\bullet -}$ from molecular oxygen and unquestionably produces H_2O_2 by the dismutation reaction, the rate of which depends on pH and O_2 concentration. H_2O_2 induces the oxidative

Fig. 12.7 Hydrogen peroxide production by reduction of molecular oxygen

$$O_2 + 2e^- + 2H^+ \longrightarrow H_2O_2$$

$$O_2^{\cdot -} + 2H^+ \xrightarrow{SOD} H_2O$$

$$O_2^{\cdot -} + e^- + 2H^+ \longrightarrow H_2O_2 + O_2$$

degradation of most biological macromolecules such as lipids, proteins or enzymes, carbohydrates and nucleic acids, through generation of the hydroxyl radical ($OH^•$), the most potent and reactive ROS. H_2O_2 in presence of Cu(I) or Fe(II) salts in-vivo produces OH radicals by using transition metal ion-catalyzed Fenton and Haber–Weiss reactions (Ozyurek et al. 2010). Besides that, the most frequently used in-vitro H_2O_2 scavenging assay is carried out on the principle that there is a decrease in absorbance of H_2O_2 upon oxidation in the presence of antioxidants. A solution of H_2O_2 prepared in phosphate buffer (pH 7.4) is used and the solution of sample extract or pure compound in phosphate buffer is added which acts as an antioxidant. The absorbance of the reaction mixture is documented spectrophotometrically at 230 nm (Ak and Gulcin 2008). The percentage H_2O_2 scavenging by standard compounds is calculated as per equation (Eq. 12.2).

$$H_2O_2 \text{ scavenging effect } (\%) = (1 - A_s/A_c) \times 100 \tag{12.2}$$

where A_c is the absorbance of the control solution and A_s is the absorbance in the presence of standard compound which acts as a scavenger.

12.11 Hypochlorous Acid (HOCl) Radical Scavenging Capacity Assay

Hypochlorous Acid (HOCl) radicalis produced in the living system by the heme-peroxidase enzyme myeloperoxidase from activated neutrophils, monocytes, and macrophages. The HOCl is the well-known strongest oxidant formed by neutrophils and is recognized as a powerful pro-inflammatory agent. Moreover, it can also react with O_2 to produce one more microbicidal species $OH^•$. These HOCl radicals may be regulated by the internal antioxidant systems such as SOD, catalase, glutathione peroxidase (GPx), and glutatione S-transferase. However, when HOCl levels exceed beyond control, they may directly react with amino acids, proteins, carbohydrates, thiol-containing antioxidants, membrane lipids, and nucleic acids. These processes cause several inflammatory diseases such as arthritis, cystic fibrosis, asthma, heart diseases, and even certain forms of cancer. Thus, the use of external antioxidants from natural sources is required for treating these oxidative stress mediated anomalies. The HOCl radical scavenging capacity assay is based on the elimination of the catalase peak due to the destruction of the heme prosthetic group by HOCl and is determined spectrophotometrically at 404 nm. The HOCl scavenging capacity of the compounds is measured by the inability of HOCl to reduce the peak in a concentration dependent manner. A solution of catalase is mixed with HOCl with increasing concentrations of sample compounds acting as HOCl radical scavengers (Arguello Garcia et al. 2010)

12.12 Reducing Power Assay

Basic principle underlying the reducing power assay is that substances with a reduction potential react with potassium ferricyanide (Fe^{3+}) to form potassium ferrocyanide (Fe^{2+}), which subsequently reacts with ferric chloride to form ferric ferrous complex exhibiting an absorption maxima at 700 nm. Antioxidants acting as electron donor reduce the Fe^{3+} complex to its Fe^{2+} (Fig. 12.8) (Kumar et al. 2014a).

Ascorbic acid (various concentrations) is routinely used as standard antioxidant compound and increased absorbance of the reaction mixture specifies increased reducing power (Jayanthi and Lalitha 2011; Kumar et al. 2013b). Conversion of the yellow coloured reaction mixture to the bluish green during assay signifies reducing action of the compound and intensity of the colour produced further depends upon the reducing ability of the test compound. Natural foods/vegetables/plant extracts show considerable reducing power at higher concentrations as shown by absorbance values signifying dose dependent response. There are several articles on dose dependent reducing power of plant extracts. Phytochemicals particularly flavonoids have been stated to function as reducing agents. So the reducing ability of plant based natural products could be ascribed to the presence of phenolic compounds which might act as reductones.

12.13 Phosphomolybdate Assay

The phosphomolybdate assay is based on the reduction of Mo(VI) at acidic pH. It is regularly used to assess the total antioxidant capacity of food supplements/plant extracts and other compounds using propyl gallate as standard (Negi and Jayaprakasha 2004). In this method molybdenum (VI) is reduced to a green coloured molybdenum (V) complex by antioxidants which shows maximum absorption at 695 nm. The total antioxidant capacity is expressed as μg propyl gallate equivalents per gram of sample. The difference in antioxidant capacity of different samples is a function of differences in their chemical composition. The antioxidant activities of the specific phenolic compounds depend on structural features namely, the number and position of phenolic hydroxyl or methoxyl groups, flavone hydroxyl, free carboxylic groups, keto groups (Kumar et al. 2014b).

$$\underset{\text{(Ferricyanide)}}{2Fe(CN)_6^{3-}} + \underset{\text{(Ascorbate)}}{C_6H_7O_6^-} \longrightarrow \underset{\text{(Ferrocyanide)}}{2Fe(CN)_6^{4-}} + \underset{\text{(Dehydroascorbate)}}{C_6H_6O_6} + H^+$$

At pH 5.6, phosphate buffer 0.2M

$$\underset{\text{(Ferric ion)}}{4\,Fe^{3+}} + \underset{\text{(Ferrocyanide)}}{3\,Fe(CN)_6^{4-}} \longrightarrow \underset{\text{(Prussian blue) } (\lambda\text{max 700nm})}{Fe_4[Fe(CN)_6]_3}$$

Fig. 12.8 Mechanism of reducing power assay

12.14 Singlet Oxygen ($^1O_2$) Scavenging Capacity Assay

Singlet oxygen ($^1O_2$) is a high energy form of oxygen which exists in the singlet state with a total quantum spin of zero. It was first predicted in 1931 as a meta-stable intermediate state of oxygen in dye-sensitized photo-oxygenations. However, in 1964, it was finally recognized that $^1O_2$ was same entity whether generated in sensitized photo-oxygenations, through radio frequency or via chemical reactions with H_2O_2/NaOCl. It is an important free radical in the living system associated with cell signaling and oxidative stress (Luqman et al. 2016). The $^1O_2$-scavenging capacity is determined by a fluorescence method that consists of monitoring the oxidation of the non-fluorescent dihydrorhodamine 123 (DHR) to fluorescent rhodamine by the reaction with $^1O_2$. The singlet oxygen species is generated by thermal breakdown of an earlier synthesized water-soluble endoperoxide [disodium 3,3-′-(1,4-naphthalene)bispropionate] at 37 °C. The results are expressed in terms of percentage inhibition of $^1O_2$-induced oxidation of DHR (Rebeiro et al. 2015).

12.15 Lipid Peroxidation Inhibition Assay

The magnitude of oxidative damage caused in a lipid based system is difficult to determine because of the short half-lives of free radicals as well as many of the products primarily formed by attack of free radical on polyunsaturated fatty acids, the electron rich substrates. Therefore, the revealing of oxidative stress depends typically on the quantification of compounds known as thiobarbituric acid reactive substances (TBARS) such as malondialdehyde (MDA). The TBARS are the byproducts resulting from the degradation of primary products of free radical attack on lipids (Antolovich et al. 2002; Kumar and Pandey 2013). The reaction between MDA and 2-thiobarbituric acid at high temperature and low pH produces a pink colored compound which is a measured spectrophotometrically at 532 nm. The intensity of the color produced varies with the degree of lipid peroxidation induced by free radicals. It is very frequently used method for lipid peroxidation assay (Oakes and Kraak 2003; Sharma et al. 2017b).

12.16 Conclusion

Interest in antioxidants, especially those intended to thwart the apparent harmful effects of free radicals in the human system along with the deterioration of lipids and other components of foodstuffs, is rising. This chapter describes a comprehensive account of the in-vitro methods commonly used for the evaluation of antioxidant potential of different food substances, plant extracts and other compounds. Most of the in-vitro models for determining antioxidant activity rely on three important mechanisms namely, free radical scavenging, reducing property and metal ion chelating abilities. Mechanisms related with different in-vitro assays have been described in specific methods. These methods are very useful for the assessment of antioxidant evaluation of vegetables, fruits and other food products, food supplements, neutraceuticals and pharmaceutical agents.

Acknowledgement RK and RG acknowledges financial support from CSIR New Delhi, India, in the form of Junior Research Fellowship, AG acknowledges financial support from University of Allahabad, Allahabad, India in the form of UGC CRET research fellowships. All the authors also acknowledge DST-FIST and UGC-SAP facility of the Department of Biochemistry, University of Allahabad, Allahabad, India.

References

Adom KK, Liu RH. Rapid peroxyl radical scavenging capacity (PSC) assay for assessing both hydrophilic and lipophilic antioxidants. J Agric Food Chem. 2005;53:6572–80.

Ak T, Gulcin I. Antioxidant and radical scavenging properties of curcumin. Chem Biol Interact. 2008;174:27–37.

Antolovich M, Prenzler PD, Patsalides E, et al. Methods for testing antioxidant activity. Analyst. 2002;127:183–98.

Arguello-Garcia R, Medina-Campos ON, Perez-Hernandez N, et al. Hypochlorous acid scavenging activities of thioallyl compounds from garlic. J Agric Food Chem. 2010;58:11226–33.

Benzie IFF, Strain JJ. The ferric reducing ability of plasma as a measure of antioxidant power: the FRAP assay. Anal Biochem. 1996;239:70–6.

Boora F, Chirisa E, Mukanganyama S. Evaluation of nitrite radical scavenging properties of selected zimbabwean plant extracts and their phytoconstituents. J Food Process. 2014;2014:918018.

Carr AC, Mccall MR, Frei B. Oxidation of LDL by myeloperoxidase and reactive nitrogen species reaction pathways and antioxidant protection. Arterioscl Throm Vas. 2000;20:1716–23.

Carter P. Spectrophotometric determination of serum ion at the submicrogram level with a new reagent (Ferrozine). Anal Biochem. 1971;40:450–8.

Chipault JR. Antioxidants for use in foods. In: Lundberg WO, editor. Autoxidation and antioxidants. 2nd ed. New York: Interscience; 1962. p. 477–542.

Cui K, Luo X, Murthy MRV. Role of oxidative stress in neurodegeneration: recent developments in assay methods for oxidative stress and nutraceutical antioxidants. Prog Neuro Psycopharma Biol Psychiatry. 2004;28:771–99.

Dinis TCP, Madeira VMC, Almeida LM. Action of phenolic derivatives (acetaminophen, salicylate, and 5-aminosalicylate) as inhibitors of membrane lipid peroxidation and as peroxyl radical scavengers. Arch Biochem Biophys. 1994;315:161–9.

El-Bahr SM. Biochemistry of free radicals and oxidative stress. Forensic Sci Int. 2013;1:111–7.

Gilbert DL. Oxygen and living processes: an interdisciplinary approach. New York: Springer; 1981.

Gomes A, Fernandes E, Lima JLFC. Use of fluorescence probes for detection of reactive nitrogen species: a review. J Fluoresc. 2006;16:119–39.

Halliwell B. How to characterize a biological antioxidant. Free Radic Res Commun. 1990;9:1–32.

Halliwell B, Gutteridge JMC. Free radicals in biology and medicine. Oxford: Oxford University Press; 1985.

Hazra B, Biswas S, Mandal N. Antioxidant and free radical scavenging activity of *Spondiaspinnata*. BMC Complement Altern Med. 2008;8:63.

Jayanthi P, Lalitha P. Reducing power of the solvent extracts of *Eichhornia Crassipes* (Mart.) Solms. Int J Pharm Pharm Sci. 2011;3:126128.
Kumar S, Pandey AK. Antioxidant, lipo-protective and antibacterial activities of phytoconstituents present in *Solanum xanthocarpum* root. Intl Rev Biophys Chem. 2012;3:42–7.
Kumar S, Pandey AK. Phenolic content, reducing power and membrane protective activities of *Solanum xanthocarpum* root extracts. Vegetos. 2013;26:301–7.
Kumar S, Pandey AK. Medicinal attributes of *Solanum xanthocarpum* fruit consumed by several tribal communities as food: an in vitro antioxidant, anticancer and anti HIV perspective. BMC Complement Altern Med. 2014;14:112.
Kumar S, Pandey AK. Free radicals: health implications and their mitigation by herbals. Br J Med Med Res. 2015;7:438–57.
Kumar S, Gupta A, Pandey AK. *Calotropis procera* root extract has the capability to combat free radical mediated damage. ISRN Pharmacol. 2013a;2013:691372.
Kumar S, Mishra A, Pandey AK. Antioxidant mediated protective effect of *Parthenium hysterophorus* against oxidative damage using in vitro models. BMC Complement Altern Med. 2013b;13:20.
Kumar S, Kumar R, Dwivedi A, Pandey AK. In vitro antioxidant, antibacterial and cytotoxic activity and in vivo effect of *Syngonium podophyllum* and *Eichhornia crassipes* leaf extracts on isoniazid induced oxidative stress and hepatic markers. Biomed Res Int. 2014a;2014:459452.
Kumar S, Pandey S, Pandey AK. In vitro antibacterial, antioxidant, and cytotoxic activities of *Parthenium hysterophorus* and characterization of extracts by LC-MS analysis. Biomed Res Int. 2014b;2014:495154.
Kunchandy E, Rao MNA. Oxygen radical scavenging activity of curcumin. Int J Pharm. 1990;58:237–40.
Leonardi C, Ambrosino P, Esposito F, et al. Antioxidative activity and carotenoid and tomatine contents in different typologies of fresh consumption tomatoes. J Agric Food Chem. 2000;48:4723–7.
Liu F, Ooi VEC, Chang ST. Free radical scavenging activities of polysaccharide extracts. Life Sci. 1997;60:763–71.
Lobo V, Phatak A, Chandra N. Free radicals and functional foods: impact on human health. Pharmacogn Rev. 2010;4:118–26.
Luqman S, Fatima K, Masood N. A high through put spectrophotometric method for singlet oxygen quenching. Protocol Exch. 2016; https://doi.org/10.1038/protex.2016.054.
Namiki M. Anatioxidants/antimutagens in food. Food Sci Nutr. 1990;29:273–300.
Negi PS, Jayaprakasha GK. Control of foodborne pathogenic and spoilage bacteria by garcinol and *Garcinia indica* extracts, and their antioxidant activity. J Food Sci. 2004;69:61–5.
Oakes KD, Kraak GJ. Utility of the TBARS assay in detecting oxidative stress in white sucker (*Catostomus commersoni*) populations exposed to pulp mill effluent. Aquat Toxicol. 2003;63:447–63.
Ozyurek M, Bektasolu B, Guclu K, et al. Hydroxyl radical scavenging assay of phenolics and flavonoids with a modified cupric reducing antioxidant capacity (CUPRAC) method using catalase for hydrogen peroxide degradation. Anal Chim Acta. 2008;616:196–206.
Ozyurek M, Bektasolu B, Guclu K, et al. Novel hydrogen peroxide scavenging assay of phenolics and flavonoids using cupric reducing antioxidant capacity (CUPRAC) methodology. J Food Compos Anal. 2010;23;689–98.
Re R, Pellegrini N, Proteggente A, et al. Antioxidant activity applying an improved ABTS radical cation decolorization assay. Free Radic Biol Med. 1999;26:1231–7.
Rebeiro AB, Berto A, Chiste RC, et al. Bioactive compounds and scavenging capacity of extracts from different parts of *Vismia cauliflora* against reactive oxygen and nitrogen species. Pharm Biol. 2015;53:1267–76.
Scalzo RL, Todaro A, Rapisarda P. Methods used to evaluate the peroxyl (ROO) radical scavenging capacities of four common antioxidants. Eur Food Res Technol. 2012;235:1141–8.

Shareef MI, Reddy PJM, Gopinath SM, et al. Superoxide anion scavenging activity of *Carthamus tinctorius* flower. Int J Inn Res Sci Eng Technol. 2014;3:10101–4.

Sharma AK, Kumar S, Pandey AK. Ferric reducing, anti-radical and cytotoxic activities of *Tinospora cordifolia* stem extracts. Biochem Anal Biochem. 2014;3:153.

Sharma UK, Sharma AK, Pandey AK. Medicinal attributes of major phenylpropanoids present in cinnamon. BMC Complement Altern Med. 2016;16:156.

Sharma AK, Sharma UK, Pandey AK. Protective effect of *Bauhinia variegate* leaf extracts against oxidative damage, cell proliferation and bacterial growth. Proc Natl Acad Sci India Sect B Biol Sci. 2017a;87:45–51.

Sharma UK, Sharma AK, Gupta A, et al. Pharmacological activities of cinnamaldehyde and eugenol: antioxidant, cytotoxic and anti-leishmanial studies. Cell Mol Biol. 2017b;63:73–8.

Sharma UK, Kumar R, Gupta A, et al. Renoprotective effect of cinnamaldehyde in food color induced toxicity. 3 Biotech. 2018;8:212.

Chapter 13
In-vitro Models in Anticancer Screening

Dhanya K. C., Aditya Menon, and Laxmi Shanker Rai

13.1 Introduction

Cancer is the abnormal growth of cells due to genome wide changes in gene expression. These alterations in gene expression lead to deregulation in the balance of cell proliferation and cell death. The genetically altered cells escape from the cell cycle checkpoints towards cell death and become immortal. They proliferate uncontrollably and ultimately evolve into cancer cells. These cells can invade nearby tissues and metastasize to distant sites, causing significant morbidity and, if untreated, death of the host. Even though it seems that cancer is a complex family of diseases, it is a disease caused by similar molecular defects resulting from alterations in genome from a molecular and cell biological point of view. Caner may be considered as a genetic disease since it results from alterations within specific genes, but in most cases, it is not an inherited disease.

13.2 Treatment Modalities of Cancer

The conventional treatment modalities for cancer are surgery, radiotherapy and chemotherapy. Several new methods for specific and targeted cancer treatment such as Targeted therapies, Immunotherapy, Hormone therapy, use of Angiogenesis inhibitors, Gene Therapy, Virotherapy, etc. are now being emerged.

Dhanya K. C.
Department of Microbiology, St. Mary's College, Thrissur, Kerala, India

A. Menon
Origin Diagnostics and Research, Karunagappally, Kerala, India

L. S. Rai (✉)
Department of Mycology, Institut Pasteur, Paris, France
e-mail: laxmi.rai@pasteur.fr

S. Kumar, C. Egbuna (eds.), *Phytochemistry: An in-silico and in-vitro Update*,
https://doi.org/10.1007/978-981-13-6920-9_13

As long as the growth of the tumor remains localized, the disease can be cured by surgical removal of the tumor. But malignant tumors tend to metastasize, that is, few cells break away from the original tumor mass, enter the lymphatic or vascular circulation, and spread to distant sites in the body where they establish lethal secondary tumors (metastases) that are no longer amenable to surgical removal.

Radiation therapy is used to eradicate cancer and as a palliative aid to relieve pain associated with metastases. Radiotherapy is the most common modality for treating human cancers. Eighty percent of cancer patients need radiotherapy at some time or other, either for curative or palliative purpose. The primary focus in radiotherapy is to increase DNA damage in tumor cells, as double strand breaks leads to cell death. To obtain optimum results, a judicious balance between the total dose of radiotherapy delivered and the threshold limit of the surrounding normal critical tissues is required.

Among the common therapeutic modalities of cancer, chemotherapy is the important one. In cancer chemo therapy, the chemotherapeutic agent exerts anticancer action through cytotoxic mechanisms (Gutman et al. 2000).

13.3 Need for the Discovery of Newer Anticancer Molecules

Most chemotherapeutic agents currently being used possess poor selectivity toward the target tissue and can harm normal cells as well as cancer cells. Therefore, it has several toxic side effects such as hematopoietic depression, gastrointestinal toxicity, hemorrhagic cystitis, hemopoetic suppression, nephrotoxicity, urotoxicity, cardiotoxicity and hepatic toxicity (Amudha et al. 2007). Cytotoxicity towards normal host tissue is the primary dose-limiting factor in chemotherapy that reduces quality of life and restricts treatment protocol. Cancer is one of the primary and most prevalent disease that cause morbidity and mortality in millions of people worldwide, there is undoubtedly an unmet need to discover novel anticancer drugs.

Thus there was a transition from cytotoxic chemotherapy to molecularly targeted cancer drugs. This has resulted in an increasing number of successful therapeutic agents. However, for many patients the therapeutic options are still limited and the process of bringing a new effective drug is still frustratingly slow.

However, the traditional process of identifying a new drug and its development is lengthy and expensive. Discovery and development of anticancer agents are the key focus of many pharmaceutical companies and government and non-government organizations, like the National Cancer Institute (NCI) in the United States, the European Organization for Research and Treatment of Cancer (EORTC), and the British Cancer Research Campaign (CRC).

13.4 Anticancer Screening: General Methodology and Conventional Methods

Initially cancer was recognized as a disease of uncontrolled cell division and thus all efforts to derive an anticancer drug were directed toward the identification of antiproliferative compounds. Regression of tumor size was the primary objective in preclinical and clinical testing. Rapidly growing murine models of cancer were developed and used for screening of new compounds in the drug discovery programs and resulted in identifying several clinically important anticancer compounds. Thus, treatment successes occurred mainly in the rapidly growing malignancies such as lymphomas, childhood leukemia, germline tumors, etc. and relatively fewer successes for the slow-growing solid-tumors such as lung, breast, and colorectal cancers (Suggitt and Bibby 2005). Thus the investigators had to modify the prescreening and screening protocols so that a variety of cell lines and tumor types were included.

Schwartsmann et al. (1988) noted that from over 600,000 compounds screened only less than 40 agents reached to the clinic. The recent advances in molecular sciences, genomics and proteomics have generated several potential new drug targets for anticancer drug screening.

Conventional anticancer drug screening focused and selected agents that had significant cytostatic or cytotoxic activity on tumor cells and caused tumor regression in murine tumor models and the drugs were discovered either by serendipity or as metabolic pathways inhibitors.

Although this strategy has achieved significant success, the drugs identified often possessed significant side effects and toxicity. The recent developments in molecular biology of cancer made researchers to come up with target-based drugs. They devised molecules pre-designed to inhibit and/or modify a selected molecular marker important in cancer growth and several target-based drugs have emerged.

Conventional screening models for anticancer agents aim to select cytotoxic drugs based on animal screening models via analyzing tumor regression and survival, while the human clinical trials are utilized to determine the dose. Toxicity and tumor-regression effects of cytotoxic agents are based on similar mechanism.

The molecularly targeted agents act on the extra cellular, transmembrane, or nonnuclear intracellular processes and a few examples are receptor tyrosine kinase inhibitors, matrix metalloproteinase (MMP) inhibitors, farnesyl transferase inhibitors, etc. (McKeage 2008). These agents often cause tumor growth inhibition, rather than regression, in animal models.

An important element of preclinical and clinical screening of molecularly targeted agents is the investigation on their solubility, stability in the solid state and solution, pH solubility and stability studies, identification of degradation pathways, absorption studies in cell culture models and animals, etc. The anticancer activity is prescreened in-vitro in cell culture models by cell growth inhibitory or clonogenic assays. Further studies are performed in murine allograft or human xenograft mouse models (Amundson et al. 2008).

Furthermore, high throughput preclinical screening methods and mathematical models to explain the mechanisms of drug activity are being adopted.

13.5 Strategies of Anticancer Activity Screening

Cancer patients need better anticancer drugs and for an efficient drug discovery process, reliable screening methods are necessary. These methods should not only detect the compounds with the higher therapeutic potential, but also predict whether such potential is optimal to deserve additional attention.

Large-scale screening using animal systems as practiced in the past is highly unethical and, particularly in Europe, strictly regulated. In majority of cases, either cellular or target-based high-throughput assays will precede in-vivo evaluation of potential anticancer drugs.

Preclinical screening is also important to prioritize compounds for further development. In the era of target- oriented molecular cancer therapeutics, screening procedures are tailored toward the desired mechanism of tumor inhibition. Thus, rational drug design or drug discovery approaches combined with novel knowledge from genome and proteome research as well as bioinformatics are promising. The drug screening and discovery pathways have evolved into an integrated approach which combines the use of cell line and tumor xenograft models that resemble very closely the patient characteristics and response (Fiebig et al. 2007). Target-driven drug development has led to the availability of many useful cell signal transduction inhibitors and antibodies targeting growth factor receptors.

Before any potential anticancer agent is subjected to human testing, the prospective drug undergoes a series of qualifying studies. Initially, primary in-vitro screening is conducted, using the NCI60 human tumor cell line anticancer drug screening protocol, whereby a potential candidate is evaluated for its ability to inhibit the growth of tumor cells in culture. This modern pharmaceutical in-vitro screening was developed by the National Cancer Institute (NCI) and comprises a panel of 60 human tumor cell lines. Currently, the NCI60 is the most commonly used system for the preliminary screening of potential anticancer drugs (Shoemaker 2006).

Once in-vitro screening gives promising anticancer activity, then it undergoes in-vivo animal testing. This phase of evaluation is critical for understanding the in-vivo antitumor activity and is mainly conducted using inbred laboratory mice (Workman et al. 2010); however, other animals such as dogs and primates are also often utilized. Mouse tumor models are utilized for conducting cancer research and screening potential anticancer compounds due to similarity of human and mouse genomes, low cost of maintenance, short gestation period, rapid reproduction rate, and the ease of growing implanted tumors.

Higher vertebrates, such as primates, are avoided in routine toxicological studies due to high cost and ethical issues.

13.5.1 In-vivo Screening: And Disadvantages

Animal models have been indispensable to conducting further research to understand the cancer biology and develop anticancer drugs. The animal models can be divided into two classes: models in which tumors are transplanted into mice, and other where tumors develop in-situ, either spontaneously or induced.

13.5.1.1 Early in-vivo Test Models

The first in-vivo test model used for the screening of anticancer compounds was the mouse leukemia models grown as ascites tumor (Teicher 2009). In Human tumor mouse xenotransplantant models human tumor cells, cultured in-vitro, are implanted to immunodeficient mice (Rygaard and Poulsen 1969). When the tumor reaches a given size, introduction of a potential anticancer drug is made and the efficacy of the drug is determined by changes in the tumor size.

Autochthonous tumors either arise spontaneously or can be induced by carcinogens or other chemical, viral, bacterial, or physical triggers (Workman et al. 2010). Genetically engineered mouse models enable testing of drugs within an immunocompetent tumor microenvironment (Damia and D'Incalci 2009).

Due to differences in the genome of mouse and human, a question raised about relevancy of using rodents in drug screening. For example most murine cells have functionally active telomerase unlike human cells and also changes in certain genes like TP53, Rb and Ras, etc. Therefore, results obtained in laboratory testing using mouse model may not be directly extrapolated to humans. A personalized medicine screening approach where biopsied tissue are used to screen antitumor substances instead of tumor cell lines will lead closer to the actual clinical conditions (Blatt et al. 2013).

13.5.2 In-vitro Screening Model for Cancer

Discovery of a new drug and its establishment is a complex, time consuming and expensive process Therefore, drug screening under in-vitro conditions are more advantageous as compared to in-vivo models. However, majority of the promising drug candidates identified during in-vitro studies failed under preclinical and clinical stage of drug development. This lack of success is due to the fact that the in-vitro models used in oncology, 2D in-vitro model of cancer cell lines (CCLs), are not able to represent the tumor complexity and also sensitive to chemotherapy.

Conventional in-vitro drug screening test for the analysis of viable cells may not distinguish tumor cells from normal cells (do not differentially analyse tumour cell/normal cell viability). Conventional viability (e.g. MTT, Alamar Blue) or cytotoxicity (e.g. LDH release) assays do not distinguish normal/accessory cells and cancer cells. Assays that can distinguish tumor cells from normal cells measure cell proliferation rather than viability and require radioactivity (e.g. 3H-thymidine incorporation, Cr release assay) or involve laborious steps (e.g. flow cytometry). Following are the in-vitro models used in various experimental setups.

1. Cell-Based Screening Assays
2. Biochemical Screening
3. Nonmammalian Model Organisms
4. In-vitro tumor models
    (i) *Transwell based assays*
    (ii) *Spheroids*
    (iii) *Hybrid models*
    (iv) *Tumour microvessel models*
5. Hollow Fiber Assay
6. Colony-Forming Assay
7. In-silico or bioinfomatic methods

13.5.2.1 Cell Based Screening Assays

The human tumor cell lines have been used in cellular screens during in-vitro cancer research. How cell viability and activity of anticancer drugs are measured is an important factor.

Several procedures are employed to determine cell growth or viability. The broadly used growth inhibition assay which was developed by Mosmann and the NCI screening staff, is the methylthiazoldiphenyl tetrazolium (MTT) assay. The yellow MTT dye is reduced by mitochondria into a purple formazan, which can be read with ultraviolet/visible light scanners (Mosmann 1983). It has a limitation of usage of large quantities of hazardous solvent, dimethyl sulfoxide. Thus the sulforhodamine B (SRB) assay was developed, SRB is a dye that stains protein. There are also fluorescence or luminescence detection assays where propidium iodide (PI) staining for DNA or use of a luciferase reporter can be utilized (Dengler et al. 1995).

Other methods for the determination of cell viability such as trypan blue dye exclusion assay and the lactate dehydrogenase assay can give false results when a time delay in scoring result or if the drug under study affects on intracellular activities (Kumar et al. 2016).

Currently one-dimensional or monolayer cultures are used to measure cell growth. These methods also have limitations for in-vitro evaluation of certain anticancer agents such as Selection of cytotoxic drugs, lack of extracellular matrix and angiogenesis, Gradients of oxygen tension, extracellular pH, nutrients, catabolites, etc. and drug penetration barriers. Cancer stem cell-targeted drugs and inhibitors are screened using cancer stem cells.

13.5.2.2 Biochemical Screening

Biochemical assays allow evaluation of large numbers of compounds (Aherne et al. 2002) in a fully automated manner. Suitable platforms currently under use are enzyme-linked immunosorbent assays (ELISA) or enzyme-based colorimetric methods, radiometric assays, fluorescence based methods, and luminescence detection methods. More recently, fluorescence resonance energy transfer (FRET) techniques which are suitable for cell-free and cellular systems are developed (Tian et al. 2007).

13.5.2.3 Nonmammalian Model Organisms for Screening

Non-mammalian organisms such as yeast *Saccharomyces cerevisiae*, the nematode *Caenorhabditis elegans*, or fruit fly *Drosophila melanogaster* were used for anti-cancer drug screening in the late 1990s since they share similar signaling and growth regulatory pathways with humans (Hartwell et al. 1997). The advantage of yeast is that the complete genome comprises only 6250 genes, and many genes involved in human tumors have homologs in this organism. For example, the p53 tumor suppressor gene has its structural homolog in RAD9, the cyclins D and E in cyclin DDm and cyclin EDm the mismatch repair genes MSH2 and MSH1 in MSH2Sc and MLH1Sc, respectively.

13.5.2.4 In-vitro Tumor Models

In-vitro tumor models have proved itself as an important tool for low-cost screening platform for anticancer drugs. Better understanding of cancer progression and treatment have been an impetus in rapid advancement in creating in-vitro models with increased accuracy and stability. This has resulted in, in-vitro models becoming progressively complex and diversification of output parameters. Advances in tumor cell biology, 3D cell culture, tissue engineering, biomaterials, microfabrication, and microfluidics have resulted in rapid development of new in-vitro models. Current model incorporates multiple cell types, extracellular matrix materials, and spatial and temporal introduction of soluble factors.

In-vitro models have been developed to provide a working insight into tumor growth, proliferation, angiogenesis and drug delivery. The variable factors of in-vitro models are cell source, biophysical properties, extracellular matrix and biochemical signals.

The molecular profiles of a large number of human cancer cell lines are available in the Cancer Cell Line Encyclopedia (Barretina et al. 2012), and these profiles can be compared to the profiles of a large number of human tumors. Using such characterised cells to develop an in-vitro model is an ideal method for recapitulating specific aspects of the tumor microenvironment.

13.5.2.4.1 Advantages of in-vitro Tumor Models

Primarily in-vitro screening is a cost effective method compared to in-vivo models. Advances in 3D cell culture, tumor cell biology, biomaterials, microfabrication, tissue engineering have increased the complexity of in-vitro tumor models. This has resulted in an increase and diversification in output parameters. The requirement of precision medicine for cancer have led to increase in adapting in-vitro models for drug screening.

13.5.2.4.2 Transwell-Based Models

This is a method to assess the ability of cancer cells to migrate and invade. This process is relevant in metastasis. Migration is the movement of cells, which may be random or influenced by matrix, soluble factors or electric field. Invasion on the other hand is the ability of the cells to traverse through an extracellular matrix (ECM) (Katt et al. 2016). There are three main adaptations of Transwell based assay,

1. Migration assays
2. Invasion assays
3. Transendothelial migration assays

Drug screening and study of migration, intravasation, extravasation and matrix remodelling can be done using transwell-based assays and this is primarily achieved by counting the cells. The applications include Studies of the influence of chemoattractants on migration and invasion, Studies of influence of other cell types (e.g., macrophages and fibroblasts), Testing the influence of knockdown, transfection, and antibody treatment on invasion and migration, Assessing drug therapies in reducing invasion (Yang et al. 2016), etc.

13.5.2.4.3 Spheroids

Spheroids are cells grown in suspension or embedded in a 3D matrix using 3D cell culture methods. Cancer cell spheroids are known as multicellular tumor spheroids (MCTS). They represent avascular tumor nodules or micro-metastases (Friedrich et al. 2009). Spheroids which bear necrotic cores mimick poorly vascularized tumors due to the establishment of oxygen and nutrient gradient. The protein and gene expressions profile of such spheroids are similar to clinical and in-vivo conditions.

There are four methods of spheroid culture

1. Suspension culture
2. Non-adherent surface methods
3. Hanging drop methods
4. Microfluidic methods

Spheroids are primarily used to study following processes.

13.5.2.4.3.1 Cell Function

The behaviour of tumor cells in an avascular tumor microenvironment include growth kinetics, tumor cell biology and composition of different types of cells. Invasion of cells can also be studied using spheroids (Hirschhaeuser et al. 2010).

13.5.2.4.3.2 Drug Screening

Studies comparing gene expression profiles of spheroids and 2D cultures revealed that spheroids closely resemble in-vivo tumor models, thus enabling it to be used in a high-throughput tool for negative selection of drug candidates to reduce animal testing (Friedrich et al. 2009).

13.5.2.4.3.3 Angiogenesis

Assessment of migration of endothelial cells into tumor spheroids or the formation of vascular networks in the spheroids (Timmins et al. 2004).

13.5.2.4.3.4 Immune Cell Response

Culturing MCTS with immune cells and observing migration and invasion. Co-culturing spheroids of tumor cells and immune cells and studying their interactions (Gottfried et al. 2006).

13.5.2.4.4 Hybrid Models

There are various types of in-vitro tumor models which cannot be classified as spheroids or transwell-based. These include,

1. Ex-vivo tumor sections
2. 3D invasion models
3. Avascular microfluidic models

The main feature of these models are that they recreate the tumor microenvironment although retaining the simplicity of in-vitro models. Various applications are as follows

1. Embedded ex-vivo Tumor Section developed from patient biopsies or resected tumor sections, can be used to probe the tumor microenvironment. This model is used for the studying fundamentals of tumor growth and invasion (Dark et al. 1997). Primary identification of individualised chemotherapeutic regiments can also be examined using this model.
2. 3D Invasion Models developed by seeding cancer cells on ECM materials reduce the complexities of tumor microenvironment. This helps in imaging of live cells to study cell morphology, cell interactions and movement (Koch et al. 2012).

3. Avascular Microfluidic Model is used to assess cancer cell migration through small channels with respect to chemotherapeutic agents (Konstantopoulos et al. 2013).

13.5.2.4.5 Tumor-Microvessel Models

Tumor vasculature is an important component of a tumor microenvironment, providing nutrients. The cells of these vessels also secrete factors promoting or suppressing tumor growth. Tumor-microvessel models recreate the complex interplay of vasculature and cancer cells under in-vitro conditions. Cancer cell movement through the vasculature can be visualized in real time and studied for various behaviors like invasion, tumor-driven angiogenesis, intravasation, and extravasation (Wang et al. 2015).

All the in-vitro models can be used for specific purposes and will have advantages and disadvantages to other models. A diagrammatic representation of in-vitro tumour culture models are given in Fig. 13.1. Advantage and disadvantage of each in-vitro model is given in Table 13.1.

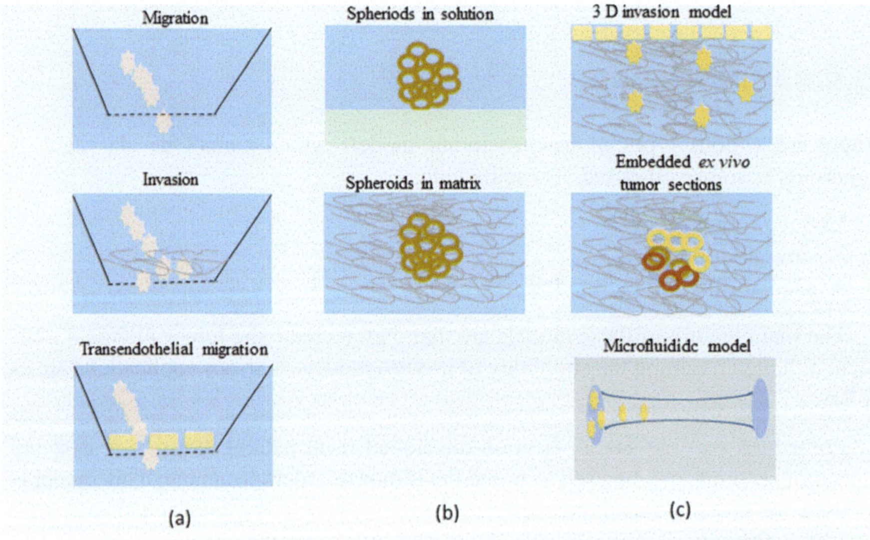

Fig. 13.1 Cartoon depict different types of transwell-based motility assays. (**a**) Migration, invasion, and transendothelial migration. (**b**) Spheroid-based assays – spheroids in media and in matrix. (**c**) Hybrid models – embedded ex-vivo tumor sections, 3D invasion models, and avascular microfluidic models

Table 13.1 Advantages and disadvantages of various in-vitro tumor models (Katt et al. 2016)

	In-vitro model	Advantages	Disadvantages
Transwell model	Migration	Easily implemented, low-cost assay	Low physiological relevance
	Invasion	High throughput	Can only assay single-cell motility
	Transendothelial migration	Can be used to compare metastatic potential of cells	Migration and invasion assays can result in conflicting data
Spheroid	Cell suspension culture	Simple	No individual compartments for spheroids
		Mass production	Cannot control uniformity (size, composition)
		Long-term culture	High shear force
	Non-adherent surfaces	Simple	Low throughput
		Better efficiency (MW)	Long-term culture difficult
		High throughput (MW)	
		Uniform spheroid size (MW)	
		Coculture (MW)	
	Hanging drop technique	Control of spheroid size	Low throughput
		Uniform spheroid size	Long-term culture difficult
		Allows coculture with defined cell types	Not efficient
	Microfluidic devices	Control of spheroid size	Difficulty collecting cells for analysis
		Control of spheroid growth parameters	
		Continuous perfusion	
		Faster spheroid	
Hybrid tumor microvessel model	Embedded ex-vivo tumor sections	Maintains tumor heterogeneity	Lacks flow through vasculature
		Patient-specific assay	
		Mimics outgrowth into surrounding tissues	
	3D invasion models	3D microenvironment	Lacks vasculature
		Allows real-time tracking of cells	Lacks tumor complexity
		Balance of complexity and experimental control	
	Avascular microfluidic	Simple migration assay	Lacks vasculature
		Easy to isolate effect of variables	Typically lacks 3D environment
		Allows real-time tracking of cells	

13.5.2.5 Hollow Fiber Assay

Hollow Fiber Assay (HF) is a combined in-vitro/in-vivo testing procedure. The HF assay was developed by Decker et al. (2004) at the NCI and is composed of 2 cm tubes consisting of polyvinylidene fluoride (PVDF) filled with tumor cell lines implanted into mice at two sites (intraperitoneal and subcutaneous). The mice undergo drug treatment and fibers are removed after 4–6 days from the animal and analyzed in-vitro for tumor cell growth.

The Hollow Fiber Assay Bridge the gap between in-vitro and in-vivo xenograft screening of anticancer compounds (Shnyder et al. 2005).

One disadvantage of HF assay is that tumor growth is inhibited by the inside diameter of the tube and another drawback is that the fiber wall is an artificial barrier between the tumor and its environment which hampers the diffusion of large biomolecules, such as DNA and antibodies.

13.5.2.6 Colony-Forming Assay

Another combined in-vitro/in-vivo testing model is the soft agar colony-forming assay, it is also known as tumor clonogenic assay (TCA). The TCA can be used for sensitivity screening of patient tumor material in-vitro for predicting clinical response and also with fresh xenograft tissue for selecting the most appropriate in-vivo model for analysis (Chumsri et al. 2007). The development of the human tumor stem cell (HTSC) assay offered an approach to determine and predict a reasonable clinical response and is disease-orientated in concept.

13.5.2.7 In-silico or Bioinfomatic Methods

Developing a new drug is a complex, expensive and time-consuming task. Many novel technologies and methodologies have been developed to increase the efficiency of the drug discovery process, and computational methodologies or bioinformatic techniques have become a crucial component of drug discovery programs. Techniques such as ligand- or structure based virtual screening are widely used for designing potential anticancer drugs and drug candidates (Grinter et al. 2011).

Inverse docking is computational docking of a specific small molecule of interest to a library of receptor structures. The technique can be used to identify new potential biological targets or to identify targets among a family of related receptors (Prada-Gracia et al. 2016).

The in-silico studies or computational drug design approaches and mathematical-/statistical-based modeling provide insights into the physicochemical properties associated with anticancer activity at the molecular level. These are classified mainly into Structural based and Ligand-based techniques. These are widely used for the in-silico screening of large chemical databases, whose aim is to detect novel bioactive ligands. Models for constructing predictive molecules and separating active from inactive ligands are developed using certain optimization methods such as neural networks, genetic algorithms, and the k-nearest neighbor algorithm (Shahaf et al. 2016, Pappalardo et al. 2017). An interesting example of structure-based pharmacophore modeling is the identification of p53 upregulated modulator of apoptosis (PUMA) inhibitors (Mustata et al. 2011).

13.6 Conclusions

The limited availability of clinical samples and high cost in performing clinical trials has made in-vitro tumor models to be the premier pre-clinical tool for cancer studies. These models are relatively cheap, easily manipulated and accumulated data from these cell lines can be used in cancer pharmacogenomics studies. In-vitro models have already proved their worth in cancer research by helping to easily identify pharmacogenomics biomarkers and enabling the testing of hypothesis. By improving the pre-clinical cell line models, the translation of clinically and pharmacologically relevant markers from the research laboratory to the hospital setup can be accelerated. 3D tumor models and other hybrid models have taken cancer research to new heights. Further advancements in in-vitro tumor models will enable in developing personalized tumor research systems and therapies, and improve their clinical translation efficiency.

Acknowledgments DKC acknowledge Department of Microbiology, St. Mary's College, Thrissur. AM acknowledges Origin Diagnostics and Research, Karunagappally. LSR acknowledge Institut Pasteur, Paris.

References

Aherne W, Garrett M, McDonald E, Workman P. Mechanism based high throughput screening for novel anticancer drug discovery. Anticancer Drug Des. 2002;249:67.

Amudha G, Josephine A, Mythili Y, Sundarapandiyan R, Varalakshmi P. Therapeutic efficacy of dl-α-lipoic acid on cyclosporine A induced renal alteration. Eur J Pharmacol. 2007;571:209–14.

Amundson SA, Do KT, Vinikoor LC, Lee RA, Koch-Paiz CA, Ahn J, Reimers M, Chen Y, Scudiero DA, Weinstein JN, Trent JM. Integrating global gene expression and radiation survival parameters across the 60 cell lines of the National Cancer Institute Anticancer Drug Screen. Cancer Res. 2008;68:415–24.

Barretina J, Caponigro G, Stransky N, Venkatesan K, Margolin AA, Kim S, Wilson CJ, Lehár J, Kryukov GV, Sonkin D, Reddy A. The Cancer Cell Line Encyclopedia enables predictive modelling of anticancer drug sensitivity. Nature. 2012;483:603.

Blatt NL, Mingaleeva RN, Khaiboullina SF, Kotlyar A, Lombardi VC, Rizvanov AA. *In vivo* screening models of anticancer drugs. Life Sci J. 2013;10:1892.

Chumsri S, Phatak P, Edelman MJ, Khakpour N, Hamburger AW, Burger AM. Cancer stem cells and individualized therapy. Cancer Genomics Proteomics. 2007;4:165–74.

Damia G, D'Incalci M. Contemporary pre-clinical development of anticancer agents–what are the optimal preclinical models? Eur J Cancer. 2009;45:2768–81.

Dark GG, Hill SA, Prise VE, Tozer GM, Pettit GR, Chaplin DJ. Combretastatin A-4, an agent that displays potent and selective toxicity toward tumor vasculature. Cancer Res. 1997;57:1829–34.

Decker S, Hollingshead M, Bonomi CA, Carter JP, Sausville EA. The hollow fibre model in cancer drug screening: the NCI experience. Eur J Cancer. 2004;40:821–6.

Dengler WA, Schulte J, Berger DP, Mertelsmann R, Fiebig HH. Development of a propidium iodide fluorescence assay for proliferation and cytotoxicity assays. Anti-Cancer Drugs. 1995;6:522–32.

Fiebig HH, Schüler J, Bausch N, Hofmann M, Metz T, Korrat A. Gene signatures developed from patient tumor explants grown in nude mice to predict tumor response to 11 cytotoxic drugs. Cancer Genomics Proteomics. 2007;4:197–209.

Friedrich J, Seidel C, Ebner R, Kunz-Schughart LA. Spheroid-based drug screen: considerations and practical approach. Nat Protoc. 2009;4:309.

Gottfried E, Kunz-Schughart LA, Andreesen R, Kreutz M. Brave little world: spheroids as an *in vitro* model to study tumor-immune-cell interactions. Cell Cycle. 2006;5:691–5.

Grinter SZ, Liang Y, Huang SY, Hyder SM, Zou X. An inverse docking approach for identifying new potential anti-cancer targets. J Mol Graph Model. 2011;29:795–9.

Gutman RL, Peacock G, Lu DR. Targeted drug delivery for brain cancer treatment. J Control Release. 2000;65:31–41.

Hartwell LH, Szankasi P, Roberts CJ, Murray AW, Friend SH. Integrating genetic approaches into the discovery of anticancer drugs. Science. 1997;278:1064–8.

Hirschhaeuser F, Menne H, Dittfeld C, West J, Mueller-Klieser W, Kunz-Schughart LA. Multicellular tumor spheroids: an underestimated tool is catching up again. J Biotechnol. 2010;148:3–15.

Katt ME, Placone AL, Wong AD, Xu ZS, Searson PC. *In vitro* tumor models: advantages, disadvantages, variables, and selecting the right platform. Front Bioeng Biotechnol. 2016;4:12.

Koch TM, Münster S, Bonakdar N, Butler JP, Fabry B. 3D traction forces in cancer cell invasion. PLoS One. 2012;7:e33476.

Konstantopoulos K, Wu PH, Wirtz D. Dimensional control of cancer cell migration. Biophys J. 2013;104:279.

Kumar S, Bajaj S, Bodla RB. Preclinical screening methods in cancer. Indian J Pharm. 2016;48:481.

McKeage MJ. The potential of DMXAA (ASA404) in combination with docetaxel in advanced prostate cancer. Expert Opin Investig Drugs. 2008;17:23–9.

Mosmann T. Rapid colorimetric assay for cellular growth and survival: application to proliferation and cytotoxicity assays. J Immunol Methods. 1983;65:55–63.

Mustata G, Li M, Zevola N, Bakan A, Zhang L, Epperly M, S Greenberger J, Yu J, Bahar I. Development of small-molecule PUMA inhibitors for mitigating radiation-induced cell death. Curr Top Med Chem. 2011;11:281–90.

Pappalardo M, Rayan M, Abu-Lafi S, Leonardi ME, Milardi D, Guccione S, Rayan A. Homology-based modeling of rhodopsin-like family members in the inactive state: structural analysis and deduction of tips for modeling and optimization. Mol Inform. 2017;36:1700014.

Prada-Gracia D, Huerta-Yépez S, Moreno-Vargas LM. Application of computational methods for anticancer drug discovery, design, and optimization. Bol Med Hosp Infant Mex (Engl Ed). 2016;73:411–23.

Rygaard J, Poulsen CO. Heterotransplantation of a human malignant tumour to "Nude" mice. Acta Pathol Microbiol Scand C. 1969;77:758–60.

Schwartsmann G, Winograd B, Pinedo HM. The main steps in the development of anticancer agents. Radiother Oncol. 1988;12:301–13.

Shahaf N, Pappalardo M, Basile L, Guccione S, Rayan A. How to choose the suitable template for homology modelling of GPCRs: 5-HT7 receptor as a test case. Mol Inform. 2016;35:414–23.

Shnyder SD, Hasan J, Cooper PA, Pilarinou E, Jubb E, Jayson GC, Bibby MC. Development of a modified hollow fibre assay for studying agents targeting the tumour neovasculature. Anticancer Res. 2005;25:1889–94.

Shoemaker RH. The NCI60 human tumour cell line anticancer drug screen. Nat Rev Cancer. 2006;6:813.

Suggitt M, Bibby MC. 50 years of preclinical anticancer drug screening: empirical to target-driven approaches. Clin Cancer Res. 2005;11:971–81.

Teicher BA. In vivo/ex vivo and in situ assays used in cancer research: a brief review. Toxicol Pathol. 2009;37:114–22.

Tian H, Ip L, Luo H, Chang DC, Luo KQ. A high throughput drug screen based on fluorescence resonance energy transfer (FRET) for anticancer activity of compounds from herbal medicine. Br J Pharmacol. 2007;150:321–34.

Timmins N, Dietmair S, Nielsen L. Hanging-drop multicellular spheroids as a model of tumour angiogenesis. Angiogenesis. 2004;7:97–103.

Wang XY, Pei Y, Xie M, Jin ZH, Xiao YS, Wang Y, Zhang LN, Li Y, Huang WH. An artificial blood vessel implanted three-dimensional microsystem for modeling transvascular migration of tumor cells. Lab Chip. 2015;15:1178–87.

Workman P, Aboagye EO, Balkwill F, Balmain A, Bruder G, Chaplin DJ, Double JA, Everitt J, Farningham DA, Glennie MJ, Kelland LR. Guidelines for the welfare and use of animals in cancer research. Br J Cancer. 2010;102:1555.

Yang F, Hu M, Lei Q, Xia Y, Zhu Y, Song X, Li Y, Jie H, Liu C, Xiong Y, Zuo Z. Nifuroxazide induces apoptosis and impairs pulmonary metastasis in breast cancer model. Cell Death Dis. 2016;6:e1701.

Chapter 14
In-vitro Techniques to Study Cell Signaling

Natasha Jaiswal

14.1 Introduction

Cellular signaling is choreography of events occurring at similar time, involving ballets of cues interacting and stimulating surrounding proteins, lipids, and ions, resulting in cytoskeletal reorganization, modulation of differentiation and induction of gene expression. It is an important mechanism via which both plant and animal cells respond to external environment; which are essential for cell development, differentiation, programmed cell Apoptosis, their behavior to biotic and abiotic stresses as in plant cell (Bari and Jones 2009) etc.

14.2 Principles of Cell Signaling Pathway

In order for the cells to interact and respond to their external environment, cell receives signals (information) from the external environment which is processed and integrated to regulate cell behavior. Intracellular signalling pathway consists of a series of event consisting of Receptor (sensors) and Cell signaling molecule (signal transducers). The basic cellular processes are common to both animals and plants. Though much attention is applied to eukaryotic cell signaling, less stress is applied to plant cell signaling despite the fact that most of the cellular responses were first known in plant cells.

N. Jaiswal (✉)
Department of Medicine, University of Pennsylvania, Philadelphia, PA, USA

S. Kumar, C. Egbuna (eds.), *Phytochemistry: An in-silico and in-vitro Update*,
https://doi.org/10.1007/978-981-13-6920-9_14

14.2.1 Receptors

Receptors are the proteins with which cells interact with the external environment through event such as a receptor–ligand interaction, cell–cell contact or cell–ECM contact. Organic substances such as Ligands, Growth hormones play a very crucial role to initiate the cell surface while Receptors act as sensors to recognize their own ligand/hormone to initiate specific cellular response. In plants there are five classical phytohormones, namely auxins, abscisic acid (ABA), cytokinin (CK), gibberellins (GAs), and ethylene, including brassinosteroids (BRs), jasmonate, salicylic acid, nitric oxide, and strigolactones (Santner and Estelle 2009), which are required in numerous physiological processes. For example Auxins regulate variety of physiological processes, such as cell enlargement, cell division, tropic responses, root initiation, and vascular tissue differentiation. In animals receptors can be categorized into four: G-protein coupled receptors, ion-channel receptors, tyrosine kinase-linked receptors and receptors with intrinsic enzymatic activity (Harvey et al. 2000). In addition to these classic classes of receptors, cells can also respond to their extracellular environment through integrin receptors or proteoglycan receptors such as the syndecan family. Whereas the largest class of enzyme-linked receptors in animals is receptor tyrosine kinases, this type of receptor is extremely rare in plants. Instead, plants feel confident on transmembrane receptor serine/threonine kinases. Plant receptors can be classified as: two component based receptors, Leucine-rich repeat (LRR) based receptors and Ubiquitination based hormone receptors.

In short, the cells respond to a number of extracellular signaling molecules by binding to the receptor present at the surface (as these molecules are too big to enter the cells) and initiate a cascade of events leading to cell behavior response. However, there are some receptors which are localized internally and they respond to mostly small molecules that can cross the membrane.

14.2.2 Signaling Pathway (Cell Response to the Stimuli)

A signaling pathway consists of activated and inactivated proteins which are turned on or off through the processes of phosphorylation, dephosphorylation, intracellular location, nucleotides (ADP/ ATP or GDP/GTP) cycles, calcium/ion levels and many more. In animal cells ligand binding to the receptor activates the pathway via activating the downstream proteins/molecules. For example, GPCR pathway, one among the primary studied pathway in animal cells, is activated in response to the activation of second messenger molecules cyclic AMP (cAMP), cyclic GMP (cGMP), diacylglycerol (DAG), phosphoinositols, and calcium (Ca^{+2}). The activation of these molecules allows them to undergo conformational change in the membrane making G-protein accessible to the receptor at their cytoplasmic domain. Similarly, Nucleoside triphosphates and diphosphates (ADP/ATP or GDP/GTP), that functions both as autocrine and paracrine signaling molecules in the animal extracellular matrix (ECM), affects a broad range of physiological processes by activating receptors specific for an exceptional operation and subsequent down stream signal transduction cascades such as neurotransmission, the regulation of blood pressure, and the immune response. Plant cells, just like animals, also make use of peptides and proteins as signals. Plant hormone such as Abscisic Acid (ABA) signals via protein kinase activation while inhibiting protein phosphatase familiar in mammalian cell signaling pathways, including phosphoprotein phosphatases, protein kinases, reactive oxygen species and calcium ions (acting as intracellular second messengers) (Hauser et al. 2011). A Similar pattern of signaling is also observed in gibberellin (GA) (Sun 2011) and auxin (Leyser 2011; Depuydt and Hardtke 2011) signal transduction pathways.

Thus, the fundamental cellular processes mediating physiological effects are common to both animals and plants and is more than a normal straight forward process as the crosstalk and integration, both in form and function, between the ECM and Cell membrane molecules (Porter and Hogg 1998; Ossowski and Aguirre-Ghiso 2000) and among the small molecules intracellularly, is required to attain a physiological phenotype. Additionally, the intracellular molecules must recognize their interacting partner. Further, since a cell is not just a target but also sends signals, a cell is constantly operating, receiving signals, integrating and sending them out as per requirement through a nicely coordinated complex network of events. Therefore, understanding such a complex series of event is important and necessary to understand how signal integration is accomplished, their physiological responses and how the defect at the levels of any of the key players leads to physiological and metabolic abnormalities, both in plant and animal cells. Progress have been made past 40 years to develop various biochemical technologies to understand this complex series of events and to develop methodologies and pharmacological approaches that allow temporal analysis of signaling events at high spatial resolution in single cells.

Some of the various in-vitro approaches used to study signal transduction in plant and animals are discussed below.

14.3 In-vitro Techniques to Study Cell Signaling

Cell signaling is an intracellular wiring of complex signaling systems that is being best represented as a network rather than a simple linear pathway and a number of specific methodologies have been invented to establish the signaling mechanisms (Schwartz and Baron 1999; Sastry and Burridge 2000; Schwartz and Shattil 2000) in both plants and animals. Both Plant and animal cells respond to a variety of external stimuli mediated via profound changes in gene expression resulting in the changes in the composition of their transcriptome, proteome, and metabolome (Pérez-Alfocea et al. 2011). The change in proteome and metabolome can be both transcriptome dependent and/or independent. Thus, a combination of molecular, genetic and proteomics approaches are used to study the complex signaling pathways at different levels as it allows direct and unbiased observance of the factors and provide the data that can be directly used to investigate the complex interaction in the cells, their metabolism and also the stress/stimuli in response to the environmental or the biological threats (insects, fungi, or other pathogens). The main biotechnological in-vitro approaches to study plant and animal responses to external stimuli are therefore categorized here into three broad categories: The Genomics, Proteomics and Metabolomics approach.

14.3.1 Genomics Approach

Genes are transcriptionally activated in response to a number of external stimuli and analysis of the function of numerous gene is beneficial in understanding the complex signaling and adaptive processes. A gene by gene approach has been typically used to understand the signaling molecules and its function. Large-scale gene function analysis is performed using high throughput technology for functional genomics and involves the interaction of gene products at cellular and organism level. Various genomic approaches such as Expressed sequence tags (ESTs) and cDNA sequences, cDNA libraries, microarray, and serial analysis of gene expression (SAGE) are a number of various methods used to analyze gene expression profiles using functional genomics approach.

14.3.1.1 Expressed Sequence Tags (ESTs), cDNA Sequences and cDNA Libraries

A cDNA (complementary DNA) sequencing is used to know the genetic profile that are actively expressed in the cells as the cDNA will only have the sequences that can be coded as it is formed from mRNA. The cDNA in the eukaryotic cells doesn't need splicing as they are produced without introns and therefore have the benefit to be directly expressed in a bacterial cell. This serves as a powerful tool to study the role of a specific gene as the products of a gene are easily identified. A collection of such a portion of actively expressed transcriptome is known as a "library" and can be expressed into host cells to form a cDNA library. These libraries are most useful in reverse genetics where the function of a protein helps to identify the gene such as in functional cloning. However, this important tool has a few drawbacks as well as they are devoid of information about noncoding regions such as enhancers, introns and other regulatory elements found in a genomic DNA library but lacks in cDNA library.

Expressed sequence tag (EST) is a small sub-sequence of a cDNA sequence (Fraire-Velázquez et al. 2011) and represent a small portion of active/expressed genes they are a part of cDNA and are derived from one-shot sequencing of a cloned cDNA (ESTs Factsheet).

14.3.1.2 Microarray for Gene Analysis

Microarray is a trending technology used to study the genetic makeup of plant and animals as it allows unraveling changes in different conditions as an acknowledgement to the environmental factors. The process is also called expression analysis or expression profiling. In this technique, diminutive DNA spots (picomoles (10–12 mol) of a specific DNA sequence, referred to as probes (or reporters or oligos), are compiled and attached to a strong surface and are then used to hybridize a CDNA or CRNA (also called anti-sense RNA) which serves as sample (called object). The whole process is performed in sternness conditions and is regularly used to measure the expression levels of large numbers of genes simultaneously. This is another powerful technique to study changes in cell signaling in response to external stimuli via genotyping varied regions of a genome.

14.3.1.3 Mutagenesis

Mutagenesis is a technique where mutations are induced in the genes for large-scale expression studies so that the specific function of the gene can be identified. Both random and site specific methods to induce mutations are known and is compliant for large-scale analysis of genes (Lukowitz et al. 2000). Gene tagging is another similar tool to study large scale gene analysis. The technique allows the understanding of the complexity of signaling process through analysis of the function of varied genes involved in particular response in both plants and animals. It allows the identification of a gene function through modification. In plants such a technique is often used to develop improved variety of stress tolerant crops.

14.3.1.4 Quantitative Trait Loci (QTL)

This molecular marker technology available since the 1980s helps to unravel the inheritance of all traits through the variation in trait method. A QTL corresponds to the variation in a phenotype and are mapped by identifying molecular markers correlating to the observed trait. It helps to understand whether a specific trait is controlled by single genes or many genes acting together. With the progress of QTL mapping, the complexity of the genetic architecture is determined and in plants has led to the development of new breeding approaches such as marker-assisted selection and breeding by design (Peleman and van der Voort 2003).

Improvements in plant and animal genomics research have unlocked new aspect to study cell signaling at the gene level and for functional analysis of genome. It proposes newer opportunities for improving animal health and enhancing the productivity in plants.

In conclusion all the genomics technologies discussed so far, along with genome sequencing, mutant screen etc. have been found beneficial to elucidate the mutagenicity of reactions.

14.3.2 Proteomics Approach

Proteomics is the large-scale study of proteins. Since proteins are directly involved in any response, proteome studies can certainly provide information to understand complex signaling pathways. The change in protein alters as per requirement, time, response to external/environmental stimuli or in response to stress that a plant or animal cell undergoes (Marques et al. 2009). Although changes at the protein level is not consistent with the changes in gene expression or vice versa, investigating proteomics has been highly important since proteins, unlike gene, are straightforward effected in response to any stimuli. Changes in proteins is a reflection of changes in metabolites as well as changes in the transcription and translation machinery, resulting in modified protein and therefore, provide a better understanding of physiological processes. In proteomics, multiple methods are used to study proteins. Generally, proteins can be detected using numerous tools like antibodies (immunoassays) or mass spectrometry. Use of specific antibody is considered to analyze proteins in complex biological samples. Methods such as quantitative dot blot analysis (qdb) and biochemical separation methods such as Electrophoresis needs to be performed before the detection step as is aids in accurate detection and quantification of the protein. Various in-vitro techniques used to study cell signaling at proteome level are as follow:

14.3.2.1 Specific Inhibitor Approach

Pharmacologic agents have been used over the past 20 years to target Signal transduction proteins in order to establish their role. Many inhibitors are available for both generalized and specific interactions with either receptors or specific kinases. This is the simplest approach to study the specific role of signal molecules with the drawbacks of having off target effect at higher doses. This is the reason for the technique being replaced by the use of RNAi, SiRNA recently. However, many efficient pharmacological inhibitors are available which are often used to target a specific signaling molecule and understand their role in signal transduction.

14.3.2.2 Cell Behavior Assays

Attachment, migration, changes in cytoskeletal elements, cell shape changes, matrix binding, and migration or differentiation are the assays used as characteristics to determine whether cells reacted differently or not. Cell behavior is assessed based on various biochemical assays by activation/deactivation of specific proteins involved in signal transduction via expressing inactive form or constitutively active form of proteins (Berrier et al. 2000) or by blocking the translational machinery with antisense oligonucleotides.

14.3.2.3 Spectrophotometry

The technique involves the use of photometers, known as spectrophotometers, to quantitate the protein concentration in the sample based on intensity of light beam in the visible region. The intensity of light produced is directly proportional to the concentration of protein.

14.3.2.4 Western Blotting

It is the most traditional biochemical assay method to study signal transduction pathway. The western blot is a tool used in molecular biology and other molecular biology disciplines to detect and analyze specific proteins in a biological sample such as tissue homogenate or cell lysates. The events can be detected using antibody against specific molecule/protein which are activated and can be identified using specific antibodies. For example, antibodies against phosphorylated forms (Ser/Thr) can help to compare the activation of proteins in cells in different states.

14.3.2.5 Immunoprecipitation

Immunoprecipitation is relatively easy and can be considered as an extension of Western blots. The proteins present in unsubstantial levels are hard to be detected using Western Blotting techniques. Such a protein is generally enriched using Beads, made from a variety of substances, including metals. In the laboratory, primary antibodies are cross-connected to a specific protein to the beads which are then used to precipitate antibody specific protein from the cell lysates. The proteins are then separated by using SDS electrophoresis and then the proteins are identified by using the procedures for Western blots. Both Western blotting and Immunoprecipitation methods are used to compare the status of a certain protein at a specific time.

14.3.2.6 GST Binding or "Pull-Down" Assays

The glutathione S-transferase (GST) binding or "pull-down" assay resembles immunoprecipitation, except it determines direct protein–protein interactions and a protein is used to bait the protein instead of the Antibody. Here the protein of interest is integrated on to GST fusion proteins using the domains that may interact with other specific proteins. This fused protein is then used to identify the specific proteins, often referred to as prey. This is a powerful tool to study protein-protein interaction and to identify the interacting partners of the protein of interest.

14.3.2.7 Visualizing Signal Transduction Events

This is an important and commonly used in-vitro technique to study the specific role of a protein in signal transduction via tagging them with fluorescent protein gene. The proteins are labeled with a green fluorescent protein (GFP) and have advantage to be used in tracer experiments.

14.3.2.8 Immunohistochemistry

Immunohistochemistry is an important in-vitro method used to visualize signaling proteins as the location and distribution of many proteins itself indicates their activity. Phosphorylation/de-phosphorylation defines the activity of many proteins and their localization. Using appropriately labelled secondary antibody against a protein can help to know the activity of a particular protein and helps to compare them in different conditions. Co-localization is the extension of the method to visualize two or more proteins at the same time and differentiate their role based on their activity and distribution. Using constructed fusion proteins—Living cells can also be observed with GFP-tagged proteins to track the expression patterns of the protein in cells, tissues, and transgenic animals. The method is also useful to determine protein–protein interactions (Michaelson et al. 2001).

14.3.2.9 Two Imaging Techniques FRET and FRAP

The discovery of two imaging techniques, Fluorescence resonance energy transfer (FRET) (Matyus 1992) and fluorescence recovery/redistribution after photobleaching (FRAP), re-appeared as useful methods and further advancement in imaging technique. FRET is the nonradioactive transfer of energy from a donor that emits a higher light wavelength when excited by a specific wavelength of light to a nearby fluorescent acceptor. Donor/ acceptor pair examples are Cy3 and Cy5 or CFP and YFP. The energy transfer is dependent on the distance between the fluorescent molecules. As the acceptor absorbs the donor fluorescence, the donor will quench and its lifetime will decrease. FRET is a modified light microscopic technique and more powerful than confocal microscopy where it allows the investigator to determine the relative distance and distribution of two proteins within 10–70 angstroms of each other. FRAP is similar to FRET except for tracing the molecule based on recovery i.e. the cells preloaded with the fluorescent molecule of interest are photo bleached in a specific region and the movement of the fluorescent molecules back into the bleached area is measured. This method allows the investigator to determine the mobility and diffusion of small molecules in the cytoplasm of living cells or record the movement of macromolecules (RNA or large proteins, drugs) into and out of cell organelles such as the nucleus. These double-labelling experiments are more advantageous as it allows the visualization of additional proteins for which anti-active antibodies are not yet available.

14.3.2.10 The Enzyme-Linked Immunosorbent Assay (ELISA)

This is a more simplified method that has been used for decades to detect and quantitate protein levels at different stages of the signaling pathways. It is a type of analytic biochemistry assay that uses a solid-phase enzyme immunoassay (EIA) to recognize the presence of an "analyte", usually an antigen, in a liquid sample or wet sample. There are three kinds of ELISA known today: Direct ELISA, Sandwich ELISA and Competitive ELISA. Using specific antibody adhered to the polystyrene coated plate, the presence (activation)/absence (deactivation) of specific protein in the sample can be detected.

14.3.2.11 Flow Cytometry

Flow cytometry is a laser- or impedance-based, biophysical technology employed in cell counting, cell sorting, biomarker detection and protein engineering, by suspending cells in a stream of fluid and passing them through an electronic detection apparatus known as flow cytometer. The method allows simultaneous multipara metric analysis of the physical and chemical characteristics of thousands of particles per second. This is a routinely used method used in the diagnosis of health disorders, in basic research, clinical practice and clinical trials.

14.3.2.12 Mass Spectrometry

Mass spectrometry (MS) is an analytical technique that ionizes chemical species and sorts the ions based on their mass-to-charge ratio. In simpler terms, a mass spectrum measures the masses within a sample. Mass spectrometry is used in many different fields and is applied to pure samples as well as complex mixtures to elucidate the chemical structures of molecules, such as peptides and is used to explore changes in proteins in response to external stimuli based on mass.

14.3.2.13 Two-Dimensional Electrophoresis

Two-dimensional electrophoresis (2DE) is a biochemical approach to separate proteins from different samples in parallel is separating them in two dimensions based on charge and mass property. The proteins are first separated in the first dimension based on their isoelectric points and are further separated perpendicularly according to their molecular mass, followed by selection and staining of differentially expressed. However, this method has its limitation because of its inability to resolve all the proteins within a sample if they fall in the same molecular range.

14.3.3 Metabolomics Approach

This is the scientific study of chemical processes involving metabolites, the small molecule intermediates and products of metabolism. It is the study of their small-molecule metabolite profiles. The metabolome represents the complete set of metabolites in a biological cell, tissue, organ or organism, which are the end products of cellular processes. Metabolomics provides a direct "functional readout of the physiological state" of an organism (Jaspers and Kangasjärvi 2010) and is intended to provide an integrated view of the functional status of an organism as changes in metabolites could help understanding many physiological processes.

14.4 Conclusion

In summary, rapid progress has been made in the development of in-vitro methodologies to understand and explore the biochemical pathway both in plants and animals. However, many of the present experimental procedures document signal transduction snapshots of individual events, without the benefit of visualizing the whole stage simultaneously. The emergence of newer assays using fluorescently tagged transfected proteins combined with live cell imaging analysis are useful to confirm and modify current perceptions of signal transduction events. Nonetheless, future experiments designed to dissect signal transduction pathways and their morphologic effects will require a combination of all of these approaches and surely new ones. The reemergence of FRET and FRAP to visualize living cells, use of transgenic models in both plant and animals has revolutionized how signal transduction events are understood. Further, innovation in the procedure utilizing custom fusion proteins, transfection of engineered proteins into embryonic tissues to view signal transduction events in whole tissues are some of the area which needs to be explored further. Also, methodologies for rapid detection of phosphorylation, dephosphorylation, protein kinase action, is essential and should always expect to uncover some newer methodologies and techniques to improve and better understand this intense network of signaling.

References

Bari R, Jones JD. Role of plant hormones in plant defence responses. Plant Mol Biol. 2009;69:473.

Berrier AL, Mastrangelo AM, Downward J, Ginsberg M, LaFlamme S. Activated R-Ras, Rac1, PI 3-kinase and PKC [small element of] can each restore cell spreading inhibited by isolated integrin b1 cytoplasmic domains. J Cell Biol. 2000;151:1549–60.

Depuydt S, Hardtke CS. Hormone signalling crosstalk in plant growth regulation. Curr Biol. 2011;21:R365–73.

ESTs Factsheet. National Center for Biotechnology Information.

Fraire-Velázquez S, Rodríguez-Guerra R, Sánchez-Calderón L. Abiotic and biotic stress response crosstalk in plants. In: Shanker A, editor. Abiotic stress response in plants—physiological, biochemical and genetic perspectives. Rijeka: InTech; 2011. p. 3–26.

Harvey L, Arnold B, S Lawrence Z, Paul M, David B, James D. Molecular Cell Biology. 4th ed. 2000.

Hauser F, Waadt R, Schroeder JI. Evolution of abscisic acid synthesis and signaling mechanisms. Curr Biol. 2011;21:R346–55.

Jaspers P, Kangasjärvi J. Reactive oxygen species in abiotic stress signaling. Physiol Plant. 2010;138:405–13.

Leyser O. Auxin, self-organisation and the colonial nature of plants. Curr Biol. 2011;21:R331–7.

Lukowitz W, Gillmor CS, Scheible WR. Positional cloning in Arabidopsis. Why it feels good to have a genome initiative working for you. Plant Physiol. 2000;23:795–805.

Marques MC, Alonso-Cantabrana H, Forment H, et al. A new set of ESTs and cDNA clones from full length and normalized libraries for gene discovery and functional characterization in citrus. BMC Genomics. 2009;10:428.

Matyus L. Fluorescence resonance energy transfer measurements on cell surfaces A spectroscopic tool for determining protein interactions. J Photochem Photobiol B. 1992;12:323–37.

Michaelson D, Silletti J, Murphy G, D'Eustachio P, Rush M, Philips MR. Differential localization of Rho GRPases in live cells: regulation by hypervariable regions and RhoGDI binding. J Cell Biol. 2001;152:111–26.

Ossowski L, Aguirre-Ghiso JA. Urokinase receptor and integrin partnership: coordination of signaling for cell adhesion, migration and growth. Curr Opin Cell Biol. 2000;12:613–20.

Peleman JD, van der Voort JR. Breeding by design. Trends Plant Sci. 2003;8:330–4.

Pérez-Alfocea F, Ghanem ME, Gómez-Cadenas A, Dodd I. Omics of root-to-shoot signaling under salt stress and water deficit. OMICs J Integr Biol. 2011;15:893–901.

Porter JC, Hogg N. Integrins take partners: cross-talk between integrins and other membrane receptors. Trends Cell Biol. 1998;8:390–6.

Santner A, Estelle M. Recent advances and emerging trends in plant hormone signalling. Nature 2009;459:1071–8.

Sastry SK, Burridge K. Focal adhesions: a nexus for intracellular signaling and cytoskeletal dynamics. Exp Cell Res. 2000;261:25–36.

Schwartz MA, Baron V. Interactions between mitogenic stimuli, or a thousand and one connections. Curr Opin Cell Biol. 1999;11:197–202.

Schwartz MA, Shattil SJ. Signaling networks linking integrins and rho family GTPases. Trends Biochem Sci. 2000;25:388–91.

Sun TP. The molecular mechanism and evolution of the GA–GID1– DELLA signalling module in plants. Curr Biol. 2011;21:R338–45.

Chapter 15
In-vitro Assays for Antimicrobial Assessment

Jaya Kurhekar, Genevieve D. Tupas, and Maria Catherine B. Otero

15.1 Introduction

The emergence of antibiotic resistance is a grave global issue that warrants great concern and attention. This phenomenon is mainly due to abuse and/or misuse of antibiotics necessitating the restriction of their use. In general, antimicrobial drugs are often used inappropriately, as chemotherapy for viral infections or prophylaxis for clean surgeries among others. Broad spectrum antimicrobial agents are used too often even in the absence of appropriate indications. Moreover some patients do not complete the treatment regimen due to financial constraints or maybe because they already felt better. In some parts of the world, antimicrobials are easily available and accessible and a general uncaring attitude exists towards rules and regulations for use of drugs. Too much of antibiotic usage and abuse have resulted in the emergence of antibiotic resistance, nosocomial infections, and drug resistant community-acquired infections (Brunton et al. 2017).

Medicinal plants may prove to be of help in this scenario. Antimicrobial agents have been derived from plants and the potential of finding new antibiotics can help combat the development of resistance (Livermore 2011)

Antimicrobial assessment of natural extracts or pure compounds from plants is an essential laboratory tool to evaluate the potential of these preparations or substances to kill or inhibit the growth of microorganisms. This can be done with the help of

J. Kurhekar (✉)
Bharati Vidyapeeth's Dr. Patangrao Kadam Mahavidyalaya, Sangli, Maharashtra, India

G. D. Tupas
College of Medicine, Department of Pharmacology, Davao Medical School Foundation, Inc., Davao City, Philippines

M. C. B. Otero
College of Medicine Research Center, Davao Medical School Foundation, Inc., Davao City, Philippines

S. Kumar, C. Egbuna (eds.), *Phytochemistry: An in-silico and in-vitro Update*,
https://doi.org/10.1007/978-981-13-6920-9_15

various in-vitro assays. These assays utilize inherent properties of anti-microbial agents and micro-organisms, which help researchers, assess the potential of certain plants as therapeutic agents.

"In-vitro" is a Latin word which means "in glass", indicating studies conducted in test tubes, flasks or petri plates, using artificial culture media or solutions. These studies make use of part or parts of a biological entity isolated from its normal biological environment. An assay can be defined as an analytical procedure used in the laboratory to assess the presence (qualitative assessment), amount (quantitative assessment), or a particular activity of an unknown substance termed as an analyte, which can be in various forms like a biochemical substance, or organic sample or a cell, or a cell of an organism(Boston and Houghton 2006; Abate and Jewell 2001). Properties of analytes are measured and expressed in proper units.

In-vitro assays can be qualitative or quantitative, using tube or plate or slide. Addition of reagents can be employed, by keeping their fixed quantities so that the quality or quantity of the test sample can be deduced by calculating the difference between the two. Older generation qualitative bioassays generally depend on measuring inhibition or dysfunction of an organism or cells in a large population. In-vitro bioassays are more complex and their interpretation may not be very easy, predictable or accurate, like chemical assay. But it provides more robust data because it can measure both potency and bioactivity of the antimicrobial substances. It can also estimate the effective dose of the candidate antimicrobial on a biological system (Cowan 1999).

A variety of methods have been developed for assessment of anti-microbial properties since the nineteenth century up to the present. These include Agar Well Diffusion and Disc Diffusion methods; Dynamic Contact assay; Thin-Layer Chromatography-Bioautography; Time-Kill assay and Biofilm assays; Microdilution methods and broth assays for Minimum Inhibitory Concentration (MIC), Minimum Bactericidal Concentration (MBC), Minimum Doubling Time/Growth Curve (MDT); Flow Cytofluorometric Assay and Bioluminescence Assay. This chapter aims to provide the readers a brief but comprehensive methodology of these in-vitro assays, along with their advantages and disadvantages. This can be used to serve as a guide for researchers in choosing the most appropriate antimicrobial assay considering the research capability and resources of their institutions.

15.2 In-vitro Assays Used for Anti-bacterial Assessment

15.2.1 Agar Diffusion Method

Agar diffusion can be defined as the spreading or movement of molecules through the gel matrix of the agar. Agar is an extracted product of seaweed, which is liquid in molten conditions which solidifies or hardens, as it cools down. The hardened polymer consists of myriad of agar strands with spaces present in between. The resulting material is permeable, through which small molecules can diffuse. If standard and controlled conditions are maintained throughout the assay procedure, the extent of movement of molecules proves to be proportional to the concentration of the molecules (www.goldbio.com).

Agar diffusion is a very popular method used for anti-microbial assays. It is basically of two types: (1) Agar Well Diffusion method, and (2) Disc Diffusion antibiotic sensitivity test, otherwise known as Kirby–Bauer Antibiotic Susceptibility Test (abbreviated as KB).

In the nineteenth century, Rideal, Walker and other co-workers worked on methods for determining the efficiency of an antimicrobial agent. Alexander Fleming was the first to use such a test successfully, which was known as ditch plate agar diffusion method. Diffusion of the antimicrobial substance through agar gel was the basis of the agar diffusion assay devised in 1944. In earlier days, these procedures were utilized by the Oxford group to check the presence of antibiotics in blood. Antibiotics were made to diffuse out of wells, cups, or reservoirs placed on the surface of solid medium (Lalitha 2004).

As the variety of antimicrobials brought into routine use increased, it was essential to carry out antimicrobial susceptibility testing (AST). In this specific method of AST, the antimicrobial agent is placed in a container, set-up in a Petri dish freshly seeded with test organisms, and allowed to diffuse out and interact with the test organisms All techniques related to antimicrobial susceptibility involve diffusion of antimicrobial agent in agar medium or diluting the antibiotic in agar or broth. All automated techniques are in one way or another, modifications or variations of these methods.

15.2.1.1 Agar Well Diffusion Method

15.2.1.1.1 Principle

In this method, the antimicrobial substance, of a pre-decided appropriate concentration is added to a well cut into the agar. It moves or diffuses away from the well, from a region of high concentration to the surrounding region of lower concentration. A bacterial mat or lawn is grown on the surface of an appropriate nutrient rich agar medium. Known concentrations of the antimicrobial material are then added in a pre-determined amount to a well dug into the agar medium on which the test organism has been spread. Agar plates are observed after a pre-defined incubation period (Bonev et al. 2008).

Agar Well Diffusion method was the first type of Zone of Inhibition (ZOI) assay used for assessing antimicrobial potency. The zone of inhibition (ZOI) assay aims at assessing the potency or activity of leachable antimicrobial agents applied to a solid substrate medium. The leachable test agent will exhibit a zone of clearance, termed as zone of inhibition, around the well if it has antimicrobial properties. Quantitatively, the potency of the antimicrobial agent can be assessed by measuring zone diameters. If the growth of culture extends up to the well containing the antimicrobial substance, it can be concluded that the organism is resistant to the substance. If there is a clearing or zone of inhibition around the well, the organisms can be concluded to be adversely affected by it. Depending on the antimicrobial efficiency of the substance against the tested microbial culture, at a particular concentration, there will be no growth of organisms in the region where the concentration of the substance in the agar is greater than or equal to the inhibitory concentration (Antimicrobial Drugs Lumen).

The inhibition zone sizes are measured and compared to standards, so as to determine if the microbial bacterial strain is sensitive to the substance. Graphically, by plotting the standard concentrations against the sizes of zones of inhibition, the potency of the sample in question can be assessed (Bonev et al. 2008).

15.2.1.1.2 Sample Procedure (Pharmaceutical Microbiology Manual 2014)

1. Double-layered plates of Assay medium are prepared by pouring 4 mL seed layer (inoculated with the desired strain of standardized microbial suspension) over a solidified 21 mL base layer of assay medium in a standard sized Petri dish, allowed to stand for 30 min for solidification.
2. Five millimeter-diameter wells are bored at three points per double-layered plates of assay medium which have been spread with the sensitive culture. One hundred microliter of each standard or test solution is pipetted out in individual wells. Generally more than two Petri dishes are used for each assay in order to test the unknown sample with each standard concentration.
3. The plates are allowed to stand for 1–4 h at low temperature so as to allow some time for pre-incubation diffusion to minimize the effects of variation in time between the applications of the different solutions, at the same time preventing the growth of the sensitive culture. The plates are then incubated at 37 °C for 24 h.
4. After the complete incubation period, the diameters (in mm) of the inhibition zones are accurately measured by a zone reader and the results are observed. Assay plates are generally tested in triplicate, for statistical significance. The average of all readings is taken as the correction point for the curve.
5. The potency of the unknown "Test" sample can be estimated from the graph drawn with concentrations of standards on X axis and logarithms of squares of zone diameters on Y axis and dropping a perpendicular line from the logarithm of the square of zone diameter of the test sample, on Y axis, corresponding to the concentration.

   Agar well diffusion assay is the least time consuming and most efficient method. It can be used for simultaneous screening of a variety of leachable antimicrobial substances. It is routinely used as a screening assay, providing data which can be directly compared with other materials.

15.2.1.2 Agar Disc Diffusion Method

15.2.1.2.1 Principle

Agar Disc Diffusion method is a very popular method used even today, for antimicrobial assays. It has varied terms; Disc Diffusion Antibiotic Sensitivity Test or Agar Diffusion Test or Kirby–Bauer Antibiotic Test or abbreviated as KB test or Zone of Inhibition (ZOI) assay.

A pure microbial culture, identified, characterized and procured from a standard culture collection centre is suspended in an appropriate buffer, its density is standardized and it is swabbed uniformly spread or swabbed on the surface of a sterile nutrient agar plate. A filter-paper disc (filter paper chosen according to the requirement) is soaked with the antimicrobial substance to be tested and placed on the agar surface of the plate. The leachable substance diffuses all around, from the filter paper onto the agar. The substance concentration is highest in the vicinity of the disc, which goes on decreasing as the distance from the disc increases (Bonev et al. 2008).

The clear space surrounding every antimicrobial disc indicates the susceptibility of that microbial culture to that particular antimicrobial substance. The extent of clear space surrounding the disc indicates the antimicrobial potency of that substance. More efficient the antimicrobial substance, wider will be the zone of no microbial growth, while an inefficient antimicrobial substance will show no inhibition of surrounding microbial growth. The effectiveness of anti-microbial substance can be estimated by measuring the zones of inhibition. The smaller the minimum inhibitory concentration (MIC) of the anti-microbial substance for the particular microbial culture the larger the zone of inhibition. Estimation of potency of the test sample is done by observing the Zone of Inhibition produced by the test sample and compared with the zone of a reference compound, of a known concentration (Balouiri et al. 2016).

15.2.1.2.2 Sample Procedure (Pharmaceutical Microbiology Manual 2014)

1. Isolated colonies of susceptible and resistant strains are grown overnight.
2. On the day of testing, a pure microbial test culture is suspended in buffer. Aseptically, a sterile swab is inserted in the broth culture of the test organism and the excess liquid is gently removed, by pressing the swab alongside the tube. Its turbidity is standardized; diluted to an OD of 0.5 McFarland turbidity standards (approximately 150 million cells per mL) at 600 nm and used for inoculation.
3. A stock solution of the antimicrobial substance to be tested and a control antimicrobial substance is prepared. Stock solutions are prepared using dilutions decided by trial and error method. Aqueous stock solution can be sterilized by filtration.

4. While choosing the medium, it was observed that best results are obtained with Mueller-Hinton agar, pH 7.2–7.4, poured 4 mm deep, in 100 mm or 150 mm size Petri plates. The pH level of the agar must be between 7.2 and 7.4.
5. Each agar plate is appropriately labelled and divided in three to four sections, depending on the number of antimicrobial substances to be tested. One hundred fifty microliter of each respective culture is spread evenly on the plate. Extra liquids are allowed to dry on the plate. Uniform growth, is obtained by spreading the plate with a glass spreader in one direction, rotating the plate by 90° and spreading the plate again in that direction, repeating this three times. The plate is allowed to dry for about 5 min.
6. Twenty microliter of the antibiotic solution to be tested is added to each of the filter paper discs on test plate and to control antimicrobial substance on the control plate. Discs should be fully dried before placing on the plates.
7. With the help of flame- sterilized forceps, one blank disc is placed in first section of each plate, gently on top of the agar and pressed gently with the forceps so that the disc is attached to the agar. Sequentially, the discs containing the various dilutions of antimicrobial substance are placed in respective sections. In the last section, the disc containing the control antimicrobial substance is placed. Discs are placed with care so that the bacterial growth in the surrounding area is not disturbed.
8. The plates are incubated overnight so as to allow complete growth, at an incubation temperature of 37 °C/98.6 °F.
9. After incubation, the area of inhibited bacterial growth is measured with the help of a ruler. Zone sizes are measured from the edge of the disk to the end of the clear zone. Generally the antimicrobial substance being tested should be equal to or slightly more than the control for the susceptible strain and roughly equal to that of the resistant strain.
10. Zones of inhibition are plotted as linear dimensions or squares of distances (Y axis) as a function of the natural logarithm of concentration of antimicrobial substances in the disks (X axis). The MIC is determined from the zero intercept of a linear regression fit using the data. The intercept depicts the logarithm of the MIC. The slope of the regression line is related to the diffusion coefficient of that particular antibiotic in the agar (Bauer et al. 1966). Potency of the antimicrobial substance under investigation can be directly interpreted from the graph.

15.2.2 Dynamic Contact Assay

Dynamic Contact Assay evaluates the resistance of antimicrobial – treated devices by exposing the test sample to known concentrations of microorganisms under dynamic conditions. There are a few types of dynamic contact assays like Preservative Efficacy Testing (PET) or Antimicrobial Effectiveness Testing (AET), which assess the performance of any preservative system in a finished product or they can be used as tools in the process of formulation or product development (ASTM E2149 – 13a 2018).

Dynamic contact assay is used to determine the efficiency of antimicrobial agents that are bonded to a substrate surface and thus cannot leach during conditions of normal use. Through a process of continuous agitation, the contact between the non-leachable antimicrobial agent and the bacterial species against which it is being tested is maximized during this test (ASTM E2149 – 13a 2018).

In this method, study organisms of high concentration of known optical density, are inoculated into the test product and mixed thoroughly, to establish a dynamic contact between them. Contact time of 1 h in an appropriate buffer solution allows microbiostasis of the microbial population. This method standardizes the conditions of growth of the test species and substrate contact times, to reduce the variable parameters associated with growth phase of the microorganism. At pre-decided time intervals, specific portions of the inoculated product are withdrawn and subjected to enumeration using standard techniques. Decrease in population can be calculated by subtracting the final inoculum level from the initial one.

This assay is recommended when non-leachable antimicrobial coatings for any type of substrate are to be developed. This is a high-throughput assay that enables performance assessment of multiple types of antimicrobial materials using various species of bacteria. This test is used for injections, emulsions, ear related products, sterile nasal products, ophthalmic products made with aqueous bases or vehicles emulsions applied to mucous membranes, oral products made with aqueous bases or vehicles, antacids made with aqueous bases or vehicles (ASTM E2149 2018).

Various factors influence the measurements of the efficiency or concentrations of specific antimicrobial substances and susceptibility of the species to particular antimicrobial substances, listed below. When designing an experiment, these factors have to be considered (Li et al. 2017);

(a) Metabolic state of the test organisms
(b) Life cycle stage cycle of the test species employed
(c) Concentration of antimicrobial substances used
(d) Water solubility of the antimicrobial substance; as water soluble substances will be easily removed from the test surface as compared to insoluble compounds.
(e) Test conditions being utilized as they will influence the process of leaching
(f) .End use of the product as it will influence the process of leaching

15.2.2.1 Sample Procedure (Pharmaceutical Microbiology Manual 2014)

1. Fresh test cultures are used from the original stock culture. *Candida albicans* (ATCC No. 10231), *Aspergillus brasiliensis* (*Aspergillus niger* ATCC No. 16404), *Escherichia coli* (ATCC No. 8739), *Pseudomonas aeruginosa* (ATCC No. 9027), *Staphylococcus aureus* (ATCC No. 6538), are test cultures of choice.
2. Appropriate agar medium, which is favourable for optimal growth of the culture under investigation, is selected for the cultivation of the test organisms.

Sabouraud's Dextrose Agar or Sabouraud's Dextrose Broth, Soybean Casein Digest Agar or Soybean Casein Digest Broth, are the common media of choice. If necessary, a suitable neutralizer or inactivator is added to the solid or liquid media, for investigating specific antimicrobial properties in the product.Solid media are used for pour plate method to determine plate count or colony forming units of the microorganisms, which need to be $\geq$70% of the calculated value of the inoculum.

3. Soybean-Casein Digest agar or broth medium is used for growth of bacteria like *Staphylococcus aureus* ATCC 6538, *Pseudomonas aeruginosa* ATCC 9027 *Escherichia coli* ATCC 8739, incubated at 32.5 ± 2.5 °C for 3–5 days while Sabouraud Dextrose medium is used for fungal cultures like *Candida albicans* ATCC 10231 (22.5 ± 2.5 °C for 3–5 days) and *Aspergillus brasiliensis* ATCC 16404 (22.5 ± 2.5 °C for 3–7 days). In special cases, as per product requirement, other microorganisms may be used for testing.
4. The cultures are harvested by washing the microbial growth with sterile saline to obtain a microbial count of about 1×10^8 CFU/mL. If cultures are grown in a liquid medium, i.e. Soybean Casein Digest Broth or Sabouraud's Dextrose Broth, they can be harvested by centrifugation, washing and suspending in sterile saline to obtain a count of about 1×10^8 Colony Forming Units (CFU) per mL, estimated turbidometrically and confirmed by plate count. The suspension should be used within 2 h, or else, stored at low temperature. Bacterial and yeast culture suspensions should be employed within 24 h after harvesting. Fungal culture can be stored at 2–8 °C for 7 days.
5. The surface of the chosen appropriate agar medium is spread with a fresh culture of test microorganisms. This test is carried out with 100–150 ml or grams of the test sample; smaller volume may be used if the product is less. Each container is inoculated with one of the prepared and standardized inoculums and mixed well. Suspension inoculum volume used should be 0.5–1.0% of the product volume. Concentration of the test preparation after inoculation needs to be 1×10^5 to 1×10^6 colony forming organisms (CFU) per mL of product.
6. Inoculated containers are incubated between 22.5 ± 2.5 °C in an incubator and sampled at periodic specific intervals, 7, 14, and 28 days or 14 and 28 days, as per the requirement. Any changes in product appearance are noted at these intervals. At each sampling interval, number of viable microorganisms per mL is determined by plate count method, using media with proper neutralizers.
7. Colony Forming Units per ml of the test microorganism, at the beginning of the test and those after 7, 14 and 28 days of the exposure will help in calculating the logarithmic reduction of the test organism. This test indicates the antimicrobial efficiency of the sample to inhibit the growth of organisms in a particular environment or product.

All plate counts are performed in duplicate per dilution in a dilution series to detect inhibited growth by the antimicrobial substance at lower dilutions. Generally testing up to 10^{-3} dilution is considered enough.

"No increase" is defined as not more than 0.5 log10 unit higher than the previous value measured. For example, in injections or emulsions, bacteria should not show less than 1.0 log reduction from the initial calculated count at 7 days, not less than 3.0 log reduction from the initial count at 14 days and no increase from the 14 day count at 28 days. For Yeast and Moulds, it must exhibit no increase from the initial calculated count at 7, 14, and 28 days (Pharmaceutical Microbiology Manual 2014).

While designing dynamic contact assays, the following points should be considered:

(a) Contact of test inoculum with the antimicrobial treated substance must be ensured.
(b) Various contact time slots, for removing of substance in contact, in order to test them, can be used.
(c) Various static conditions can be tried out.
(d) Different conditions of sensitivity can be used.
(e) Variety of reproducibility parameters can be tested.
(f) Different types of treated substrates can be evaluated.
(g) A wide range of microorganisms can be tested against various substances.
(h) Treated substrates employed for testing can be exposed to a wide variety of physical/chemical stresses like investigating the effect of contaminating substances like chemicals, salts in water, blood, various types of proteins etc.
(i) Surface antimicrobial activity can be calculated by comparing results of controls and test samples.
(j) Neutralization test can be performed after the contact reaction to check the presence of an antimicrobial substance.
(k) SI units are used as standard units.

(General Information/Reference Standards 2017)

The dynamic contact assay evaluates antimicrobial substances which are not leached out from the aqueous surface of the medium. This test can be used for routine quality control and screening tests for evaluation of substrate-bound antimicrobial substances. It can also be used to evaluate modified specimens like radiation sterilization, steam sterilization, exposure to UV, manipulation of solvents, susceptibility to temperature, physical and chemical manipulation, by using appropriate controls (Ristić et al. 2011).

Dynamic contact assay can also assess the performance of any preservative system in a finished product. It can be employed in the manufacturing or development of pharmaceutical and cosmetic products, and any other antimicrobial product (ASTM E2149 2018).Microbial contamination in various finished products is prevented by addition of antimicrobial preservatives, which are generally toxic. Hence, its concentration in the final product requires being below the prescribed toxic level. Antimicrobial preservatives are added to inhibit microbial growth that may probably be introduced during manufacturing, after manufacturing or during daily use of the product by the end consumer (ASTM E2149 2018).

15.2.3 *Thin Layer Chromatography: Bioautography*

In Thin-Layer Chromatography-Bioautography, or Direct Bioautography, chromatographically separated components are assayed for antimicrobial activity against known bacterial suspensions (Balouiri et al. 2016). Coupled with the power of Thin-Layer Chromatography (TLC) and spectroscopic detection, phytochemical fractions can directly be observed for antimicrobial activity by inhibiting microbial growth in areas where there is direct contact (Choma and Jesionek 2015).

15.2.3.1 Sample Procedure (Dewanjee et al. 2015; Choma and Jesionek 2015)

1. Prepare fresh broth cultures (10^6 CFU/mL) of *Bacillus subtilis*, *Staphylococcus aureus*and *Escherichia coli*. The use of 90:10 Mueller-Hinton broth:Agar is recommended to have a medium that can sufficiently disperse microbes while still being viscous enough to adhere to the developed TLC plates (Okusa et al. 2010).
2. Dip the developed TLC plate and incubate for 48 h at 25 °C under humid conditions.
3. After incubation, spray tetrazolium salts on the TLC plate.
4. Reincubate the plates at 37 °C for 3–4 h. White or cream-colored spots in a background of purple formazan indicate areas of growth inhibition.

Recent modifications further pair TLC-Bioautography with bioluminescent assays using either *Photobacteriumphosphoreum* or *Aliivibriofischeri* (Choma and Jesionek 2015). Antimicrobial substances interfere with the metabolism of these bioluminescent bacteria, and this result in the quenching of their inherent luminescence. Right after immersing the developed TLC plates in the microbial broth, bioluminescence is observed at 490 nm using a cooled CCD camera. Emissions that are either darker or brighter than the normal blue-green light indicates that the antimicrobial agent in question has bacteria-toxic substances (Horvath et al. 2010). Downstream quantitative analyses can also be performed (Baumgartner and Schwack 2010).

TLC – Bioautography provides a faster and direct assay for antimicrobial potential even before the sample's components are isolated and purified (Ncube et al. 2008; Urbain and Simoes-Pires 2014). As such, this method is ideal for high-throughput antimicrobial screening, and for bioassay-guided isolation of the antimicrobial components of interest (Urbain and Simoes-Pires 2014).

15.2.4 *Time-Kill Assay*

Once the antimicrobial potential of a natural product extract is established, its specific inhibitory or bactericidal characteristics in relation to an identified microorganism must be further investigated. In Time-Kill assay, a microbial isolate is exposed to a pre-determined concentration of the antimicrobial agent of interest, and the rate at which the microbes are killed is measured (Forbes et al. 2007). Ideally, Minimum Inhibitory Concentration (MIC) and Minimum Bactericidal Concentration (MBC) must already be measured before Time-Kill Assays are undertaken.

15.2.5 Sample Procedure (Forbes et al. 2007)

1. Broth medium is inoculated with an antimicrobial agent or extract of known concentration.
2. At regular time intervals (e.g. 0, 4, 8, 12, 16, 20, 24 h), an aliquot from the broth culture will be inoculated in an agar medium.
3. Following incubation, the number of Colony Forming Units (or CFUs) will be counted. CFU counts versus time will be plotted to determine the killing rate of the samples or extracts. A reduction in CFU by 1000-fold after 24 h of incubation compared to the initial number of viable cells is ass1igned as evidence of bactericidal activity.

A modification of the above procedure is described in the CLSI M26-A document, and is the procedure of choice in Time-Kill Assays in well-equipped microbiology laboratories. At least three bacterial suspensions with 5×10^5 CFU/mL are required: 1 as positive growth control, 1 with $0.25 \times$ MIC of the test sample, and 1 containing $1 \times$ MIC. Growth is measured at time intervals via agar plate count method, as described above. For this method, the bactericidal activity of the extracts or samples is defined as 90% lethality after 6 h of incubation, or 99.9% lethality after 24 h (Balouiri et al. 2016).

Although labor-intensive and highly technical (Forbes et al. 2007), Time-Kill Assays provide insights on the interactions between the microbial strain and the antimicrobial agent or extracts used. It also allows investigations on the synergism or antagonism in different formulations or drug combinations (Pfaller et al. 2004; Balouiri et al. 2016).

15.2.6 Biofilm Assay

Biofilms are micro-colonies of bacteria surrounded by an extracellular polymeric matrix (Forbes et al. 2007). Pathogenic biofilm-associated microbes differ from their planktonic counterparts in terms of growth, gene expression, and consequently, drug resistance (Donlan 2002). The behaviour of cells in biofilms is influence mainly by quorum sensing. Thus, inhibitors of quorum sensing can control biofilm formation (Buommino et al. 2014). For this reason, biofilms assays are given special attention in order to screen for quorum sensing inhibitory (QSI) activity in natural products.

In recent years, microplate-based test for investigating activity against biofilms have been developed. Biofilm-producing bacteria, such as *Staphylococcus aureus, Escherichia coli, Pseudomonas aeruginosa*, certain enterococci and fungi, are grown in nutrient-rich medium in a microplate with or without pegs (O'Toole 2011; Dalecki et al. 2016). After washing the biofilms produced, growth will be challenged with different dilutions of antimicrobial test samples. After another round of incubation and washing, the biofilm will be added with fluorescent dyes, such as resazurin, to visually or photometrically check for viability (Dalecki et al. 2016). In the sample procedure below, solubilisation of biofilm-associated dye was used as a proxy for measuring amount of biofilm produced by test microorganism (O'Toole 2011). Although rudimentary, the method allows rapid screening of biofilm production as affected by candidate QSI agents.

15.2.6.1 Sample Procedure (O'Toole 2011)

1. Grow *Pseudomonas aeruginosa* in Lauria broth overnight.
2. Prepare 1:100 dilution of the overnight culture with fresh biofilm assay culture (commonly M63 minimal medium added with $MgSO_4$ and arginine).
3. Add 100 μL of each test compound in a 96-well microplate.
4. Place 100 μL of the diluted culture in the 96-well microplate, with 4 up to 8 replicate wells for each treatment. Incubate the plate at 37 °C for 4 up to 24 h.
5. After the incubation period, dispose the unattached cells and extra media in the plates by turning it over and shaking.
6. Wash the plates in sterile water twice, and shake out excess liquid.
7. Place 125 μL 0.1% crystal violet solution (in water) to all wells. Then, incubate the plates for 10 up to 15 min at room temperature.
8. Discard the stain and wash the microplate thrice, as in step 5.
9. Leave the microplate to dry in a sterile environment for up to 24 h.
10. Next, place 125 μL 30% acetic acid (in water) to all wells. This step will solubilize crystal violet trapped in the biofilm. Leave the plate to stand for 10–15 min at RT.
11. Get 125 μL of the solubilized CV and place in a fresh flat-bottom microplate. Measure its absorbance at 550 nm, with 30% acetic acid as blank.

More sophisticated biofilm assays have been developed recently that allow direct observation of biofilm-associated cells, excluding planktonic cells from the analysis. In this format, biofilm assays can be scaled for medium-throughput testing of antimicrobial activity against biofilm-producing drug-resistant pathogens (Dalecki et al. 2016).

15.2.7 Microdilution Methods

Certain antimicrobial activity tests can be adapted to microplate format. For example, the Minimum Inhibitory Concentration (MIC) and Minimum Doubling Time (MDT) tests are traditionally performed using broth cultures in test tubes. To reduce preparation time, and volumes of test samples, these tests are down-scaled from milliliters to microliters in microplates (Balouiri et al. 2016). Biofilm assays and bioluminescent assays are also easier and faster to do using micro-volumes. Results are visualized with the aid of colorimetric reactions, or with Optical Density, using a spectrophotometer.

15.2.7.1 Minimum Inhibitory Concentration (MIC) Test

The Minimum Inhibitory Concentration or MIC is the smallest antimicrobial agent concentration that can completely inhibit visible bacterial growth (Forbes et al. 2007). A doubling dilution series of the antimicrobial substance is prepared, and mixed with an appropriate culture medium. For example, you can start with 16 μg/mL, and diluted down twofold until you reach 0.25 μg/mL. A known concentration of test microorganism is then added, and growth is checked after incubation. Sterility or negative control, and a growth or positive control must both be used in all MIC determinations. Depending on the scale of analysis, measurement of microbial growth can be done visually or using absorbance readings in a spectrophotometer (Balouiri et al. 2016).

One of the earliest methods of determining MIC in antimicrobial susceptibility test is through Agar Diffusion. In this method, a pre-determined concentration of an antimicrobial agent is mixed to agar medium. Only one concentration of the antimicrobial agent can be added in one Petri plate. After inoculation and incubation, colony formation is observed (Balouiri et al. 2016).

Another common way for MIC determination is broth macro dilution. Instead of agar medium, at least 1 mL of broth medium is used. Growth can be recorded as turbidity, and sediment and pellicle formation. A recent modification of this is called broth micro-dilution, performed in microplate format with broth volumes between 50 and 100 μL (Jorgensen and Ferraro 2009). Microbial growth is colorimetrically measured, using dye reagents such as tetrazolium salts, MTT, XTT, and resazurin. Although commercial systems for both broth and agar dilution are available for clinical use, no commercially produced panel is available for natural products screening.

In all variations of the MIC test, the lowest drug concentration without optically measurable growth is the Minimum Inhibitory Concentration. MIC results can be further translated into categories: resistant, intermediate, or susceptible (Forbes et al. 2007). However, for screening natural products for antimicrobial activity, MIC values are more useful. The advantages and disadvantages of the different MIC methods in use, have been depicted in Table 15.1.

Table 15.1 Advantages and disadvantages of the different MIC methods

Parameters	Agar dilution	Broth microdilution	Broth macrodilution
Specialized equipment and reagents required	No	No	Yes
Reduced medium and test volumes	No	Yes, at least 1 mL	Yes, microvolumes
Preparation time	Longer	Longer	Shorter due to microvolumes
Number of test organisms	Up to 4 per plate	1 only	1 only
Number of test samples	1 per plate	1 per tube	Up to 8, with 10 dilutions

15.2.7.1.1 Sample Protocol for Broth Microdilution for Aerobic Bacteria (CLSI Document M07- A9 2012)

1. A 1/10 dilution of the plant extracts (reconstituted in DMSO) in Mueller-Hinton Broth will be prepared.
2. Then, 100 uL of this dilution will be dispensed in the first well of a 96-well plate.
3. Next, 50 uL of MHB will be dispensed in wells 2–12. Twofold dilutions will be prepared using the plant extracts by transferring 50 uL successively from the first well to the 11th well.
4. Bacterial suspensions of *Staphylococcus aureus, Pseudomonas aeruginosa, Escherichia coli,* and *Enterococcus faecalis* adjusted to 0.5 MacFarland will be diluted 1/150 with MHB to a final concentration of 1×10^6 CFU/mL. Then, 50 uL of the inoculum will be added to all wells.
5. The microplate will be incubated at 35 °C for 18 h.
6. Next, 20 μl of the redox indicator Resazurin in concentration 0.2 mg/ml will be added to each well. Then the plate will be incubated for the next 2 h at 37 °C.
7. Lastly, bacterial respiration will be measured using a microplate reader, as a difference in absorption of reduced form of Resazurin (pink-coloured) at 570 nm and the oxidized form of Resazurin (blue-coloured) at 600 nm. Alternatively, plates can be inspected visually, where pink wells denote the presence of living cells, and blue wells denote inhibited bacterial growth.

15.2.7.2 Minimum Bactericidal Concentration (MBC) Test

The Minimum Bactericidal Concentration (MBC), also called the Minimum Lethal Concentration (MLS), is the concentration of an antimicrobial agent that gives 99.9% reduction of CFU/mL in a microbial culture (Forbes et al. 2007). It is one of the most frequent estimation method for determining fungicidal or bactericidal activity. MBC takes off from MIC results of broth dilution (CLSI Document M26-A 1998).

15.2.7.2.1 Sample Procedure (Balouiri et al. 2016)

1. Aliquots from all tubes or wells without visible growth are sub-cultured in enrichment media, commonly with sheep blood agar.
2. After 24-h incubation, the cultures are inspected, and the CFU per aliquot volume will be computed.
3. The smallest concentration of an antimicrobial agent or plant extract that kills 99.9% of microbes is reported as the MBC.

It is of note that theoretical MBC may not be fully achieved in the clinical or animal study setting, especially if the MBC is greater than the amount of antimicrobial in serum in the test organism. Tolerance in the host or animal is also an issue, and this manifests as MBC that is greater by 32 times or more than the MIC (Forbes et al. 2007). In this case, instead of killing the test microbes, growth is only inhibited.

15.2.7.3 Minimum Doubling Time (MDT)/Growth Curve Test

The Minimum Doubling Time (MDT), or the generation time, is defined as the time needed to double the population of microbes in a broth culture (Forbes et al. 2007). The assumption is that all requirements for bacterial propagation (such as oxygen, nutrients, pH, and temperature) are present in optimal amounts. Once the growth curve of the test microorganism is known, the effect of an antimicrobial agent or plant extract on the growth of the microorganism can be observed.

Optical Density of a broth culture can be proportional to the concentration of the cell population in a given sample, when properly calibrated (Stevenson et al. 2016). Adopting the microplate format for MDT, changes in OD over a given set of time points can be plotted to determine the minimum doubling time.

15.2.7.3.1 Sample Procedure (Othman 2016)

1. Bacterial isolates are grown in MH broth (usually 18-h cultures, or comparable to 0.5 MacFarland standard).
2. An inoculum of 50 μL broth culture and 50 μL test sample is added to 100 μL fresh MH broth. The control tube should contain only MH broth and inoculum.
3. Microplates are incubated for 17 h at 37 °C.
4. The Optical Density at wavelength 600 nm (OD_{600}) of the broth cultures are then collected at specific time points, depending on the known growth curve of the test microorganism (for *Escherichia coli,* every 2 h from 0 to 18 h).

The wavelength used to measure OD may be changed, as long all downstream measurements are in the same wavelength. Broth culture without inoculum is used as blank for OD measurements. When a spectrophotometer or photometer is unavailable, a known volume of aliquots at the different time points can be placed in a microscopic counting chamber.

To determine MDT, the natural logarithm of the OD (ln OD) should be graphed against the different time points. Using 2 points from the linear portion of the graph (corresponding to the exponential growth stage), the rate of growth can be computed (Gaupp et al. 2015) as per the formula (Eq. 15.1) given below;

$$r = \frac{(\ln [OD1/OD2])}{(T2 - T1)} \tag{15.1}$$

Minimum Doubling Time is equal to ln 2/r.

It is recommended to screen for antimicrobial potential must first be demonstrated before using MDT (Othman 2016). After MDT, plant extracts that delay, decrease, or inhibit bacterial growth can then be further characterized for specific antimicrobial mechanisms of action. However, test microorganism that is not culturable in broth medium cannot be evaluated using the MDT described above. Additionally, interaction of plant extracts with media components is not observable.

15.2.8 Flow Cytofluorometric Assay

Flow cytofluorometry allows determination of cell viability, quantity, size, and structure via light scattering, and type and intensity of fluorescence (Davey 2004). This is important in unravelling the mechanisms of action of antimicrobial agents. For example, if a microbicidal plant extract induces cell membrane damage, dyes like propidium iodide (PI) can be used. PI cannot penetrate live or viable cells. It can also intercalate in double-stranded DNA (Balouiri et al. 2016; Davey 2004). Thus, red fluorescence with PI indicates that the cell membrane is compromised after treatment, and reveals where the double stranded DNA molecules are localized. Dose-response studies using flow cytofluorometrycan pinpoint the optimal dose for the expected outcome.

Other dyes used in flow cytofluorometry include Calcafluor white, pyronin Y, fluorescein isothiocyanate (FITC), phycoerythrin (PE), Hoechst, and 4′,6-diamidino-2-phenylindole (DAPI) (Balouiri et al. 2016). Flow cytofluorometry has the added advantage of detecting multiple target cells or structures. Certain models of flow cytometers permit analysis of five up to ten cellular parameters. Thus, differential fluorescence in one cell type can be achieved. When surface markers or fluorescent proteins have to be examined at the same time, cross-linking fixatives have to be added to enable efficient analysis (Davey 2004).

15.2.8.1 Sample Protocol (Nuding and Zabel 2013)

1. Grow bacteria in broth, and adjust to 0.5 McFarland at log phase.
2. Incubate bacterial broth culture with test samples for 120 mins at 37 °C. Use commercial antibiotics for the control set-up.

3. Add 0.5 μg/mL DiBAC4 dye [bis-(1, 3-dibutylbarbituric acid) trimethineoxonol], and stand for another 120 mins at RT.
4. Centrifuge the broth culture at 420 × g for 10 mins, RT.
5. Resuspend bacterial pellet in PBS, and analyse via flow cytometry. DiBAC4 accumulates in the cytoplasm and binds intracellular membranes of proteins in damaged cell, showing green fluorescence at 516 nm.

Commercial systems have been developed over the years. Despite the resolution and power of flow cytofluorometry in antimicrobial susceptibility testing, it has not yet been fully utilized in this area. One of the factors that lead to this under utilization is the inaccessibility of flow cytometry to many natural products laboratories and the tedious validation of methods for AST. Moreover, technically, flow cytometry cannot distinguish between bactericidal and bacteriostatic substances (Van Belkum and Dunne 2013).

15.2.9 Bioluminescence Assay

Antimicrobial activity of natural products be assessed using Bioluminescence Assays. This type of assay is anchored on the coupling of luciferase-mediated light emission with a target parameter, such as cell viability (Balouiri et al. 2016)

One example of the bioluminescent assay is the ATP Bioluminescence Assay. This assay takes advantage of the adenosine triphosphate (ATP) that is produced by metabolically bacteria and fungi. When ATP is present, D-luciferin is converted by luciferase into the light-emitting oxyluciferin. Emitted light is measured in Relative Light Units (RLU) in a luminometer, and then transformed into RLU/mole of ATP. Microbes have a defined amount of ATP inside each living cell. Thus, by measuring the total amount of ATP, the population of viable microbes in a given medium can be estimated (Balouiri et al. 2016).

15.2.9.1 Sample Procedure (Cai et al. 2015)

1. Prepare fresh bacterial broth cultures.
2. Dilute broth culture with pre-warmed cation-adjusted MHB to yield 1–-5 × 10^5 CFU/mL final concentrations.
3. Transfer 50 μL of diluted culture to a 96-well microplate (flat-bottom).
4. Add 50 μL of test antimicrobial agent to the wells. Prepare replicate wells and controls.
5. Cover the microplate with sealing film and incubate it with agitation for 24 h at 35 °C.
6. After 24 h, determine total ATP content in each well with the BacTiter Glo microbial viability assay (Promega, Madison, WI), following manufacturer's recommendations. Add 100 μl BacTiter-Glo assay reagent to 100 μl test sample, and incubate at room temperature for 15 min.

7. Record bioluminescent light (in RLU) in each test well. Background RLU values can be determined using the negative control.

ATP bioluminescence assay used in testing antibacterial and antifungal activity gives quicker results than the traditional dilution methods, and is adaptable to high-throughput formats. It can also be used for in-situ and in-vivo antimicrobial testing (Balouiri et al. 2016)

15.3 Conclusion

Microbial infections remain to be a formidable threat to mankind; hence, the need to continue to look for ways to control them will always be a priority. The use of in-vitro antimicrobial assays is among the initial steps taken in order to come up with potential leads towards finding a novel and effective antimicrobial agent. This chapter discusses the different methods of in-vitro antimicrobial testing, their advantages and disadvantages. Researchers are always faced with a problem as to which method to use for their studies especially in relation to antimicrobial assessment. To help researchers decide, we came up with the following recommendations:

For preliminary screening of antimicrobial activity, the Disk Diffusion Assay remains to be the most commonly used method. However, TLC – Direct Bioautography shortens the process by pinpointing the specific chromatographically separated fractions which are bioactive against the test microorganisms. When candidate substances are identified, researchers can proceed to the next-level ASTs, such as MIC determination, MBC, MDT, Time-Kill assays and biofilm assays. Flow cytofluorometry and bioluminescence assays provide more information on the mechanism of action of candidate antimicrobials. Dynamic contact assays evaluate bioactivity during the formulation and at the later stage production of antimicrobials.

In deciding which in-vitro antimicrobial assay to utilize, researchers should also consider the following factors: availability of the test and its reagents, the type of microorganism to be tested, reliability of the procedure, individual circumstances, other resources at hand, allotted time and budgetary constraints. It is possible that more than one assay may be needed to achieve a specific outcome and such a case may be tried out by trial and error. This comprehensive write-up may prove to be a guiding light in such endeavours!

References

Abate F, Jewell E. The new Oxford American dictionary. 1st ed. Oxford: Oxford University Press; 2001.

Antimicrobial Drugs. Testing the effectiveness of antimicrobials, lumen, microbiology.

ASTM E2149 – 13a. Standard test method for determining the antimicrobial activity of antimicrobial agents under dynamic contact conditions, active standard ASTM E2149 | Developed by Subcommittee: E35.15, Book of Standards Volume: 11.08.2018.

Balouiri M, Sadiki M, Ibnsouda SK. Methods for in vitro evaluating antimicrobial activity: a review. J Pharm Anal. 2016;6(2):71–9.

Bauer AW, Kirby WM, Sherris JC, Turck M. Antibiotic susceptibility testing by a standardized single disk method. Tech Bull Regist Med Technol. 1966;36(3):49–52.

Baumgartner V, Schwack W. Enhanced quantitative evaluation of the HPTLC-bioluminescence detection. J Liq Chromatogr Relat Technol. 2010;33(7–8):980–95.

Bonev B, Hooper J, Parisot J. Principles of assessing bacterial susceptibility to antibiotics using the agar diffusion method. J Antimicrob Chemother. 2008;61(6):1295–301.

Boston MA, Houghton M. The American heritage dictionary of the English language 4th ed.; 2006.

Brunton LL, Chabner B, Knollmann BC. Goodman and Gilman's the pharmacological basis of therapeutics. 12th ed. New York: Mc Graw Hill Medical; 2017.

Buommino E, Scognamiglio M, Donnarumma G, Fiorentino A, D'Abrosca B. Recent advances in natural product-based anti-biofilm approaches to control infections. Mini-Rev Med Chem. 2014;14(14):1169–82.

Cai Y, Leck H, Lim TP, Teo J, Lee W, Hsu LY, Kwa ALH. Using an adenosine triphosphate bioluminescent assay to determine effective antibiotic combinations against carbapenem-resistant gram negative bacteria within 24 hours. PLoS One. 2015;10(10):e0140446.

Choma IM, Jesionek W. TLC-direct bioautography as a high throughput method for detection of antimicrobials in plants. Chromatography. 2015;2(2):225–38.

CLSI. CLSI document M26-A. Wayne: Clinical and Laboratory Standards Institute; 1998.

CLSI. CLSI document M07-A9. Wayne: Clinical and Laboratory Standards Institute; 2012.

Cowan MM. Plant products as antimicrobial agents. Clin Microbiol Rev. 1999;12(4):564–82.

Dalecki AG, Crawford CL, Wolschendorf F. Targeting biofilm associated *Staphylococcus aureus* using resazurin based drug-susceptibility assay. J Vis Exp 2016; (111).

Davey H. Flow cytometry for clinical microbiology. CLI; (2–3), 12. 2004.

Dewanjee S, Gangopadhyay M, Bhattacharya N, Khanra R, Dua TK. Bioautography and its scope in the field of natural product chemistry. J Pharm Anal. 2015;5(2):75–84.

Donlan RM. Biofilms: microbial life on surfaces. Emerg Infect Dis. 2002;8(9):881.

Forbes BA, Sahm DF, Weissfeld AS. Bailey and Scott's. Diagnostic microbiology. 12th ed. St Louis: Mosby Elsevier; 2007.

Gaupp R, Lei S, Reed JM, Peisker H, Boyle-Vavra S, Bayer AS, Somerville GA. *Staphylococcus aureus* metabolic adaptations during the transition from a daptomycin susceptibility phenotype to a daptomycin non susceptibility phenotype. Antimicrob Agents Chemother. 2015;59 (7):4226–38.

General Information/Reference Standards. JP XVII. 2017.

Horváth G, Jámbor N, Végh A, Böszörményi A, Lemberkovics É, Héthelyi É, Kocsis B. Antimicrobial activity of essential oils: the possibilities of TLC–bioautography. Flavour Fragr J. 2010;25(3):178–82.

Jorgensen JH, Ferraro MJ. Antimicrobial susceptibility testing: a review of general principles and contemporary practices. Clin Infect Dis. 2009;49:1749–55.

Lalitha MK. Manual on antimicrobial susceptibility testing. Indian Association of Medical Microbiologists. 2004.

Li J, Shuyu X, Saeed A, Funan W, Yufeng G, Chaonan Z, Ximan C, Yalan W, Jinxia C, Guyue C. Antimicrobial activity and resistance: influencing factors. Front Pharmacol. 2017;8:364.

Livermore DM. Discovery research: the scientific challenge of finding new antibiotic. J Antimicrob Chemother. 2011;66(9):1941–4.

Ncube NS, Afolayan AJ, Okoh AI. Assessment techniques of antimicrobial properties of natural compounds of plant origin: current methods and future trends. Afr J Biotechnol. 2008;7 (12):1797–806.

Nuding S, Zabel LT. Detection, identification, and susceptibility testing of bacteria by flow cytometry. J Bacteriol Parasitol. 2013;S5:005. https://doi.org/10.4172/2155-9597.

O'Toole GA. Microtiter dish biofilm formation assay. J Vis Exp; 2011; (47).

Okusa P, Stevigny C, Devleeschouwer M, Duez P. Optimization of the culture medium used for direct TLC-bioautography. Application to the detection of antimicrobial compounds from Cordia gilletii De Wild (Boraginaceae). JPC-J Planar Chromatogr-Mod TLC. 2010;23 (4):245–9.
Othman AS. Determination of the antibacterial effect of some natural products against some gram-positive and gram-negative bacteria. Egypt Pharm J. 2016;15(1):10.
Pfaller MA, Sheehan DJ, Rex JH. Determination of fungicidal activities against yeasts and molds: lessons learned from bactericidal testing and the need for standardization. Clin Microbiol Rev. 2004;17(2):268–80.
Pharmaceutical Microbiology Manual. ORA.007, Version 1.2 DATE: 03-30-2015 Chapter 1: antimicrobial effectiveness testing, Office of Regulatory Affairs Office of Regulatory Science Medical Products and Tobacco Scientific Staff. 2014.
Ristić T, Lidija FZ, Monika N, Marjetka KK, Silva S, Nina GC, Simona S. Science against microbial pathogens: communicating current research and technological advances A. Méndez-Vilas (Ed.), Antimicrobial efficiency of functionalized cellulose fibres as potential medical textiles, ©FORMATEX; 2011. 36–51.
Stevenson K, McVey AF, Clark IB, Swain PS, Pilizota T. General calibration of microbial growth in microplate readers. Sci Rep. 2016;6:38828.
Urbain A, Simões-Pires CA. Thin-layer chromatography of plants, with chemical and biological detection methods. Encyclopedia of Analytical Chemistry. 2014.
Van Belkum A, Dunne WM. Next-generation antimicrobial susceptibility testing. J Clin Microbiol. 2013;51(7):2018–24.

Chapter 16
In-vitro Assays for Determining Anti-HIV Potential of Phytochemicals

Pascaline Fonteh, Walter Chingwaru, Debra Meyer, and Petrina Kapewangolo

16.1 Introduction

The HIV epidemic continues to affect people worldwide. According to the UNAIDS fact sheet of 2018, approximately 36.9 million people were living with HIV in 2017 and since the start of the epidemic, approximately 35.4 million people had died from AIDS related illnesses (UNAIDS 2018).

Antiviral treatment currently involves the use of combination antiretroviral therapy abbreviated as cART (Cihlar and Fordyce 2016). cART usually consists of two nucleotide or nucleoside reverse transcriptase inhibitors and a third drug from another class (Cihlar and Fordyce 2016). Such a combination is known to significantly suppress virus replication leading to substantial improvement in the clinical management of HIV infection by delaying disease progression, prolonging survival and improving the overall quality of life (Cihlar and Fordyce 2016). A total of 25 anti-HIV drugs from 6 different mechanistic classes have reportedly been developed for use since the approval of zidovudine (AZT) in 1987 (Cihlar and Fordyce 2016).

Limitations of existing treatment such as toxicity to the host, development of drug resistant viruses together with the persistence of latent pools of the virus makes lifelong treatment a necessity (Tang and Shafer 2012). AZT and other HIV-drugs originated from anti-cancer investigations of phytochemicals, the latter known for

P. Fonteh
Department of Surgery, University of the Witwatersrand, Johannesburg, South Africa

W. Chingwaru
Biological Sciences Department, Bindura University Science Education, Bindura, Zimbabwe

D. Meyer
Department of Biochemistry, University of Johannesburg, Johannesburg, South Africa

P. Kapewangolo (✉)
Department of Chemistry & Biochemistry, University of Namibia, Windhoek, Namibia
e-mail: pkapewangolo@unam.na

S. Kumar, C. Egbuna (eds.), *Phytochemistry: An in-silico and in-vitro Update*,
https://doi.org/10.1007/978-981-13-6920-9_16

chemical novelty that is higher in natural products than in any other sources (Jadaun et al. 2016). There is ongoing research to support the role of phytochemicals as anti-HIV agents. This chapter focuses on various in-vitro screening assays that are used in the development of phytomedicines for HIV management. Some of the medicinal plants and phytochemicals with potential for use in HIV management are described followed by a description of the replication cycle of the virus as it relates to treatment and some of the associated anti-viral assays that have been used in testing these phytochemicals for HIV enzyme inhibition. The various in-vitro anti-HIV assays that are enzyme and replication based are then discussed.

16.2 Medicinal Plants in HIV Management

The globe is estimated to have approximately 61,000 tree/shrub species (Qian et al. 2018). Of these, approximately 23,400 species are found in Southern Africa alone (Chingwaru et al. 2015) – making the flora of the region one of the richest in the world. The rich plant diversity in Southern Africa coincides with an unequivocally high HIV burden in the region compared to the rest of the world. The region is also affected by high levels of poverty (Mbirimtengerenji 2007). The Eastern and Southern African region is also home to more than half (53%) of the world's HIV-infected population (UNAIDS 2018). Despite the fact that conventional HIV drugs have become widely available around the world, their high cost and side-effects/drug resistance have negatively impacted their uptake among the poorer communities particularly those in Southern Africa and parts of Asia (Institute of Medicine 2011). Medicinal plants are popularly used in traditional medicinal practices in communities around the world (Pan et al. 2014). Empirical evidence shows that a number of these medicinal plants help alleviate symptoms of HIV/AIDS (Chingwaru et al. 2015). While Southern Africa is endowed with a rich floral diversity, only a handful of the plants growing in the region have been assessed for anti-HIV activities. The use of plant based medicines in the management of HIV disease is also reported in Indian and Chinese traditional medicine.

Two plants have dominated the traditional management of HIV in Southern Africa, namely *Hypoxis hemerocallidea* (common name: African potato) and *Sutherlandia frutescens* (Maroyi 2014). Other plants with a history of use in the management of HIV in Southern Africa and most with in-vitro anti-HIV evidence, include *Plectranthus barbatus*, *Siphonochilus natalensis* (Maroyi 2014; Chingwaru et al. 2015), *Abrus precatorius* (Ma et al. 1998), *Erythrina abyssinica* (Mohammed et al. 2012), *Cassia fistula* L. bark (Xu et al. 1996), *Bulbine alooides* (L.) Willd, *Hypoxis sobolifera* var. *sobolifera* (Jacq.) Nel., *Leonotis leonurus* (L.) R.Br. (Klos et al. 2009) and, *Dodonaea angustifolia* L.f. (Asres et al. 2001).

In India, HIV patients reportedly use plants such as *Boxus sempervirens* (Boxwood), *Andrographis paniculata* (Andrographolide), *Azadirachta indica* (Neem), *Rasagandhi mezhuga, Amukkara chooranum* and *Nellikkai lehyam* in the management of HIV (Fritts et al. 2008).

The gaps that are apparent in the discovery of plant based medicines for use in HIV management require concerted research efforts to ensure safety in the traditional management of the disease. In-vitro assays are an essential platform to improve our understanding of the mechanisms of action of phytomedicines and their levels of safety.

16.3 Anti-HIV Phytochemicals

A number of plant derived compounds have been identified as promising anti-HIV agents including alkaloids, flavonoids, coumarins, tannins, saponins, phenolics, quinones, and lignans (Kurapati et al. 2016). The remainder of this section provides examples of phytochemicals with anti-HIV activity.

16.3.1 Alkaloids

Michellamines (A, B and C) are atropisomeric naphthylisoquinoline alkaloid dimers that were isolated from a Cameroonian plant, *Ancistrocladus korupensis* (Boyd et al. 1994). These alkaloids inhibited HIV-induced cell killing and viral replication in a variety of human cell lines (Boyd et al. 1994). Anti-HIV studies showed that michellamine B was active against a panel of biologically diverse laboratory and clinical strains of HIV-1, including the AZT resistant strain G910-6 and the pyridinone-resistant strain A17; the compound also inhibited several strains of HIV-2 (Boyd et al. 1994). Michellamine D, a homolog of michellamines A-C underwent extensive preclinical evaluation as a potential HIV drug but was found to be highly toxic (Singh et al. 2011). Cepharanthine is a biscoclaurine alkaloid isolated from *Stephania cepharantha* and was found to suppress HIV-1 LTR-driven gene expression through the inhibition of NF-kB activation (Okamoto et al. 1998). These are just a few of this class of compounds, a lot more alkaloids are reported in the literature with potential inhibitory activity against HIV-1 (Singh et al. 2011).

16.3.2 Flavonoids

Flavonoids are one of the groups of plant based compounds with known anti-HIV activities, particularly known to inhibit all three HIV enzymes; Reverse transcriptase (RT), protease (PR) and integrase (IN) (Singh et al. 2011). Flavonoids have also been shown to inhibit the attachment of viral gp120 to CD4+ cells as well as reducing viral replication in-vitro (Singh et al. 2011). Flavonoids with potent anti-HIV activity include chalcones, hydroxypanduratin A, quercetin, chrysin, epigallocatechin, thalassionlins A-C, and taxifolin (Singh et al. 2011).

16.3.3 Coumarins

Calanolide is a popular coumarin which inhibits HIV-1 RT. This compound was first isolated from a tropical tree (*Calophyllum lanigerum*) in Malaysia (César et al. 2011). The safety and pharmacokinetics of this compound have already been evaluated and it is one of the few plant derived compounds undergoing HIV clinical trials (Saklani and Kutty 2008). Other types of coumarins with reported anti-HIV activity include khellolactone, furanocoumarins, immperatorin, and heraclenol (Singh et al. 2011).

16.3.4 Terpenes

This is a large group of medicinal compounds that mainly falls under three classes; diterpenes, triterpenes and sesquiterpenes (Singh et al. 2011). These cyclic compounds have been reported to mostly inhibit HIV RT as well as prevent viral replication in cells. Betulinic acid, also known as PA-457 or bevirimat, potently inhibits HIV-1 replication by specifically blocking CA-SP1 cleavage resulting in antimaturation activity (Stoddart et al. 2007). Even though clinical trials of berivimat indicated a significant and clinical reduction of the viral load in infected individuals, a high baseline drug resistance was also revealed (Dang et al. 2013). The high baseline drug resistance poses serious limitations to the clinical potential and further development of this terpene as a potential HIV drug (Dang et al. 2013). New berivimat derivatives are being developed to overcome the drug resistance of the parent compound (Dang et al. 2013). Other terpenes with anti-HIV properties include butenolide-3-epi-litsenolide D_2, lactone, lanicilacton C, and limonoid (Singh et al. 2011).

16.3.5 Phenolic Acids

Several phenolic acids have been highlighted to have anti-HIV properties. Two of these phenolics, dicaffeoyl quinic acid and dicaffeoyl-tartaric, have been identified as potent and selective HIV-1 IN inhibitors (Mcdougall et al. 1998). Other phenolics with anti-HIV activity are phloroglucinol alpha pyrone arzanol which inhibits HIV-1 replication in-vitro (Appendino et al. 2007), and polyphenols geraniin and corilagin that reportedly inhibit HIV-1 RT (Notka et al. 2003).

16.3.6 Chlorophyll Derivatives

Chlorophyll is a naturally occurring plant pigment known to be highly unstable (Humphrey 2004). The progressive degradation of chlorophyll leads to the formation of similarly colored derivatives (Lanfer-Marquez et al. 2005). Commonly reported chlorophyll derivatives include chlorophylls a and b, pheophytins a and b, and pheophorbide a and b (Lanfer-Marquez et al. 2005). Pheophytin a reportedly inhibits HIV PR in-vitro and also suppressed HIV replication in a chronically infected cell model (Kapewangolo et al. 2017a). Pheophorbide-a has been found to possess anti-HIV effect in-vitro (Zhang et al. 2003).

16.4 HIV Life Cycle and Phytochemical Targets for Therapy

HIV infects cells of the immune system, mostly CD4+ T helper cells (Martínez-Bonet et al. 2015), and the infection ends up destroying the host's ability to effectively fight off the virus due to a compromised defense system. The replication cycle of HIV, commonly referred to as the life cycle of HIV, consist of two main steps. These are; (1) the entry and integration step, and (2) translation and budding step. Understanding the life cycle of HIV has led to the development and improvement of antiretroviral drugs with potential for controlling or completely eliminating the virus (Kirchhoff 2013). Key aspects of the life cycle that have been explored for treatment are virus fusion, reverse transcription, integration steps consisting of 3′ processing and the strand transfer (ST) steps as well as protein maturation steps (Pommier et al. 2005). Current HIV therapy targets different steps of the HIV cycle, however, complete eradication of the virus is compromised by the existence of reservoirs that harbor latent or dormant HIV (Schwartz et al. 2017) which is reactivated upon treatment interruption (Martínez-Bonet et al. 2015). The different stages of the HIV replication cycle are explained below.

16.4.1 Viral Entry and Integration

This step occurs at the surface of the cell and involves the virion bearing two copies of ribonucleic acid (RNA) binding to the CD4+ receptor and chemokine co-receptors, CCR5 or CXCR4. This is mediated by gp120/gp41 protein complex found on the surface of the viral envelop (Meanwell and Kadow 2003). A series of processes involving conformational changes and eventually fusion, with the viral core being inserted into the host cell followed by uncoating and the release of viral contents within the cytoplasm of these cells then follows (Chan and Kim 1998). The viral RNA is reverse transcribed by reverse transcriptase to a complementary deoxyribonucleic acid (cDNA) strand, which is subsequently transported into the nucleus as the pre-integration complex (PIC). Within the nucleus, the cDNA is integrated into the host DNA catalysed by the viral enzyme, integrase. Integration of the viral DNA into the host allows for the production of HIV proteins during the normal gene expression process.

16.4.2 Translation and Budding

The transcription step of the viral DNA results in viral genomic RNA and subsequently translation of viral protein which are processed and assembled in the cytoplasm by HIV protease. HIV protease further catalyses the maturation of viral particles via proteolytic processing into infectious virions. The virions consist of viral proteins and of two single stranded unspliced viral RNA which then bud off from the cells. As the virus replicates and makes new copies of itself, the course of infection in the infected individual is the gradual loss and destruction of naive and memory CD4+ T cells leading to AIDS which is the final stages of the infection course (Vidya Vijayan et al. 2017).

16.4.3 HIV Phytochemical Targets

Phytochemicals with anti-HIV properties are widely reported. Classes of phytochemicals targeting different steps of the HIV cycle include alkaloids, flavonoids, coumarins, saponins, terpenes and phenolics (Kurapati et al. 2016). Unique anti-HIV agents of plant origin exist and the most notable one is possibly Bevirimat, a novel inhibitor of HIV isolated from a Chinese herb, *Syzygium claviflorum*, which targets maturation of newly formed HIV particles. The bioactivity of anti-HIV plant extracts could be justified as promising therapy for HIV since they have a high potential of targeting more than one stage in the HIV cycle due to the presence of various phytochemicals in the extract.

16.5 In-vitro HIV Enzyme Models for Screening Phytochemicals for Anti-HIV Activity

Specific HIV enzyme-based assays have been established to screen for potential anti-HIV activity in-vitro. The principles behind these assays are described here. It should be noted that there are no techniques specific for investigating the anti-HIV potential of natural products and assays described here have been used to determine in-vitro anti-HIV activity of both synthetic and natural drug agents.

16.5.1 HIV Reverse Transcriptase Activity Assays

The RT enzyme of HIV is a heterodimer consisting of 66- and 51-kDa subunits (Fields et al. 1996) and it is involved in converting viral RNA to cDNA. This multifunctional enzyme is involved in RNA-dependent and DNA-dependent polymerisation, strand displacement synthesis and strand transfer, and degrades the RNA strand in the RNA/DNA hybrid (Schultz and Champoux 2008). It performs these functions through its polymerase function (for which there are two classes of inhibitors, the nucleoside reverse transcriptase and non-nucleoside reverse transcriptase inhibitors) and an RNase H function which is unique to the C terminus of the p66 subunit (Su et al. 2010). The polymerase function requires either RNA or DNA as the template making use of a host transfer RNA (tRNA) as primer ($tRNA_{lys3}$) (Sarafianos et al. 2009). The RNase H activity is required for processing the tRNA primer to begin minus-strand DNA synthesis and degradation of viral RNA during synthesis followed by preparation of the polypurine tract DNA-RNA hybrid which serves as the primer for positive strand DNA synthesis (Schultz and Champoux 2008). All these processes result in the copying of a single stranded RNA to a double stranded DNA (Sarafianos et al. 2009; Schultz and Champoux 2008).

16.5.1.1 Targeting the Polymerase Domain

The assay provides a quantitative measure of RT activity and takes advantage of the ability of RT to synthesise DNA, starting from the template/primer hybrid poly (A) x oligo $(dT)_{15}$. Digoxigenin and biotin labelled nucleotides are incorporated into the same DNA molecule as it is freshly synthesized by RT. The detection and quantification of newly synthesized DNA follows a sandwich ELISA protocol where the biotin-labelled DNA binds to the surface of streptavidin-coated microplate modules. An antibody to digoxigenin, conjugated to peroxidase (anti-DIG- POD) is then added and binds to the digoxigenin-labelled nucleotides. Finally, a peroxidase substrate, 2, 2′-azino-di-[3-ethylbenzthiazoline sulfonate (6)] diammonium salt crystals (ABTS), is added. The peroxidase enzyme catalyzes the cleavage of the substrate to produce a coloured reaction product. The absorbance of the samples is determined using a microplate (ELISA) reader, and is directly correlated to the level of RT activity in the sample (Kapewangolo et al. 2017a). Other assay variations which make use of the same or similar principles have been reported (Suzuki et al. 1993). These assays have moved from the previous unsafe isotope-based assays which required special equipment and tedious procedures (Kuno et al. 1999). Screening for HIV anti-RT activity using the method described here can be done both as direct enzyme assays or using compound treated cells supernatant in a replication assay. Examples of plant products that have reportedly shown anti HIV activity as a result of inhibiting the polymerase domain of HIV include the flavonoid myricetin (Pasetto et al. 2014) and crude plant extracts of *Hoodia gordonii* (Kapewangolo et al. 2016a) and *Sceletium tortuosum* (Kapewangolo et al. 2016b).

16.5.1.2 RNAse H Activity Assays

Drugs that target the RNase H function of RT are important since they offer alternatives which focus on the RNase H binding function of RT. Unfortunately, assays that target RNase H are cumbersome and costly, not sensitive and make use of radiolabelled substrates, as such, this target has not been widely explored (Wu et al. 2017). An isotope based anti- RNase H activity study on the effect of nine South African medicinal plants against HIV-1 was assessed by measuring the degree of degradation of the 3H-labelled RNA strand in a RNA/DNA hybrid by RT in the presence of the test substance (Bessong et al. 2005). Of the nine plants, *Combretum molle* demonstrated the highest inhibition effect on the RNAse activity (Bessong et al. 2005). Wu et al. (2017) recently reported a label free fluorescence assay for rapid detection of RNase H activity based on Tb^{3+}-induced G-quadruplex conjugates (Wu et al. 2017). The assay makes use of an RNA:DNA hybrid probe which in the presence of RNase H, can be cleaved to produce a single-strand G-rich oligonucleotide. The ssG rich oligonucleotide is then subsequently induced by Tb^{3+} to form a stable G-quadruplex structure leading to an enhancement in Tb^{3+} fluorescence. This quick, rapid and much safer method is a positive move from the traditional techniques.

16.5.2 HIV Protease Activity Assays

These assays make use of quantitative techniques such as fluorescence and high performance liquid chromatography. One such fluorescence technique involves the use of a fluorogenic HIV PR substrate 1; Arg-glu-(EDANS)-Ser-Gln-Asn-Tyr-Pro-Ile-Val-Gln-Lys-(DABCYL)-Arg (Sigma Aldrich, Missouri USA). This substrate is a synthetic peptide Tyr-Pro cleavage site for HIV PR as well as two covalently modified amino acids for the detection of cleavage (Matayoshi et al. 1990). One of the modifications involves the attachment of the fluorophore 5-(2-aminoethylamino)-1-naphthalene sulfonate (EDANS) to the glutamic residue. The other modification is the addition of an acceptor chromophore 4-′-dimethylaminoazobenzene-4-caboxylate (DABCYL) to the lysine residue. The modified amino acids are on opposite sides of the cleavage site. Spatial orientation and overlap of the DABCYL absorbance with the EDANS emission permits resonance energy transfer between the two moieties and quenching of the EDANS fluorescence at 490 nm occurs. However, when HIV PR cleaves the peptide, the DABCYL group is no longer proximal to the fluorophore and emission at 490 nm cannot be detected. A drug that inhibits HIV PR therefore prevents this cleavage thus allowing quenching to occur such that the EDANS fluorescence signal is diminished. Models of such an assay are used for high throughput screening in 96-well format. Medicinal plants with inhibition effect on HIV-1 PR include *Ocimum labiatum* and the isolated bioactive compound pheophytin a (Kapewangolo et al. 2017a), a popular commercial plant *Hoodia gordonii* (Apocynaceae) (Kapewangolo et al. 2016a), *Sceletium tortuosum* (Kapewangolo et al. 2016b) and *Bulbine frutescens* (Shikalepo et al. 2018). Polyphenols have diverse multi-target against HIV which includes inhibition of HIV PR (Andrae-marobela et al. 2013).

16.5.3 HIV Integrase Activity Assays

Unlike the other two HIV enzymes, RT and PR, HIV IN doesn't have a mammalian equivalent making it a suitable target for HIV therapy. HIV IN integrates the viral DNA into the host genome in two steps. The first step involves the removal of two nucleotides from the viral cDNA and the second step which is referred to as DNA strand transfer involves the integration of the cDNA into the host genome (Goldgur et al. 1999). Direct HIV IN assays have been developed with inhibitors such as phytochemicals/crude plant extracts targeting the 3′-processing (removal of nucleotides) and subsequent DNA strand transfer recombination reaction. One such method is a non-radioactive assay used to quantitatively measure the effects of anti-viral agents on HIV-1 IN activity in a 96-well plate format (Kapewangolo et al. 2016a). In this assay, a full-length recombinant HIV-1 integrase protein is loaded onto the HIV-1 LTR U5 donor substrate (DS) DNA containing an end-labeled biotin. Inhibitors are added to the enzyme reaction with subsequent addition of a different double stranded target substrate (TS) DNA containing a 3′-end modification. The HIV-1 integrase cleaves the terminal two bases from the exposed 3′-end of the HIV-1 LTR DS DNA and then catalyzes a strand-transfer recombination reaction to integrate the DS DNA into the TS DNA. Detection of the reaction products is done colorimetrically using an HRP-labeled antibody directed against the TS 3′-end modification. In addition to other anti-HIV properties, flavonols such as quercetin and kaempferol reportedly inhibit HIV IN (Andrae-marobela et al. 2013). Medicinal plants tested for HIV IN inhibition include *Bulbine frutescens* (Shikalepo et al. 2018) and *Sceletium tortuosum* (Kapewangolo et al. 2016b).

16.6 HIV Replication Assay Models for in-vitro Anti-HIV Screening

HIV primarily infects CD4+ T cells and causes AIDS as a result of the depletion of this subpopulation of cells, which is crucial for an effective immune system (Rambaut et al. 2004). Most importantly, the virus infects cells that express the CD4 receptor and chemokine co-receptors, CCR5 and CXCR4, allowing for viral entry (Alkhatib 2009).

Some anti-HIV assays aimed at studying HIV replication focusing on inhibition of replication by potential drugs including phytochemicals are the simple prescreening methods such as the tetrazolium dye based assays, 3-[4,5-dimethylthiazol-2-yl]-2,5 diphenyl tetrazolium bromide (MTT) and 2,3-bis [2-methoxy-4-nitro-5-sulfophenyl]-2H-tetrazolium-5-carboxanilide (XTT), the p24 expression assay, syncytium formation assay and luciferase activity detection assays. These assays are performed on human T cells including both primary peripheral blood mononuclear cells (PBMCs) and continuous cell lines such as CEM-SS, MT4, H9 and C8166 (Rege et al. 2015). These inexpensive 96 well format assays facilitate the estimation of the 50% cytotoxic concentrations (CC_{50}s) and effective concentrations (EC_{50}) of potential drugs albeit with limitations such as no information on the mechanism of action of the potential compound.

16.6.1 Tetrazolium Dye Based Assays

The MTT and XTT assays involve dyes whose reduction to formazan is used to estimate cell viability and proliferation by spectrophotometric measurements. Considering that several T cell lines are susceptible to the lytic effect of HIV replication, these dyes have been employed in measuring both this cytolytic effect and hence anti-HIV activity as well as the cytotoxic effect of potential drugs (Rege et al. 2015).

In the case of MTT (developed by Mosmann 1983), an insoluble blue product is produced and requires a solubilisation step using solvents (Mosmann 1983). Solvents such as acidified isopropanol in a 1:9 ratio (1 part 1 M HCl and 9 part of isopropanol) and DMSO have been used. The product obtained from XTT reduction is water soluble and brownish. An additional coupling reagent, phenazine methosulfate (PMS) is used in the XTT assays to enhance formazan formation. For both assays, spectrophotometric readings are taken within 4–24 h of treatment at one of different wavelengths of the absorbance spectrum; MTT (540–600 nm and 650–690 background) and XTT (450–500 nm and 630–690 nm background). MTT has reportedly been used in assessing the antiviral and cytotoxic effect of phytochemicals (Liu et al. 2018). The selectivity index of these products for HIV is measured from dividing the CC_{50} by the EC_{50}.

The use of the dynamic impedance based real time, label-free cell electronic sensing analyser for measuring cell proliferation, cytotoxicity start time, cell recovery, and cell response patterns (Fonteh et al. 2015), is a useful alternative to MTT and XTT albeit much more expensive. The system is applicable in research applications in drug development, toxicology, cancer, medical microbiology and virology. Other cell viability and proliferation assays such as the annexin/PI apoptosis assay and the caboxyfluoreiscenin succinimidyl ester assays which are flow cytometry based are alternatives which are more specific but with low throughput and more costly. These are not discussed here.

16.6.2 Syncytium Formation Cell Counting Assay

Syncytium results from cell fusion induced by HIV. Microscopy has been used to determine syncytium formation by counting the giant multinucleated cells under an inverted microscope at 100x magnification (Rashed et al. 2013). This method does not allow for quantification of syncytium nor does it provide an estimate of the number of cells involved in cell fusion (Wünschmann and Stapleton 2000). Wunchsmann and Stapleton (2000) described a technique that involved staining of the DNA with propidium iodide to determine cell size and fluorescence parameters using flow cytometry. Using this technique, the authors were able to distinguish between syncytium forming and non syncytium forming cells. The assay showed that a decrease in cell numbers in HIV infected MT-2 cell lines was as a result of decrease in syncytium formation than viral cell death making it an inexpensive assay for use in place of p24 ELISA (Wünschmann and Stapleton 2000). Syncytium inhibition studies have been reported for flavonoids such as scutellarin (Zhang et al. 2005).

16.6.3 p24 Antigen Assay

The p24 antigen is the main structural protein of HIV whose detection serves as an early diagnosis for HIV infection (Xia et al. 2015). The assay is performed by collecting cell free supernatant from compounds treated cells following incubation for a desired number of period usually 4 days post treatment which is optimal for viral production (Rege et al. 2015). An ELISA assay is then performed for detection of p24 activity. Plant derived compounds have been shown to be activators of latent virus, resulting in p24 increase in culture supernatant (Kapewangolo et al. 2017b) as well as inhibitors of the HIV replicative process resulting in decreases in p24 antigen production (Pasetto et al. 2014).

16.6.4 Luciferase Activity Assay for Viral Infectivity Measuremen

This is an HIV replication assay performed using cells such as the TZM-bl cell line. These cells were generated from JC.53 cells by introducing separate integrated copies of the luciferase and β-galactosidase genes under the control of the HIV-1 promoter (Platt et al. 1998) and are highly sensitive to infection with diverse isolates of HIV-1. These assays make use of molecularly cloned pseudoviruses designed to undergo a single round of infection readily detectable in genetically engineered cell lines that contain a *tat*-responsive reporter gene such as luciferase e.g. the TZM-bl cell line (Montefiori 2004).

In assays that involve TZM-bl cell line, the cells are seeded overnight followed by the addition of virus pretreated with varying concentration of phytochemicals and incubated for 4 h. Cells are then washed to remove unbound virus and further incubated for 48 h. The cells are washed with PBS, lysed and the lysate treated with a substrate followed by measurement of luciferase activity using a fluorimeter. The results are expressed as percentage inhibition and IC_{50} is calculated. Flavonoids have been shown to inhibit HIV activity as a result of inhibiting the luciferase activity (Pasetto et al. 2014).

16.7 Conclusion

Constituents of medicinal plants continue to hold promise as potential anti-HIV drug agents. In-vitro anti-HIV assays are important to improve our understanding of the biology of HIV, the mechanisms of action of plant based compounds, biochemical reactions with implications towards inhibition of HIV and safety of such medicines. While the in-vitro assays are important, they cannot be used alone without the need for animal/clinical studies. The current drive for marketing phytochemicals is drawn from the ongoing popular use of medicinal plants in the traditional management of HIV/AIDS.

References

Alkhatib G. The biology of CCR5 and CXCR4. Curr Opin HIV AIDS. 2009;4:96–103. https://doi.org/10.1097/COH.0b013e328324bbec.

Andrae-marobela K, Ghislain FW, Okatch H, Majinda RRT. Polyphenols: a diverse class of multi-target anti-HIV-1 agents. Curr Drug Metab. 2013;14:392–413.

Appendino G, Ottino M, Marquez N, et al. Arzanol, an anti-inflammatory and anti-HIV-1 phloroglucinol alpha-Pyrone from Helichrysum italicum ssp. microphyllum. J Nat Prod. 2007;70:608–12. https://doi.org/10.1021/np060581r.

Asres K, Bucar F, Kartnig T, et al. Antiviral activity against human immunodeficiency virus type 1 (HIV-1) and type 2 (HIV-2) of ethnobotanically selected Ethiopian medicinal plants. Phyther Res. 2001;15:62–9. https://doi.org/10.1002/1099-1573(200102)15:1<62::AID-PTR956>3.0.CO;2-X.

Bessong PO, Obi CL, Andréola ML, et al. Evaluation of selected South African medicinal plants for inhibitory properties against human immunodeficiency virus type 1 reverse transcriptase and integrase. J Ethnopharmacol. 2005;99:83–91. https://doi.org/10.1016/j.jep.2005.01.056.

Boyd MR, Hallock YF, Cardellina JH, et al. Anti-HIV mechallamines from Ancistrocladus korupensis. J Med Chem. 1994;37:1740–5.

César G-ZJ, Alfonso M-GG, Marius M-M, et al. Inhibition of HIV-1 reverse transcriptase, toxicological and chemical profile of *Calophyllum brasiliense* extracts from Chiapas, Mexico. Fitoterapia. 2011;82:1027–34. https://doi.org/10.1016/j.fitote.2011.06.006.

Chan DC, Kim PS. HIV entry and its inhibition. Cell. 1998;93:681–4.

Chingwaru W, Vidmar J, Kapewangolo PT. The potential of Sub-Saharan African plants in the management of human immunodeficiency virus infections: a review. Phytother Res. 2015; https://doi.org/10.1002/ptr.5433.
Cihlar T, Fordyce M. Current status and prospects of HIV treatment. Curr Opin Virol. 2016;18:50–6. https://doi.org/10.1016/j.coviro.2016.03.004.
Dang Z, Ho P, Zhu L, et al. New betulinic acid derivatives for bevirimat-resistant human immunodeficiency virus type-1. J Med Chem. 2013;56:2029–37. https://doi.org/10.1021/jm3016969.
Fields BN, Knipe DM, Howley PM. Fields virology. 3rd ed. Philadelphia: Lippincott-Raven Publishers; 1996.
Fonteh P, Elkhadir A, Omondi B, et al. Impedance technology reveals correlations between cytotoxicity and lipophilicity of mono and bimetallic phosphine complexes. Biometals. 2015;28:653–67.
Fritts M, Crawford CC, Quibell D, et al. Traditional Indian medicine and homeopathy for HIV/AIDS: a review of the literature. AIDS Res Ther. 2008;5:1–9. https://doi.org/10.1186/1742-6405-5-25.
Goldgur Y, Craigie R, Cohen GH, et al. Structure of the HIV-1 integrase catalytic domain complexed with an inhibitor: a platform for antiviral drug design. Proc Natl Acad Sci. 1999;96:13040–3. https://doi.org/10.1073/pnas.96.23.13040.
Humphrey A. Chlorophyll as a color and functional ingredient. J Food Sci. 2004;69:C422–5.
Institute of Medicine (US). Committee on envisioning a strategy for the long-term burden of HIV/AIDS: African needs and U.S. interests. Preparing for the future of HIV/AIDS in Africa: a shared responsibility, The Burden of HIV/AIDS: Implications for African States and Societies, vol. 4. Washington, DC: National Academies Press (US); 2011.. Available from: https://www.ncbi.nlm.nih.gov/books/NBK209743/. Accessed on 19.08.2018
Jadaun P, Khopkar P, Kulkarni S. Repurposing phytochemicals as anti-HIV agents. J Antivirals Antiretrovirals. 2016;8:139–41.
Kapewangolo P, Knott M, Shithigona REK, et al. In vitro anti-HIV and antioxidant activity of *Hoodia gordonii* (Apocynaceae), a commercial plant product. BMC Complement Altern Med. 2016a;16:114. https://doi.org/10.1186/s12906-016-1403-7.
Kapewangolo P, Tawha T, Nawinda T, et al. Sceletium tortuosum demonstrates in vitro anti-HIV and free radical scavenging activity. S Afr J Bot. 2016b;106:140–3. https://doi.org/10.1016/j.sajb.2016.06.009.
Kapewangolo P, Kandawa-Schulz M, Meyer D. Anti-HIV activity of Ocimum labiatum extract and isolated pheophytin-a. Molecules. 2017a;22:1763. https://doi.org/10.3390/molecules22111763.
Kapewangolo P, Omolo J, Fonteh P, et al. Triterpenoids from Ocimum labiatum activates latent HIV-1 expression in vitro: potential for use in adjuvant therapy. Molecules. 2017b;22:1703. https://doi.org/10.3390/molecules22101703.
Kirchhoff F. HIV life cycle: overview. Encyclopedia of AIDS. 2013. p. 1–9.
Klos M, Van de Venter M, Milne PJ, Traore HN, Meyer D, Oosthuizen V. In vitro anti-HIV activity of five selected South African medicinal plant extracts. J Ethnopharmacol. 2009;124:182–8.
Kuno H, Ikeda H, Takeuchi M, Yoshida T. A simple and rapid reverse transcriptase assay for the detection of retroviruses in cell cultures. Cytotechnology. 1999;29:221–7.
Kurapati KRV, Atluri VS, Samikkannu T, et al. Natural products as anti-HIV agents and role in HIV-associated neurocognitive disorders (HAND): a brief overview. Front Microbiol. 2016;6:1 14. https://doi.org/10.3389/fmicb.2015.01444.
Lanfer-Marquez UM, Barros RMC, Sinnecker P. Antioxidant activity of chlorophylls and their derivatives. Food Res Int. 2005;38:885–91. https://doi.org/10.1016/j.foodres.2005.02.012.
Liu Y-P, Zhao W-H, Feng X-Y, et al. Novel tetrahydrofuran derivatives from Trigonostemon howii with their potential anti-HIV-1 activities. Bioorg Chem. 2018;79:111–4.
Ma CM, Nakamura N, Hattori M. Saponins and C-glycosyl flavones from the seeds of *Abrus precatorius*. Chem Pharm Bull. 1998;46(6):982–7.
Maroyi A. Alternative medicines for HIV/AIDS in resource-poor settings: insight from traditional medicines use in Sub- Saharan Africa. Trop J Pharm Res. 2014;13:1527–36

Martínez-Bonet M, Puertas MC, Fortuny C, et al. Establishment and replenishment of the viral reservoir in perinatally HIV-1-infected children initiating very early antiretroviral therapy. Clin Infect Dis. 2015;61:1169–78. https://doi.org/10.1093/cid/civ456.

Matayoshi ED, Wang GT, Krafft GA, Erickson J. Novel fluorogenic substrates for assaying retroviral proteases by resonance energy transfer. Science. 1990;247(80):954–8.

Mbirimtengerenji ND. Is HIV/AIDS epidemic outcome of poverty in sub-saharan Africa? Croat Med J. 2007;48(5):605.

Mcdougall B, King PJ, Wu BW, et al. Dicaffeoylquinic and dicaffeoyltartaric acids are selective inhibitors of human immunodeficiency virus type 1 integrase. Antimicrob Agents Chemother. 1998;42:140–6.

Meanwell N, Kadow J. Inhibitors of the entry of HIV into host cells. Curr Opin Drug Discov Devel. 2003;6:451–61.

Mohammed MM, Ibrahim NA, Awad NE, Matloub AA, Mohamed-Ali AG, Barakat EE, Mohamed AE, Colla PL. Anti HIV-1 and cytotoxicity of the alkaloids of *Erythrina abyssinica* Lam. growing in Sudan. Nat Prod Res. 2012;26(17):1565–175.

Montefiori DC. Evaluating neutralizing antibodies against HIV, SIV, and SHIV in luciferase reporter gene assays. Curr Protoc Immunol. 2004;64:12.11. 11–7.

Mosmann T. Rapid colorimetric assay for cellular growth and survival: application to proliferation and cytotoxicity assays. J Immunol Methods. 1983;65:55–63.

Notka F, Meier G, Wagner R. Inhibition of wild-type human immunodeficiency virus and reverse transcriptase inhibitor-resistant variants by *Phyllanthus amarus*. Antivir Res. 2003;58:175–86. https://doi.org/10.1016/S0166-3542(02)00213-9.

Okamoto M, Ono M, Baba M. Potent inhibition of HIV type 1 replication by an antiinflammatory alkaloid, cepharanthine, in chronically infected monocytic cells. AIDS Res Hum Retrovir. 1998;14:1239–45.

Pan SY, Litscher G, Chan K, Yu ZL, Chen HQ, Ko KM. Traditional medicines in the world: where to go next? Evid-Based Compl Altern. 2014;2014:525340.

Pasetto S, Pardi V, Murata RM. Anti-HIV-1 activity of flavonoid myricetin on HIV-1 infection in a dual-chamber in vitro model. PLoS One. 2014;9:1–18. https://doi.org/10.1371/journal.pone.0115323.

Platt EJ, Wehrly K, Kuhmann SE, et al. Effects of CCR5 and CD4 cell surface concentrations on infections by macrophagetropic isolates of human immunodeficiency virus type 1. J Virol. 1998;72:2855–64.

Pommier Y, Johnson AA, Marchand C. Integrase inhibitors to treat HIV/AIDS. Nat Rev Drug Discov. 2005;4:236.

Qian H, Deng T, Sun H (2018) Global and regional tree species diversity. J Plant Ecol 1–6. https://doi.org/10.1093/jpe/rty013.

Rambaut A, Posada D, Crandall KA, Holmes EC. The causes and consequences of HIV evolution. Nat Rev Genet. 2004;5:52.

Rashed K, Luo M-T, Zhang L-T, Zheng Y-T. Phytochemical content and anti-HIV-1 activity of *Sapindus Saponaria* (L). J Forest Prod Ind. 2013;2:22–6.

Rege A, Dahake R, Roy S, Chowdhary A. Screening of natural products for anti-HIV potential: an in vitro approach. J Virol Curr Res. 2015;1:7.

Saklani A, Kutty SK. Plant-derived compounds in clinical trials. Drug Discov Today. 2008;13:161–71. https://doi.org/10.1016/j.drudis.2007.10.010.

Sarafianos SG, Marchand B, Das K, et al. Structure and function of HIV-1 reverse transcriptase: molecular mechanisms of polymerization and inhibition. J Mol Biol. 2009;385:693–713.

Schultz SJ, Champoux JJ. RNase H activity: structure, specificity, and function in reverse transcription. Virus Res. 2008;134:86–103.

Schwartz C, Bouchat S, Marban C, et al. On the way to find a cure: purging latent HIV-1 reservoirs. Biochem Pharmacol. 2017;146:10–22. https://doi.org/10.1016/J.BCP.2017.07.001.

Shikalepo R, Mukakalisa C, Kandawa-Schulz M, et al. In vitro anti-HIV and antioxidant potential of Bulbine frutescens (Asphodelaceae). J Herb Med. 2018;12:73–8. https://doi.org/10.1016/j.hermed.2017.09.007.

Singh J, Kumar A, Kumar V. Plant derived anti-human immunodeficiency virus (HIV) compounds. Int J Nat Prod Sci. 2011;2:1–9.

Stoddart CA, Joshi P, Sloan B, et al. Potent activity of the HIV-1 maturation inhibitor bevirimat in SCID-hu Thy/Liv mice. PLoS One. 2007;2:e1251. https://doi.org/10.1371/journal.pone.0001251.

Su H-P, Yan Y, Prasad GS, et al. Structural basis for the inhibition of RNase H activity of HIV-1 reverse transcriptase by RNase H active site-directed inhibitors. J Virol. 2010;84:7625–33.

Suzuki K, Craddock BP, Kano T, Steigbigel RT. Colorimetric reverse transcriptase assay for HIV-1. J Virol Methods. 1993;41:21–8.

Tang MW, Shafer RW. HIV-1 antiretroviral resistance. Drugs. 2012;72:e1–e25.

UNAIDS. Fact sheet – latest global and regional statistics on the status of the AIDS epidemic. 2018.

Vidya Vijayan K, Karthigeyan KP, Tripathi SP, Hanna LE. Pathophysiology of CD4+ T-cell depletion in HIV-1 and HIV-2 infections. Front Immunol. 2017;8:580.

Wu K, Ma C, Liu H, et al. Label-free fluorescence assay for rapid detection of RNase H activity based on Tb 3+-induced G-quadruplex conjugates. Anal Methods-UK. 2017;9:3055–60.

Wünschmann S, Stapleton JT. Fluorescence-based quantitative methods for detecting human immunodeficiency virus type 1-induced syncytia. J Clin Microbiol. 2000;38:3055–60.

Xia L, Zhang J, Cui C, et al. In vitro affinity maturation and characterization of anti-P24 antibody for HIV diagnostic assay. J Biochem. 2015;158:531–8.

Xu HX, Wan M, Loh BN, Kon OL, Chow PW, Sim KY. Screening of traditional medicines for their inhibitory activity against HIV-1 protease. Phytother Res. 1996;10:207–10.

Zhang H-J, Tan GT, Hoang VD, et al. Natural anti-HIV agents. Part IV. Anti-HIV constituents from Vatica cinerea. J Nat Prod. 2003;66:263–8. https://doi.org/10.1021/np020379y.

Zhang G, Wang Q, Chen J, et al. The anti-HIV-1 effect of scutellarin. Biochem Biophys Res Commun. 2005;334:812–6. https://doi.org/10.1016/j.bbrc.2005.06.166.

Chapter 17
Nanoencapsulation of Phytochemicals and in-vitro Applications

Jordy Kim Ung Ling, Yiik Siang Hii, Jaison Jeevanandam, Yen San Chan, and Michael K. Danquah

17.1 Introduction

The compounds that are obtained from biological organisms are called biological compounds. These compounds are highly beneficial in variety of applications. Biological compounds are attained from the crude metabolites that are extracted from microbes such as bacteria, algae, fungi and also from various parts of plants. Biological compounds from microbes are difficult to extract as it takes a longer time and undergo tedious downstream separation processes. However, biological compounds from plants are relatively easy to extract and have easier scalability due to the wide availability. Thus, plant-based biological compounds are highly recommended for various applications, especially in pharmaceutical and nutraceutical industries. These plant-based biological compounds are called phytochemicals that are extracted and separated from the crude plant metabolites. These phytochemicals are mainly responsible for the flavor and the color of the plants by being an integral part of their immune system. It can be noted that highly colored plants have the potential of containing high phytochemicals in their extracted metabolites. There are numerous phytochemicals that can be obtained from plants such as pyrimidines, flavonoids, phenols, carotenes, allicin, steroids, terpenes and nucleic acid which possess potential bioactivity in human cells. The natural chemical compounds that are present in phytochemicals are highly beneficial for the physiological and immunological systems of humans for sickness and ailments prevention. Currently, there are several studies that show the efficiency of phytochemicals in the treatment of common to deadly ailments. It was listed in these studies that about 900 phytochemicals are identified from plants which

J. K. U. Ling · Y. S. Hii · J. Jeevanandam · Y. S. Chan
Department of Chemical Engineering, Curtin University Malaysia, Miri, Malaysia

M. K. Danquah (✉)
Chemical Engineering Department, University of Tennessee, Chattanooga, TN, USA
e-mail: michael-danquah@utc.edu

S. Kumar, C. Egbuna (eds.), *Phytochemistry: An in-silico and in-vitro Update*,
https://doi.org/10.1007/978-981-13-6920-9_17

can be used in diagnosis and treatment of various diseases. However, high solubility and low absorption in body fluids, inability to reach targeted site and poor bioavailability are some of the drawbacks in the extracted free phytochemicals. Thus, there is a need for an alternate to address these challenges posed by phytochemicals for bioactivity enhancement in pharmaceutical application (Barros and Ferreira 2017; Jeevanandam et al. 2017).

Nanoencapsulation is a method of formulating and loading a substance in nanosized carriers that has the ability to hold, to carry and to deliver the molecule to the targeted site. The substance can be nanoencapsulated using dendrimers via divergent and convergent growth, polymeric nanoparticles via dispersion of preformed polymers, supercritical fluid technology and polymerization, liposomes via reverse phase evaporation, detergent removal, high-pressure extrusion, nanoemulsions via dispersion, ultrasonication, phase inversion temperature, spontaneous and high energy emulsification, micelles via direct dissolution, film casting, dialysis and oil in water. These nanoformulation strategies have high potential in encapsulating phytochemicals which can be eventually useful in phytochemical delivery applications (Jeevanandam et al. 2016b). Several studies done in recent times that showed the ability of nanoencapsulation to load phytochemicals and their controlled release towards target cells are listed in this chapter. Additionally, the cytotoxicity of nanoencapsulated phytochemicals and their ability to enhance drug delivery efficiency in cancer treatment and antimicrobial property are also discussed.

17.2 Nanoencapsulated Phytochemicals

Various types of nanocarriers were developed to overcome the issues associated with the stability and bioavailability of phytochemicals. The potential nanocarriers such as liposomes, dendrimers, polymers, emulsions, micelles, solid lipid nanoparticles and hydrogels as displayed in Fig. 17.1 that are used to encapsulate, load and deliver phytochemicals are described in this section.

17.2.1 Liposomes

Liposomes possess the structure of lipid bilayers which are composed of an internal aqueous core and an outer phospholipids bilayer which serves as a versatile carrier for both hydrophilic and hydrophobic bioactive phytocompounds (Xie et al. 2016). Liposomes possess numerous advantages such as biocompatibility, reduced toxicity, biodegradability, non-immunogenicity, enhanced solubility, efficacy and bioavailability of poor soluble phytochemicals (Li et al. 2017). Despite these advantages, the applications of liposomes are hindered due to their poor stability. This occurs when

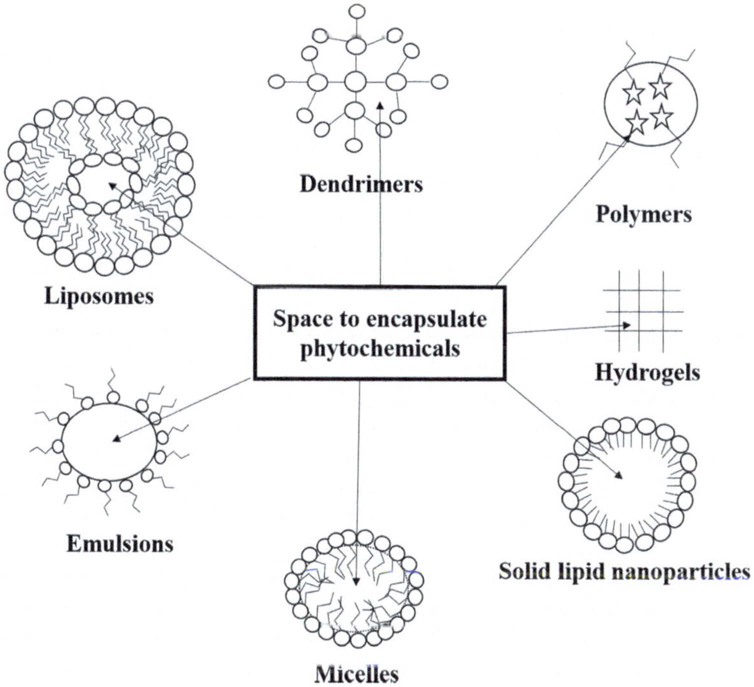

Fig. 17.1 Schematics of encapsulation structures for phytochemical formulation

they are exposed to adverse conditions in the gastrointestinal tract and consequently restrict the bioavailability of bioactive phytocompounds. Thus, researchers have shifted the focus from the conventional liposomes to nanoliposomes which provides larger interfacial area of bioactive phytocompounds towards target cells which extensively increases their bioavailability.

Dutta and Bhattacharjee (2017) studied the ability of nanoliposomes in protecting piperine (rich pepper extract) from degradation during the storage stage. Their study revealed that nanoliposomes are able to protect piperine from degradation and increased its stability to 2.4 times at 4 ± 1 °C, compared to the free piperine. Likewise, Aliakbari et al. (2018) investigated the stability of nanoliposome encapsulated baicalein which is a flavonoid used to treat neurodegenerative disorders. The results showed that the stability of baicalein was increased by 27% which proved the efficiency of nanoliposome formulation as a potential phytochemical encapsulation agent. Moreover, Amjadi et al. (2018) have reported in a recent study that the gummy candy with nanoliposome encapsulated betanin increases its stability and antioxidant activity significantly than the gummy candy with free betanin, after a 60-days storage.

17.2.2 Dendrimers

Dendrimers are monodispersed synthetic polymers with a well-defined three-dimensional globular architecture (Jeevanandam et al. 2016a). They have unique properties such as uniform nanosize, well-defined structure, high degree of branching and possess copious internal cavities. With these unique properties, dendrimers have become one of the major materials of research interest for the encapsulation of phytochemicals. The bioactive phytocompounds are either entrapped within the internal core or covalently attached with the outer terminal functional groups. Madaan et al. (2016) incorporated quercetin in polyamidoamine (PAMAM) dendrimer and showed a significant improvement in quercetin solubility in targeted site which facilitates its delivery. In the in-vitro study, the quercetin-loaded dendrimer provided a biphasic release profile, indicating their ability in enhancing the bioavailability of poorly water-soluble phytochemicals. Thus, the study also investigated the potential of PAMAM dendrimer to entrap the curcumin to enhance its solubility and stability. The results showed that the solubility of curcumin-loaded dendrimers increased up to 450 times, as compared to free curcumin. Furthermore, a prolonged release profile of curcumin was obtained, indicating an enhancement in their bioavailability.

17.2.3 Polymers

Polymeric nanocarriers that are developed from biocompatible natural or synthetic polymers are an interesting option to augment the bioavailability of phytochemicals. Synthetic polymers undergo chemical and enzymatic polymerization which usually possess higher sustainability and high reproducibility than natural polymers. Various synthetic polymers such as polyglycolides (PGAs), poly (butylene succinate), poly (ethylene glycol) (PEG), poly (lactic acids) (PLA), poly (lactic-co-glycolic acids) (PLGA) and polyester amides (PEAs) have been exploited for pharmaceutical and nutraceutical delivery of phytochemicals (Kamil et al. 2015; Xiao et al. 2017). On the other hand, natural polymers, mainly polysaccharides and proteins are macromolecular compounds that are synthesized by the living microorganisms. Natural polymers exhibited several remarkable properties such as complete biodegradability and biocompatibility, and thus, are often used as components of nanoformulations in various industries (Balaji et al. 2018).

Sechi et al. (2016) encapsulated poorly water-soluble flavonoid namely fisetin using poly- (ε-caprolactone) (PCL) and PLGA-PEG-COOH to increase their oral bioavailability. The results revealed that the polymeric nanoparticle can preserve fisetin with only less than 15% of release in 2 h and a controlled release profile for fisetin was observed under simulated gastric condition. Similarly, Gorbunova et al. (2018) alginate nanocapsules were used to encapsulate phytochemicals such as betacyanins, betaxanthin and polyphenols that are extracted from the beet green extracts. It was observed that the alginate nanocapsules are intact in simulated gastric conditions which indicates their potential in preserving the phytochemicals. Likewise, Rahaiee et al. (2017) also reported the encapsulation of a water-soluble carotenoid namely crocin in chitosan-alginate biopolymer nanoparticles which showed a controlled release profile up to 48 h.

17.2.4 Emulsions

An emulsion is a colloidal dispersion of two immiscible liquids such as oil and water which can enhance the bioavailability and stability of phytochemicals. This can also directly protect them against degradation. Nano-sized emulsions have become a more popular approach in the protection of phytochemicals that exhibit a better stability to particle aggregation compared to conventional emulsion due to their small droplet sizes (Gupta et al. 2016). The ability of nanoemulsion in increasing the stability of phytochemicals compounds was revealed by Shu et al. (2018) by encapsulating a lipid-soluble carotenoid namely astaxanthin in oil-in- water nanoemulsion and stabilized by a natural emulsifier called ginseng saponins. The nanoemulsified astaxanthin experienced only a 2.5% loss, compared to free astaxanthin, after 15 days of storage at 5 °C, indicating the potential of nanoemulsion based system in preserving the phytochemicals. Likewise, Rabelo et al. (2018) produced a water-in-oil nanoemulsion to protect the anthocyanin which is unstable and susceptible to deterioration. Their study successfully proved the feasibility of nano-emulsion in protecting unstable anthocyanins against degradation by retaining their superior antioxidant activity after 30 days of storage period.

17.2.5 Micelles

Micelles are different from liposomes which are made up of monolayers of phospholipids formed from amphiphilic molecules with a hydrophobic tail and hydrophilic head and a hydrophobic lipid core. A hydrophobic phytochemical is usually encapsulated in the micelle core, while the hydrophilic shell provides sufficient colloidal stability and prevents aggregation for controlled release and improves bioavailability of bioactive phytocompounds (Joseph et al. 2017). Feng et al. (2017) enhanced the water solubility of curcumin by encapsulating them into glycyrrhetinic acid-PEG-PCL copolymeric micelle system. The results showed that the water solubility of curcumin in micelle formulation is 1.87 mg/ml which is 1.70×10^5 higher than the crude curcumin. Similarly, Desai and Patravale (2018) incorporated curcumin in a micellar nanocarrier for nose-to-brain delivery and revealed a sixfold increment of curcumin bioavailability, compared to free curcumin upon nasal administration. Besides that, Yashaswini et al. (2017) enhanced the bioavailability of an antioxidant compound namely sesamol from sesame by incorporating them in phosphatidylcholine mixed micelles which showed a 1.2-fold better cellular uptake, compared to free sesamol.

17.2.6 Solid Lipid Nanoparticles (SLN)

Solid lipid nanoparticles are a class of emulsion which contain a solid or semi-solid lipid core structure that exhibits several advantages such as biocompatibility, high encapsulation efficiency, solvent-free nature, enhanced stability, ability to restrict degradation and ensure sustained release (Gao and McClements 2016). Specifically, SLNs are relatively suitable for encapsulating phytochemicals due to their excellent biocompatibility. For instance, Mehrad et al. (2018) improved the physiochemical stability of β-carotene by encapsulating them into SLN and stabilized via whey protein. It was observed that there was no β-carotene degradation, when the SLN formulation is stored for 18 days under high temperature. Similarly, Baek and Cho (2017) formulated N-carboxymethyl chitosan in SLN to initiate a rapid release of curcumin, when exposed to simulated gastric condition. Additionally, Hamishehkar et al. (2018) prepared sclareol-loaded SLN to improve the water solubility of the phytochemical to increase its therapeutic efficacy. The study showed that the sclareol-SLN gave an initial burst, followed by a prolonged release for 1 week, whereas free sclareol experienced a release of 90% within first 12 h which shows the efficiency of SLN in phytochemical formulation.

17.2.7 Hydrogel

Hydrogel beads are polymer networks that can absorb water by maintaining their three-dimensional structure. They play a critical role in bioactive phytochemical delivery system as they possess great potential for improving its oral bioavailability (McClements 2017). Zhang et al. (2016) incorporated β-carotene in alginate hydrogel beads and discovered that the hydrogel significantly enhanced the chemical stability of the phytochemical. Moreover, Patel et al. (2018) incorporated a pentacyclic triterpene namely lupeol found in vegetables in chitosan-gelatin hydrogel film and proved a rapid and constant biphasic release pattern of lupeol with approximately 91% after 24 h via in-vitro release study. Meanwhile, Lei et al. (2017) entrapped nobiletin using alginate hydrogel system to avoid phytochemical crystallization during digestion process. The study showed that the hydrogel system has limited the release of nobiletin (<27%) at pH 1.2 and allowed a sustained release at pH 7.4 (>80%) whereas no nobiletin crystal was found in the gastrointestinal region. This reveals that the filled-hydrogel system is effective in eliminating the phytochemical precipitation during digestion and improves their bioavailability. Thus, all these recent studies as listed in Table 17.1 provided ample reasons to use

Table 17.1 List of phytochemicals that are nanoencapsulated in recent studies

Nanoformulation	Phytochemicals	References
Liposomes	Piperine (rich pepper extract)	Dutta and Bhattacharjee (2017)
	Baicalein	Aliakbari et al. (2018)
	Betanin in gummy candy	Amjadi et al. (2018)
Dendrimers	Quercetin	Madaan et al. (2016)
Polymers	Fisetin	Sechi et al. (2016)
	Betacyanins, betaxanthin and polyphenols	Gorbunova et al. (2018)
	Water soluble carotenoids	Rahaiee et al. (2017)
Emulsions	Astaxanthin	Shu et al. (2018)
	Anthocyanin	Rabelo et al. (2018)
	Curcumin	Feng et al. (2017)
Micelles	Curcumin	Desai and Patravale (2018)
	Sesamol	Yashaswini et al. (2017)
	β-carotene	Mehrad et al. (2018)
Solid lipid nanoparticles	Curcumin	Baek and Cho (2017)
	Sclareol	Hamishehkar et al. (2018)
	β-carotene	Zhang et al. (2016)
Hydrogel	Lupeol	Patel et al. (2018)
	Nobiletin	Lei et al. (2017)

nanoencapsulated phytochemicals to improve their bioavailability and prolong their release in in-vivo condition for pharmaceutical and nutraceutical applications.

17.3 In-vitro Studies of Nanoencapsulated Phytochemicals

In-vitro studies will help to identify the efficiency of nanoencapsulated phytochemicals in a cellular model. Microbes such as bacteria, fungi and algae are used as microbial models to study the antimicrobial activity and various synthetic cells from mouse, human and monkey are utilized as cell models to understand the behavior of biological degradation, drug loading, carrying and delivering efficacy of nanoencapsulated phytochemicals.

17.3.1 Antimicrobial Analysis

Generally, several phytochemicals have proven to possess antimicrobial activity and nanoencapsulation of these antimicrobial phytochemicals have shown to increase the microbial inhibiting capability. Pereira et al. (2018) nanoencapsulated phytochemical compounds from guabiroba extract (GE) using poly (DL-lactide-co-glycolide) (PLGA) and studied the antimicrobial activity against *Listeria innocua*. The results showed that a low antimicrobial activity was observed in free GE with minimum inhibitory concentration (MIC) value of 8107 μg/ml due to their low water solubility. However, a higher antimicrobial activity was observed in nanoencapsulated GE with MIC values ranging from 2251 to 2670 μg/ml by increasing their water solubility. Similarly, Lee et al. (2017) nanoencapsulated silymarin extracted from *Silybum marianum* with water soluble chitosan and poly-γ-glutamic acid (γ-PGA) to study their antibacterial activity against *Bacillus subtilis* and *Staphylococcus aureus*. The MIC values of free silymarin and nanoencapsulated silymarin against *B. subtilis* was presented as 750 μg/ml and 375 μg/ml, respectively, whereas 375 μg/ml and 188 μg/ml of MIC values were observed against *S. aureus*. The result revealed that nanoencapsulation enhances the water solubility of silymarin by 7.7-fold which leads to higher antimicrobial activity. In addition, Oliveira et al. (2017) investigated the antimicrobial activity of nanoencapsulated extracts of passion fruit by-products and its antimicrobial activity was tested against *Escherichia coli* and *Listeria innocua*. It was demonstrated that the MIC values of free extracts for both *E. coli* and *L. innocua* were greater than 4000 μg/ml while PLGA encapsulation decreased its MIC value from 188 to 585 μg/ml for both bacterial cultures. This study indicates that nanoencapsulation significantly reduces the extracts concentration needed for bacterial growth inhibition.

Recently, Benjemaa et al. (2018) investigated the antibacterial efficiency of nanoemulsified *Thymus capitatus* essential oil (EO) and free EO against food-borne pathogens such as *E. coli* and *B. subtilis*. The results showed that the nanoemulsified EO showed higher antibacterial activity against Gram-positive bacteria (*B. subtilis*) than Gram-negative bacteria (*E. coli*) due to the present of carvacrol in EO. Similarly, Silva Gündel et al. (2018) also studied the antimicrobial effect of nanoemulsified essential oil from *Cymbopogon flexuosus* against *S. aureus*, *Candida albicans*, *Cryptococcus grubii*, and *Pseudomonas aeruginosa* and reported that nanoemulsified EO shows better antibacterial activity than free EO. Remarkably, Rossi et al. (2017) studied the antimicrobial activity of nanoemulsified EO from *C. flexuosus* against rapidly growing mycobacteria (RGM) such as *Mycobacterium fortuitum*, *Mycobacterium massiliense* and *Mycobacterium abscessus*. The free EO showed inhibitory effect whereas the nanoemulsified EO possess 2.5 to 10-fold higher antimicrobial potential against all three strains. All these studies showed that nanoencapsulated phytochemicals possess higher potential in inhibiting microbial growth via controlled delivery than free phytochemicals.

17.3.2 Cytotoxicity of Nanoencapsulated Phytochemicals

Cytotoxic analysis of nanoencapsulated phytochemicals plays a major role in recommending less toxic nano-sized formulations for nutraceutical and pharmaceutical applications. Recently, Chittasupho and Athikomkulchai (2018) investigated the cytotoxicity of nanoencapsulated phytochemicals from *Combretum quadrangulare* leaf extracts (CQLE) against adenocarcinomic human alveolar basal epithelial (A549) lung cancer cells using 3-(4, 5-dimethylthiazol-2-yl)-2, 5 – diphenyltetrazolium bromide (MTT) assay. The results showed that nanoencapsulated CQLE induced cell death of A549 cells, comparatively higher than free CQLE and suggested that nanoencapsulation was capable of lowering the toxicity of CQLE by controlling their cellular release. Besides, Nejat et al. (2017) studied the cytotoxicity of nanoencapsulated cardamom extract (CE) against human glioblastoma (U87MG) brain tumor cell lines via MTT assay. The study indicated that nanoencapsulated CE are highly toxic to U87MG cells, compared to free CE and a minimal cytotoxic effect was observed towards noncancerous cells.

Likewise, Ibrahim et al. (2018) nanoencapsulated curcumin (CUR) from *Curcuma longa* in a marine lipid-based liposome called marinosomes which consists of high polyunsaturated fatty acids ratio and their cytotoxicity was evaluated in lung cancer cells (A549) and human umbilical vein endothelial cells (HUVECs) via water soluble tetrazolium salts (WST) assay. From the results, it can be concluded that nanoencapsulated CUR are highly toxic towards cancer cells than free CUR due to their passive diffusions. Meanwhile, Jayaprakasha et al. (2016) also nanoencapsulated CUR using whey protein against colon cancer (SW480) and prostate cancer (LNCaP) cells. The MTT cytotoxic analysis results of these nanoencapsulated CUR showed comparatively higher cytotoxicity than free CUR after 72 h of incubation time. Recently, a core-shell nanocarrier was synthesized by Sedeky et al. (2018) for encapsulating a major alkaloid named piperine (PIP) extracted from *Piper Nigrum L* and to evaluate their cytotoxicity against human brain cancer (Hs683) cell line using MTT assay. The results proved that the core-shell encapsulation of PIP boosted their cytotoxicity towards cancer cells and enhances their solubility to improve PIP release at targeted site.

Several studies such as chitosan nanoencapsulation of CUR and PIP against neuroblastoma (SH-SY5Y) cell line by Baspinar et al. (2018), nanoencapsulation of CUR and chrysin using mesoporous silica against olfactory bulb neuroblastoma (OBFG400) cell line by Lungare et al. (2016), nanoencapsulated resveratrol against murine macrophage (RAW 264.7) cell line by Vankayala et al. (2018), resveratrol-loaded nanocapsules against murine melanoma tumor growth (B16F10) cell lines by Carletto et al. (2016), solid lipid nanoencapsulated lycopene synthesized by Jain et al. (2017), chitosan-alginate nanoencapsulated crocin from *Crocus sativus L.* flowers fabricated by Rahaiee et al. (2017) against breast cancer (MCF-7) cell lines and nanoencapsulated quercetin (Q) via PLGA, chitosan (CS) and polyethylene glycol (PEG) against liver cancer (HepG2) and colorectal cancer (HCT 116) cell lines by Abd-Rabou and Ahmed (2017) showed that nanoencapsulations efficiently increase the release of phytochemicals into the targeted cancer cells and inhibits their growth. Meanwhile, the controlled release of phytochemicals in targeted cells helps to improve their bioavailability and bypass blood-brain barrier which is highly essential for pharmaceutical applications. Table 17.2 is the summarized list of recent nanoencapsulated phytochemicals and their cytotoxic study in in-vitro models.

Table 17.2 List of recent nanoencapsulated phytochemicals and their cytotoxic study in in-vitro models

Phytochemicals	Nanoencapsulations	Models for in-vitro studies	References
Guabiroba extract	PLGA polymer	*Listeria innocua*	Pereira et al. (2018)
Silymarin	Chitosan and γ-PGA polymer	*Bacillus subtilis*, *Staphylococcus aureus*	Lee et al. (2017)
Passion fruit extract by-products	PLGA polymer	*Escherichia coli*, *Listeria innocua*	Oliveira et al. (2017)
Thymus capitatus essential oil	Nanoemulsion	*Escherichia coli*, *Bacillus subtilis*	Benjemaa et al. (2018)
Cymbopogon flexuosus essential oil	Nanoemulsion	*Staphylococcus aureus*, *Candida albicans*, *Cryptococcus grubii*, *Pseudomonas aeruginosa*	da Silva Gündel et al. (2018)
Cymbopogon flexuosus essential oil	Nanoemulsion	*Mycobacterium fortuitum*, *Mycobacterium massiliense*, *Mycobacterium abscessus*	Rossi et al. (2017)
Combretum quadrangulare leaf extracts	Leaf extract nanoparticle	A549 lung cancer cells	Chittasupho and Athikomkulchai (2018)
Cardamom extract	Gelatin nanoparticle	U87MG brain tumor cell lines	Nejat et al. (2017)
Curcumin from *Curcuma longa*	Marinosomes	A549 lung cancer cells, HUVECs human umbilical vein endothelial cells	Ibrahim et al. (2018)
Curcumin	Whey protein	SW480 colon cancer, LNCaP prostate cancer	Jayaprakasha et al. (2016)
Piperine from *Piper Nigrum L.*	Core-shell nanocarrier	Hs683 human brain cancer cell line	Sedeky et al. (2018)
Curcumin and piperine	Chitosan	SH-SY5Y neuroblastoma cell line	Baspinar et al. (2018)
Curcumin and chrysin	Mesoporous silica	OBFG400 olfactory bulb neuroblastoma cell line	Lungare et al. (2016)
Resveratrol	Nanovesicles	RAW 264.7 murine macrophage cell line	Vankayala et al. (2018)
Resveratrol	Nanocapsules	B16F10 murine melanoma tumor growth cell lines	Carletto et al. (2016)
Lycopene	Solid lipid nanocapsules	MCF-7 breast cancer cell lines	Jain et al. (2017)
Crocin from *Crocus sativus L.* flowers	Chitosan-alginate	MCF-7 breast cancer cell lines	Rahaiee et al. (2017)
Quercetin	PLGA, chitosan, polyethylene glycol	HCT116 colorectal and HepG2 liver cancer cells	Abd-Rabou and Ahmed (2017)

17.4 Recent Pharmaceutical Applications of Nanoencapsulated Phytochemicals

Recently, nanoencapsulated phytochemicals have been the topic of interest in disease treatment applications for controlled delivery in the targeted cell or organ. Until now, the potential of nanoencapsulated phytochemicals have been investigated in cancer treatment and antimicrobial activity. Wang et al. (2014) reported the application of nanoencapsulated diet-derived phytochemicals with improved bioavailability and used them in the treatment of diseases (Wang et al. 2014). Polyphenols such as curcumin and gallic acid derivatives possess anticancer activity against colon-rectal cancer. These polyphenols shown to enhance anticancer activity when encapsulated using lipid based colloids (Santos et al. 2013). Even though, nanoformulations are recommended for encapsulation of phytochemicals to enhance efficacy for a long time, their utilization in disease managements have not yet been explored and widely investigated in recent times. Several nanoformulations that are made up of liposomes, lipids, polymers and metals in phytochemical encapsulation for active and passive target applications against lung cancer (Aw-Yong et al. 2018). These nanoencapsulated phytochemicals are used to protect the phytochemical from body fluids, bypass blood-brain barrier and inhibit the growth of cancer cells. Thus, nanoformulation is regarded as a potential drug candidate to control and reduce cancer growth as well as other diseases via phytochemicals without side effects.

Apart from disease management, nanoencapsulated phytochemicals are extensively used in antioxidant and antimicrobial applications. Antioxidant and antimicrobial oleoresin from black pepper was nanoencapsulated using hydroxypropyl beta-cyclodextrin through kneading technology to increase their stability and bioavailability. It can be noted that the nanoformulated oleoresin retained the total phenolic content and enhanced the antioxidant activity. The antimicrobial analysis of nanoformulated oleoresin against *Escherichia coli* K12 and *Salmonella enterica serovar* Typhymurium LT2 revealed that the nanoencapsulated phytochemical inhibits the growth of bacterial colonies at lower concentrations compared to compared to free extracts (Teixeira et al. 2013). Similarly, hydrophobic polyphenols and carotenoids present in *Campomanesia xanthocarpa* O. Berg fruit extract shown to possess antimicrobial and antioxidant properties. Poly ($_{\text{DL}}$-lactide-co-glycolide) (PLGA) was used to encapsulate the hydrophobic phytochemicals which helps to improve their stability and bioavailability to be used in food and pharmaceutical applications. The results showed that the nanoformulation releases carotenoids with an initial burst are followed by a controlled release of the phytochemicals. The nanoencapsulated hydrophobic phytochemicals inhibits the growth of *Listeria innocua* and antioxidant activity at low concentration than free extract which showed that nanoformulated phytochemicals are useful in controlled phytochemical release which reduces their lethal concentration (Pereira et al. 2015).

Prakash et al. (2018) listed the nanomaterials that are used as carriers for essential oils from plants in food systems. Essential oils from plants are widely used as antimicrobial agents as they are less toxic compared to synthetic antimicrobials.

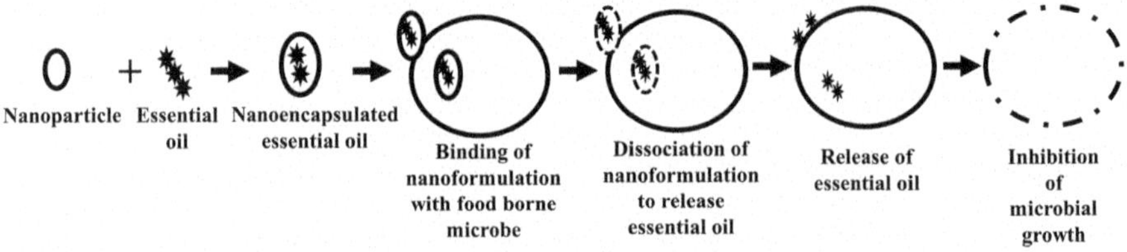

Fig. 17.2 Mechanism of microbial inhibition using nanoencapsulated phytochemical (essential oil)

However, low volatility, stability, water solubility and bioavailability of phytochemicals make them unreliable for biomedical applications. Thus, nanoencapsulations such as nanocarriers, liposomes, nanoemulsions and solid-lipid nanoparticles were used to formulate essential oils to enhance their essential antimicrobial properties. The work also postulated a possible mechanism of nanoencapsulated plant essential oils against food borne microbes (Prakash et al. 2018) as shown in Fig. 17.2. Likewise, flavonoids were encapsulated using chitosan nanoparticles to improve their quorum sensing and inhibitory potential of *E. coli* biofilm formation (Omwenga et al. 2018). These nanoencapsulations are useful as nano-drugs in loading the bioactive phytochemicals and helps in the controlled and targeted release of the drugs. Thus, these novel phytochemicals as drugs can prevent several microbes-mediated diseases by inhibiting the growth and served as a curative agent against several diseases in which nervous system, lung, liver, kidney, cardiovascular system and other organs can be weaken. In future, the potential of phytochemicals can be further explored through the nanoencapsulated phytochemicals in curing almost all the diseases including deadly rare diseases.

17.5 Conclusion

Plant metabolites in the form of phytochemicals are highly beneficial in pharmaceutical and nutraceutical applications due to their improved bioactivity and efficacy in curing wide variety of diseases. However, free and raw phytochemical possess several challenges including high solubility in body fluid which leads to the search for a perfect formulation to protect them. Nanoformulation of phytochemicals via encapsulation provides opportunities for phytocompounds to be delivered in the target site which eventually increases their bioavailability and efficacy in diseases treatments. The advancements in the field of pharmacology and nanotechnology, helps in the discovery of novel phytochemical drugs which can be enhanced by combining both technologies in the form of nanoencapsulations. Nanoencapsulated phytochemicals enhance the ability of phytocompounds and further reduces toxicity which is highly valuable, compared to conventional chemical drugs which cause side effects. Thus, these nanoencapsulated phytochemicals are expected to capture the market of phytochemical drugs and conventional synthetic drugs to cure deadly diseases without any side effects in future.

References

Abd-Rabou AA, Ahmed HH. CS-PEG decorated PLGA nano-prototype for delivery of bioactive compounds: a novel approach for induction of apoptosis in HepG2 cell line. Adv Med Sci. 2017;62:357–67. https://doi.org/10.1016/j.advms.2017.01.003.

Aliakbari F, et al. Formulation and anti-neurotoxic activity of baicalein-incorporating neutral nanoliposome. Colloids Surf B. 2018;161:578–87. https://doi.org/10.1016/j.colsurfb.2017.11.023.

Amjadi S, Ghorbani M, Hamishehkar H, Roufegarinejad L. Improvement in the stability of betanin by liposomal nanocarriers: its application in gummy candy as a food model. Food Chem. 2018;256:156–62. https://doi.org/10.1016/j.foodchem.2018.02.114.

Aw-Yong PY, Gan PH, Sasmita AO, Mak ST, Ling APK. Nanoparticles as carriers of phytochemicals: recent applications against lung cancer. Int J Res Biomed Biotech. 2018;7:1–11.

Baek J-S, Cho C-W. Surface modification of solid lipid nanoparticles for oral delivery of curcumin: improvement of bioavailability through enhanced cellular uptake, and lymphatic uptake. Eur J Pharm Biopharm. 2017;117:132–40. https://doi.org/10.1016/j.ejpb.2017.04.013.

Balaji AB, Pakalapati H, Khalid M, Walvekar R, Siddiqui H. Natural and synthetic biocompatible and biodegradable polymers. In: Engineering WPSiCSa, editor. Biodegradable and biocompatible polymer composites: processing, properties and applications. Duxford: Woodhead Publishing; 2018. p. 3–32.

Barros L, Ferreira IC. Phytochemicals and their effects on human health. Curr Pharm Des. 2017;23:2695–6.

Baspinar Y, Ustundas M, Bayraktar O, Sezgin C. Curcumin and piperine loaded zein-chitosan nanoparticles: development and in-vitro characterisation. Saudi Pharm J. 2018;26:323–34. https://doi.org/10.1016/j.jsps.2018.01.010.

Benjemaa M, Neves MA, Falleh H, Isoda H, Ksouri R, Nakajima M. Nanoencapsulation of Thymus capitatus essential oil: formulation process, physical stability characterization and antibacterial efficiency monitoring. Ind Crop Prod. 2018;113:414–21. https://doi.org/10.1016/j.indcrop.2018.01.062.

Carletto B, et al. Resveratrol-loaded nanocapsules inhibit murine melanoma tumor growth. Colloids Surf B. 2016;144:65–72. https://doi.org/10.1016/j.colsurfb.2016.04.001.

Chittasupho C, Athikomkulchai S. Nanoparticles of Combretum quadrangulare leaf extract induce cytotoxicity, apoptosis, cell cycle arrest and anti-migration in lung cancer cells. J Drug Deliv Sci Technol. 2018;45:378–87. https://doi.org/10.1016/j.jddst.2018.04.003.

da Silva Gündel S, et al. Nanoemulsions containing Cymbopogon flexuosus essential oil: development, characterization, stability study and evaluation of antimicrobial and antibiofilm activities. Microb Pathog. 2018;118:268–76. https://doi.org/10.1016/j.micpath.2018.03.043.

Desai PP, Patravale VB. Curcumin cocrystal micelles – multifunctional nanocomposites for management of neurodegenerative ailments. J Pharm Sci. 2018;107:1143–56. https://doi.org/10.1016/j.xphs.2017.11.014.

Dutta S, Bhattacharjee P. Nanoliposomal encapsulates of piperine-rich black pepper extract obtained by enzyme-assisted supercritical carbon dioxide extraction. J Food Eng. 2017;201:49–56. https://doi.org/10.1016/j.jfoodeng.2017.01.006.

Feng R, Deng P, Song Z, Chu W, Zhu W, Teng F, Zhou F. Glycyrrhetinic acid-modified PEG-PCL copolymeric micelles for the delivery of curcumin. React Funct Polym. 2017;111:30–7. https://doi.org/10.1016/j.reactfunctpolym.2016.12.011.

Gao S, McClements DJ. Formation and stability of solid lipid nanoparticles fabricated using phase inversion temperature method. Colloids Surf A. 2016;499:79–87. https://doi.org/10.1016/j.colsurfa.2016.03.065.

Gorbunova N, Bannikova A, Evteev A, Evdokimov I, Kasapis S. Alginate-based encapsulation of extracts from *beta Vulgaris* cv. beet greens: stability and controlled release under simulated gastrointestinal conditions. LWT Ed. 2018;93:442–9. https://doi.org/10.1016/j.lwt.2018.03.075.

Gupta A, Eral HB, Hatton TA, Doyle PS. Nanoemulsions: formation, properties and applications. Soft Matter. 2016;12:2826–41.

Hamishehkar H, Bahadori MB, Vandghanooni S, Eskandani M, Nakhlband A, Eskandani M. Preparation, characterization and anti-proliferative effects of sclareol-loaded solid lipid nanoparticles on A549 human lung epithelial cancer cells. J Drug Deliv Sci Technol. 2018;45:272–80. https://doi.org/10.1016/j.jddst.2018.02.017.

Ibrahim S, Tagami T, Kishi T, Ozeki T. Curcumin marinosomes as promising nano-drug delivery system for lung cancer. Int J Pharm. 2018;540:40–9. https://doi.org/10.1016/j.ijpharm.2018.01.051.

Jain A, et al. Fabrication and functional attributes of lipidic nanoconstructs of lycopene: an innovative endeavour for enhanced cytotoxicity in MCF-7 breast cancer cells. Colloids Surf B. 2017;152:482–91. https://doi.org/10.1016/j.colsurfb.2017.01.050.

Jayaprakasha GK, Chidambara Murthy KN, Patil BS. Enhanced colon cancer chemoprevention of curcumin by nanoencapsulation with whey protein. Eur J Pharmacol. 2016;789:291–300. https://doi.org/10.1016/j.ejphar.2016.07.017.

Jeevanandam J, Chan YS, Danquah MK. Nano-formulations of drugs: recent developments, impact and challenges. Biochimie. 2016a;128–129:99–112. https://doi.org/10.1016/j.biochi.2016.07.008.

Jeevanandam J, San Chan Y, Danquah MK. Nano-formulations of drugs: recent developments, impact and challenges. Biochimie. 2016b;128:99–112.

Jeevanandam J, Aing YS, Chan YS, Pan S, Danquah MK. Nanoformulation and application of phytochemicals as antimicrobial agents. In: Antimicrobial nanoarchitectonics. Amsterdam: Elsevier; 2017. p. 61–82.

Joseph M, Trinh HM, Mitra AK. Chapter 7 – Peptide and protein-based therapeutic agents∗. In: Emerging nanotechnologies for diagnostics, drug delivery and medical devices. Boston: Elsevier; 2017. p. 145–67. https://doi.org/10.1016/B978-0-323-42978-8.00007-3.

Kamil A, Chen CYO, Blumberg JB. The application of nanoencapsulation to enhance the bioavailability and distribution of polyphenols. In: Nanotechnol funct foods. Chichester: Wiley; 2015. p. 158–74.

Lee J-S, Kim ES, Lee HG. Improving the water solubility and antimicrobial activity of silymarin by nanoencapsulation. Colloids Surf B. 2017;154:171–7.

Lei L, Zhang Y, He L, Wu S, Li B, Li Y. Fabrication of nanoemulsion-filled alginate hydrogel to control the digestion behavior of hydrophobic nobiletin. LWT Food Sci Technol. 2017;82:260–7. https://doi.org/10.1016/j.lwt.2017.04.051.

Li R, et al. Liposomes coated with thiolated chitosan as drug carriers of curcumin. Mater Sci Eng. 2017;80:156–64. https://doi.org/10.1016/j.msec.2017.05.136.

Lungare S, Hallam K, Badhan RKS. Phytochemical-loaded mesoporous silica nanoparticles for nose-to-brain olfactory drug delivery. Int J Pharm. 2016;513:280–93. https://doi.org/10.1016/j.ijpharm.2016.09.042.

Madaan K, Lather V, Pandita D. Evaluation of polyamidoamine dendrimers as potential carriers for quercetin, a versatile flavonoid. Drug Deliv. 2016;23:254–62.

McClements DJ. Recent progress in hydrogel delivery systems for improving nutraceutical bioavailability. Food Hydrocoll. 2017;68:238–45. https://doi.org/10.1016/j.foodhyd.2016.05.037.

Mehrad B, Ravanfar R, Licker J, Regenstein JM, Abbaspourrad A. Enhancing the physicochemical stability of β-carotene solid lipid nanoparticle (SLNP) using whey protein isolate. Food Res Int. 2018;105:962–9. https://doi.org/10.1016/j.foodres.2017.12.036.

Nejat H, et al. Preparation and characterization of cardamom extract-loaded gelatin nanoparticles as effective targeted drug delivery system to treat glioblastoma. React Funct Polym. 2017;120:46–56. https://doi.org/10.1016/j.reactfunctpolym.2017.09.008.

Oliveira DA, Angonese M, Ferreira SRS, Gomes CL. Nanoencapsulation of passion fruit by-products extracts for enhanced antimicrobial activity. Food Bioprod Process. 2017;104:137–46. https://doi.org/10.1016/j.fbp.2017.05.009.

Omwenga E, Hensel A, Shitandi A, Goycoolea F. Chitosan nanoencapsulation of flavonoids enhances their quorum sensing and biofilm formation inhibitory activities against an *E. coli* top 10 biosensor. Colloids Surf B. 2018;164:125–33.

Patel S, Srivastava S, Singh MR, Singh D. Preparation and optimization of chitosan-gelatin films for sustained delivery of lupeol for wound healing. Int J Biol Macromol. 2018;107:1888–97. https://doi.org/10.1016/j.ijbiomac.2017.10.056.

Pereira MC, Hill LE, Zambiazi RC, Mertens-Talcott S, Talcott S, Gomes CL. Nanoencapsulation of hydrophobic phytochemicals using poly (DL-lactide-co-glycolide)(PLGA) for antioxidant and antimicrobial delivery applications: guabiroba fruit (Campomanesia xanthocarpa O. Berg) study. LWT Food Sci Technol. 2015;63:100–7.

Pereira MC, et al. Effect of nanoencapsulation using PLGA on antioxidant and antimicrobial activities of guabiroba fruit phenolic extract. Food Chem. 2018;240:396–404.

Prakash B, Kujur A, Yadav A, Kumar A, Singh PP, Dubey N. Nanoencapsulation: an efficient technology to boost the antimicrobial potential of plant essential oils in food system. Food Control. 2018;89:1–11.

Rabelo CAS, Taarji N, Khalid N, Kobayashi I, Nakajima M, Neves MA. Formulation and characterization of water-in-oil nanoemulsions loaded with açaí berry anthocyanins: insights of degradation kinetics and stability evaluation of anthocyanins and nanoemulsions. Food Res Int. 2018; https://doi.org/10.1016/j.foodres.2018.01.017.

Rahaiee S, Hashemi M, Shojaosadati SA, Moini S, Razavi SH. Nanoparticles based on crocin loaded chitosan-alginate biopolymers: antioxidant activities, bioavailability and anticancer properties. Int J Biol Macromol. 2017;99:401–8. https://doi.org/10.1016/j.ijbiomac.2017.02.095.

Rossi GG, et al. Antibiofilm activity of nanoemulsions of Cymbopogon flexuosus against rapidly growing mycobacteria. Microb Pathog. 2017;113:335–41. https://doi.org/10.1016/j.micpath.2017.11.002.

Santos IS, Ponte BM, Boonme P, Silva AM, Souto EB. Nanoencapsulation of polyphenols for protective effect against colon–rectal cancer. Biotechnol Adv. 2013;31:514–23.

Sechi M, et al. Nanoencapsulation of dietary flavonoid fisetin: formulation and in vitro antioxidant and α-glucosidase inhibition activities. Mater Sci Eng. 2016;68:594–602. https://doi.org/10.1016/j.msec.2016.06.042.

Sedeky AS, Khalil IA, Hefnawy A, El-Sherbiny IM. Development of core-shell nanocarrier system for augmenting piperine cytotoxic activity against human brain cancer cell line. Eur J Pharm Sci. 2018;118:103–12. https://doi.org/10.1016/j.ejps.2018.03.030.

Shu G, Khalid N, Chen Z, Neves MA, Barrow CJ, Nakajima M. Formulation and characterization of astaxanthin-enriched nanoemulsions stabilized using ginseng saponins as natural emulsifiers. Food Chem. 2018;255:67–74. https://doi.org/10.1016/j.foodchem.2018.02.062.

Teixeira BN, Ozdemir N, Hill LE, Gomes CL. Synthesis and characterization of nano-encapsulated black pepper oleoresin using hydroxypropyl beta-cyclodextrin for antioxidant and antimicrobial applications. J Food Sci. 2013;78:1913–20.

Vankayala JS, Battula SN, Kandasamy R, Mariya GA, Franklin MEE, Pushpadass HA, Naik LN. Surfactants and fatty alcohol based novel nanovesicles for resveratrol: process optimization, characterization and evaluation of functional properties in RAW 264.7 macrophage cells. J Mol Liq. 2018;261:387–96. https://doi.org/10.1016/j.molliq.2018.04.058.

Wang S, Su R, Nie S, Sun M, Zhang J, Wu D, Moustaid-Moussa N. Application of nanotechnology in improving bioavailability and bioactivity of diet-derived phytochemicals. J Nutr Biochem. 2014;25:363–76.

Xiao J, Cao Y, Huang Q. Edible nanoencapsulation vehicles for oral delivery of phytochemicals: a perspective paper. J Agric Food Chem. 2017;65:6727–35.

Xie J, Yang Z, Zhou C, Zhu J, Lee RJ, Teng L. Nanotechnology for the delivery of phytochemicals in cancer therapy. Biotechnol Adv. 2016;34:343–53. https://doi.org/10.1016/j.biotechadv.2016.04.002.

Yashaswini PS, Kurrey NK, Singh SA. Encapsulation of sesamol in phosphatidyl choline micelles: enhanced bioavailability and anti-inflammatory activity. Food Chem. 2017;228:330–7. https://doi.org/10.1016/j.foodchem.2017.02.002.

Zhang Z, Zhang R, McClements DJ. Encapsulation of β-carotene in alginate-based hydrogel beads: impact on physicochemical stability and bioaccessibility. Food Hydrocoll. 2016;61:1–10. https://doi.org/10.1016/j.foodhyd.2016.04.036.

Chapter 18
Phytotherapy in Inflammatory Lung Diseases: An Emerging Therapeutic Interventional Approach

Ridhima Wadhwa, Shakti Dhar Shukla, Dinesh Kumar Chellappan, Gaurav Gupta, Trudi Collet, Nicole Hansbro, Brian Oliver, Kylie Williams, Philip Michael Hansbro, Kamal Dua, and Pawan K. Maurya

18.1 Introduction

Inflammation is the defense by the body to uncertain response including allergens or tissue injury responsible to several inflammatory diseases such as allergies, respiratory distress, cancers, metabolic disorders, auto-immune diseases. Respiratory diseases are amongst the most common inflammatory diseases affecting millions all

R. Wadhwa
Faculty of Life Sciences and Biotechnology, South Asian University, Chanakyapuri, New Delhi, India

S. D. Shukla
Priority Research Centre for Healthy Lungs, University of Newcastle & Hunter Medical Research Institute, Callaghan, NSW, Australia

D. K. Chellappan
Department of Life Sciences, School of Pharmacy, International Medical University, Kuala Lumpur, Malaysia

G. Gupta
School of Pharmaceutical Sciences, Jaipur National University, Jaipur, India

T. Collet
Innovative Medicines Group, Institute of Health & Biomedical Innovation, School of Clinical Sciences, Queensland University of Technology, Brisbane, Australia

N. Hansbro
Priority Research Centre for Healthy Lungs, University of Newcastle & Hunter Medical Research Institute, Callaghan, NSW, Australia

School of Life Sciences, Faculty of Science, University of Technology Sydney (UTS), Ultimo, Australia

B. Oliver
School of Life Sciences, Faculty of Science, University of Technology Sydney (UTS), Ultimo, Australia

S. Kumar, C. Egbuna (eds.), *Phytochemistry: An in-silico and in-vitro Update*,
https://doi.org/10.1007/978-981-13-6920-9_18

over the world. With the advancement in research, therapeutics for prevention and cure has improved. According to World Health Organization (WHO), 328 million people have been estimated suffering from acute to severe chronic obstructive pulmonary disease (COPD), out of which 3 million deaths occurs each year and is the third leading cause of mortality worldwide (Cruz 2007). While WHO reports, 235 million people suffer from asthma and is the most common disease in children, 250,000 people died due to asthma (Asher and Pearce 2014). Inflammatory distress impairs the quality of physical as well as physiological well-being of the patients. It also implies an economic and social setback in terms of cost associated with treatment, hospitalization and productivity.

Inhaled and oral anti-inflammatory and bronchodilators have been used as therapeutic since past three decades for the treatment of COPD and asthma. The essential clinical consequence of drugs corresponds with adverse effects. Therefore, there is urge for other effective and safe therapeutics including phytomedicine which could be used for treatment, cure or complement the prevailing treatment corticosteroids and short and long-acting β2-adrenergic agonists (LABA; 5) bronchodilators overuse in case of COPD or asthma specifically in children effects bone growth, increased morbidity and delayed response when improper inhalation method is adopted (Food and Administration 2010; Bender 2002). Thus the use of these inflammation controlling or suppressing steroids and non-steroids. The use of natural anti-inflammatory compounds can be used as therapeutic by elevating pharmaceutical response with low adverse effects. The medicinal plants contain chemical entities or secondary metabolites similar to a synthetic drug which can be substituted as therapeutics. The aim of this chapter is to elucidate plant-based molecule having more desired anti-inflammatory properties compared to bronchodilators and glucocorticoids with a minimal systemic side effect.

K. Williams
Discipline of Pharmacy, Graduate School of Health, University of Technology Sydney, Ultimo, NSW, Australia

P. M. Hansbro
Priority Research Centre for Healthy Lungs, University of Newcastle & Hunter Medical Research Institute, Callaghan, NSW, Australia

School of Life Sciences, Faculty of Science, University of Technology Sydney (UTS), Ultimo, Australia

Centre for Inflammation, Centenary Institute, Sydney, NSW 2050, Australia

K. Dua
Priority Research Centre for Healthy Lungs, University of Newcastle & Hunter Medical Research Institute, Callaghan, NSW, Australia

Discipline of Pharmacy, Graduate School of Health, University of Technology Sydney, Ultimo, NSW, Australia

Centre for Inflammation, Centenary Institute, Sydney, NSW 2050, Australia

P. K. Maurya (✉)
Department of Biochemistry, Central University of Haryana, Mahendergarh, Haryana, India
e-mail: pkmaurya@cuh.ac.in

18.2 Overview of Inflammatory Lung Diseases

18.2.1 Pathophysiology of COPD

COPD is determined by slow progression in airflow limitation with poor reversibility and is related to lung inflammation due to noxious particles and gases (GOLD 2019). The Global Burden of Obstructive Lung Disease (GOLD) defines the disease as; "COPD is a common, preventable and treatable disease that is characterized by persistent respiratory symptoms and airflow limitation that is due to airway and/or alveolar abnormalities, usually caused by significant exposure to noxious particles or gases" (GOLD 2019).

The major symptoms of the disease include a chronic cough, mucus hypersecretion and progressive dyspnea. COPD is a complex and heterogeneous disease that includes several pathologies such as chronic bronchiolitis, fibrosis, small airways obstruction, emphysema, enlarged airspaces and loss of lung elasticity. These pathologies greatly vary between COPD patients and hence, the standard treatment may not benefit all the patients alike. Cigarette smoke accounts significantly in COPD especially in developed countries while air pollution, poor diet and occupational exposure also contribute significantly to the development and progression of the disease (Barnes 2003). The natural history of COPD involves obliteration of ~50% of small airways even before any symptoms appear, or any changes in the lung functions are detectable (Koo et al. 2018). This is in combination to or followed by classic symptoms of mucus hypersecretion in the large airways, that manifests as chronic cough and breathlessness. The pathology of both small and large airways could be further compounded with the presence of emphysema, that usually occurs in around half of the COPD patients. Collectively, these pathophysiological processes result in significantly decreased lung function and increased risk of frequent microbial (bacterial/viral) infections, which is termed as acute exacerbations (AE) (Agustí and Celli 2017). Narrowing of small airways is attributed to increased lymphoid follicle formation and collagen deposition in the outer wall of airways which restricts airways.

18.2.2 Pathophysiology of Asthma

Asthma is a chronic inflammatory disease distinguished by airways hyper-responsiveness causing sporadic, repeated wheezing, breathlessness, tightness of chest and continuous coughing. Asthma reciprocates between predisposed genes and environmental factors. There occurs infiltration in airways by the inflammatory cells leading to vasodilation. Thereby, several structural changes include hypertrophy of smooth muscles, fibrosis, elastic fiber destruction, hyperplasia of goblet cells and glandular submucosa, edema and desquamation of epithelium in airways. Eosinophilic infilteration is the characteristic property of asthma which allows distinguishing from other inflammatory diseases (Foster et al. 2008). Thus, major basic protein (MBP) and eosinophilic cationic protein (ECP) present in the eosinophilic granules can be detected in the broncholavage fluids (BALF) (Kay and Menzies-Gow 2003). A number of mast cells in the mucosal layer does not vary greatly but their number significantly increases in asthmatics. Similarly, T cell count is notable during the symptomatic phase which corresponds to Th2 cells (Barnes et al. 2009). Deposition of granular proteins is responsible for desquamation and destruction of the airways epithelium. In chronic asthma, fibronectin and Type III and V collagen deposit below the basement membrane especially in the central and peripheral bronchi (Davies 2009). However, eosinophil based TGF-ß and PDGF-B are critical in airways remodeling (Ohno et al. 1995).

18.2.3 Cellular and Molecular Mediators of Inflammatory Lung Diseases

Inflammation is crucial in maintaining homeostasis by providing protection against mechanical injury to host cells and tissues, as well as limiting the infection by a variety of microbes, including important pathogens. Inflammatory mediators also play a key role in tissue repair post injury/infection. In the context of lungs, the inflammation is primarily induced for removal of inspired foreign particles (microorganisms, particles) that reach distal airways and alveoli (Reynolds 1987). However, in several diseases, such as chronic inflammatory lung diseases (such as asthma and COPD), the inflammatory response often occur in an uncontrolled, non-resolving manner, thus resulting in a vicious cycle of tissue damage and remodeling that continues throughout the course of the disease (Barnes 2011) as shown in Fig. 18.1.

A myriad of immune cells, cytokines, chemokines and pro-/anti-/regulatory-inflammation genes are often involved in chronic inflammatory lung diseases. Moreover, inflammatory signals that regulate both key inflammatory signaling transduction pathways and target proteins involved in airway and lung inflammation, play a crucial role in the persistence of inflammation in chronic lung inflammatory diseases, which is a hallmark feature of these diseases (Lee and Yang 2013). Importantly, the inflammatory profiles between asthma and COPD differ substantially, which is attributed to differences in the mechanisms involved in the pathogenesis of asthma and COPD. Recent research further indicates the presence of distinct end types (i.e., patients exhibiting a specific inflammatory profile) within both asthmatics and patients with COPD. However, in severe asthma or patients with asthma who are smokers, the inflammatory profiles resemble patients with COPD. Given the increasing global burden of both asthma and COPD, there is an urgent need to understand the complex pathways perpetuating inflammatory process in the lung that may lead to the development of novel therapies for these diseases that are currently not fully treatable and/or difficult and expensive to manage.

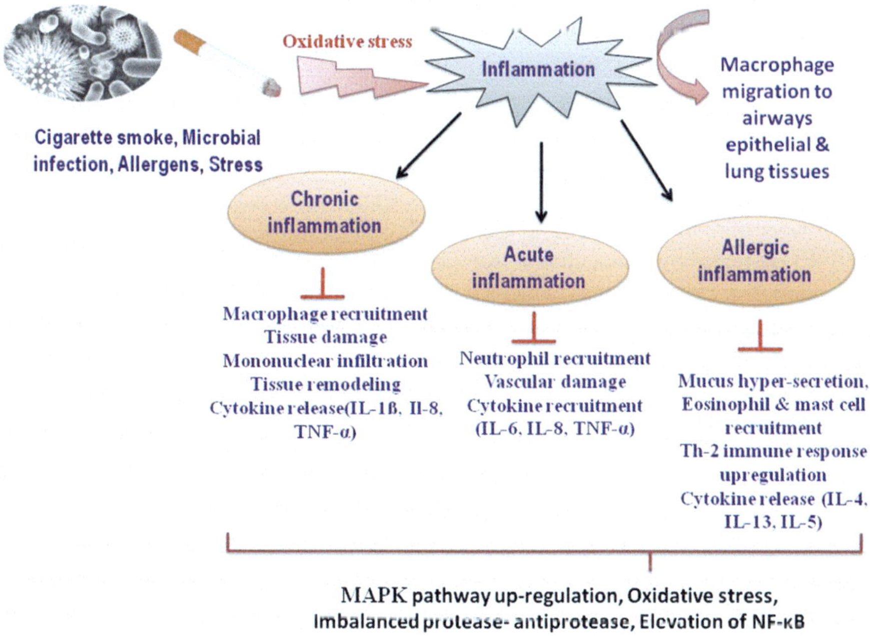

Fig. 18.1 Possible mechanism involved in pulmonary inflammation in respiratory diseases. In pulmonary inflammation elevate transcription of NF- B to the nucleus, thus develop inflammatory processes triggered by aan llergen, cigarette smoke, virus, or bacteria. Therefore, lead to over-expression of inflammatory cytokines and oxidative stress. These effects together culminate with the decline of lung function and with the increase of pulmonary inflammation

18.2.4 Inflammatory Mechanisms in COPD

COPD is a complex disease with variable inflammatory profiles, however, the key inflammatory cell types are macrophages, neutrophils and T cells (Cosio et al. 2009; O'Donnell et al. 2006), with various other immune cell types also playing a role in the progression of the disease (O'Donnell et al. 2006). In addition, the epithelial and lung structural cells play a crucial role in driving inflammatory processes in the COPD lung (Comer et al. 2013). The inflammation in COPD is chronic and involves components of both innate and adaptive immune system, which primarily affects the broncho-alveolar walls of both large and small airways (Suissa et al. 2012). The number of infiltrating macrophages in COPD airways is increased to a larger extent when compared to asthma (O'Donnell et al. 2006). Alveolar macrophages (AM) increase inflammation through the release of chemokines that then attract neutrophils, monocytes and T cells (O'Donnell et al. 2006). AMs also release proteases, especially MMP 9, an elastolytic endopeptidase, that further causes inflammation and associated pathology in COPD (Atkinson et al. 2011). In addition, macrophages also generate and release local ROS burden in the lungs (Boukhenouna et al. 2018). Notably, increased number of pro-inflammatory M1-macrophages and decreased number of M2-macrophages is reported in the small airways of healthy smokers without COPD and COPD patients compared to healthy, non-smoking controls (Eapen et al. 2017). In contrast, the opposite trend was observed in luminal macrophages, whichexhibited a dominant M2-macrophage phenotype in both healthy smokers and COPD subjects (Eapen et al. 2017). This implies that both the number and specific phenotype is important in determining the inflammatory profile in at least cigarette smoke-induced COPD.

Airways fibrosis occurs due to elevated levels of transforming growth factor-ß (TNF-ß) by releasing connective tissue growth factor (CTGF) and causes collagen deposition (Ihn 2002). Chronic mucus hypersecretion also attributes to inflammation around the submucosal glands. In COPD condition, neutrophils, macrophages, CD8-T and Th17 lymphocytes are increased in epithelial cells and fibroblast of alveoli and small airways of lungs. These inflammatory cells stimulate the release of leukotriene (LT) B4, interleukin (IL)-8, tumor necrosis factor-α (TNF-α), Interferon-γ (INF-γ), transforming growth factor beta (TGF- β), chemokines like CC (cysteine cysteine) and CXC (two N terminal cysteines separated by one aminoacid), neutrophil elastase (NE), matrix metalloproteinase (MMP)-2, 9 and 12 responsible for lung damage (Larsson 2007). Cigarette smoke and noxious gases stimulate neutrophils and macrophages alongwith, induces reactive oxygen species (ROS)/reactive nitrogen species (RNS) which further stimulates NF-κB. NF-κB activates cascade involving inflammatory genes. Oxidative stress (OS) causes the production of H_2O_2, this has been elucidated from increased concentration of H_2O_2 in the exhaled breath of COPD patients as well as increased level of carbon monoxide (CO) production in the airways (Nowak et al. 1999). The concentration of 8- isoprostaglandin $F_{2\alpha}$ also increase, its increased levels are estimated in breath and urine of COPD patients (PraticÒ et al. 1998). 4-hydoxy-2-nonenal results in adducts with basic amino acids causing an elevation in lipid peroxidation in lungs (Rahman et al. 2002). Exposure to cigarette smoke inactivates α1-antitrypsin develops early onset of COPD and emphysema (Barnes et al. 2003). Neutrophil elastase causes MUC5AC expression in ROS dependent manner (Fischer and Voynow 2002). Corticosteroids have been extensively used as therapeutic which regulate inflammatory genes by hiring HDAC2 at transcription site and deacetylates hyperacetylated histones. But in diseased state HDAC2 levels are reduced, down-regulates anti-inflammatory response against COPD (Ito et al. 2001). Moreover, chemokines associated with monocytic and neutrophilic inflammation (such as CXCL8, CXCL1 and CCL2) is also detected in COPD patients. An increase in reactive oxygen species (ROS), reactive nitrogen species (RNS) and a number of proteases may further complicate the immunopathological profile of COPD patients (Boukhenouna et al. 2018). Importantly, this inflammatory profile is not restricted to lungs only and may spill over systemically that potentially manifests as co-morbidities such as cardiovascular diseases, lung cancer and arthritis.

18.2.5 Inflammatory Mechanisms in Asthma

Asthma is primarily a chronic disease of the proximal airways, characterized by periodic symptoms, variable airflow obstruction and most importantly, airway inflammation (Bateman et al. 2008). The airway inflammation in asthma is heterogeneous and complex (Barnes 2008), and asthma subtypes have been proposed based on the prevalence of predominant inflammatory cells in the respiratory samples, including induced sputum and endobronchial biopsies (Simpson et al. 2006; Wenzel et al. 1999). The distinct inflammatory profiles in asthmatics, referred to as endotypes, are only now beginning to gain popularity in making clinical decisions while managing the disease. Notably, airway inflammation in asthma could be categorized into four inflammatory subtypes, i.e., neutrophilic, eosinophilic, mixed granulocytic and paucigranulocytic (Simpson et al. 2006). Moreover, sputum color has been proposed as a cost-effective way to identify asthmatics with neutrophilic inflammation (Pabreja et al. 2017). In addition, basophils in sputum and airways are increased in eosinophilic asthma compared with non-eosinophilic asthma phenotypes (Brooks et al. 2017; Suzuki et al. 2017). Basophils have been proposed to play particularly crucial roles in acute allergic reactions, often presented by patients with asthma (Korosec et al. 2018). Similarly, Type 2 innate lymphoid cells (ILC2) has been shown to be significantly increased in eosinophilic mild to moderate asthma (Liu et al. 2015). Importantly, infiltration of airway smooth muscle by tryptase-positive mast cells has been linked with airway hyperresponsiveness in asthmatics (Brightling et al. 2002). A number of immune cells, including dendritic cells, natural killer cells, mast cells, macrophages, and more recently have also been shown to play critical roles in immunopathological mechanisms in asthma onset/ pathogenesis (Barnes 2008).

Conventional adaptive immune mechanisms in asthma are categorized as either type I or type II responses. Type 1 immune response is defined as delayed hypersensitivity and increased production of interleukin (IL)-2 and interferon (IFN)-γ. Type 2 responses primarily depend upon B-cell mediated humoral immunity and overproduction of IL-4, IL-5, IL-9 and IL 13 (Mosmann and Coffman 1989). These inflammatory mediators act through activation of the transcription factors GATA-binding protein 3 (GATA3) and nuclear factor of activated T cells (NFAT) (Barnes 2008). Notably, the "Th2 hypothesis" proposed an upregulation of Th2 response (confirmed with an increase in IL-3, IL-4, IL-5 andgranulocyte-macrophage-colony-stimulating factor [GM-CSF]) and downregulation of Th1 responses in patients with atopic asthma (Robinson et al. 1992). In severe asthma, the disease pathology also involves peripheral airways and is characterized by increased infiltration of neutrophils and CD8+ T cells (Barnes 2008).

In addition to the immune cells, increases in the levels of inflammatory mediators that are implicated in broncho-constriction (such as histamine and cysteinyl leukotrienes) are observed (Laidlaw and Boyce 2012). Moreover, epithelial-derived cytokines, also known as alarmins, such as thymic stromal lymphopoietin (TSLP), a member of the IL-7 family, induces type 2 inflammation by activating dendritic cells and releasing chemokines specific for Th2 cells and ILC2s (Mitchell and O'Byrne 2017). Moreover, other alarmins, such as IL-33 and IL-25/IL-17E, have been shown to play crucial roles in various asthma pathologies, including airway fibrosis, ILC2 biology and mediating interaction between mast cells and neutrophils (Divekar and Kita 2015).

18.3 Drawbacks of Current Therapeutic Strategies

β-adrenoceptor agonists (β-agonists), anticholinergics, theophylline, anti-inflammatory drugs and glucocorticoids reverse broncho-constriction have been used as a treatment therapy for both COPD and asthma. However, pharmacodynamic side effects are associated due to the aging process, mainly due to receptor changes, post-receptor modifications and altered adaptive homeostatic response (McLean and Le Couteur 2004). Adverse effects potentially arise from co-morbidities like heart diseases, muscle problems, metabolic disorders prevail during lung obstruction (Gupta and O'Mahony 2008). The focus of treatment in inflammatory respiratory disease patients is lung function and proper oxygenation, thus osteoporosis is left undiagnosed which further attributes to impaired respiratory function in case of the patients with vertebral compressions and loss of height due to bone degradation (Masala et al. 2014). Glucocorticoid treatment for respiratory obstruction increases the chances of fractures due to osteoporosis, responsible for disability and motility in COPD patients (De Vries et al. 2007). They have successfully used for treating inflammatory diseases. The mechanism of action involves interference of most inflammatory pathways and suppression of inflammation in disease. As of now, glucocorticoids are the most effective therapeutic but are limited by systemic side effects (Barnes 1998). Inhaled glucocorticoids are highly lipophilic, which quickly invade and bind cytosolic receptors and rapidly moves in the nucleus and induces gene transcription resulting in therapeutic effect (Patel and Savjani 2015). Alongwith, the anti-inflammatory and immunosuppressant effects glucocorticoids cause severe partially reversible and non-reversible side effects like hypothalamic-pituitary-adrenal axis suppression, diabetes mellitus, peptic ulcer, Cushing's syndrome, osteoporosis, skin atrophy, psychosis, glaucoma etc. (Patel and Savjani 2015). Therefore, there use is limited by these side effects, and there is an urgent need for the development of compounds exhibiting anti-inflammatory potential of standard glucocorticoids with reduced side effects.

Use of ß-agonist broncho-constrictors can decrease forced expiratory volume (FEV) by 20% especially with the increase in age and salbutamol consumption (Barisione et al. 2010). This phenomenon has been reported in 8% of patients as reported by Barisione, Giovanni, et al. (2010). This condition tends to get severe due to nebulization and metered dose inhaler (MDI) as it can cause death. Regular use of ß-agonists can reduce bronchodilator response due to down-regulation of ß-adrenoceptor (Barisione et al. 2010). Thus, they can increase asthma exacerbations and deaths due to asthma upto two to fourfolds. Anti-cholinergics are widely associated to treat COPD. Inhalation of anti-cholinergics have a broad spectrum of tolerance and are poorly absorbed. Some of the side effects caused are dry mouth, unpleasant taste, occasional cough, paradoxical broncho-constriction, mild congtive decline (Gross 2006).

Theophylline is another therapeutic with major limitation due to its side effects including vomiting, gastric problems and headache (Barr et al. 2003). It is a narrow spectrum drug with altered pharmacokinetic and drug interaction, so it s risky for use of older population. In a randomized study with 116 patient serum having >30 mg/L theophylline concentration, it was found that 7% had life-threatening toxicity, 50% with moderate toxicity and 38% with mild toxicity (Sessler 1990). It has been reported to cause cardiac problems including myocardial ischaemia, tachycardias, arrhythmias and sudden death. Prolonged use of theophylline has been reported to activate central A_1 receptors and causes tremors, headaches, nervousness, depression and behavioral changes (Gupta and O'Mahony 2008). However, the dose concentration should be closely monitored; poly-pharmacy or alternative therapeutics should be considered for treating COPD and asthma.

18.4 Emerging Need for Alternative Therapy

Human beings, from time immemorial, have been using various substances from edible roots, bitter tonics, animal parts to inorganic minerals as sources of medicinal and therapeutic substances. These also included practices that ranged from blood-letting, piercing, manipulating bodily organs to extreme rituals. Drug therapy took different dimensions through the ages from spiritual, astronomical, ritual and astrological forms. Healers and priests were highly sought after, as they were considered gods. The rise and fall of great empires and kingdoms resulted in the spread of such therapies far and wide. Nevertheless, there came a time when such primitive practices and beliefs took a beating when man invented newer and advanced machines (PBRN 2011). Technologies and equipment never known to mankind started emerging, and widespread interest in developing newer and safer therapies took root. Chemicals and more potent substances were developed for diseases even as safety issues of such products became a monumental concern. However, as time changed, and as the number of adverse events and untoward effects of drugs multiplied, there arose a renewed interest in traditional methods and therapies. Such therapies were together classified and were called as alternative therapies. The National Center for Complementary and Alternative Medicine (NCCAM) defines such therapies as alternative therapies which complement the conventional system of medicine. These are generally followed as traditional therapies which are non-orthodox in nature (Kayne 2002). These diverse medical systems, according to the World Health Organization (WHO) are not generally adopted into an existing healthcare system (Renzella et al. 2018). There are several reasons contributed towards the current renewed interest for phytotherapy. Firstly, plant kingdom consists of a vast army of different plant species of which only a minor fraction of these have been scientifically studied. A huge amount of potent therapeutic principles still lies untapped. Secondly, the occurrence of life-threatening adverse effects reported with the use of conventional medicine has paved the way for the re-emergence of traditional and alternative therapies. Among several alternative therapies, phytotherapy has gained much importance in the recent decades, as plants have an unlimited arsenal of potent untapped therapeutic principles within them (Freeman and Lawlis 2004). A separate study carried out in 2003 reported that a large proportion of adults in developed countries namely the US, the UK, and Australia have turned back to alternative therapies, primarily phytotherapy (Ismail and Chan 2004) to manage illnesses. The National Health Interview Survey (NHIS) reported that more than 35% of adults in America use alternative therapy to treat diseases and health concerns (Kendall 2013). Thirdly, there is a belief that the symptoms of diseases and ailments are not completely removed by the drugs in the conventional system. On the other hand, plants and plant products used in several systems of medicine have evidence of curing the symptoms of diseases completely (Nahin et al. 2010). Fourthly, phytotherapy is believed to be safe, as it primarily involves plants and plant products of natural origin. Nearly 70% of the world population consumes herbal products, consisting either fresh or dried plant parts, as reported by the WHO (Renzella et al. 2018). Although there are several reports published in the scientific literature on the harmful effects of herbal products, the belief still exists that phytopharmaceuticals are safe and has negligible side effects (Merican 2002; Hamilton et al. 2016). Lastly, the cost of herbal products is much cheaper compared to conventional drugs. Majority of the world's population lives in rural areas. Primarily, people residing in rural areas find the cost of conventional therapy unaffordable. All these factors have highlighted the need for alternative therapy.

18.5 Herbal Products: Experimental and Clinical Approaches

The incidence of pulmonary diseases has been increasing and affects millions of peoples all around the world. The most common types of pulmonary diseases are asthma and COPD which are linked with a high rate of mortality and morbidity. In both of these pulmonary diseases, pulmonary inflammation is common that might be acute or chronic and there many inflammatory mediators are involved in pathophysiology of both diseases. At present lots of bronchodilator, oral and/or inhaled anti-inflammatory medications are available for the treatment of both conditions. Even though available medications have a potent efficacy but therapeutic options are limited in number and not without side effects (May and Li 2015; Cukic et al. 2012; Qureshi et al. 2014; Montuschi 2006). Hence it is required to search new therapeutic options including herbal products that might be used as alternative therapeutic options for asthma and COPD.

The plant can be defined as a biosynthetic laboratory which produces a number of secondary metabolites including flavonoids, alkaloids, saponins, glycosides, tannins, resins etc. (Senguttuvan et al. 2014). These compounds are responsible for various types of medicinal property.

18.5.1 Evidence of Herbal Products in the Treatment of Asthma

Asthma is a chronic pulmonary inflammatory disease characterized by recurrent episodes of wheezing, chest tightness, and breathlessness that are usually related with variable airflow obstruction (Kaplan et al. 2009; Ukena et al. 2008).

As per earlier reported literature flavonoids isolated from the various plant extracts have shown strong anti-inflammatory effects (Panche et al. 2016). Sakuranetin, a flavonoid isolated from Baccharisretusa has decreased the number of inflammatory cells in asthma animal model. According to the Toledo et al., sakuranetin was significantly reduced the T2cytokines like RANTES, IL-5, and Eotaxintreated experimental animal (Toledo et al. 2013). Other flavonoids such as kuwanon G, kaempferol-3-O-rhamnoside, kaempferol, wogonin, Naringin, Curcumin (Fig. 18.2) etc. were also reported for anti inflammatory effects in asthmatic animal models. They were exhibited anti-inflammatory effect by inhibiting an increasing in the number of various types of cytokines such as L-4, IL-5, and IL-13. Some flavonoids act through reduces airway allergic responses, such as eosinophilia, specific IgE, lympho-cytein filtration (Santana et al. 2016; Wang et al. 2018; Rho et al. 2011).

18.5.2 Evidence of Herbal Products in the Treatment of COPD

COPD is majorly responsible for chronic respiratory mortality and morbidity in all around the world and it is characterized by persistent limitation of airflow which is associated with an increased inflammatory response to toxic gases or particles. This type of inflammatory response generally promotes destructions of parenchymal cells which further leads to emphysema and in some cases also causing a small airways fibrosis. Various types of environmental pollutants or cigarette smoking stimulates the macrophages in the epithelial cells of the lung to generate a ROS and reactive nitrogen species in more amount which causes an imbalance in the system. The imbalance protease/antiprotease is recognized as the main mechanism involved in the development of emphysema (van Eeden and Sin 2008; Barnes 2007; King 2015).

Numerous antioxidative agents including polyphenols, thiols obtained from the herbal products and many other secondary natural compounds like resveratrol,

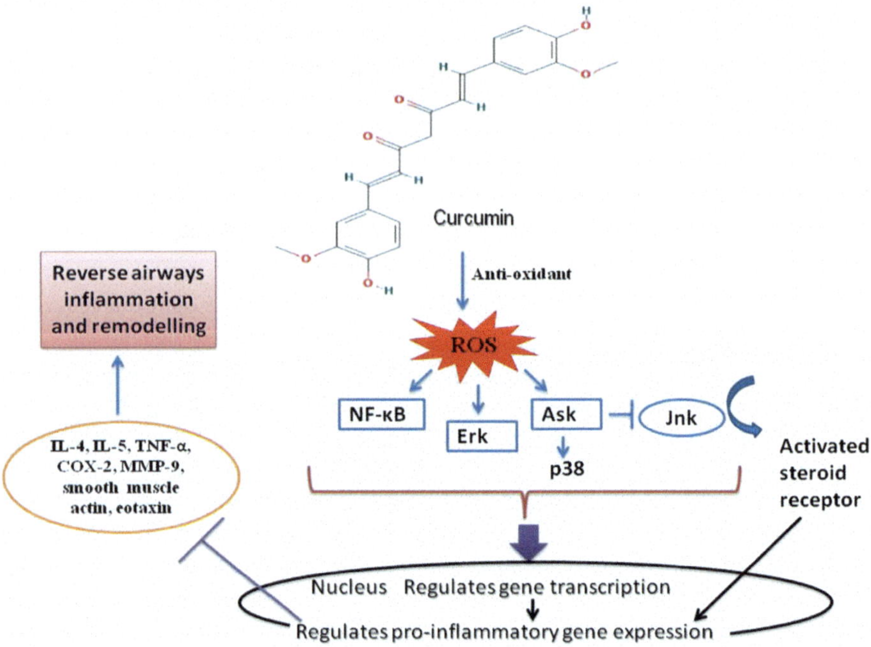

Fig. 18.2 Curcumin as the mediator for the regulation of pro-inflammatory gene expression in asthma. Except for pro-inflammatory gene suppression, it also suppresses secondary genes involved in inflammatory response. It down-regulates the expression of pro-inflammatory mediators, such as matrix meatalloproteinase-9 (MMP-9), tumor necrosis factor (TNF-α), chemokines. It also modulates kinase signaling pathways such as Jun N-terminal kinases (JNK), p38, serine/threonine-specific protein kinase (AKT), Janus kinase (JAK), extracellular signal-regulated kinases (ERK). Thereby, responds to airways inflammation

quercetin, and curcumin have been investigated in emphysema models (Rahman 2006; Niedzwiecki et al. 2016). Considering the pathophysiology of COPD, substances that reduce oxidative stress and imbalance in protease/antiprotease should be considered to be valuable in COPD (Pandey et al. 2017; Lomas 2016). In viewing of herbal medicines, some studies have examined the effect of these compounds in the development of emphysema. As per an earlier study published by Lee et al., herbal formulation PM014 significantly decreases IL-6 and TNF-α levels as well as inflammatory cells in an elastase + LPS-induced emphysema model (Lee et al. 2012). In one of the study, Lee et al. have observed the effect of traditional herb, *Callicarpa japonica* in the treatment of inflammatory diseases. The outcome of the study reported that extract of *Callicarpa japonica* successfully decreased oxidative stress, productions of cytokines including TNF-α and IL-6 as well as decreasedinfiltration of neutrophil in the cigarette smoke animal model (Lee et al. 2015).

18.6 Conclusion

The current therapies for inflammatory respiratory diseases (COPD and Asthma) are not sufficient. Phytomedicine is one of the essential resources from ancient times which is evident from the traditional medicine system including Ayurveda, Siddha of India and the Chinese medicine system. Several of these plants have bronchodilatory, relaxant, anti-cholinergic, antitussive, anti-spasmodic and mucociliary clearance action. Chronic diseases have been treated by using crude extracts in a traditional system which provides better therapeutic effect compared to an isolated constituent, due to synergic activity with other constituents. For the clinical use, the isolated constituent or the crude extract needs extensive screening at molecular (IL-6, IL-8, NFκB, ROS) and cellular (macrophages, eosinophils, andneutrophils) level in-vitro and in-vivo. Furthermore, the toxicity should also be considered and for better delivery of the lead, molecule can be associated with nanotechnology for better therapeutic effects due to high surface area ratio and reduced toxic effects.

References

Agustí A, Celli B. Natural history of COPD: gaps and opportunities. ERJ Open Res. 2017;3(4). https://doi.org/10.1183/23120541.00117-2017.

Asher I, Pearce N. Global burden of asthma among children. Int J Tuberc Lung Dis. 2014;18 (11):1269–78.

Atkinson JJ, Lutey BA, Suzuki Y, Toennies HM, Kelley DG, Kobayashi DK, Ijem WG, Deslee G, Moore CH, Jacobs ME, Conradi SH, Gierada DS, Pierce RA, Betsuyaku T, Senior RM. The role of matrix metalloproteinase-9 in cigarette smoke-induced emphysema. Am J Respir Crit Care Med. 2011;183(7):876–84. https://doi.org/10.1164/rccm.201005-0718OC.

Barisione G, Baroffio M, Crimi E, Brusasco V. Beta-adrenergic agonists. Pharmaceuticals. 2010;3 (4):1016–44.

Barnes PJ. Anti-inflammatory actions of glucocorticoids: molecular mechanisms. Clin Sci. 1998;94 (6):557–72.

Barnes PJ. New concepts in chronic obstructive pulmonary disease. Annu Rev Med. 2003;54 (1):113–29.

Barnes PJ. Chronic obstructive pulmonary disease: a growing but neglected global epidemic. PLoS Med. 2007;4(5):e112. https://doi.org/10.1371/journal.pmed.0040112.

Barnes PJ. Immunology of asthma and chronic obstructive pulmonary disease. Nat Rev Immunol. 2008;8:183. https://doi.org/10.1038/nri2254.

Barnes PJ. Similarities and differences in inflammatory mechanisms of asthma and COPD. Breathe. 2011;7(3):229–38. https://doi.org/10.1183/20734735.026410.

Barnes PJ, Shapiro S, Pauwels R. Chronic obstructive pulmonary disease: molecular and cellular mechanisms. Eur Respir J. 2003;22(4):672–88.

Barnes PJ, Drazen JM, Rennard SI, Thomson NC. Asthma and COPD: basic mechanisms and clinical management. Amsterdam: Elsevier; 2009.

Barr RG, Rowe BH, Camargo CA. Methylxanthines for exacerbations of chronic obstructive pulmonary disease. Cochrane Database Syst Rev 2003; 2: CD002168. https://doi.org/10.1002/14651858.CD002168.

Bateman ED, Hurd SS, Barnes PJ, Bousquet J, Drazen JM, FitzGerald JM, Gibson P, Ohta K, O'Byrne P, Pedersen SE, Pizzichini E, Sullivan SD, Wenzel SE, Zar HJ. Global strategy for asthma management and prevention: GINA executive summary. Eur Respir J. 2008;31(1):143–78. https://doi.org/10.1183/09031936.00138707.

Bender BG. Overcoming barriers to nonadherence in asthma treatment. J Allergy Clin Immunol. 2002;109(6):S554–9.

Boukhenouna S, Wilson MA, Bahmed K, Kosmider B. Reactive oxygen species in chronic obstructive pulmonary disease. Oxidative Med Cell Longev. 2018;2018:5730395. https://doi.org/10.1155/2018/5730395.

Brightling CE, Bradding P, Symon FA, Holgate ST, Wardlaw AJ, Pavord ID. Mast-cell infiltration of airway smooth muscle in asthma. N Engl J Med. 2002;346(22):1699–705. https://doi.org/10.1056/NEJMoa012705.

Brooks CR, van Dalen CJ, Hermans IF, Gibson PG, Simpson JL, Douwes J. Sputum basophils are increased in eosinophilic asthma compared with non-eosinophilic asthma phenotypes. Allergy. 2017;72(10):1583–6. https://doi.org/10.1111/all.13185.

Comer DM, Kidney JC, Ennis M, Elborn JS. Airway epithelial cell apoptosis and inflammation in COPD, smokers and nonsmokers. Eur Respir J. 2013;41(5):1058–67. https://doi.org/10.1183/09031936.00063112.

Cosio MG, Saetta M, Agusti A. Immunologic aspects of chronic obstructive pulmonary disease. N Engl J Med. 2009;360(23):2445–54. https://doi.org/10.1056/NEJMra0804752.

Cruz AA. Global surveillance, prevention and control of chronic respiratory diseases: a comprehensive approach. Geneva: World Health Organization; 2007.

Cukic V, Lovre V, Dragisic D, Ustamujic A. Asthma and chronic obstructive pulmonary disease (COPD) – differences and similarities. Mater Sociomed. 2012;24(2):100–5. https://doi.org/10.5455/msm.2012.24.100-105.

Davies DE. The role of the epithelium in airway remodeling in asthma. Proc Am Thorac Soc. 2009;6(8):678–82.

De Vries F, Bracke M, Leufkens HG, Lammers JWJ, Cooper C, Van Staa TP. Fracture risk with intermittent high-dose oral glucocorticoid therapy. Arthritis Rheum. 2007;56(1):208–14.

Divekar R, Kita H. Recent advances in epithelium-derived cytokines (IL-33, IL-25, and thymic stromal lymphopoietin) and allergic inflammation. Curr Opin Allergy Clin Immunol. 2015;15(1):98–103. https://doi.org/10.1097/aci.0000000000000133.

Eapen MS, Hansbro PM, McAlinden K, Kim RY, Ward C, Hackett TL, Walters EH, Sohal SS. Abnormal M1/M2 macrophage phenotype profiles in the small airway wall and lumen in smokers and chronic obstructive pulmonary disease (COPD). Sci Rep. 2017;7(1):13392. https://doi.org/10.1038/s41598-017-13888-x.

Fischer BM, Voynow JA. Neutrophil elastase induces MUC 5AC gene expression in airway epithelium via a pathway involving reactive oxygen species. Am J Respir Cell Mol Biol. 2002;26(4):447–52.

Foster PS, Rosenberg HF, Asquith KL, Kumar RK. Targeting eosinophils in asthma. Curr Mol Med. 2008;8(6):585–90.

Freeman L, Lawlis G. Exercise as an alternative therapy. Mosby's complementary and alternative medicine: a research-based approach. 2nd ed. St Louis: Mosby; 2004. p. 481–518.

GOLD. Global initiative for chronic obstructive lung disease-global strategy for the diagnosis, management, and prevention of COPD. 2019.

Gross NJ. Anticholinergic agents in asthma and COPD. Eur J Pharmacol. 2006;533(1–3):36–9.

Gupta P, O'Mahony MS. Potential adverse effects of bronchodilators in the treatment of airways obstruction in older people. Drugs Aging. 2008;25(5):415–43.

Hamilton WL, Doyle C, Halliwell-Ewen M, Lambert G. Public health interventions to protect against falsified medicines: a systematic review of international, national and local policies. Health Policy Plan. 2016;31(10):1448–66.

Ihn H. Pathogenesis of fibrosis: role of TGF-β and CTGF. Curr Opin Rheumatol. 2002;14(6):681–5.

Ismail I, Chan S. Knowledge and practice of complementary medicine amongst public primary care clinic doctors in Kinta district, Perak. Med J Malaysia. 2004;59(1):4–10.

Ito K, Watanabe S, Kharitonov S, Hanazawa T, Adcock I, Barnes P. Histone deacetylase activity and gene expression in COPD patients. Eur Respir J. 2001;18:316S.

Kaplan AG, Balter MS, Bell AD, Kim H, McIvor RA. Diagnosis of asthma in adults. CMAJ Can Med Ass J. 2009;181(10):E210–20. https://doi.org/10.1503/cmaj.080006.

Kay AB, Menzies-Gow A. Eosinophils and interleukin-5: the debate continues. Am J Respir Crit Care Med. 2003;167:1586–7.

Kayne SB. Complementary therapies for pharmacists. London: Pharmaceutical Press; 2002.

Kendall DE. National Center for Complementary and Alternative Medicine (NCCAM) introduction to acupuncture with disclaimer. Schweiz Z Ganzheitsmedizin Swiss J Integr Med. 2013;25 (4):241–6.

King PT. Inflammation in chronic obstructive pulmonary disease and its role in cardiovascular disease and lung cancer. Clin Transl Med. 2015;4(1):68. https://doi.org/10.1186/s40169-015-0068-z.

Koo HK, Vasilescu DM, Booth S, Hsieh A, Katsamenis OL, Fishbane N, Elliott WM, Kirby M, Lackie P, Sinclair I, Warner JA, Cooper JD, Coxson HO, Pare PD, Hogg JC, Hackett TL. Small airways disease in mild and moderate chronic obstructive pulmonary disease: a cross-sectional study. Lancet Respir Med. 2018;6(8):591–602. https://doi.org/10.1016/s2213-2600(18)30196-6.

Korosec P, Gibbs BF, Rijavec M, Custovic A, Turner PJ. Important and specific role for basophils in acute allergic reactions. Clin Exp Allergy. 2018;48(5):502–12. https://doi.org/10.1111/cea.13117.

Laidlaw TM, Boyce JA. Cysteinyl leukotriene receptors, old and new; implications for asthma. Clin Exp Allergy. 2012;42(9):1313–20. https://doi.org/10.1111/j.1365-2222.2012.03982.x.

Larsson K. Aspects on pathophysiological mechanisms in COPD. J Intern Med. 2007;262 (3):311–40.

Lee I-T, Yang C-M. Inflammatory signalings involved in airway and pulmonary diseases. Mediat Inflamm. 2013;2013:12. https://doi.org/10.1155/2013/791231.

Lee H, Kim Y, Kim HJ, Park S, Jang YP, Jung S, Jung H, Bae H. Herbal formula, PM014, attenuates lung inflammation in a murine model of chronic obstructive pulmonary disease. Evid Based Complement Alternat Med: eCAM. 2012;2012:769830. https://doi.org/10.1155/2012/769830.

Lee JW, Shin NR, Park JW, Park SY, Kwon OK, Lee HS, Hee Kim J, Lee HJ, Lee J, Zhang ZY, Oh SR, Ahn KS. *Callicarpa japonica* Thunb. attenuates cigarette smoke-induced neutrophil inflammation and mucus secretion. J Ethnopharmacol. 2015;175:1–8. https://doi.org/10.1016/j.jep.2015.08.056.

Liu T, Wu J, Zhao J, Wang J, Zhang Y, Liu L, Cao L, Liu Y, Dong L. Type 2 innate lymphoid cells: a novel biomarker of eosinophilic airway inflammation in patients with mild to moderate asthma. Respir Med. 2015;109(11):1391–6. https://doi.org/10.1016/j.rmed.2015.09.016.

Lomas DA. Does protease-antiprotease imbalance explain chronic obstructive pulmonary disease? Ann Am Thorac Soc. 2016;13(Suppl 2):S130–7. https://doi.org/10.1513/AnnalsATS.201504-196KV.

Masala S, Magrini A, Taglieri A, Nano G, Chiaravalloti A, Calabria E, Di Trapano R, Pietroiusti A, Simonetti G. Chronic obstructive pulmonary disease (COPD) patients with osteoporotic vertebral compression fractures (OVCFs): improvement of pulmonary function after percutaneous vertebroplasty (VTP). Eur Radiol. 2014;24(7):1577–85.

May SM, Li JTC. Burden of chronic obstructive pulmonary disease: healthcare costs and beyond. Allergy Asthma Proc. 2015;36(1):4–10. https://doi.org/10.2500/aap.2015.36.3812.

McLean AJ, Le Couteur DG. Aging biology and geriatric clinical pharmacology. Pharmacol Rev. 2004;56(2):163–84.

Merican I. Traditional/complementary medicine: the way ahead. Med J Malays. 2002;57(3):261–5.

Mitchell PD, O'Byrne PM. Epithelial-derived cytokines in asthma. Chest. 2017;151(6):1338–44. https://doi.org/10.1016/j.chest.2016.10.042.

Montuschi P. Pharmacological treatment of chronic obstructive pulmonary disease. Int J Chron Obstruct Pulmon Dis. 2006;1(4):409–23.

Mosmann TR, Coffman RL. TH1 and TH2 cells: different patterns of lymphokine secretion lead to different functional properties. Annu Rev Immunol. 1989;7:145–73. https://doi.org/10.1146/annurev.iy.07.040189.001045.

Nahin RL, Dahlhamer JM, Stussman BJ. Health need and the use of alternative medicine among adults who do not use conventional medicine. BMC Health Serv Res. 2010;10(1):220.

Niedzwiecki A, Roomi MW, Kalinovsky T, Rath M. Anticancer efficacy of polyphenols and their combinations. Nutrients. 2016;8(9):552. https://doi.org/10.3390/nu8090552.

Nowak D, Kasielski M, Antczak A, Pietras T, Bialasiewicz P. Increased content of thiobarbituric acid-reactive substances and hydrogen peroxide in the expired breath condensate of patients with stable chronic obstructive pulmonary disease: no significant effect of cigarette smoking. Respir Med. 1999;93(6):389–96.

O'Donnell R, Breen D, Wilson S, Djukanovic R. Inflammatory cells in the airways in COPD. Thorax. 2006;61(5):448–54. https://doi.org/10.1136/thx.2004.024463.

Ohno I, Nitta Y, Yamauchi K, Hoshi H, Honma M, Woolley K, O'Byrne P, Dolovich J, Jordana M, Tamura G. Eosinophils as a potential source of platelet-derived growth factor B-chain (PDGF-B) in nasal polyposis and bronchial asthma. Am J Respir Cell Mol Biol. 1995;13(6):639–47.

Pabreja K, Gibson P, Lochrin AJ, Wood L, Baines KJ, Simpson JL. Sputum colour can identify patients with neutrophilic inflammation in asthma. BMJ Open Respir Res. 2017;4(1):e000236. https://doi.org/10.1136/bmjresp-2017-000236.

Panche AN, Diwan AD, Chandra SR. Flavonoids: an overview. J Nutr Sci. 2016;5:e47. https://doi.org/10.1017/jns.2016.41.

Pandey KC, De S, Mishra PK. Role of proteases in chronic obstructive pulmonary disease. Front Pharmacol. 2017;8:512. https://doi.org/10.3389/fphar.2017.00512.

Patel SS, Savjani JK. Systematic review of plant steroids as potential antiinflammatory agents: current status and future perspectives. J Phytopharmacol. 2015;4(2):121–5.

PBRN A. Pharmacotherapy: a pathophysiologic approach. 2011.

PraticÒ D, Basili S, Vieri M, Cordova C, Violi F, Fitzgerald GA. Chronic obstructive pulmonary disease is associated with an increase in urinary levels of isoprostane F2 α-III, an index of oxidant stress. Am J Respir Crit Care Med. 1998;158(6):1709–14.

Qureshi H, Sharafkhaneh A, Hanania NA. Chronic obstructive pulmonary disease exacerbations: latest evidence and clinical implications. Ther Adv Chronic Dis. 2014;5(5):212–27. https://doi.org/10.1177/2040622314532862.

Rahman I. Antioxidant therapies in COPD. Int J Chron Obstruct Pulmon Dis. 2006;1(1):15–29.

Rahman I, van Schadewijk AA, Crowther AJ, Hiemstra PS, Stolk J, MacNee W, De Boer WI. 4-Hydroxy-2-nonenal, a specific lipid peroxidation product, is elevated in lungs of patients with chronic obstructive pulmonary disease. Am J Respir Crit Care Med. 2002;166(4):490–5.

Renzella J, Townsend N, Jewell J, Breda J, Roberts N, Rayner M, Wickramasinghe K. What national and subnational interventions and policies based on Mediterranean and Nordic diets are recommended or implemented in the WHO European Region, and is there evidence of effectiveness in reducing noncommunicable diseases? 2018.

Reynolds HY. Lung inflammation: normal host defense or a complication of some diseases? Annu Rev Med. 1987;38:295–323. https://doi.org/10.1146/annurev.me.38.020187.001455.

Rho HS, Ghimeray AK, Yoo DS, Ahn SM, Kwon SS, Lee KH, Cho DH, Cho JY. Kaempferol and kaempferol rhamnosides with depigmenting and anti-inflammatory properties. Molecules (Basel, Switzerland). 2011;16(4):3338–44. https://doi.org/10.3390/molecules16043338.

Robinson DS, Hamid Q, Ying S, Tsicopoulos A, Barkans J, Bentley AM, Corrigan C, Durham SR, Kay AB. Predominant TH2-like bronchoalveolar T-lymphocyte population in atopic asthma. N Engl J Med. 1992;326(5):298–304. https://doi.org/10.1056/nejm199201303260504.

Santana FPR, Pinheiro NM, Mernak MIB, Righetti RF, Martins MA, Lago JHG, Lopes FDTQDS, Tibério IFLC, Prado CM. Evidences of herbal medicine-derived natural products effects in inflammatory lung diseases. Mediat Inflamm. 2016;2016:2348968. https://doi.org/10.1155/2016/2348968.

Senguttuvan J, Paulsamy S, Karthika K. Phytochemical analysis and evaluation of leaf and root parts of the medicinal herb, *Hypochaeris radicata* L. for in vitro antioxidant activities. Asian Pac J Trop Biomed. 2014;4(Suppl 1):S359–67. https://doi.org/10.12980/APJTB.4.2014C1030.

Sessler CN. Theophylline toxicity: clinical features of 116 consecutive cases. Am J Med. 1990;88(6):567–76.

Simpson JL, Scott R, Boyle MJ, Gibson PG. Inflammatory subtypes in asthma: assessment and identification using induced sputum. Respirology (Carlton, Vic). 2006;11(1):54–61. https://doi.org/10.1111/j.1440-1843.2006.00784.x.

Suissa S, Dell'Aniello S, Ernst P. Long-term natural history of chronic obstructive pulmonary disease: severe exacerbations and mortality. Thorax. 2012;67(11):957–63. https://doi.org/10.1136/thoraxjnl-2011-201518.

Suzuki Y, Wakahara K, Nishio T, Ito S, Hasegawa Y. Airway basophils are increased and activated in eosinophilic asthma. Allergy. 2017;72(10):1532–9. https://doi.org/10.1111/all.13197.

Toledo AC, Sakoda CPP, Perini A, Pinheiro NM, Magalhães RM, Grecco S, Tibério IFLC, Câmara NO, Martins MA, Lago JHG, Prado CM. Flavonone treatment reverses airway inflammation and remodelling in an asthma murine model. Br J Pharmacol. 2013;168(7):1736–49. https://doi.org/10.1111/bph.12062.

U.S. Food and Drug Administration. FDA Drug Safety Communication: drug labels now contain updated recommendations on the appropriate use of long-acting inhaled asthma medications called Long Acting Beta Agonists (LABAs). 2010. Retrieved from URL.

Ukena D, Fishman L, Niebling W-B. Bronchial asthma: diagnosis and long-term treatment in adults. Deutsches Arzteblatt Int. 2008;105(21):385–94. https://doi.org/10.3238/arztebl.2008.0385.

van Eeden SF, Sin DD. Chronic obstructive pulmonary disease: a chronic systemic inflammatory disease. Respir Int Rev Thorac Dis. 2008;75(2):224–38. https://doi.org/10.1159/000111820.

Wang J, Fang X, Ge L, Cao F, Zhao L, Wang Z, Xiao W. Antitumor, antioxidant and anti-inflammatory activities of kaempferol and its corresponding glycosides and the enzymatic preparation of kaempferol. PLoS One. 2018;13(5):–e0197563. https://doi.org/10.1371/journal.pone.0197563.

Wenzel SE, Schwartz LB, Langmack EL, Halliday JL, Trudeau JB, Gibbs RL, Chu HW. Evidence that severe asthma can be divided pathologically into two inflammatory subtypes with distinct physiologic and clinical characteristics. Am J Respir Crit Care Med. 1999;160(3):1001–8. https://doi.org/10.1164/ajrccm.160.3.9812110.

Part II
Phytochemicals and in-vivo Updates

Chapter 19
In-silico Tools in Phytochemical Research

Ajeet Singh, Shafaque Zahra, and Shailesh Kumar

19.1 Introduction

Phytochemicals (Greek, *phyton* means plants) are chemical compounds procured from plants. They are essential for the overall metabolic functioning of the plants, and defend them against microbial infection, infestation by pests, herbivores as well as changes in environmental conditions, thus they have disease preventive and healing properties. Apart from affecting the physiology of plants, phytochemicals show different types of effects on other organisms exposed to these chemicals. They can be beneficial or deleterious to other organisms including human beings. Although they are not essential nutrients, many plants are exploited for preparation of traditional herbal medicines in less developed countries (Davidson-Hunt 2000) since ancient times for treating numerous chronic diseases like diabetes, blood pressure anomalies, central nervous system disorders, cancer etc. Presently, jillions of phytochemicals have been structurally and functionally characterized. Some phytochemicals abundantly found in common plants are flavonoids, anthocyanins, and lycopene. Phytochemicals are grouped in different classes on the basis chemical constitution viz. alkaloids, catechins, stilbenes, flavones, flavonols, phenolic acids, tannins, terpenes, anthocyanins, essential oils, and steroids (Hamuel 2012). The study of phytochemicals includes their separation, extraction, purification, identification, structural as well as functional analyses. The frequently used techniques used for phytochemical research involve different chromatographic techniques along with mass spectrometry as well as nuclear magnetic resonance.

There are different effects of individual phytochemicals such as antioxidant activity, hormonal action, stimulatory effect on enzymes, anti-bacterial, hindering cellular machinery or acting as a physical barrier. Due to their pharmacological

A. Singh · S. Zahra · S. Kumar (✉)
National Institute of Plant Genome Research (NIPGR), New Delhi, India
e-mail: shailesh@nipgr.ac.in

S. Kumar, C. Egbuna (eds.), *Phytochemistry: An in-silico and in-vitro Update*,
https://doi.org/10.1007/978-981-13-6920-9_19

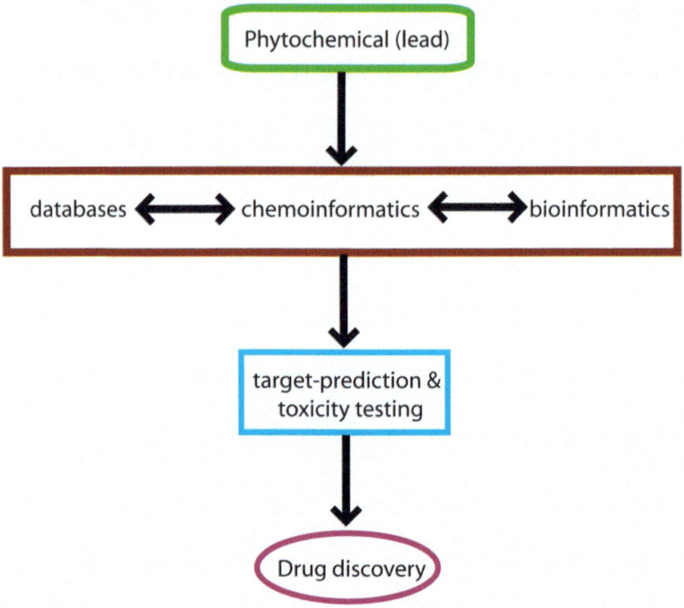

Fig. 19.1 Flow-chart representing the use of phytochemical in drug discovery using chemo- and bioinformatics approach

action as well as their capability of showing pleiotropic effect (multiple targeting potentials), they can be a boon in drug development domain leading to cost-efficient approach with lesser side effects for synthesizing burgeons of new medicines for multiple ailments.

Yet, most of the phytochemicals are unexplored because their biological effects are not known. In-silico methods could be exploited for discovering the actions of un-investigated phytochemicals by identification of their molecular targets using an amalgamation of chemical informatics and bioinformatics along with systems biology approaches (Fig. 19.1), hence advantageous for drug discovery (Lagunin et al. 2014). These approaches make simultaneous use of publicly accessible databases and various software tools for screening of potential chemicals, pattern recognition, and analyzing their binding interactions with their probable targets. The omics tools are categorized based on their respective functions like descriptor and pharmacophore generation, molecular docking, biological activity prediction, quantitative or qualitative structure (QSAR) modeling (Barlow et al. 2012). This chapter sheds light on these software tools, which are elaborately described here along with brief information on available databases and their mode of operation.

19.1.1 Database

Databases storing valuable information about the phytochemicals obtained from a variety of plants are available, which are helpful in studying structures as well as functional aspects of a range of chemicals extracted from plants. Most of the databases contain information like:

(a) Name of the plant (including the local and vernacular name of plants)
(b) Phytochemicals found in those plants (tissues)
(c) Chemical structure and stereochemistry of those phytoconstituents
(d) Pharmacological activities
(e) Toxicity, if present

Medicinal plants have a reservoir of diverse phytochemicals are tabulated for development of databases of therapeutic plants. Many famous and popularly used databases are mentioned below, depending upon the following features and advantages: ease of accessibility, abundance of useful data and coverage of information, curability, peer-reviewed in scientific publications, contain 2D and 3D structure of phytochemicals, should contain information about known or predicted targets, as well as toxicity levels and the databases, must be live and updated.

Although a single database cannot exploited to fetch information regarding all the existing, known phytochemicals, few of the most extensively used, and comprehensive databases that can utilized for in-silico-based approaches for drug discovery are mentioned in this chapter. Cheminformatics databases like Dictionary of Natural Products (http://dnp.chempnetbase.com) and Traditional Chinese Medicine Integrated Database (TCMID) (Xue et al. 2012), KNApSAcK Core DB (Afendi et al. 2012), Medicinal Plants Database for Drug Discovery (MPD3), AfroDb (Ntie-Kang et al. 2013), Indian Medicinal Plants, Phytochemistry And Therapeutics (IMPPAT) (Mohanraj et al. 2018), Phytochemica (Pathania et al. 2015), Herbalog (Wang et al. 2016), Super Natural II (Banerjee et al. 2015), and PubChem Substance Database (Kim et al. 2016) are available and can access from a single link: https://omictools.com.

19.2 Functional Classification of Software and Tools

Along with the databases, different types of software and programming languages are utilized for phytochemical-based drug discovery as categorized below.

19.2.1 Virtual Screening Programs

Virtual screening employs computational techniques for searching libraries of phytochemical molecules to find out the most suitable ligand for binding to the specific target. The target can be either enzymes or receptor proteins, which will ultimately pave a way for discovering therapeutic drugs (Rollinger et al. 2008). There are two methods of virtual screening:

19.2.1.1 Ligand-Based

Using this method, from a given set of ligands with known structures, binding to a specific receptor, a model of that receptor can be designed from collective information of the ligands which is called pharmacophore model. Then a candidate ligand can be compared with the generated pharmacophore model and checked for their binding and interactions. One more approach to ligand-based virtual screening is the use of 2D chemical similarity analysis procedure which is a ligand-based approach to skim a database of molecules against many ligand structures (Willett et al. 1998). Further, an alternative for ligand-based screening is based on finding molecules with an identical shape to that of well-known ones, with anticipation that these molecules will fit the binding site of prey or target protein, and ultimately binds to the target.

19.2.1.2 Structure-Based

This approach involves docking of candidate ligands to a target protein and scoring function is applied for estimating the likelihood of high-affinity binding to the ligand to protein (Kroemer 2007). The algorithms and computational tools used for virtual screening make use of classification/regression trees (including Random Forest), linear discriminant analysis, artificial neural networks, as well as support vector machines.

19.2.2 Descriptors Generator and Pharmacophore Programs

Generation of molecular descriptors of phytochemical molecules is achieved by giving mathematical treatment to chemical information contained within these molecules using algorithms giving discrete and different values to individual chemical entities. Programs developed for descriptor generation (Rudik et al. 2014) include AFGen, CODESSA, DRAGON, MOLGEN, and PaDEL (Yap 2011).

A pharmacophore is the description of atoms and molecules responsible for molecular recognition of a ligand by a biological macromolecule. Pharmacophores are also used as starting points for developing 3D- QSAR models. Both ligand and structure-based programs for pharmacophore generation are available for identifying putative ligands or targets. Some representative examples of such packages are LigandScout (Wolber and Langer 2005), Accelrys, Discovery- Studio (http://accelrys.com), Schrödinger's small-molecule drug discovery suite (https://www.schrodinger.com/suites/small-molecule-drug-discovery-suite).

19.2.3 QSAR Modeling

Quantitative or qualitative structure-activity relationship (QSAR) interlinks biological activities with physicochemical properties. QSAR analysis statistically connects one or multiple molecular descriptors with the molecular activity. The statistical models hence created are exploited for prediction of the biological activity of newly discovered chemical compounds yet to be tested experimentally. It is also very important to the study of ADME/T (absorption, distribution, metabolism, excretion, and toxicity) properties of phytochemicals used for drug development. There are many software packages available with appropriate QSAR models for the evaluation of ADME/T properties for chemicals including phytochemicals based on their structures. The step-wise processes involved in the development of a QSAR model (Vilar et al. 2008) are:

I. Selection of the database of compounds with known functions
II. Calculation of molecular descriptors
III. Statistical model designing for relating the activity with calculated descriptors
IV. The assessment of the generated model with a mock set

QSAR modeling tools (Rudik et al. 2014) include Accelrys', Discovery Studio, SYBYL®-X, ChemBench, KNIME, StarDrop™, GUSAR, ADMET Predictor™.

19.2.4 Docking Programs

Docking is the computational simulation method by which most preferred orientation of one molecule with another molecule is predicted when they are bound together as stable complex minimizing their free energy (Lengauer and Rarey 1996). It is the most frequently used method in structure-based drug designing because of its ability to predict the binding conformation of small ligand molecules to the appropriate target molecule binding site. For performing docking, the 3D structure of target protein should be available which is used to ‘dock’ candidate ligands into the active site of target proteins for optimization of their conformational and chemical complementarity and assigning scoring functions for their binding affinity. Inverse docking (Chen and Zhi 2001) (INVDOC) and target fishing (Li et al. 2006) (TarFishDoc) are other in-silico screening techniques for identifying protein targets.

Various software available for docking are: AutoDock (Morris et al. 2009), FlexX (https://www.biosolveit.de/FlexX/), Glide (Friesner et al. 2004), GOLD (Verdonk et al. 2003), LigandScout (Wolber and Langer 2005), Molegro Virtual Docker (MVD) (Rudik et al. 2014).

19.2.5 Pattern Recognition, Statistics, and Machine Learning Software

Pattern recognition, machine learning, and artificial intelligence approaches play an increasingly important role in rational drug design, screening and identification of candidate molecules and studies on quantitative structure-activity relationships (QSAR). There is a rapid growth of network analysis methods for discovering drug targets. Network pharmacology is an emerging model in drug discovery which was first proposed by Hopkins in 2007 (Hopkins 2007). For understanding the mechanisms of action of drugs and their efficiency, it is necessary to explore the chemistry of these molecules and their relationship with biological networks.

Various language and software packages like R (Stowell 2012), MATLAB (Caballero and Fernandez 2008), PSPP, SAS®/STAT, SIMCA, SPSS software, STATISTICA (Scotti et al. 2012) and WEKA are utilized for statistical as well as network analyses, and machine learning based applications, for drug designing and discovery.

All the above-mentioned categories of software utilized for drug discovery are dependent on each other and functionally inter-related. Some of the widely used and freely accessible software packages are mention in the preceding sections.

19.3 SwissSimilarity

SwissSimilarity, a web interface, virtually screen the small to large libraries of small molecules on ligand base. It includes many query compounds including drugs, bioactive and commercial molecules, and 205 million virtual compounds. Small-molecule libraries in SwissSimilarity are (Zoete et al. 2016):

I. Drug Collection
One thousand and five hundred compounds (approved), 4800 experimental compounds, 500 compounds from the investigation, 160 compounds (withdrawn drugs), 170 illicit molecules, and 78 nutraceuticals.
II. Bioactive Small Molecules
Nineteen thousand and five hundred molecules from PDB structural entries, 177,000 highly curated ChEMBEL molecules with an activity lower than 10 μM against a well-defined target, 28,000 ChEBI compounds, 480,000 kinase and GPCR inhibitors from ChEMBL and GLASS (GPCR-ligand association database) and 39,000 metabolomes from HMDB (Human Metabolome Database).
III. Molecule collections from the ZINC database.
IV. Two hundred five million virtual compounds that can be synthesized from commercially available reagents, and filtered for stability, non-toxicity, and lack of promiscuous character.

19.3.1 *Working of SwissSimilarity*

19.3.1.1 Input

The user can provide the query molecule either by drawing it or by pasting the SMILES in the devoted content box. Both are synchronized (Fig. 19.2a). Radio button checking in the list provides the user to screen the chemical library and the ligand-based virtual screening method (Fig. 19.2b).

19.3.1.2 Output

Figure 19.2b shows an output page obtained by screening the FDA approved drugs against Diclofenac with FP2 fingerprints. The molecules similar to the reference compound are displayed, along with the similarity score and a reference for corresponding entries in their databases of origin (e.g. DrugBank, ChEMBL, etc.).

19.3.1.3 Result Prediction

The similarity score ranges from 0 to 1; where 0 for different molecules and 1 for identical compounds.

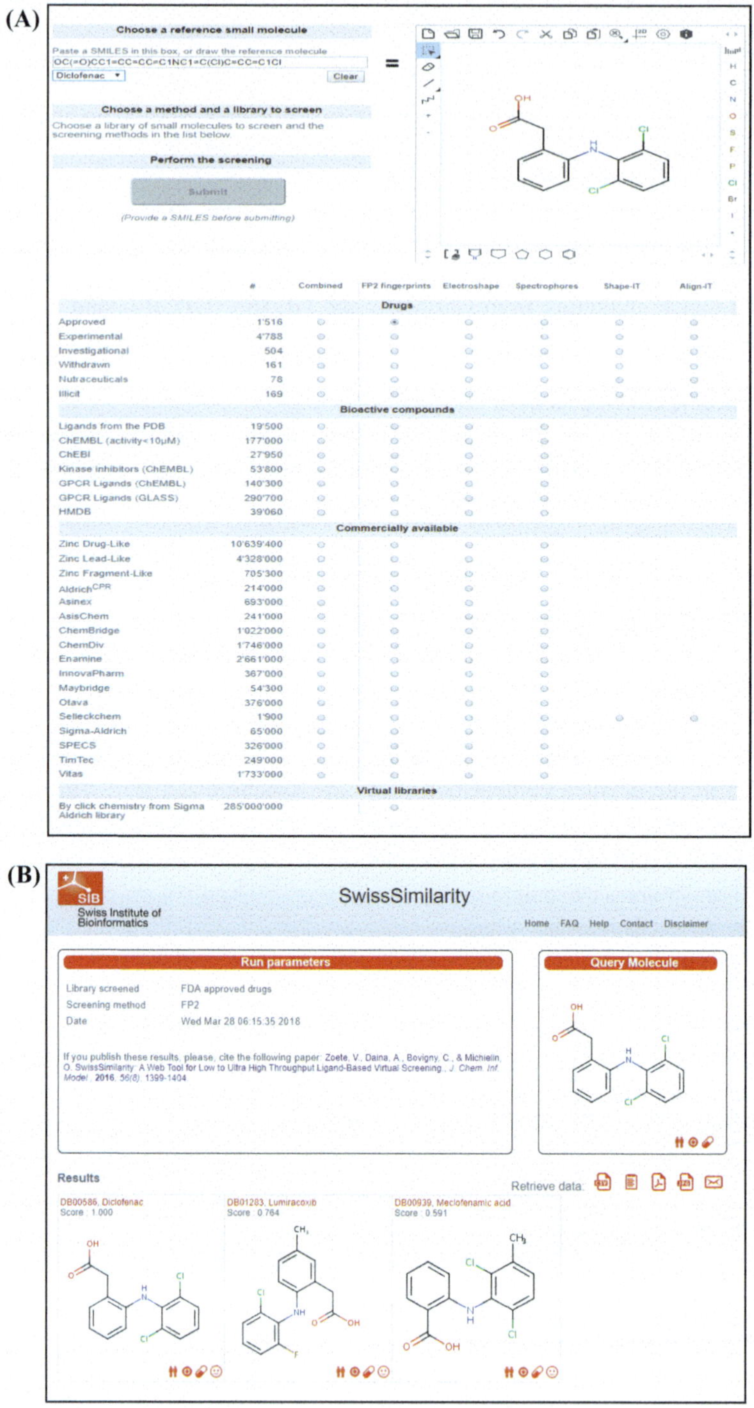

Fig. 19.2 Web interface of SwissSimilarity. (**a**) Input of molecule (ex. Diclofenac) with a selection of parameters in SwissSimilarity. (**b**) Result output with run parameters

Notes

1. Outputs are put away on the server for no less than 1 week, amid which results can be retrieved or shared without having to download them.
2. After screening the virtual library, the proposed organic reactions to synthesize the virtual hit compound are retrieved by clicking on the molecule identifier.
3. A single click on a reagent's name opens the corresponding entry in the website of the provider, where availability and purchase information can be checked.
4. An E-mail containing the URL of the output page can also be sent automatically by the user by clicking on a dedicated icon.

19.4 AutoDock

AutoDock predicts the interaction of ligands with biomacromolecule targets in an automated procedural way. The current version is 4.2.6. To make accurate and potent, AutoDock use: (a) rapid grid-based energy estimation, and (b) efficient search of torsional freedom. (Morris et al. 2009).

AutoDock 4.2.6 utilizes the Lamarckian Genetic Algorithm and factual free energy scoring function, afterward gives reproducible docking results for ligands with ~10 pliable bonds. AutoDock GUI can be accessed by AutoDockTools (ADT).

19.4.1 Working of AutoDock

Several steps for AutoDock calculations are as follows:

19.4.1.1 Preparation of Coordinate File

A protein-ligand model (polar H includes; C-H excludes) is used by AutoDock4.2. This tool's force field uses many atom types for common atoms which includes: different types of aromatic and aliphatic carbon atoms; different types of hydrogen bond forming and without bond polar atoms. Info regarding the torsional degree of freedom is also included in PDB files (PDBQT). Also, if there is a specific side chain in protein that treated as flexible than for that coordinates a different PDBQT file is created.

19.4.1.2 AutoGrid Calculation

Enhanced energy evaluation in AutoGrid procedure includes the pre-calculation of atomic affinity for docking atom types in ligand molecules. In this procedure, a probe atom is placed at each point of protein embedded 3-D grid and energy interaction for singlet atom with protein is allocated to the grid point. Affinity grids are calculated for each atom in the ligand (generally carbon, hydrogen, nitrogen, and oxygen) and also for de-solvation and electrostatic potential.

19.4.1.3 Docking

AutoDock run many times for a consistent result with several docked conformations and energy prediction, so that best solution become identified. It includes methods like Lamarckian genetic algorithm, traditional algorithms, and simulated annealing also accessible for choice-based docking.

19.4.1.4 Analysis Using ADT

As ADT provides a graphical user interface to the user, so it becomes simple to analyze the docking output or results. User analyzes the docking simulation, visualize conformations, similarities, ligand-protein interactions, and many other docking tasks (as user explore ADT).

19.4.2 AutoDock Vina and AutoDock 4

AutoDock Vina is an open source program for performing molecular docking. It utilizes an easier scoring function that permits a quicker pursuit technique and gives reproducible outcomes to bigger frameworks with upwards of 20 flexible bonds (Trott and Olson 2009).

AutoDock Vina and AutoDock 4: Similarities and comparison

(a) Both are used PDB file (PDBQT) format. This makes it easy to use Vina with the existing auxiliary software developed for AutoDock, such as ADT, for preparing the files, choosing the search space, and viewing the results.
(b) Moreover, manually picking the atom type for grid maps, computing grid map delineate with AutoGrid, selection of 'search parameters' and clustering the outcomes after docking is not any more important, as Vina ascertains its own grid maps rapidly and automatically and does not store them on disk. It additionally clusters and positions the outcomes without presenting the user to intermediate details.

(c) A much of the time experienced issue for AutoDock 4 users is to do with the fixed data structure sizes in the program i.e. maximum grid map size, number of atoms, number of rotatable bonds etc. These points of confinement are settled at the aggregate time and setting them higher may squander time and memory. To change these breakpoints to address their issues, the users are required to change them in the source code and then recompile the software. This is error-prone and intimidating for many users. While Vina does not have such constraints, it automatically adapts to input.

19.4.2.1 Working Steps

The steps for molecular docking using AutoDock-Vina are illustrated below

19.4.2.1.1 Target File Preparation

Download protein file from PDB.

1. Open in WordPad/Notepad and remove HEATM and save again PDB file.
2. Open AutoDock tool, go to *File* and click *Read Molecule* for open PDB file.
   (a) *Edit* → Delete water.
   (b) Hydrogens → Add (to polar only)
   (c) Charges → Add Kollman charges
   (d) Grid → Macromolecules → Choose → Select Molecule → Save as file.pdbqt
   (e) Grid Box → Set grid box to the protein site; Save orientation values in CONF. txt file.

19.4.2.1.2 Ligand File Preparation

1. Ligand → Input → Open (open ligand file)
2. Torsion Tree → Detect Root
3. Output → Save as PDBQT (save as filename.pdbqt)
4. Quit AutoDock Tools

19.4.2.1.3 Run Vina

1. Put both. pdbqt files (Protein and Ligand) in Vina Folder
2. Write the name(s) of ligand(s) in ligands.txt and save in the same folder/directory
3. Open Command Line Prompt (Windows+R → cmd → enter)

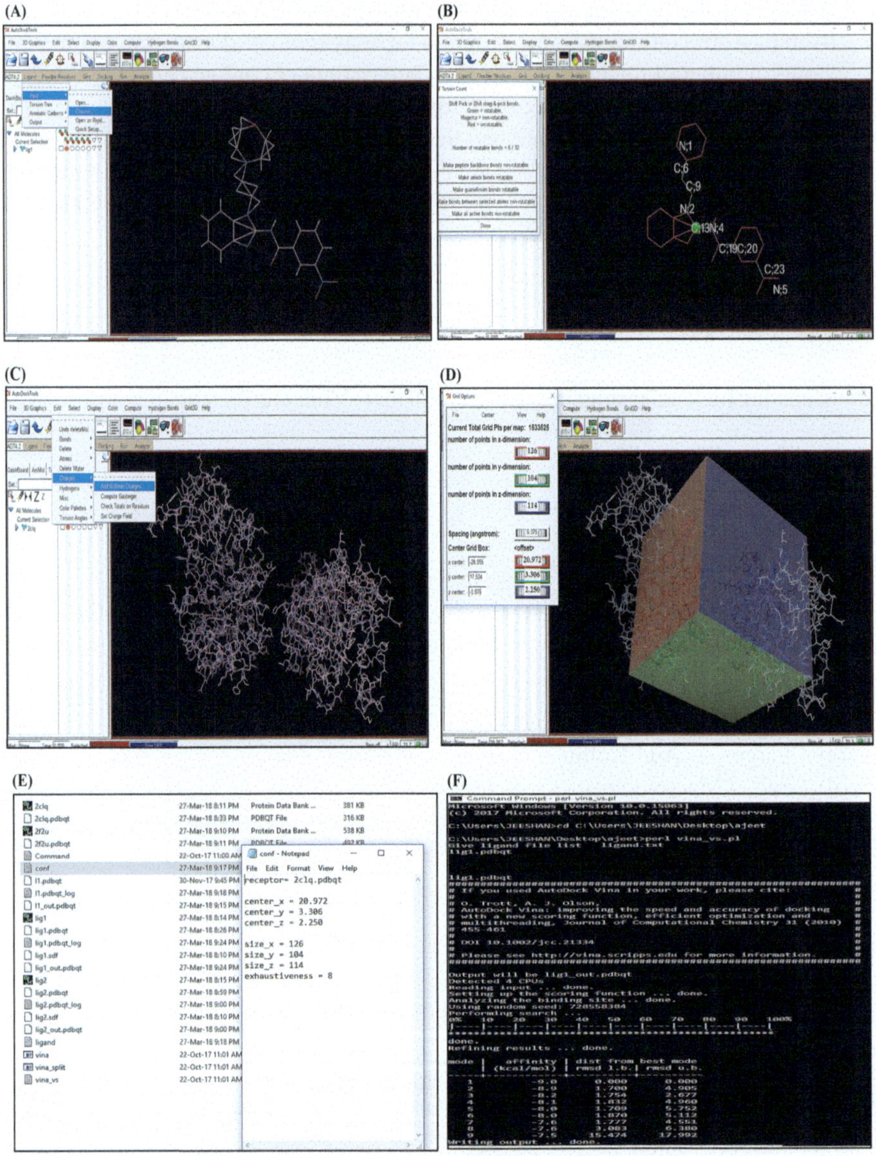

Fig. 19.3 Autodock GUI with AutoDock-Vina by ADT. (**a**) Schematic representation of ADT with input ligand (IRAK-1/4 inhibitor). (**b**) ADT window showing Torsion Count parameter with root (green ball in structure). (**c**) Schematic representation of ADT showing add Kollman charges in target protein (2CLQ- mitogen-activated protein kinase). (**d**) A window showing Grid-box parameter setting in AutoDock by ADT. (**e**) Text file with Grid-box parameter for AutoDock Vina. (**f**) Command prompt window showing AutoDock vina result

4. Set the directory (Address) of Vina Folder and run the following commands-
   (a) perl vina_vs.pl
   (b) ligands.txt
   (c) ligandfilename_out.pdbqt containing docking conformations and ligandfilename.log containing binding affinities, files will be appearing in Vina folder for further analysis.
   (d) Pymol can be used for visualization purpose.
   (e) Running of AutoDock-Vina by using example molecules IRAK-1/4 inhibitor as ligand and 2CLQ- mitogen-activated protein kinase as target protein is demonstrated in Fig. 19.3a–f).

19.5 SWISSDOCK

SwissDock is a docking web server, based on EADocks DSS (docking software) that predicts the molecular interactions between a small molecule and target protein. Docking calculations are run on the server side, so it does not take computation power from the user side. After running, results interpretations are facilitated by visualizing it on a web browser using UCSF Chimera molecular viewer. This is launched by clicking the link we get after query completion (Grosdidier et al. 2011).

19.5.1 The Methodology of EADock DSS Algorithm

1. In local docking binding modes are generated in the box and in blind docking binding, modules are generated near all target cavities.
2. Then on the grid, CHARMM energies are estimated simultaneously.
3. With FACTS the binding modes with the most favorable energies are evaluated and clustered.
4. The most favorable clusters can be visualized and downloaded by the user.

19.5.2 Working of SwissDock

19.5.2.1 Inputs

There is need to check three parameters to start a docking assay through the web interface of SwissDock: users must define a protein structure, docking parameters and one or several putative ligands (Fig. 19.4b). Several sample files are supplied to users and can be directly uploaded into the form simply by clicking on a link. The corresponding sample output files are also provided.

(A)
(B)

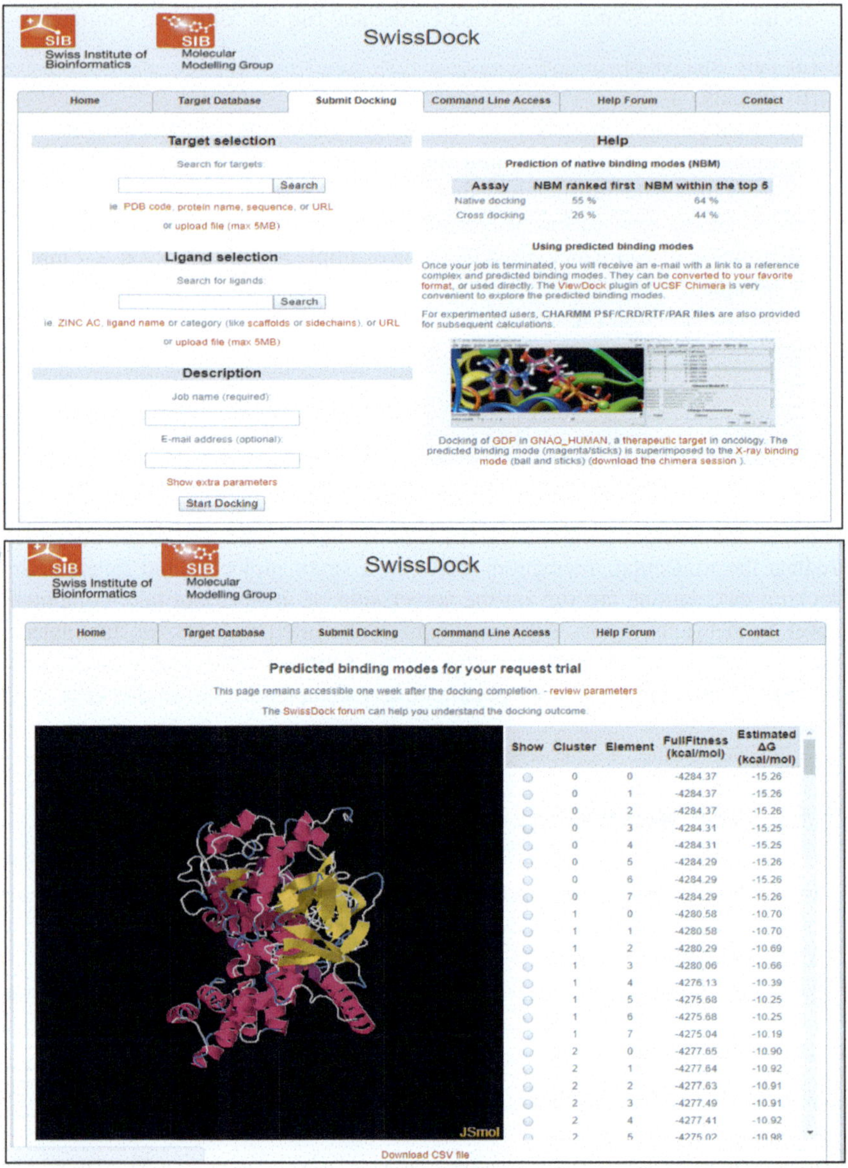

Fig. 19.4 SwissDock web interface for query submission. (**a**) SwissDock query submission portal view. (**b**) An example showing SwissDock result output with parameters

19.5.2.2 Target Selection

The identifier specification from PDB or uploaded structure will determine the target protein structure. Here PDB option is for the users who are not intimate with 3-D structure files. If there are a lot of options in PDB for the target, then choose that one which is similar to ligand and highly resolute.

19.5.2.3 Ligand Selection

The user can upload the structure file for ligand and can also select it from the ZINC database. This is because of wide acceptability of SwissDock, as for target protein, besides Mol2 format, it also supports direct uploading of CHARMM files. The ligand CHARMM files have PDB coordinate, docking parameters, and extra topology description.

19.5.2.4 Outputs

A URL is provided for tracking the query submission. If an (optional) email address has been specified in the submission form, this URL is also sent to the user by email, as well as a link to the docking result web page once the docking is completed. A Jmol applet is featured in result web page for result description with parameters. SwissDock result output is represented by various parameters in Fig. 19.4b.

Note
After 7 days, docking predictions with uploaded files are automatically deleted. For statistical purposes, other variables are kept unremarkably.

19.6 Molinspiration

Molinspiration, a Java-based software that can execute on any Windows, Mac, UNIX or LINUX, is distributed in a form of engines, which have the option to be utilized as stand-alone computational engines, used to power web-based tools, or effortlessly consolidated into larger in-house Java applications. This has nearly 3000 citations. There is ten chemInformatics software provided by Molinspiration; their respective uses are listed in Table 19.1.

Molinspiration additionally gives free electronic administrations to ascertaining logP, polar surface zone, number of hydrogen bond contributors and acceptors and

Table 19.1 Molinspiration software products and their uses (www.molinspiration.com)

Tools name	Uses
mib engine	It calculates important molecular properties, molecular processing (SMILES canonicalization, normalization of charges), conversion between SMILES and SD files, SMILES depiction, generation of molecular images
misearch engine	Flexible molecular database supporting substructure, similarity, and pharmacophore similarity searches
miscreen engine	Fragment-based virtual screening engine enabling development of pharmacophore models, validation, and screening of large molecular libraries
Galaxy	Generates 3D structures from SMILES
Molinspiration clusterer	Identify groups (clusters) of similar structures in molecular datasets. Two types of clustering are currently supported: similarity-based clustering and scaffold tree clustering
Molinspiration property calculator	Calculate molecular properties and generate QSAR tables in an interactive way
Molinspiration molecule viewer	Visualization of large sets molecules
Molinspiration data viewer	Visualization of QSAR datasets with interactive molecule display.
Molinspiration depiction	The high-quality depiction of molecules encoded as SMILES of MDL Molfile
Database of bioactive substituents and linkers	Database of 49,000 linkers and 21,000 substituents extracted from bioactive molecules includes calculated properties also

other sub-atomic properties, and also the expectation of bioactivity score for the most critical drug targets (GPCR ligands, kinase inhibitors, particle channel modulators, atomic receptors).

Using an example molecule (Fig. 19.5a), cyclohexane, molecular properties and bioactivity score are calculated using web services provided by Molinspiration. The respective output of query completion is depicted in Fig. 19.5b–d.

19.7 Toxicity Estimation Software Tool (TEST)

This tool, which is based on CDK, estimates the toxicity of chemicals using QSAR methodologies. The latest version is 4.2.1. The TEST contains a model for several physical properties like normal boiling point, surface tension, viscosity, density, water solubility, melting point and so on.

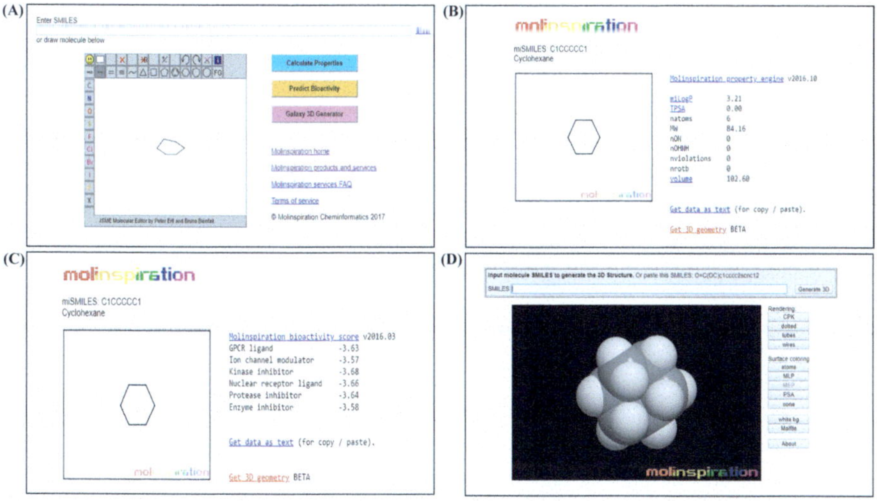

Fig. 19.5 Free web tools by Molispiration. Taking an example of cyclohexane molecule, molecular properties and bioactivity score are calculated. (**a**) The open online window for calculating molecular properties and bioactivity score. (**b**) Calculated molecular properties for cyclohexane. (**c**) Results of bioactivity scores for cyclohexane. (**d**) 3-D structure of cyclohexane generated by Molinspiration Galaxy 3D Generator

19.7.1 QSAR Methodologies (Epa et al. 2016)

There are several methods uses in the TEST for QSAR, which are

1. Hierarchical method: It uses many different models which are obtained by applying Ward's method to portioning the training set in structurally similar class series. In the hierarchical method, models are generated earlier to runtime
2. FDA method: In this method models are generated at runtime. A new model is used in this for prediction (for each test chemical) that is to be fitted with the chemicals which are similar to the test compound.
3. Single model method and Group contribution method: Predictions are made utilizing a multilinear regression model that is fit to the preparation set (single model- using molecular descriptors as independent variables; group contribution- using molecular fragment counts as autonomous factors). In both, the regression models are generated prior to runtime.
4. Nearest neighbor method: In this method, the average of most test similar three chemicals in the training set is used for the estimation of predicted toxicity.

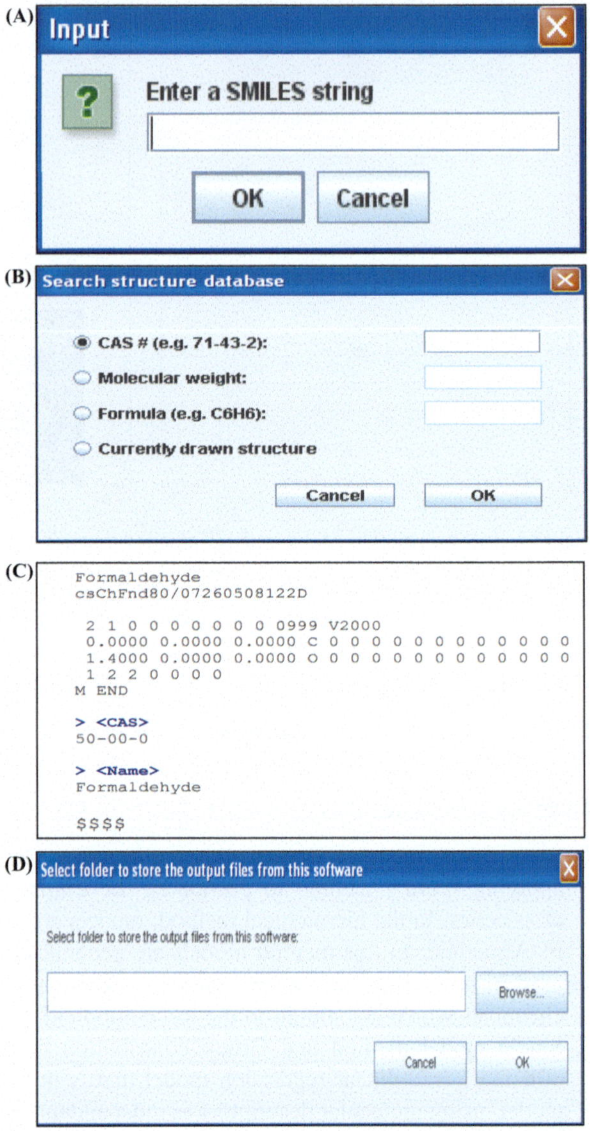

Fig. 19.6 Workflow of toxicity estimation software tool. (**a**) The window for entering smiles string in T.E.S.T. (**b**) Schematic representation of window for import structure from the database. (**c**) An example of SDfile (formaldehyde). (**d**) The window for select folder (for storing output files)

19.7.2 Working Flow of TEST

The usage of the TEST package is very user-friendly beginning with the import of the chemical structure files to perform toxicity predications. The software also facilitates simultaneous import of multiple compounds (batch import).

1. The single compound can be imported in the following ways
   (a) By drawing a structure in the molecular structure drawing tool.
   (b) By importing from MDL molfile.
   (c) Importing from a SMILES string (Fig. 19.6a). For example, to import benzene enter c1ccccc1 as the SMILES string.
   (d) The user can also import compound from the included structure database (Fig. 19.6b).
2. **Import multiple compounds**:
   (a) Importing from MDL SDfile (an example of SDfile including formaldehyde – Fig. 19.6c).
   (b) By importing from a list of SMILES strings.
   (c) Importing from a list of CAS numbers.
3. **Performing toxicity predictions**

   To perform the toxicity prediction, first, enter the molecule ID/CAS number, and then select a toxicity endpoint from the drop-down list. Then using the provided drop-down list, choose the QSAR toxicity estimation method (default is hierarchical clustering method). Finally, click the *Calculate* for toxicity prediction. But here, before running the calculation, select the location for storage of output (Fig. 19.6d).
4. **Interpretation of results**

   After performing the toxicity estimation calculations, a web page is generated which displays the results. The predicted toxicity using the consensus method shows the average of predicted toxicities from all QSAR methods (in TEST).

19.8 R Language

R is a computer programming language and environment for statistical computation for drug development, which is exploited for experimental data analysis and experimental design. With R, one can design new trials, perform exploratory and confirmatory analysis of data in those trials (e.g. compare therapies, assess their ability to produce intended result, safety, toxicity, evaluate drugs by performing pharmacokinetic/pharmaco-dynamic analyses), perform data validation against errors, types, and frauds, transform data between various formats, tools, and databases.

An excellent example is that of the package, **ezqsar**, for QSAR modeling using R. This package is capable of developing Multiple Linear Regression (MLR) QSAR models from 2D or 3D structures and their corresponding activities via a single line command.

The descriptor generation in 'ezqsar' is done using the CDK library. It computes 2D and 3D descriptors. They are classified into five groups "topological", "geometrical" "hybrid", "constitutional", and "electronic". If the input structures are in 3D coordinates, the 3D descriptors will be calculated otherwise, the value for the 3D

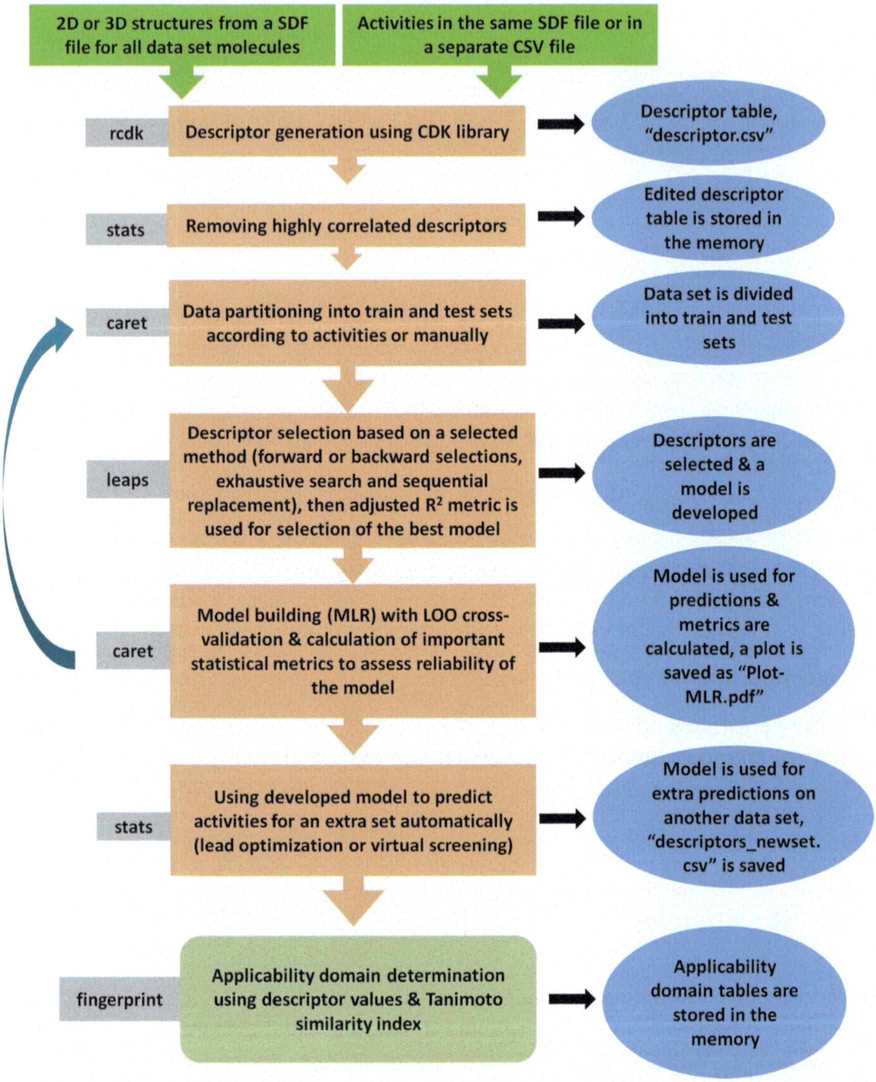

Fig. 19.7 The workflow of ezqsar_f function from ezqsar package

descriptors would be zero. 'ezqsar' only accepts SDF file as an input. The workflow of ezqsar_f function from ezqsar package is represented in Fig. 19.7 (Shamsara 2017).

Important links for tools

1. www.swisssimilarity.ch
2. www.swissdock.ch
3. www.molinspiration.com
4. www.epa.gov/chemical-research/toxicity-estimation-software-tool-test

References

Afendi FM, Okada T, Yamazaki M, et al. KNApSAcK family databases: integrated metabolite–plant species databases for multifaceted plant research. Plant Cell Physiol. 2012;53:e1. https://doi.org/10.1093/pcp/pcr165.

Banerjee P, Erehman J, Gohlke B-O, et al. Super natural II—a database of natural products. Nucleic Acids Res. 2015;43:D935–9. https://doi.org/10.1093/nar/gku886.

Barlow DJ, Buriani A, Ehrman T, et al. In-silico studies in Chinese herbal medicines' research: evaluation of in-silico methodologies and phytochemical data sources, and a review of research to date. J Ethnopharmacol. 2012;140:526–34. https://doi.org/10.1016/J.JEP.2012.01.041.

Caballero J, Fernandez M. Artificial neural networks from MATLAB® in medicinal chemistry. Bayesian-regularized genetic neural networks (BRGNN): application to the prediction of the antagonistic activity against human platelet thrombin receptor (PAR-1). Curr Top Med Chem. 2008;8:1580–605. https://doi.org/10.2174/156802608786786570.

Chen YZ, Zhi DG. Ligand-protein inverse docking and its potential use in the computer search of protein targets of a small molecule. Proteins. 2001;43:217–26.

Davidson-Hunt I. Ecological ethnobotany: stumbling toward new practices and paradigms. MASA J. 2000;16:1–13.

Epa U, of Research O, Technology Division S. User's guide for T.E.S.T. (version 4.2) (toxicity estimation software tool): a program to estimate toxicity from molecular structure. 2016.

Friesner RA, Banks JL, Murphy RB, et al. Glide: a new approach for rapid, accurate docking and scoring. 1. Method and assessment of docking accuracy. J Med Chem. 2004;47:1739–49. https://doi.org/10.1021/jm0306430.

Grosdidier A, Zoete V, Michielin O. SwissDock, a protein-small molecule docking web service based on EADock DSS. Nucleic Acids Res. 2011;39:W270–7. https://doi.org/10.1093/nar/gkr366.

Hamuel J. Phytochemicals: extraction methods, basic structures and mode of action as potential chemotherapeutic agents. Phytochemicals – a global perspective of their role in nutrition and health. InTechOpen, UK; 2012.

Hopkins AL. Network pharmacology. Nat Biotechnol. 2007;25:1110–1. https://doi.org/10.1038/nbt1007-1110.

Kim S, Thiessen PA, Bolton EE, et al. PubChem substance and compound databases. Nucleic Acids Res. 2016;44:D1202–13. https://doi.org/10.1093/nar/gkv951.

Kroemer RT. Structure-based drug design: docking and scoring. Curr Protein Pept Sci. 2007;8:312–28. https://doi.org/10.2174/138920307781369382.

Lagunin AA, Goel RK, Gawande DY, et al. Chemo- and bioinformatics resources for in silico drug discovery from medicinal plants beyond their traditional use: a critical review. Nat Prod Rep. 2014;31:1585–611. https://doi.org/10.1039/C4NP00068D.

Lengauer T, Rarey M. Computational methods for biomolecular docking. Curr Opin Struct Biol. 1996;6:402–6. https://doi.org/10.1016/S0959-440X(96)80061-3.

Li H, Gao Z, Kang L, et al. TarFisDock: a web server for identifying drug targets with docking approach. Nucleic Acids Res. 2006;34:W219–24. https://doi.org/10.1093/nar/gkl114.

Mohanraj K, Karthikeyan BS, Vivek-Ananth RP, et al. IMPPAT: a curated database of Indian medicinal plants, phytochemistry and therapeutics. Sci Rep. 2018 Mar 12;8(1):4329. https://doi.org/10.1038/s41598-018-22631-z.

Morris GM, Huey R, Lindstrom W, et al. AutoDock4 and AutoDockTools4: automated docking with selective receptor flexibility. J Comput Chem. 2009;30:2785–91. https://doi.org/10.1002/jcc.21256.

Ntie-Kang F, Zofou D, Babiaka SB, et al. AfroDb: a select highly potent and diverse natural product library from African medicinal plants. PLoS One. 2013;8:e78085. https://doi.org/10.1371/journal.pone.0078085.

Pathania S, Ramakrishnan SM, Bagler G. Phytochemica: a platform to explore phytochemicals of medicinal plants. Database J Biol Databases Curation. 2015;2015:bav075. https://doi.org/10.1093/database/bav075.

Rollinger JM, Stuppner H, Langer T. Virtual screening for the discovery of bioactive natural products. Prog Drug Res. 2008;65(211):213–49.

Rudik AV, Konova VI, Pogodin PV, et al. Drug discovery from medicinal plants beyond their traditional use: a critical review. Nat Prod Rep. 2014;00:1–27. https://doi.org/10.1039/C4NP00068D.

Scotti L, Bezerra Mendonça Junior FJ, Magalhaes Moreira DR, et al. SAR, QSAR and docking of anticancer flavonoids and variants: a review. Curr Top Med Chem. 2012;12:2785–809.

Shamsara J. Ezqsar: an R package for developing QSAR models directly from structures. Open Med Chem J. 2017;11:212–21. https://doi.org/10.2174/1874104501711010212.

Stowell S. Instant R: an introduction to R for statistical analysis. New York: Jotunheim Publishing; 2012. isbn:978-0-957-46490-2.

Trott O, Olson AJ. AutoDock Vina: improving the speed and accuracy of docking with a new scoring function, efficient optimization, and multithreading. J Comput Chem. 2009;31:455–61. https://doi.org/10.1002/jcc.21334.

Verdonk ML, Cole JC, Hartshorn MJ, et al. Improved protein-ligand docking using GOLD. Protein Struct Funct Bioinform. 2003;52:609–23. https://doi.org/10.1002/prot.10465.

Vilar S, Cozza G, Moro S. Medicinal chemistry and the molecular operating environment (MOE): application of QSAR and molecular docking to drug discovery. Curr Top Med Chem. 2008;8:1555–72. https://doi.org/10.2174/156802608786786624.

Wang Y, Hu J-S, Lin H-Q, et al. Herbalog: a tool for target-based identification of herbal drug efficacy through molecular docking. Phytomedicine. 2016;23:1469–74. https://doi.org/10.1016/j.phymed.2016.08.008.

Willett P, Barnard JM, Downs GM. Chemical similarity searching. 1998. doi:https://doi.org/10.1021/CI9800211.

Wolber G, Langer T. LigandScout: 3-D pharmacophores derived from protein-bound ligands and their use as virtual screening filters. J Chem Inf Model. 2005;45:160–9. https://doi.org/10.1021/ci049885e.

Xue R, Fang Z, Zhang M, et al. TCMID: traditional Chinese medicine integrative database for herb molecular mechanism analysis. Nucleic Acids Res. 2012;41:D1089–95. https://doi.org/10.1093/nar/gks1100.

Yap CW. PaDEL-descriptor: an open source software to calculate molecular descriptors and fingerprints. J Comput Chem. 2011;32:1466–74. https://doi.org/10.1002/jcc.21707.

Zoete V, Daina A, Bovigny C, Michielin O. SwissSimilarity: a web tool for low to ultra high throughput ligand-based virtual screening. J Chem Inf Model. 2016;56:1399–404. https://doi.org/10.1021/acs.jcim.6b00174.

Chapter 20
In-silico Approach to Target Cancer Cell DNA Repair Pathway

Azra Yasmin, Zarrin Basharat, and Naila Safdar

20.1 Introduction

Every creature living on earth faces different type of stresses. These stresses can be endogenous, resulting from the internal cellular conditions or exogenous, resulting from the environmental exposure. Point of satisfaction is that all living organisms develop different types of defense mechanisms in response to internal or external pressures. These defense mechanisms operate at different levels from restriction enzymes working in microbial population to the secondary metabolites production in plants. Mammals have developed defense mechanisms at very large scale varying from the immune systems functioning to the DNA repair mechanisms.

Though DNA is a highly stable and versatile molecule, but in all living organisms it suffers damage either due to the usual cellular activities or due to certain physical, chemical or biological agents. DNA damage can interfere with the replication or transcription processes thereby limiting the viability of the cell. Specific DNA injury can cause mutations resulting in acute or chronic tumorigenesis/carcinogenesis. The perpetuation of genetic material from generation to generation depends upon keeping the rates of mutation at low level. Fortunately, the cell has developed a highly coordinated DNA damage response (DDR) to combat the DNA injuries as the DNA is the vital hereditary material of the living organisms and its protection with integrity is prelude to cellular functioning. DDR may include transient cell cycle arrest, apoptosis activation or the DNA repair systems depending upon the cellular

A. Yasmin (✉) · N. Safdar
Microbiology & Biotechnology Research Lab, Department of Biotechnology, Fatima Jinnah Women University, Rawalpindi, Pakistan
e-mail: azrayasmin@fjwu.edu.pk

Z. Basharat
Jamil-ur-Rahman Center for Genome Research, Dr. Panjwani Center for Molecular Medicine and Drug Research, International Center for Chemical and Biological Sciences, University of Karachi, Karachi, Pakistan

S. Kumar, C. Egbuna (eds.), *Phytochemistry: An in-silico and in-vitro Update*,
https://doi.org/10.1007/978-981-13-6920-9_20

conditions (Connor 2015). DNA repair system comprises of five major repair pathways depending upon the type of damage DNA has suffered. The importance of repairing pathways can be implicated by investigating the neurological and carcinogenic disorders where these well-known repairing pathways or repairing proteins get incompetent. Before understanding the repairing pathways, different DNA damage products should be well known as the repairing pathways are specific for the specific type of damages product DNA produces.

20.2 Upshots of DNA Damage

DNA has the capability to undergo chemical reactions either through endogenous sources or the exogenous agents. Diverse types of DNA damage products are produced through the usual endogenous reactions like hydrolytic, deamination, oxidation and alkylation that react with the sugar or the base of the DNA to form diverse products. Some of the notable DNA damage products produced through these reactions include hypoxanthine, xanthine, uracil, Apurinic or Apyrimidinic sites, oxidative and alkylated DNA adducts like 8-oxo guanine, O-6 methyl guanine, 7-methyl guanine. Single stranded or double stranded scission may also occur in DNA during the said events (Krokan et al. 2002; Yonekura et al. 2009; Ciccia and Elledge 2010). Introduction of occasional insertions and deletions, mistakenly incorporation of nucleotides, error in the proof reading capability of DNA replicative enzymes and error in the DNA repair processes can also generate injured DNA (McClendon and Osheroff 2007; Giglia et al. 2011; Vermeij et al. 2016).

Along with the endogenous sources, exogenous sources (environment) also contribute to the DNA damage products. Physical agents like Ultraviolet light, ionizing radiations (X-rays, cosmic rays, gamma rays, radiotherapies) and chemical agents like nitrosoamines, polycyclic aromatic hydrocarbons (PAH), heterocyclic amines, benzopyrene aflatoxins troubles the cellular DNA. Bulky adducts like cyclobutane pyrimidine dimers, 6,4 pyrimidone are produced when these agents come in contact with DNA. Enhanced Production of ROS and double stranded breaks are also the outcome of these stressful agents (Sordet et al. 2009; Nambiar and Raghavan 2011). Figure 20.1 illustrates different sources responsible for DNA damage with their respective products and the DNA repair systems working in accordance with the DNA damage products.

20.3 Five Major DNA Repair Systems

DNA repair pathways allow the cell to maintain the stability and integrity of the master molecule DNA; thus overcoming the cellular and DNA damage. Mammalian systems usually consist of five major DNA repair pathways with the simplest error free direct reversal mechanism. Photolyase and O-6-methyl guanine/O-6-alkyl

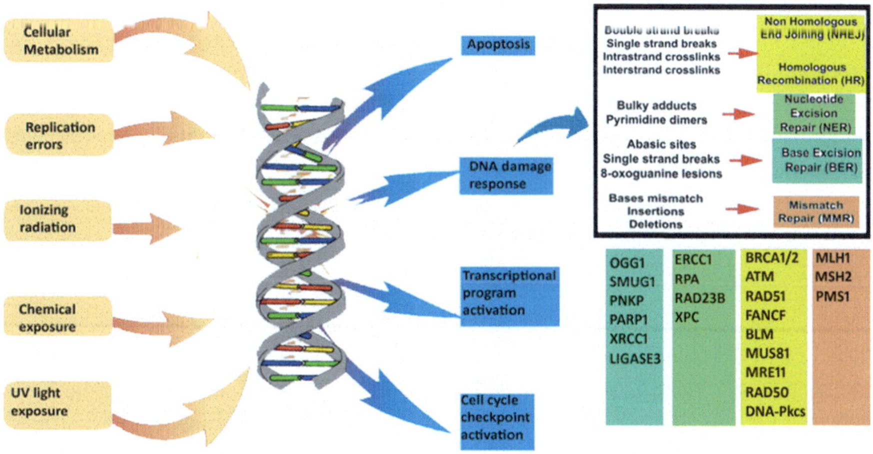

Fig. 20.1 DNA damage causing agents lead to changes in the cell pathway including various lesions such as strand breaks, adducts, insertion/deletion mutations. DNA repair pathways then try to rectify the change by repairing the damage. Repair depends on cell cycle stage as well as the inflicted lesion. Some of the key genes relevant to specific repair pathway shown in same colored boxes beneath the repair mechanism

guanine methyl transferase (MGT/AGT) enzymes are involved in direct repairing of DNA lesions. Complex repair mechanisms involve five major types briefly discussed one by one below.

20.3.1 Base Excision Repair System (BER)

BER is the leading DNA repair mechanism among the other repairing systems. As the name indicates, it works where the base gets damaged and the overall helical structure of DNA does not get distorted significantly. This pathway involves series of enzymes working in a coordinated and systematic manner with the DNA glycosylases being the first player of this pathway. Twelve specific DNA glycosylases work on specific base lesions creating Apurinic or Apyrimidinic (Abasic) site followed by APE1 enzyme repairing the Abasic sites (Abbotts and Madhusudan 2010). APE1 has an intrinsic property of endonuclease and 3′phosphodiesterase for restoration of the 3′ hydroxyl ends; thereby removing the blocking lesions. Tyrosyl-DNA phosphodiesterase 1 (Tdp1) and aprataxin (APTX) are the end processing enzymes involved in repairing of the termini produced during DNA damage. Base excision repair can be of short DNA patch or long DNA patch. DNA polymerase and DNA ligase successfully fill the nucleotide space and ligate it to the remaining strand if the short DNA strand is damaged. DNA polymerase δ or ε, PCNA (proliferating cell nuclear antigen), RFC (replication factor-C), FEN1 (flap endonuclease-1), and DNA ligase I are involved in the repairing of the long patch BER. Besides the above mentioned proteins, X-ray repair cross-complementing protein 1 (XRCC1) and poly (ADP-ribose) polymerase 1 (PARP1) act as facilitative proteins in the BER pathway where they recruit and interact with other BER repair proteins; thereby allowing their interaction to the DNA damaged sites (Krokan and Bjoras 2013). On the whole, contributions of these sequential proteins in the BER pathway formulate effortless identification and repairing of the damaged DNA bases.

20.3.2 Nucleotide Excision Repair System (NER)

NER is a well-known complex pathway for removing bulky lesions from the DNA damaged by UV radiations. Thirty different kinds of proteins participate in the NER system at various steps involving sensing DNA damage, unwinding the DNA around the lesion, removing damaged part of the DNA, repairing and ligation. These entire multisteps require complex proteins' family working together in a coordinated manner. NER system defective cells demonstrates Xeroderma pigmentosum, trichothiodystrophy and Cockayne syndrome in humans depicting the biological importance of this repairing system (Hass and Lam 2012; Dietlein et al. 2014). NER has been categorized in two subtypes GG-NER (Global genome NER) and TC-NER (Transcription coupled NER). In GG type, NER removes DNA lesions throughout the genome whereas TC-NER repair lesions from the coding strand of the actively transcribed gene. Almost similar proteins work in both these types, the only difference exists between the initial steps where specific proteins have to recognize the damage DNA. In the former type XPC/HR23B/CEN2 act as a DNA damaging recognition factor. HR23B and CEN2 recruits XPC and increases the specifity and binding of the XPC to the DNA lesion whereas CSA and CSB (Cockayne syndrome proteins A and B) work in the later type where they dislocate the RNA polymerase II halted at the DNA lesion and allowing other NER proteins to reach the damaged site (Barakat and Tuszynski 2013). Following the damage recognition ten, different protein complex and multi- transcription factors work in a coordinated manner for the repair process. XPA, RPA, XPG, XPF, ERCC1, DNA polymerase, DNA ligase are all the NER proteins commonly working in both the subtypes.

20.3.3 Mismatch Repair System (MMR)

As the name indicates, this system repairs the in-borne replication errors produced by DNA polymerase that escapes the proof reading activity and mismatched nucleotides get incorporated. MMR system also repairs IDL (Insertion/deletion loops) produced by polymerase slippage during replication. Microsatellite instability and increased mutation tendency is reported in MMR deficient cells. If MMR mutation happens in germ lines, then different types of cancer like Lunch syndrome are known to appear. Therefore, this repair system is indispensable for the normal functioning of cellular life (Martin et al. 2010; Fishel 2015). Two MUT family proteins participate in MMR system, i.e. MutS and MutL. In humans, MutS is heterodimer with the combination of two proteins MSH2-MSH6 and MSH2 and MSH3. MSH2/6 recognizes mismatched bases while MSH2/3 identifies larger Insertion deletion loops. Like MutS, MutL also exist in heterodimers with three types MLH1/PMS2 as first heterodimer followed by MLH1/MLH2 and MLH1/MLH3 as a second and third heterodimer respectively. The former heterodimer type participates in repairing of the DNA process, while the rest two types has minor roles observed in the current system (Fukui 2010). PCNA, Replication factor C, Exo 1, Replication protein A, DNA ligase I are the well-known proteins participating in the MMR system for DNA repair.

20.3.4 Homologous Recombination (HR)

HR repairs double stranded DNA damages which are most hazardous and lethal damages of the cell if not repaired. Homologous recombination utilizes the homologous sister chromatid for the replication information and correcting the damaged or lost DNA, thereby, becoming an error free system. This system chiefly works in S and G2 phase of cell cycle. HR consists of mainly the following steps.

1. DNA ends resection for the production of extruding ssDNA
2. Filament forming at the single stranded overhang in order to find homology with the sister chromatid
3. Filament invasion into the homologous region
4. DNA heteroduplex (D loop) formation
5. Reading the homology area through D loop
6. Holliday junction formation
7. Migration and restoration of the repaired DNA strand (Krejci et al. 2012).

Number of proteins play an important role in the HR repair system but RPA complex and Rad51 are known as the two decisive players in repairing process. Both these proteins work in the early and later stages of the repair process. Mutations in any of these proteins can halt the HR repairing system. Rad51 complexes with other proteins and exchanges the strand forming the filaments that could read the homology and synthesize the right nucleotide. Balance production of both proteins help the HR pathway work efficiently (Yan et al. 2011).

20.3.5 Non-homologous End Joining (NHEJ)

NHEJ is an error prone mechanism mediated by relatively many small proteins that does not require sequence homology and efficiently works in the G0 and G1 of the cell cycle. Ku70/Ku80 heterodimer is the initial protein working in NHEJ recognizing and binding to the exposed DNA termini forming a complete ring shape around the DNA; thus encircling it. DNA-PKCS (DNA dependent protein kinases) exhibiting the protein kinase activity is then recruited by the DNA-Ku complex to the double stranded breaks where synapsis of the DNA molecules is promoted (Budke et al. 2012). DNA polymerase μ and DNA polymerase λ belongs to polymerase family resynthesizing the missing nucleotides. APE1, Tdp1, PNKP, Exo1, WRN AND Artemis are the NHEJ proteins participating in the exo and endonuclease activity of DNA; thus processing DNA termini. Once the DNA termini have been processed appropriately, then DNA ligase IV and XRCC4 performs the function of ligation. XLF, another protein factor encourages DNA ligation by interacting with the XRCC4-DNA ligase complex (Deriano and Roth 2013).

20.4 Promising DNA Repair Proteins and Specific Cancer Types

Role of DNA repair proteins in the specific repair systems have been discussed in the previous section. These repair proteins tend to protect and repair the DNA from endogenous and exogenous damage by products. Damaging DNA is the key event in the carcinogenic conditions; though number of cancer treatments is available like chemotherapy, radiotherapy, surgery etc., but this dilemma cannot be completely cured. Chemo and radiotherapy can make the cancer cells resistant to drugs and the current research has now moved and focused on overcoming this resistance problem. One potential approach is to block the DNA repairing pathways in cancerous cells which becomes activated during cancer chemo and radiotherapies. Sensitizing the cancerous cells to anti-cancerous drugs can make them an efficient target in comparison to normal healthy cells. DNA repairing proteins present the prospective target for inhibiting the DNA repairing pathways in cancerous cells specifically; thus blocking their cell proliferation. Literature suggests that direct and BER repair pathways contain the proteins that can potentially be targeted for the cancer therapeutics. Other DNA repair proteins involved in pathways like MMR and NER have not shown profound clinical results (Tell and Wilson 2010). Anyhow, future investigations of DNA repair proteins working in other pathways are still on their way for the possible role in carcinogenic systems. Promising DNA repair proteins as anti-carcinogenic targets with their functions in specific cancer types have been discussed below.

20.4.1 O-6-Alkylguanine DNA Methyltransferase (AGT)

AGT is a typical enzyme involved in direct repair of damaged DNA. Literature suggests potential role of AGT in tumour suppression (Weller et al. 2010). Inhibition of this protein in tumorogenic cells renders them sensitivity against the chemotherapeutic agents. One research study demonstrated the potential role of AGT in brain tumour and malignant melanomas. A chemo drug temozolomide (TMZ) methylates N7 AND N3 of guanine and adenine respectively generation DNA lesions in brain tumor cells. These lesions would be usually repaired by the AGT and BER pathway proteins. Inhibition of the AGT and the BER proteins represents the successful target for increasing the sensitization of the tumor cells towards drug, alternatively leading to tumor cell death. AGT has been successfully inhibited in combination with other chemo drugs like dacarbazine, streptozotocin, procarbazine, BCNU, CCNU (1-2-chloroethyl-3-cyclohexyl-1-nitrosourea) thus presenting a promising target in cancer treatment. Number of clinical trials is on a way assessing the impact of AGT inhibition with chemodrug in glioma, colon cancer, sarcoma and lymphomas (Fan et al. 2013; Kelley and Fishel 2013).

20.4.2 APE1/Ref-1

APE1/Ref-1 is an indispensable enzyme in BER pathway performing versatile functions such as

1. Repairing of DNA damaged by oxidation and alkylation
2. Maintaining replicative transcription factors in active state due to its redox property
3. Stimulation of Transcription factors involved in cancer development to bind their target DNA

Fung and Demple (2005) showed that siRNA synthesized against APE/Ref-1 gene results in an increased addition of AP sites and cellular apoptosis emphasizing its importance in cell survival and propagation. This protein has shown altered expressions in different cancer types like ovarian, prostrate, colon, rhabdomyosarcoma and cervical. APE-1/Ref-1 works in both SP- AND LP-BER and coordinates with numerous proteins as stated before like XRCC1, PCNA, FEN1, β Polymerase; thus offering the potency as a suitable candidate for targeting and enhancing the cytotoxic potential of drugs. Understanding the coordination mechanisms and functional regulations in normal and cancerous cells would be of extreme importance in developing effective inhibitors/drugs against the selected protein. Tell et al. (2009) and Cardoso et al. (2012) have tried to review and analyse the multifunctional roles and post translational modifications of APE-1/Ref-1 protein for potent designing of selective inhibitors against it but still more understanding is required for its significant and consequential clinical interventions.

20.4.3 PARP-1

PARP-1 (poly (ADP) ribosepolymerase-1) is not directly involved in DNA damage removal or repair but it serves as a recruitment factor of interacting with many other BER proteins like β-polymerase, XRCC1, DNA ligase III so that the breaks in the DNA strand could be sensed and repaired by bringing all repairing proteins to the target sites. PARP-1 uses NAD as a substrate for the addition of ADP ribose units to histones and different repairing proteins; thus signalling for various cellular processes like replication, transcription, post translational modifications (Sonnenblick et al. 2015). PARP is a super family of 17 different isoforms with PARP-1 being the first member. Otto et al. (2005) knocked out PARP-I and PARP-2 in mouse embryo and found dead embryos the next week, highlighting the importance of these proteins in embryo development. PARP-1 deficient cells have ineffective BER system and represent a potential target for sensitizing the cancerous cells to chemo and alkylating drugs. PARP inhibitors are found tremendously effectual in BRCA1 and BRCA 2 (Breast cancer susceptible proteins) deficient cell lines, most probably due to lack of PARP signalling for repairing DNA breaks (Venkitaraman 2014). This suggests high sensitivity and killing of BRCA1 and BRCA 2 deficient cells using PARP inhibitors. PARP proteins present promising set of chemotherapeutic agents that will prove effective against a range of cancers.

20.4.4 Double Strand Break (DSB) Repair Proteins

Double stand breaks are the most lethal breaks/damage faced by DNA and is known for triggering the carcinogenesis. These double strand breaks (if not repaired) can inhibit the replication and transcription, eventually inhibiting the cellular division and leading to cellular death. HR and NHEJ repairs the double stranded breaks by recruiting specific proteins to damaged site. MRN complex senses double stranded breaks and recruits ATM protein- the regulatory protein controlling HR pathway (Chapman et al. 2012). Studies have found increased sensitivity to chemo-drugs for ATM defective cells. Different types of drugs/inhibitors have been tested in ATM deficient cells but their side effects like inhibiting other family member proteins poses problem. Combination of various chemo-drugs is being tested in animal models for identifying the specific ATM inhibitor with no side effects (Batey et al. 2013). Other double stranded DNA repair proteins include Ligase 4, Artemis, the Ku proteins, Ku70 and Ku80, XRCC4 and DNA-protein kinase (DNA-PK) functioning in NHEJ. Among all these known proteins, DNA-PK is the ideal candidate in targeting cancer treatment. Over expression of this protein is known to enhance the tumorogenesis resulting in poor cell survival. Contrary, decreased expression of DNA-PK was found to increase the radio sensitization. Clinical trials are underway where inhibitors against DNA-PK are being under test in head, thoracic and neck

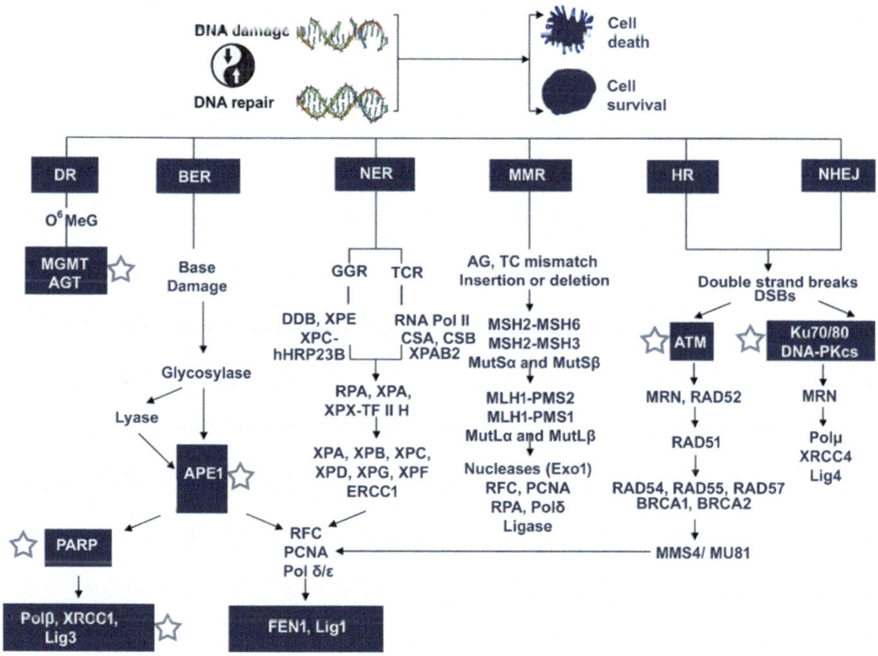

Fig. 20.2 DNA repair systems with specific anti-cancerous target proteins. Boxes with stars represent promising anticancer targets. Availability of other DNA repair proteins as anticancer targets requires further testing. Abbreviations: *DR* Direct Reversal, *BER* Base Excision Repair, *NER* Nucleotide Excision Repair, *GGR* Global Genome Repair, *TCR* Transcription Coupled Repair, *MMR* Mismatch Repair, *HR* Homologous Recombination, *MRN* Mre11–Rad50–Nbs1 complex, *NHEJ* Non-Homologous End Joining, *O6MeG* O6-methylguanine. (Adapted from Tell and Wilson 2010)

cancers; checking their sensitivity together with radio and chemotherapy in cancerous cells (Sishc and Davis 2017).

Inhibition of MMR and NER pathway proteins has not shown successful relationship with cancer sensitization. Though increased resistance of tumor cells to cisplatin owing to the presence of malfunctioned MMR has been reported. This suggests NER or MMR enzyme inhibition may not be an appropriate choice for cancer chemotherapeutics. However, recent investigations into these pathways will open the door for the possible manipulation of these repair proteins. One recent study found ERCC2 NER protein mutated in 15% bladder cancer patients. ERCC2 mutations increased the sensitivity of Cisplatin therapy. Another NER protein, ERCC1 also showed progressive response to Cisplatin when expressed in little cellular amounts (Mouw 2017). Future investigations into NER and MMR proteins can explore their possible role in cancer treatments.

Promising DNA repair proteins as anticancer targets are highlighted in Fig. 20.2 with their specific DNA repair systems. Specific drugs or inhibitors that have been

Table 20.1 Exploring druggable DNA repair proteins

Targeted DNA repair proteins	Drugs/inhibitors under clinical trials
PARP	Olaparib, Veliparib, Niraparib, Talazoparip, Rucaparip, NMS-P118, Iniparip, Methoxy amine, CEP-9722
BER proteins APE1, FEN1, Pol β, XRCC1	E3330, APEI, Formusosin, NSC-124854 (tested for in-vitro cell studies, no drug under clinical trial so far)
NER proteins	NER101, X-80, TDRL/505/551 (tested for in-vitro cell studies)
RPA	Trabecdetin, UCN-01 (under clinical trials)
HR proteins	Mirin, KU-55993, KU-60019 (For in-vitro cell studies)
ATM, and Rad 51	No drug directly under clinical trials
NHEJ proteins DNA-PK, PNKP and Ligase IV	Wortmannin, NU7026, NU7441, L189, SCR7 (compounds under cell studies; no clinical trial under investigation)

Translation of the successful DNA repair protein inhibitors/drugs into future clinical trials are required (Data taken from Kelley et al. 2014)

explored and tested against these target DNA repair proteins are also highlighted in Table 20.1.

20.5 Phytochemicals as Therapy Sensitizers

Phytochemicals as the name suggests compose of plant derived chemical constituents. FDA (Food and drug administration) has approved 30–50% plant based therapies suggesting the importance of plants phytochemicals in different medical treatments (Tuorkey 2015). Phytoconstituents have distinctive well known biological and pharmacological properties like antimicrobial, anti-inflammatory, anticancerous, antioxidants. Around 10,000 of phytochemicals have been identified with anticancerous properties. Currently, phytochemicals are employed synergistically with the cancer chemodrugs/DNA repair inhibitors so as to increase the sensitization of the carcinogenic cells resulting in their cell death (Siddique et al. 2010), but this study needs more clinical trials and results for validation. Though many reviews are available in literature where phytochemicals have been known to show anticancerous properties modulating different signaling pathways and molecular mechanisms in carcinogenic systems (Orlikova and Diederich 2012; Mollakhalili Meybodi et al. 2017; Radhiga et al. 2016). Table 20.1 represents different well known molecular pathways altered by diverse phytoconstituents for the treatment of carcinogenic systems. However, new target approach has been utilized where phytochemicals are evaluated as a potent DNA repair inhibitors in synergism to chemodrugs for increased sensitization of cancerous cells (Dietlein et al. 2014) (Table 20.2).

Table 20.2 Well-known modern phytochemicals with their modulated mechanisms in carcinogenic treatments

Phytochemicals	Mechanisms	References
Arctiin	Modulation of Wnt and MAPK signaling pathways via mRNA expression	Lee et al. (2014)
Ellagic acid	Modulation of Nrf2-Keap1 signalling	Hseu et al. (2012)
Glycyrrhizic Acid	Prevention of photoaging via modulation of NFκB and MMP	Afnan et al. (2012)
Luteolin	Modulation of MMP1 expression via MAPKs and AP1-dependent signalling	Wölfle et al. (2012)
Magnolol	Induces apoptosis via an increased cleavage of caspase-8	Chilampalli et al. (2011)
Myricetin	Modulation of central kinases such as MEK, Janus kinase 1, Akt, and mitogen activated protein kinase 4 kinase by direct binding	Kang et al. (2011)
Plumbagin	Activation of AMPK directly phosphorylates Raptor to inhibit mTOR complex 1 (mTORC1) activation and Bcl-2 expression	Chen et. (2013)
Rottlerin	Inhibiting Wnt/β-catenin and mTORC1 signalling promoted LRP6 degradation and the low density lipoprotein receptor-related protein-6.	Lu et al. (2014)
Rutin	Inhibition of COX2 and iNOS by the potential targeting of p38, MAPK and JNK	Choi et al. (2014)
6-Shogaol	Modulatory effect on mRNA levels of chemokine, cytokine and apoptosis regulatory genes (Il-7, CClS, BAX, BCl2, p21, and p27)	Saha et al. (2014)

Examples extracted from Radhiga et al. (2016) and Tuorkey (2015)

Recent investigations have initiated using phytochemicals either alone or in combination with chemodrugs for possible cancer treatment. One such compound curcumin has shown significant results sensitizing the tumor cells. Curcumin is a well-known compound extracted from *Curcuma longa* and is known to inhibit HDAC activity (Histone deacetylase) which in turn suppresses DNA damage response and DNA repair. In one study, Wang et al. (2015) treated yeast cells (DSB induction) with curcumin and found increased sensitization of yeast cells to DNA lesions. Curcumin inactivated Mec1-DNA checkpoint and Rad52 recombinase resulting in sustaining double stranded breaks and imperfect repair followed by apoptosis. Chapman et al. (2012) identified catechol and 3-hydroxypyridones (a natural polyphenols) as the inhibitor of ERCC1-XPF proteins and augmenter of cisplatin drug effect. Ko et al. (2010) investigated Emodin, a natural anthraquinone as a chemosensitizer and found down regulation of repair protein ERCC1 in lung cancer alone as well as with cisplatin.

20.6 DNA Replication Proteins as Targets of Phytochemicals

DNA damage response and repair protein mutations make one predisposed to cancer. Tumors also accumulate such mutations, leading to defective repair of DNA. Scientists have applied the idea of synthetic lethality in such cases and exploited several drugs against defective proteins to treat malignancies (Brown et al. 2017). Typically, synthetic drugs have been used for the treatment but use of phytochemicals or their derivatives is being recognized as a potent complementary treatment against cancer. Due to their pharmacological properties and inhibition of oxidative stress, they are an asset for use as components of cancer vaccines and adjuvants against cancer (Yin et al. 2017).

Genomic instability due to DNA replication and damage response provides foundation for stipulated targeting of tumors or cancerous cells (over normal cells). Targeting of DNA damage and response protein repair as well as epigenetic players with subsequent role in DNA repair enhances the efficacy of therapeutic agents (including chemotherapeutic compounds, ionizing as well as ultraviolet radiation and radiomimetic chemicals).

Inhibition of these defective proteins via phytochemicals, could cause the death of cancer cells due to enhanced DNA damage response and inhibition of DNA repair mechanism (Rajendran et al. 2011). In-vivo therapeutic efficacy of such phytochemical compounds has been reviewed in detail previously (Link et al. 2010; Rajendran et al. 2011) and advocates that effective concentrations can be attained which alter DNA damage and repair to achieve desired response against tumors. Phytochemicals with pleiotropic properties may also modulate DNA damage and repair through alteration of epigenetic mechanisms, e.g. via microRNAs and DNA methylation in cancer. Better understanding of such auxiliary mechanisms could provide enhanced understanding and strong foundation for coalescing dietary compounds and other therapies like ionizing radiation or chemotherapy, thereby achieving better treatment and response in the clinical setting.

DNA response and repair regulating protein defects have been attempted to be treated by phytochemicals. Some of the examples are mentioned here. Chirnomas et al. (2006) reported that curcumin inhibits the fanconi anemia/BRCA (FA/BRCA) pathway. It is involved in downregulation of MGMT (O6-methylguanine-DNA methyltransferase), DNAPK, Ku70, Ku80, and ERCC-1 protein (Dhandapani et al. 2007). Resveratrol also inhibits DNA repair via activation of ATM pathway (Leone et al. 2010). Quercetin inhibits DNA repair via competitive inhibition of DNAPK, a repair protein involved in NHEJ (Izzard et al. 1999). Roy et al. (2015) reported that thymoquinone from *Nigella sativa* acts on DNA-PKcs. Ruiz et al. (2018) have demonstrated sensitization of cancer stem cells to resveratrol upon binsing RAD51, decrease in viability and expression of the protein and ultimately, apoptosis. Curcumin has also been reported to induce DNA damage by acting on RAD51 and sensitizing the tumor to DNA damage agents as well as PARP1 and DNA-PK inhibitors (Zhao et al. 2018). Oxidative stress in tumor microenvironment is also curbed by phytochemicals such as curcumin, artemisinin, resveratrol, piperine, paclitaxel, berberine, isothiocyanate, noscapine and its derivatives. Apart from testing in mouse models, these have also been evaluated in clinical trials (Cheng et al. 2016).

20.6.1 PARP Inhibition with Phytochemicals

PARP (Poly (adenosine diphosphate [ADP]–ribose) polymerase) inhibition is a therapeutic strategy for the treatment of breast cancer with DNA-repair problem arising in patients with germline mutations of BRCA1/BRCA2. PARP belongs to multifunctional enzyme family and plays a major role in DNA single-strand break repair. Therefore, its inhibition can cause single-strand breaks, which can further lead to DNA double-strand breaks. Usually, such breaks are repaired by means of accurate homologous-recombination of double-stranded DNA. BRCA1 and BRCA2 are major components of this repair pathway but defective alleles of these genes lead to several type of cancers, including breast, prostate and ovarian cancer. PARP inhibitors are cytotoxic to tumor, cause DNA single strand breaks which further lead to double-strand breaks and collapse the replication fork. Since the normal tissue comprises of cells having heterozygous BRCA mutations, they retain accurate homologous-recombination function and are sensitive to inhibitors of PARP. This results in a high therapeutic impact via PARP inhibition in such patients. Fong et al. (2009) have demonstrated via in-vitro assays that BRCA1/2 deficient cells show increased sensitivity (~1000-fold) to PARP inhibitors as compared to wild-type cells. Growth inhibition of ovarian tumor was also seen in BRCA2-deficient xenografts upon PARP and other DNA replication pathway proteins i.e. CHK1 and ATR (George et al. 2017).

PARP1 inhibition has been exploited by several researchers and some of the compounds are in clinical trials. Some scientists have virtually screened phytochemicals against PARP1 protein as well. Mukherjee et al. (2017) reported binding of 12 out of 27 compounds from *Withania somnifera*, screened against PARP1. Five of these compounds i.e. Withaferin A, Withacnistin, Stigmasterol, Withanolide B and Withanolide G showed a remarkable binding affinity to the catalytic domain of PARP1, even greater than an anticancer drug Talazoparib, by Pfizer and currently in trial for breast cancer. Another group screened 16 chemicals from *Lycopersicon esculentum* against PARP1 and found a compound lupeol to have high binding affinity for the protein (Oche et al. 2018).

20.7 Case Study

20.7.1 PARP1 Screening Against Ayurvedic Compounds

In this study we attempted the virtual screening and docking of phytochemicals with the DNA replication pathway protein PARP1 against a library of ~2000 ayurvedic pharmacopia compounds from more than 50 medicinal plants (Langunin et al. 2015). Ayurveda is traditional Indian medicine and dates back to ancient times yet still practiced. Even though it has been imparting good results, scientific testing and validity to the claims needs to be provided (Patwardhan et al. 2005). The ayurvedic compound library, consisting of more than 2000 compounds was obtained from Indian-Russian joint project website http://ayurveda.pharmaexpert.ru/in 3D .sdf form. To the best of authors knowledge, this is the largest library of phytochemicals screened against this protein to date. PARP1 protein structure was obtained from PDB with ID: 5XSR (deposited recently by Chen et al. (2018) during screening of inhibitors against PARP1). Preparation of protein and ligand was carried out as described previously (Basharat et al. 2018). Docking was then carried out with MOE and drug likeliness was also evaluated using Lipinski's drug-like test option in the 2D descriptors in QuaSAR descriptor module of MOE. Only those compounds were selected as top ones which qualified the criteria of drug-like compounds in addition to being potent binder. Top compound with a binding S-value of −7.6 was 1-(2,4-dihydroxyphenyl)-3-hydroxy-3-(4-hydroxyphenyl)-1-propanone (-β-hydroxy-DHP) and shown in Fig. 20.3. Rafi et al. (2002) has reported isolation of this compound from Licorice root (*Glycrrhiza glabra*). It successfully initiated apoptosis and arrest of G2/M cell cycle in both breast cancer and colorectal cancer cell lines. Binding analysis with PARP1 further clarified the phenomenon. Other compounds showing good binding affinity after β-hydroxy-DHP were

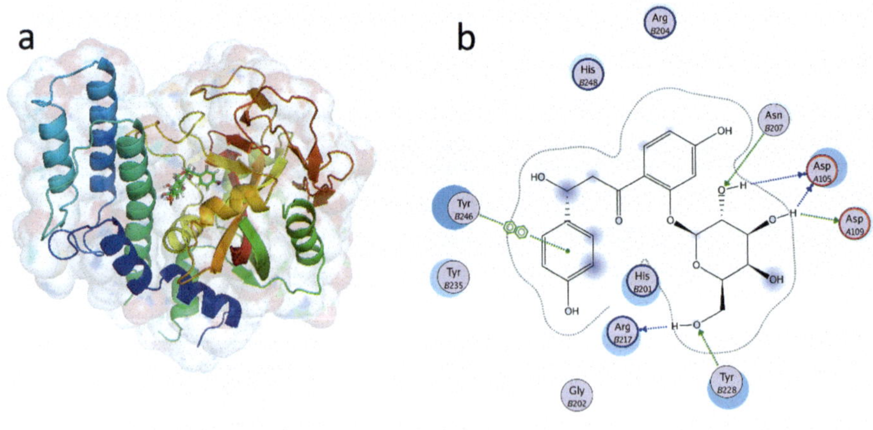

Fig. 20.3 (**a**) 3D representation of docked complex of PARP1 and 1-(2,4-dihydroxyphenyl)-3-hydroxy-3-(4-hydroxyphenyl)-1-propanone and (**b**) 2D representation of docked complex

2′,4,4′-Trihydroxy-3′-prenylchalcone, Bavacoumestan A, (4-Hydroxybenzyl) thiocarbamic acid, Skyrin, Boeravinone B and 4,4′,7,9-Tetrahydroxy-(8->9′)-neolign-7′-ene.

20.7.2 PARP1 Screening against African Natural Product Database

PARP1 was also screened against North African natural product database (Ntie-Kang et al. 2017) available at http://african-compounds.org/nanpdb/ (with ~6000 compounds). North African natural product database is largest annotated compounds from the North African region. Screening was done using the same procedure mentioned before (Basharat et al. 2018). Oleoside 11-methyl ester was the top binding compound with S-value of −10.55 (Fig. 20.4), followed by apigenin 4′-(6-″-methylglucuronide) with an S-value of −10.51. Oleoside 11-methyl ester is a constituent of *Jasminum sambac* (Arabian jasmine) and usually found in tea, olive, and herbs and spices (Rothwell et al. 2013). It is also a constituent of floral buds of *Syringa patula* and shown anti-inflammatory and cytotoxic properties (El-Desouky and Gamal-Eldeen 2009). Apigenins are flavonoids found in the flowers of *Bellis perennis* L (Nazaruk and Gadej 2000). Subsequent to these compounds in binding energy were antirrhinoside, hispidulin-7-sulphate, kaempferol, 3-O-beta-D-glucopyranoside.

Docking results and draggability analysis showed that these compounds have a potential to be tested against cancer and need to be experimentally tested further. If they pass ADME and toxicity filter and enhance effect of other synthetic drugs currently being used for cancer treatment, they could be implicated for cancer cure.

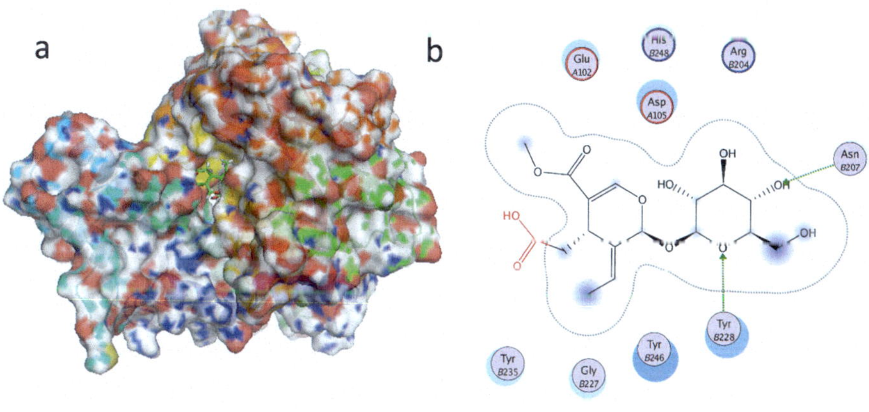

Fig. 20.4 (**a**) 3D depiction of the PARP1 oleoside 11-methyl ester binding. The ligand is visible in the deep pocket (**b**) 2D depiction of the binding interaction of PARP1 with oleoside 11-methyl ester

20.8 Perspective

Although in-silico studies are quick but they only depict the binding potency and some parameters for safety. Lab testing and clinical trials need to be conducted before use of screened compounds. Use of dietary phytochemicals, herbal medicinal products as well as supplements is increasing, and it is a swift approach to computationally screen large libraries and use them for further testing. Many phytoconstituents have been in clinical trials (Yang et al. 2012) but majority of these remain untested in large populace. According to an estimate, around four billion people from developing countries primarily depend on herbal and plant medicinal products for healthcare (Ekor 2014). This is because majority perceives natural compounds as potentially safe, which is both untrue and misleading. Numerous studies have reported adverse reactions and even death upon administration to the patients (Ernst 2002). An Aussie as well as people from Japan and Taiwan died from kidney failure after using aristolochic acid containing herbal medicine (prepared from *Aristolochia fangchi* root and marketed for psoriasis) (Yang et al. 2002; Hong et al. 2006; Chau et al. 2011). Use of this acid containing Chinese herbs has also revealed risk of urothelial cancer (Chen et al. 2012). Another study reported development of liver fibrosis in people using traditional herbal drugs in Uganda (Auerbach et al. 2012). Herbal constituents of *Anacardium occidentalis* and *Entandrophragma utile*, marketed for blood purification by an African company were tested in mice. Investigation revealed that even though the plant extracts and potion evidenced safety during acute toxicity study, testing for chronic toxicity showed enlargement of spleen in several mice and tumor formation in the lung of one (John et al. 1997). Adverse reactions by use of herbal and phytochemical products has been reviewed elsewhere (Ekor 2014). Apart from inadequate knowledge about phytochemicals mode of action, contraindications and possible adverse reactions, interaction with food and other drugs being taken also remains a mystery. There is a need of intense testing, safety establishment and regulatory policy making for phytochemicals to be safe in order to introduce them in the clinics and prevent mishaps of adverse reactions.

References

Abbotts R, Madhusudan S. Human AP endonuclease 1 (APE1): from mechanistic insights to druggable target in cancer. Cancer Treat Rev. 2010;36(5):425–35.

Afnan Q, Adil MD, Nissar-Ul A, Rafiq AR, Amir HF, Kaiser P, et al. Glycyrrhizic acid (GA), a triterpenoid saponin glycoside alleviates ultraviolet-B irradiation-induced photoaging in human dermal fibroblasts. Phytomedicine. 2012;19:658–64.

Auerbach BJ, Reynolds SJ, Lamorde M, Merry C, Kukunda-Byobona C, Ocama P, Semeere AS, Ndyanabo A, Boaz I, Kiggundu V, Nalugoda F. Traditional herbal medicine use associated with liver fibrosis in rural Rakai, Uganda. PLoS One. 2012;7(11):e41737.

Barakat KH, Tuszynski JA. Nucleotide excision repair inhibitors: still a long way to go. In: Chen C, editor. New research directions in DNA repair. Arizona: InTech; 2013. p. 559–76.

Basharat Z, Zaib S, Yasmin A, Tong Y. Drug target identification and virtual screening in pursuit of phytochemical intervention of *Mycobacterium chelonae*. bioRxiv. 2018;315408.

Batey MA, Zhao Y, Kyle S, Richardson C, Slade A, Martin NM, et al. Preclinical evaluation of a novel ATM inhibitor, KU59403, in vitro and *in vivo* inp53 functional and dysfunctional models of human cancer. Mol Cancer Ther. 2013;12:959–67.

Brown JS, O'Carrigan B, Jackson SP, Yap TA. Targeting DNA repair in cancer: beyond PARP inhibitors. Cancer Discov. 2017;7(1):20–37.

Budke B, Logan HL, Kalin JH. RI-1: a chemical inhibitor of RAD51 that disrupts homologous recombination in human cells. Nucleic Acids Res. 2012;40(15):7347–57.

Cardoso AA, Jiang Y, Luo M, et al. APE1/Ref-1 regulates STAT3 transcriptional activity and APE1/Ref-1-STAT3 dual-targeting effectively inhibits pancreatic cancer cell survival. PLoS One. 2012;7(10):e47462.

Chapman JR, Taylor MR, Boulton SJ. Playing the end game: DNA doublestrand break repair pathway choice. Mol Cell. 2012;47:497–510.

Chau W, Ross R, Li JY, Yong TY, Klebe S, Barbara JA. Nephropathy associated with use of a Chinese herbal product containing aristolochic acid. Med J Aust. 2011;194:367–8.

Chen CH, Dickman KG, Moriya M, Zavadil J, Sidorenko VS, Edwards KL, Gnatenko DV, Wu L, Turesky RJ, Wu XR, Pu YS. Aristolochic acid-associated urothelial cancer in Taiwan. Proc Natl Acad Sci. 2012;109(21):8241–6.

Chen MB, Zhang Y, Wei MX, et al. Activation of AMP-activated protein kinase (AMPK) mediates plumbagin-induced apoptosis and growth inhibition in cultured human colon cancer cells. Cell Signal. 2013;25:1993–2002.

Chen X, Huan X, Liu Q, Wang Y, He Q, Tan C, Chen Y, Ding J, Xu Y, Miao Z, Yang C. Design and synthesis of 2-(4, 5, 6, 7-tetrahydrothienopyridin-2-yl)-benzoimidazole carboxamides as novel orally efficacious Poly (ADP-ribose) polymerase (PARP) inhibitors. Eur J Med Chem. 2018;145:389–403.

Cheng YT, Yang CC, Shyur LF. Phytomedicine – modulating oxidative stress and the tumor microenvironment for cancer therapy. Pharmacol Res. 2016;114:128–43.

Chilampalli C, Guillermo R, Zhang X, Kaushik RS, Young A, Zeman D, et al. Effects of magnolol on UVB-induced skin cancer development in mice and its possible mechanism of action. BMC Cancer. 2011;11:456.

Chirnomas D, Taniguchi T, de la Vega M, Vaidya AP, Vasserman M, Hartman AR, Kennedy R, Foster R, Mahoney J, Seiden MV, D'Andrea AD. Chemosensitization to cisplatin by inhibitors of the Fanconi anemia/BRCA pathway. Mol Cancer Ther. 2006;5(4):952–61.

Choi KS, Kundu JK, Chun KS, Na HK, Surh YJ. Rutin inhibits UVB radiation-induced expression of COX-2 and iNOS in hairless mouse skin: p38 MAP kinase and JNK as potential targets. Arch Biochem Biophys. 2014;559:38–45.

Ciccia A, Elledge SJ. The DNA damage response: making it safe to play with knives. Mol Cell. 2010;40:179–204.

Deriano L, Roth DB. Modernizing the nonhomologous end-joining repertoire: alternative and classical NHEJ share the stage. Annu Rev Genet. 2013;47:433–55.

Dhandapani KM, Mahesh VB, Brann DW. Curcumin suppresses growth and chemoresistance of human glioblastoma cells via AP-1 and NFκB transcription factors. J Neurochem. 2007;102 (2):522–38.

Dietlein F, Thelen L, Reinhardt HC. Cancer-specific defects in DNA repair pathways as targets for personalized therapeutic approaches. Trends Genet. 2014;30(8):326–39.

Ekor. The growing use of herbal medicines: issues relating to adverse reactions and challenges in monitoring safety. Front Pharmacol. 2014;4:177.

El-Desouky SK, Gamal-Eldeen AM. Cytotoxic and anti-inflammatory activities of some constituents from the floral buds of *Syringa patula*. Pharm Biol. 2009;47(9):872–7.

Ernst E. Toxic heavy metals and undeclared drugs in Asian herbal medicines. Trends Pharmacol Sci. 2002;23:136–9.

Fan CH, Liu WL, Cao H, Wen C, Chen L, Jiang G. O(6)-methylguanine DNA methyltransferase as a promising target for the treatment of temozolomide-resistant gliomas. Cell Death Dis. 2013;4:876.

Fishel R. Mismatch repair. J Biol Chem. 2015;290(44):26395–403.

Fong PC, Boss DS, Yap TA, Tutt A, Wu P, Mergui-Roelvink M, Mortimer P, Swaisland H, Lau A, O'connor MJ, Ashworth A. Inhibition of poly (ADP-ribose) polymerase in tumors from BRCA mutation carriers. N Engl J Med. 2009;361(2):123–34.

Fukui K. DNA mismatch repair in eukaryotes and bacteria. J Nucleic Acids. 2010;2010:1–6.

Fung H, Demple B. A vital role for Ape1/Ref1 protein in repairing spontaneous DNA damage in human cells. Mol Cell. 2005;17(3):463–70.

George E, Kim H, Krepler C, Wenz B, Makvandi M, Tanyi JL, Brown E, Zhang R, Brafford P, Jean S, Mach RH. A patient-derived-xenograft platform to study BRCA-deficient ovarian cancers. JCI Insight. 2017;2(1):e89760.

Giglia-Mari G, Zotter A, Vermeulen W. DNA damage response. Cold Spring Harb Perspect Biol. 2011;3:000745.

Hass CS, Lam K. Repair-specific functions of replication protein A. J Biol Chem. 2012;287 (6):3908–18.

Hong YT, Fu LS, Chung LH, Hung SC, Huang YT, Chi CS. Fanconi's syndrome, interstitial fibrosis and renal failure by aristolochic acid in Chinese herbs. Pediatr Nephrol. 2006;21:577–9.

Hseu YC, Chou CW, Senthil-Kumar KJ, Fu KT, Wang HM, Hsu LS, et al. Ellagic acid protects human keratinocyte (HaCaT) cells against UVA-induced oxidative stress and apoptosis through the up regulation of the HO-1 and Nrf-2 antioxidant genes. Food Chem Toxicol. 2012;50:1245–55.

Izzard RA, Jackson SP, Smith GC. Competitive and noncompetitive inhibition of the DNA-dependent protein kinase. Cancer Res. 1999;59(11):2581–6.

John TA, Okoli CC, Ekor M. Some phytochemical, safety, antimicrobial and haematological studies of super b and seven keys to power blood purifier herbal tonics. Nig J Med Res. 1997;1:37–43.

Kang NJ, Jung SK, Lee KW, Lee HJ. Myricetin is a potent chemo preventive phytochemical in skin carcinogenesis. Ann N Y Acad Sci. 2011;1229:124–32.

Kelley MR, Fishel ML. DNA repair and cancer therapeutics. In: Tan D, Lynch HT, editors. Principles of molecular diagnostics and personalized cancer medicine. Pennsylvania: Lippincott Williams & Wilkins; 2013. p. 566–79.

Kelley MR, Logsdon D, Fishel ML. Targeting DNA repair pathways for cancer treatment: what's new? Future Oncol. 2014;10(7):1215–37.

Ko JC, Su YJ, Lin ST, Jhan JY, Ciou SC, Cheng CM, Chiu YF, Kuo YH, Tsai MS, Lin YW. Emodin enhances cisplatin induced cytotoxicity via down-regulation of ERCC1 and inactivation of ERK1/2. Lung Cancer. 2010;69:155–64.

Krejci L, Altmannova V, Spirek M, Zhao X. Homologous recombination and its regulation. Nucleic Acids Res. 2012;40(13):5795–818.

Krokan HE, Bjoras M. Base excision repair. Cold Spring Harb Perspect Biol. 2013;5:a012583.

Krokan HE, Drablos F, Slupphaug G. Uracil in DNA–occurrence, consequences and repair. Oncogene. 2002;21(58):8935–48.

Lagunin AA, Druzhilovsky DS, Rudik AV, Filimonov DA, Gawande D, Suresh K, Goel R, Poroikov VV. Computer evaluation of hidden potential of phytochemicals of medicinal plants of the traditional Indian ayurvedic medicine. Biomeditsinskaia Khimiia. 2015;61(2):286–97.

Lee GT, Cha HJ, Lee KS, Lee KK, Hong JT, Ahn KJ, et al. Arctiin induces an UVB protective effect in human dermal fibroblast cells through microRNA expression changes. Int J Mol Med. 2014;33:640–8.

Leone S, Cornetta T, Basso E, Cozzi R. Resveratrol induces DNA double-strand breaks through human topoisomerase II interaction. Cancer Lett. 2010;295(2):167–72.

Link A, Balaguer F, Goel A. Cancer chemoprevention by dietary polyphenols: promising role for epigenetics. Biochem Pharmacol. 2010;80(12):1771–92.

Lu W, Lin C, Li Y. Rottlerin induces Wnt co-receptor LRP6 degradation and suppresses both Wnt/beta-catenin and mTORC1 signaling in prostate and breast cancer cells. Cell Signal. 2014;26:1303–9.

Martin SA, Lord CJ, Ashworth A. Therapeutic targeting of the DNA mismatch repair pathway. Clin Cancer Res. 2010;16(21):5107–13.

McClendon AK, Osheroff N. DNA topoisomerase II, genotoxicity, and cancer. Mutat Res. 2007;623(1–2):83–97.

Mollakhalili Meybodi N, Mortazavian AM, Monfared AB, Sohrabvandi S, Meybodi FA. Phytochemicals in Cancer prevention: a review of the evidence. Iranian J Cancer Preven. 2017;10(1): e7219.

Mouw KW. DNA repair pathway alterations in bladder cancer. Cancer. 2017;9:28.

Mukherjee S, Kumar G, Patnaik R. Identification of potential inhibitors of PARP-1, a regulator of caspase-independent cell death pathway, from *Withania somnifera* phytochemicals for combating neurotoxicity: a structure-based in-silico study. J Theor Comput Chem. 2017;16(07):1750062.

Nambiar M, Raghavan SC. How does DNA break during chromosomal translocations? Nucleic Acids Res. 2011;39:5813–25.

Nazaruk JOLANTA, Gudej J. Apigenin glycosides from the flowers of *Bellis perennis* L. Acta Pol Pharm. 2000;57(2):129–30.

Ntie-Kang F, Telukunta KK, Döring K, Simoben CV, Moumbock A, F A, Malange YI, Njume LE, Yong JN, Sippl W, Günther S. NANPDB: a resource for natural products from northern African sources. J Nat Prod. 2017;80(7):2067–76.

O'Connor MJ. Targeting the DNA damage response in cancer. Mol Cell. 2015;60:547–60.

Oche GA, Eniafe Gabriel O, Olumide Iyang K, Olanrewaju John A, Omotuyi Olaposi I. Lupeol from *Lycopersicon esculentum* as a potential antagonist of the poly (ADP-ribose) polymerase-1 (PARP-1) for cancer chemotherapy: an *in silico* Study. Int J Sci Eng Res. 2018;9(2), 534–552.

Orlikova B, Diederich M. Power from the garden: plant compounds as inhibitors of the hallmarks of cancer. Curr Med Chem. 2012;19:2061–87.

Otto H, Reche PA, Bazan F, Dittmar K, Haag F, Koch-Nolte F. BMC Genomics. 2005;6:139.

Patwardhan B, Warude D, Pushpangadan P, Bhatt N. Ayurveda and traditional Chinese medicine: a comparative overview. Evid Based Complement Alternat Med. 2005;2(4):465–73.

Radhiga T, Agilan B, Muzaffer U, Karthikeyan R, Kanimozhi G, Paul VI, et al. Phytochemicals as modulators of ultraviolet-b radiation induced cellular and molecular events: a review. J Radiat Cancer Res. 2016;7:2–12.

Rafi MM, Vastano BC, Zhu N, Ho CT, Ghai G, Rosen RT, Gallo MA, DiPaola RS. Novel polyphenol molecule isolated from licorice root (*Glycrrhiza glabra*) induces apoptosis, G2/M cell cycle arrest, and Bcl-2 phosphorylation in tumor cell lines. J Agric Food Chem. 2002;50(4):677–84.

Rajendran P, Ho E, Williams DE, Dashwood RH. Dietary phytochemicals, HDAC inhibition, and DNA damage/repair defects in cancer cells. Clin Epigenetics. 2011;3(1):4.

Rothwell JA, Pérez-Jiménez J, Neveu V, Medina-Ramon A, M'Hiri N, Garcia Lobato P, Manach C, Knox K, Eisner R, Wishart D, Scalbert A. Phenol-explorer 3.0: a major update of the phenol-explorer database to incorporate data on the effects of food processing on polyphenol content. Database. 2013;2013:bat070.

Roy M, Kunnumakkara AB, Mukherjee A, Sarkar R, Mukherjee S, Biswas J. Repair activity impaired by Arsenic: recovery by phytochemicals. Int J Curr Microbiol App Sci. 2015;4(3):578–87.

Ruíz G, Valencia-González HA, León-Galicia I, García-Villa E, García-Carrancá A, Gariglio P. Inhibition of RAD51 by siRNA and resveratrol sensitizes cancer stem cells derived from HeLa cell cultures to apoptosis. Stem Cells Int. 2018;2018:1–11.

Saha A, Blando J, Silver E, et al. 6-Shogaol from dried ginger inhibits growth of prostate cancer cells both in vitro and in vivo through inhibition of STAT3 and NF-kappaB signaling. Cancer Prev Res (Phila). 2014;7:627–38.

Siddique YH, Ara G, Beg T, et al. Assessment of cell viability, lipid peroxidation and quantification of DNA fragmentation after the treatment of anticancerous drug mitomycin C and curcumin in cultured human blood lymphocytes. Exp Toxicol Pathol. 2010;62:503–8.

Sishc BJ, Davis AJ. The role of the core non-homologous end joining factors in carcinogenesis and cancer. Cancer. 2017;9:81.

Sonnenblick A, de Azambuja E, Azim HA Jr, Piccart M. An update on PARP inhibitors – moving to the adjuvant setting. Nat Rev Clin Oncol. 2015;12(1):27.

Sordet O, Redon CE, Guirouilh-Barbat J, Smith S, Solier S, Douarre C, Conti C, Nakamura AJ, Das BB, Nicolas E, et al. Ataxia telangiectasia mutated activation by transcription- and topoisomerase I-induced DNA double-strand breaks. EMBO Rep. 2009;10:887–93.

Tell G, Quadrifoglio F, Tiribelli C, Kelley MR. The many functions of APE1/Ref-1: not only a DNA repair enzyme. Antioxid Redox Signal. 2009;11:601–20.

Tell G, Wilson DM. Targeting DNA repair proteins for cancer treatment. Cellul Mol Life Sci. 2010;67(21):3569–72.

Tuorkey MJ. Cancer therapy with phytochemicals: present and future perspectives. Biomed Environ Sci. 2015;28(11):808–19.

Venkitaraman AR. Cancer suppression by the chromosome custodians, BRCA1 and BRCA2. Science. 2014;343:1470–5.

Vermeij WP, Hoeijmakers JH, Pothof J. Genome integrity in aging: human syndromes, mouse models, and therapeutic options. Annu Rev Pharmacol Toxicol. 2016;56:427–45.

Wang SH, Lin PY, Chiu YC, Huang JS, Kuo YT, Wu JC, et al. Curcumin-mediated HDAC inhibition suppresses the DNA damage response and contributes to increased DNA damage sensitivity. PLoS One. 2015;10(7):e0134110.

Weller M, Stupp R, Reifenberger G, Brandes AA, van den Bent MJ, Wick W, Hegi ME. MGMT promoter methylation in malignant gliomas: ready for personalized medicine. Nat Rev Neurosci. 2010;6:39–51.

Wölfle U, Heinemann A, Esser PR, Haarhaus B, Martin SF, Schempp CM. Luteolin prevents solar radiation-induced matrix metalloproteinase-1 activation in human fibroblasts: a role for p38 mitogen-activated protein kinase and interleukin-20 released from keratinocytes. Rejuvenation Res. 2012;15:466–75.

Yan H, Toczylowski T, McCane J, Chen C, Liao S. Replication protein A promotes $5' \rightarrow 3'$ end processing during homology-dependent DNA double-strand break repair. J Cell Biol. 2011;192 (2):251–61.

Yang SS, Chu P, Lin YF, Chen A, Lin SH. Aristolochic acid-induced Fanconi's syndrome and nephropathy presenting as hypokalemic paralysis. Am J Kidney Dis. 2002;39:E14.

Yang Z, Kulkarni K, Zhu W, Hu M. Bioavailability and pharmacokinetics of genistein: mechanistic studies on its ADME. Anti Cancer Agents Med Chem. 2012;12(10):1264–80.

Yin SY, Yang NS, Lin TJ. Phytochemicals approach for developing cancer immunotherapeutics. Front Pharmacol. 2017;8:386.

Yonekura S, Nakamura N, Yonei S, Zhang-Akiyama QM. Generation, biological consequences and repair mechanisms of cytosine deamination in DNA. J Radiat Res. 2009;50(1):19–26.

Zhao Q, Guan J, Qin Y, Ren P, Zhang Z, Lv J, Sun S, Zhang C, Mao W. Curcumin sensitizes lymphoma cells to DNA damage agents through regulating Rad51-dependent homologous recombination. Biomed Pharmacother. 2018;97:115–9.

Chapter 21
In-silico Targets in Immune Response

Amit Bhattacharya, Nidhi Sharma, Neetu Bhattacharya, and Sabyasachi Senapati

21.1 Introduction

Immune system defines the overall natural protection of any living organism against known or unknown antigens. Common antigens include microbes, toxins or could be any synthetic chemicals or environmental particles. Stringent specification, functional hierarchy and molecular compartmentalization contribute to the sophistication of immune system in higher organisms such as in vertebrates, especially in mammals. Single cell prokaryotes and early metazoans have immune system which is primitive in organization but highly robust and adaptive.

Scientific observation on immune system began as early as around 425–430 BC, when Thucydides documented superior immunity status of plague survivors after Peloponnesian War in the Athens. It was the first instance when attack of possible contagious agent on human was predicted based on scientific observation (Retief and Cilliers 1998; Littman 2009). Further, basic or advanced immunological studies were performed either to cure or prevent an epidemic or endemic human diseases due to pathogen infection. Thereafter, along with rapid technological advancements holistic knowledge of the immune system and its functionalities have been excelled exponentially in last few decades.

Integration of computational sciences in biology, and bioinformatics based algorithms implemented in the 1990s helped in the generation of many prediction based data and analyzing experimental results. Development of several open end databases, and databanks further created a niche for more scope of in-silico studies. A

A. Bhattacharya
Department of Zoology, Ramjas College, Delhi University, New Delhi, India

N. Sharma · S. Senapati (✉)
Department of Human Genetics and Molecular Medicine, Central University of Punjab, Bathinda, Punjab, India

N. Bhattacharya
Department of Zoology, Dyal Singh College, Delhi University, New Delhi, India

S. Kumar, C. Egbuna (eds.), *Phytochemistry: An in-silico and in-vitro Update*,
https://doi.org/10.1007/978-981-13-6920-9_21

recent field of Immunoinformatics that uses bioinformatics approach to study immune biology targets the critical issues related to immune system functionality has been very successful in analyzing the diversity and complexity of the human immune system (Wendelsdorf et al. 2012). Use of in-silico study for biological research is advantageous as it is cost effective, time saving and eliminate the need for ethical clearance (Siqueira-Batista 2014). Immunoinformatics facilitate holistic prediction about mechanisms of immune action and thus helps in efficient diagnosis, drug designing and disease prognosis monitoring.

21.2 Computational Tools in Various Immune Responses

Homeostasis is a physiological phenomenon of the body to maintain a stable internal environment through interaction with components of external environment. This concept was given by W.B. Cannon an American Biologist in 1929 (Taniguchi et al. 2009). Immune system senses the external antigens and initiates an immune response to eliminate these foreign antigens and get back to its supervision mode after executing immune action (Husband 1995). The characteristic feature of immune system is to discriminate between host antigen and foreign antigen. This feature is used by immune system as a protective medium to selectively eliminate the foreign pathogen without causing any harm to its own tissue or niche (Chaplin 2010).

21.2.1 Multi-agent System/AutoSimmune

Multi-agent system used to analyze immune system comprises of basic immune simulator (BIS) used to evaluate the interaction between innate and adaptive immune system. *AutoSimmune* is specialized computational system that helps in examining the immune modulations (Siqueira-Batista 2014).

It is an open source system generally used in an agent based modeling where agents are the components of immune system i.e. antigen, antibody, T-cell, B-cell, cytokines, macrophages, mast cells, NK-cells etc. This system has been used to explore the role of mast cell in inflammation and in immune response after infection. Besides this it has been used to study Malaria, Chagas disease and Sepsis. An example of this system is a computational model used for neutrophils. In *AutoSimmune* model neutrophil agents represents the neutrophils and imitate their phagocytic action. After inserting these agents in the bone marrow zone, neutrophils get associated with the different zones after receiving the signal from pre-inflammatory cytokine. Neutrophil agents are capable of identifying the pathogens via showing their affinity (affinity threshold parameter) towards the cells showing PK1 (a cellular stress signaling substance) (Siqueira-Batista 2014) (Fig. 21.1).

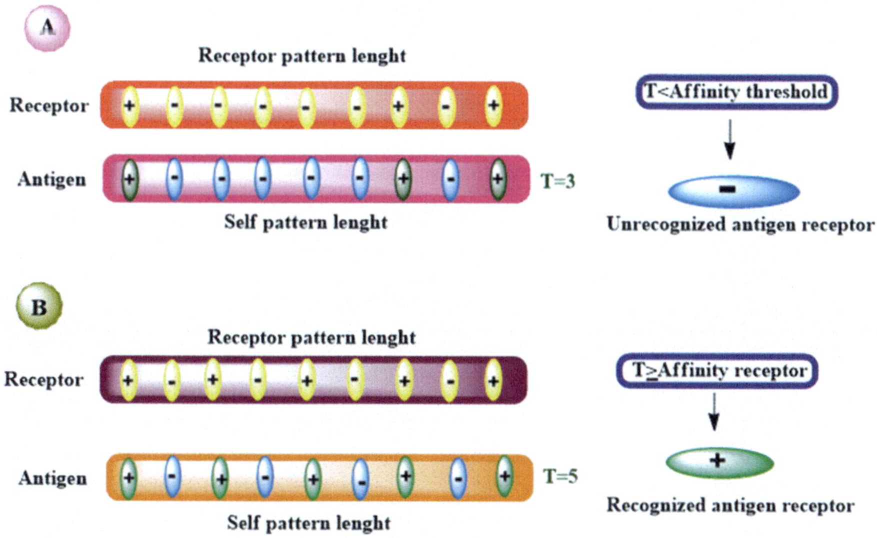

Fig. 21.1 Degree of affinity of antigen for its receptor: T represents the no. of recognized receptors which are 3 in part A and 5 in part B. Green color indicates the recognition/matching of receptor while blue color indicates the unrecognized/mismatch receptor

21.2.2 In-silico Tool for Intestinal Immune Response

ENISI elaborated as an Enteric Immunity stimulator is an in-silico system developed to initiate the mucosal immune response of the gut system. Enteric system is related to the gastrointestinal tract. By using ENISI researchers can test and design hypothesis for disease related to enteric system and would be able to propose an experiment via causing infection in the gut. Advantages of this computer based system is that,

1. It initiates the cellular components (T-cell, B-cell, dendritic cell) of the regulatory network that are involved in maintaining the homeostasis of the gut system
2. It enables detailed representation of activated (through different pathways) macrophages, $CD4^{+}T$ cells and dendritic cell is a characteristic feature of ENISI
3. It allows integration of cellular interaction such as movement, proliferation and differentiation with tissue level phenotype and can be represented using ENISI with high resolution. (Wendelsdorf et al. 2012)

21.2.3 In-silico Model in Cancer Vaccine

SimTriplex model an agent based system is an example of this category which models cancer vaccine. It imitates the reaction of "Triplex"- a cancer preventive vaccine in Her-2/neu transgenic mice, susceptible to mammary tumor. A successful vaccine schedule is important to inhibit the formation of solid tumor for the lifetime. SimTriplex model along with optimized techniques permit us to identify the vaccine schedule, based on immune system functionality and guided action (Hattotuwagama et al. 2007).

21.3 Strategies for Immune Modulation: Recent Updates

21.3.1 Adaptive Immune System

With the arrival of next-generation sequencing (NGS) and advance immunoinformatics tools based on computational methods, it has become feasible to predict various immunological functions/responses. Tumor exome and transcriptome sequencing help to decipher information for cancer immunotherapy such as mutated genes, over/under-expressed genes and aberrant splicing (Schmidt et al. 2017; Vita et al. 2010). Different in-silico prediction softwares are available for peptide predictions (Artificial Neural Network, IEDB), proteasomal cleavage predictions, peptide prediction accuracy, MHC ligand prediction algorithms, known HLA alleles (International ImMunoGeneTics information system database, IMGT) etc. (Backert and Kohlbacher 2015). One of the oldest databases is SYFPEITHI, which contains information on peptide sequences, anchor positions, MHC specificity, source proteins, source organisms and references. IMGT, created in 1989, is a high-quality immunogenetics and immunoinformatics resource covering immunoglobulins (IG) or antibodies, T cell receptors (TCR), Major histocompatibility complexes (MHC) of human and other vertebrate species, and the immunoglobulin superfamily (IgSF), MH superfamily (MhSF) and related proteins of the immune system (RPI) of vertebrates and invertebrates. The IPD-IMGT/HLA Database, part of the IGMT, provides a special repertoire for sequences of the human major histocompatibility complex (MHC) and comprises the official sequences named by the WHO Nomenclature Committee for Factors of the HLA System. DFRMLI database links the gap between immunological data and computer science/machine learning applications. AntiJen is a thermodynamic and cellular database that also includes peptide library, copy numbers and diffusion coefficient data along with MHC Ligand molecules, MHC Ligand kinetics, T Cell Epitope, TAP, B Cell Epitope molecules, Protein-Protein interactions and Protein complexes. Mathematical models and computer simulations have been extensively used also to study immune responses (such as adaptive immunity) to viral infections (Human Immunodeficiency Virus type 1, Simian immunodeficiency viruses, Lymphocytic choriomeningitis virus, and influenza A virus) (Lee et al. 2009). This two-compartment model is a simplification of the complex biological process that can compute the interaction between viral replication and adaptive immunity and various parameters affecting the production of CD8 T-cell effectors. Multi agent system (MAS)- based immune system modelling approaches have been effectively useful to simulate many complex systems and testing of hypotheses about a disease (such as poststreptococcal glomerulonephritis, PSGN) (Bastos et al. 2016). The model primarily focused its operation in the Repast Simphony framework1, version 1.2 (Possi et al. 2011). This immune system simulator (AutoSimmune) had the representation of numerous essentials for immune system simulation such as space, time, and representations of cells, tissues, and substances. The following parameters were used for simulations: Initial number of Pyogenes (initial number of bacteria at the start of simulation), Pyogenes Virulency (the ability of the bacteria to replicate) and Pyogenes Latency. The standard version of AutoSimmune contains several components relating to immune response cells such as Natural killer (NK) cells, cytotoxic T lymphocytes, B lymphocytes, macrophages, and dendritic cells.

Newer Immunoinformatics algorithms for identifying T-cell epitopes have been developed since the initial ones by Berzofsky, Margalit, and DeLisi (in the 1980s). The 'immunogenicity score' of a protein from such algorithms helps to evaluate and predict protein therapeutics in the preclinical phase and immunogenicity to provoke an immune response. One such example is the 'EpiMatrix', a T-cell epitope mapping algorithm and computational tool. In EpiMatrix, eight most common HLA class II alleles and six super-type HLA class I alleles are considered. Peptide scoring above 1.64 (on the EpiMatrix Z scale) and above 2.32 indicate a significant binding and extremely likely to bind to the desired MHC molecule respectively. However, precise prediction of B-cell antigenic epitopes (regions of the antigen protein surface preferentially recognised by B cells) is helpful for designing vaccines and immunodiagnostic kits. Nevertheless, novel computational tools for B-cell epitopes (continuous or discontinuous) prediction remain elusive due to conformational dependence of antigen-antibody interactions, computational complexity and limited number of known antibody-antigen complex structures. Some prediction algorithms for discontinuous epitope prediction are CEP, DiscoTope, BEpro (PEPITO), ElliPro, SEPPA, EPITOPIA, and EPCES, EPSVR, EPMeta, and Bpredictor (reviewed in Yao et al. 2013). The B-cell epitope prediction tools 3D-Epitope-Explorer (3DEX) algorithm tool helps to allocate the conformational epitopes in 3D protein structures considering the physicochemical environs of C(alpha)- or C(beta)-atoms of each amino acids. These different prediction algorithms and freely accessible web servers or downloadable software packages have augmented the in-silico prediction studies to identify antigenic or immunogenic determinants on protein antigens (Table 21.1).

21.3.2 Innate Immune System

The immune response is a complex mechanism that involves understanding the function of the host, infectious agents, regulated variations in the pathogen virulence and interplay with the host microbiome (Clark and Kupper 2005). For example, a significant difference has been recorded in the macrophage responses to viable pathogens organisms (such as *Mycobacterium tuberculosis*) as compared with the response to inactivated bacteria. The innate immune response is essential for

Table 21.1 List of some immunoinformatics databases

Database	Contents available	Links
SYFPEITHI	Peptide sequences known to bind class I and class II MHC molecules	http://www.syfpeithi.de/
IMGT	Antibodies, T-cell receptors	http://www.imgt.org/
IEDB (The Immune Epitope Database and Analysis Resource)	Epitopes, epitope-MHC/B-cell receptor complexes	https://www.iedb.org/
MHCBN 4.0	Major histocompatibility complex (MHC) binding, non-binding peptides and T-cell epitopes	http://crdd.osdd.net/raghava/mhcbn/
IPD-IMGT/HLA	Human leukocyte antigens (HLA) alleles	https://www.ebi.ac.uk/ipd/imgt/hla/
Dana-Farber Repository for Machine Learning in Immunology (DFRMLI)	MHC ligands	http://projects.met-hilab.org/DFRMLI/
AntiJen	MHC ligands, T cell receptor (TCR)-MHC complexes, T-cell epitopes, TAP, B-cell epitopes, protein-protein interactions	http://www.ddg-pharmfac.net/antijen/AntiJen/antijenhomepage.htm
EpiMatrix	T-cell epitopes	http://i-cubed.org/tools/ivax/ivax-tool-kit/epimatrix/

protection against bacterial, fungal and parasitic infections while the adaptive immune response is responsible for combating disease pathogens and budding into the memory response mechanism. The two immune systems share several receptors, regulatory systems, signal transduction passageway and effector machinery. The advance biomedical research has helped to decode the innate immune response, often regarded as human body's first line of defence against invading pathogens, and complex communication between the innate and adaptive immune systems to improve our understanding of their significance in protection against a range of pathogens (Gardy et al. 2009). As a result, novel 'in-silico' research tools are now used to supplement the traditional research studies to decode the biocomplexity. Over the years, several bioinformatics platforms have been developed to understand innate and acquired immune system interactions and reasons for maladaptive interactions between these two systems that lead to harmful immunologic responses (such as allergy, autoimmunity, and allograft rejection) (Table 21.2).

Transcriptomics, genome-wide meta-analysis of the immune response, is enabling researchers to unearth processes and pathways of host innate immune system to a range of pathogens. Microarrays and new technologies (such as next-generation sequencing and exon and microRNA arrays) are giving new insights into innate immunity (Ricciardi-Castagnoli and Granucci 2002; Wang et al. 2009). Meta-analysis resources such as Array Express and Gene Expression Omnibus (GEO) are

Table 21.2 List of web-accessible bioinformatics resources

Bioinformatics resource	Database name	Web-link (URL)
Immunology-specific resources	Immune Response In Silico (IRIS)	http://share.gene.com/clark.iris.2004/iris/iris.html
	The Immunological Genome Project	http://www.immgen.org
	The Immunome Database	http://bioinf.uta.fi/Immunome/
	Innate Immunity Database	http://db.systemsbiology.net/IIDB
	InnateDB	http://www.innatedb.ca
	The Immunology Database and Analysis Portal (ImmPort)	http://www.immport.org/
	Reference Database of Immune Cells (RefDIC)	http://refdic.rcai.riken.jp
Pathway databases	Integrating Network Objects with Hierarchies (INOH)	http://www.inoh.org/
	Kyoto Encyclopedia of Genes and Genomes (KEGG)	http://www.genome.ad.jp/kegg/
	NCI-Nature Pathway Interaction Database (PID)	http://pid.nci.nih.gov
	NetPath	http://www.netpath.org
	Pathguide	http://www.pathguide.org/
	Reactome	http://www.reactome.org/
	Integrating Network Objects with Hierarchies (INOH)	http://www.inoh.org/
Interaction databases	Biomolecular Interaction Network Database (BIND)	http://bond.unleashedinformatics.com
	Database of Interacting Proteins (DIP)	http://dip.doe-mbi.ucla.edu/
	Molecular Interaction Database (MINT)	http://mint.bio.uniroma2.it/mint/
	Pathogen Interaction Gateway (PIG)	http://molvis.vbi.vt.edu/pig/
	The Biological General Repository for Interaction Datasets (BioGRID)	http://www.thebiogrid.org/
	VirusMINT	http://mint.bio.uniroma2.it/virusmint/
	Biomolecular Interaction Network Database (BIND)	http://bond.unleashedinformatics.com
Network analysis platforms	CEll REgion-Based Rendering And Layout (CEREBRAL)	http://www.pathogenomics.ca/cerebral/
	Cytoscape	http://www.cytoscape.org/
	HUB oBjects Anlyzer (HUBBA)	http://hub.iis.sinica.edu.tw/Hubba/
	Network Analysis Tools (NEAT)	http://rsat.ulb.ac.be/rsat/index_neat.html

(continued)

Table 21.2 (continued)

Bioinformatics resource	Database name	Web-link (URL)
Gene expression repository	Array Express	http://www.ebi.ac.uk/microarray-as/ae/
	Gene Expression Omnibus	http://www.ncbi.nlm.nih.gov/geo/
	International HapMap Project	http://www.hapmap.org/
	miRBase	http://microrna.sanger.ac.uk

public repository for transcriptomics data from all species while International HapMap Project describes the patterns of genetic variation in humans. Several research groups have created online publicly available databases of innate and adaptive immunology-relevant transcriptomics datasets. Some of these resources are: Reference Database of Immune Cells (RefDIC; Database of immune cells and tissues mRNS and protein profiles), Immune Response In Silico database (IRIS; Microarray expression datasets), Innate Immunity Database (Genomic annotations from microarray experiments), The Immunological Genome Project (Genome-wide gene expression datasets in mouse immune cells) and The Immunome Database (Database of Human immunity related proteins) etc. Jenner and Young (2005) by transcriptional profiling of 32 in-vitro human studies identified a cluster of 511 genes demonstrating host response to bacteria, viruses and selected other pathogens. However understating genome-wide genetic regulatory networks novel promising high-throughput approaches (chromatin-immunoprecipitation (ChIP) array, ChIP-chip and ChIP-seq technology) are developed. Kyoto Encyclopedia of Genes and Genomes (KEGG) and NCI-Nature Pathway Interaction Database (PID) are curated database of biological pathways. The Basic Immune Simulator (BIS), an agent-based model with Graphical User Interface (GUI), was developed to study the innate and adaptive interactions of the immune response to pathogens (Folcik et al. 2007). Agent-based modelling (also known as individual-based modelling, bottom-up modelling, and pattern-oriented modelling) are used to study the non-linear behaviour of complex systems. In such computer-simulated environment, the basic elements consist of agents and signals while the interactions between agents are also shown. The BIS was shaped using the open-source software Recursive Porus Agent Simulation Toolkit (RepastJ) library and consists of 03 virtual spaces indicating parenchymal tissue, secondary lymphoid tissue and the lymphatic/humoral circulation. These in-silico platforms methodically identify potential targets for more efficient cure for diseases such as hypersensitivity reactions (allergies, asthma), autoimmunity and cancer.

21.4 Conclusion

Significant advancement has been made in the emerging field of immunoinformatics and in-silico immune response studies. Different mathematical, computational and statistical models have been successfully applied to understand the immune modulation upon epitope recognition. It has also been successfully applied to identify novel immune interactions and antigen epitope prediction. Most of these tools are on the open access and thus provided ample scope for further improvement. Newer robust and sensitive tools are warranted to make this approach scientifically more successful.

References

Backert L, Kohlbacher O. Immunoinformatics and epitope prediction in the age of genomic medicine. Genome Med. 2015;7:119.

Bastos CA, Oliveira A, Cerqueira FR et al. Multiagent-based simulation of the human immune system: a study of the immune response and antimicrobial therapy in post-streptococcal glomerulonephritis. In: FLAIRS Conference; 2016. p. 374–9.

Chaplin DD. Overview of the immune response. J Allergy Clin Immunol. 2010;125(2):S3–S23.

Clark R, Kupper T. Old meets new: the interaction between innate and adaptive immunity. J Invest Dermatol. 2005;125:629–37.

Folcik VA, An GC, Orosz CG. The basic immune simulator: an agent-based model to study the interactions between innate and adaptive immunity. Theor Biol Med Model. 2007;4:39.

Gardy JL, Lynn DJ, Brinkman FSL, et al. Enabling a systems biology approach to immunology: focus on innate immunity. Trends Immunol. 2009;30:249–62.

Hattotuwagama CK, Guan P, Davies M, et al. Empirical, AI, and QSAR approaches to peptide-MHC binding prediction. In: Flower D, Timmis J, editors. In silico immunology. Boston: Springer; 2007. p. 139–75.

Husband AJ. The immune system and integrated homeostasis. Immunol Cell Biol. 1995;73(4):377.

Jenner RG, Young RA. Insights into host responses against pathogens from transcriptional profiling. Nat Rev Microbiol. 2005;3:281–94.

Lee HY, Topham DJ, Park SY, et al. Simulation and prediction of the adaptive immune response to influenza A virus infection. J Virol. 2009;83(14):7151–65.

Littman RJ. The plague of Athens: epidemiology and paleopathology. Mt Sinai J Med. 2009;76 (5):456–67.

Possi MA, Oliveira AP., Chaves C et al. An *in-silico* immune system model for investigating human autoimmune diseases. In XXXVII Conferencia Latinoamericana en Informatica CLEI. 2011.

Retief FP, Cilliers L. The epidemic of Athens, 430–426 BC. Afr Med J. 1998;88(1):50–3.

Ricciardi-Castagnoli P, Granucci F. Interpretation of the complexity of innate immune responses by functional genomics. Nat Rev Immunol. 2002;2(11):881.

Schmidt J, Guillaume P, Dojcinovic D, et al. *In silico* and cell-based analyses reveal strong divergence between prediction and observation of T-cell–recognized tumor antigen T-cell epitopes. J Biol Chem. 2017;292:11840–9.

Siqueira-Batista R. The artificial neutrophil and a proposal of an in silico research of the immune response in human bacterial diseases. Abakós. 2014;2(2):79–91.

Taniguchi Y, Yoshioka N, Nakata K, et al. Mechanism for maintaining homeostasis in the immune system of the intestine. Anticancer Res. 2009;29(11):4855–60.

Vita R, Zarebski L, Greenbaum JA, et al. The immune epitope database 2.0. Nucleic Acids Res. 2010;38:D854–62.

Wang Z, Gerstein M, Snyder M. RNA-Seq: a revolutionary tool for transcriptomics. Nat Rev Genet. 2009;10(1):57.

Wendelsdorf KV, Alam M, Bassaganya-Riera J, et al. ENteric immunity SImulator: a tool for in silico study of gastroenteric infections. IEEE Trans Nanobiosci. 2012;11(3):273–88.

Yao B, Zheng D, Liang S, et al. Conformational B-cell epitope prediction on antigen protein structures: a review of current algorithms and comparison with common binding site prediction methods. PLoS One. 2013;8(4):e62249.

Chapter 22
In-silico Targets in Neurodegenerative Disorders

Aniko Nagy and Timea Polgar

22.1 Neurodegenerative Diseases

Amyotrophic lateral sclerosis (ALS), Cerebellar ataxia (CA), Parkinson's disease (PD), Alzheimer's disease (AD), and Motor Neuron disease (MND) are known as the most frequently occurring human neurodegenerative diseases. All of them are related to aging, and they are progressive disorders affecting certain classes of neurons found in the central nervous system (CNS). For example, PD is described by loss of nigrostriatal dopaminergic neurons in the substantia nigra pars compacta (SNc), which causes reduction in dopamine levels and cytoskeletal inclusion of Lewy bodies (Pandey et al. 2018).

Of all neurodegenerative diseases, AD is the most common, which was first identified as a primary cause of death for people suffering from mental illness (Katzman 1976). There are several hypotheses about the underlying reasons that can be classified into the following three categories (Herrup 2015): cellular, genetic, and molecular imbalances (Thai et al. 2017).

Its etiology is mainly unknown. It has already been demonstrated that oxidative stress can play a key role in the etiology of late onset neurodegenerative diseases (Flynn and Melov 2013). Oxidative stress was found to be relevant in a range of neurodegenerative diseases, and emerging evidence from in-vitro and in-vivo disease models demonstrates that oxidative stress can be important in general disease pathogenesis (Pandey et al. 2018).

A. Nagy
NaturAI Inc, Grandville, MI, USA
http://www.naturai.net

T. Polgar (✉)
NaturAI Inc, Grandville, MI, USA

Globe Program Association, Fort Collins, CO, USA
http://www.globe pa.org

S. Kumar, C. Egbuna (eds.), *Phytochemistry: An in-silico and in-vitro Update*,
https://doi.org/10.1007/978-981-13-6920-9_22

Pharmacological treatment of neurodegenerative disorders has not been solved; therefore, there is a need to investigate new drug candidates. Modification of endogenous chemicals can offer new drug candidates with better potencies and/or further functions that may be of therapeutic benefit. Computational sciences can support the discovery of such candidates and can further improve the execution speed, decrease cost and the usage of test animals (Wenzel and Klegeris 2018). Computational biology offers a way for discovery of novel pharmacological targets and can pinpoint the genetic background of such diseases. While computational chemical approaches involve multiple methods that can support the discovery of new chemical entities. These methods for example include cheminformatics (Pandey et al. 2018), docking and molecular dynamics modeling (Thai et al. 2017).

Ziprasidone is an example of a recent drug candidate, which serves the basis of currently applied therapies in many countries for the treatment of bipolar disorder or psychosis. This dual-target antipsychotic treatment combines and optimizes the following pharmacophores: dopamine and naphthylpiperazine, a serotonin receptor ligand. Ziprasidone was shown to have less serious side effects compared to other antipsychotics (Wenzel and Klegeris 2018).

In this chapter, we would like to discuss the possible targets of the two most common neurological diseases, AD and PD. The known and the possible new targets are shown and their therapeutic importance is also detailed. In addition, the methods of their discovery is highlighted demonstrating the importance of the in-silico discovery of new targets in neurological diseases.

22.2 Alzheimer's Disease (AD)

AD is the most common type of dementia in the population over age 65 (Dias Viegas et al. 2018). According to World Alzheimer's Report, in 2015, more than 35 million people suffered from AD, and it is predicted to grow exponentially in the future (Sang et al. 2017).

AD is described by progressive loss in cognition, decrease in functional and motor capacity, damage in behavioral, social autonomy and finally death (Dias Viegas et al. 2018). AD is a result of a neurodegeneration, which is caused by extracellular β-amyloid plaques, while intracellular neurofibrillary tangles cause neurotoxicity and synaptic loss (Agatonovic-Kustrin et al. 2018).

Current clinical treatment is only palliative and restricted to five pharmaceuticals approved by the US FDA. These are donepezil, galantamine, hupersine A and rivastigmine, which are reversible or competitive acetylcholinesterase inhibitors (AChEI), and memantine, which is a non-competitive N-methyl-D-aspartate (NMDA) receptor antagonist (Agatonovic-Kustrin et al. 2018).

None of the approved drugs demonstrated adequate efficacy, tolerability in most patients. Current therapy only provides symptomatic cures and temporarily delay cognitive decline in about half of the patients, while having only a minor effect in severe, advanced cases (Blennow et al. 2006). Furthermore, they have some adverse

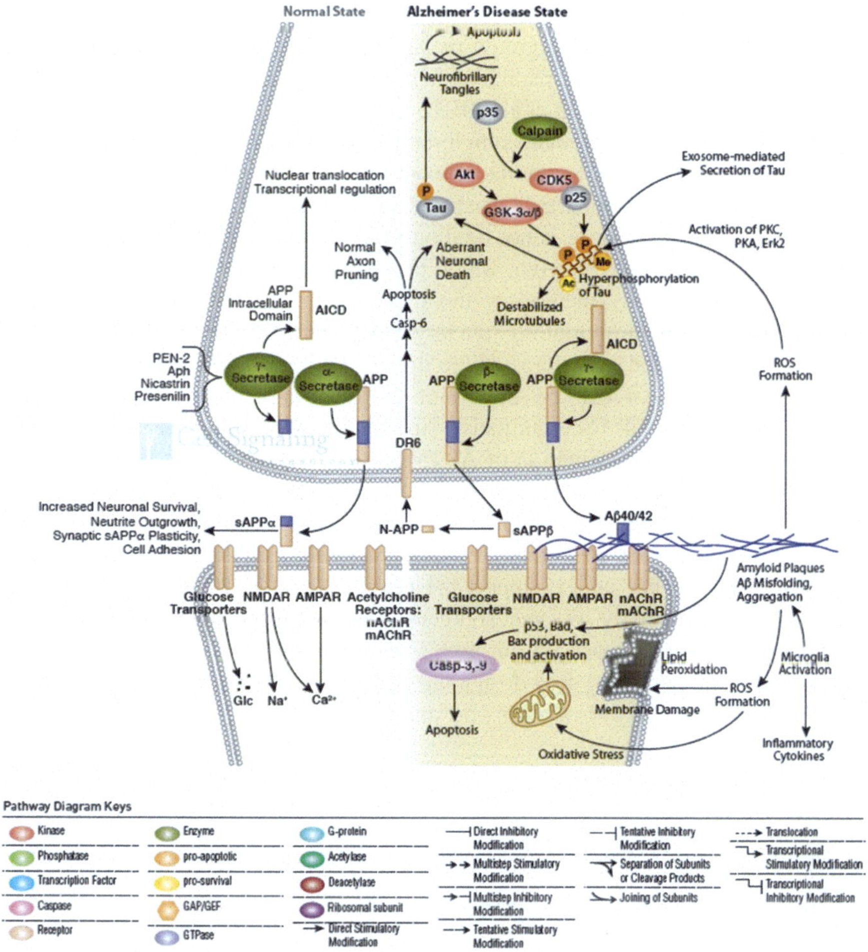

Fig. 22.1 Amyloid plaque formation in Alzheimer's disease signaling interactive pathway. (Illustration reproduced courtesy of Cell Signaling Technology, Inc. (www.cellsignal.com))

effects such as diarrhea, dizziness, nausea, vomiting, because of their poor bioavailability, hepatotoxicity and non-selectivity (Agatonovic-Kustrin et al. 2018).

Current medication cannot even stop the progression of the disease. AD was already described by multiple physiological and biochemical malfunctions involving simultaneously operating chemical mediators and numerous protein targets (Dias Viegas et al. 2018). Some of the concerned factors are low levels of acetylcholine, beta amyloid (Aβ) deposits, tau protein hyperphosphorylation and aggregation, oxidative stress, neuroinflammation and dyshomeostasis of bio-metals (Jalili-Baleh et al. 2018).

Several therapeutic strategies have been under exploration to increase older population functionally and life expectancy, like using acetylcholinesterase

(AChE) and butyrylcholinesterase (BuChE) inhibitors to elevate cholinergic neurotransmission. Furthermore, prevention of Aβ deposition and oxidative damages in the brain are also an important parameter to decrease the pathological complications of AD (Fig. 22.1) (Jalili-Baleh et al. 2018).

Important to note that, many single-target candidates for neurodegenerative diseases that are effective in animal models finally have not been effective in human clinical trials (Zimmerman et al. 2007). The reasons for these failures can be for example, that the drug is not binding to its in-vivo human target or it is not reaching the human target in-vivo, which problems could be solved by modifying the structure or delivery method of the candidates. Another, more complicated problem can be, if the drug binding to a target does not lead to enough of an impact to reduce the signs and symptoms of the disease (Wenzel and Klegeris 2018). To develop new drugs, it can be important to use computational chemistry and cheminformatics to look for quantitative structure-activity relationships to identify molecules to inhibit the formation of Aβ (Nastase and Boyd 2012). Many in-silico approaches have been broadly applied for 'single-target' anti-AD drug research based on the conception of 'therapeutic target' including docking, molecular dynamics, quantum mechanics, and other methods that could result in quantitative structure. Novel approaches could support the development of 'multi-target' anti-AD drugs with integration of diverse sources of information, such as microarrays, literature mining-data or protein– protein interactions (PPIs) (Sun et al. 2013). In conclusion, computer-aided drug design has been one of the main tools used in drug discovery projects used to lower the cost and process time.

22.2.1 Pharmacological Targets

22.2.1.1 Acetylcholinesterase (AchE)

Disturbance in the acetylcholine (ACh) releasing cholinergic neurotransmission was shown to likely be coupled with the memory loss in the brain of AD patients. Several types of learning and cortex plasticity are dependent on Ach production. The cholinesterase enzymes lower the concentration of ACh through hydrolyzing it. Cholinesterase blockers can bind to these enzymes leading to higher levels of ACh in synapses (Kumar et al. 2016).

The relation between the malfunctions of neurotransmissions and AD severity revealed a rationale for the application of AchE inhibitors, like donepezil, rivastigmine and galantamine. Cholinesterase blockers can reversibly bind to cholinesterase enzyme, which is responsible for degradation of acetylcholine within the synaptic cleft and increases cholinergic transmission between neurons. Randomized controlled trials have showed significant proof of benefit of AchE inhibitors on cognitive and functional function, although these benefits tend to have only insignificant clinical effects (Epperly et al. 2017).

AchE inhibitors recommended to be used in association with other drugs for the most serious cases of the disease. This also demonstrates the importance of new multi-target (polypharmacological), anti-Alzheimer's drug candidates (Qian et al. 2018). For instance, novel donepezil-derived *N*-acyl-aryl-hydrazone ligands with multifunctional purposes can be promising novel drug candidates for therapy. Donepezil's AChE inhibitory activity as a reference drug for mild and moderate stages of the disorder and previous research studies on *N*-acylhydrazone moiety further support this idea. It can act as pharmacophore or auxophore subunit, with a broad range of bioactivity profile, depending on other functionalities present in the molecular structure. Pharmacological evaluation demonstrated that these derivative molecules showed significant AChE inhibitory activity (Silva et al. 2016) and antioxidant capacity with metal chelate ability (Turnaturi et al. 2016).

With these activities of basic chemicals, donepezil-derived *N*-acyl-aryl-hydrazone could have an effect through cholinergic signaling, excitotoxic mechanisms and the amyloid cascade, microglial cells and neuroinflammation, which play as key targets in the polypharmacology-derived concept of multi-target directed pharmaceuticals for AD (Dias Viegas et al. 2018).

22.2.2 *Butyrylcholinesterase (BuChE)*

The brain contains two known cholinesterases, acetylcholinesterase (AChE) and butyrylcholinesterase (BuChE). BuChE can be found at high concentrations in the plasma. Until today, its physiological role has not been understood yet. It can have a facilitating role in influencing synaptic transmission and act as a support for AChE. Normally, BuChE can hydrolyze ACh in case of total AChE inhibition. Moreover, high levels of BuChE in plasma also leads to protection for AChE by becoming a scavenger of cholinergic toxins (Agatonovic-Kustrin et al. 2018). Therefore, the inhibition of BuChE might be a successful therapeutic method for AD as well (Jalili-Baleh et al. 2018).

22.2.3 *N-Methyl-D-Aspartate Receptor (NMDAR)*

The NMDAR is an ionotropic glutamate receptor mediating glutamatergic transmission and affecting many functions in the central nervous system, such as plasticity, neurodevelopment, learning, synaptic and memory formation. Their over-activation leads to pathological excitotoxicity that can be associated with several neurodegenerative diseases, including AD. Memantine, the only noncompetitive NMDAR antagonist drug, inhibits the NMDAR ion channel (Deora et al. 2017).

22.2.4 β-Amyloid Precursor Protein Cleaving Enzyme 1 (BACE-1)

Aβ accumulation leads to extracellular and intracellular formation of plaques and neurofibrillary tangles in the brain, and causes neuronal death (Haghighijoo et al. 2017).

Amyloid beta peptides are produced by proteolytic cleavage of APP (amyloid precursor protein) by BACE-1 enzyme as a first step, then by γ-secretases as a second step (Thai et al. 2017). Therefore, BACE1 blockers that reduce the production of all forms of Aβ can serve as potential drug candidates for AD therapy (Haghighijoo et al. 2017).

Haghighijoo and coworkers designed new 3-methylquinazolin-4(3H)-one hydrazones as potential nonpeptide BACE1 blockers, like 2-(2-(2,3-dichlorobenzylidene) hydrazinyl)-3-methylquinazolin-4(3H)-one, which had an acceptable blood brain barrier (BBB) crossing, Caco-2 permeability and logS value, in contrast with peptidic compounds with low blood brain barrier (BBB) crossing, poor oral bioavailability and susceptibility to P-glycoprotein (Pgp) transport (Haghighijoo et al. 2017).

22.2.5 Peroxisome Proliferator-Activated Receptor γ (PPARγ)

The peroxisome proliferator-activated receptor γ (PPARγ) is a ligand-inducible transcription factor that regulates genes involved in inflammation control. For instance, it blocks proinflammatory gene expression in microglia by suppressing the action of NF-κB. Moreover, PPARγ regulates genes with role in lipid and carbohydrate metabolism, which is important, because number of studies show that higher cellular cholesterol levels trigger high amyloid beta production. Therefore, PPARγ agonists have been advanced as a novel disease altering approach to AD treatment (Thai et al. 2017; Combarros et al. 2011).

22.2.6 The Liver X Receptor (LXR)

LXR is in close relation with PPARs, thus it can be considered as a target for AD treatment. It has two isoforms LXRα and LXRβ. According to a study with AD mouse model, treatment with LXRα agonist T090131 can lower Aβ formation (Koldamova et al. 2005), however, T090131 was shown to enhance plasma and liver triglycerides, underlining that this compound is not a good drug candidate. These results confirm the importance of LXR agonists, and highlight the role of drug discovery to find a good drug candidate without side effects (Mandrekar-Colucci and Landreth 2011).

22.2.7 Cathepsin B (CatB)

CatB can be a potential therapeutic target to decrease neuroinflammation that adds to the progression of neurodegenerative disorders. In patients with AD, CatB activity is increased in the central nervous system and blood serum, while its activity decreases in monocytes, lymphocytes compared to normal, same age people. Blocking CatB bioactivity can possibly slow down the progression of AD (Wenzel and Klegeris 2018).

22.2.8 Monoacylglycerol Lipases (MAGLs)

MAGLs being part of the endocannabinoid system, modulating pain, cognition and several immune functions (Wenzel and Klegeris 2018). MAGL belongs to the lipid-metabolizing enzymes in the serine hydrolase superfamily. Inhibition of MAGL was shown to downregulate proinflammatory cytokines and prostaglandins. Moreover, it can increase glial cell-derived neurotrophic factor and transforming growth factor, that promote the survival of neurons and glial cells (Fernández-Suárez et al. 2014). Another important factor, inhibition of MAGL does not lead to gastric hemorrhages, in contrast to anti-inflammatory therapies that modify prostaglandin synthesis. However, blocking MAGL can raise microglial activation marker ionized calcium binding adaptor molecule 1 (Fernández-Suárez et al. 2014). Accordingly, blocking MAGL could serve as an important strategy for designing drugs to hinder the progression of AD (Wenzel and Klegeris 2018).

22.2.9 Dual Specificity Phosphatase 2 (DUSP 2)

DUSP 2 enzyme inactivates via dephosphorylation of the phospho-amino acid residues on mitogen-activated protein kinases (MAPKs). These kinases play an important role in mitogenic signal transduction, stress responses, cell survival, apoptosis. That is why, application of DUSP 2 inhibitors in drug design could be a successful way for modulating the inflammatory cascades of neurodegenerative disorders (Wenzel and Klegeris 2018).

22.2.10 Multi-target Therapeutic Methods to Neurodegeneration

Animal models of neurodegenerative disorders can show effective single-target drugs, however most of them have not been effective in human clinical studies of AD. The reasons behind these failures involve (a) the drug does not bind to the in-vivo human target; (b) does not reach the human target in-vivo, (c) the drug-target interaction is compensated, (d) the drug administered to patients at an late stage of the disease, and (e) variability in the endpoints can mask the beneficial effect of the drug (Wenzel and Klegeris 2018).

Ligands (MTDLs) affecting multiple targets with more than two complementary bioactivities in a single drug molecule can show relevant advantages for the treatment of complex diseases like AD (Jalili-Baleh et al. 2018). This single-molecule approach can decrease side effects and drug interactions of MTDLs compared to combinations of single-target drug (Wenzel and Klegeris 2018).

AP2469 is one of the well-known coumarin-based MTDL with AChE and beta-secretase blocking activities, and having anti-Aβ aggregating, antioxidant, and neuroprotective characteristics. Therefore, coumarin-based MTDLs can be successful drug candidates for AD (Tarozzi et al. 2014).

MTDLs act by creating a combinatory effect on multiple proteins in the biological network simultaneously, which may lower the therapeutic dose, and improving therapeutic efficacy, preventing drug resistance, and reducing side effects (Kalash et al. 2017).

22.3 Importance of Novel in-silico Ligands

Over the past few years, several small molecules or bioproducts have been under development as a therapy for Alzheimer's disease, such as small molecule drugs and antibodies. There are some successful natural product based approaches, such as the *Ginkgo biloba* extract EGb761 and *Salvia officinalis* extracts have been shown to have anti-oxidative, anti-apoptosis, neuroprotective and neuromodulatory effects. *Ginkgo biloba* could improve attention and memory performance in patients as observed during a phase III clinical study (Sun et al. 2013).

Huperzia serrata was also studied in clinical trials focusing on the component named Huperzine A (a blood-brain barrier permeable alkaloid) The study showed significant improvements (Sun et al. 2013).

Nordihydroguaiaretic acid from Larrea tridentates was shown to block cholinesterase similarly to the marketed drugs and even has further antiaggregation effect on Aβ. Flavonols and flavones, myricetin and quercetin demonstrated very good BACE1 inhibitory effect (Kumar et al. 2016).

Sarsasapogenin, from the rhizomes of Anemarrhena asphodeloides Bunge, and its derivatives was shown to have neuroprotective activities. Its piperazine derivatives were active against cell injury induced by hydrogen peroxide, lipopolysaccharide and beta amyloid (Yang et al. 2018).

Actinomycetales are considered as prolific producers of a wide variety of bioactive secondary metabolites, such as antibiotics (Mohammadipanah et al. 2012), enzyme inhibitors (Imada 2005), anti-angiogenic and proangiogenic agents (Azarakhsh et al. 2017).

Research for natural anti-AChE metabolites from marine sources, found Actinobacteria, Streptomyces sp. UTMC 1334. Its extract and its high percentage of pyrrole derivatives revealed both anti-acetylcholinesterase and anti-oxidant activities, that might be potentially useful as an adjuvant candidate in the treatment of cognitive disorders (Almasi et al. 2018).

Apelin is a neuropeptide from bovine stomach extracts. It is an endogenous ligand of the APJ receptor (Tatemoto et al. 1998). The apelin/APJ system has various important functions, such as cardiac contractility, blood pressure regulation, glucose metabolism, immunity, water homeostasis, cell proliferation, angiogenesis, neuroprotection (Wu et al. 2017). A study showed that serum levels of apelin-13 decreased in patients with AD. Apelin can lower the production of Aβ by reducing the amount of APP and blocking the activity of β-secretase, and by raising the levels of ABCA1 and increasing the activity of Neprilysin, can lead to degradation of Aβ and lower its accumulation. Apelin may be able to decrease phosphorylation and accumulation of tau protein (Masoumi et al. 2018).

To explore more effective drugs for the treatment of AD, various molecules against known and novel targets of AD could be designed using computational approaches. Dual or multiple inhibitors that block two or more targets of AD can also be studied. Currently there is no treatment to prevent or cure AD but several approved drugs treat several of the symptoms and improve quality of life. Targeting the direct cause of neuronal degeneration could offer a rational strategy and potentially further novel prospects for the treatment of AD (Kumar et al. 2016).

22.4 Parkinson's Disease (PD)

PD was first described by James Parkinson, in 1817, as a neurological disturbance showing symptoms like resting tremor and a distinctive form of the progressive motor disorder, designated as shaking palsy or paralysis agitans (Parkinson 2002). Today, it is considered the most common neurodegenerative disorder after Alzheimer's Disease (AD). PD is generally defined as a progressive, irreversible, chronic neurological disorder characterized by increasingly disabling motor symptoms. These are associated to impaired coordinated movements involving bradykinesia, resting tremor, cogwheel rigidity, postural instability, and gait disorders (Dauer and Przedborski 2003). The majority of PD patients suffer from motor disabilities and numerous non-motor symptoms as well that finally lead to decreased the quality of life (Poewe 2008).

Mechanisms, neuropathology, pathophysiology features involve neuronal loss in specific areas of the substantia nigra and widespread intracellular α-synuclein protein accumulation. These two main pathologies are specific for a definitive diagnosis of idiopathic type PD. In the early-stage disorder, loss of dopaminergic neurons is limited to the ventrolateral substantia nigra with relative sparing of dopaminergic neurons in other areas. The dramatic loss of these dopaminergic neurons even in the early phase reveals that the degeneration here begins before the onset of motor symptoms.

The other characteristic neuropathology is the abnormal occurence of α-synuclein (Lewy bodies) in the cytoplasm. In patients suffering from AD, there is a different pattern of α-synuclein pathology, that is characteristic for limbic brain regions. Despite heritable forms of PD only occur 5–10% in all patients, they provided important clues to the mechanisms. Some of the PD associated gene-coded proteins are related to a set of signaling pathways that may trigger such a neuropathology that resembles sporadic type PD. Moreover, large genome-wide association studies (GWAS) showed that some of these genes are also affected in sporadic PD. Among the examples are: α-synuclein proteostasis, oxidative stress, calcium homeostasis, mitochondrial function, axonal transport, neuroinflammation. Interesting enough that several lines of evidence showed mitochondrial dysfunction as a key element in the pathogenesis of PD. A new theory is one of a vicious cycle in which α-synuclein aggregation and mitochondrial dysfunction facilitate each other, which could provide an explanation why these cellular modifications occur together in degenerating neurons in PD (Poewe et al. 2017).

22.4.1 Pharmacological Targets

22.4.1.1 Dopaminergic Pharmacological Targets

Losing dopaminergic neurons in the substantia nigra pars compacta causes striatal dopamine depletion, which is the underlying mechanism in the cardinal motor symptoms. The systemic administration of the dopamine precursor amino acid, the LDOPA showed a breakthrough in the therapy a 50 years ago. Since then, other relevant advances in the understanding of the molecular pharmacology have demonstrated further targets for presynaptic or postsynaptic dopaminergic therapies (PD Med Collaborative Group et al. 2014).

22.4.1.2 Catechol-O-Methyltransferase Targets

Current LDOPA preparations include blockers of aromatic amino acid decarboxylase enzyme (AADC) like carbidopa or benserazide. These prevent peripheral metabolism of dopamine and support better bioavailability. The peripheral metabolism of LDOPA is done via a secondary metabolic pathway involving ortho-methylation of l-DOPA through catechol-O-methyltransferase (COMT) activity. Therefore, inhibition of COMT can improve bioavailability and the half-life of l-DOPA. Increasing the half life of l-DOPA doses by the application of COMT inhibitors became a first-line therapy in many patients (Ferreira et al. 2016).

22.4.1.3 Monoamine Oxidase Type B (MAOB) Inhibitors

Oxidation by MAOB in glial cells is an important mechanism for clearance of synoptically excreted dopamine, in addition to presynaptic reuptake by the dopamine transporter. Blocking MAOB raises synaptic dopamine levels, leads to symptomatic efficacy. Recent studies established the antiparkinsonian efficacy of monotherapy using selegiline and another novel MAOB inhibitor rasagiline, which was shown to be effective in combination with l-DOPA in patients with motor fluctuations. While selegiline, rasagiline are irreversible blockers, there is a novel and promising reversible inhibitor, safinamide, on the market (Schapira 2011).

22.4.1.4 Dopamine Agonists

The dopamine activity on striatal medium spiny neurons is regulated by two classes of dopamine receptors. Dopaminomimetics have direct activity to dopamine receptors (dopamine receptor agonist) targeting the D2 receptor family. These were applied as PD treatment first in the 1970s with bromocriptine and have since become an important treatment method for motor symptoms. First drugs belonging to this class were ergoline derivatives, that also activate 5-hydroxytryptamine (5-HT) receptors, such as 5-HT2B subtype. Later, these became linked to pleuropulmonary and cardiac valvular fibrosis, raising important safety concerns (Connolly and Lang 2014).

22.4.1.5 Non-dopaminergic Pharmacological Targets

Despite the relevant effect of dopaminergic therapy on PD, there is still a clear need for further therapies targeting other pharmacological routes. The symptom that need to be addressed by such therapies include the complications of l-DOPA treatment, such as motor fluctuations and l-DOPA-induced dyskinesia, l-DOPA-resistant ('non-dopaminergic') motor features like treatment-resistant tremor, postural instability, falls, swallowing, speech disturbances. Today, the only effective therapy for l-DOPA-induced dyskinesia is amantadine, an N-methyl-d-aspartate receptor antagonist (Kalia et al. 2013).

The breakthrough for DBS as a therapy for PD arrived in 1993 when novel concepts of the basal ganglia circuitry led to the identification of the subthalamic nucleus as a new target for DBS. DBS is based on the observation that high-frequency (100–200 Hz) electrical stimulation of specific brain targets can mimic the effect of a region without the need for damaging brain tissue, and it involves the implantation of an electrode in brain tissue (Chandra et al. 2018).

22.5 Nano-particle Mediated Inhibition of Parkinson's Disease Using Computational Biology Approach

Research for the development of such approaches that can accelerate intracerebral dopamine concentrations and/or stimulates central dopamine receptors was carried out by the application of nanoparticles (NPs). It was shown to be a revolutionary treatment because of their site directed target delivery and ability to penetrate through the blood-brain barrier (BBB). For example, NPs surface linked with peptidomimetic antibodies was demonstrated as a molecular Trojan horse that can transport bulky molecules such as drugs and genes across the BBB. Moreover, biocompatible gold nanoparticles (AuNPs) were also shown to induce a strong α-synuclein aggregation. Other NPs, like graphene and superparamagnetic iron-oxide nanoparticles (SPIONs) demonstrated that they can block the Aβ fibrillation process (Seppi et al. 2011).

Phosphodiesterases (PDEs), found in several regions of the human brain, represent a class of enzymes that selectively cleave cyclic adenosine monophosphate (cAMP) and/or cyclic guanosine monophosphate (cGMP). They are well-known as therapeutic applications for erectile dysfunction and pulmonary hypertension (PDE5 inhibitors) and severe Chronic Obstructive Pulmonary Disease (PDE4 inhibitors), there are other different isoforms of PDEs, which are captivating the interest of scientists looking for new targets for neurodegenerative diseases. Blocking PDE4, PDE7 and PDE10 seems to be involved in protecting dopaminergic neurons. A study focused on a class of derivatives like berberine (isolated from Berberis sp.), and other plants because in traditional Chinese medicine berberine and its analogues have long been reported to have several bioactive and therapeutic roles. These are for example, anticancer, endocrine, cardiovascular, anti-inflammatory and immunomodulatory activities. The activity of berberine on animal models is, while berberine was shown to block PDE activity, a study on its potential effect on other targets that can be included in the anti-PD activity was proven. A combined in-silico evaluation study revealed that berberine and its synthetic derivatives can play a role in the onset and the progression of the neurodegeneration (Zanforlin et al. 2017).

22.6 Importance of in-silico Target Discovery

Until today, four signaling pathways were shown to be implicated in familial parkinsonism. These are synaptic neurotransmission, endosomal trafficking, lysosomal autophagy, mitochondrial metabolism. Although these processes can be separate, they employ very similar signalling pathways, therefore they may be temporally and functionally related. For instance, synaptic dysfunction, resulting from or leading to alpha-synucleinopathy, can do harm on balancing exo- and endocytosis, neurotransmission or early endosomal receptor recycling. These changes will influence the flux through the endosomal pathway and ultimately lead to autophagy and lysosomal fusion with multivesicular bodies (Gao et al. 2013).

As a conclusion, further genomic, genetic investigations should be done in the future. Knowledge of the pathogenic signaling pathways underlying the etiology and ontology of PD is needed for therapeutic development. The understanding of the normal physiology of certain neuronal populations, signaling pathways and the role of individual proteins, is essential. Understanding parkinsonism and genetic insights can be unbiased and unequivocal but those insights should be carefully considered. Genetic association studies do not provide clear picture and further work is required to fully understand the specific contribution of GWAS, let alone to translate the information from such studies into novel therapeutic methods for idiopathic PD. When interpreting linkage and exome studies, the phenotypes and families have to be also considered. Several neuroscience studies on PD was done using model systems with toxin administration, and these may not accurately reflect the human condition. Why the substantia nigra pars compacta is selectively lost in PD stays still questionable, but through human genetics we now have relevant molecular targets and research tools to study this. With such advances, therapeutic prospects for disease modification (neuroprotection) should be treated with true optimism (Lin and Farrer 2014).

22.7 Conclusion

One of the most common neurodegenerative disease is Alzheimer's, which is characterized by progressive loss in cognition, decrease in motor and functional capacity, impairment in behavioral and social autonomy and death (Dias Viegas et al. 2018). Current clinical treatment of AD is only palliative and restricted to five prescription pharmaceutical drugs approved by the US FDA, which are not demonstrated good efficacy and tolerability over a wide range of patients (Blennow et al. 2006). Therefore, there is a need to investigate new drug candidates. Some of the FDA approved drugs are listed in Table 22.1. There are more potential pharmacological targets, which are studied in different projects using computational science, like acetylcholinesterase, butyrylcholinesterase, N-methyl-D-aspartate receptor, β-amyloid precursor protein cleaving enzyme 1, peroxisome proliferator-activated receptor γ, liver X receptor, cathepsin B monoacylglycerol lipases and Dual Specificity Phosphatase 2. Animal models of neurodegenerative diseases can show effective single-target drugs, but most of them have not been effective in human clinical trials of AD. Therefore, researchers need to focus on multitarget-directed ligands with two or more complementary biological activities in a single drug molecule to display significant advantages against complex diseases like AD (Kalash et al. 2017).

Parkinson's disease is the second recurrent neurodegenerative disorder after AD. It is progressive, irreversible and characterized by increasingly disabling motor symptoms. As it's pharmacological treatment has not been solved; therefore, there is a need to investigate and discover new drug candidates and targets, next to the known ones (dopaminergic pharmacological targets, catechol-O-methyltransferase, monoamine oxidase type B inhibitors, dopamine agonists and non-dopaminergic pharmacological targets).

Table 22.1 FDA approved drugs in Alzheimer's and Parkinson's disease (Alzheimer's association 2017; Houghton et al. 2018)

Target	Ligand	Pharmacological effect	Side effects
Cholinesterase inhibitors	Tacrine, Donepezil, Rivastigmine, Galantamine	Mild to moderate AD	Vomiting, nausea, loss of appetite, frequent bowel movements
N-methyl-D-aspartate antagonists	Memantine	Moderate to severe AD	Headache, confusion, constipation, dizziness
Levodopa	Carbidopa	Mono-, combination therapy: stiffness, slowness, tremor (PD)	Low blood pressure, confusion, nausea, dyskinesia
Dopamine agonists	Ropinirole, Pramipexole, Rotigotine, Apomorphine	Mono-, combination therapy: stiffness, slowness, tremor (PD)	Low blood pressure, leg swelling, nausea, discoloration, sleep attacks, confusion, compulsive behaviors
MAO-B inhibitors	Selegiline, Rasagiline, Zydis selegiline HCL Oral disintegrating	Mono-, combination therapy: stiffness, slowness, tremor (PD)	Nausea, dry mouth, light-headedness, constipation; worse dyskinesia;
COMT-inhibitors	Entacapone, Tolcapone	Levodopa-combination therapy: motor fluctuations	Diarrhea, colored urine, enhancing adverse effects of levodopa: dyskinesia and confusion
Anticholinergics	Trihexyphenidyl, Benztropine	Mono-, or combination therapy: tremor (young patients with PD); not suggested for elderly patients	Confusion, memory issues, dry mouth, hallucinations, blurry vision, urinary retention
Other Antiparkinson Medications	Amantadine	Monotherapy: tremor, slowness, stiffness; Levodopa combination therapy for induced motor fluctuations; helpful for reducing dyskinesia (PD)	Nausea, leg discoloration, confusion, mild anti-cholinergic effects

To develop new drugs, it is necessary to use computational chemistry and cheminformatics to search for quantitative structure-activity relationships to identify, for example, inhibitors of key enzymes (Nastase and Boyd 2012). New approaches could facilitate the development of 'multi-target' anti-AD drugs with integration of various sources of information, such as microarrays, literature mining-data, protein–protein interactions (Sun et al. 2013).

In conclusion, computer-aided drug design has been one of the major tools applied in drug discovery programs used to reduce the cost and process time.

References

Agatonovic-Kustrin S, Kettle C, Morton DW. A molecular approach in drug development for Alzheimer's disease. Biomed Pharmacother. 2018;106:553–65.

Almasi F, Mohammadipanah F, Adhami HR, Hamedi J. Introduction of marine-derived *Streptomyces* sp. UTMC 1334 as a source of pyrrole derivatives with anti-acetylcholinesterase activity. J Appl Microbiol. 2018;5:1370–82.

Alzheimer's association. FDA approved treatments for Alzheimer's 800.272.3900 alz.org®. 2017.

Azarakhsh Y, Mohammadipanah F, Nassiri SM, Siavashi V, Hamedi J. Isolation and screening of proangiogenic and antiangiogenic metabolites producing rare actinobacteria from soil. J Appl Microbiol. 2017;122:1595–602.

Blennow K, de Leon MJ, Zetterberg H. Alzheimer's disease. Lancet (London, UK). 2006;368:387–403.

Cell Signaling Technology, Inc. (www.cellsignal.com).

Chandra KA, Bharadwaj S, Kumar S, Wei DQ. Nano-particle mediated inhibition of Parkinson's disease using computational biology approach. PMC Sci Rep. 2018;8:9169.

Combarros O, Rodríguez-Rodríguez E, Mateo I, Vázquez-Higuera JL, Infante J, Berciano J, Sánchez-Juan P. APOE dependent-association of PPAR- genetic variants with Alzheimer's disease risk. Neurobiol Aging. 2011;(3):547.e1–6.

Connolly BS, Lang AE. Pharmacological treatment of Parkinson disease: a review. JAMA. 2014;311:1670–83.

Dauer W, Przedborski S. Parkinson's disease, mechanisms and models. Neuron. 2003;39:889–909.

Deora GS, Kantham S, Chan S, Dighe SN, Veliyath SK, McColl G, Parat MO, McGeary RP, Ross BP. Multifunctional analogs of kynurenic acid for the treatment of Alzheimer's disease: synthesis, pharmacology and molecular modeling studies. ACS Chem Neurosci. 2017;8 (12):2667–75.

Dias Viegas FP, de Freitas Silva M, Divino da Rocha M, Castelli MR, Riquiel MM, Machado RP, et al. Design, synthesis and pharmacological evaluation of N-benzyl-piperidinyl-aryl-acylhydrazonederivatives as donepezil hybrids: discovery of novel multi-target anti-alzheimer prototype drug candidates. Eur J Med Chem. 2018;147:48–65.

Epperly T, Dunay MA, Boice JL. Alzheimer disease: pharmacologic and nonpharmacologic therapies for cognitive and functional symptoms. Am Fam Physician. 2017;95:771–8.

Fernández-Suárez D, Celorrio M, Riezu-Boj JI, Ugarte A, Pacheco R, González H, Oyarzabal J, Hillard CJ, Franco R, Aymerich MS. The monoacylglycerol lipase inhibitor JZL184 is neuroprotective and alters glial cell phenotype in the chronic MPTP mouse model. Neurobiol Aging. 2014;35:2603–16.

Ferreira JJ, Lees A, Rocha JF, Poewe W, Rascol O, Soares-da-Silva P, Bi-Park 1 Investigators. Opicapone as an adjunct to levodopa in patients with Parkinson's disease and end-of-dose motor fluctuations: a randomised, doubleblind, controlled trial. Lancet Neurol. 2016;15:154–65.

Flynn JM, Melov S. SOD2 in mitochondrial dysfunction and neurodegeneration. Free Radic Biol Med. 2013;62:4–12.

Gao L, Fang JS, Bai XY, Zhou D, Wang YT, Liu AL, Du GH. In silico target fishing for the potential targets and molecular mechanisms of baicalein as an antiparkinsonian agent: discovery of the protective effects on NMDA receptor-mediated neurotoxicity. Chem Biol Drug Des. 2013;81:675–87.

Haghighijoo Z, Firuzi O, Hemmateenejad B, Emami S, Edraki N, Miri R. Synthesis and biological evaluation of quinazolinone-based hydrazones with potential use in Alzheimer's disease. Bioorg Chem. 2017;74:126–33.

Herrup K. The case for rejecting the amyloid cascade hypothesis. Nat Neurosci. 2015;18:794–9.

Houghton D, Hurtig H, Metz S, Giroux M, Petzinger G, Fisher B, Hawthorne L, Jakowec M. Parkinson's disease medications. National Parkinson Foundation. 2018. www.parkinson.org.

Imada C. Enzyme inhibitors and other bioactive compounds from marine actinomycetes. Antonie Van Leeuwenhoek. 2005;87:59–63.

Jalili-Baleh L, Forootanfar H, Küçükkılınç TT, Nadri H, Abdolahi Z, Ameri A, Jafari M, Ayazgok B, Baeeri M, Rahimifard M, Abbas Bukhari SN, Abdollahi M, Ganjali MR, Emami S, Khoobi M, Foroumadi A. Design, synthesis and evaluation of novel multi-target-directed ligands for treatment of Alzheimer's disease based on coumarin and lipoic acid scaffolds. Eur J Med Chem. 2018;152:600–14.

Kalash L, Val C, Azuaje J, Loza MI, Svensson F, Zoufir A, Mervin L, Ladds G, Brea J, Glen R, Sotelo E, Bender A. Computer-aided design of multi-target ligands at A_1R, A_2AR and PDE10A, key proteins in neurodegenerative diseases. J Cheminform. 2017;30:67.

Kalia LV, Brotchie JM, Fox SH. Novel nondopaminergic targets for motor features of Parkinson's disease: review of recent trials. Mov Disord. 2013;28:131–44.

Katzman R. The prevalence and malignancy of Alzheimer disease: a major killer. Arch Neurol. 1976;33:217–8.

Koldamova RP, Lefterov IM, Staufenbiel M, Wolfe D, Huang S, Glorioso JC, et al. The liver X receptor ligand T0901317 decreases amyloid β production in vitro and in a mouse model of Alzheimer's disease. J Biol Chem. 2005;280:4079–88.

Kumar A, Nisha CM, Silakari C, Sharma I, Anusha K, Gupta N, Nair P, Tripathi T, Kumar A. Current and novel therapeutic molecules and targets in Alzheimer's disease. J Formos Med Assoc. 2016;115:3–10.

Lin MK, Farrer MJ. Genetics and genomics of Parkinson's disease. Genome Med. 2014;6:48.

Mandrekar-Colucci S, Landreth GE. Nuclear receptors as therapeutic targets for Alzheimer's disease. Expert Opin Ther Targets. 2011;15:1085–97.

Masoumi J, Abbasloui M, Parvan R, Mohammadnejad D, Pavon-Djavid G, Barzegari A, Abdolalizadeh J. Apelin, a promising target for Alzheimer disease prevention and treatment. Neuropeptides. 2018;70:76–86.

Mohammadipanah F, Matasyoh J, Hamedi J, Klenk HP, Laatsch H. Persipeptides A and B, two cyclic peptides from Streptomyces sp. UTMC 1154. Bioorg Med Chem. 2012;20:335–9.

Nastase AF, Boyd DB. Simple structure-based approach for predicting the activity of inhibitors of beta-secretase (BACE1) associated with Alzheimer's disease. J Chem Inf Model. 2012;52:3302–7.

Pandey S, Singh B, Yadav SK, Mahdi AA. Novel biomarker for neurodegenerative diseases- motor neuron disease (MND), cerebellar ataxia (CA) and Parkinson's disease (PD). Clin Chim Acta. 2018;485:258–61.

Parkinson J. An essay on the shaking palsy. 1817. J Neuropsychiatr Clin Neurosci. 2002;14:223–36.

PD Med Collaborative Group, Gray R, Ives N, Rick C, Patel S, Gray A, Jenkinson C, McIntosh E, Wheatley K, Williams A, Clarke CE. Long-term effectiveness of dopamine agonists and monoamine oxidase B inhibitors compared with levodopa as initial treatment for Parkinson's disease (PD MED): a large, open-label, pragmatic randomised trial. Lancet. 2014;384:1196–205.

Poewe W. Non-motor symptoms in Parkinson's disease. Eur J Neurol Suppl. 2008;1:14–20.

Poewe W, Seppi K, Tanner CM, Halliday GM, Brundin P, Volkmann J, Schrag AE, Lang AE. Parkinson disease. Nat Rev Dis Prim. 2017;3:17013.

Qian S, He L, Wang Q, Wong YC, Mak M, Ho CY, Han Y, Zuo Z. Intranasal delivery of a novel acetylcholinesterase inhibitor HLS-3 for treatment of Alzheimer's disease. Life Sci. 2018;207:428–35.

Sang Z, Wang K, Wang H, Wang H, Ma Q, Han X, Ye M, Yu L, Liu W. Design, synthesis and biological evaluation of 2-acetyl-5-O-(aminoalkyl) phenol derivatives as multifunctional agents for the treatment of Alzheimer's disease. Bioorg Med Chem Lett. 2017;27:5046–52.

Schapira AH. Monoamine oxidase B inhibitors for the treatment of Parkinson's disease: a review of symptomatic and potential disease-modifying effects. CNS Drugs. 2011;25:1061–71.

Seppi K, Weintraub D, Coelho M, Perez-Lloret S, Fox SH, Katzenschlager R, Hametner EM, Poewe W, Rascol O, Goetz CG, Sampaio C. The Movement Disorder Society evidence-based medicine review update: treatments for the non-motor symptoms of Parkinson's disease. Mov Disord. 2011;26:S42–80.

Silva GS, Figueiró M, Tormena CF, Coelho F, Almeida WP. Effects of novel acylhydrazones derived from 4-quinolone on the acetylcholinesterase activity and Aβ 42 peptide fibrils formation. J Enzyme Inhib Med Chem. 2016;6366:1–7.

Sun Y, Zhu R, Ye H, Tang K, Zhao J, Chen Y, Liu Q, Cao Z. Towards a bioinformatics analysis of anti-Alzheimer's herbal medicines from a target network perspective. Brief Bioinform. 2013;14:327–43.

Tarozzi A, Bartolini M, Piazzi L, Valgimigli L, Amorati R, Bolondi C, et al. From the dual function lead AP2238 to AP2469, a multi-target-directed ligand for the treatment of Alzheimer's disease. Pharmacol Res Perspect. 2014;2:e00023.

Tatemoto K, Hosoya M, Habata Y, Fujii R, Kakegawa T, Zou MX, Kawamata Y, Fukusumi S, Hinuma S, Kitada C. Isolation and characterization of a novel endogenous peptide ligand for the human APJ receptor. Biochem Biophys Res Commun. 1998;251:471–6.

Thai NQ, Nguyen HL, Linh HQ, Li MS. Protocol for fast screening of multi-target drug candidates: application to Alzheimer's disease. J Mol Graph Model. 2017;77:121–9.

Turnaturi R, Oliveri V, Vecchio G. Biotin-8-hydroxyquinoline conjugates and their metal complexes: exploring the chemical properties and the antioxidant activity. Polyhedron. 2016;110:254–60.

Wenzel TJ, Klegeris A. Novel multi-target directed ligand based strategies for reducing neuroinflammation in Alzheimer's disease. Life Sci. 2018;207:314–22.

Wu L, Chen L, Li L. Apelin/APJ system: a novel promising therapy target for pathological angiogenesis. Clin Chim Acta. 2017;466:78–84.

Yang GX, Ge SL, Wu Y, Huang J, Li SL, Wang R, Ma L. Design, synthesis and biological evaluation of 3-piperazinecarboxylate sarsasapogenin derivatives as potential multifunctional anti-Alzheimer agents. Eur J Med Chem. 2018;156:206–15.

Zanforlin E, Zagotto G, Ribaudo G. The medicinal chemistry of natural and semisynthetic compounds against Parkinson's and Huntington's diseases. ACS Chem Neurosci. 2017;11:2356–68.

Zimmerman G, Lehar J, Keith C. Multi-target therapeutics: when the whole is greater than the sum of the parts. Drug Discov Today. 2007;12:34–42.

Chapter 23
Phytochemicals, Cancer and miRNAs: An in-silico Approach

Neha Kashyap, Prem P. Kushwaha, Atul K. Singh, Santosh Maurya, Ashish K. Sahoo, and Shashank Kumar

23.1 Introduction

Cancer involves the process of carcinogenesis which turns normal cells into cancerous cells. The process includes three major steps of initiation, promotion and progression. In order to design therapeutics and drugs against cancer development one need to understand these major steps.

Initiation is the first step which is a fast process comprising the interaction of pre malignant cells with a carcinogenic agent. The second step of promotion is a lengthy phase where abnormal cells persists, replicates and originate into pre neoplastic cells. Progression is the final step of conversion of premalignant cells into cancerous cells having metastasis potential with a new blood vessel formation called angiogenesis. In the first step of cancer development, epigenetic changes take place where target cell DNA is damaged. Hence, most promising step to prevent carcinogenesis is to scavenge the Reactive Oxygen Species (ROS) or by inducing the phase-II conjugating enzymes (glutathione-S-transferases (GST), glucuronidases and sulphotransferases) to promote the detoxification of the carcinogenic agent. In the step of promotion the cell cycle check points and apoptosis need to be checked out to restore the disturbance between cell division and apoptosis. In the last step (progression) interruption of angiogenesis or the prevention of invasion of malignant cells to from one place to another could be targeted (Adams et al. 2010; Aggarwal and Shishodia 2006; Aithal and Rajeswari 2013; Alam et al. 2018).

During the process of carcinogenesis there are several factors which participate in the process out of which the different signalling pathways like AKT, phosphatidylinositol kinase (PI3K), MAPKs (mitogen activated protein kinase), and nuclear factor kappa B (NF-KB) play an important role (Manson 2003; Ramos 2007; Surh 2003). Different stages work under regulation of key proteins related to cellular

N. Kashyap · P. P. Kushwaha · A. K. Singh · S. Maurya · A. K. Sahoo · S. Kumar (✉)
Department of Biochemistry and Microbial Sciences, Central University of Punjab, Bathinda, Punjab, India

S. Kumar, C. Egbuna (eds.), *Phytochemistry: An in-silico and in-vitro Update*,
https://doi.org/10.1007/978-981-13-6920-9_23

antioxidant defences, cellular proliferation and survival transduction pathways. The proteins like Growth factor receptors (GFR), antiapoptic members of bcl-2 family genes, signalling pathways like AKT, PI3K, and NF-KB are up regulated and caspases, proapoptic members of bcl-2 family genes are down regulated.

The wide range of phytochemicals extracted from plants have the capacity to up regulate and down regulate the key proteins involved in several signalling pathways and affect different stages of cancer cell development like cell proliferation, cell invasion or metastasis, angiogenesis, apoptosis resulting in a protein beneficial effect (Manson 2003; Ramos 2007; Surh 2003). Polyphenols, a class of phytochemicals have shown higher cytotoxicity in cancer cells when compared to normal cells hence proves to be more sensitive and beneficial (Ahmad et al. 2000; Park et al. 2005). It becomes important to understand that variety of phytochemicals exists with their individual properties and hence consequently clarifying the mechanisms at molecular level by which phytochemicals will become an effectual agent in exerting a potential anti-carcinogenic effect on cancer cells.

MicroRNAs (miRNA) are group of short ribonucleotide strands of around 22–25 nucleotides which are excellent gene regulators and show abnormal expressions in different kinds of human diseases including the most deadly one, cancer. As predicted, about 60% of mRNA is under direct control of miRNA; hence this regulation appears to be the most abundant form of posttranscriptional regulation (Bartel 2009; Fabian and Sonenberg 2012; Saj and Lai 2011). MicroRNA plays a crucial role in targeting the double stranded mRNA and silencing the translation of proteins. They modulate the cellular pathways of cancer cells and are responsible for tumor progression and formation in the body. The 'undruggable' proteins can be targeted through their miRNA gene regulators and their expression can be controlled which will turn out to prove a good chance to cure number of diseases which seems impossible to get treated. A therapeutic strategy can be followed by correcting the deficiencies of miRNA by either up regulating or down regulating the miRNA function. With the advent of study in the field of miRNAs it has been concluded that miRNAs links to various pathways like Notch, Hedgehog, JAK/STAT, Akt/mTOR which corresponds or targets the phenomenon of metastasis, inflammation, angiogenesis and uncontrollable proliferation of cells making it cancerous. Hence miRNAs are turning out to be an efficient biomarkers and potential tool for therapeutics and drug targets in clinics. MicroRNAs can be oncogenic and simultaneously it can also suppress the tumor making it difficult to study the link between miRNAs and cancer development. Examples include different function shown by a single miRNA like miR-29 is responsible for tumor suppressor of lung tumors whereas in breast cancer it has oncogenic properties (Fabbri et al. 2007). The approach of this chapter is to view the action of bioactive compounds extracted from plants, known as phytochemicals in diminishing the recurrence of cancer (Le Marchand et al. 2000; Nakachi et al. 1998). Majority of phytochemicals are polyphenols which upregulate or downregulate miRNAs which are linked to several signaling pathways in humans.

The returning challenge in studying the link between miRNAs and cancer development is the level of usefulness of miRNA when the disease is in progression.

The miRNAs which is commonly regulated by phytochemicals and also modulates cancer hallmarks like angiogenesis, metastatic, apoptosis are studied and further analyzing their link with associated pathways. All these approaches including new chemicals which we call now drug-like will nevertheless go in vain and successful results of the different approaches targeting miRNAs will surely generate a new era of useful drugs which can be helpful in eradicating diseases which are incurable at present scenario (Bhattacharyya et al. 2011; Camps et al. 2008; Caporali and Emanueli 2011; Chan and Loscalzo 2010; Chan et al. 2011; Cheng et al. 2013; Chung et al. 2013).

23.2 Cancer Cellular Signaling Pathways

23.2.1 Notch Signaling

Notch signalling pathway is the most discussed pathway among different types of signaling pathways occurring in cells of vertebrates and invertebrates. Notch receptor participating in the cell-cell communication is a single transmembrane receptor protein with a large extracellular domain and short intracellular domain (Li et al. 2017a, b, c). Notch signalling pathway helps in transporting signals between adjacent cells. It maintains homeostasis of the tissue and regulates neuronal, immune, cardiac and endocrine development of the body. The signaling takes place when the receptor binds the ligand. Ligand binding with the extracellular domain of the notch receptor induces proteolytic cleavage which results in the release of intracellular domain travelling inside the cell nucleus to modulate the gene expression (Braune and Lendahl 2016) and this process was first identified in *Drosophila melanogaster* (Struhl and Greenwald 1999).

The notch signalling pathways studied till now is of two different types, the canonical and the non-canonical pathway (Kovall and Blacklow 2010). The canonical model in which the notch receptor binds to Delta-serrate-LAG2 (DSL) ligands. γ-secretase mediates the release of notch intracellular domain by the process of cleavage out of plasma membrane. Non-canonical notch pathway gets activated by proteins by Delta like-1(DLK-1), Delta and Notch like Epidermal Growth factor (DNER) in which there is absence of DSL motifs (Hori et al. 2013). The notch signaling pathway is of great importance as it is linked to human diseases. It has been found that notch signaling gets deregulated in many cancers like T-cell acute lymphoblastic leukaemia and other diseases like Multiple Sclerosis, Teratology, and Alagille Syndrome (Yamamoto et al. 2014). Notch signaling seems to get mutated in cancer cells and hence the notch receptor could be changed by replacing the intracellular and extracellular domain of our choice. This would allow researchers to investigate the genes which are up regulated in response and detection of ligands and hence facilitating new ways of both basic and applied research in cell-cell signaling. This pathway will account for broad range of therapeutics and pharmacological interventions.

23.2.2 *Hedgehog Signaling*

Hedgehog signalling pathway is one of the crucial signal transduction pathways that help in the transfer of information to embryonic cells for cell differentiation (Gorojankina 2016). Diseases resulting due to malfunctioning of this pathway include cancer. This pathway is one of the important pathways in developmental biology processes in mammals and arthropods. The pathway is responsible for development of many important portions of our body like brain, lungs, heart, and GI tract and muscle cells (Bai et al. 2016). Hedgehog (Hh), the signalling molecule is a chemical mediator secreted by the larva during its development. It is a segment polarity gene or protein secreted during larval development of drosophila. It designates which side of the larva will be anterior and posterior part and helps in segmentation of the body.

Three hedgehogs have been studied: Desert (DHH), Indian (IHH), and Sonic (SHH) of which sonic is the most common. Sonic interacts with Wingless signal transduction pathway resulting in sequential segmentation of the larva. Hence this pathway connects two different pathway and known as intercellular signalling pathway. Different cells crosstalk among them and helps in formation of heart and lungs.

Two different proteins patch and smoothen participate in the signalling. Patch is embedded on the cell surface and smoothen is present in cytosol of the cell. Gli proteins are target genes present in the cytoplasm of the cell which when gets activated by the Smoothen protein, travel inside the nucleus inducing the transcription of the genes for cell proliferation and differentiation. Smoothen protein only gets activated when the Hedgehog molecule binds with the Patch protein. In the absence of the hedgehog molecule there will be no transcription of the genes (Petricci and Manetti 2015). This pathway has shown its impact in development of cancer (Gonnissen et al. 2015). The recent pharmaceutical companies are designing drugs targeting this pathway to modulate the body's important process like cell differentiation, cell proliferation which gets disturbed during the development of these diseases.

23.2.3 *JAK/STAT Signaling*

JAK/STAT signaling pathway is the most simplest and conserved pathways studied in metazoans. JAK/STATs signaling pathway is activated by cytokines and growth factors and is responsible for transcription activation in cells. It mediates many cellular responses such as cell differentiation, proliferation, cell survival, migration and apoptosis. JAK/STATs signaling are necessary to control numerous developmental and homeostatic processes which include stem cell maintenance, immune cell development, haematopoiesis, organismal growth and mammary gland development (Villarino et al. 2015).

JAK/STATs signaling are made of three proteins: cell surface receptor (gp 130), Janus Kinase (JAK), and signaling transducer and activator of transcription proteins (STATs). The signaling pathway is a cascade of events. It starts from the binding of a cytokine (IL-6) to the cytokine receptor present on cell surface which dimerizes the receptors and results in the phosphorylation of the JAK which in turn phosphorylates the tyrosine residues on the receptor creating activation loops for STATs possessing SH2 domains. STATs then bind to the phosphorylated tyrosine on the receptor through their SH2 domains and then they are tyrosine-phosphorylated by JAKs, resulting in the dissociation of STATs from the receptor. The STATs dimerizes and enter nucleus activating their target genes for transcription (Wagner and Siddiqui 2012). Inhibitor of JAK/STATs signaling pathway is known as SOCS (suppressors of cytokine signaling). It contains SOCS box of 40 amino acids which can interact with number of proteins forming a protein complex which can cause breakdown of JAK and the receptor, inhibiting the JAK/STATs signaling pathway. This pathway is of clinical significance because it is responsible for many fundamental processes like apoptosis and inflammation. Alteration in pathway can lead to deadly cancer development. It can be concluded that excessive JAK/STATs signaling pathway is responsible for uncontrolled cell growth in any part of the body (Amoyel et al. 2014). It allows transcription of genes such as BCL2 and c-Myc which are responsible for cell division. We can use drugs which inhibits excessive JAK/STATs signalling for the treatment of cancer hence increasing the survival of patients with cancer (Goropevšek et al. 2017).

23.2.4 NF-KB Signaling

The nuclear factor kappa b (Nf-κb) pathway is a multi-component signaling pathway that regulates the expression of hundreds of genes which are involved in cell proliferation, cell survival, and cellular stress response in innate immunity and inflammation.

NF-κB molecules are typically dimers. A normal structure of Nf-κB is P50-P65 dimer (Nf-κB1/RelA). The dimer formation is fundamental for DNA binding; two NF-kB monomers bind to DNA as a dimer. The N-terminal regions of dimer are accountable for precise DNA contact. The C-terminal areas are often totally conserved; they're accountable for dimerization and nonspecific DNA phosphate contact. The entire NF-kB molecular is just like a pliers vies on the DNA chain and function as a transcription aspect (Mitchell et al. 2016)

These inhibitor proteins, which incorporate IκBa, IκBb and IκBg, contain 6–7 ankyrin repeats that mediate binding to the RHD (Rel homology domain). These repeats are also present in the C-terminal halves of the Nf-κB2/p100 and Nf-κB1/p105 precursors, which also operate as IκBs and retain their partners, the Rel proteins, in the cytoplasm. There are two principal signaling pathways result in the Iκ-B protein inhibitor dissociation from NF-κB dimer and let the translocation of NF-κB dimers from the cytoplasm into the nucleus. The binding to a certain inhibitors referred to as Iκ-B protein, which could bind to the Rel homology domain (RHD) of NF-κB. There are two pathways involved the canonical and the non-canonical pathway. The inhibitors of this pathway are IRFD1. In cancer development the proteins that control this pathway are mutated or wrongly expressed. Malfunctioning of this pathway increases the number of cells leading to apoptosis. NF-κB signaling pathway inhibition methods are seeking clinical trials in laboratories as it is of utmost importance in eradicating inflammatory diseases and cancer (Dai et al. 2016).

23.2.5 PI3K Signaling

The PI3K signaling pathway is also termed as AKT and mTOR pathway. This pathway regulates the cell cycle, cell proliferation, cell death and its longevity. The pathway decides when the cell has to live or undergo programmed cell death. It comprises of the signaling molecules and the signaling receptors. The pathway is complex to understand as it has many loops and branches. More occurrence rates of this pathway in the body induce more cell proliferation than apoptosis leading to cancer development.

The PI3K/AKT pathway starts with the activation of phosphotidylinositol3 or PI3 kinase. This pathway can be accomplished by three different pathways (Lin et al. 2014). Two pathways begin with the activation of receptor belonging to tyrosine kinase receptor family by an extracellular growth factor. Binding of extracellular growth factor leads to the dimerization of the receptor monomers and heterologous auto-phosphorylation of the monomers. Depending on the receptor different proteins may bind to phosphorylated domain. Insulin receptor substrate 1 (IRS-1) binds to the activated IGF-1 receptor dimer. Receptor bound IRS-1 serves as a binding and activation site for PI3K. In addition PI3K may bind directly to a phosphorylated receptor tyrosine kinase.

A completely different mechanism of PI3K activation begins with a small membrane bound GTP and Ras. By binding to active GTP bound Ras PI3 kinase is activated. At the second level of pathway phosphatidylinositol (3, 4, 5)-triphosphate (PIP3) is formed. This leads to the activation of serine threonine kinase (AKT). The activated PI3K migrates to the inner membrane and binds to phosphatidylinositol (3, 4)-diphosphate (PIP2) which is a regular component of the membrane and is anchored by its two fatty acids in the lipid layer of the membrane. PI3K phosphorylates PIP2 to PIP3. PIP3 can activate the kinase AKT which is also called phosphorylated kinase B and was named after its homologous protein in retrovirus AKT-8. There are many downstream effects of AKT. The serine-threonine kinase is a proto-onco protein with many substrates and effects. It initiates protein synthesis or translation and responsible for inhibition of apoptosis. In the absence of AKT mitochondrial holes created by BAX leads to apoptosis by caspase cascade. AKT prevents tumour suppressor protein FOXO from inhibiting proliferation of cells (Wang et al. 2010).

23.3 Phytochemicals and Cancer Signaling Pathways

Phytochemicals are non-nutritive plant products that proved to be chemo-preventive and have anti-cancerous properties. Phytochemicals, plant by-products also termed as secondary metabolites seems to range around thousand types. Some of the well-known phytochemicals are lycopene in tomatoes, flavonoids in fruits, isoflavones in soy, genistein in soyabean, catechins in green tea, 6-gingerol from ginger, eugenol from cloves etc. These exhibit different properties such as antioxidant activity, anti-atherosclerotic, anti-carcinogenic, anti-inflammatory, spasmolytic, hepatoprotective, anti-viral, anti-microbial, anti-allergic, anti-diarrheal and oestrogenic activity. They also affect many cell processes including proliferation, apoptosis, cell-cycle control, angiogenesis, inflammation and DNA repair. These plants natural by-products have emerged as a need to solve the various problems relating to the disease cancer. In addition to various useful activities they are chemo preventive and can act as modulators of miRNA and hence their target genes (Lançon et al. 2013). Dietary phytochemicals have been reported to inhibit the progression of cancer and proves to be anti-metastatic and apoptotic. Including dietary phenols in diet has been linked to reduction in the rate of metastatic cancer in many preclinical animal models and human epidemiological studies. Dietary phytochemicals play a significant role in modulating various biological events including epigenetic events like non-coding micro-RNAs, histone modification and DNA methylation and multiple signalling transduction pathways like1 Wnt/β-catenin, notch, Hedgehog, COX-2, EGFR, MAPK-ERK,JAK/STAT, Akt/PI3k/mTor, Nf-κB, AP-1 etc. which plays a key role in regulation of metastasis cascade (Parasramka et al. 2012). Phytochemicals work differently from each other and dietary polyphenols have significant inhibitory activity at nearby each step of the metastatic cascade. Dietary polyphenols have anticancer effects by including apoptosis and by inhibiting cell growth, migration, invasion and angiogenesis. Most phytochemicals have antioxidant activity and protect our cells against oxidative damage. Many of these phytochemicals include terpenoids, carotenoids, phenolic compounds like flavonoids, isoflavanoids, aurones, chalconoids, flavonolignans, lignans, stilbenoids, curcuminoids, tannins, aromatic acids, phenylethanoids, glucosinolates, betalains, naringenin etc. Isoflavones, a group of phytochemicals regulate hormonal level of oestrogen; some phytochemicals called terpenes interfere with enzymes, and are protease inhibitors. Dietary polyphenols are natural agents which have potential to increase the efficiency of chemotherapy and radiotherapy through the regulation of multiple signalling pathways (Petric et al. 2015). They participate in molecular mechanisms of regulating metastatic cascade and highlight the potential of these dietary polyphenols as promising therapeutic inhibitors of cancer. A wide range of phytochem icals distinguished as polyphenols show their inhibiting effects on the cell proliferation, cell migration, and cell differentiation of cancer cells by acting on the activation of different signalling pathways like AKT pathway, NF-KB pathway, PI3K pathway and other pathways involved in the formation of cancerous cells and tumour formation. As cancer tends to rise globally we are attempting to find and learn inexpensive ways in which inhibitory signalling pathways are in effect of these phytochemicals to improve the health worldwide. A review of the history and latest clinical trials has been in study of the role of polyphenols such as curcumin, EGCG, quercetin and other antioxidants against the cancer development. Dietary intervention is a new field of strategy to stop the progression of cancer which is increasing at an alarming rate (Sarkar et al. 2009). Chemoprevention of cancer is the upcoming recent advanced topic which provides a practical approach to identify potential useful inhibitors of cancer development and also the study of formation of tumour. Dietary polyphenols from vegetables, fruits, medicinal plants and spices have attracted many researchers as well as the normal public due to their excellent demonstration to inhibit the mechanism of cancer development.

23.4 Phytochemical Anti-metastatic Potential and Epigenetic Modification

Metastasis is the key step which relates to the occurrence of cancer. Metastasis results from several interlinked processes including cell multiplication, angiogenesis, cell adhesion, cell migration and invasion into the surrounding tissue. Researcher proved that regular intake of dietary phytochemicals decreases the progression of metastasis when studied in a population and offers significant promise for reducing the incidence and mortality of cancer (Raffoul et al. 2006). Phytochemicals extracted from plants can be used as ancient traditional medicines. These plants are not only source of phytochemicals but also an excellent source of fibre and minerals and also bioactive molecules such as polyphenols, terpenoids, alkaloids, vitamins, pigments which will serve more basic nutrition function. Dietary polyphenols helps in suppression of metastasis phenomenon in cancer cells (Sarkar 2010; Aggarwal et al. 2006). They have therefore become one of the most extensively studied inhibitors of both cancer cell growth and metastasis and have the potential to be of substantial clinical importance to patients with various types of cancer.

Epigenetic events are associated with tumor formation in its every step. DPs are used in recognition to affect DNA integrity this is in turn can impact on chromatin integrity and epigenetic profile (Hadi et al. 2007). Dr. S.M. Hadi with his group were first to investigate transition metal-mediated oxidative DNA damage-dependent anticancer mechanism of DPs in human cancer cell lines and also concluded that this property of DPs could be further used for exploring anticancer agents with higher therapeutic indices (Nazeem et al. 2009; Bhat et al. 2006, 2007; Azmi et al. 2005, 2006; Uddin et al. 2004). Copper, a key metal found in chromatin, was observed to be elevated in a number of malignancies (Linder 2012). Dietary polyphenols including quercetin, reservatol and other induces oxidative DNA damage in the presence of copper ions (Hadi et al. 2000, 2007; Azmi et al. 2006). This innovative antitumor mechanism of copper bounded chromatin and the consequent pro-oxidant action (Ullah et al. 2011).

Epigenetic refers partially reversible and somatically inheritable DNA sequence that modulates and regulates gene expression, structure of chromatin, and cell cycle regulation and apoptosis induction. Non-coding miR expression, DNA methylation and histone modifications are major crucial events of epigenetics. DNA hyper methylation and changes in the level of key histone modifications are major crucial events of epigenetics. DNA hyper methylation and changes in the levels of key histone configurations important for activation and repressor of gene transcript have also been frequently found in metastatic tumors and in the blood of cancer patients (Fraga et al. 2005).

23.4.1 Regulation of microRNA Expression by Phytochemicals

miRNA are a family of small, non-coding RNAs 18–24 nucleotides in length that regulate gene expression in a sequence specific manner. miRNA have been studied in linkage with tumorigenesis by functioning as tumour metastasis suppressors. Hence miRNAs modulate the hallmarks of cancer progression. Several methods involving the ways to modulate miRs have been identified with the potential to target different signalling pathways deregulated in variety of cancers (Jansson and Lund 2012; Kumar et al. 2017; Lei et al. 2009; Li et al. 2010; Liu et al. 2011; Madanecki et al. 2013; Marí-Alexandre et al. 2015; Mathew et al. 2014; Neergheen et al. 2014; Ramón et al. 2012; Ramos 2008; Ribas and Lupold 2010; Sabatel et al. 2011; Sasahira et al. 2012; Schmidt 2014; Singh et al. 2014; Sun et al. 2012; Ueda et al. 2009; Xu et al. 2011; Zheng et al. 2013; Zhou et al. 2013; Zhu et al. 2013; Farazi et al. 2011; Sethi et al. 2013). Additionally, miRs regulate the expression of numerous functional proteins by regulating miR stability and translation in cancer progression. Aberrant miR expression affects signalling pathways to enhance tumorigenesis. Recent investigation further suggests that DPs can restore tumour suppressor activity of miRs, with successful suppression of metastasis cascade which include induction of cell death and inhibition of cell proliferation. Recently, it has been studied that curcumin is responsible for induction of apoptosis in A549/DDP multidrug-resistant human lung adenocarcinoma cells which is associated with upregulation of miR-186 (>2.5-fold) (Zhang et al. 2010a, b). Curcumin activity even suppressed the expression of miR-196 (an oncogenic miR) and induced miR-22 expression (a tumor suppressor miR) in gastric cancers (Zhang et al. 2010a, b). Resveratrol addition to monocytic cells inhibited AP-1 expression by upregulating miR-663 and simultaneously reducing the expression of oncogenic miR-155 (Tili et al. 2010a, b). 1,3-di-O-galloyl-4,6-(s)-HHDP-b-D-glucopyranose, an ellagitannin, was found to regulate the expression of let-7 family members, miR-370, miR-373, and miR-526b, thereby inhibiting metastasis events such as proliferation and differentiation in HepG2 cells (Wen et al. 2009). Indole-3-carbinol (I3C) restores the miRNAs which inhibits the proliferation of tumor cells such as miR-146 (NF-κB activation); miR-192 (Ras activation); let-7a, let-7c (cell proliferation, angiogenesis, Ras activation); and miR-123, miR-222, (angiogenesis, cell proliferation) and miR-99b (apoptosis) (Izzotti et al. 2010). The recent findings advocate that microRNA expression is often deregulated in cancer and these have uncovered a wholly new horizon of molecular objectives upstream of gene expression, which enables new investigations to further elucidate their precise function in malignancy. Studies done in-vitro and in-vivo reveal that phytochemicals prevent human cancer by regulating miRNA expression. DPs also modulate different signalling pathways hence proving that miRNA expression is controlled by phytochemicals or dietary polyphenols.

23.4.2 DNA Hypermethylation

DNA methylation is the process of addition of a methyl group covalently to the fifth carbon position in the pyrimidine ring of cytosine in eukaryotic DNA. DNA methylation occurs through the help of CpG dinucleotides (Ren et al. 2011). DNA methyltransferase enzyme (DNMTs) protein regulates the phenomenon of DNA methylation. Several account of studies show that hyper methylation of CpG dinucleotides leading to abnormal gene silencing is an important cellular modification system which allows the cancer cells to survive in such toxic surroundings and hence taking advantage of gene mutation (Jones and Baylin 2007). This epigenetic regulation of DNA hyper methylation has become targets for chemo preventive phytochemicals which will result in suppression of activity of DNA methyl transferase and reactivation of glutathione-S-transferase (Pandey et al. 2010).

Dietary phytochemicals like green tea EGCG (Fang et al. 2003), genistein (Majid et al. 2009), apigenin (Fang et al. 2007), curcumin (Shu et al. 2011), sulforaphane (SFN) (Shu et al. 2011; Meeran et al. 2010a, b) and reservatol (Papoutsis et al. 2010) have been demonstrated to inhibit DNA hypermethylation. EGCG inhibited DNMT1 activity and induces activation of several genes like $p16^{INK4a}$, retinoic acid receptor β (RAR β), MGMT and hLMH1 of human oesophageal KYSE 510 and 150 cells (Fang et al. 2004). It also induces promoter demethylation of WIF-1 in lung cancer cells (Gao et al. 2009) and EGCG can block the entry of the main nucleotide cytosine into the active site of DNMT by hydrogen bonds and thus prevent DNA methylation. Human telomerase reverse transcriptase (hTERT), a catalytic subunit of telomerase is deactivated by EGCG (Berletch et al. 2008). EGCG treatment to oral carcinoma cells has shown to induce epigenetic reactivation of RECK leading to suppression of matrix metalloproteinases (MMPS) and inhibition of tumour invasion, angiogenesis and metastasis (Kato et al. 2008). Hypomethylation is also reported in variety of cancers including pancreas, liver, colon, breast, etc. (Meeran et al. 2010a, b). Indirect inhibition of DNA methylation pattern is also measured with treatment of chlorogenic acid and caffeic acid (Lee and Zhu 2006). This happened due to decrease in S-adenosyl L-methionine. In breast cancer cells, treatment with SFN inhibited the expression of hTERT (Meeran et al.

2010a, b). Resesrvatol was found to partially prevent silencing of BRCA-1 in MCF-7 cells which was related to the inhibition of DNMT-1 activity, MBD2, and enrichment of monomethylated H3K9 (Papoutsis et al. 2010). Hence dietary polyphenols show their effect on DNA methylation suggesting their role in decreasing cancer metastasis and as potential therapeutic target.

23.4.3 Histone Modifications

Dietary polyphenols alters the posttranscriptional events like aceylation and deaceylation by regulating the activities of the enzymes. Histone acetyltransferase (HAT) and histone deacetyltransferase (HDAC) which causes aceylation and deaceylation (Kang et al. 2006).

Curcumin has been found to decrease aceylation of H_3 and H_4 by inhibiting HAT activity, associated with caspases-dependent cellular apoptosis in brain cancer cells (Kang et al. 2006). It also restores the UV-hyperaceylated inflammatory related genes: COX-2, ATF-3 and MKP1 in human keratinocytes (Pollack et al. 2009). Curcumin is also found to downregulate H3 hyperaceylation, NF-kB binding, and p300 and H3S10 phosphorylation (Chiu et al. 2009). Treatment with SFN to human colorectal cancer cells (HCT116) increases aceylation at the $p21^{WAF1/CIPI}$ promoter through down regulation of HDAC activity (Myzak et al. 2004), thereby inducing growth arrest apoptosis and anti-metastatic events in prostate cancer cells. EGCG strongly abrogated p300-induced p65 acetylation in-vitro and in-vivo which is associated with the inhibition of HAT activity (Choi et al. 2009). Recently, Pandey and his coworkers demonstrated that GIPS at 1–10 μg/ml showed inhibitory potential on HDAC1-3 expression and increased the levels of aceylated histone H3 (Lys H9/18) and H4 levels, thereby reactivating GSTP1 gene which is an important hallmark in prostate tumorigenesis (Nihal et al. 2010). Soy genistein has been also reported to increase acetylation of histones at the $p21^{WAF1/CIPI}$ and $p16^{INK4a}$ transcription start sites by inducing HAT activity followed by cell cycle arrest and cyclin and caspase-dependent apoptosis (Majid et al. 2008). Reservatol has been found to induce activation of SIRT1 and p300 (type III HDAC inhibitors) in multiple in-vitro and in-vivo models, thereby regulating cell proliferation through inhibition of surviving expression, because of SIRT1 deacetylase activity (Kaeberlein et al. 2005). Many other phytochemicals like cynadins, baicalein, apigenin, rosmarinic and have been reported to have antimetastatic activity through inhibition of DNMT activity or histone modifications (Fang et al. 2007) (Table 23.1).

Table 23.1 Modulation of signaling pathways modulated by phytochemicals by regulation of miRNAs

Phytochemical	miRNAs	Pathway	References
Resveratrol	miR-21 ↓, miR-30a-5p ↓, miR-19↓, miR-34a↑, miR-663↑, miR-17↓, miR-21↓, miR-23a↓, miR-23b↓, miR-25↓, miR-29c↓, miR-92-a-2↓, miR-103-1↓, miR-103-2↓, miR-146a↓	AKT, β-catenin, STAT-3, COX-2, NF-κB, PI3k/mTOR, MAPK/Erk1/2,TGFβ	Campagnolo et al. (2015), Wang et al. (2015a, b), Kumazaki et al. (2013), and Tili et al. (2010a, b)
Quercetin	Let-7c↑, miR-155↓, miR-28↑	Notch, Nrf2, JAK/STAT, STAT3	Nwaeburu et al. (2016) and Jiang et al. (2010)
Curcumin	miR-21↓, miR-186↓, miR-15a↑, miR-16↑, miR-22↑	Akt, NF-κB	Sun et al. (2008), Zhang et al. (2010a, b), Mudduluru et al. (2011), Ali et al. (2010), and Yang et al. (2010)
Genistein	miR-200↑, miR-27a↓, miR-221↓, miR-222↓, miR-146a↑, let-7↑	SMAD 4 ↓, SMAD 3↓	Li et al. (2009), Chen et al. (2011), and Sun et al. 2009
EGCG	miR-20a↓, miR-182↑ miR-498↑, Let-7↑, miR-16↑, miR-330↑, miR-20a↓, miR-125b↓, miR-27a↓, miR-100↓, miR-224↓, miR-23a↓, miR-24↓, miR-31↓, miR-93↓, miR99b↓, miR-15b↓, miR-27b↓, miR-767-5p↓, miR-21↓	Bcl 2↓, Bclxl↓	Ahn et al. (2010), Tsang and Kwok (2010), and Siddiqui et al. (2011)

↓-down regulation; ↑-up regulation

23.5 Therapeutic Effects of Phytochemicals on Cancer Hallmarks

23.5.1 Inflammation

NF-KB, tumour necrosis factor (TNF) and cyclooxygenase -2(Cox-2) are involved in the process of inflammation (Lu et al. 2006). Cancer and inflammation are associated with each other in their functions and the above mentioned key players are involved not only in inflammation but also cell proliferation, anti-apoptotic activity, metastasis and angiogenesis. COX-2 isoenzyme is highly overexpressed in premalignant and malignant conditions in colon, liver, pancreas, breast, lung, skin, bladder, stomach, head, and neck and oesophagus cancer cells (Subbaramaiah and Dannenberg 2003). Therefore to inhibit the activity of COX-2 phytochemicals are to be used which will play an important role in slow progress of cancer cell development.

Genistein treatment to these cancer cells inhibited the COX-2 activity without increasing the apoptosis process (Ye et al. 2004). Treatment with flavone induced apoptosis in human colon carcinoma cells through changes in mRNA levels of cell cycle and apoptosis-related genes including COX-2 and NF-KB (Wenzel et al. 2000). EGCG treatment led to a down regulation of AKT pathways and inhibition of COX-2 promoter activity by inhibiting NF-KB activity (Peng et al. 2006). Some natural phenolic compounds inhibit the activity of TNF-α and lower its level such as geraniin, corilagin and EGCG (Fujiki et al. 2003).

23.5.2 Angiogenesis

Angiogenesis is key step in tumour growth and cancer development. Many polyphenols like ellagic acid, EGCG, genistein and anthocyanin -rich berry extracts through down regulation of vascular endothelial factor (VEGF), VEGF receptor-2, PDGF, PDGF receptor, hypoxia inducible factor 1α (HIF-1α) and matrix metalloproteases (MMPs), also inhibits phosphorylation of EGFR,VEGFR and PDGFR (Annabi et al. 2002; Yamakawa et al. 2004; Labrecque et al. 2005; Bagchi et al. 2004).

Quercetin, myricetin and galanin were able to suppress the VEGF stimulated HUVEC tubular structure formation and to inhibit the activated U937 monocyctic cell adhesion to HUVEC cells, hence playing an important role in the prevention of angiogenesis (Kim et al. 2006). A polyphenol delphinidin inhibits the VEGF-induced migration and proliferation through the blockade of cell cycle in G_0/G_1 phase (Favot et al. 2003).

23.5.3 Metastasis

Dietary polyphenols exist with properties of inhibition of metastasis and invasion of cancer cells. The process of metastasis is interlinked among extracellular matrix (ECM) degradation, proteolysis, cell adhesion, cell migration, angiogenesis and invasion (Woodhouse et al. 1997).

Some dietary polyphenols like Baicalein, Epicatechin and EGCG inhibited cell shedding and invasion by a decreased ROS generation and down regulation of MMP-9 expression (Gunther et al. 2007). Hydrolysable tannins also inhibited the MMP-9 and -2 activities without suppressing the ERK-MAPK and PI3K/AKT pathways (Tanimura et al. 2005). EGCG prevented metastasis and invasion through its inhibitory effects on activation of MMP and focal adhesion kinase (FAK) (Annabi et al. 2002; Yamakawa et al. 2004; Zhen et al. 2006; Hung et al. 2005).

EGCG unfortunately activated proMMP-7 induction and production in HT-29 human colorectal cancer cells through ROS formation and activation of JNK, c-fos, AP-1 but not p38 MAPK or ERK (Kim et al. 2005). Quercetin induces cytoskeletal alterations in microtubule polymerization through directly binding to tubulin. Quercetin and luteolin inactivated the EGFR (endothelial growth factor) tyrosine kinase activity, which stimulates cell migration and downstream signalling pathways and to reduce MMP secretion (MMP-2 and -9) (Lee et al. 2004; Huang et al. 1999; Vijayababu et al. 2006). Curcumin has also resulted in down regulation of MMP-1 and -2 due to the inhibition of NF-KB/AP-1 mediated mRNA transcription which explains reduced invasion and metastasis (Bachmeier et al. 2007). Polyphenols also disturbs the movement of cancer cells through different mechanisms. EGCG breaks the stress fibres and decreases the phosphorylation rate of myosin-II regulatory light chain which is important for cell division and contractile ring formation (Umeda et al. 2005). It also disrupts the structure and framework of the cellular space and leading to inhibition of cell proliferation (Ermakova et al. 2005). Flavones play an important role in expression and affinity of integrin α2β1, FAK phosphorylation, actin cytoskeleton, MMPS activities, migration and network formation on 3-D Matrigel, suggesting a stopping effect on tumour cell behaviour (Hung et al. 2005). The miRNAs that stimulates metastasis and invasion of cancer cells are miR-101, miR-146b, miR-10b, miR-520c, miR-21, miR-206, mir-155, miR-335, miR-200, miR-378, miR-126, let-7. These miRNAs are downregulated by phytochemicals like EGCG, flavones, curcumin and anthocyanins and hence proving antimetastatic.

23.6 Phytochemicals Efficacy on miRNA Modulation and Associated Pathways: An in-silico Approach

Normal functioning of the various signaling pathways is involved in the maintenance of normal physiological status of the cell. During diseases such as cancer, aberration in the signaling function has been reported. Aberration in signaling pathways (Notch, Wnt, JAK-STAT and IP3) are responsible for anti-apoptosis, angiogenesis, cell cycle check point bypass, migration, invasion and resistance in cancer cells. Literature revealed that miRNAs are involved in the regulation of these signaling pathways. Phytochemicals are known to regulate miRNA expression levels and thereby affect the activity of genes involved in pathogenesis of various diseases such as cancer. Epi-gallocatechin-3-gallate, curcumin, resveratrol and isoflavone are among some of the most studied compounds for their miRNA expression modulation potential. Screening of the effect of phytochemicals on miRNA expression profile and associated signaling pathways in a cancer cell is an expensive and time taking process. Thorough literature search on the miRNA expression modulation potential of the phytochemicals and prediction of genes involved in the signaling pathway together with the use of some computer based application is a tool to identify the relation between phytochemicals and miRNAs associated with the particular disease. In the preceding sections of the chapter we will discuss how to use computer based application to identify the miRNA modulatory potential of phytochemicals. Following is the step by step methodology to identify the interrelation between phytochemicals, affected miRNAs and associated signaling pathways (Carotenuto et al. 2016):

23.6.1 Literature Based Identification of miRNAs Involved in the Cancer Pathogenesis

Initially, a literature search should be performed using Google Scholar and/or PubMed databases to identify miRNAs involved in the pathophysiology of cancer. Date and year of the literature search as well as exact keywords should be reported. Prepare the list of miRNAs in a tabular format. If lots of literature is available on the particular keyword, we should choose only most appropriate and exhaustive reviews available on the topic. For example, we have taken literature based identification of miRNAs involved in the cancer cell angiogenesis (Tables 23.2 and 23.3).

Table 23.2 Micro RNAs involved in angiogenesis promotion

miRNAs	References
miR-210, miR-182, miR-55, miR-519c, miR-494, miR-93, miR-34, miR-20a, miR-20b, miR-424, miR-130a, miR-130b, mir-200b, miR-200c, miR-429, miR-17, miR-31, miR-146a, miR-21,let-7c, miR-296, miR-378, miR-27b, miR-126, miR-382, miR-26a, miR-194, miR-17~92, miR-467, miR-21, miR-29c, miR-27a	Yang et al. (2016), Chiang et al. (2016), Kong et al. (2014), Cha et al. (2010), Mao et al. (2015); Lin et al. (2012, 2014), Saito et al. (2011); Lee et al. (2007, 2010), Maroof et al. (2014), Szade et al. (2015), Yin et al. (2015), Kuehbacher et al. (2007), Wurdinger et al. (2008), Urbich et al. (2012), Wang et al. (2008), Seok et al. (2014), Chai et al. (2013), Sundaram et al. (2011), Ramon et al. (2011), Bhattacharyya et al. (2012), Zhang et al. (2012), and Ohlsson et al. (2010)

Table 23.3 MicroRNAs involved in angiogenesis inhibition

miRNAs	References
miR-15b, miR-1228, miR-107, miR-17~92, miR-218, miR-206, miR-7, miR-100, miR-140-5p, miR-20b, miR-29b, miR-200c, miR-377, miR-126, miR-135a, miR-361-5p, miR-497, miR-134, miR-144-3p, miR-200b, miR-20a, mir-20b, miR-199a, mir-155, miR-16, miR-222, miR-221, miR-92a, miR-424, miR-320a, miR-217, miR-503, miR-34, miR-214, miR-21, miR-18a, miR-145, miR-519c, miR-22, miR-126, miR-202-3p, miR-17-5p, miR-125a	Nikaki et al. (2012), Jia et al. (2017), Yamakuchi et al. (2010), Ma et al. (2016), Guan et al. (2017), Liang et al. (2016), Beyer et al. (2017), Pakravan et al. (2017), Hu et al. (2017), Cascio et al. (2010), Li et al. (2017a, b, c), Shi et al. (2013), Kong et al. (2016), Wang et al. (2014), Zhou et al. (2017), Kanitz et al. (2012), Tu et al. (2015), Zhang et al. (2018), Wu et al. (2017), Sinha et al. (2015), Hua et al. (2006), Rane et al. (2009), Bruning et al. (2011), and Yin et al. (2015)

Table 23.4 Phytochemical mediated up and down regulated miRNAs (Lançon et al. 2013)

Phytochemicals	miRNAs
	Down-regulated miRNAs
Curcumin	miR-199a, miR-510, miR-196a, miR-7, miR-15b, miR-195, miR-74, miR-98
Isoflavone	miR-34c, miR-376a, miR-196a, miR-320a, miR-654, miR-34c, miR-196
EGCG	miR-10a, miR-18a, miR-19a, miR-26b, miR-29b, miR-34b, miR-98, miR-129, miR-181d
Resveratrol	miR-589, miR-20b, miR-291b-5p, miR-550-1, miR-326, miR-532, miR-128a, miR-17, let-7g, miR-132, miR-20b, miR-133b, miR-292-5p, miR-342, miR-466, miR-181b-1, miR-684, miR-129-5p, miR-14, miR-497, miR-17-92, miR-106a, miR-106b, miR-21, miR-155
	Up-regulated miRNAs
Curcumin	miR-22, miR-34a, miR-24, miR-181a, miR-21, miR-181b, miR-27a
Isoflavone	miR-200b, miR-200c, let-7b, let-7c, let-7d, let-7e, miR-663, miR-146a, miR-374b
EGCG	let-7c, miR-16, miR-18b, miR-20a, miR-25, miR-92, miR-93, miR-221, miR-320a
Resveratrol	miR-199a-5p, miR-339, miR-629, miR-712-5p, miR-9-5p, miR-27b, miR-688, miR-702, miR-20a, miR-676-3p, miR-694,miR-692, miR-125b, miR-21, miR-34b, let-7a-2, miR-200b, miR-184, miR-199a-3p, miR-16-1, miR-125b2, miR-155, miR-138-1, miR-122-5p, miR-542-3p, miR-16, miR-141, miR-143, miR-200c, miR-663, miR-34c, miR-299-5p, miR-194, miR-338-3p, miR-582-3p, miR-92a-2,miR-137, miR-15a, miR-539-5p

23.6.2 Literature Based Identification of miRNAs Modulated by Test Phytochemicals

Tools and methodology used for the search of associated miRNAs may repeat for the literature based identification of phytochemical modulated miRNAs in a particular disease. Prepare the list of miRNAs in a tabular format. Next identify the list of miRNAs modulated by phytochemicals. For example, we have taken literature based identification of miRNAs modulated by Epi-gallocatechin-3-gallate, curcumin, resveratrol and isoflavone (Table 23.4).

23.6.3 Identification of Phytochemical Modulated miRNAs Involved in Angiogenesis

The miRNAs involved in the cancer pathogenesis (Tables 23.2 and 23.3) and phytochemical modulated (Table 23.4) are used to build the Venn diagram (http://bioinfogp.cnb.csic.es/tools/venny/). This diagram predicts the group of miRNAs modulated by all phytochemicals, single phytochemical, two set of phytochemicals, three set of phytochemicals, and n-1 set of phytochemical (Figs. 23.1 and 23.2). Where 'n' is the maximum number of phytochemicals involved in the study. This diagram also predicts the miRNAs that have not been modulated by any of the study phytochemicals.

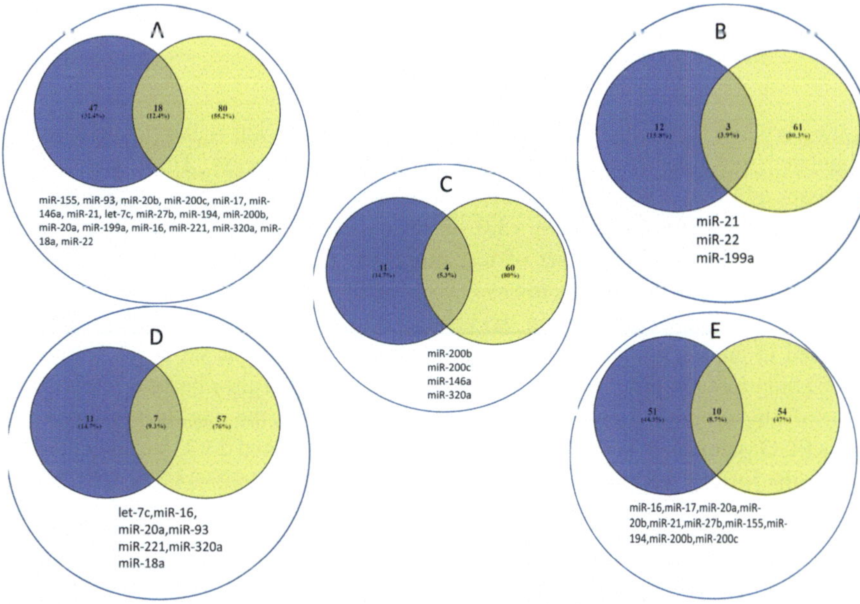

Fig. 23.1 Venn diagram showing relation between (**a**) Phytochemical modulated miRNAs and miRNAs involved in angiogenesis (**b**) Curcumin modulated miRNAs and miRNAs involved in angiogenesis (**c**) EGCG modulated miRNAs and miRNAs involved in angiogenesis (**d**) Isoflavone modulated miRNAs and miRNAs involved in angiogenesis and (**e**) Resveratrol modulated miRNAs and miRNAs involved in angiogenesis

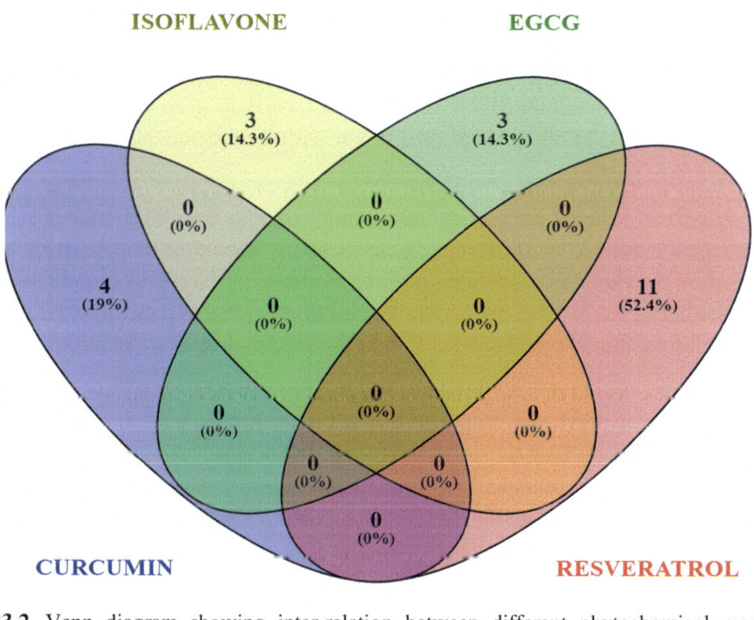

Fig. 23.2 Venn diagram showing inter-relation between different phytochemical modulated miRNAs involved in angiogenesis

23.6.4 Pathway Prediction of Phytochemical Modulated miRNAs Involved in Angiogenesis

In the above discussion and Venn diagram, we came to know the phytochemical modulated miRNAs associated with the angiogenesis process. There are various databases and online tools are available to predict the pathways associated with the test miRNAs. DIANA-miRPath v3.0, is an online tool that predicts the KEGG pathways modulated by the set of test miRNAs. For example, we used DIANA tool to predict the KEGG pathways associated with the phytochemical (Eepigallocatechin-3-gallate, curcumin, resveratrol and isoflavone) modulated miRNAs involved in angiogenesis. The target prediction threshold was set at 0.85. p-value <0.05. Step by step process is given in subsequent part of the chapter. The association of phytochemical modulated miRNAs involved in the angiogenesis process and KEGG pathway is shown in Tables 23.4, 23.5, 23.6, and 23.7. Visualization of data in the form of heat map is depicted in Figs. 23.9, 23.10, 23.11, and 23.12.

Step 1 need to go on DIANA mirPath v.3 tool page on internet (http://snf-515788.vm.okeanos.grnet.gr/). The page (Fig. 23.3) will appear on your screen showing the KEGG pathway analysis for miRNAs.

Step 2 includes the selection of species (mouse, *D.melanogaster* etc.), before entering the miRNAs into the box. The species is set as human by default (Fig. 23.4).

Step 3 involves addition of miRNA. In the column "Add miRNAs" type the test miRNAs and click on add button. Click add button after the entry of each test miRNA. The box just beside the microRNA entry needs to be set as micro T-CDS (miRNA database) for the KEGG pathway analysis (Fig. 23.5).

Step 4 involves the analysis of KEGG pathway. After addition of all the miRNAs, the default P-value threshold (0.05) and MicroT threshold (0.8) should be used. The list of miRNAs added will be displayed down with the associated number of genes (Fig 23.6).

Step 5 involves visualization of pathway analysis. To see the associated KEGG pathways of the added test miRNAs, click on pathways union tab. A list of KEGG pathway with p-value and associated number of genes and number of pathways will be generated (Fig 23.7). The list can be pasted into a word file or the results can be downloaded in a excel spread sheet file by clicking on download results.

Step 6 involves visualization of data in Heat map format. By clicking on pathways union we obtain a third box stated as Enrichment Analysis Method which is by default Fisher's Exact Test (Hypergeometric Distribution). And for generating the absence or presence of statistical significant pathways one need to select Targeted Pathways clusters/Heat map (Fig. 23.8). The miRNA cluster dendrogram and heat map is generated and it can be downloaded easily.

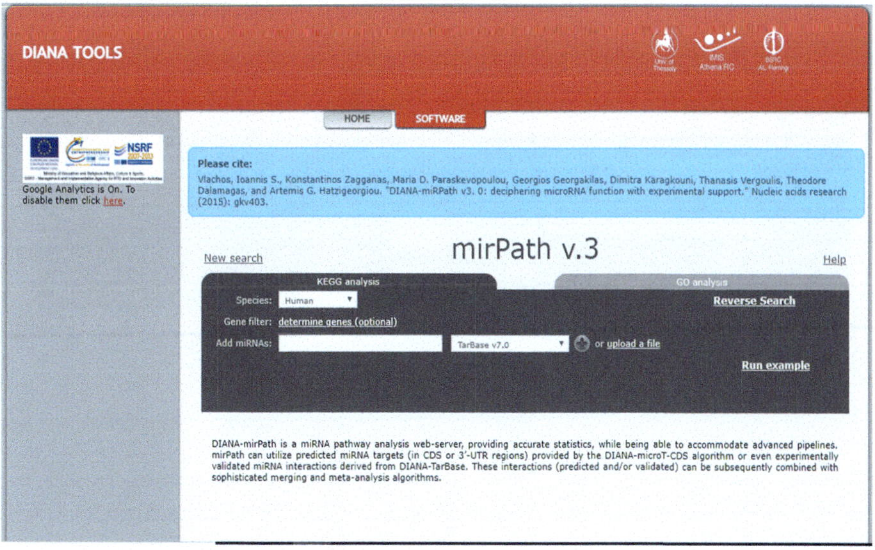

Fig. 23.3 Webpage of DIANA mirPath v.3 tool

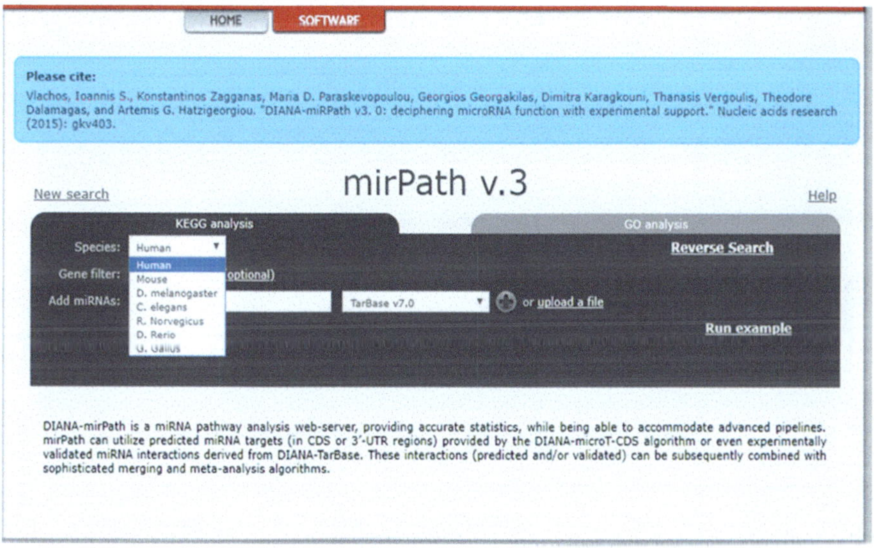

Fig. 23.4 Selection of species in DIANA mirPath v.3 tool

Based upon the above mentioned methodology, we predicted the significantly associated KEGG pathways with the phytochemical (epi-gallocatechin-3-gallate, curcumin, resveratrol and isoflavone) modulated miRNAs involved in angiogenesis. Moreover the significantly associated KEGG pathways influenced by curcumin, epi-gallocatechin-3-gallate, isoflavone and resveratrol modulated miRNAs

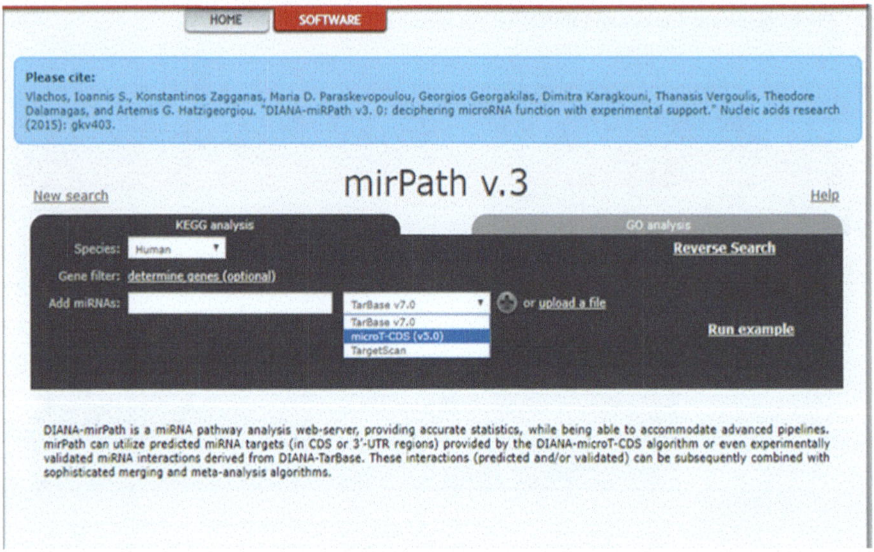

Fig. 23.5 Addition of miRNAs and selection of database in DIANA mirPath v.3 tool for KEGG analysis

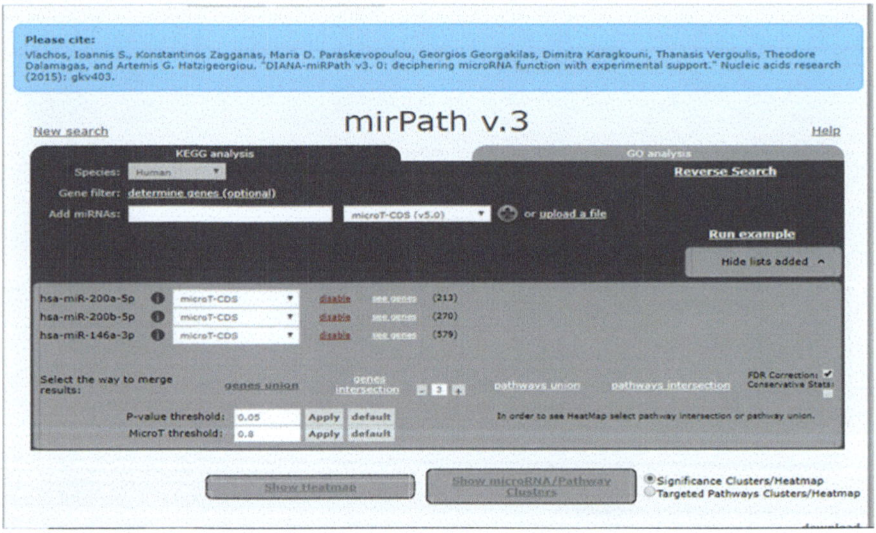

Fig 23.6 Selection of appropriate statistics in DIANA mirPath v.3 tool for KEGG analysis

individually involved in angiogenesis were also predicted (Tables 23.5, 23.6, 23.7, and 23.8) respectively. Heat map generated by the DIANA tool v.3mirPath for these data is depicted in Figs. 23.9, 23.10, 23.11, 23.12, and 23.13 respectively as per discussed.

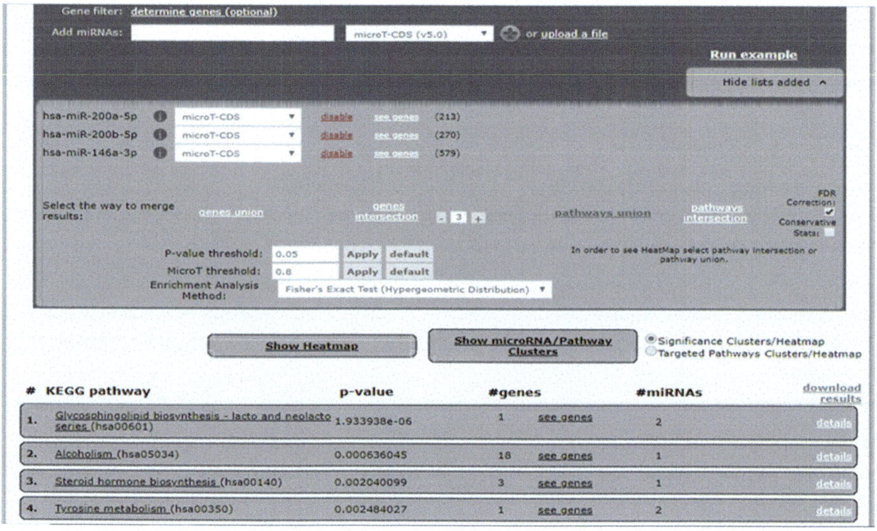

Fig. 23.7 Selection of appropriate statistics in DIANA mirPath v.3 tool for KEGG analysis

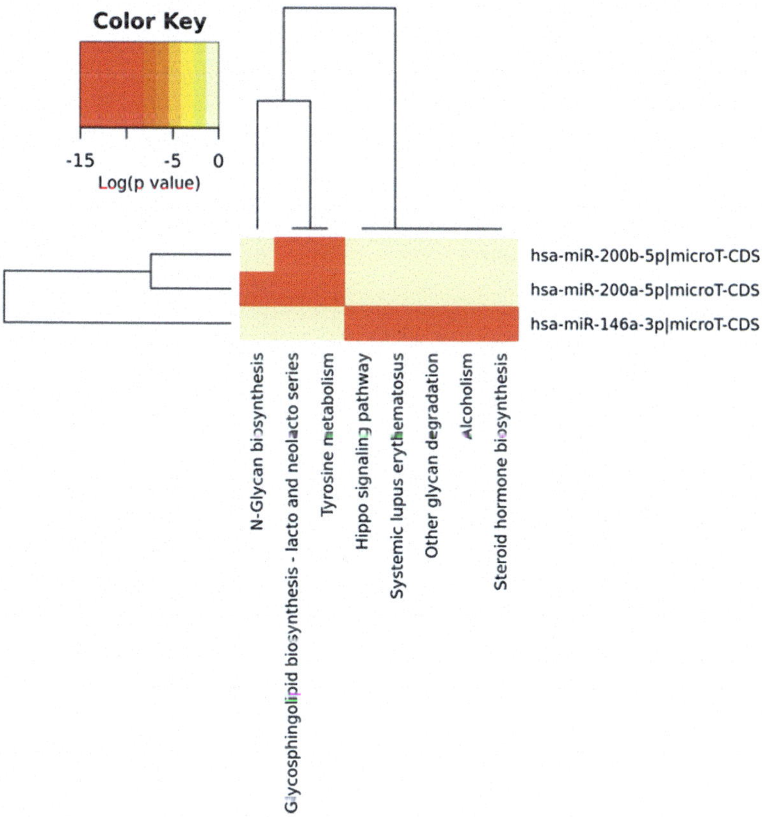

Fig. 23.8 Production and visualization of KEGG pathway analysis data in heat map format in DIANA mirPath v.3 tool

Table 23.5 Enriched KEGG pathways for predicted curcumin modulated microRNAs involved in angiogenesis

KEGG pathway	p-value	Genes	miRNAs
Hippo signaling pathway (hsa04390)	0.00046147	8	2
Cytokine-cytokine receptor interaction (hsa04060)	0.0004863493	11	1
2-Oxocarboxylic acid metabolism (hsa01210)	0.03547975	3	1

Table 23.6 Enriched KEGG pathways for predicted EGCG modulated microRNAs involved in angiogenesis

KEGG pathway	p-value	Genes	miRNAs
Fatty acid biosynthesis (hsa00061)	<1e-325	4	1
Signaling pathways regulating pluripotency of stem cells (hsa04550)	1.320676e-09	54	4
Glioma (hsa05214)	7.253241e-06	24	4
TGF-beta signaling pathway (hsa04350)	1.180666e-05	18	2
Prostate cancer (hsa05215)	1.627159e-05	39	4
Fatty acid metabolism (hsa01212)	3.594566e-05	7	1
Proteoglycans in cancer (hsa05205)	0.0001045923	55	4
Lysine degradation (hsa00310)	0.0001073424	11	2
Hippo signaling pathway (hsa04390)	0.0001722163	40	3
Non-small cell lung cancer (hsa05223)	0.0004484873	18	3
FoxO signaling pathway (hsa04068)	0.0007460847	42	3
Melanoma (hsa05218)	0.0009623208	26	3
Pathways in cancer (hsa05200)	0.001242614	67	2
Estrogen signaling pathway (hsa04915)	0.002697365	16	2
Axon guidance (hsa04360)	0.007773555	19	1
Long-term depression (hsa04730)	0.009130671	13	2
Glycosaminoglycan biosynthesis – heparan sulfate/heparin (hsa00534)	0.01036648	4	2
AMPK signaling pathway (hsa04152)	0.01263703	35	3
Endometrial cancer (hsa05213)	0.01472444	15	2
Renal cell carcinoma (hsa05211)	0.01711761	18	2
Circadian rhythm (hsa04710)	0.01783856	13	2
Thyroid hormone signalling	0.02786344	15	1
MAPK signaling pathway (hsa04010)	0.03234459	29	1
Colorectal cancer (hsa05210)	0.03265998	15	2
Endocytosis (hsa04144)	0.03343939	36	2
Morphine addiction (hsa05032)	0.03653941	3	1
Pancreatic cancer (hsa05212)	0.03684178	15	2
Rap1 signaling pathway (hsa04015)	0.04671777	32	2

Table 23.7 Enriched KEGG pathways for predicted isoflavone modulated microRNAs involved in angiogenesis

KEGG pathway	p-value	Genes	miRNAs
Glioma (hsa05214)	0.001082791	13	2
Hippo signaling pathway (hsa04390)	0.001093889	23	2
Alcoholism (hsa05034)	0.001591541	18	1
Steroid hormone biosynthesis (hsa00140)	0.007862145	3	1
Glycosphingolipid biosynthesis – lacto and neolacto series (hsa00601)	0.01671999	1	1
Other glycan degradation (hsa00511)	0.01681244	2	1
Systemic lupus erythematosus (hsa05322)	0.01681244	17	1
Estrogen signaling pathway (hsa04915)	0.02080689	11	1
Prostate cancer (hsa05215)	0.03075329	16	1
Proteoglycans in cancer (hsa05205)	0.0422731	18	1
Circadian rhythm (hsa04710)	0.0486534	10	1

Table 23.8 Enriched KEGG pathways for predicted resveratrol modulated microRNAs involved in angiogenesis

KEGG pathway	p-value	Genes	miRNAs
Fatty acid biosynthesis (hsa00061)	<1e-325	4	1
Signaling pathways regulating pluripotency of stem cells (hsa04550)	2.847193e-05	52	4
Glioma (hsa05214)	0.0009141194	19	3
Fatty acid metabolism (hsa01212)	0.001721963	7	1
Hippo signaling pathway (hsa04390)	0.002795897	27	3
Pathways in cancer (hsa05200)	0.003248205	67	2
Proteoglycans in cancer (hsa05205)	0.005258917	38	2
Prostate cancer (hsa05215)	0.007910257	27	2
TGF-beta signaling pathway (hsa04350)	0.01052044	20	2
Axon guidance (hsa04360)	0.01361044	19	1
Hepatitis B (hsa05161)	0.01499133	16	1
FoxO signaling pathway (hsa04068)	0.01671007	31	2
ErbB signaling pathway (hsa04012)	0.01836461	14	1
Renal cell carcinoma (hsa05211)	0.01869079	1	1
Prolactin signaling pathway (hsa04917)	0.02083174	16	2
Neurotrophin signaling pathway (hsa04722)	0.02095697	17	1
MAPK signaling pathway (hsa04010)	0.02151759	29	1
Cytokine-cytokine receptor interaction (hsa04060)	0.02607704	11	2

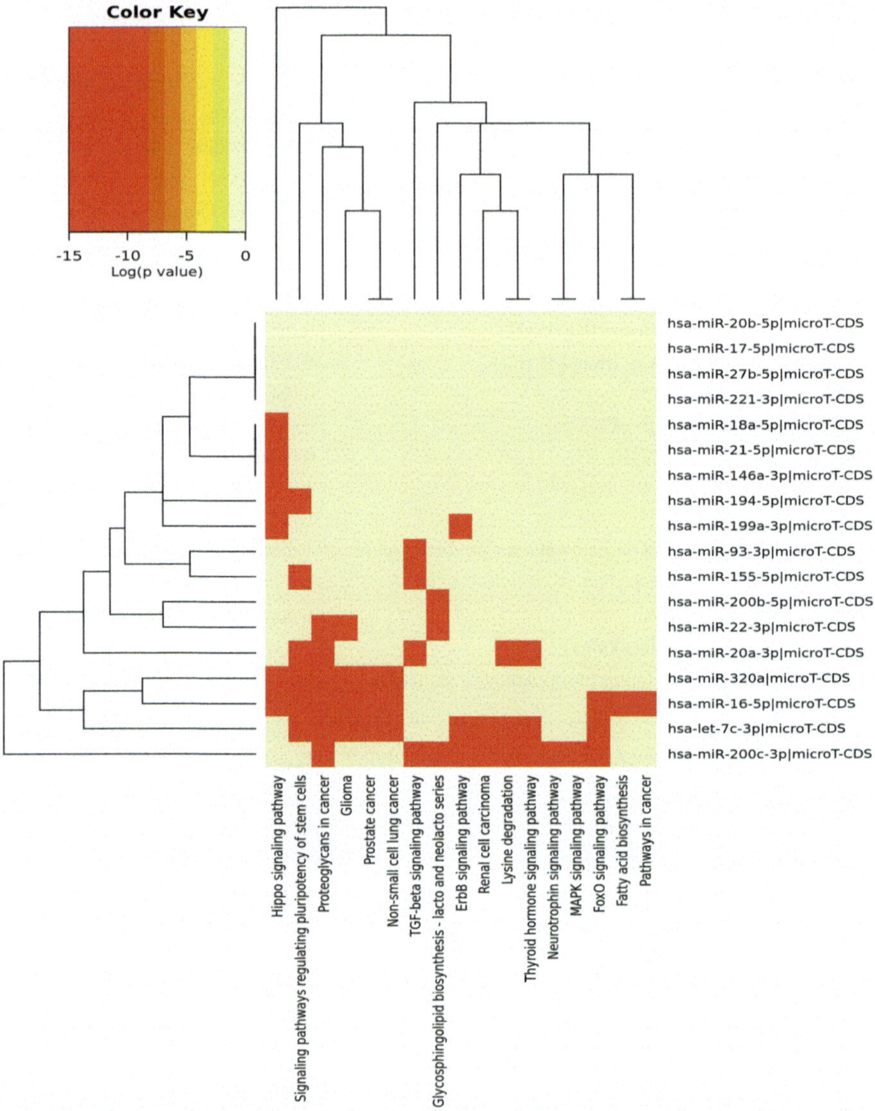

Fig. 23.9 Binary heat map of predicted phytochemical (curcumin, Isoflavone, EGCG, and Resveratrol) modulated microRNAs involved in angiogenesis. All significantly targeted pathways are marked with deep red

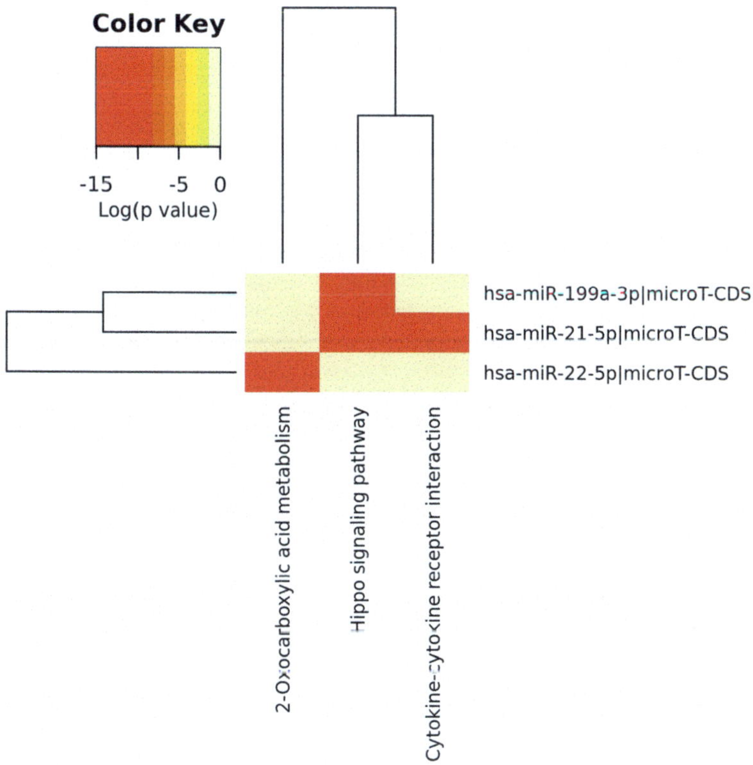

Fig. 23.10 Binary heat map of predicted curcumin modulated microRNAs involved in angiogenesis. All significantly targeted pathways are marked with deep red

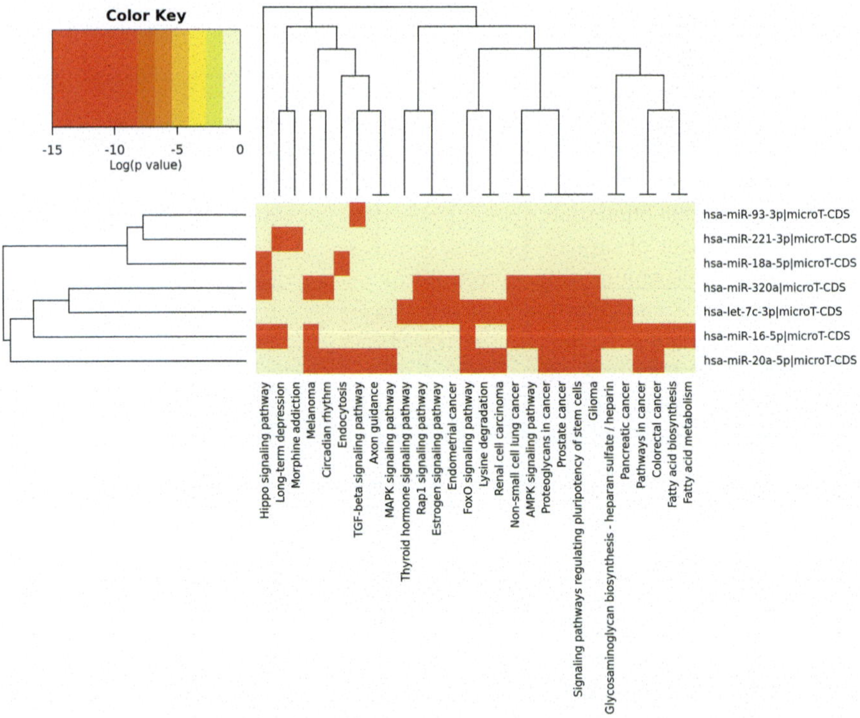

Fig 23.11 Binary heat map of predicted EGCG modulated microRNAs involved in angiogenesis. All significantly targeted pathways are marked with deep red

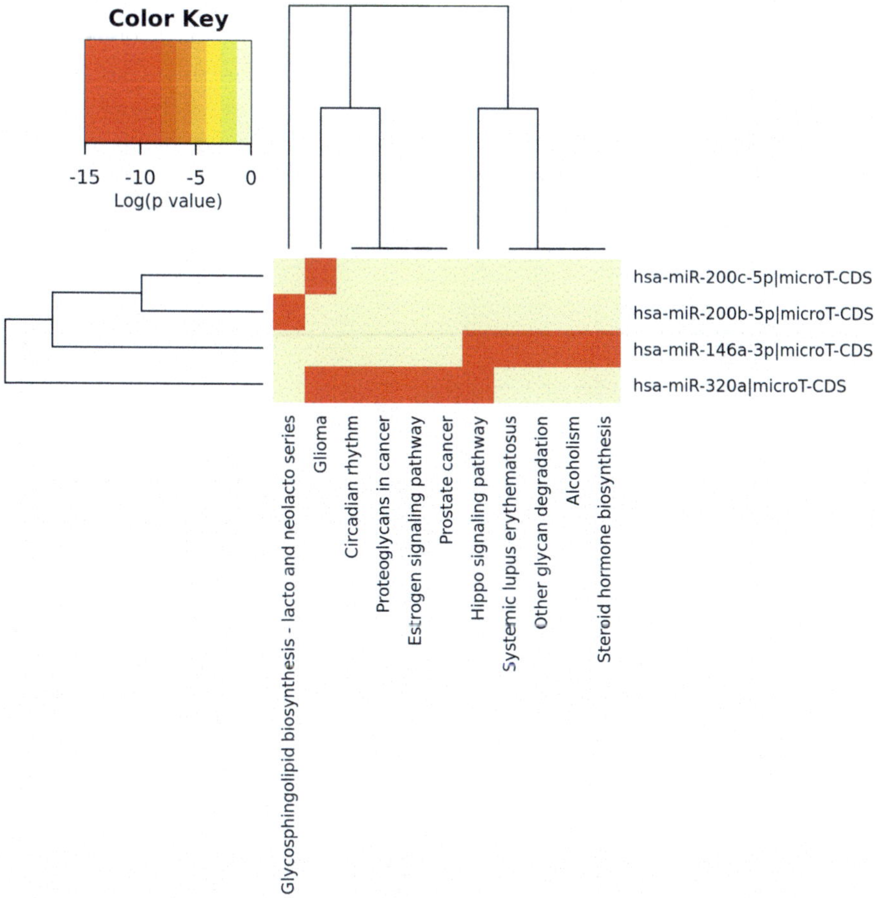

Fig. 23.12 Binary heat map of predicted isoflavone modulated microRNAs involved in angiogenesis. All significantly targeted pathways are marked with deep red

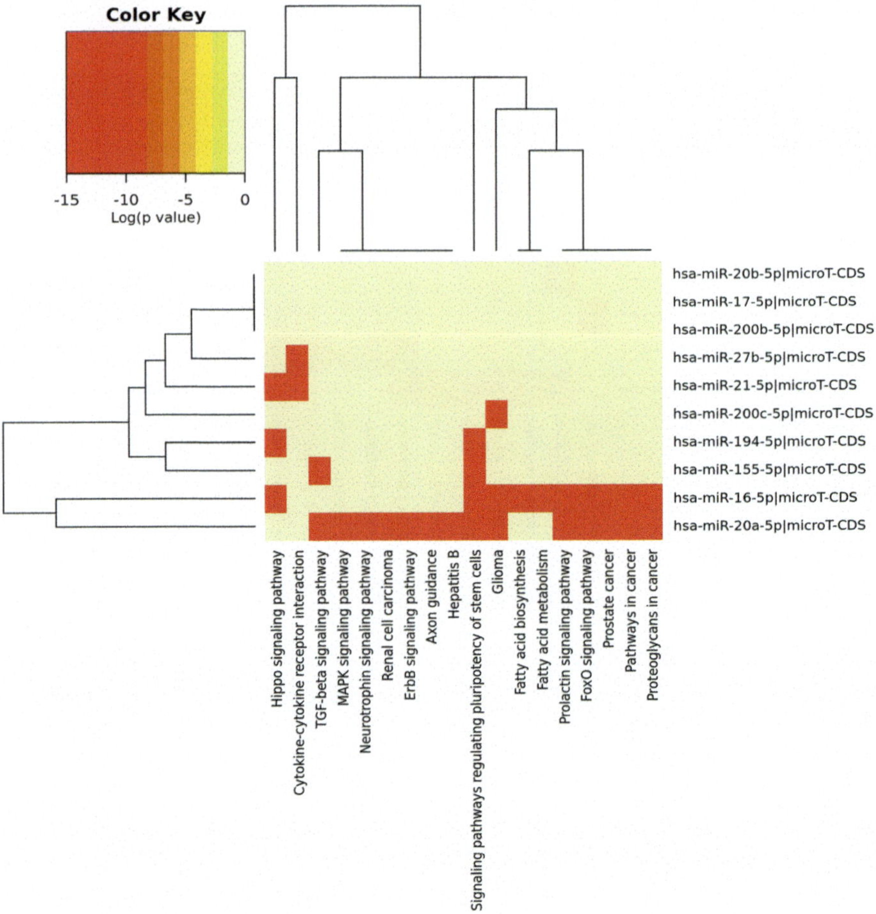

Fig 23.13 Binary heat map of predicted resveratrol modulated microRNAs involved in angiogenesis. All significantly targeted pathways are marked with deep red

23.7 Conclusion

The study highlights the importance of phytochemicals in regulating different cellular signalling pathways. Epigenetic changes for chemoprevention and therapy of various forms of cancers is also controlled by phytochemicals by modulating genetic and epigenetic targets. Clinical trials are required to confirm the effects of these agents for the prevention of cancer development and its treatment. Many phytochemicals should be introduced in clinical trials and their applications can be checked out. Dietary phytochemicals proves to be an efficient agent in chemoprevention of cancer development. It should be noted that clinical application of these agents can only be measured when they have passed preclinical toxicity-profiling, pharmacokinetic profiling and other profiling in advanced animal model systems.

Acknowledgements NK acknowledges Central University of Punjab for providing infrastructure facility. PPK and AKS acknowledge financial support from University Grants Commission, India in the form of CSIR-UGC Senior/Junior Research fellowship. SK acknowledges Department of Science and Technology, India, for providing financial support in the form of DST-SERB Grant [EEQ/2016/000350].

References

Adams LS, Phung S, Yee N, Seeram NP, Li L, Chen S. Blueberry phytochemicals inhibit growth and metastatic potential of MDA-MB-231 breast cancer cells through modulation of the phosphatidylinositol 3-kinase pathway. Cancer Res. 2010;70:3594–605.

Aggarwal BB, Shishodia S. Molecular targets of dietary agents for prevention and therapy of cancer. Biochem Pharmacol. 2006;71(10):1397–421.

Aggarwal BB, Ichikawa H, Garodia P, Weerasinghe P, Sethi G, Bhatt ID, Pandey MK, Shishodia S, Nair MG. From traditional Ayurvedic medicine to modern medicine: identification of therapeutic targets for suppression of inflammation and cancer. Expert Opin Ther Targets. 2006;10 (1):87–118.

Ahmad N, Gupta S, Mukhtar H. Green tea polyphenol epigallocatechin-3-gallate differentially modulates nuclear factor κB in cancer cells versus normal cells. Arch Biochem Biophys. 2000;376(2):338–46.

Ahn JI, Jeong KJ, Ko MJ, Shin HJ, Kim HS, Chung HJ, Jeong HS. Changes of miRNA and mRNA expression in HepG2 cells treated by epigallocatechin gallate. Mol Cell Toxicol. 2010;6 (2):169–77.

Aithal MG, Rajeswari N. Role of notch signalling pathway in cancer and its association with DNA methylation. J Genet. 2013;92(3):667–75.

Alam MN, Almoyad M, Huq F. Polyphenols in colorectal cancer: current state of knowledge including clinical trials and molecular mechanism of action. Biomed Res Int. 2018;2018:4154185.

Ali S, Ahmad A, Banerjee S, Padhye S, Dominiak K, Schaffert JM, Wang Z, Philip PA, Sarkar FH. Gemcitabine sensitivity can be induced in pancreatic cancer cells through modulation of miR-200 and miR-21 expression by curcumin or its analogue CDF. Cancer Res. 2010;13:0008–5472.

Amoyel M, Anderson AM, Bach EA. JAK/STAT pathway dysregulation in tumors: a drosophila perspective. Semin Cell Dev Biol. 2014;28:96–103. Academic

Annabi B, Lachambre MP, Bousquet-Gagnon N, Pagé M, Gingras D, Bëliveau R. Green tea polyphenol (−)-epigallocatechin 3-gallate inhibits MMP-2 secretion and MT1-MMP-driven migration in glioblastoma cells. Biochim Biophys Acta (BBA)-Mol Cell Res. 2002;1542 (1–3):209–20.

Azmi AS, Bhat SH, Hadi SM. Resveratrol–Cu (II) induced DNA breakage in human peripheral lymphocytes: implications for anticancer properties. FEBS Lett. 2005;579(14):3131–5.

Azmi AS, Bhat SH, Hanif S, Hadi SM. Plant polyphenols mobilize endogenous copper in human peripheral lymphocytes leading to oxidative DNA breakage: a putative mechanism for anticancer properties. FEBS Lett. 2006;580(2):533–8.

Bachmeier B, Nerlich A, Iancu C, Cilli M, Schleicher E, Vené R, Dell'Eva R, Jochum M, Albini A, Pfeffer U. The chemopreventive polyphenol curcumin prevents hematogenous breast cancer metastases in immunodeficient mice. Cell Physiol Biochem. 2007;19(1–4):137–52.

Bagchi D, Sen CK, Bagchi M, Atalay M. Anti-angiogenic, antioxidant, and anti-carcinogenic properties of a novel anthocyanin-rich berry extract formula. Biochemistry (Moscow). 2004;69(1):75–80.

Bai Y, Bai Y, Dong J, Li Q, Jin Y, Chen B, Zhou M. Hedgehog signaling in pancreatic fibrosis and cancer. Medicine. 2016;95(10):e2996.

Bartel DP. MicroRNAs: target recognition and regulatory functions. Cell. 2009;136(2):215–33.

Berletch JB, Liu C, Love WK, Andrews LG, Katiyar SK, Tollefsbol TO. Epigenetic and genetic mechanisms contribute to telomerase inhibition by EGCG. J Cell Biochem. 2008;103 (2):509–19.

Beyer S, Fleming J, Meng W, Singh R, Haque SJ, Chakravarti A. The role of miRNAs in angiogenesis, invasion and metabolism and their therapeutic implications in gliomas. Cancers. 2017;9(7):85.

Bhat SH, Azmi AS, Hanif S, Hadi SM. Ascorbic acid mobilizes endogenous copper in human peripheral lymphocytes leading to oxidative DNA breakage: a putative mechanism for anticancer properties. Int J Biochem Cell Biol. 2006;38(12):2074–81.

Bhat SH, Azmi AS, Hadi SM. Prooxidant DNA breakage induced by caffeic acid in human peripheral lymphocytes: involvement of endogenous copper and a putative mechanism for anticancer properties. Toxicol Appl Pharmacol. 2007;218(3):249–55.

Bhattacharyya S, Balakathiresan NS, Dalgard C, Gutti U, Armistead D, Jozwik C, Srivastava M, Pollard HB, Biswas R. Elevated miR-155 promotes inflammation in cystic fibrosis by driving hyper-expression of interleukin-8. J Biol Chem. 2011;286:11604–15.

Bhattacharyya S, Sul K, Krukovets I, Nestor C, Li J, Adognravi OS. Novel tissue-specific mechanism of regulation of angiogenesis and cancer growth in response to hyperglycemia. J Am Heart Assoc. 2012;1(6):e005967.

Braune EB, Lendahl U. Notch—a goldilocks signaling pathway in disease and cancer therapy. Discov Med. 2016;21(115):189–96.

Bruning U, Cerone L, Neufeld Z, Fitzpatrick SF, Cheong A, Scholz CC, Simpson DA, Leonard MO, Tambuwala MM, Cummins EP, Taylor CT. MicroRNA-155 promotes resolution of hypoxia-inducible factor-1α activity during prolonged hypoxia. Mol Cell Biol. 2011;31:4087–96. MCB-01276

Campagnolo P, Hong X, Di Bernardini E, Smyrnias I, Hu Y, Xu Q. Resveratrol-induced vascular progenitor differentiation towards endothelial lineage via MiR-21/Akt/β-catenin is protective in vessel graft models. PLoS One. 2015;10(5):e0125122.

Camps C, Buffa FM, Colella S, Moore J, Sotiriou C, Sheldon H, Harris AL, Gleadle JM, Ragoussis J. hsa-miR-210 Is induced by hypoxia and is an independent prognostic factor in breast cancer. Clin Cancer Res. 2008;14(5):1340–8.

Caporali A, Emanueli C. MicroRNA regulation in angiogenesis. Vasc Pharmacol. 2011;55 (4):79–86.

Carotenuto F, Albertini MC, Coletti D, Vilmercati A, Campanella L, Darzynkiewicz Z, Teodori L. How diet intervention via modulation of DNA damage response through microRNAs may have an effect on cancer prevention and aging, an in silico study. Int J Mol Sci. 2016;17(5):752.

Cascio S, D'Andrea A, Ferla R, Surmacz E, Gulotta E, Amodeo V, Bazan V, Gebbia N, Russo A. miR-20b modulates VEGF expression by targeting HIF-1α and STAT3 in MCF-7 breast cancer cells. J Cell Physiol. 2010;224(1):242–9.

Cha ST, Chen PS, Johansson G, Chu CY, Wang MY, Jeng YM, Yu SL, Chen JS, Chang KJ, Jee SH, Tan CT. MicroRNA-519c suppresses hypoxia-inducible factor-1α expression and tumor angiogenesis. Cancer Res. 2010;16:0008–5472.

Chai ZT, Kong J, Zhu XD, Zhang YY, Lu L, Zhou JM, Wang LR, Zhang KZ, Zhang QB, Ao JY, Wang M. MicroRNA-26a inhibits angiogenesis by down-regulating VEGFA through the PIK3C2α/Akt/HIF-1α pathway in hepatocellular carcinoma. PLoS One. 2013;8(10):e77957.

Chan SY, Loscalzo J. MicroRNA-210: a unique and pleiotropic hypoxamir. Cell Cycle. 2010;9 (6):1072–83.

Chan YC, Khanna S, Roy S, Sen CK. miR-200b targets Ets-1 and is down-regulated by hypoxia to induce angiogenic response of endothelial cells. J Biol Chem. 2011;286(3):2047–56.

Chen Y, Zaman MS, Deng G, Majid S, Saini S, Liu J, Tanaka Y, Dahiya R. MicroRNAs 221/222 and genistein-mediated regulation of ARHI tumor suppressor gene in prostate cancer. Cancer Prev Res. 2011;4(1):76–86.

Cheng HS, Sivachandran N, Lau A, Boudreau E, Zhao JL, Baltimore D, Delgado-Olguin P, Cybulsky MI, Fish JE. MicroRNA-146 represses endothelial activation by inhibiting pro-inflammatory pathways. EMBO Mol Med. 2013;5(7):1017–34.

Chiang CH, Chu PY, Hou MF, Hung WC. MiR-182 promotes proliferation and invasion and elevates the HIF-1α-VEGF-A axis in breast cancer cells by targeting FBXW7. Am J Cancer Res. 2016;6(8):1785.

Chiu J, Khan ZA, Farhangkhoee H, Chakrabarti S. Curcumin prevents diabetes-associated abnormalities in the kidneys by inhibiting p300 and nuclear factor-κB. Nutrition. 2009;25(9):964–72.

Choi KC, Jung MG, Lee YH, Yoon JC, Kwon SH, Kang HB, Kim MJ, Cha JH, Kim YJ, Jun WJ, Lee JM. Epigallocatechin-3-gallate, a histone acetyltransferase inhibitor, inhibits EBV-induced B lymphocyte transformation via suppression of RelA acetylation. Cancer Res. 2009;69 (2):583–92.

Chung MY, Lim TG, Lee KW. Molecular mechanisms of chemopreventive phytochemicals against gastroenterological cancer development. World J Gastroenterol: WJG. 2013;19(7):984.

Dai C, Li B, Zhou Y, Li D, Zhang S, Li H, Xiao X, Tang S. Curcumin attenuates quinocetone induced apoptosis and inflammation via the opposite modulation of Nrf2/HO-1 and NF-kB pathway in human hepatocyte L02 cells. Food Chem Toxicol. 2016;95:52–63.

Ermakova S, Choi BY, Choi HS, Kang BS, Bode AM, Dong Z. The intermediate filament protein vimentin is a new target for epigallocatechin gallate. J Biol Chem. 2005;280(17):16882–90.

Fabbri M, Garzon R, Cimmino A, Liu Z, Zanesi N, Callegari E, Liu S, Alder H, Costinean S, Fernandez-Cymering C, Volinia S. MicroRNA-29 family reverts aberrant methylation in lung cancer by targeting DNA methyltransferases 3A and 3B. Proc Natl Acad Sci. 2007;104 (40):15805–10.

Fabian MR, Sonenberg N. The mechanics of miRNA-mediated gene silencing: a look under the hood of miRISC. Nat Struct Mol Biol. 2012;19(6):586.

Fang MZ, Wang Y, Ai N, Hou Z, Sun Y, Lu H, Welsh W, Yang CS. Tea polyphenol (−)-epigallocatechin-3-gallate inhibits DNA methyltransferase and reactivates methylation-silenced genes in cancer cell lines. Cancer Res. 2003;63(22):7563–70.

Fang MZ, Wang Y, Ai N, Hou Z, Sun Y, Lu H, Welsh W, Yang CS. Tea polyphenol (−)-epigallocatechin-3-gallate inhibits DNA methyltransferase and reactivates methylation-silenced genes in cancer cell lines. Cancer Res. 2004;63(22):7563–70.

Fang M, Chen D, Yang CS. Dietary polyphenols may affect DNA methylation. J Nutr. 2007;137 (1):223S–8S.

Farazi TA, Spitzer JI, Morozov P, Tuschl T. miRNAs in human cancer. J Pathol. 2011;223 (2):102–15.

Favot L, Martin S, Keravis T, Andriantsitohaina R, Lugnier C. Involvement of cyclin-dependent pathway in the inhibitory effect of delphinidin on angiogenesis. Cardiovasc Res. 2003;59 (2):479–87.

Fraga MF, Ballestar E, Villar-Garea A, Boix-Chornet M, Espada J, Schotta G, Bonaldi T, Haydon C, Ropero S, Petrie K, Iyer NG. Loss of acetylation at Lys16 and trimethylation at Lys20 of histone H4 is a common hallmark of human cancer. Nat Genet. 2005;37(4):391.

Fujiki H, Suganuma M, Kurusu M, Okabe S, Imayoshi Y, Taniguchi S, Yoshida T. New TNF-α releasing inhibitors as cancer preventive agents from traditional herbal medicine and combination cancer prevention study with EGCG and sulindac or tamoxifen. Mutat Rese/Fundam Mol Mech Mutagen. 2003;523:119–25.

Gao ZH, Xu Z, Hung MS, Lin YC, Wang T, Gong MI, Zhi X, Jablon DM, You L. Promoter demethylation of WIF-1 by epigallocatechin-3-gallate in lung cancer cells. Anticancer Res. 2009;29(6):2025–30.

Gonnissen A, Isebaert S, Haustermans K. Targeting the hedgehog signaling pathway in cancer: beyond smoothened. Oncotarget. 2015;6(16):13899.

Gorojankina T. Hedgehog signaling pathway: a novel model and molecular mechanisms of signal transduction. Cell Mol Life Sci. 2016;73(7):1317–32.

Goropevšek A, Holcar M, Avčin T. The role of STAT signaling pathways in the pathogenesis of systemic lupus erythematosus. Clin Rev Allergy Immunol. 2017;52(2):164–81.

Guan B, Wu K, Zeng J, Xu S, Mu L, Gao Y, Wang K, Ma Z, Tian J, Shi Q, Guo P. Tumor-suppressive microRNA-218 inhibits tumor angiogenesis via targeting the mTOR component RICTOR in prostate cancer. Oncotarget. 2017;8(5):8162.

Gunther S, Ruhe C, Derikito MG, Böse G, Sauer H, Wartenberg M. Polyphenols prevent cell shedding from mouse mammary cancer spheroids and inhibit cancer cell invasion in confrontation cultures derived from embryonic stem cells. Cancer Lett. 2007;250(1):25–35.

Hadi SM, Asad SF, Singh S, Ahmad A. Putative mechanism for anticancer and apoptosis-inducing properties of plant-derived polyphenolic compounds. IUBMB Life. 2000;50(3):167–71.

Hadi SM, Bhat SH, Azmi AS, Hanif S, Shamim U, Ullah MF. Oxidative breakage of cellular DNA by plant polyphenols: a putative mechanism for anticancer properties. Semin Cancer Biol. 2007;17(5):370–6. Academic

Hori K, Sen A, Artavanis-Tsakonas S. Notch signaling at a glance. J Cell Sci. 2013;126:2135–40.

Hu Y, Li Y, Wu C, Zhou L, Han X, Wang Q, Xie X, Zhou Y, Du Z. MicroRNA-140-5p inhibits cell proliferation and invasion by regulating VEGFA/MMP2 signaling in glioma. Tumor Biol. 2017;39(4):1010428317697558.

Hua Z, Lv Q, Ye W, Wong CK, Cai G, Gu D, Ji Y, Zhao C, Wang J, Yang BB, Zhang Y. MiRNA-directed regulation of VEGF and other angiogenic factors under hypoxia. PLoS One. 2006;1(1): e116.

Huang YT, Hwang JJ, Lee PP, Ke FC, Huang JH, Huang CJ, Kandaswami C, Middleton E, Lee MT. Effects of luteolin and quercetin, inhibitors of tyrosine kinase, on cell growth and metastasis-associated properties in A431 cells overexpressing epidermal growth factor receptor. Br J Pharmacol. 1999;128(5):999–1010.

Hung CF, Huang TF, Chiang HS, Wu WB. (−)-Epigallocatechin-3-gallate, a polyphenolic compound from green tea, inhibits fibroblast adhesion and migration through multiple mechanisms. J Cell Biochem. 2005;96(1):183–97.

Izzotti A, Calin GA, Steele VE, Cartiglia C, Longobardi M, Croce CM, De Flora S. Chemoprevention of cigarette smoke–induced alterations of microRNA expression in rat lungs. Cancer Prev Res. 2010;3(1):62–72.

Jansson MD, Lund AH. MicroRNA and cancer. Mol Oncol. 2012;6(6):590–610.

Jia L, Chen J, Xie C, Shao L, Xu Z, Zhang L. microRNA-1228□ impairs the pro-angiogenic activity of gastric cancer cells by targeting macrophage migration inhibitory factor. Life Sci. 2017;180:9–16.

Jiang S, Zhang HW, Lu MH, He XH, Li Y, Gu H, Liu MF, Wang ED. MicroRNA-155 functions as an OncomiR in breast cancer by targeting the suppressor of cytokine signaling 1 gene. Cancer Res. 2010;30:0008–5472.

Jones PA, Baylin SB. The epigenomics of cancer. Cell. 2007;128(4):683–92.

Kaeberlein M, McDonagh T, Heltweg B, Hixon J, Westman EA, Caldwell S, Napper A, Curtis R, Di Stefano PS, Fields S, Bedalov A. Substrate specific activation of sirtuins by resveratrol. J Biol Chem. 2005;280:17038–45.

Kang SK, Cha SH, Jeon HG. Curcumin-induced histone hypoacetylation enhances caspase-3-dependent glioma cell death and neurogenesis of neural progenitor cells. Stem Cells Dev. 2006;15(2):165–74.

Kanitz A, Imig J, Dziunycz PJ, Primorac A, Galgano A, Hofbauer GF, Gerber AP, Detmar M. The expression levels of microRNA-361-5p and its target VEGFA are inversely correlated in human cutaneous squamous cell carcinoma. PLoS One. 2012;7(11):e49568.

Kato K, Long NK, Makita H, Toida M, Yamashita T, Hatakeyama D, Hara A, Mori H, Shibata T. Effects of green tea polyphenol on methylation status of RECK gene and cancer cell invasion in oral squamous cell carcinoma cells. Br J Cancer. 2008;99(4):647.

Kim WK, Bang MH, Kim ES, Kang NE, Jung KC, Cho HJ, Park JH. Quercetin decreases the expression of ErbB2 and ErbB3 proteins in HT-29 human colon cancer cells. J Nutr Biochem. 2005;16(3):155–62.

Kim JD, Liu L, Guo W, Meydani M. Chemical structure of flavonols in relation to modulation of angiogenesis and immune-endothelial cell adhesion. J Nutr Biochem. 2006;17(3):165–76.

Kong W, He L, Richards EJ, Challa S, Xu CX, Permuth-Wey J, Lancaster JM, Coppola D, Sellers TA, Djeu JY, Cheng JQ. Upregulation of miRNA-155 promotes tumour angiogenesis by targeting VHL and is associated with poor prognosis and triple-negative breast cancer. Oncogene. 2014;33(6):679.

Kong R, Ma Y, Feng J, Li S, Zhang W, Jiang J, Zhang J, Qiao Z, Yang X, Zhou B. The crucial role of miR-126 on suppressing progression of esophageal cancer by targeting VEGF-A. Cell Mol Biol Lett. 2016;21(1):3.

Kovall RA, Blacklow SC. Mechanistic insights into Notch receptor signaling from structural and biochemical studies. Curr Top Dev Biol. 2010;92:31–71. Academic Press

Kuehbacher A, Urbich C, Zeiher AM, Dimmeler S. Role of dicer and Drosha for endothelial microRNA expression and angiogenesis. Circ Res. 2007;101(1):59–68.

Kumar A, Rimando AM, Levenson AS. Resveratrol and pterostilbene as a microRNA-mediated chemopreventive and therapeutic strategy in prostate cancer. Ann N Y Acad Sci. 2017;1403 (1):15–26.

Kumazaki M, Noguchi S, Yasui Y, Iwasaki J, Shinohara H, Yamada N, Akao Y. Anti-cancer effects of naturally occurring compounds through modulation of signal transduction and miRNA expression in human colon cancer cells. J Nutr Biochem. 2013;24(11):1849–58.

Labrecque L, Lamy S, Chapus A, Mihoubi S, Durocher Y, Cass B, Bojanowski MW, Gingras D, Béliveau R. Combined inhibition of PDGF and VEGF receptors by ellagic acid, a dietary-derived phenolic compound. Carcinogenesis. 2005;26(4):821–6.

Lançon A, Michaille JJ, Latruffe N. Effects of dietary phytophenols on the expression of microRNAs involved in mammalian cell homeostasis. J Sci Food Agric. 2013;93(13):3155–64.

Le Marchand L, Murphy SP, Hankin JH, Wilkens LR, Kolonel LN. Intake of flavonoids and lung cancer. J Natl Cancer Inst. 2000;92(2):154–60.

Lee LT, Huang YT, Hwang JJ, Lee AY, Ke FC, Huang CJ, Kandaswami C, Lee PP, Lee MT. Transinactivation of the epidermal growth factor receptor tyrosine kinase and focal adhesion kinase phosphorylation by dietary flavonoids: effect on invasive potential of human carcinoma cells. Biochem Pharmacol. 2004;67(11):2103–14.

Lee DY, Deng Z, Wang CH, Yang BB. MicroRNA-378 promotes cell survival, tumor growth, and angiogenesis by targeting SuFu and Fus-1 expression. Proc Natl Acad Sci U S A. 2007;104 (51):20350–5.

Lee ST, Chu K, Jung KH, Yoon HJ, Jeon D, Kang KM, Park KH, Bae EK, Kim M, Lee SK, Roh JK. MicroRNAs induced during ischemic preconditioning. Stroke. 2010;41(8):1646–51.

Lee WJ, Zhu BT. Inhibition of DNA methylation by caffeic acid and chlorogenic acid, two common catechol-containing coffee polyphenols. Carcinogenesis. 2006;27(2):269–77.

Lei Z, Li BO, Yang Z, Fang H, Zhang GM, Feng ZH, Huang B. Regulation of HIF-1α and VEGF by miR-20b tunes tumor cells to adapt to the alteration of oxygen concentration. PLoS One. 2009;4 (10):e7629.

Li Y, VandenBoom TG, Kong D, Wang Z, Ali S, Philip PA, Sarkar FH. Up-regulation of miR-200 and let-7 by natural agents leads to the reversal of epithelial-to-mesenchymal transition in gemcitabine-resistant pancreatic cancer cells. Cancer Res. 2009;69(16):6704–12.

Li Y, VandenBoom TG, Wang Z, Kong D, Ali S, Philip PA, Sarkar FH. miR-146a suppresses invasion of pancreatic cancer cells. Cancer Res. 2010;70:1486–95.

Li F, Liang X, Chen Y, Li S, Liu J. Role of microRNA-93 in regulation of angiogenesis. Tumour Biol. 2014;35(11):10609–13.

Li Y, Cai B, Shen L, Dong Y, Lu Q, Sun S, Liu S, Ma S, Ma PX, Chen J. MiRNA-29b suppresses tumor growth through simultaneously inhibiting angiogenesis and tumorigenesis by targeting Akt3. Cancer Lett. 2017a;397:111–9.

Li B, Xu WW, Han L, Chan KT, Tsao SW, Lee NP, Law S, Xu LY, Li EM, Chan KW, Qin YR. MicroRNA-377 suppresses initiation and progression of esophageal cancer by inhibiting CD133 and VEGF. Oncogene. 2017b;36(28):3986.

Li L, Tang P, Li S, Qin X, Yang H, Wu C, Liu Y. Notch signaling pathway networks in cancer metastasis: a new target for cancer therapy. Med Oncol. 2017c;34(10):180.

Liang Z, Bian X, Shim H. Downregulation of microRNA-206 promotes invasion and angiogenesis of triple negative breast cancer. Biochem Biophys Res Commun. 2016;477(3):461–6.

Lin SC, Wang CC, Wu MH, Yang SH, Li YH, Tsai SJ. Hypoxia-induced microRNA-20a expression increases ERK phosphorylation and angiogenic gene expression in endometriotic stromal cells. J Clin Endocrinol Metabol. 2012;97(8):E1515–23.

Lin HY, Yang SH, Tang HY, Cheng GY, Davis PJ, Grasso P. Biologically active leptin-related synthetic peptides activate STAT3 via phosphorylation of ERK1/2 and PI-3K. Peptides. 2014;57:95–100.

Linder MC. The relationship of copper to DNA damage and damage prevention in humans. Mutat Res-Fund Mol M. 2012;733(1–2):83–91.

Liu LZ, Li C, Chen Q, Jing Y, Carpenter R, Jiang Y, Kung HF, Lai L, Jiang BH. MiR-21 induced angiogenesis through AKT and ERK activation and HIF-1α expression. PLoS One. 2011;6(4): e19139.

Lu H, Ouyang W, Huang C. Inflammation, a key event in cancer development. Mol Cancer Res. 2006;4(4):221–33.

Ma H, Pan JS, Jin LX, Wu J, Ren YD, Chen P, Xiao C, Han J. MicroRNA-17~ 92 inhibits colorectal cancer progression by targeting angiogenesis. Cancer Lett. 2016;376(2):293–302.

Madanecki P, Kapoor N, Bebok Z, Ochocka R, Collawn JF, Bartoszewski R. Regulation of angiogenesis by hypoxia: the role of microRNA. Cell Mol Biol Lett. 2013;18(1):47.

Majid S, Kikuno N, Nelles J, Noonan E, Tanaka Y, Kawamoto K, Hirata H, Li LC, Zhao H, Okino ST, Place RF. Genistein induces the p21WAF1/CIP1 and p16INK4a tumor suppressor genes in prostate cancer cells by epigenetic mechanisms involving active chromatin modification. Cancer Res. 2008;68(8):2736–44.

Majid S, Dar AA, Ahmad AE, Hirata H, Kawakami K, Shahryari V, Saini S, Tanaka Y, Dahiya AV, Khatri G, Dahiya R. BTG3 tumor suppressor gene promoter demethylation, histone modification and cell cycle arrest by genistein in renal cancer. Carcinogenesis. 2009;30(4):662–70.

Manson MM. Cancer prevention–the potential for diet to modulate molecular signalling. Trends Mol Med. 2003;9(1):11–8.

Mao G, Liu Y, Fang X, Liu Y, Fang L, Lin L, Liu X, Wang N. Tumor-derived microRNA-494 promotes angiogenesis in non-small cell lung cancer. Angiogenesis. 2015;18(3):373–82.

Marí-Alexandre J, García-Oms J, Barceló-Molina M, Gilabert-Aguilar J, Estellés A, Braza-Boïls A, Gilabert Estellés J. MicroRNAs and angiogenesis in endometriosis. Thromb Res. 2015;135: S38–40.
Maroof H, Salajegheh A, Smith RA, Lam AK. Role of microRNA-34 family in cancer with particular reference to cancer angiogenesis. Exp Mol Pathol. 2014;97(2):298–304.
Mathew LK, Skuli N, Mucaj V, Lee SS, Zinn PO, Sathyan P, Imtiyaz HZ, Zhang Z, Davuluri RV, Rao S, Venneti S. miR-218 opposes a critical RTK-HIF pathway in mesenchymal glioblastoma. Proc Natl Acad Sci. 2014;111(1):291–6.
Meeran SM, Patel SN, Tollefsbol TO. Sulforaphane causes epigenetic repression of hTERT expression in human breast cancer cell lines. PLoS One. 2010a;5(7):e11457.
Meeran SM, Ahmed A, Tollefsbol TO. Epigenetic targets of bioactive dietary components for cancer prevention and therapy. Clin Epigenetics. 2010b;1(3):101.
Mitchell S, Vargas J, Hoffmann A. Signaling via the NFκB system. Wiley Interdiscip Rev Syst Biol Med. 2016;8(3):227–41.
Mudduluru G, George-William JN, Muppala S, Asangani IA, Kumarswamy R, Nelson LD, Allgayer H. Curcumin regulates miR-21 expression and inhibits invasion and metastasis in colorectal cancer. Biosci Rep. 2011;31(3):185–97.
Myzak MC, Karplus PA, Chung FL, Dashwood RH. A novel mechanism of chemoprotection by sulforaphane: inhibition of histone deacetylase. Cancer Res. 2004;64(16):5767–74.
Nakachi K, Suemasu K, Suga K, Takeo T, Imai K, Higashi Y. Influence of drinking green tea on breast cancer malignancy among Japanese patients. Jpn J Cancer Res. 1998;89(3):254–61.
Nazeem S, Azmi AS, Hanif S, Ahmad A, Mohammad RM, Hadi SM, Kumar KS. Plumbagin induces cell death through a copper-redox cycle mechanism in human cancer cells. Mutagenesis. 2009;24(5):413–8.
Neergheen VS, Bahorun T, Taylor EW, Jen LS, Aruoma OI. Targeting specific cell signaling transduction pathways by dietary and medicinal phytochemicals in cancer chemoprevention. Toxicology. 2014;278(2):229–41.
Nihal M, Roelke CT, Wood GS. Anti-melanoma effects of vorinostat in combination with polyphenolic antioxidant (−)-epigallocatechin-3-gallate (EGCG). Pharm Res. 2010;27(6):1103–14.
Nikaki A, Piperi C, Papavassiliou AG. Role of microRNAs in gliomagenesis: targeting miRNAs in glioblastoma multiforme therapy. Expert Opin Investig Drugs. 2012;21(10):1475–88.
Nwaeburu CC, Bauer N, Zhao Z, Abukiwan A, Gladkich J, Benner A, Herr I. Up-regulation of microRNA Let-7c by quercetin inhibits pancreatic cancer progression by activation of Numbl. Oncotarget. 2016;7(36):58367.
Ohlsson Teague EM, Print CG, Hull ML. The role of microRNAs in endometriosis and associated reproductive conditions. Hum Reprod Update. 2010;16(2):142–65.
Pakravan K, Babashah S, Sadeghizadeh M, Mowla SJ, Mossahebi-Mohammadi M, Ataei F, Dana N, Javan M. MicroRNA-100 shuttled by mesenchymal stem cell-derived exosomes suppresses in vitro angiogenesis through modulating the mTOR/HIF-1α/VEGF signaling axis in breast cancer cells. Cell Oncol. 2017;40(5):457–70.
Pandey M, Shukla S, Gupta S. Promoter demethylation and chromatin remodeling by green tea polyphenols leads to re-expression of GSTP1 in human prostate cancer cells. Int J Cancer. 2010;126(11):2520–33.
Papoutsis AJ, Lamore SD, Wondrak GT, Selmin OI, Romagnolo DF. Resveratrol prevents epigenetic silencing of BRCA-1 by the aromatic hydrocarbon receptor in human breast cancer cells. J Nutr. 2010;140(9):1607–14.
Parasramka MA, Ho E, Williams DE, Dashwood RH. MicroRNAs, diet, and cancer: new mechanistic insights on the epigenetic actions of phytochemicals. Mol Carcinog. 2012;51(3):213–30.
Park HK, Han DW, Park YH, Park JC. Differential biological responses of green tea polyphenol in normal cells vs. cancer cells. Curr Appl Phys. 2005;5(5):449–52.
Peng G, Dixon DA, Muga SJ, Smith TJ, Wargovich MJ. Green tea polyphenol (−)-epigallocatechin-3-gallate inhibits cyclooxygenase-2 expression in colon carcinogenesis. Mol Carcinog. 2006;45(5):309–19.

Petric RC, Braicu C, Raduly L, Zanoaga O, Dragos N, Monroig P, Dumitrascu D, Berindan-Neagoe I. Phytochemicals modulate carcinogenic signaling pathways in breast and hormone-related cancers. Onco Targets Ther. 2015;8:2053.

Petricci E, Manetti F. Targeting the hedgehog signaling pathway with small molecules from natural sources. Curr Med Chem. 2015;22(35):4058–90.

Pollack BP, Sapkota B, Boss JM. Ultraviolet radiation-induced transcription is associated with gene-specific histone acetylation. Photochem Photobiol. 2009;85(3):652–62.

Raffoul JJ, Wang Y, Kucuk O, Forman JD, Sarkar FH, Hillman GG. Genistein inhibits radiation-induced activation of NF-κB in prostate cancer cells promoting apoptosis and G 2/M cell cycle arrest. BMC Cancer. 2006;6(1):107.

Ramon LA, Braza-Boïls A, Gilabert-Estellés J, Gilabert J, España F, Chirivella M, Estellés A. microRNAs expression in endometriosis and their relation to angiogenic factors. Hum Reprod. 2011;26(5):1082–90.

Ramón LA, Braza-Boïls A, Gilabert J, Chirivella M, Espana F, Estellés A, Gilabert-Estelles J. microRNAs related to angiogenesis are dysregulated in endometrioid endometrial cancer. Hum Reprod. 2012;27(10):3036–45.

Ramos S. Effects of dietary flavonoids on apoptotic pathways related to cancer chemoprevention. J Nutr Biochem. 2007;18(7):427–42.

Ramos S. Cancer chemoprevention and chemotherapy: dietary polyphenols and signalling pathways. Mol Nutr Food Res. 2008;52(5):507–26.

Rane S, He M, Sayed D, Vashistha H, Malhotra A, Sadoshima J, Vatner DE, Vatner SF, Abdellatif M. Downregulation of miR-199a derepresses hypoxia-inducible factor-1α and Sirtuin 1 and recapitulates hypoxia preconditioning in cardiac myocytes. Circ Res. 2009;104(7):879–86.

Ren J, Singh BN, Huang Q, Li Z, Gao Y, Mishra P, Hwa YL, Li J, Dowdy SC, Jiang SW. DNA hypermethylation as a chemotherapy target. Cell Signal. 2011;23(7):1082–93.

Ribas J, Lupold E. The transcriptional regulation of miR-21, its multiple transcripts and their implication in prostate cancer. Cell Cycle. 2010;9(5):923–9.

Sabatel C, Malvaux L, Bovy N, Deroanne C, Lambert V, Gonzalez ML, Colige A, Rakic JM, Noël A, Martial JA, Struman I. MicroRNA-21 exhibits antiangiogenic function by targeting RhoB expression in endothelial cells. PLoS One. 2011;6(2):e16979.

Saito K, Kondo E, Matsushita M. MicroRNA 130 family regulates the hypoxia response signal through the P-body protein DDX6. Nucleic Acids Res. 2011;39(14):6086–99.

Saj A, Lai EC. Control of microRNA biogenesis and transcription by cell signaling pathways. Curr Opin Genet Dev. 2011;21(4):504–10.

Sarkar FH. Nutraceuticals and cancer. Preface. Cancer Metastasis Rev. 2010;29(3):381–2.

Sarkar FH, Li Y, Wang Z, Kong D. Cellular signaling perturbation by natural products. Cell Signal. 2009;21(11):1541–7.

Sasahira T, Kurihara M, Bhawal UK, Ueda N, Shimomoto T, Yamamoto K, Kirita T, Kuniyasu H. Downregulation of miR-126 induces angiogenesis and lymphangiogenesis by activation of VEGF-A in oral cancer. Br J Cancer. 2012;107(4):700.

Schmidt MF. Drug target miRNAs: chances and challenges. Trends Biotechnol. 2014;32 (11):578–85.

Seok JK, Lee SH, Kim MJ, Lee YM. MicroRNA-382 induced by HIF-1α is an angiogenic miR targeting the tumor suppressor phosphatase and tensin homolog. Nucleic Acids Res. 2014;42 (12):8062–72.

Sethi S, Kong D, Land S, Dyson G, Sakr WA, Sarkar FH. Comprehensive molecular oncogenomic profiling and miRNA analysis of prostate cancer. Am J Transl Res. 2013;5(2):200.

Shi L, Zhang S, Wu H, Zhang L, Dai X, Hu J, Xue J, Liu T, Liang Y, Wu G. MiR-200c increases the radiosensitivity of non-small-cell lung cancer cell line A549 by targeting VEGF-VEGFR2 pathway. PLoS One. 2013;8(10):e78344.

Shu L, Khor TO, Lee JH, Boyanapalli SS, Huang Y, Wu TY, Saw CL, Cheung KL, Kong AN. Epigenetic CpG demethylation of the promoter and reactivation of the expression of Neurog1 by curcumin in prostate LNCaP cells. AAPS J. 2011;13(4):606–14.

Siddiqui IA, Asim M, Hafeez BB, Adhami VM, Tarapore RS, Mukhtar H. Green tea polyphenol EGCG blunts androgen receptor function in prostate cancer. FASEB J. 2011;25(4):1198–207.
Singh BN, Singh HB, Singh A, Naqvi AH, Singh BR. Dietary phytochemicals alter epigenetic events and signaling pathways for inhibition of metastasis cascade. Cancer Metastasis Rev. 2014;33(1):41–85.
Sinha M, Ghatak S, Roy S, Sen CK. microRNA–200b as a switch for inducible adult angiogenesis. Antioxid Redox Signal. 2015;22(14):1257–72.
Struhl G, Greenwald I. Presenilin is required for activity and nuclear access of notch in drosophila. Nature. 1999;398(6727):522.
Subbaramaiah K, Dannenberg AJ. Cyclooxygenase 2: a molecular target for cancer prevention and treatment. Trends Pharmacol Sci. 2003;24(2):96–102.
Sun M, Estrov Z, Ji Y, Coombes KR, Harris DH, Kurzrock R. Curcumin (diferuloylmethane) alters the expression profiles of microRNAs in human pancreatic cancer cells. Mol Cancer Ther. 2008;7(3):464–73.
Sun Q, Cong R, Yan H, Gu H, Zeng Y, Liu N, Chen J, Wang B. Genistein inhibits growth of human uveal melanoma cells and affects microRNA-27a and target gene expression. Oncol Rep. 2009;22(3):563–7.
Sun X, Icli B, Wara AK, Belkin N, He S, Kobzik L, Hunninghake GM, Vera MP, Blackwell TS, Baron RM, Feinberg MW. MicroRNA-181b regulates NF-κB–mediated vascular inflammation. J Clin Invest. 2012;122(6):1973–90.
Sundaram P, Hultine S, Smith LM, Dews M, Fox JL, Biyashev D, Schelter JM, Huang Q, Cleary MA, Volpert OV, Thomas-Tikhonenko A. p53-responsive miR-194 inhibits thrombospondin-1 and promotes angiogenesis in colon cancers. Cancer Res. 2011;71:7490–501. canres-1124
Surh YJ. Cancer chemoprevention with dietary phytochemicals. Nat Rev Cancer. 2003;3(10):768
Szade A, Grochot-Przeczek A, Florczyk U, Jozkowicz A, Dulak J. Cellular and molecular mechanisms of inflammation-induced angiogenesis. IUBMB Life. 2015;67(3):145–59.
Tanimura S, Kadomoto R, Tanaka T, Zhang YJ, Kouno I, Kohno M. Suppression of tumor cell invasiveness by hydrolyzable tannins (plant polyphenols) via the inhibition of matrix metalloproteinase-2/-9 activity. Biochem Biophys Res Commun. 2005;330(4):1306–13.
Tili E, Michaille JJ, Alder H, Volinia S, Delmas D, Latruffe N, Croce CM. Resveratrol modulates the levels of microRNAs targeting genes encoding tumor-suppressors and effectors of TGFβ signaling pathway in SW480 cells. Biochem Pharmacol. 2010a;80(12):2057–65.
Tili E, Michaille JJ, Adair B, Alder H, Limagne E, Taccioli C, Ferracin M, Delmas D, Latruffe N, Croce CM. Resveratrol decreases the levels of miR-155 by upregulating miR-663, a microRNA targeting JunB and JunD. Carcinogenesis. 2010b;31(9):1561–6.
Tsang WP, Kwok TT. Epigallocatechin gallate up-regulation of miR-16 and induction of apoptosis in human cancer cells. J Nutr Biochem. 2010;21(2):140–6.
Tu Y, Liu L, Zhao D, Liu Y, Ma X, Fan Y, Wan L, Huang T, Cheng Z, Shen B. Overexpression of miRNA-497 inhibits tumor angiogenesis by targeting VEGFR2. Sci Rep. 2015;5:13827.
Uddin Q, Malik A, Azam S, Hadi N, Azmi AS, Parveen N, et al. The biflavonoid, amentoflavone degrades DNA in the presence of copper ions. Toxicol In Vitro. 2004;18(4):435–40.
Ueda R, Kohanbash G, Sasaki K, Fujita M, Zhu X, Kastenhuber ER, McDonald HA, Potter DM, Hamilton RL, Lotze MT, Khan SA. Dicer-regulated microRNAs 222 and 339 promote resistance of cancer cells to cytotoxic T-lymphocytes by down-regulation of ICAM-1. Proc Natl Acad Sci. 2009;106(26):10746–51.
Ullah MF, Khan HY, Zubair H, Shamim U, Hadi SM. The antioxidant ascorbic acid mobilizes nuclear copper leading to a prooxidant breakage of cellular DNA: implications for chemotherapeutic action against cancer. Cancer Chemother Pharmacol. 2011;67(1):103–10.
Umeda D, Tachibana H, Yamada K. Epigallocatechin-3O-gallate disrupts stress fibers and the contractile ring by reducing myosin regulatory light chain phosphorylation mediated through the target molecule 67 k Da laminin receptor. Biochem Biophys Res Comm. 2005;333:628–35.

Urbich C, Kaluza D, Frömel T, Knau A, Bennewitz K, Boon RA, Bonauer A, Doebele C, Boeckel JN, Hergenreider E, Zeiher AM. MicroRNA-27a/b controls endothelial cell repulsion and angiogenesis by targeting semaphorin 6A. Blood. 2012;119(6):1607–16.

Vijayababu MR, Arunkumar A, Kanagaraj P, Venkataraman P, Krishnamoorthy G, Arunakaran J. Quercetin downregulates matrix metalloproteinases 2 and 9 proteins expression in prostate cancer cells (PC-3). Mol Cell Biochem. 2006;287(1–2):109–16.

Villarino AV, Kanno Y, Ferdinand JR, O'Shea JJ. Mechanisms of Jak/STAT signaling in immunity and disease. J Immunol. 2015;194(1):21–7.

Wagner MA, Siddiqui MA. The JAK-STAT pathway in hypertrophic stress signaling and genomic stress response. Jak-Stat. 2012;1(2):131–41.

Wang S, Aurora AB, Johnson BA, Qi X, McAnally J, Hill JA, Richardson JA, Bassel-Duby R, Olson EN. The endothelial-specific microRNA miR-126 governs vascular integrity and angiogenesis. Dev Cell. 2008;15(2):261–71.

Wang LX, Wang YP, Chen Z, Liu XY, Liu XH, Liu L, Chen WJ, Liu LB. Exendin-4 protects murine pancreatic β-cells from dexamethasone-induced apoptosis through PKA and PI-3K signaling. Diabetes Res Clin Pract. 2010;90(3):297–304.

Wang W, Ren F, Wu Q, Jiang D, Li H, Shi H. MicroRNA-497 suppresses angiogenesis by targeting vascular endothelial growth factor A through the PI3K/AKT and MAPK/ERK pathways in ovarian cancer. Oncol Rep. 2014;32(5):2127–33.

Wang W, Zhang E, Lin C. MicroRNAs in tumor angiogenesis. Life Sci. 2015a;136:28–35.

Wang G, Dai F, Yu K, Jia Z, Zhang A, Huang Q, Kang C, Jiang H, Pu P. Resveratrol inhibits glioma cell growth via targeting oncogenic microRNAs and multiple signaling pathways. Int J Oncol. 2015b;46(4):1739–47.

Wen XY, Wu SY, Li ZQ, Liu ZQ, Zhang JJ, Wang GF, Jiang ZH, Wu SG. Ellagitannin (BJA3121), an anti-proliferative natural polyphenol compound, can regulate the expression of MiRNAs in HepG2 cancer cells. Phytother Res: Int J Devoted Pharmacol Toxicol Eval Nat Prod Deriv. 2009;23(6):778–84.

Wenzel U, Kuntz S, Brendel MD, Daniel H. Dietary flavone is a potent apoptosis inducer in human colon carcinoma cells. Cancer Res. 2000;60(14):3823–31.

Woodhouse EC, Chuaqui RF, Liotta LA. General mechanisms of metastasis. Cancer: Interdiscip Int J Am Cancer Soc. 1997;80(S8):1529–37.

Wu M, Huang C, Huang X, Liang R, Feng Y, Luo X. MicroRNA-144-3p suppresses tumor growth and angiogenesis by targeting SGK3 in hepatocellular carcinoma. Oncol Rep. 2017;38(4):2173–81.

Wurdinger T, Tannous BA, Saydam O, Skog J, Grau S, Soutschek J, Weissleder R, Breakefield XO, Krichevsky AM. miR-296 regulates growth factor receptor overexpression in angiogenic endothelial cells. Cancer Cell. 2008;14(5):382–93.

Xu Q, Liu LZ, Qian X, Chen Q, Jiang Y, Li D, Lai L, Jiang BH. MiR-145 directly targets p70S6K1 in cancer cells to inhibit tumor growth and angiogenesis. Nucleic Acids Res. 2011;40(2):761–74.

Yamakawa S, Asai T, Uchida T, Matsukawa M, Akizawa T, Oku N. (−)-Epigallocatechin gallate inhibits membrane-type 1 matrix metalloproteinase, MT1-MMP, and tumor angiogenesis. Cancer Lett. 2004;210(1):47–55.

Yamakuchi M, Lotterman CD, Bao C, Hruban RH, Karim B, Mendell JT, Huso D, Lowenstein CJ. P53-induced microRNA-107 inhibits HIF-1 and tumor angiogenesis. Proc Natl Acad Sci. 2010;107(14):6334–9.

Yamamoto S, Schulze KL, Bellen HJ. Introduction to Notch signaling. In: Notch signaling. New York: Humana Press; 2014. p. 1–14.

Yang J, Cao Y, Sun J, Zhang Y. Curcumin reduces the expression of Bcl-2 by upregulating miR-15a and miR-16 in MCF-7 cells. Med Oncol. 2010;27(4):1114–8.

Yang Y, Zhang J, Xia T, Li G, Tian T, Wang M, Wang R, Zhao L, Yang Y, Lan K, Zhou W. MicroRNA-210 promotes cancer angiogenesis by targeting fibroblast growth factor receptor-like 1 in hepatocellular carcinoma. Oncol Rep. 2016;36(5):2553–62.

Ye F, Wu J, Dunn T, Yi J, Tong X, Zhang D. Inhibition of cyclooxygenase-2 activity in head and neck cancer cells by genistein. Cancer Lett. 2004;211(1):39–46.

Yin KJ, Hamblin M, Eugene Chen Y. Angiogenesis-regulating microRNAs and ischemic stroke. Curr Vasc Pharmacol. 2015;13(3):352–65.

Zhang J, Zhang T, Ti X, Shi J, Wu C, Ren X, Yin H. Curcumin promotes apoptosis in A549/DDP multidrug-resistant human lung adenocarcinoma cells through an miRNA signaling pathway. Biochem Biophys Res Commun. 2010a;399(1):1–6.

Zhang J, Du Y, Wu C, Ren X, Ti X, Shi J, Zhao F, Yin H. Curcumin promotes apoptosis in human lung adenocarcinoma cells through miR-186* signaling pathway. Oncol Rep. 2010b;24 (5):1217–23.

Zhang X, Ng WL, Wang P, Tian L, Werner E, Wang H, Doetsch P, Wang Y. MicroRNA-21 modulates the levels of reactive oxygen species by targeting SOD3 and TNFα. Cancer Res. 2012;72:4707–13.

Zhang L, Lv Z, Xu J, Chen C, Ge Q, Li P, Wei D, Wu Z, Sun X. Micro RNA-134 inhibits osteosarcoma angiogenesis and proliferation by targeting the VEGFA/VEGFR 1 pathway. FEBS J. 2018;285(7):1359–71.

Zhen MC, Huang XH, Wang Q, Sun K, Liu YJ, Li W, Zhang LJ, Cao LQ, Chen XL. Green tea polyphenol epigallocatechin-3-gallate suppresses rat hepatic stellate cell invasion by inhibition of MMP-2 expression and its activation. Acta Pharmacol Sin. 2006;27(12):1600.

Zheng Y, Li S, Ding Y, Wang Q, Luo H, Shi Q, Hao Z, Xiao G, Tong S. The role of miR-18a in gastric cancer angiogenesis. Hepato-Gastroenterology. 2013;60(127):1809–13.

Zhou B, Ma R, Si W, Li S, Xu Y, Tu X, Wang Q. MicroRNA-503 targets FGF2 and VEGFA and inhibits tumor angiogenesis and growth. Cancer Lett. 2013;333(2):159–69.

Zhou Y, Li S, Li J, Wang D, Li Q. Effect of microRNA 135a on cell proliferation, migration, invasion, apoptosis and tumor angiogenesis through the IGF-1/PI3K/Akt signaling pathway in non-small cell lung cancer. Cell Physiol Biochem. 2017;42(4):1431–46.

Zhu K, Pan Q, Zhang X, Kong LQ, Fan J, Dai Z, Wang L, Yang XR, Hu J, Wan JL, Zhao YM. MiR-146a enhances angiogenic activity of endothelial cells in hepatocellular carcinoma by promoting PDGFRA expression. Carcinogenesis. 2013;34(9):2071–9.

Chapter 24
Protein Homology Modeling in Phytochemical Research

Aditya Narayan, Ajeet Singh, and Shailesh Kumar

24.1 Introduction

24.1.1 Phytochemicals

In the broadest sense, we may define phytochemicals as chemical compounds produced by plants with very sharp, and distinctive properties such as taste, odor and so on. More significantly, they may also play key roles in metabolism, defense mechanisms and other aspects of plant physiology. Although they are non-essential nutrients, intake of some phytochemicals has been shown to provide protective and disease preventive properties. There are burgeons of phytochemicals belonging to different classes with multipartite functions. Some of the functions include antioxidant action (well-known examples include carotenoids and flavonoids), influencing hormone release (isoflavones in soy imitate estrogen), enzyme stimulation (indoles in cabbage), impairment of cell replication (saponins), DNA protective mechanisms (capsaicin) and protection from pathogens by physically inhibiting access to cell walls (Hamuel 2012).

Absolutely critical to the study of phytochemistry is the basic understanding of the 3D structures of the vast array of existing chemicals. Experimental validation of these structures is time-consuming and inefficient, thereby limiting the knowledge of the vast majority of protein structures. Till now, for exploring the biochemical role of a number of phytochemicals at any level, it becomes necessary to employ in-silico methods of structure modeling.

A. Narayan
University of Virginia, Charlottesville, VA, USA

A. Singh · S. Kumar (✉)
National Institute of Plant Genome Research (NIPGR), New Delhi, India
e-mail: shailesh@nipgr.ac.in

S. Kumar, C. Egbuna (eds.), *Phytochemistry: An in-silico and in-vitro Update*,
https://doi.org/10.1007/978-981-13-6920-9_24

The utility of such tools may be assessed by how the homology modeling and molecular docking analysis are used on phytochemicals for better understanding and designing solutions for disease-causing mechanisms along with their prevention. Some supportive examples include screening phytochemicals against the ligand binding sites of the E1 protein of chikungunya virus responsible for host-binding, docking phytoconstituents in psoriasis-causing protein corneodesmosin for better understanding of its active site and testing flavonoid inhibitors against the hyperactive 26S proteasome subunit in cancer cells (Vasavi et al. 2010; Panda et al. 2014; Salomi et al. 2016). Homology modeling may even be applied to whole plant proteomes in order to understand the set of phytoconstituents contributing to antioxidant activity in spinach (Sahay and Shakya 2010).

24.1.1.1 Molecular Modeling

Molecular modeling comprises of the wide range of theoretical, computational, and biochemical techniques employed to generate and study the structure of different molecules. As stated previously, the most commonly applied method for molecular modeling is known as homology modeling (also referred to as comparative modeling or template-based modeling), which employs template structures which are evolutionarily related to the structure of interest to serve as a model from which the structure can be predicted. Although not as empirically accurate as certain techniques such as x-ray crystallography, homology modeling may yield large volumes of structural information that are critical in the study of molecular structure, dynamics, and protein-ligand interactions. Thus, the development and use of these tools bear wide-reaching industrial and medical implications. In particular, given the experimental difficulties related to structural determination, faster methods of protein structure prediction yielding accurate models have significantly emerged nowadays.

24.1.2 Homology Modeling Steps

The process of homology modeling can be further categorized into a number of steps which are largely identical across the variety of in-silico homology modeling tools (Fig. 24.1).

24.1.2.1 Template Recognition and Initial Alignment

Homology modeling hinges on the fact that 3D structures among evolutionarily related protein structures are highly analogous. A variety of databases, particularly

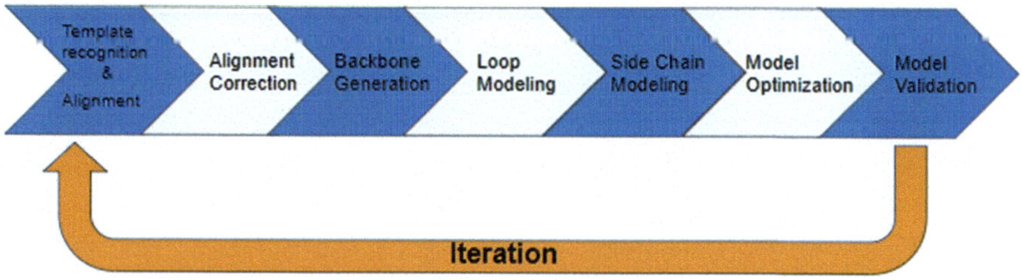

Fig. 24.1 Workflow/methodology of protein homology modeling tools

the Protein Data Bank (http://www.wwpdb.org/), and database scanning tools exist which allow users to enter query sequences/structural information that will be used to enquire appropriate template homologs. The best-known tool for this purpose is BLAST. Once a template is found, it is necessary to create a multiple sequence alignment of the template, the structure of interest and other related proteins to gain an insight into related domains, motifs, features within the protein family and other relevant structural information.

The discussion regarding the "twilight zone" of sequence similarity for proteins is demonstrated graphically by Sivakumar (http://biosiva.50webs.org/alignment.htm). In comparing sequences, the similarity with respect to the residue/size/charge/hydrophobicity is measured as sequence identity. When sequence identity is high, generally defined as greater than ~25%, it is possible to argue the protein's evolutionary relatedness (homology) with some degree of confidence. Below this limit, the sequence identity is said to fall within the "twilight zone", in which homologous sequences mix with randomly aligned sequences. Below 20%, homologous relationships may not be reliably determined – the "midnight zone." Sequence length must also be accounted for as shorter sequences will have a greater chance that alignments are a result of random chance. Irrespective of the sequence, however, there are many situations in which proteins falling in the "twilight zone" of sequence identity maintain similar folds. However, such situations are rare and thus the goal of modeling tools is to identify templates with sufficient sequence homology that they may be used to generate accurate structures.

24.1.2.2 Alignment Correction

Residue properties must be taken into consideration in aligning multiple sequences such that analogous core residues are sufficiently hydrophobic while the outermost residues may be more variable. It may become necessary to introduce small insertions and deletions in the non-conserved regions of the structure by hand to correct for sequence disparities. If sequence fragments are not present in the target but are found in the template, they may be deleted from the template. If there is an insertion in the target, then the template will contain a gap. This ultimately improves the overall quality of the target-template alignment.

24.1.2.3 Backbone Generation

The backbone of the target structure is generated from analogous coordinates in the template given the conservation of 3D structures. This applies also to the conserved side chains.

24.1.2.4 Loop Modeling

Due to deletions and insertions in the alignment as a result of alignment correction, the template may contain gaps in the sequence. This prevents modeling of the target and such sequences must be constructed as flexible surface loops. These may be generated *ab initio* through identification of similar sequences in the PDB which bear similar residues/environments. Various tools may be used to determine the most accurate loop model which is then added to the target structure.

24.1.2.5 Side-Chain Modeling

Side chains of conserved residues are easily modeled in the target structure. However, non-conserved residues must be added by accounting for a number of constraints. Torsional angle is often conserved across homologs for the majority of residues and certain rotamers are optimal for certain backbone structures/steric constraints etc. Libraries of side chain structures under such constraints are analyzed to choose the optimal rotamers.

24.1.2.6 Model Optimization

In order to account for backbone and side chain changes over the course of modeling, each must be adjusted in accordance with the other as constraints are altered. Molecular dynamics simulations are performed within a force-field modeling tool such that protein folding may be modeled. Ultimately the side chain rotamers and backbone will be adjusted until the potential energy of the structure is at a minimum.

24.1.2.7 Model Validation

A number of model validation tools may be employed to determine if the final structure produced from the previous steps abides by acceptable ranges with respect to bond angles, torsion, bond lengths, distribution of side chains/side chain properties and how the model folds/active site compare to homologs.

24.2 Homology Modeling Tools

In order to determine the most effective homology modeling tools, it is common to refer to international blind trials of protein structure prediction methods known as the Critical Assessment of protein Structure Prediction or CASP. For majority of the commonly used protein modeling tools, CASP trials indicate relatively little difference in accuracy – particularly in situations when homology is not readily inferred with known structures (Kelley et al. 2015). However, here we will discuss differences in methodology, algorithms applied, input, output, and speed among commonly applied homology modeling tools. The overview of advantages and disadvantages of each tool discussed in the chapter is highlighted in Table 24.1.

24.2.1 Modeller

24.2.1.1 Outline/Methodology

MODELLER is the most commonly used tool used in generating 3D structures via homology modeling. This tool is applied to create a model through analysis of spatial restraints as well as statistical assessments of homologous structures in the PDB and it was conceived in order to minimize alignment errors in the comparative modeling

Table 24.1 Overview of each tool's advantages and disadvantages

Tools name	Pros	Cons
MODELLER	Fast, high quality, free, reliable	Command-line only interface, poor core and side-chain modelling
I-Tasser	Fast, high quality, free	Unreliable model selection, poor domain splitting
Swiss-Model	Fast, high quality, good core modeling, good stereochemistry	Unreliable crashes frequently
Phyre2	Fast, reliable, ease of use	Low accuracy
HHPred	Very fast, ease of use	Low accuracy
Robetta	High quality, ease of use	Computationally demanding

process to produce the most accurate possible target structure. It does so through the application of a genetic algorithm to optimize alignment, after which a comparative modeling procedure is used to create the 3D structure. Application of a genetic algorithm allows for individuals within the population to undergo recombination and random mutation (alignment changes) under the guidance of a fitness function to converge on an optimal model as evaluated by an assessment score. In summary, the MODELLER outputs a list of comparative models after undergoing an iterative process of alignment using a genetic algorithm, model building, and assessment until the top model is output (as calculated by composite modeling score) (John and Sali 2003).

24.2.1.2 Pros/Cons

MODELLER has the advantage of being able to create high-quality figures relatively quickly and for free though it may be improved with respect to core and side chain modeling (Wallner and Elofsson 2005). However, its availability as command-line only interface may be difficult to use for the inexperienced users.

24.2.1.3 Input

(a) Multiple sequence alignment of the target (PIR format) with homologs.
(b) One or more template structures.
(c) Template profile of multiple structural alignments of the templates with close homologs.

24.2.1.4 Output

Target–template alignments alongside comparative models for the sequence ranked in order of the model score.

24.2.1.5 Working Steps (Basic Modeling)

I. Searching for structures related to the target.

   (a) Put the target sequence into PIR format (. ali) in which the first line contains ">P1; code". The second line contains information of the structure file with the term "sequence" and the model file name separated by a colon. The sequence follows with "*" at the end.

```
>> Summary of successfully produced models:
Filename                        molpdf      DOPE score    GA341 score
----------------------------------------------------------------------
TvLDH.B99990001.pdb         1763.56104   -38079.76172        1.00000
TvLDH.B99990002.pdb         1560.93396   -38515.98047        1.00000
TvLDH.B99990003.pdb         1712.44104   -37984.30859        1.00000
TvLDH.B99990004.pdb         1720.70801   -37869.91406        1.00000
TvLDH.B99990005.pdb         1840.91772   -38052.00781        1.00000
```

Fig. 24.2 MODELLER output log displaying a summary of models built

(b) The profile. Build (sequence.ali) MODELLER command is used to identify homologs.

II. Selecting template.

(a) The output of the profile. build command, "build_profile.log" contains PDB sequences for identified templates, the corresponding sequence identity, and e-value. The greatest similarity sequences (identifiable by sequence identity and e-value) are used as an argument in the alignment. compare_structures command. Within this command, structural and sequence similarity with templates are iteratively compared with the input. The results are evaluated and presented in a tabular format (compare.log file).

(b) The output table presents relevant pairwise sequence distances as well as a template clustering tree to illustrate the distinction between templates. From this result, the final template may be selected. Note that it is possible to consider crystallographic R-factor in making the selection.

III. Template alignment.

(a) The align2D command accepts the selected template and is used to generate the sequence alignment in PIR and PAP formats (model-selected_template. ali and model-selected_template.pap).

IV. Model building.

(a) The auto model class is automatically applied by MODELLER. This process loads the auto model class, generates an auto model object, names the file containing the target-template alignment with alnfile, defines the template structure with the known function, obtains alignment scores with assess methods, defines the target sequence name with sequence, defines the number of calculated models with starting model and ending model, and finally applies the make command.

(b) The most significant output file is "model-single.log" which contains a summary of models built with associated confidence scores (Fig. 24.2).

V. Model Evaluation.

(a) It is possible to evaluate the validity of the model with external programs or within MODELLER by using the MODELLER objective function, DOPE/SOAP assessment scores or GA341 assessment score.

Notes

1. PSI-BLAST may also be used to identify target and template profiles.
2. Additional features of MODELLER include de-novo modeling of loops, optimization of the structure through user-defined functions, multiple sequence/structure alignment, and clustering.

24.2.2 *I-TASSER*

24.2.2.1 Outline/Methodology

I-Tasser is a Unix based tool utilised for fully automated protein structure prediction and functional annotation. It does so by assembling the target model from fragments of threading templates. The process begins by threading the target sequence through a PDB library with LOMETS while generating alignments through a variety of tools (HMM models, Psi-BLAST, Needleman-Wunsch, Smith-Waterman). Consistent/aligned fragments are applied to the model while non-aligned regions are constructed through *ab initio* modeling by replica-exchange Monte Carlo simulations. Structural decoys are clustered iteratively through SPICKER and the lowest energy structures of the clusters matching the desired properties are selected. REMO and FG-MD are used for the final model and refinement with the top structures rated by I-TASSER confidence score (C-Score). Structural annotation is performed with COFACTOR and COACH (Zhang 2008).

24.2.2.2 Pros/Cons

I-TASSER, similar to MODELLER, provides a relatively fast structural prediction for free and based on CASP trials has outperformed other tools with respect to accuracy. Minor issues include poor domain splitting and a propensity for unreliable model selection that greatly skews results.

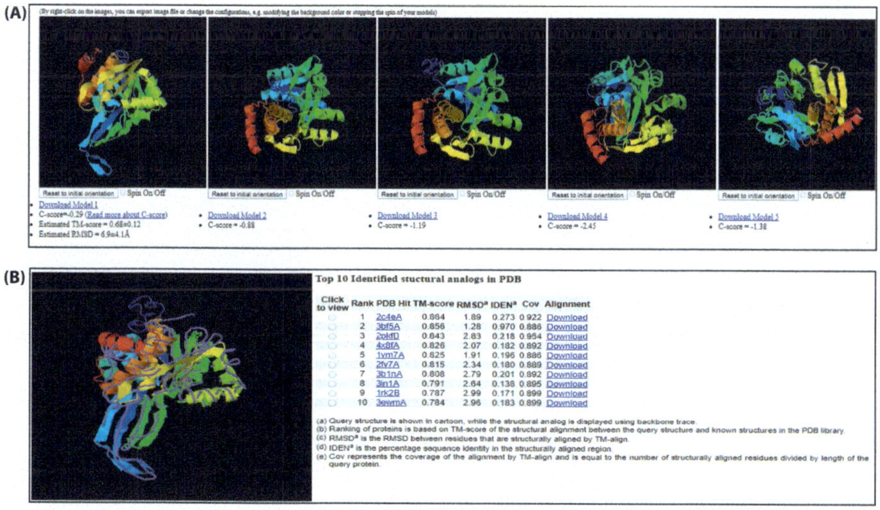

Fig. 24.3 Demonstration of I-Tasser overall output for a given query sequence. (**A**) Top 5 modeling results displayed in a browser window from decoy clustering. (**B**) Top 10 structurally similar homologs to the given target

24.2.2.3 Input

The primary amino acid sequence of protein query in the FASTA format.

24.2.2.4 Output

The output page includes information on the sequence, secondary structure/structural feature prediction, predicted normalized B factor (thermal mobility), top structural models with associated accuracy estimation from TM-Align (Fig. 24.3a), top threading alignments (Fig. 24.3b), solvent accessibility, functional annotation, EC number/GO term and editable models exportable as gif, jpg, or png.

24.2.2.5 Working Steps

I. Submit protein amino acid sequence using the form on I-TASSER site or upload directly and press the Run button.
II. Results will be provided via input email.

Notes

The I-Tasser suite contains a variety of tools including PSSpred, LOMETS, SPICKER (decoy clustering), ModRefiner, ResQ, COFACTOR, COACH, and NW-align.

24.2.3 Swiss-Model

24.2.3.1 Outline/Methodology

SWISS-MODEL, a web server was created initially with the goal of creating a flexible, easy-to-use protein modeling tool for structural analysis. This flexibility is evident in the multiple interfaces available including a "first approach mode" which minimizes complications for the user by only allowing an amino acid sequence input. It is also possible to use an "alignment mode" which accepts a sequence alignment as well as a "project mode" allowing manual control of a wide variety of parameters. SWISS-MODEL functions by first selecting a template through template structure library analysis (specifically the Swiss-Model Template Library derived from PDB), after which a structural alignment is constructed by iteratively superimposing structures and removing incompatible templates. If no candidate is found, a conformational space search is applied. The backbone structure is constructed as an average of templates with loops created with *ab initio* modeling. Finally, side chains are chosen from a rotamer library and the structures are energy minimized via the CHARMM27 force field and the optimal structures output (Biasini et al. 2014).

24.2.3.2 Pros/Cons

SWISS-MODEL has shown to provide highly accurate structures, particularly in the protein core and with respect to stereochemistry, at high speed. However, the program has been shown to crash, particularly when only low sequence identity is available (Wallner and Elofsson 2005).

24.2.3.3 Input

There are a number of potential inputs for SWISS-MODEL within the web interface. The simplest one is an amino acid sequence in FASTA format, with one letter sequence or through UniProt accession code. Additional inputs include a target-template MSA or a Deep View project file.

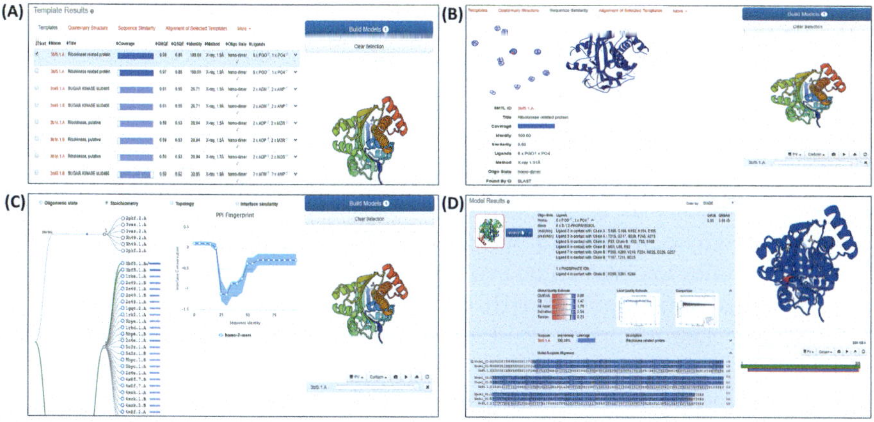

Fig. 24.4 Representation of the working modules of the tool SwissModel. (**A**) List of top alignments found by SwissModel and model constructed from the respective alignments on the right. (**B**) Sequence similarity of alignments visualized. The target protein is shown in red and each template as a blue circle with the N terminus of the target protein shown at the top of the circle and the C terminal at the clockwise end. The distance between circles represents similarity. Clicking on a circle takes the user to template-specific information. (**C**) Phylogenetic tree displaying the relationship between template homologs. (**D**) Modeling results for a specific template with relevant information including ligand binding sites, model-template alignments, and quality scores

24.2.3.4 Template Search Output

The output of the template search includes template results (Fig. 24.4a), relationships between detected templates and the query. It is possible to manually select the template within the browser interface through inspection of visual representations of the alignments (Fig. 24.4b). Alternatively, the templates may be selected automatically. Results are presented both graphically (interactive 3D and 2D formats), as a relationship tree (Fig. 24.4c), and also in tabular format alongside relevant information about each template.

From each of the selected templates, the program provides model coordinates and structure, predicted accuracy and similar relevant information (Fig. 24.4d). This information can be downloaded from the website for future use as zip files containing models in PNG, PDB, and JSON file formats as well as information on templates.

24.2.3.5 Working Steps

I. Submit protein the amino acid sequence through the form on the Swiss-Model site into the "Target Sequence" box.
II. Press "Build Model" and results will be presented within the web interface.

24.2.4 *Phyre2 (ProteinHomology/Analogy Recognition Engine)*

24.2.4.1 Outline/Methodology

Phyre2, a web tool, which distinguished by its ease of use and speed while providing advanced tools such as batch submission, Backphyre (detects homology across genomes) and Phyre Investigator, which allows for detailed model analysis. Phyre2 first scans libraries to construct an evolutionary profile, or statistical distribution of residues across homologs, with HHBlits that is then used in conjunction with a secondary structure prediction (Psi-Pred) to convert the query to an HMM. This HMM is scanned and aligned against HMMs in a fold library using HHsearch, to generate the protein fold while loops are modeled from a fragment (2–15 amino acid length) library or *ab initio* modeling to correct for insertions/deletions. Finally, side chains are modeled from a rotamer library with the R3 protocol (Kelley et al. 2015).

24.2.4.2 Pros/Cons

Phyre2 provides relatively high ease of use for those lacking expertise in protein modeling while sacrificing little in the terms of accuracy. The limitations of Phyre2 are not unique to the program and include difficulty in producing accurate structures when no homologs of known structure are detected, difficulty predicting the structural effect of mutations and inability to model multimeric proteins.

24.2.4.3 Input

The initial input for the standard protocol used to model a single sequence is the primary amino acid sequence in one letter code (FASTA format).

24.2.4.4 Output

Once complete, the user will be sent via an email, several pieces of information about the models produced including confidence in the models and a list of the highest scoring models. A web page of results will be attached containing the interactive views of all models produced with confidence scores, secondary structure/function predictions and summary tables including information on homologs and ligand binding predictions (Fig. 24.5a, b). Models may be downloaded in the PDB format or analyzed through the built-in Phyre Investigator tool to explore mutations, model quality, and functional assignments.

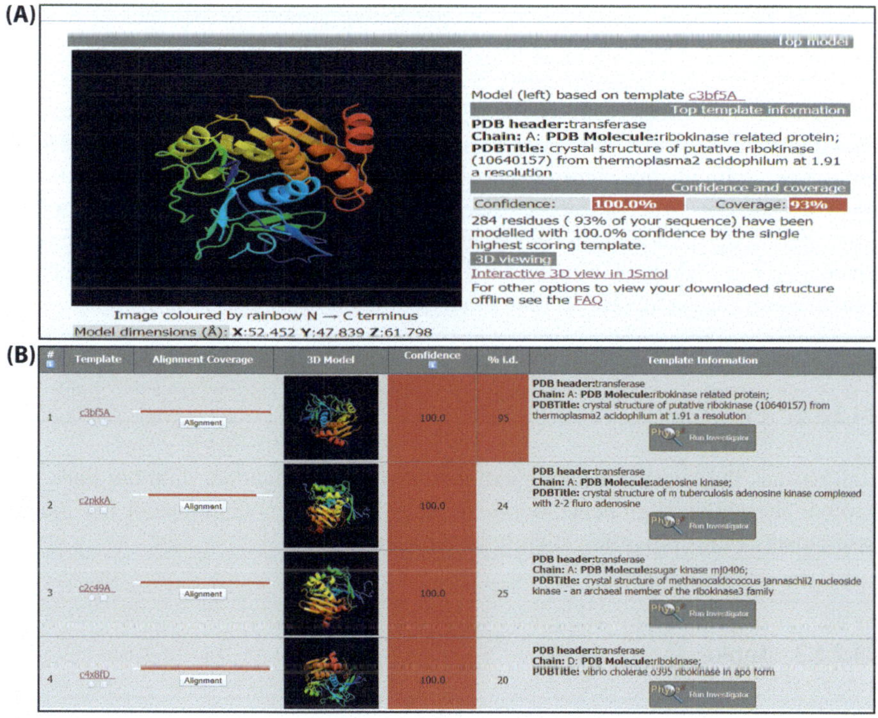

Fig. 24.5 The Phyre2 Output displaying the generation of 3D model. (**A**) Summary of information for top model generated as well as confidence score and download link for 3D model. (**B**) List of top models with alignments applied to generate them

24.2.4.5 Working Steps

I. In the Phyre2 homepage, enter relevant information such as email, job name etc.
II. Enter the amino acid sequence in the form provided.
III. Click the "Phyre search" button which directs one to the job monitoring page.
IV. On completion, an email update will be sent containing summary information and a job identifier used to explore models within the website.

24.2.5 HHPred

24.2.5.1 Outline/Methodology

The HHPred server was designed as an intermediary between fast, but lower accuracy homology search programs such as BLAST and high accuracy but slow protein prediction programs with the ultimate goal of predicting protein structure and function. The program proceeds by first generating a multiple sequence alignment through iterative PSI-BLAST searches by applying the HHSearch software. An HMM is constructed from the query and used to search for homologous templates in alignment databases such as Pfam via HMM-HMM alignment with a neural network used to then re-rank the potential templates. Distinct from other modeling programs, HHPred then generates multiple alignments with progressively lower diversity for the target sequence. Each alignment receives a TM-score for the structural models, allowing for template selection based on the optimal alignment diversity for each domain. Finally, MODELLER is run with the optimal template to generate a 3D structure (Hildebrand et al. 2009).

24.2.5.2 Pros/Cons

HHPred has the benefit of taking extremely low computational time but does not explore alternative alignments, optimize the side chain, model loops or use model assessments - thus providing slightly lower accuracy.

24.2.5.3 Input

Protein sequence or MSA (FASTA/Clustal/A3M format) in the HHPred form online. The user may provide additional parameters such as databases to search.

24.2.5.4 Output

Predicted alignments with visualization of overlap for top hits as well as NCBI reference sequences and accompanying confidence scores displayed within the browser window (Fig. 24.6a, b). Additional information on each protein may be identified on the NCBI website.

24.2.5.5 Working Steps

I. Submit protein amino acid sequence in an approved format using the form available on the HHPred site.
II. Press “Submit Job” and results will be displayed within the web interface.

(A)

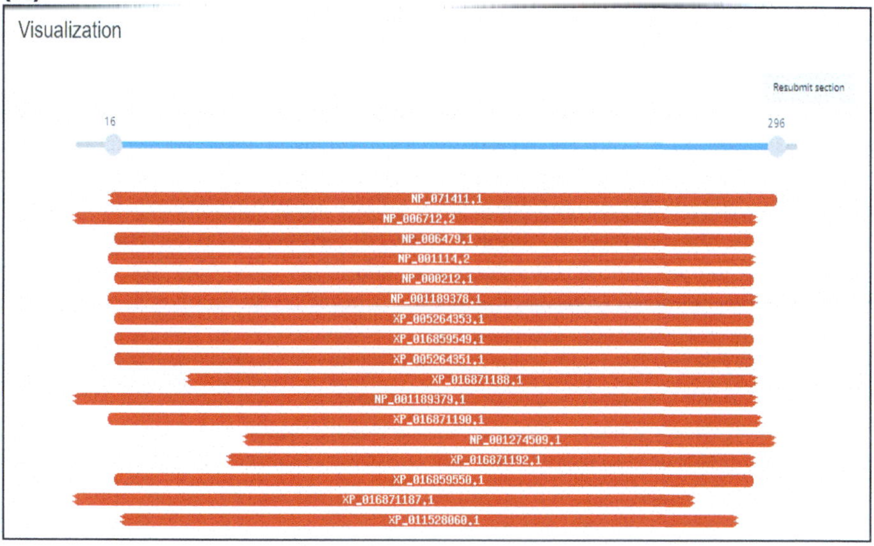

(B)

Hitlist

Show 25 entries

Nr	Hit	Name	Probability	E-value	SS	Cols	Target Length
1	NP_071411.1	ribokinase isoform 1 [Homo sapiens]	100	2.3e-31	30.4	277	322
2	NP_006712.2	adenosine kinase isoform b [Homo sapiens]	99.96	2e-28	34.5	286	362
3	NP_006479.1	ketohexokinase isoform b [Homo sapiens]	99.95	2e-28	32.9	264	298
4	NP_001114.2	adenosine kinase isoform a [Homo sapiens]	99.95	2.5e-28	32.8	271	345
5	NP_000212.1	ketohexokinase isoform a [Homo sapiens]	99.95	2.4e-28	31.5	264	298
6	NP_001189378.1	adenosine kinase isoform c [Homo sapiens]	99.95	1.1e-28	28.1	272	327
7	XP_005264353.1	PREDICTED: ketohexokinase isoform X8 [Homo sapiens]	99.94	3.3e-27	31.8	264	299
8	XP_016859549.1	PREDICTED: ketohexokinase isoform X4 [Homo sapiens]	99.93	1.1e-25	33.7	264	343
9	XP_005264351.1	PREDICTED: ketohexokinase isoform X3 [Homo sapiens]	99.93	1.4e-26	27.7	269	344

Fig. 24.6 Visualization of working of the HHPred and model generation. (**A**) HHPred alignment overlap visualization. (**B**) HHPred hitlist displaying aligned proteins

24.2.6 *ROBETTA*

24.2.6.1 Outline/Methodology

Robetta is an automated web tool which functions through application of the integrated ROSETTA software which applies a fragment insertion method. Input sequences are separated into domains through the "Ginzu" method, a hierarchical screening procedure that assigns domains. If a PDB homolog is found with BLAST or HHSearch, the template is aligned through methods such as HHSearch or Compass, comparative models are generated, and loops assembled from fragments and finally are optimized to fit the aligned structure. If no homolog is available, the ROSETTA *de novo* fragment insertion method (Bonneau) is applied by assembling fragment libraries that are used to generate the model. To produce the final structure, an iterative domain assembly protocol is applied by introducing linker regions between successive domains. Notable advancements include the ability to apply experimental NMR data to more accurately construct the model as well as prediction of the effect of mutagenesis on protein-protein interactions (Kim et al. 2004).

24.2.6.2 Pros/Cons

Robetta is a significantly demanding tool, requiring 4–6 h to run a 150 residues query and the *de novo* algorithm is largely optimized for small single domain proteins (making the domain assignment step highly significant) although the program performs well compared to other servers.

24.2.6.3 Input

The user inputs the primary amino acid sequence (FASTA) into the form displayed within the Robetta website, specifying a domain identification or structure prediction job.

24.2.6.4 Output

Results are presented within the web interface and are accessible by clicking the job ID from the queue. The ten (10) full structure predictions are shown with accompanying relevant information including secondary structure prediction, disordered region prediction, domain prediction and top PSI-BLAST analogs with annotations (Fig. 24.7a–c).

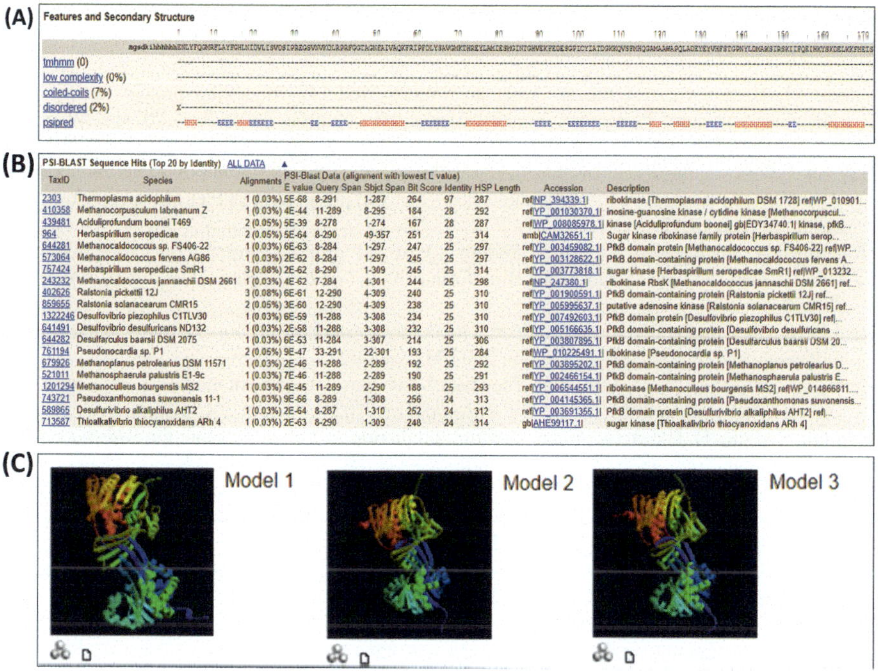

TaxID	Species	Alignments	PSI-Blast Data (alignment with lowest E value) E value	Query Span	Sbjct Span	Bit Score	Identity	HSP Length	Accession	Description
2303	Thermoplasma acidophilum	1 (0.03%)	5E-68	8-291	1-287	264	97	287	ref|NP_394339.1|	ribokinase [Thermoplasma acidophilum DSM 1728] ref|WP_010901...
410358	Methanocorpusculum labreanum Z	1 (0.03%)	4E-44	11-289	8-295	184	28	292	ref|YP_001030370.1|	inosine-guanosine kinase / cytidine kinase [Methanocorpuscul...
439481	Aciduliprofundum boonei T469	2 (0.05%)	5E-39	8-278	1-274	167	28	287	ref|WP_008085978.1|	kinase [Aciduliprofundum boonei] gb|EDY34740.1| kinase, pfkB...
964	Herbaspirillum seropedicae	2 (0.05%)	5E-64	8-290	49-357	251	25	314	emb|CAM32651.1|	Sugar kinase ribokinase family protein [Herbaspirillum serop...
644281	Methanocaldococcus sp. FS406-22	1 (0.03%)	6E-63	8-284	1-297	247	25	297	ref|YP_003459082.1|	PfkB domain protein [Methanocaldococcus sp. FS406-22] ref|WP...
573064	Methanocaldococcus fervens AG86	1 (0.03%)	2E-62	8-284	1-297	245	25	297	ref|YP_003128622.1|	PfkB domain-containing protein [Methanocaldococcus fervens A...
757424	Herbaspirillum seropedicae SmR1	3 (0.08%)	2E-62	8-290	1-309	245	25	314	ref|YP_003773818.1|	sugar kinase [Herbaspirillum seropedicae SmR1] ref|WP_013232...
243232	Methanocaldococcus jannaschii DSM 2661	1 (0.03%)	4E-62	7-284	4-301	244	25	298	ref|NP_247380.1|	ribokinase RbsK [Methanocaldococcus jannaschii DSM 2661] ref...
402626	Ralstonia pickettii 12J	3 (0.08%)	6E-61	12-290	4-309	240	25	310	ref|YP_001900591.1|	PfkB domain-containing protein [Ralstonia pickettii 12J] ref...
859655	Ralstonia solanacearum CMR15	2 (0.05%)	3E-60	12-290	4-309	238	25	310	ref|YP_005995637.1|	putative adenosine kinase [Ralstonia solanacearum CMR15] ref...
1322246	Desulfovibrio piezophilus C1TLV30	1 (0.03%)	6E-59	11-288	3-308	234	25	310	ref|YP_007492603.1|	PfkB domain protein [Desulfovibrio piezophilus C1TLV30] ref|...
641491	Desulfovibrio desulfuricans ND132	1 (0.03%)	2E-58	11-288	3-308	232	25	310	ref|YP_005166635.1|	PfkB domain-containing protein [Desulfovibrio desulfuricans ...
644282	Desulfarculus baarsii DSM 2075	1 (0.03%)	6E-53	11-284	3-307	214	25	306	ref|YP_003807895.1|	PfkB domain-containing protein [Desulfarculus baarsii DSM 20...
761194	Pseudonocardia sp. P1	2 (0.05%)	9E-47	33-291	22-301	193	25	284	ref|WP_010225491.1|	ribokinase [Pseudonocardia sp. P1]
679926	Methanoplanus petrolearius DSM 11571	1 (0.03%)	2E-46	11-288	2-289	192	25	292	ref|YP_003895202.1|	PfkB domain-containing protein [Methanoplanus petrolearius D...
521011	Methanosphaerula palustris E1-9c	1 (0.03%)	7E-46	11-288	2-289	190	25	291	ref|YP_002466154.1|	PfkB domain-containing protein [Methanosphaerula palustris E...
1201294	Methanoculleus bourgensis MS2	1 (0.03%)	4E-45	11-289	2-290	188	25	293	ref|YP_006544551.1|	ribokinase [Methanoculleus bourgensis MS2] ref|WP_014866811...
743721	Pseudoxanthomonas suwonensis 11-1	1 (0.03%)	9E-66	8-289	1-308	256	24	313	ref|YP_004145365.1|	PfkB domain-containing protein [Pseudoxanthomonas suwonensis...
589865	Desulfurivibrio alkaliphilus AHT2	1 (0.03%)	2E-64	8-287	1-310	252	24	312	ref|YP_003691355.1|	PfkB domain protein [Desulfurivibrio alkaliphilus AHT2] ref|...
713587	Thioalkalivibrio thiocyanoxidans ARh 4	1 (0.03%)	2E-63	8-290	1-309	248	24	314	gb|AHE99117.1|	sugar kinase [Thioalkalivibrio thiocyanoxidans ARh 4]

Fig. 24.7 Overall features and working of the modeling tool Robetta alongwith model generation. (**A**) Features and secondary structures of the query structure. TMHMM is applied to detect transmembrane regions, SEG to identify regions of low complexity, COILS to identify coiled-coil regions, DISOPRED to identify disordered regions, PSIPRED for secondary structures (*H* Helix, *E* Strand, *C* Coil). (**B**) PSI-BLAST hits (Top 20 by identity) displaying potential template homologs. (**C**) Top 3 generated models

24.2.6.5 Working Steps

I. From the Robetta site, select "Submit" under the section titled "Domain Parsing & 3-D Modeling."
II. Input the target name, FASTA amino acid sequence, and relevant contact information.
III. Results of modeling will be provided within the web interface.
IV. In order to generate models, from the results, one must click on "Predict domain structure with comparative modeling" under the respective Ginzu domain prediction.
V. Final models will be provided via email and within the web interface.

References

Biasini M, Bienert S, Waterhouse A, et al. SWISS-MODEL: modelling protein tertiary and quaternary structure using evolutionary information. Nucleic Acids Res. 2014;42:W252–8. https://doi.org/10.1093/nar/gku340.

Hamuel J. Phytochemicals: extraction methods, basic structures and mode of action as potential chemotherapeutic agents. In: Phytochemicals – a global perspective of their role in nutrition and health. InTech. 2012.

Hildebrand A, Remmert M, Biegert A, Söding J. Fast and accurate automatic structure prediction with HHpred. Proteins: Struct Funct Bioinforma. 2009;77:128–32. https://doi.org/10.1002/prot.22499.

John B, Sali A. Comparative protein structure modeling by iterative alignment, model building and model assessment. Nucleic Acids Res. 2003;31:3982–92.

Kelley LA, Mezulis S, Yates CM, et al. The Phyre2 web portal for protein modeling, prediction and analysis. Nat Protoc. 2015;10:845–58. https://doi.org/10.1038/nprot.2015.053.

Kim DE, Chivian D, Baker D. Protein structure prediction and analysis using the Robetta server. Nucleic Acids Res. 2004;32:W526–31. https://doi.org/10.1093/nar/gkh468.

Panda PK, Ibrahim D, Kumar P, Gupta P. Computational modeling and analysis of theoretical structure of Corneodesmosin receptor protein with existing phytochemicals in psoriasis | request PDF. Indian J Fundam Appl Life Sci. 2014;4(4):346–55.

Sahay A, Shakya M. In silico analysis and homology modelling of antioxidant proteins of spinach. J Proteomics Bioinforma. 2010;03:148–54. https://doi.org/10.4172/jpb.1000134.

Salomi MV, Manimekalai R, Shoba G. Comparative in silico docking analysis of phytochemicals from Murraya koenigii and commonly used drugs in the treatment of cancer. Int J Curr Res Acta Rev. 2016;3:119–33.

Vasavi CS, Saptharshi, Radhika Devi R, et al. Homology modeling and protein ligand interaction to identify potential inhibitor for E1 protein of chikungunya. Berlin: Springer; 2010. p. 510–3.

Wallner B, Elofsson A. All are not equal: a benchmark of different homology modeling programs. Protein Sci: Publ Protein Soc. 2005;14:1315–27. https://doi.org/10.1110/ps.041253405.

Zhang Y. I-TASSER server for protein 3D structure prediction. BMC Bioinforma. 2008;9:40. https://doi.org/10.1186/1471-2105-9-40.

Chapter 25
In-silico Approaches to Study Therapeutic Efficacy of Nutraceuticals

Ramesh Kumar, Amit Kumar Singh, Ashutosh Gupta, and Abhay K. Pandey

25.1 Introduction

Nowadays, human life is difficult without Drugs. Even though drug development has been propelled by diversity of technology advancements that can also speed up the growth and development of medicinal agents from herbal world, drug development still remains a prolonged process with a lower success rate and enormous capital investment. Generally about 10–15 years are taken for a newly synthesized molecule to be converted into a saleable therapeutic drug. Big Pharma in 2005 exhausted about US$ 50 billion, which is very huge amount spent in 1996 (Barden and Weaver 2010). Worldwide the medical industries are classified among the major support industries as indicator for the economic development. In 2006, international expenditure on recommended drugs exceeded 640 billion US dollar and the USA accounted for approximately half of the universal pharma sell (Wu et al. 2011). Complementary and alternative medicine therapies (CAMTs), which are achieving recognition globally, are categorized into drug and nondrug-based CAMT (Pan et al. 2012). Amid a variety of drug-based CAMTs, medicines frequently practiced are of herbal origin. Humans have advanced from the herbivorous animals to omnivorous. Use of herbal ingredients for encouraging human health has long history, and the scientific details are accessible for research related with current drug discovery. The most ancient facts of human utilization of plants for medicinal usage can be traced to the Neanderthal period (Winslow and Kroll 1998). Plant derived herbal compounds are beneficial to human health. Fascinatingly, it has been seen that non-human vertebrates would consume specific herbs for medication under diseased states (Krief et al. 2008; Fowler et al. 2007). Not amazingly, the interpretation of serendipitous eating behaviour in animals has led to the finding of plants having medicinal efficacy. The study of self-medication in animals paved the approach to drug discovery for humans. About three fourth of the world's population still depends

R. Kumar · A. K. Singh · A. Gupta · A. K. Pandey (✉)
Department of Biochemistry, University of Allahabad, Allahabad, India

S. Kumar, C. Egbuna (eds.), *Phytochemistry: An in-silico and in-vitro Update*,
https://doi.org/10.1007/978-981-13-6920-9_25

on herbs for health care system. World Health Organization report suggested that about one third of all plant species have been used for medicinal purposes (Schippmann et al. 2002).

In recent times, with greater understanding of structural and functional properties of compounds, new methods and techniques have been applied for drug discovery programmes (Yi et al. 2016). A quick and convenient method needs to be established for accurate prediction of enormous chemical compounds followed by in-vitro and in-vivo pharmacological investigations for validation will considerably improve the effectiveness of assessing the biological activities of medicinal plants. One of the famous antimalarial drug artemisinin was discovered through large-scale herbal screening in 1950s. Subsequently many FDA approved drugs of natural origin have been discovered using in-silico approaches (Yi et al. 2016). Computer machinery have been made excellent impact on effectiveness, producing precise data for society through communications, and market requirement estimating, to shielding public safety. Computational approach are executed as a device for generating novelty for search, design and optimization of new drugs planned to treat human diseases. However, in-silico technique assist diminish drug erosion rates during use of pharmacovigilance computational data mining techniques, therapeutic bioinformatics approaches and prognostic computer models for recognize possible drug toxicity (Valerio Jr. 2012). Currently in-silico modelling has become a necessary tool for drug discovery. Mostly molecular docking algorithms presume the proteins as inflexible objects which show improper correlation with the docking results. No docking algorithm is known that can accurately forecast the binding affinities of ligand in molecular interface (Lakshmi Ranganatha et al. 2013). A chief benefit of in-silico methodologies is that they assist pace the rate of manufacture and screening of drug based on scrutiny of designed possessions and forecast models for drug remedial targets and recognition of protective responsibility While all the minimizing the requirement for costly and time-consuming animal and in-vitro assay laboratory work.

25.2 Software and Database of Medicinal Plant Compounds

There are several software/websites available in commercial either free or access restricted which can be used to construct a databases of medicinal plant compounds of therapeutic application. A database of complete medicinal nutraceuticals must provide following descriptions: (1) Compound's storage number; (2) Name and CAS/Pubchem ID of the compound; (3) Information regarding plant sources (Latin name and extract fraction); and (4) structural insight of the compound. Besides this many software also compute relative molecular mass of the compound, the number of bonds showing rotation bonds, lipid partition coefficient, the number of hydrogen donors/acceptors and several other physicochemical attributes of medicinal agents. This set of information can make continuation work needing transparency and data processing more user-friendly. Table 25.1 enlists the most common databases of medicinal plant compound having therapeutic properties (Yi et al. 2018).

Table 25.1 Common molecular databases of plant secondary metabolites and its therapeutic applications

Databases	Developed by	URL/websites	Description
NAPRALERT	University of Illinois at Chicago, USA	www.napralert.org	A natural product database based on more than 200,000 bibliographical records including pharmacological and biochemical information
NCI	National Cancer Institute, USA	www.cactus.nci.nih.gov	Data archive containing the chemical, physical and biological properties as well as associated activity prediction value
TCM	China Medical University, Taiwan	www.tcm.cmu.edu.tw	Compendium of many Traditional Chinese medicines along with 20,000 Compounds and their physical, chemical and structural (3D) properties
Alkamid	Ghent University, Belgium	www.alkamid.ugent.be	Details of plants' sources, biosynthesis and pharmacological information related withof *N*-alkyl amides
Asian anti-cancer materia	Institute of East-West Medicine, USA	http://www.asiancancerherb.info/herbList/list_s.htm	Details about 700 types of anti-cancer drugs of Asian origin
TCMSP	Centre for Bioinformatics, College of Life Science, Northwest A&F University, China	www.lsp.nwsuaf.edu.cn/tcmsp.php	Information about 500 Chinese pharmacopoeia
Timtec	Timtec, Russia	www.timtec.net	Database of more than 10,000 natural products and their derivatives
TradiMed (Traditional Chinese medicine D	TradiMed, Korea	www.tradimed.com	Record of prescriptions in Korean and Chinese traditional medicines and structural features
ZINC	University of California, USA	www.zinc.docking.org	Computational database of millions of compounds and 3D structure for docking
Chinese National Compound Library	National Health and Family Planning Commission of the People's Republic of China	www.app.cncl.org.cn	A library of small molecule compounds
CHMIS-C (A Comprehensive Herbal Medicine Information System for Cancer)	University of Michigan Medical School, USA	www.sw16.im.med.umich.edu/chmis-c	Database of many anticancer herbal prescriptions, components and small molecules for treatment of variety of cancer

25.3 Analysis of Drug Like Properties of Medicinal Compounds

There are large numbers of compounds obtained from medicinal plants; however, the majority of them do not have pharmacological efficacy. In order to increase screening efficiency, the first and foremost step is to eliminate the non-potential compounds to improve the desired compounds activity. This is known as pre-treatment of selected medicinal compounds. Drug-like feature is a qualitative trait employed in drug design for knowing a compound's effectiveness with respect to factors such as bioavailability, which is a function of molecular structural attributes (Kerns and Li 2008). There are certain structural characteristic which show that a compound has a higher possibility of becoming an effective drug. Earlier research on these drug characteristics was one of the most vital components of downstream drug development. Currently, it has become imperative to integrate the study of drug like properties during the early stages of drug discovery programme to assess their drug-likeness properties. Pharmacologists are looking for the following characteristics in drugs: firstly, the structural features such as shape, relative molecular weight, lipophilicity, hydrogen bonding capacity, polar surface area, as well as acid dissociation constant; secondly its physical and chemical properties i.e. solubility in solvents, pH value, membrane permeability and chemical stability; thirdly, its property inside cell, i.e., its metabolism, binding affinity with proteins and transport ability; and fourthly its pharmacokinetics and toxicity, half-life, bioavailability, drug interactions and median lethal dose (Yi et al. 2018).

According to Lipinski's rule of five (RO5), a small molecule needs to have following characteristics to have drug like properties. (1) A maximum of five hydrogen bond donors (the total sum of N–H and O–H bonds) should be present; (2) it should not have more than ten hydrogen bond acceptor atoms (total oxygen or nitrogen atoms); (3) Its molecular weight must be <500 Daltons; and (4) an octanol–water partition coefficient (log P) should not be greater than five (Lipinski et al. 2001). Noticeable thing is that all the numbers are multiple of five reason for the name RO5. Any small molecule which is orally active does not shows the violation of more than one criteria of RO5. Molecules satisfying the RO5 rule generally have greater bioavailability in the metabolic processes of the organisms and hence are prospective candidates to become oral therapeutic agent. Veber et al. (2002) presented a different set of requirements for the oral administration of drugs. However, exceptions are for the screening of special drugs for example narcotic drug, which cannot exclude compounds of small molecular weight, and those anti-cancerous drugs which cannot rule out metalo-organic compounds. Currently, some softwares associated with molecular docking and virtual screening studies are of great importance in drug discovery and development. Instant J Chem from Chem Axon Company which calculate the Lipinski rule, and the Discovery Studio software have potential to predict both the Lipinski and Veber's rules (Eldehna et al. 2016).

25.3.1 ADME/T (Absorption, Distribution, Metabolism, Excretion and Toxicity) Selection

ADME/T selection provides an important insight towards the drug filtering when after the establishment of drug-likeness from the analyses of the physiochemical properties and structural features of desired drug candidates. It is essential to forecast the situation and movement of a compounds inside human body while the designing of the drug molecule. ADME/T properties of a drug needs to evaluate during early stage of drug design (Van de Waterbeemd and Gifford 2003). Absorption is the process of drug transport into the human circulatory system. Distribution is the infiltration of the drug through the cell membrane barrier into the various tissues, organs or body fluids. As metabolism proceeds, the initial compound is transformed to new compounds called metabolites. Cytochrome P450 enzymes, present in liver cell catalyses this interconversion. However, excretion is the elimination of the drug either in initial form or as metabolites from the human body. The toxicity of the drug also affects the functioning of human body. Many commercial software programs are designed to have ADME/T prediction component for drug molecules. For instance, Discovery Studio software provides an ADME/T descriptor module for candidate drug screening, which covers the following features: prediction of solubility of each compound in aqueous solvent; crossing blood brain barrier, to predict the ability of compounds to enter into the brain; and CYP2D6 enzyme binding capacity (Dhiman et al. 2017). The CYP2D6 enzyme is an important member of the family CYP450 enzyme and do participates with the CYP3A4, CYP1A2, CYP2C9, and CYP2C19 enzymes in drug metabolism. These five major CYPs leads to metabolism of more than 95% of the drug metabolism in animals. The CYP2D6 enzyme normally accounts for approximately 2% of the CYP total but approximately 30% of the total drug metabolism. High binding capacity of enzyme CYP2D6 with drug reflect its excellent metabolism ability inside human body. The relevant criteria involve hepatotoxicity, the dose-dependent liver toxicity of the drug molecules; intestinal absorption, the absorption of drug molecules in the human body after oral administration; and plasma protein binding, the ability of the compound to bind to the carrier protein in the bloodstream. Suitable pharmacokinetic properties and low toxicity during body absorption, distribution, metabolism and excretion are the key factors in successfully passing the clinical trials. Prior to drug design and performing the screening process, ADME/T prediction done to reduce the drug development cost as well as increases the success rate. For ADME/T properties rational predictive have been developed on mechanistic explanations of the underlying biophysical processes. Some softwares can perform the ADME/T such as Simulations plus ADME/T predictor, PK Map the Discovery Studio from Accerlary Company (Willmann et al. 2005). Although it is considered that natural products have more preferable ADME/T properties in comparison with synthetic chemicals, the study of ADME/T properties for phytochemicals is also very important in drug development (Dhiman et al. 2017).

25.4 Tools for Analysis of Biological Activity of Medicinal Plant

With every passing year, the database of the biological activity of the medicinal compounds as well as their target macromolecules' number related with therapeutic effects is increasing. At the same time, the data pool regarding the compositions of medicinal plants has also increased. Consequently, the necessity to use bioinformatics tools to assess the biological activity of medicinal plants is inevitable.

25.4.1 QSAR (Quantitative or Qualitative "Structure-Activity" Relationships) Modelling and Virtual Screening

QSAR molecular modelling and virtual screening methods find extensive application in drug discovery related with synthetic compounds. These methods may be utilized for unravelling the biological efficacy of medicinal plants provided the information related with the structures of natural compounds is known. All these methods rely on the notion that the structure of compound defines its activity. QSAR models consist of three major components:

1. Structures and biological activity related non-contradictory data of the compounds of interest.
2. Structural description of the compounds (structural fragments, fingerprints, topological, electro-topological, constitutional, physicochemical and quantum-chemical properties)
3. Machine learning methods (multiple linear regressions, neural networks, support vector machines, random forest, similarity etc.) for identification of the association amongst descriptors, which are conventionally being used as independent variables, and biological activity.

QSAR models are of two types, local and global. The models created using heterogeneous data set is called global QSAR models while homogenous data models are called as local models. Global models have wider utility domain and may be used for virtual screening, biological activity prediction as well as target fishing and are being applied for the optimisation of hit/lead compounds. For creation of pharmacophore, local 3D-QSAR models may be used. A pharmacophore is defined as a group of atoms present in the molecule which are thought to be accountable for a pharmacological action (Lagunin et al. 2014).

Docking is a procedure for predicting the favoured orientation of one molecule vis-a-vis another molecule when they are associated with each other to produce a stable complex (Lengauer and Rarey 1996). This is commonly being used for prediction of the binding orientation of small drug-like molecules with their protein targets to predict the binding affinity and biological activity of the molecule. The three dimensional structures of target molecules are essential for the docking procedure. They may be obtained from the protein data bank (PDB) database (http://www.pdb.org) or it can be generated by molecular modelling using ligand-based and target based strategies (Chen et al. 2003). Table 25.2 depicts commercially and freely available software for these studies (Lagunin et al. 2014). For reconnoitring biological activity and the prospective targets of a new phytochemical, the computational tools stated in Table 25.2 can be used. The benefits of docking include the application of data only for targets and that it does not entail information about active compounds. Because of limited number of 3D structures of targets molecules docking suffers disadvantage in determining the docking results (Vilar et al. 2008).

25.5 Application of QSAR Modelling and Docking to Study Therapeutic Efficacy of Nutraceuticals

Molecular docking and QSAR modelling studies were performed to investigate the immuno-modulatory activity of natural coumarinolignoid derivatives extracted from *Cleome viscosa* seeds. Immuno stimulatory activity was analysed via the QSAR model generated by forward stepwise multiple linear regression and physico-chemical descriptors from Scigress Explorer (Hardy et al. 2010). Docking studies showed the likely binding affinity of coumarinolignoids to diverse immuno-modulatory receptors such as COX-1, COX-2, TLR-4, iNOS, CD14, CD86 and IKK b (Yadav et al. 2010; Meena et al. 2011). Similarly, anticancer/cytotoxic activity of analogues of glycyrrhetinic acid against the human lung cancer cell line (A-549) and immunomodulatory/anti-inflammatory activity of gallic acid derivatives were predicted (Yadav et al. 2012). Root of *Glycyrrhiza glabra* (liquorice) possesses glycyrrhizic acid which on hydrolysis produces glycyrrhetinic acid, a pentacyclic triterpenoid derivative of β-amyrin. The docking results demonstrated the high binding affinity of the proposed active compounds with the EGFR, the lung cancer target. Docking studies of gallic acid derivatives revealed that the compounds had high binding affinities for INFα-2, IL-4, and IL-6 receptors.

Maurya and coworkers (2012) reported the immunomodulatory and anti-inflammatory activity of ursolic acid and lupeol obtained from *Gentiana kurroo* and *Eucalyptus tereticornis* by QSAR modelling and docking studies using Scigress Explorer. The results proposed that the triterpenoids used in the study exhibited anti-inflammatory and immunomodulatory action because of their higher binding affinity towards human targets (receptors and enzymes): COX-2, p52, TNF-α, and NF-kβ p50. Earlier the lupane group triterpenoids were analysed using PASS software which resulted in the prediction of hepatoprotective, antigestagenic, and other biological activities (Sharma et al. 2012). QSAR modelling was used in unravelling the antimalarial activity of artemisinin derivatives from *Artemisia annua* along with docking to plasmepsins (Plm-II) using Scigress Explorer (Qidwai et al. 2012). Chemically synthesis of one of the expected active compounds was done and it was tested in-vivo in mice infected with a multidrug-resistant strain of *Plasmodium*

Table 25.2 Software for analysing biological activity, QSAR modelling and virtual screening (docking) (Lagunin et al. 2014)

Website	Software name	Description
www.acdlabs.com/products/percepta	ACD/Percepta	Prediction of ADME/T and physico-chemical properties
www.simulations-plus.com	ADMET Predictor	Prediction of ADME/T Property
www.accelrys.com/products/discovery-studio	Discovery Studio	QSAR modelling and pharmacophore generation
www.way2drug.com/GUSAR	GUSAR	QSAR modelling, antitarget interactions
www.chemcomp.com/softwarechem.htm	Molecular Operating Environment (MOE)	MOE includes tools for the creation of QSAR/QSPR models. Also used for descriptors including topological index, E-state index, structural keys, topological polar surface area, physical properties
www.vlifesciences.com/products/QSARpro/product_QSARpro.php	QSARpro	QSAR modelling
http://www.tripos.com	SYBYL®-X Suite	Pharmacophore hypothesis generation, QSAR modelling, conformational searching, molecular alignment, ADME prediction, virtual screening and docking
www.way2drug.com/GE	DIGEP-Pred	Information regarding drug-induced alterations in the gene expression based on the structural details of drug-like compounds
www.way2drug.com/GUSAR	GUSAR (web-service)	Knowledge about acute rodent toxicity, ecotoxicity endpoints and anti-targets interactions
www.farma.ku.dk/smartcyp	SMARTCyp	Susceptible site prediction in molecules that are affected by cytochrome P450 mediated metabolism
www.autodock.scripps.edu/	AutoDock	Molecular modelling simulation software including protein–ligand docking
www.schrodinger.com/productpage/14/5/21	Glide	Schrodinger's ligand–protein docking software
www.ccdc.cam.ac.uk/Solutions/GoldSuite/Pages/GOLD.aspx	GOLD	Docking, virtual screening, hit optimisation, and identification of the precise binding mode of active molecules
www.molegro.com/mvd-product.php	Molegro Virtual Docker (MVD)	Prediction of protein–ligand interactions
www.eyesopen.com/oedocking	OEDocking	Molecular docking tools and their associated workflows
www.accelrys.com/insight/affinity	Affinity	Accuracy in calculation due to simulation of molecular pairs of procedures

yoeliinigeriensis. The results revealed the potent antimalarial activity of the synthesized compound. Table 25.3 provides information about some of the docking studies, their targets and the mechanism of action (Lagunin et al. 2014).

Table 25.3 Docking studies of some phytoconstituents their target and biological application

Compound	Source	Biological effect	Target	Software used	Reference
Capsaicin	Chilli	Antibiotic	NorA efflux pump	SiteMap module of Schrodinger, extra precision scoring function of Glide	Puppala et al. (2012)
Catechol alkenyls/	*Semecarpusanacardium*	Alzheimer's disease treatment	Acetylcholinesterase	GOLD 3.1	Adhami et al. (2012)
Withanolides	*W. somnifera*	Anticancer	Mortalin, p53, p21, Nrf2	AutoDock 4.2	Vaishnavi et al. (2012)
Aqueous root extracts	*Gentiana lutea*	Inhibition of vascular smooth muscle cell proliferation	MEK1	AutoDock 4	Kesavan et al. (2013)
Withanolide derivatives	*W. somnifera*	Antimycobacterial	Protein kinase G	Glide	Santhi and aishwarya (2011)
Saponins	*Parthenium hysterophorus*	Anti-inflammatory	TNF-α	Cerius2,LigandFit, Glide	Shah et al. (2009)
Epsilon viniferins	*Vitis vinifera*	9-proteins including PDE4	700 proteins	Selnergy™	Do et al. (2005)

25.6 Conclusion

During twenty-first century the use of in-silico methods has increased extensively for herbal drug discovery programmes. With the advent of new chemo- and bioinformatics tools, in addition to the accumulating OMICs data as well as information on phytometabolite structures has paved the way for interpretation of biological and pharmacological activities of plant products. Since the number and structural characteristics of reported phytochemicals are limited in comparison to the existing total diversity of phytoconstituents on the planet, future will see the discovery of many new phytochemicals. Therefore, the application of currently available chemo- and bioinformatics resources viz., QSAR modelling, virtual screening, molecular docking, ADME/T prediction and many more approaches provide important facts for the discovery of novel drug leads and their pharmacological applications beyond their traditional use.

Acknowledgments Ramesh Kumar and Amit Kumar Singh acknowledges financial support from CSIR New Delhi, India, in the form of Junior Research Fellowship and Ashutosh Gupta acknowledges financial support from UGC-CRET New Delhi, India. All the authors also acknowledge DST-FIST and UGC-SAP facility of Department of Biochemistry University of Allahabad, Allahabad India.

References

Adhami HR, Linder T, Kaehlig H. Catechol alkenyls from Semecarpus anacardium: acetylcholinesterase inhibition and binding mode predictions. J Ethnopharmacol. 2012;139:142.

Barden CJ, Weaver DJ. The rise of micropharma. Drug Discov Today. 2010;15:84–7.

Chen X, Ung CY, Chen Y. Can an in silico drug-target search method be used to probe potential mechanisms of medicinal plant ingredients? Nat Prod Rep. 2003;20:432.

Dhiman V, Singh DK, Ladumor MK. Characterization of stress degradation products of amodiaquinedihydrochloride by liquid chromatography with high-resolution mass spectrometry and prediction of their properties by using ADMET predictor. J Sep Sci. 2017;40:4530–40.

Do QT, Renimel I, Andre P. Reverse pharmacognosy: application of selnergy, a new tool for lead discovery. The example of epsilon-viniferin. Curr Drug Discov Technol. 2005;2:161–7.

Eldehna WM, Fares M, Ibrahim HS. Synthesis and cytotoxic activity of biphenylurea derivatives containing indolin-2-one moieties. Molecules. 2016;21:E762.

Fowler AK, Outsioni Y, Sommer V. Leaf-swallowing in Nigerian chimpanzees: evidence for assumed self-medication. Primates. 2007;48:73–6.

Hardy B, Douglas N, Helma C. Collaborative development of predictive toxicology applications. J Cheminform. 2010;2:7.

Kerns EH, Li D. Drug-like properties: concepts, structure design and methods. Oxford: Elsevier Ltd.; 2008. p. 125–36.

Kesavan R, Potunuru UR, Nastasijevi B. Inhibition of vascular smooth muscle cell proliferation by *Gentiana lutea* root extracts. PLoS One. 2013;8:61393.

Krief S, Jamart A, Mahe S. Clinical and pathologic manifestation of oesophagostomosis in African great apes: does self-medication in wild apes influence disease progression? J Med Primate. 2008;37:188–95.

Lagunin AA, Goel RK, Gawande DY. Chemo- and bioinformatics resources for in silico drug discovery from medicinal plants beyond their traditional use: a critical review. Nat Prod Rep. 2014;31:1585–611.

Lakshmi Ranganatha V, Zameer F, Meghashri S. Design, synthesis and anticancer properties of novel benzophenone conjugated coumarinanalogs. Arch Pharm. 2013;346:901–11.

Lengauer T, Rarey M. Computational methods for biomolecular docking. Curr Opin Struct Biol. 1996;6:402–6.

Lipinski CA, Lombardo F, Dominy BW, Feeney PJ. Experimental and computational approaches to estimate solubility and permeability in drug discovery and development settings. Adv Drug Deliv Rev. 2001;46:3–26.

Maurya A, Khan F, Bawankule DU, et al. QSAR, docking and in vivo studies for immunomodulatory activity of isolated triterpenoids from *Eucalyptus tereticornis* and *Gentiana kurroo*. Eur J Pharm Sci. 2012;47:152–61.

Meena A, Yadav DK, Srivastava A. In silico exploration of anti-inflammatory activity of natural coumarinolignoids. Chem Biol Drug Des. 2011;78:567–79.

Pan SY, Gao SH, Zhou SF, et al. New perspectives on complementary and alternative medicine: an overview and alternative therapy. Altern Ther Health Med. 2012;18:20–36.

Puppala M, Ponder J, Suryanarayana P. The isolation and characterization of β-glucogallin as a novel aldose reductase inhibitor from *Emblica officinalis*. PLoS One. 2012;7:31399.

Qidwai T, Yadav DK, Khan F. QSAR, docking and ADMET studies of artemisinin derivatives for antimalarial activity targeting plasmepsin II, a hemoglobin-degrading enzyme from *P. falciparum*. Curr Pharm Des. 2012;18:6133–55.

Santhi N, Aishwarya S. Insights from the molecular docking of with anolide derivatives to the target protein PknG from *Mycobacterium tuberculosis*. Bioinformation. 2011;7:1–4.

Schippmann U, Cunningham AB, Leaman DJ. Impact of cultivation and gathering of medicinal plants on biodiversity: global trends and issues. Rome: FAO; 2002. p. 142–67.

Shah BA, Chib R, Gupta P. Saponins as novel TNF-α inhibitors: isolation of saponins and a nor-pseudoguaianolide from *Parthenium hysterophorus*. Org Biomol Chem. 2009;7:3230–5.

Sharma S, Chattopadhyay SK, Yadav DK. QSAR, docking and in vitro studies for anti-inflammatory activity of cleomiscosin A methyl ether derivatives. Eur J Pharm Sci. 2012;47:952.

Vaishnavi K, Saxena N, Shah N. Differential activities of the two closely related withanolides, Withaferin A and Withanone: bioinformatics and experimental evidences. PLoS One. 2012;7:44419.

Valerio LG Jr. Application of advanced in-silico methods for predictive modeling and information integration. Exp Opin Drug Metab Toxicol. 2012;8:395–8.

Van de Waterbeemd H, Gifford E. ADMET in silico modelling: towards prediction paradise? Nat Rev Drug Discov. 2003;2:192–204.

Veber DF, Johnson SR, Cheng HY, et al. Molecular properties that influence the oral bioavailability of drug candidates. J Med Chem. 2002;45:2615–23.

Vilar S, Cozza G, Moro S. Medicinal chemistry and the molecular operating environment (MOE): application of QSAR and molecular docking to drug discovery. Curr Top Med Chem. 2008;8:1555–72.

Willmann S, Lippert J, Schmitt W. From physicochemistry to absorptionand distribution: predictive mechanistic modelling and computational tools. Exp Opin Drug Metab Toxicol. 2005;1:159–68.

Winslow LC, Kroll DJ. Herbs as medicines. Arch Intern Med. 1998;158:2192–9.

Wu M, Atchley D, Greer L, et al. Dosed without prescription: preventing pharmaceutical contamination of our nation's drinking water. Environ Sci Technol. 2011;15:366–7.

Yadav DK, Meena A, Srivastava A, et al. Development of QSAR model for immunomodulatory activity of natural coumarinolignoids. Drug Des Dev Ther. 2010;4:173–86.

Yadav DK, Khan F, Negi AS. Pharmacophore modeling, molecular docking, QSAR, and in silico ADMET studies of gallic acid derivatives for immunomodulatory activity. J Mol Model. 2012;18:2513–25.

Yi F, Sun L, Xu LJ, et al. In silico approach for anti-thrombosis drug discovery: P2Y1R structure-based TCMs screening. Front Pharmacol. 2016;7:531.

Yi F, Li L, Xu LJ, et al. In silico approach in reveal traditional medicine plants pharmacological material basis. Chin Med. 2018;13:33.

Chapter 26
Bulbine frutescens Phytochemicals as a Promising Anti-cancer Drug Discovery Source: A Computational Study

Prem P. Kushwaha, Ajay Kumar, Santosh Maurya, Atul K. Singh, Amit Kumar Sharma, and Shashank Kumar

26.1 Introduction

B. frutescens is a flowering plant belongs to *Asphodelaceae* family. It is commonly used as a traditional medicine in South Africa for the treatment of skin wounds and burns (Pinar 1998). Recently in 2014, for the first time, *B. frutescens* potent anti-proliferative property has been reported (Tambama et al. 2014). There are only nine PubMed indexed articles on the *B. frutescens* species which includes phytochemical isolation and their biological activity. Aqueous extracts of *B. frutescens* have the ability to increase glucose utilization activity in-vitro (Van Huyssteen et al. 2011). *B. frutescens* phytochemical (isofuranonapthoquinone 1) are known as a potent inhibitor of Glutathione transferase P1-1 (GSTP1-1) protein, involved anticancer drug detoxification mediated drug resistance (Mukanganyama et al. 2011). Beside that HIV-1 protease inhibition by aqueous and ethanolic extracts of Bulbine genus gives credence to the use of this plant for the treatment of HIV-related skin infections.

Cancer is one of the chief health anxieties for people in present and will be for the predictable future also. Other than surgery and radiation, chemotherapy is an important clinical method to treat different cancer successfully (Kalibcrov and Buchsbaum 2012). Most of the signaling cascade such as Janus-activated kinase/ signal transducer and activator of transcription, Hedgehog, Wnt, phosphatidylinositol 3-kinase/phosphatase, tensin homolog, nuclear factor-κB facilitates many of the properties like cell fate decisions, survival, proliferation, and

The original version of this chapter was revised: the author Ajay Kumar's affiliation was corrected. The correction to this chapter is available at https://doi.org/10.1007/978-981-13-6920-9_32

P. P. Kushwaha · A. Kumar · S. Maurya · A. K. Singh · S. Kumar (✉)
Department of Biochemistry and Microbial Sciences, Central University of Punjab, Bathinda, Punjab, India

A. K. Sharma
Department of Agriculture, Uttaranchal University, Dehradun, Uttrakhand, India

S. Kumar, C. Egbuna (eds.), *Phytochemistry: An in-silico and in-vitro Update*,
https://doi.org/10.1007/978-981-13-6920-9_26

differentiation (Matsui 2016). Studies reported that overactive or abnormal signaling among these pathways may contribute to the survival of cancer (Matsui 2016). People have been using plant-derived medicines for several centuries throughout the world. Various rewards such as reduce risk of side effects, effective with chronic conditions, cost-effective and widespread availability are allied with using natural medicines. Due to these points, currently, natural compounds are widely used in cancer chemotherapy and have a promising future (Kinghorn et al. 2009). Several phytochemicals such as epigallocatechin gallate (EGCG), curcumin, resveratrol, apigenin etc. have been reported to inhibit signaling pathways involved in cancer cells and cancer stem cell initiation, progression, anti-apoptotic pathways and drug resistance (Duvoix et al. 2003; Mouli et al. 2009). The present scenario demands potent natural compounds as potential anticancer and anti-drug resistant agents to treat the devastating disease. Keeping this in our mind, we screened the anticancer potential of *B. frutescens* phytochemicals against cancer signaling pathway proteins [β-catenin (1LUJ), PI3K (3APF), TGF-β (1PY5), NF-kB (1NFK), SMO (4JKV), MEK1 (3PP1) and JAK2 (3Q32)]. Furthermore the drug likness, ADME/T, interaction with xenobiotic metaolising enzymes (CYP450) and biological activites were also studied with the use of computational softwares.

26.2 Materials and Methods

26.2.1 Screening, Retrieval and Preparation of Phytochemicals

The NCBI PubChem compounds database and kinds of literature facilitated to prepare a list of *B. frutescens* phytochemicals (Table 26.1) (Boltan et al. 2008). The 3D or 2D structure of phytochemicals and their respective reported inhibitors of target proteins were retrieved from NCBI PubChem in .sdf format. Open Babel

Table 26.1 Phytochemicals present in the *Bulbine frutescens*

S. No.	Phytochemical	References
1	Isofuranonaphthoquinone	Tambama et al. (2014) and Mukanganyama et al. (2011)
2	Jozikniphololones A, Joziknipholones B	Bringmann et al. (2008)
3	Gaboroquinones A, Gaboroquinones B, 4′-*O*-Demethylknipholone-4′-*O*-β-D-glucopyranoside, Knipholone, Bulbine-knipholone, 4′-Demethylknipholone, 3-Methyl-1,8-bis[(trimethylsilyl) oxy]anthra-9,10-quinone, Isoknipholone, Bulbineloneside D	Abegaz et al. (2002)
4	Chrysophanol	van Staden and Drewes (1994)
5	Chrysophanol Dimethyl Ether, Chrysophanol-9-anthrone, 1,8-Dihydroxy-3-methyl-anthraquinone, Chrysophanol Glucuronide, Pulmatin, Knipholone anthrone, Chrysothrone, Knipholone 6′-methyl ether	Abdissa et al. (2014)

molecule format converter (O'Boyle et al. 2011) performed conversion of 2D to 3D conformation, Marvin Sketch 15.9.7.0 performed conversion from .sdf to .pdb (for docking) and .mol (for molecular properties prediction) file. Ligands energy was minimized by applying mmff94 force field and conjugate gradients optimization algorithm using PyRx-Python prescription 0.8 for 200 steps (Dallakyan 2010).

26.2.2 Retrieval of Receptor

The 3D structure of various cancerous signaling proteins such as β-catenin (1LUJ), PI3K (3APF), TGF-β (1PY5), NF-KB (1NFK), SMO (4JKV), MEK1 (3PP1) and JAK2 (3Q32) were downloaded from RCSB-protein data bank (PDB) (Berman et al. 2006). The resolutions of the retrieved structure were between 2.1 and 2.8 Å.

26.2.3 Receptor Preparation

The different structures of the various cancer signaling proteins were loaded into UCSF Chimera for molecular docking preparation (Pettersen et al. 2004). Protein models were cleaned and optimized by removing ligands as well other heteroatoms such as water. After that energy minimization of protein structures was performed by steepest descent method has 100 steps (step size 0.02 Å), and conjugate gradient method has ten steps (step size 0.02 Å) by using UCSF Chimera.

26.2.4 Molecular Drug Properties Predictions

For docking experiments, the proteins and the ligands were loaded into Auto Dock Tools 1.5.6 (ADT) (Trott and Olson 2010). Gestgeiger partial charges assigned after merging non polar hydrogen and torsions applied to the ligands by rotating all rotatable bonds. Docking calculations carried out on the protein models. Polar hydrogen atoms, Koll man charges, and solvation parameters were added with the aid of Auto Dock tools. Auto Dock 4.2 offers the option of three search algorithms to explore the space of active binding with different efficacy. We used the Lamarckian genetic algorithm (LGA) in this study. The grid box includes the entire binding site of the particular proteins and provides enough space for the ligands translational and rotational walk. For each of the 30 independent runs, a maximum number of 27,000 GA operations generated on a single population of 150 individuals. Operator weights for the rate of crossover, rate of gene mutation, and elitism was defaulting parameters (0.80, 0.02, and 1, respectively). Thereafter, LigPlot$^+$ (v.1.4.5) and UCSF Chimera (v.1.10.2) used for visualization of the interaction pattern in the protein–ligand complex (Laskowski and Swindells 2011).

26.2.5 ADME, Toxicity and Drug Likeness Property Prediction

All ligand molecules examined for their ADME profile such as Blood Brain Barrier (BBB) penetration, CACO2 permeability, Human Intestinal Absorption (HIA), Madin-Darby Canine Kidney cells (MDCK), Pgp inhibition and Plasma Binding Protein (PPB) (http://molsoft.com/mprop/), (http://www.cbligand.org/BBB/predictor.php). Toxicity profile such as mutagenicity (Ames test), Carcino Rat and hERG Inhibition were further examined using PreADMET server (Lee et al. 2003). Drug likeness properties such as CMC like rule (CLR) and rule of five (RF).

26.3 Results

26.3.1 Molecular Drug Properties

Total 21 phytochemicals were selected for the present study based on their suitability for Lipinski's rule of five. The Lipinski's rule of five distinguishes between the drug like and non-drug like molecules. These phytochemicals include Isofuranonapthoquinone, Jozikniphololone A, Joziknipholone B, Chrysophanol, Chrysothrone, Chrysophanol Dimethyl Ether, 4′- Demethylknipholone, Pulmatin, Knipholone, Gaboroquinone B, AC1NSTKE, Palmidin C, Bulbineloneside D, Gaboroquinone A, Knipholone Anthrone, Knipholone 6′-methyl ether, Bulbine-knipholone, Isoknipholone, Chrysophanol Glucuronide, 1,8-Dihydroxy-3-methyl-anthraquinone, 4′-Demethylknipholone 2′-β-D-glucopyranoside. The drug-likeness (Lipinski's rule of Five), mutagenicity (Ames test), Human Intestinal Absorption (HIA) and BBB penetration prediction of seven inhibitors such as Dehydroxymethylepoxyquinomicin, XL147, Sulumetinib, Ruxolitinib, FH535, Erismodegib, S4317 and 21 *B. frutescens* phytochemicals presented in the Tables 26.2 and 26.3 Ames test is a method to test mutagenicity of a compound as suggested by Dr. Ames. Twenty one phytochemicals predicted as Mutagen along with six standard inhibitors. Fourteen ligands are predicted as BBB^- (unable to cross the Blood Brain Barrier) and six ligands as BBB^+ (able to cross the Blood Brain Barrier). DAPT and Dehydroxymethylepoxyquinomicin predicted as BBB^- (unable to cross the Blood Brain Barrier). Rest of the standard inhibitor predicted as BBB^+ (able to cross the Blood Brain Barrier).

26.3.2 Binding Mode Analysis of Ligands

Interaction of *B. frutescens* phytochemicals with different signaling proteins involved in cancer pathophysiology was studied by using docking technique. Tested

Table 26.2 ADME properties of the ligands and inhibitors

S.N.	Standard	PubChem ID	$BBB^{+/-}$	CACO2	HIA	MDCK	Pgp I	PPB
ADME properties of the standard inhibitor								
1	XL147	1893730	BBB^{+}	1.9768	95.65104	0.04825	N	100
2	FH535	3463933	BBB^{+}	0.44076	96.88135	0.82480	I	100
3	Dehydroxymethylepoxyquinomicin	9881652	BBB^{-}	18.8991	80.39258	0.72509	N	46.57691
4	Sulumetinib	10127622	BBB^{+}	20.6749	94.4464	0.03349	N	86.03106
5	S4317	16079008	BBB^{+}	11.6061	91.4681	0.43038	N	88.64964
6	Erismodegib	24775005	BBB^{+}	27.4794	96.55341	0.04877	I	91.8343
7	Ruxolitinib	25126798	BBB^{+}	13.7647	92.38298	34.3729	N	84.27869
ADME properties of the *B. frutescens* phytoconstituents								
1	Isofuranonapthoquinone	–	BBB^{+}	20.9882	76.64688	4.63348	N	82.34574
2	Jozknipholone A	–	BBB^{-}	19.0586	80.6972	0.04341	I	100
3	Jozknipholone B	–	BBB^{-}	18.7948	81.9672	0.04341	I	100
4	Chrysophanol	10208	BBB^{+}	16.3361	93.74017	43.0544	N	100
5	Chrysothrone	68111	BBB^{+}	16.8755	92.21604	52.9248	N	94.27973
6	Chrysophanol dimethyl ether	189763	BBB^{+}	27.4432	97.44726	38.4694	I	94.07198
7	4′- Demethylknipholone	438991	BBB^{-}	17.1424	76.78456	1.065	I	100
8	Pulmatin	442731	BBB^{-}	14.8656	78.17364	0.79829	N	68.45597
9	Knipholone	442753	BBB^{-}	18.551	86.32806	0.46471	I	93.65815
10	Gaboroquinone B	636652	BBB^{+}	19.3247	75.08605	4.06663	N	89.45054
11	ACONSTKE	5315852	BBB^{-}	15.4312	53.51182	0.04500	N	90.81391
12	Palmidin C	5320386	BBB^{+}	17.3142	88.34972	59.8017	I	100
13	Bulbineloneside D	10008440	BBB^{-}	14.2695	21.16991	0.06095	N	83.07959
14	Gaboroquinone A	10072444	BBB^{+}	18.6484	75.08921	4.06663	N	90.07599
15	Knipholone Anthrone	10093576	BBB^{-}	15.9181	87.61283	0.19259	I	95.36165
16	Knipholone 6′-methyl ether	11190093	BBB^{-}	18.3378	92.10849	0.24011	I	90.58104

(continued)

Table 26.2 (continued)

S.N.	Standard	PubChem ID	$BBB^{+/-}$	CACO2	HIA	MDCK	Pgp I	PPB
17	Bulbine-knipholone	11729754	BBB^{-}	17.4956	86.32846	0.46471	I	93.82751
18	Isoknipholone	12178518	BBB^{-}	17.4969	86.33084	0.46471	I	94.24969
19	Chrysophanol glucuronide	25033784	BBB^{-}	13.6239	69.25678	0.08650	N	72.79434
20	1,8-Dihydroxy-3-methyl-anthraquinone	91932460	BBB^{-}	16.0663	87.90467	35.4713	N	100
21	4′-Demethylknipholone 2′-β-D-glucopyranoside	102378267	BBB^{-}	13.9643	21.17161	0.04468	N	83.41376

MDCK Madin-Darby Canine Kidney cells; *PgpI* P-Glycoprotein inhibition, *BBB* Blood Brain Barrier, *BBB*$^{+}$ Penetrable to Blood Brain Barrier, *BBB*$^{-}$ Not penetrable to Blood Brain Barrier, *I* Inhibitor, *N* Non inhibitor

Table 26.3 Drug likeness and toxicity properties of the ligands and inhibitors

S. N.	Standard	PubChem ID	Drug likeness properties		Toxicity properties		
			CLR	RF	AT	CR	hERG I
Drug likeness and toxicity properties of the standard inhibitor							
1	XL147	1893730	Q	S	NM	N	A
2	FH535	3463933	Q	S	M	P	MR
3	Dehydroxymethylepoxyquinomicin	9881652	NQ	S	M	N	MR
4	Sulumetinib	10127622	Q	S	M	P	MR
5	S4317	16079008	Q	S	M	P	MR
6	Erismodegib	24775005	NQ	S	M	N	MR
7	Ruxolitinib	25126798	Q	S	M	N	MR
Drug likeness and toxicity properties of the *B. frutescens* phytoconstituents							
1	Isofuranonapthoquinone	–	Q	S	M	P	LR
2	Jozikniphölone A	–	NQ	V	M	N	HR
3	Joziknipholone B	–	NQ	V	M	N	HR
4	Chrysophanol	10208	Q	S	M	P	MR
5	Chrysothrone	68111	Q	S	M	N	MR
6	Chrysophanol Dimethyl Ether	189763	Q	S	M	P	MR
7	4′- Demethylknipholone	438991	Q	S	M	N	MR
8	Pulmatin	442731	Q	S	M	N	HR
9	Knipholone	442753	Q	S	M	N	MR
10	Gaboroquinone B	636652	Q	S	M	N	HR
11	AC1NSTKE	5315852	NQ	S	M	N	HR
12	Palmidin C	5320386	NQ	S	M	P	MR
13	Bulbineloneside D	10008440	NQ	V	M	N	HR
14	Gaboroquinone A	10072444	Q	S	M	N	HR
15	Knipholone Anthrone	10093576	Q	S	M	N	MR
16	Knipholone 6′-methyl ether	11190093	Q	S	M	P	MR
17	Bulbine-knipholone	11729751	Q	S	M	N	MR
18	Isoknipholone	12178518	Q	S	M	N	MR
19	Chrysophanol Glucuronide	25033784	Q	S	M	N	HR
20	1,8-Dihydroxy-3-methyl-anthraquinone	91932460	Q	S	M	P	MR
21	4′-Demethylknipholone2’-β-D-glucopyranoside	102378267	NQ	V	M	N	HR

CLR CMC like rule, *RF* Rule of five, *AT* Ames test, *hERGI* hERG inhibition, *CR* Carcino Rat, *HR* High risk, *LR* Low risk, *MR* Medium risk, *NM* Non mutagen, *M* Mutagen, *NQ* Not qualified, *Q* Qualified, *F* Failed, *A* Ambiguous, *S* Suitable, *V* Violated, *N* Negative, *P* Positive

ligands having lowest binding energy against proteins presented in Figs. 26.1 and 26.2. Amino acid residues of targeted proteins involved in the hydrogen bonding and hydrophobic interaction are summarized in Table 26.4. Comparative study for the lowest binding energy of all ligands indicated that out of 23 phytochemicals, most of

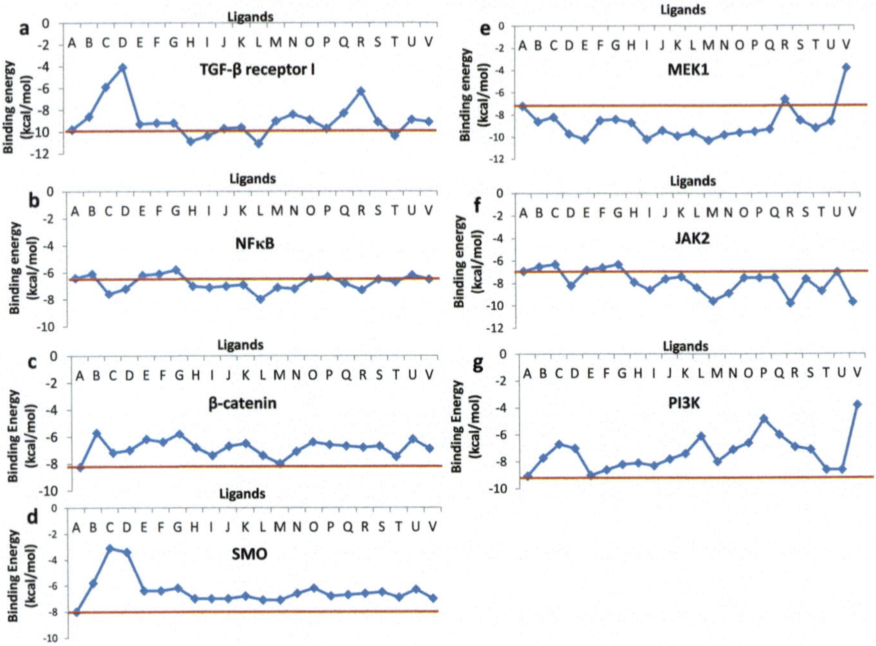

Fig. 26.1 Representation of binding energy score of the ligand-protein interaction (**a–h**). Ligand A represents standard inhibitor of the respective protein while "B" to "V" represents *B. frutescens* phytochemicals. The standard inhibitors (A of 'a' to 'g' figure,), S4317_SIGMA, Dehydroxymethylepoxyquinomicin, FH535, Erismodegib, Sulumetinib, Ruxolitinib, and XL147 were used against TGF-β receptor 1, NF-kB, β-catenin, SMO, MEK1, JAK2, and PI3K proteins respectively. B=Isofuranonapthoquinone; C=Jozikniphology A; D=Joziknipholone B; E=Chrysophanol; F=Chrysothrone; G=Chrysophanol Dimethyl Ether; H=4′-Demethylknipholone; I=Pulmatin; J=Knipholone; K=Gaboroquinone B; L=AC1NSTKE; M=Palmidin C; N=Bulbineloneside D; O=Gaboroquinone A; P=Knipholone Anthrone; Q=Knipholone 6′-methyl ether; R=Bulbine-Knipholone; S=Isoknipholone; T=Chrysophanol Glucuronide; U=1,8-Dihydroxy-3-methyl-anthraquinone; V=4′-Demethylknipholone 2′-β-D-glucopyranoside

the molecules have higher affinity for different proteins as compare to standard inhibitors. *B. frutescens* phytochemicals (Bulbine-knipholone and 4-′-Demethylknipholone 2′-β-D-glucopyranoside) showed excellent binding with Mitogen activated protein kinase kinase 1 (Mek1) than the standard inhibitor. However, 15 *B. frutescens* phytochemicals (Joziknipholone A, Joziknipholone B, 4′- Demethylknipholone, Pulmatin, Knipholone, Gaboroquinone B, AC1NSTKE, Palmidin C, Bulbineloneside D, Gaboroquinone A, Knipholone 6′-methyl ether, Bulbine-knipholone, Isoknipholone, Chrysophanol Glucuronide, 4-′-Demethylknipholone 2′-β-D-glucopyranoside) showed excellent binding with NFκB than the standard inhibitor.

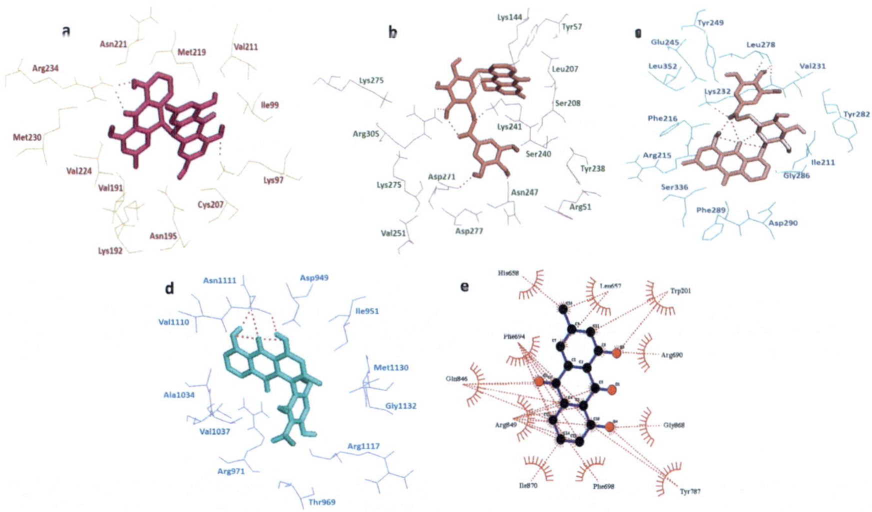

Fig. 26.2 Schematic representation of interaction between lead *B. frutescens* phytochemicals and various signaling proteins. (**a**) MEK1 and Palmidin C (**b**) NF**K**B and AC1NSTKE (**c**) TGF-β Receptor I and AC1NSTKE (**d**) JAK2 and Bulbine-knipholone (**e**) PI3K and chrysophanol

26.3.3 Interaction of B. frutescens *Phytochemicals with CYPs*

Interaction (as inhibitor or substrate) of *B. frutescens* phytoconstituents with several xenobiotic metabolizing enzymes such as CYP2D6, CYP2C19, CYP2C9 and CYP3A4 was studied and results are depicted in Fig. 26.3.

Results showed that all the *B. frutescens* phytoconstituents are non-inhibitor and non-substrate of CYP2D6 enzyme, and act as substrate for CYP2D6 enzyme. However, all the *B. frutescens* phytoconstituents act as an inhibitor of CYP2C19 and CYP2C9 enzyme. Except Pulmatin all the *B. frutescens* phytoconstituents showed inhibition potential against CYP3A4 enzyme. Isofuranonapthoquinone, Chrysophanol, Palmidin C, Isoknipholone and 4′-Demethylknipholone 2′-β-D-glucopyranoside showing as a non-substrate, Joziknipholone A, Chrysophanol Dimethyl Ether, Gaboroquinone B, Gaboroquinone A and Chrysophanol Glucuronide sowing as a substrate and Joziknipholone B, Chrysothrone, 4′-Demethylknipholone, Pulmatin, Knipholone, AC1NSTKE, Bulbineloneside D, Knipholone Anthrone, Knipholone 6′-methyl ether, Bulbine-knipholone and 1,8-Dihydroxy-3-methyl-anthraquinone showed weak substrate potential for CYP3A4 enzyme.

Table 26.4 Amino acid residues of targeted signaling protein involved in the hydrogen bonding and hydrophobic interaction with *B. frutescens* phytochemicals

Phytochemicals and protein	Residue involved in H-bonding	Residue involved in hydrophobic bonding
Isofuranonapthoquinone		
MEK1	M146, Q153	L197, M143, A95, L74, E149, V182, K97
Joziknipholone A		
NF-kB	D239, H141, K241, R305, Q306	S205, A243, P243, F307, K373, Y57
MEK1	K192, V224, K97, S194, S150	D152, K156, R227, M230, S228, Y229, N78, F223, D190, G77, D208, A76, Q153
Jozikniphololone B		
NF-kB	R54, K272	R305, F307, D239, H141, K241, Y57, E60, C59
JAK2	P1114, A1034, D949, M1130	V1037, H950, I951, K1005, E1006, D1004, A1131, V1033, S1115, R1113
MEK1	K192, Y229, N78, L97, G80	R227, N195, D152, G77, A76, S194, D208, I99, D190, S222, M230, F223, V224
Chrysophanol (10208)		
PI3K	–	H658, L657, Q846, R849, I870, G868, F698, F694, Y787, R690, W201
MEK1	M146	L197, V82, G77, A76, Q153, L74, E144, G149, A95
Chrysothrone (68111)		
MEK1	–	A76, K97, G77, V82, D208, M143, L197, M146, A95, L74
Chrysophanol Dimethyl Ether (189763)		
MEK1	S150	L74, L197, M146, M143, A95, K97, V82, D208, D152, S194, Q153, G149
4′- Demethylknipholone (438991)		
TGF-β	I211, H283, D351, E245	G286, G212, K213, V219, K232, L278, S280, L260, A350, A230, L340, D281
NF-kB	R305, S246	Q306, F307, K241, D271, N247, K272,
JAK2	D1118, R1113, D949, E1006	P1095, H950, N1111, S1115, M1130, V1033, I951
MEK1	S194, N195, S150, D152	D208, K97, D190, G80, G77, A76, G153, M143, V182, L197
Pulmatin (442731)		
TGF-β	D351, E245, W249, L278,	V279, A230, L260, S280, K232, V231, V219, I211, L340, G212, K213, K337, A350
NF-kB	L207, D3206, S208, D239, H141	A242, K144, Y57, R54, Q60, K241, P243, N244
JAK2	V1110, R1113, P1114, N1111, D949, S1115	V1033, A1034, M1130, N1129, R1117, E1097, P1095, D1118, H950
MEK1	S194, N195, D152, S150	K97, D190, D208, G77, G80, A76, Q153, L197, V182, M143

(continued)

Table 26.4 (continued)

Phytochemicals and protein	Residue involved in H-bonding	Residue involved in hydrophobic bonding
Knipholone (442753)		
NF-kB	R305	Q306, F307, K241, S246, N247, K272, D271,
JAK2	D1118, R1113, D949	M1130, S1115, V1033, E1006, I951, N1111, H950, P1095
MEK1	N195, S194, K97, D152, S150	G80, N78, G77, A76, L197, Q153, K192, M143, D190, V82, D208,
Gaboroquinone B (636652)		
NF-kB	R305, R54, S246	F307, Q306, K241, S240, N247, K272, D271
JAK2	D1118, R1113, D949	I961, P1095, M1130, V1033, E1006, N1111, H960, S1116
MEK1	K97, N195, S194, S150, D152	V82, G80, N78, G77, D190, D208, A76, Q153, L197, M143,
AC1NSTKE (5315852)		
TGF-β	L278, K232	S280, Y249, E245, V279, V231, A230, L340, W282, I211, D290, S287, K337, K335, N338, R215, G214, D351, V219, L260
NF-kB	D271, N247, K241, R305	K272, S246, S240, Y57, D239, H141, S208, P243, R54
SMO	V378, Y399, Q380, N396,	G304, R398, K204, L303, K395, Q380, Y399
JAK2	V1110, R1113, D949, N1111, M1130	S1115, E1006, S1032, A1034, V1033, H950
MEK1	D245, H188, D208, N78, N195	R109, M219, M230, D190, Y229, F223, R234, V224, K97, G77, I99, G210, F189
Palmidin C (5320386)		
β-Catenin	D390, N426, R386, N387,	S425, G422, D459, P463, E462
NF-kB	K282	Q306, R305, F307, K272
SMO	Y399, LK204	R398, A379, V378, G304, N396, K395, Q380,
JAK2	N1111, R1113, M1130, E1006	V1110, D949, H950, V1033, S1115
MEK1	R234, K97	R189, L215, I216, M219, D208, F209, G80, N78, V224, F223, D190, M230
Bulbineloneside D (10008440)		
NF-kB	D239, K241, H141, K145	Y57, C59, T143, K144, S208, L207
JAK2	E1097, D1118, S1115, F1116, R1117	P1095, R1113, N1111, A1034, V1110, E1006, V1033, E965, N1129
MEK1	K97, N195, F209	G210, D208, M219, G79, G80, A76, K192, G77, N78, D190, R189

(continued)

Table 26.4 (continued)

Phytochemicals and protein	Residue involved in H-bonding	Residue involved in hydrophobic bonding
Gaboroquinone A (10072444)		
NF-kB	R305, S246, K272, K241,	Q306, F307, D271, S240,
JAK2	D1118, R1113, A1034, V1033, E1006, D949	P1114, P1095, H950, M1130, S1115, G968, S1032
MEK1	K97, G80, V211	I99, D208, N195, G77, G79, N78, D190, F223, M219, L215, I2016, G210, I141, F209
Knipholone Anthrone (10093576)		
JAK2	–	M1130, I951, H950, E1006, V1110, N1111, A1034, V1033, F1116, E965, R1117, S1115
MEK1	S150, S194, N195	D152, D208, A76, D190, N78, K192, G77, G80, V82, K97, M43, L197, Q153
Knipholone 6′-methyl ether (11190093)		
NF-kB	R305, K272	S246, D271, N247, K241, F307, Q306
JAK2	–	R1117, E965, F1116, H950, V1033, A1034, V1110, E1006, N1111, R1113, I951, M1130, S1115,
MEK1	S150, D152, S194, N195, K97	G77, N78, G80, A76, D208, K192, D190, Q153, L197, M143, V82
Bulbine-knipholone (11729754)		
NF-kB	R305, S246	Q306, F307, K241, D271, K272, N247,
JAK2	N1111	H950, S1115, F1116, G968, V1033, I951, S1032, E1006, A1034, V1110
Isoknipholone (12178518)		
NF-kB	R305, K272, S246, K241	Q306, D271, F307, S240
JAK2	V1110	M1130, S1115, I951, H950, N1111, A1034, R1113, D949, V1033, F1116, E965
MEK1	V211, K97	I216, L215, M219, I99, N78, D190, G80, G77, G79, I41, N195, D208, F209, G210
Chrysophanol glucuronide (25033784)		
TGF-β	E245, W249, D351, K337,	S280, L278, V279, V231, K232, A230, V219, I211, L340, K213, G212, A350, L260,
NF-kB	L207, D239, H141, Y57,	S208, P243, K241, E60,
JAK2	R1113, V1110, H950, S1115, D1118	A1034, P1095, M1130, N1129, E1097, R1117, D949, N1111, V1033
MEK1	L74, Q153, K156	G149, L197, S150, E144, A95, M146, K97, V82, S194, D208, D152,
1,8-Dihydroxy-3-methyl-anthraquinone (91932460)		
JAK2	V1033, A1034	S1032, E1006, D949, N1111, H950, S1115, R1113, P1114
MEK1	Q153	G149, L74, M146, H145, V82, A95, G77, A76, G75

(continued)

Table 26.4 (continued)

Phytochemicals and protein	Residue involved in H-bonding	Residue involved in hydrophobic bonding
4′-Demethylknipholone 2′-β-D-glucopyranoside (102378267)		
NF-kB	Y57, H141, D239, L207, K144	S208, T143, P243, K241
JAK2	R1113, P1114, S1115, F1116, R1117, M1130, N1111, V1110	V1033, G968, I951, D949, E1006, A1034

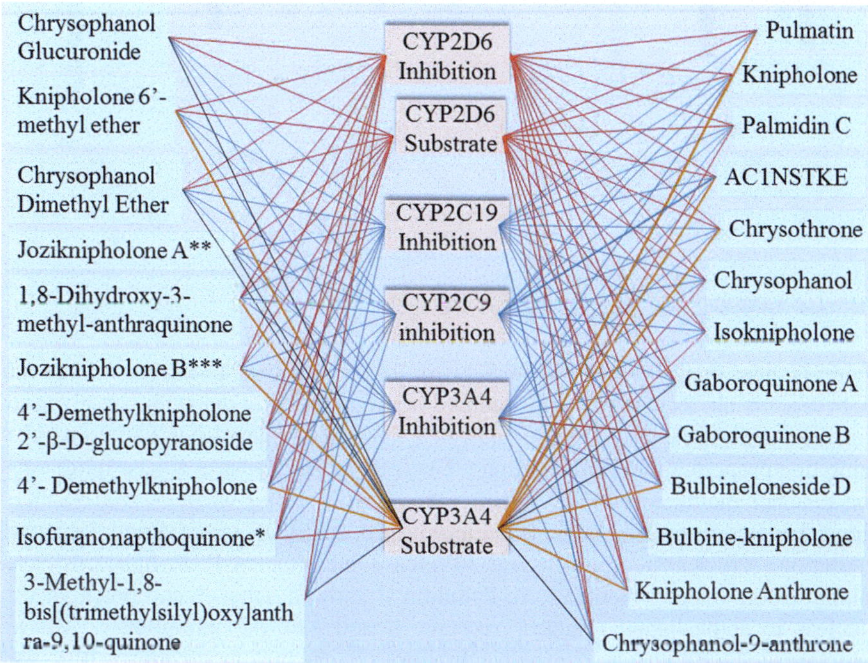

Fig. 26.3 *B. frutescens* phytochemicals having different properties against CYPs. Red line shows non-substrate and non-inhibitor interaction with CYPs. Blue lines show inhibitory interaction with respective CYPs. Phytochemicals associated with brown and black lines show weak substrate and strong substrate of CYP3A4 enzyme respectively

26.3.4 *Bioactivities of* B. frutescens

Bioactivity score of *B. frutescens* phytochemicals were calculated for different target like GPCR (G-protein coupled receptor), ion channel modulator, kinase inhibitor, nuclear receptor ligand, protease inhibitor and enzymes inhibitor. A molecule having bioactivity score more than 0.00 is most likely to exhibit considerable biological activities, while values −0.50 to 0.00 are expected to be moderately active and if

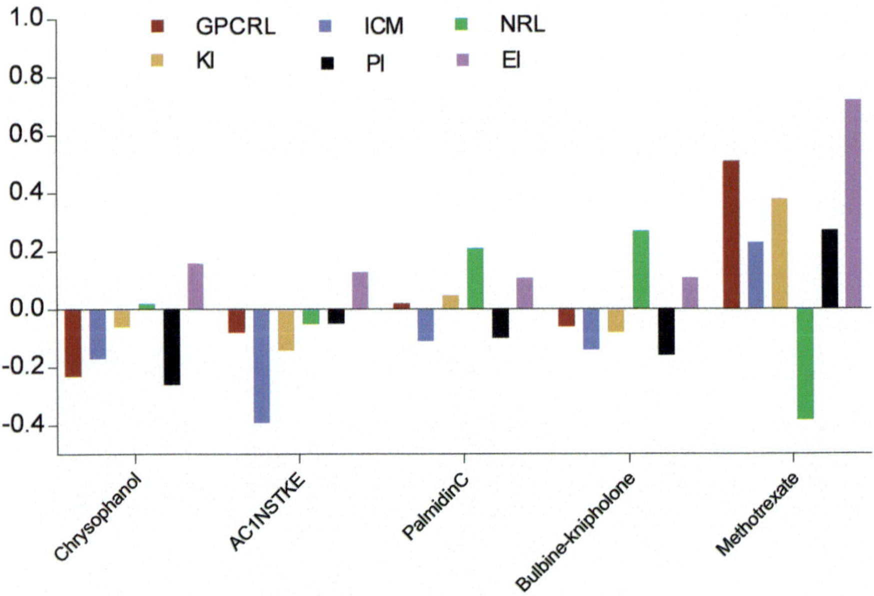

Fig. 26.4 Bioactivities *B. frutescens* lead phytochemicals

score is less than −0.50 it is presumed to be inactive. Bioactivity score of *B. frutescens* phytochemicals and standard inhibitors used in the study are given in Fig. 26.4 and Table 26.5. Result indicated that chrysophanol has better enzyme inhibition property (enzyme inhibition > Nuclear receptor ligand> Kinase inhibitor> Ion channel modulator>GPCR ligand >Protease inhibitor) whereas AC1NSTKE showed better ligand for enzyme inhibition property than other targets (enzyme inhibition> Nuclear receptor ligand >Protease inhibitor >GPCR ligand>Kinase inhibitor>Ion channel modulation). Palmidin C showed better ligand potential for Nuclear receptor ligand (Nuclear receptor ligand>Enzyme inhibition > Kinase inhibitor> GPCR ligand> Protease inhibitor>Ion channel modulation). Bulbine-knipholone showed better nuclear receptor ligand property than other predicted biological activity (Nuclear receptor ligand>enzyme inhibition>Protease inhibitor>Ion channel modulator>Kinase inhibitor>GPCR ligand).

26.4 Discussions

Based on the *B. frutescens* phytochemicals biological activity survey (Table 26.6), we prepared a list of phytochemicals and studied there in-silico anticancer and anti-drug resistant target. The study identified some novel lead anticancer compounds such as Palmidin C and, AC1NSTKE from the *B. frutescens*.

Table 26.5 Bioactivity score of *B. frutescens* phytochemicals and standard inhibitors

Bioactivity score of the compounds								
S N.	Ligand Name	PubChem ID	GPCRL	ICM	KI	NRL	PI	EI
1	Isofuranonapthoquinone	–	−0.28	−0.05	0.12	0.02	−0.32	0.11
2	Jozikniphholone A	–	−3.08	−3.61	3.52	−3.31	−2.56	−3.17
3	Jozikniphholone B	–	−3.08	−3.61	3.52	−3.31	−2.56	−3.17
4	Chrysophanol	10208	−0.23	−0.17	0.06	0.02	−0.26	0.16
5	Chrysothrone	68111	−0.08	−0.04	0.03	0.26	−0.36	0.25
6	Chrysophanol Dimethyl Ether	189763	−0.19	−0.26	0.03	−0.05	−0.23	0.03
7	4′- Demethylknipholone	438991	−0.05	−0.12	0.07	0.29	−0.13	0.11
8	Pulmatin	442731	0.06	−0.04	0.05	0.15	0.06	0.39
9	Knipholone	442753	−0.07	−0.15	0.09	0.25	−0.17	0.07
10	Gaboroquinone B	636652	0.06	0.01	0.01	0.37	−0.01	0.24
11	AC1NSTKE	5315852	−0.08	−0.39	0.14	−0.05	−0.05	0.13
12	Palmidin C	5320386	0.02	−0.11	0.05	0.21	−0.10	0.11
13	Bulbineloneside D	10008440	−0.01	−0.43	0.23	0.05	−0.05	0.10
14	Gaboroquinone A	10072444	0.13	−0.01	0.10	0.45	0.06	0.28
15	Knipholone Anthrone	10093576	0.07	−0.06	0.10	0.42	−0.14	0.15
16	Knipholone 6′-methyl ether	11190093	−0.05	−0.13	0.10	0.26	−0.17	0.10
17	Bulbine-knipholone	11729754	−0.06	−0.14	0.08	0.27	−0.16	0.11
18	Isoknipholone	12178518	0.00	−0.17	0.02	0.34	−0.10	0.11
19	Chrysophanol Glucuronide	25033784	0.08	−0.06	0.10	0.29	0.07	0.40
20	1,8-Dihydroxy-3-methyl-anthraquinone	91932460	−0.09	0.02	0.16	−0.01	−0.32	0.21
21	4′-Demethylknipholone 2′-β-D-glucopyranoside	102378267	0.04	−0.45	0.14	0.12	0.01	0.13

GPCRL G-protein coupled receptor ligand, *ICM* Ion channel modulator, *KI* Kinase inhibitor, *NRL* Nuclear receptor ligand, *PI* Protein inhibitor, *EI* Enzyme inhibitor

Table 26.6 *B. frutescens* phytochemicals and their different activity

Phytochemicals	Activities	References
B. frutescens		
	Anticancer activity of isofuranonaphthoquinone isolated from the plant	Tambama et al. (2014)
	Wound healing property in leaf	Pather et al. (2011) and Pather and Kramer (2012)
	Induced glucose utilization activity	van Huyssteen et al. (2011)
	Glutathione transferase P1-1 inhibition in cancer cells	Mukanganyama et al. (2011)
	Antimalarial and anticancer activity of Joziknipholones A and B isolated from the plant	Bringmann et al. (2008)
	Antiplasmodial and antitrypanosomal activities	Abegaz et al. (2002)
	Scar management	Widgerow et al. (2000)
B. frutescens lead phytochemicals in present study		
Palmidin C	–	–
AC1NSTKE	–	–
Bulbine-knipholone	No study has been reported yet. But on knipholone alone antimalarial, antiplasmodial, anticancer and anti-inflammatory activity has been reported	–
4′-Demethyl knipholone 2′-β-D-glucopyranoside	–	–
Chrysophanol	Anticancer activity	Lu et al. (2017)

Distortion in signaling networks are the fuel factors involved in cancer progression such as change in tumor microenvironment, angiogenesis and inflammation. Our study explores the potential of *B. frutescens* phytoconstituents that inhibits various cancer signaling proteins. Multi-targeted strategy would be more fruitful than the single one. *B. frutescens* phytoconstituent, AC1NSTKE inhibit TGF-β Receptor I kinase activity which might inhibit the PI3K mediated downstream signaling. Auto phosphorylation of the TGF-β Receptor I is the key step of the TGF- β signaling initiation. Inactivation of TGF-β ReceptorI will not able to auto phosphorylate themselves which leads to no further activation of PI3K kinases. A study demonstrated that TGF-β Receptor I inhibition by SB-431542 inhibits TGF-beta-induced transcription, gene expression, apoptosis and growth suppression. SB-431542 attenuates the tumor-promoting effects of TGF-beta including TGF-beta-induced EMT, cell motility, migration and invasion and vascular endothelial growth factor secretion in human cancer cell lines (Halder et al. 2005). Apart from it, PI3K can also be gaining their kinase activity by JAK2 or IRS1 mediated pathway. Brevilin A, a novel natural product, inhibits janus kinase activity and blocks STAT3 signaling in cancer cells (Chen et al. 2013). *B. frutescens* also have potential to inhibit both JAK2 and IRS1 mediated activation of PI3K kinases. Bulbine-Knipholone interfere with JAK2 protein residues Asn1111 and might down regulate their activity which results to loosing JAK2 potential to activate the PI3K kinase. Indole-3-carbinol, 3,3′-diindolylmethane and their derivatives are able to inhibit PI3K signaling pathway as well as the downstream NF-κB, which helps explain their ability to inhibit invasion and angiogenesis and the reversal of epithelial-to-mesenchymal transition (EMT) phenotype and drug resistance (Wu et al. 2014). Many growth factors stimulate tyrosine kinase receptor which activates PI3K mediated cellular signaling. Another *B. frutescens* isolated chrysophanol also have capability to directly inhibit the PI3K kinase. Direct inhibition of PI3K by chrysophanol might hamper both RAS and ERK/MAPK mediated downstream signaling (Fig. 26.5). Cancer cells mostly uses RAS-MEK1 pathway for their regular

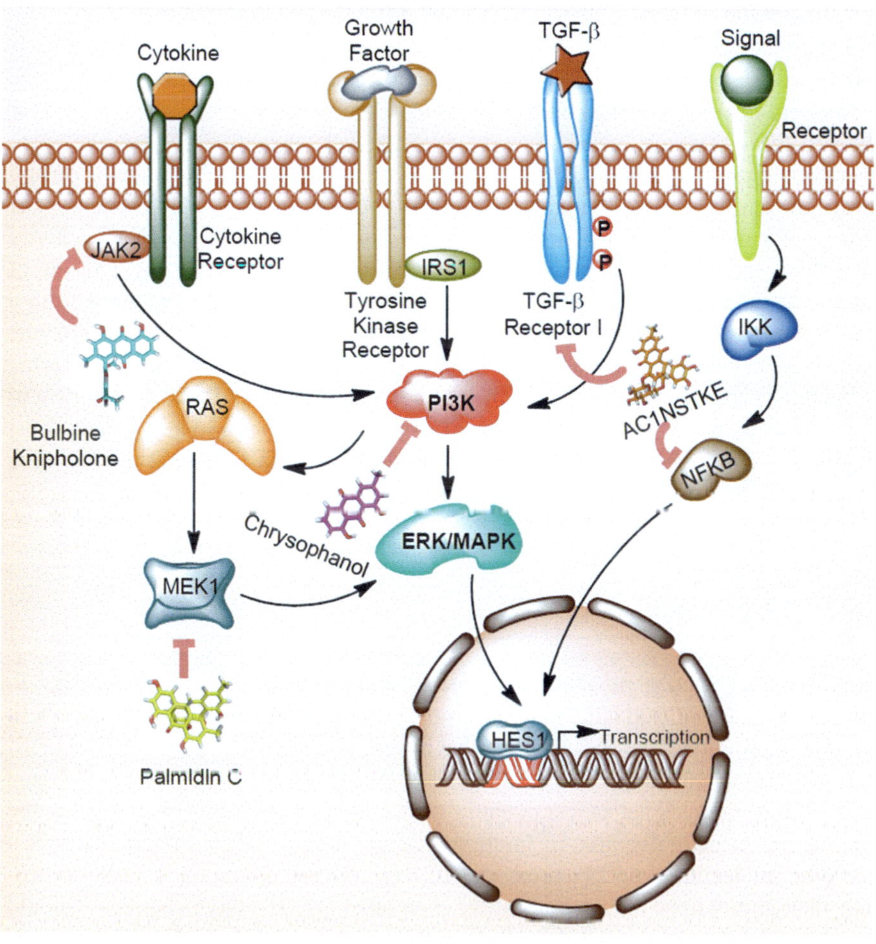

Fig. 26.5 Cancer signaling pathway proteins as target of *B. frutescens* phytochemicals

proliferation. Cancer cells have some capability to modify the RAS proteins. So we hypothesize to inhibit MEK1, the next of the RAS protein. It has been reported that myricetin strongly inhibited MEK1 kinase activity and suppressed TPA- or EGF-induced phosphorylation of extracellular signal-regulated kinase (ERK) or p90 ribosomal S6 kinase, downstream targets of MEK (Lee et al. 2007). *B. frutescens* phytochemical Palmidin C obstruct the MEK1 protein with greater strength (−10.3 kcal/mol) than the standard inhibitor (−7.2 kcal/mol). NFkB signaling protein are mostly overexpressed in cancer cells and thereby leads to increased expression of respective responsive genes. Altogether *B. frutescens* phytoconstituents have potential to inhibit different cancer signaling pathway proteins.

In the present study we reported some novel phytochemicals as anticancer agents. There was no published study on some of the reported phytochemicals (Table 26.6). Thus we studied the drug likeness, ADME/T, xenobiotic metabolizing enzyme interaction, and biological property prediction of the *B. frutescens* phytochemicals. Administration of drug in the body follows different processes such as absorption, distribution, metabolism and elimination or excretion (ADME). Pharmacokinetics is the way to study the association of body and drug once it is administered. It is the measure of the rate (kinetics) of ADME. Drug-likeness rules are set of guidelines for the structural properties of compounds, used for fast calculation of drug-like properties of a molecule. The cytochrome P450 (CYP) enzyme (xenobiotic metabolising enzymes) family plays a dominant role in the biotransformation of a vast number of structurally diverse drugs. There are several factors that influence CYP activity directly or at enzyme regulation level. Many drug interactions are a result of inhibition or induction of CYP enzymes. Inhibition based drug interactions form a major part of clinically significant drug interactions. Isoenzymes of the CYP enzyme system are mainly involved in metabolism of most of the drugs. CYP3A4 isoenzyme is the most predominant isoenzyme in the liver and is involved in the metabolism of approximately 30–40% of drugs (Xu et al. 2005). Results showed that *B. frutescens* phytochemicals depicted suitable pharmacological potentials to being used as drug molecule.

26.5 Conclusion

Multi-targeted strategy to eliminate cancer cells seems to be a good strategy to overcome cancer related ailments. Lead molecule such as Palmidin C, AC1NSTKE, Bulbine-knipholone and chrysophanol have shown potential binding ability against targeted cancer signaling pathway proteins, NFkB, TGF-β receptor I, PI3K and JAK2. Multi-targeted anticancer potential of *B. frutescens* phytochemicals makes the plant as novel natural source of future anticancer drug. Predicted pharmacological and other molecular drug like properties of *B. frutescens* compounds further potentiate their ability to be included in cancer chemotherapy. The phytochemicals should be studied in-vitro and in-vivo for further anticancer potential validation.

Acknowledgements PPK and SK acknowledges financial support from UGC, India and Central University of Punjab Bathinda, Punjab, India for providing financial assistance in the form of UGC-CSIR Senior Research Fellowship and UGC-BSR Research Start-Up-Grant, [No.F.30-372/2017(BSR)] respectively.

Conflict of Interest The authors declare that they have no conflict of interest.

References

Abdissa N, Heydenreich M, Midiwo JO, Ndakala A, Majer Z, Neumann B, Stammler HG, Sewald N, Yenesew A. A xanthone and a phenylanthraquinone from the roots of *Bulbine frutescens*, and the revision of six seco-anthraquinones into xanthones. Phytochem Lett. 2014;9:67–73.

Abegaz BM, Bezabih M, Msuta T, Brun R, Menche D, Muhlbacher J, Bringmann G. Gaboroquinones A and B and 4′-O-demethylknipholone-4′-O-beta-D-glucopyranoside, phenylanthraquinones from the roots of *Bulbine frutescens*. J Nat Prod. 2002;65:1117–21.

Berman H, Henrick K, Nakamura H, Markley JL. The worldwide Protein Data Bank (wwpdb): ensuring a single, uniform archive of PDB data. Nucleic Acids Res. 2006;35:301–3.

Bolton EE, Wang Y, Thiessen PA, Bryant SH. PubChem integrated platform of small molecules and biological activities. Annu Rep Comput Chem. 2008;4:217–41.

Bringmann G, Mutanyatta-Comar J, Maksimenka K, Wanjohi JM, Heydenreich M, Brun R, Müller WE, Peter MG, Midiwo JO, Yenesew A. Joziknipholones A and B: the first dimeric phenylanthraquinones, from the roots of *Bulbine frutescens*. Chemical. 2008;14:1420–9.

Chen X, Du Y, Nan J, Zhang X, Qin X, Wang Y, Hou J, Wang Q, Yang J. Brevilin A, a novel natural product, inhibits janus kinase activity and blocks STAT3 signaling in cancer cells. PLoS One. 2013;8:e63697.

Dallakyan S. PyRx-python prescription v. 0.8. La Jolla: The Scripps Research Institute; 2010.

Duvoix A, Morceau F, Delhalle S, Schmitz M, Schnekenburger M, Galteau MM, Dicato M, Diederich M. Induction of apoptosis by curcumin: mediation by glutathione S-transferase P1-1 inhibition. Biochem Pharmacol. 2003;66:1475–83.

Halder SK, Beauchamp RD, Datta PK. A specific inhibitor of TGF-beta receptor kinase, SB-431542, as a potent antitumor agent for human cancers. Neoplasia. 2005;7:509–21.

Kaliberov SA, Buchsbaum DJ. Cancer treatment with gene therapy and radiation therapy. Adv Cancer Res. 2012;115:221–63.

Kinghorn AD, Chin YW, Swanson SM. Discovery of natural product anticancer agents from biodiverse organisms. Curr Opin Drug Dis Dev. 2009;12:189–96.

Laskowski RA, Swindells MB. LigPlot+: multiple ligand–protein interaction diagrams for drug discovery. J Chem Inf Model. 2011;51:2778–86.

Lee SK, Lee IH, Kim HJ, Chang GS, Chung JE, No KT. The PreADME approach: web-based program for rapid prediction of physico-chemical, drug absorption and drug-like properties. EuroQSAR 2002 Designing Drugs and Crop Protectants: processes, problems and solutions. 2003. p. 418–20.

Lee KW, Kang NJ, Rogozin EA, Kim HG, Cho YY, Bode AM, Lee HJ, Surh YJ, Bowden GT, Dong Z. Myricetin is a novel natural inhibitor of neoplastic cell transformation and MEK1. Carcinogenesis. 2007;28:1918–27.

Lu L, Li K, Mao YH, Qu H, Yao B, Zhong WW, Ma B, Wang ZY. Gold-chrysophanol nanoparticles suppress human prostate cancer progression through inactivating AKT expression and inducing apoptosis and ROS generation in vitro and in vivo. Int J Oncol. 2017;51:1089–103.

Matsui WH. Cancer stem cell signaling pathways. Medicine (Baltimore). 2016;95:S8–S19.

Mouli KC, Vijaya T, Rao SD. Phytoresources as potential therapeutic agents for cancer treatment and prevention. J Global Pharma Tech. 2009;1:4–18.

Mukanganyama S, Bezabih M, Robert M, Ngadjui BT, Kapche GF, Ngandeu F, Abegaz B. The evaluation of novel natural products as inhibitors of human glutathione transferase P1-1. J Enzyme Inhib Med Chem. 2011;26:460–7.

O'Boyle NM, Banck M, James CA, Morley C, Vandermeersch T, Hutchison GR. Open babel: an open chemical toolbox. J Cheminform. 2011;3:33.

Pather N, Kramer B. *Bulbine natalensis* and *Bulbine frutescens* promote cutaneous wound healing. J Ethnopharmacol. 2012;144:523–32.

Pather N, Viljoen AM, Kramer B. A biochemical comparison of the in vivo effects of *Bulbine frutescens* and *Bulbine natalensis* on cutaneous wound healing. J Ethnopharmacol. 2011;133:364–70.

Pettersen EF, Goddard TD, Huang CC, Couch GS, Greenblatt DM, Meng EC, Ferrin TE. UCSF Chimera- a visualization system for exploratory research and analysis. J Comput Chem. 2004;13:1605–12.

Pinar R. To investigate knowledge of Turkish people about cancer. Cancer Agenda. 1998;12:66–73.

Tambama P, Abegaz B, Mukanganyama S. Antiproliferative activity of the isofuranonaphthoquinone isolated from *Bulbine frutescens* against jurkat T cells. Biomed Res Int. 2014;2014:1–14.

Trott O, Olson AJ. AutoDock Vina: improving the speed and accuracy of docking with a new scoring function, efficient optimization, and multithreading. J Comput Chem. 2010;31:455–61.

van Huyssteen M, Milne PJ, Campbell EE, van de Venter M. Antidiabetic and cytotoxicity screening of five medicinal plants used by traditional African health practitioners in the Nelson Mandela Metropole, South Africa. Afr J Tradit Complement Altern Med. 2011;8:150–08.

van Staden LF, Drewes SE. Knipholone from *Bulbine latifolia* and *Bulbine frutescens*. Phytochemistry. 1994;35:685–6.

Widgerow AD, Chait LA, Stals R, Stals PJ. New innovations in scar management. Aesthet Plast Surg. 2000;24:227–34.

Wu T, Chen C, Li F, Chen Z, Xu Y, Xiao B, Tao Z. 3,3′-Diindolylmethane inhibits the invasion and metastasis of nasopharyngeal carcinoma cells *in vitro* and *in vivo* by regulation of epithelial mesenchymal transition. Exp Ther Med. 2014;7:1635–8.

Xu C, Li CYT, Kong ANT. Induction of phase I, II and III drug metabolism/transport by xenobiotics. Arch Pharm Res. 2005;28:249.

Chapter 27
PtRFdb: Plant tRNA-Derived Fragments Database

Shafaque Zahra and Shailesh Kumar

27.1 Introduction

The emergence of fast, accurate, and cost-effective high throughput Next Generation Sequencing (NGS) technology has significantly paced up the exploration of the immense repertoire of small non-coding RNA molecules (Vickers et al. 2015). The rich plethora of small RNA molecules with diverse biological activities are produced in varied organisms, in various tissues, under different conditions, and during varied stages of development. Amongst them, microRNAs (miRNAs) and small interfering RNAs (siRNAs) have been extensively studied and proved to play a significant role in different aspects of gene expression together with gene regulation. Transfer RNA-derived fragments or tRFs with length ranging from 15 to 28 nucleotides (nt), being similar to miRNAs, both structurally and functionally, has been detected in diverse species ranging from the most primitive Archaebacteria to the most evolved human beings (Keam and Hutvagner 2015). It is reported that tRFs are present in abundance and stand next to miRNAs in the small RNA pool.

Apart from the canonical role of as adapter molecules of tRNAs during protein translation, it has been revealed that tRNAs are also shown to be involved in the regulation of cellular functionalities and metabolism (Orioli 2017). This has paved way for exploring new dimensions in tRNA biology and their implications in cellular physiology and disease. With the augmentation of the omics era, decreased cost of sequencing, and abundance of sequenced transcriptomic data, an insight into tRNA dynamics has revealed that tRNAs are also involved in generation of a novel class of small non-coding RNAs by undergoing endonucleolytic cleavage at specific positions, called tRNA-derived fragments or tRFs (Lee et al. 2009). These novel cleavage products with length ranging from 15 to 28 nucleotides have been detected in both prokaryotes and eukaryotes (Sablok et al. 2017). Some specific tRFs

S. Zahra · S. Kumar (✉)
National Institute of Plant Genome Research (NIPGR), New Delhi, India
e-mail: shailesh@nipgr.ac.in

S. Kumar, C. Egbuna (eds.), *Phytochemistry: An in-silico and in-vitro Update*,
https://doi.org/10.1007/978-981-13-6920-9_27

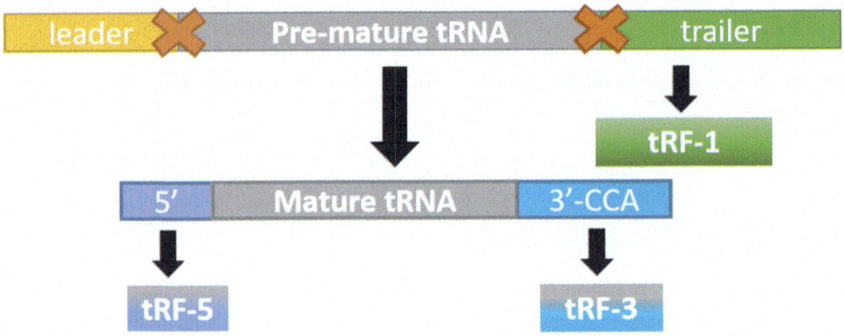

Fig. 27.1 Cleavage of pre-mature and mature tRNA for tRFs generation

overexpression in different human has unveiled their role in cancer progression and altering cellular dynamics in other organisms as well under different stress conditions (Sun et al. 2018). Although well studied in humans and other organisms, some reports in plants also confirm that tRFs are associated with Argonaute (AGO) proteins and involved in gene expression regulation under various abiotic as well as biotic stresses (Loss-Morais et al. 2013). Being functionally similar to microRNAs, recently the study of tRFs has gained momentum in plants as some tRFs have been proven to be differentially over-expressed during abiotic and biotic stresses.

In diverse organisms including plants, tRFs are synthesized by enzymatic cleavage of tRNAs in a position-specific fashion and depending on the cleavage position of tRNAs, they are classified into three types: tRF-5, tRF-3, and tRF-1 (Lee et al. 2009). The tRF-5s and tRF-3s are generated from mature tRNA from 5′ and 3′ ends respectively, while tRF-1s are synthesized from 3′-trailer sequences of pre-mature tRNAs (Fig. 27.1), which can be mediated by Dicer-like (DCL) proteins or by some DCL-independent process which is yet to be elucidated. The ribonucleases involved in tRF generation are not established yet but RNS1 is speculated to be responsible for tRF synthesis in the model plant, *Arabidopsis thaliana* (Alves et al. 2017). There is also evidence of the existence of organellar (mitochondrial and plastidial) tRFs from the previous literature but their potential functions have not been revealed yet (Cognat et al. 2017).

It has now been established that tRFs have a differential accumulation in different plants under various abiotic and abiotic stresses. They are believed to be acting as stress-related gene expression regulators. Specific tRFs viz. Ala^{AGC}, Arg^{CCT}, Arg^{TCG}, and Gly^{TCC} were reported to be overexpressed in drought and salt stress, Val^{CAC}, Tyr^{UGU}, Thr^{GUA}, and Ser^{UGA} during heat stress, Asp^{GTC} and Gly^{TCC} during phosphate deficiency, Arg^{CCT} in cold stress, Arg^{TCG} and Tyr^{GTA} during oxidative stress, Ile^{AAT}, Arg^{ACG}, and Ala^{CGC} during pathogen infection (Hsieh et al. 2009; Loss-Morais et al. 2013; Wang et al. 2016a, b; Alves et al. 2017).

tRFs are also been associated with some AGO proteins like AGO1, AGO2, AGO4, and AGO7 during stress conditions (Sablok et al. 2017). It has been discovered that tRFs also interfere with ribosomal proteins and affect translational activity. Apart from translational repression, the tRFs affect genome stability by governing the post-transcriptional activity of retrotransposons (Martinez et al. 2017). Some recent research suggests that tRFs facilitate in the root nodule formation as well as aid arbuscular mycorrhiza growth in leguminous plants (Jin et al. 2018).

For further exploration of the domain of tRFs, some currently available web-based portals are MINTbase (Pliatsika et al. 2016), tRFdb (Kumar et al. 2015), tRex (Thompson et al. 2018) and tRF2Cancer (Zheng et al. 2016). However, very little information is available for plant tRFs with the exception of tRex which is web-portal holding information about tRFs detected in *Arabidopsis* only. This chapter discusses about our recently developed database, 'PtRFdb' (www.nipgr.res.in/PtRFdb) containing complete information of tRFs detected in 10 evolutionarily dissimilar and diversified plant species (Gupta et al. 2018). This database is believed to be highly resourceful for gaining numerous useful information about different tRF types in diverse plants species. PtRFdb will be useful to elucide new pathways of gene expression regulation in plant genomics and better comprehensive understanding of the cross-talks between other small non-coding RNAs and their downstream target molecules.

27.2 Materials

The tRNA genes of the ten plant species viz. *Physcomitrella patens* (Version 1.1), *Brachypodium distachyon* (JGI v1.0 8X), *Populus trichocarpa* (January 2010 Version 2.0), *O. sativa* (v7.0), *Sorghum bicolor* (Version 1.0), *M. truncatula* (March 2009 Version 3.0), *A. thaliana* (TAIR10 February 2011), *Glycine max* (Wm82.a2), *Vitis vinifera* (Grapevine 12X) and *Zea mays* (Version 5b.60) were downloaded from GtRNAdb (Chan and Lowe 2009). Additionally, the reference genomes of the plants were also fetched from their respective genomic portals. Thus, in the FASTA format, the sequences of tRNA genes for each of the considered plant species were extracted as per the strand information. For generating pre-tRNAs, we extracted the sequences 40 nt upstream and downstream at the terminal ends of the mature tRNA genes. As mature RNA contains 'CCA' at the $3'$-end, CCA was added to tRNA sequences obtained from tRNAscan (Lowe and Eddy 1997). By combining pre-mature and mature tRNA sequences, a reference database was created for each plant species using the option 'makeblastdb' script of BLAST (Basic Local Alignment Search Tool) (Altschul et al. 1990). This reference database was utilized for the prediction of three different tRF types (tRF-5s, tRF-3s, and tRF-1s).

27.3 Methods

27.3.1 Data Procurement

Datasets constituting small RNA sequencing reads and small RNA/microRNA data comprising of unique sequences with clonal frequency were downloaded from

NCBI-SRA (https://www.ncbi.nlm.nih.gov/sra) and NCBI-GEO (http://www.ncbi.nlm.nih.gov/geo/) respectively. This data was further processed for the identification of tRFs.

27.3.2 tRFs Identification

The fragments with the length ranging from 15 to 28 nt, with a clonal frequency greater than 9 (>9) were selected and BLASTN was performed against reference database aforementioned. Only those reads were considered that mapped along 100% length to the database for further study. Similarly, raw reads were processed by using stringent filters as described in our published research cited earlier. Reads were also filtered using tDRmapper software (Selitsky and Sethupathy 2015) and only reads with a quality score of >28 were accepted. Further, for eliminating false positive reads, reads with identity equal to 100% and without a gap (0%) were selected. The reads length ranging from 15 to 28 nt were selected for incorporation in PtRFdb.

27.3.3 The Web Interface of PtRFdb

After the collection and compilation of all the information, our PtRFdb database was developed on an Apache Hypertext Transfer Protocol (HTTP) Server together with MySQL at the backend, providing commands for data storage and retrieval into the database. Hypertext Markup Language or HTML, JavaScript, in addition to Hypertext Pre-processor (PHP) for front-end of web interface designing. PHP and PERL languages were used for writing in-built scripts. The home page of PtRFdb is represented in Fig. 27.2.

27.3.4 PtRFdb Features and Tools

PtRFdb holds detailed information related to tRFs identified in different plants. It holds data at two levels- primary and secondary level. At the primary level, basic information pertaining to each tRF entry like tRF type, tRNA name, gene coordinates, plant, tissue, PubMed ID, anticodon of corresponding tRNA and GSM number are provided. At the secondary level, sequence length, mapping position, relevent frequency, publication and sequencing study was incorporated. For ease in the retrieval of information, we have provided user friendly search modules: 'Basic search', 'Advanced search' and 'Browse' (Figs. 27.3, 27.4, and 27.5). For each query, up to a maximum of ten different fields can be displayed. The columns for 'GSM number', 'Sequence', and 'PMID' are again linked with their parallel information related to experimental details, research publications, and sequence details as highlighted in the search result of Fig. 27.4.

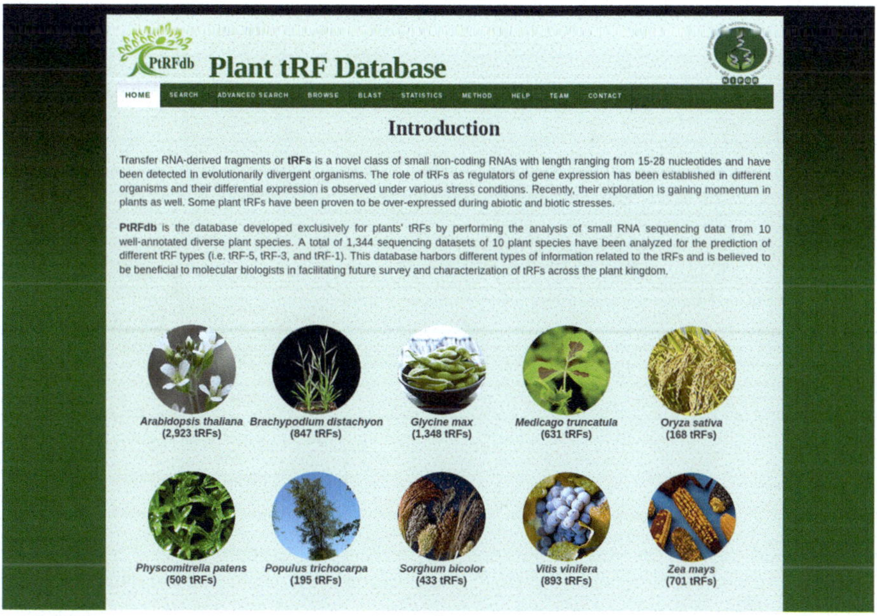

Fig. 27.2 Homepage of PtRFdb (http://www.nipgr.res.in/PtRFdb/)

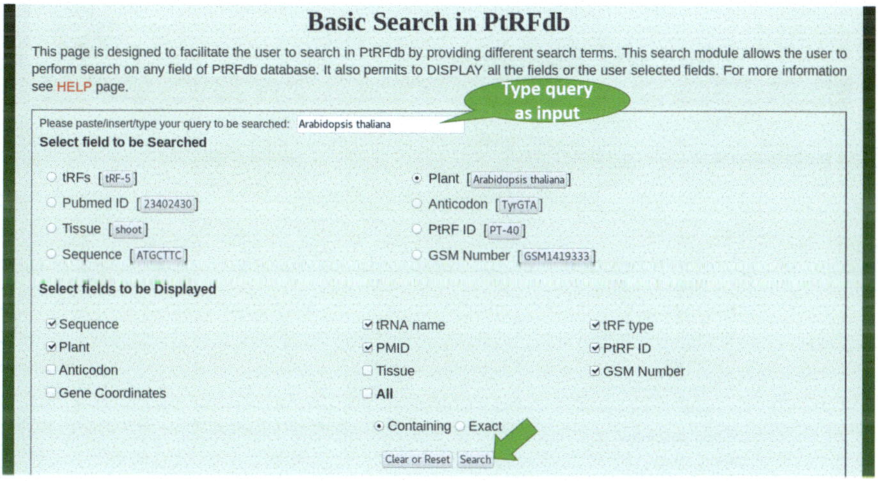

Fig. 27.3 Basic search page of PtRFdb (http://www.nipgr.res.in/PtRFdb/search.php)

Advanced search is supported with conditional and Boolean operators for user-built customized search. The 'Browse' section of PtRFdb facilitates the user to browse in three different manners: with respect to individual plants, by tRF types (i.e. tRF-5, 3, and 1) and also by anticodon type.

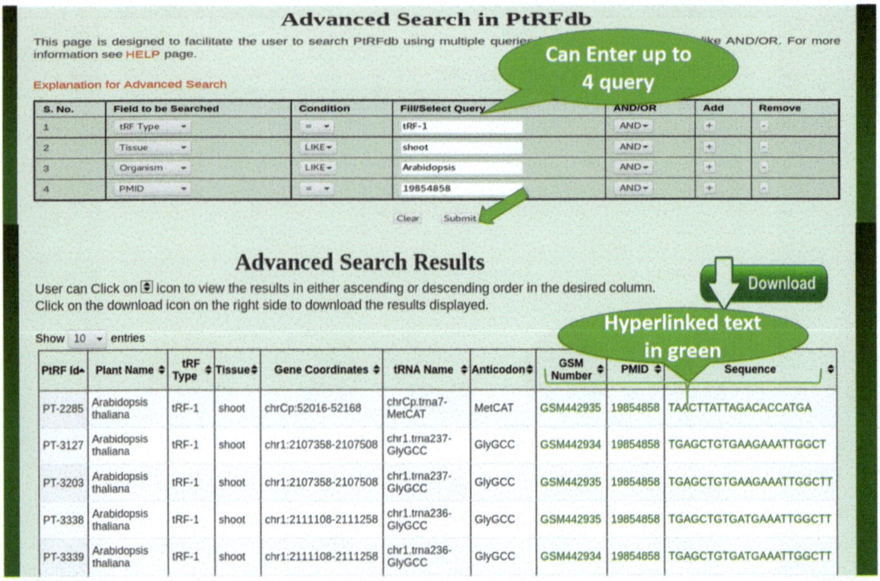

Fig. 27.4 Advanced search and its corresponding result page in PtRFdb (http://www.nipgr.res.in/PtRFdb/cond.php)

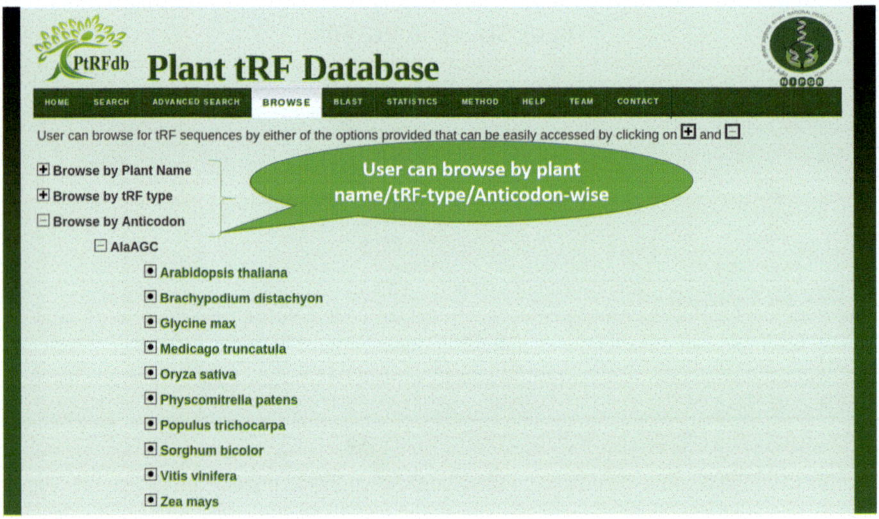

Fig. 27.5 Browse section of PtRFdb (http://www.nipgr.res.in/PtRFdb/browse.php)

In the 'BLAST' page of the PtRFdb (Fig. 27.6), BLASTN of any query nucleotide sequence can be performed against a particular plant species or over entire available datasets which can be selected as per the user's requirement.

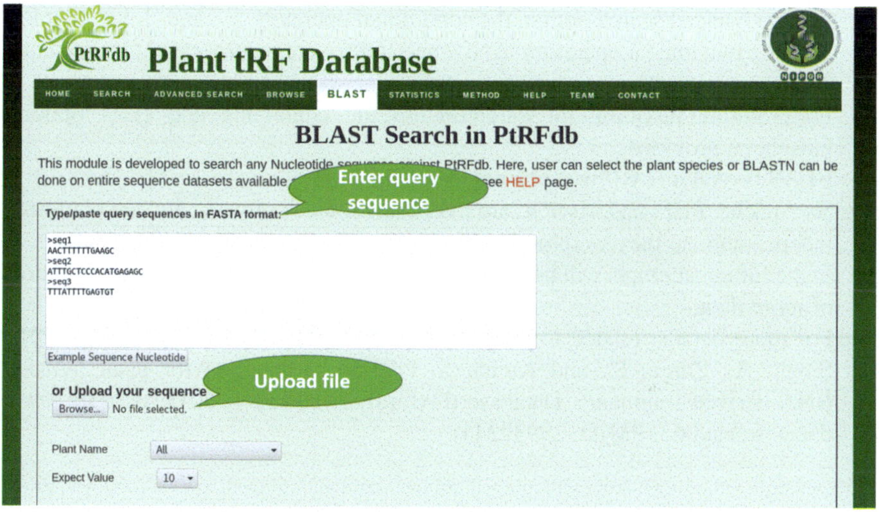

Fig. 27.6 Blast page of PtRFdb (http://www.nipgr.res.in/PtRFdb/blast.php)

To know the significance of the BLAST match, different 'Expect value or E value' ranging from 0.001 to 100 can be selected. The 'Method' section highlights sequencing steps for identification of tRFs in our study right from downloading raw datasets till tRFs prediction. In the 'Statistics' page, graphical and tabular representation of the overall distribution of different tRF types in individual plant species is provided. Lastly, the 'Help' section guides the users for simply understanding and navigating different modules of PtRFdb.

27.4 Notes

1. The total of 1344 sequencing datasets of ten plant species were used for the identification of tRFs.
2. Information associated with all analyzed GEO samples in our study was fetched by using the 'SRAdb' as well as 'GEOmetadb' libraries of the Bioconductor software package (http://www.bioconductor. org) and united for each entry of our database, PtRFdb.
3. For extraction of the mapping coordinates, in house PERL scripts were used.
4. This database holds information about 487,765 entries of tRFs (258,439 tRF-5s, 225,380 tRF-3s and 3946 tRF-1s).
5. The total number of 5607 unique tRFs sequences is incorporated in PtRFdb (2580 tRF-5s, 2269 tRF-3s, 758 tRF-1s).
6. The majority of the tRFs had a length in the range of 18–24 nt and tRF-5 were most abundant of all tRFs.

7. In the advanced search option, the conditional operators '=' and 'Like' coupled with the two logical operators 'AND' and 'OR' were incorporated for user-built customized search.
8. For providing flexibility for search options, the 'containing' and 'exact' options have been provided.
9. BLAST-version 2.6.0 was utilized in the PtRFdb.
10. As Apache, PHP, and MySQL are free, open-source software, and are platform-independent, so they were preferably utilized for our database development.
11. In the future, attempts will be made for updating our database by further addition of more data.
12. For more details, related to PtRFdb, refer to our published paper (Gupta, N., Singh, A., Zahra, S., and Kumar S. PtRFdb: a database for plant transfer RNA-derived fragments. Database (Oxford). 2018 Jan 1; 2018. doi: 10.1093/database/bay063. PMID: 29939244).

References

Altschul SF, Gish W, Miller W, et al. Basic local alignment search tool. J Mol Biol. 1990;215:403–10. https://doi.org/10.1016/S0022-2836(05)80360-2.

Alves CS, Vicentini R, Duarte GT, et al. Genome-wide identification and characterization of tRNA-derived RNA fragments in land plants. Plant Mol Biol. 2017;93:35–48. https://doi.org/10.1007/s11103-016-0545-9.

Chan PP, Lowe TM. GtRNAdb: a database of transfer RNA genes detected in genomic sequence. Nucleic Acids Res. 2009;37:D93–7. https://doi.org/10.1093/nar/gkn787.

Cognat V, Morelle G, Megel C, et al. The nuclear and organellar tRNA-derived RNA fragment population in Arabidopsis thaliana is highly dynamic. Nucleic Acids Res. 2017;45:3460–72. https://doi.org/10.1093/nar/gkw1122.

Gupta N, Singh A, Zahra S, Kumar S. PtRFdb: a database for plant transfer RNA-derived fragments. Database. 2018;2018:63.

Hsieh L-C, Lin S-I, Shih AC-C, et al. Uncovering small RNA-mediated responses to phosphate deficiency in Arabidopsis by deep sequencing. Plant Physiol. 2009;151:2120–32. https://doi.org/10.1104/pp.109.147280.

Jin D, Meng X, Wang Y, et al. Computational investigation of small RNAs in the establishment of root nodules and arbuscular mycorrhiza in leguminous plants. Sci China Life Sci. 2018;61(6):706–17. https://doi.org/10.1007/s11427-017-9203-7.

Keam SP, Hutvagner G. tRNA-derived fragments (tRFs): emerging new roles for an ancient RNA in the regulation of gene expression. Life (Basel, Switzerland). 2015;5:1638–51. https://doi.org/10.3390/life5041638.

Kumar P, Mudunuri SB, Anaya J, Dutta A. tRFdb: a database for transfer RNA fragments. Nucleic Acids Res. 2015;43:D141–5. https://doi.org/10.1093/nar/gku1138.

Lee YS, Shibata Y, Malhotra A, Dutta A. A novel class of small RNAs: tRNA-derived RNA fragments (tRFs). Genes Dev. 2009;23:2639–49. https://doi.org/10.1101/gad.1837609.

Loss-Morais G, Waterhouse PM, Margis R. Description of plant tRNA-derived RNA fragments (tRFs) associated with argonaute and identification of their putative targets. Biol Direct. 2013;8:6. https://doi.org/10.1186/1745-6150-8-6.

Lowe TM, Eddy SR. tRNAscan-SE: a program for improved detection of transfer RNA genes in genomic sequence. Nucleic Acids Res. 1997;25:955–64.

Martinez G, Choudury SG, Slotkin RK. TRNA-derived small RNAs target transposable element transcripts. Nucleic Acids Res. 2017;45:5142–52. https://doi.org/10.1093/nar/gkx103.
Orioli A. tRNA biology in the omics era: stress signalling dynamics and cancer progression. BioEssays. 2017;39:1600158. https://doi.org/10.1002/bies.201600158.
Pliatsika V, Loher P, Telonis AG, Rigoutsos I. MINTbase: a framework for the interactive exploration of mitochondrial and nuclear tRNA fragments. Bioinformatics. 2016;32:2481–9. https://doi.org/10.1093/bioinformatics/btw194.
Sablok G, Yang K, Chen R, Wen X. tRNA derived smallRNAs: smallRNAs repertoire has yet to be decoded in plants. Front Plant Sci. 2017;8:1–4. https://doi.org/10.3389/fpls.2017.01167.
Selitsky SR, Sethupathy P. tDRmapper: challenges and solutions to mapping, naming, and quantifying tRNA-derived RNAs from human small RNA-sequencing data. BMC Bioinforma. 2015;16:354. https://doi.org/10.1186/s12859-015-0800-0.
Sun C, Fu Z, Wang S, et al. Roles of tRNA-derived fragments in human cancers. Cancer Lett. 2018;414:16–25. https://doi.org/10.1016/J.CANLET.2017.10.031.
Thompson A, Zielezinski A, Plewka P, et al. tRex: a web portal for exploration of tRNA-derived fragments in Arabidopsis thaliana. Plant Cell Physiol. 2018;59:e1. https://doi.org/10.1093/pcp/pcx173.
Vickers KC, Roteta LA, Hucheson-Dilks H, et al. Mining diverse small RNA species in the deep transcriptome. Trends Biochem Sci. 2015;40:4–7. https://doi.org/10.1016/J.TIBS.2014.10.009.
Wang Q, Li T, Xu K, et al. The tRNA-derived small RNAs regulate gene expression through triggering sequence-specific degradation of target transcripts in the oomycete pathogen Phytophthora sojae. Front Plant Sci. 2016a;07:1938. https://doi.org/10.3389/fpls.2016.01938.
Wang Y, Li H, Sun Q, Yao Y. Characterization of small RNAs derived from tRNAs, rRNAs and snoRNAs and their response to heat stress in wheat seedlings. PLoS One. 2016b;11:e0150933. https://doi.org/10.1371/journal.pone.0150933.
Zheng L-L, Xu W-L, Liu S, et al. tRF2Cancer: a web server to detect tRNA-derived small RNA fragments (tRFs) and their expression in multiple cancers. Nucleic Acids Res. 2016;44:W185–93. https://doi.org/10.1093/nar/gkw414.

Chapter 28
In-silico Methods of Drug Design: Molecular Simulations and Free Energy Calculations

Fortunatus Chidolue Ezebuo, Prem P. Kushwaha, Atul K. Singh, Shashank Kumar, and Pushpendra Singh

28.1 Introduction

The term in-silico refers to "computer-aided". The term was given in 1989 as equivalence to the Latin terms in-vivo, in-vitro, and in-situ. So in-silico drug design refers to rational design through which drugs are designed or discovered via computational methods (Singh et al. 2017). It is important to note that serendipity played a vital role in the past in discovering novel drugs but the present-day trend in drug discovery has moved from discovering to designing. In-silico drug design approaches can leverage understanding of the biochemistry of diseases, pathways, identification of disease causative proteins to design compounds which are capable of modifying the characters of the proteins (Raut et al. 2015). In-silico drug design strategies include (a) Computer-based systems to get further productive drug discovery and advancement methods. (b) Creation of chemical and biological databases regarding ligands and targets/proteins to distinguish new drugs. (c) Developing in-silico methods to find out pharmacokinetic or drug-likeness characteristics for substances before screening to facilitate early identification of compounds that are going to fail in clinical phases (Barlow et al. 2009; Ferreira et al. 2015; Klabunde and Hessler 2002; Kuntz 1992). Cheminformatics Bioinformatics and are two key disciplines for in-silico drug design method that have an influence on modern drug discovery practice and accelerate drug designing. Bioinformatic techniques can help in drug target identification, validation of drug targets, protein modeling,

F. C. Ezebuo
Department of Biochemistry, College of Natural and Applied Sciences, Chrisland University, Abeokuta, Ogun State, Nigeria

P. P. Kushwaha · A. K. Singh · S. Kumar
Department of Biochemistry and Microbial Sciences, Central University of Punjab, Bathinda, Punjab, India

P. Singh (✉)
Tumor Biology Laboratory, National Institute of Pathology, New Delhi, India

S. Kumar, C. Egbuna (eds.), *Phytochemistry: An in-silico and in-vitro Update*,
https://doi.org/10.1007/978-981-13-6920-9_28

understanding of drug targets including their evolution and phylogeny. Cheminformatic techniques can be exploited for managing storage and maintenance of information relating to chemicals and related features, identification of novel bioactive compounds, optimization of leads, in-silico ADME (Absorption, Distribution, Metabolism and Elimination) forecast and other concerns which assist in reducing the last phase failure of substances (Bleicher et al. 2003; Li 2001; Singh et al. 2017).

28.2 Molecular Simulations

New approaches of in-silico drug designing come in two major groups (Jorgensen 2004; Meng et al. 2011): ligand-based and receptor-based approaches. Ligand-based method comprises QSAR (quantitative structure-activity relationship), various pharmacophore assignment/mapping, database searching or mining, The structure-based methods, that comprises molecular docking and modern molecular simulations (Examples: classical molecular dynamics simulations, QM/MM molecular dynamics simulations etc), require structural information about drug targets which are available from nuclear magnetic resonance (NMR), X-ray crystallography methods, or through protein model building on the basis of homology (Jaworska et al. 2005; Senn and Thiel 2009; Verma et al. 2010). One can obtain drug target (s) of interest through (1) bioinformatics mining of different databases and repositories (Example Protein DataBank, tdr,) (2) comparative homology modeling by using software (Example MODELLER) or online resource (Examples: Swiss-Model, ModBase etc) (Borhani and Shaw 2012; Durrant and McCammon 2011; Eswar et al. 2006; Kitchen et al. 2004). Molecular simulations as it relates to drug design requires one to have compounds library from where potential drugs will be discovered and drug target(s) of interest. There are different flexible approaches to design compound libraries of interest for drug design/discovery (Fig. 28.1).

28.3 Design of Libraries

One may design library of compounds from where a potential drug will be discovered through literature mining and generation of the structures with a molecular graphics software (like ChemOffice, ChemDraw), generate electronic structures of phytochemicals from structural elucidation studies to form a library, perform structural modifications of compounds of interest with molecular graphics software, Collect compounds from different public domains, databases and repositories (Examples: ZINC Database) and filter them according to lipisink rule of fiveto get lead library or according to rule of 3 (R03) proposed by Astex to obtain fragment library. It is of interest to know that lead molecule is commonly characterized as a small molecule which has molecular weight (MW) of nearly 500 Da, and can bind its target via H-bonds with approximately five hydrogen bond donors and ten hydrogen

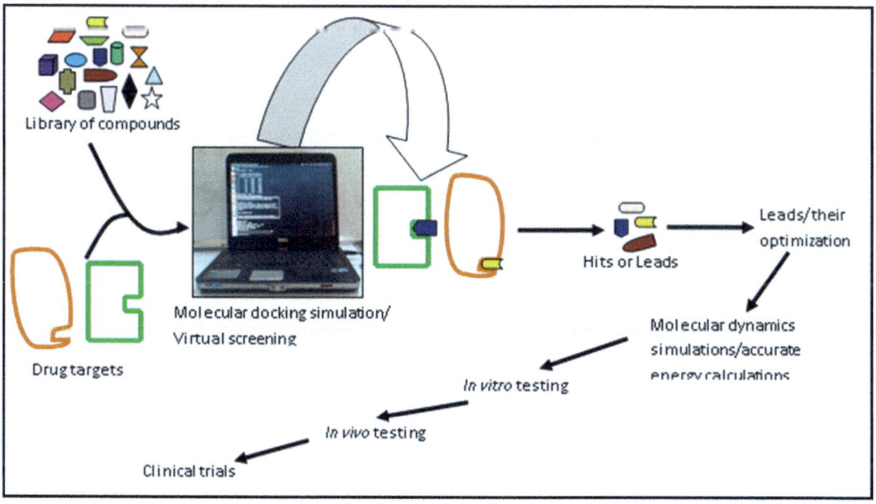

Fig. 28.1 Schematic representation of structure-based drug design method

bond acceptors, is enough with rotatable bonds to permit binding to the target, and satisfactorily lipophilic having partition coefficient (cLogP, a measure of hydrophobicity) not more than five (Buntrock 2002). General fragment libraries, invented for the purpose of screening of a wide range of targets, are different collections of compounds having high pharmacophore heterogeneity or physicochemical characteristics like molecular mass, lipophilicity etc. (Dixon et al. 2006; Guner 2000). The molecules are investigated to find out functional groups which might add to extra chemical reactivity, toxicity, and incorrect positives. Lead library fulfills "rule of five" proposed by Lipinski, that on occasion imposes understanding of the disposition properties (absorption, distribution, metabolism, and excretion, ADME) to get effective inhibitors (Lipinski 2004; Zhang and Wilkinson 2007). Through analogy to the Lipinski's "rule of five", molecules present in libraries accommodated to fragment-based screening follow the following rules (1) molecular weight of $\leq$300 Da; (2) hydrophilicity value of, clogP$\leq$3; (3) quantity of hydrogen bond donors and acceptors $\leq$3; (4) quantity of rotatable bonds $\leq$3; and, (5) to a minor extent, molecular polar surface area of $\leq$60 Å2. Once the library is created, one can perform molecular docking simulations with them after validation of docking protocols (Singh and Bast 2014, 2015b, c).

28.4 Molecular Docking Simulation

Molecular docking simulation is an automatic computer algorithm which finds out how ligands (molecules in library of interest) will attach with their binding site in drug-target (Alonso et al. 2006). This includes determination of the alignment of the ligand, its conformational geometry, and scoring. The scoring can be free energy, binding energy, or can be a qualitative numerical measure. Each docking simulation algorithm keeps the ligand into diverse alignments and conformations at the binding site and calculates a score for each one. Two key segments in molecular docking are absolute pose prediction and accurate binding free energy estimation that can be applied to grade the order of the docking poses. The promising candidates from molecular docking simulation analysis can be processed for molecular dynamics simulations after docking (Alonso et al. 2006; Thomsen and Christensen 2006). Molecular docking methodologies are of great importance in the planning and design of new drugs. Behind the advancement of the first algorithms in the 1980s, molecular docking developed as a vital apparatus for drug discovery (Medina-Franco et al. 2011). It is applied at many stages in drug discovery like estimation of the docked structure of the ligand-receptor complex and also to grade ligands depending on their score. Docking procedures help in explanation of energetically suitable binding pose of ligands with their receptor (Iman et al. 2015). However, they are less accurate than molecular dynamics simulations which are more computationally expensive but more accurate in predicting receptor-ligand interactions. Some molecular docking simulation programs include AutoDockVina®, DOCK, AutoDock, HADDOCK, FlexX, GOLD, and GLIDE among others (Singh and Bast 2014, 2015a, c; Singh et al. 2016; Alonso et al. 2006). Molecular docking simulations are not only applicable in the study of target-ligand interactions (i.e. target-ligand docking) but can also be applied in understanding protein-protein interactions (protein-protein docking).

28.5 Molecular Dynamics Simulation

Molecular dynamics (MD) investigations are the time-based development of coordinates of complex molecular modules as a purpose of time. It has grown as the main method in the collection of ways to create novel bioactive molecules and could aid to logically understand their method of action and advance chemical structures with respect to biological effect. Their foremost benefit is in explicitly handling structural flexibility and entropic effects. This provides an additionally precise approximation of the thermodynamics and kinetics in relation to drug-target recognizing and binding, as improved algorithms and hardware constructions increase their application. Classical MD simulations nowadays permit implementation of structure-based drug design approaches which fully explains structural flexibility of the overall drug-target model arrangement (Durrant and McCammon 2011; Harvey and De Fabritiis 2012) Certainly, now it is publicly acknowledged that two main drug-binding models (induced-fit and conformational selection) have outdated Emil Fischer's rigid lock-and-key binding paradigm (Boehr et al. 2009; Changeux and Edelstein 2011; Vogt and Di Cera 2012). Researchers have lately illustrated the supremacy of these approaches for investigating protein-ligand binding and determining the associated free energy and kinetics. Receptor and ligand flexibility is essential to precisely predict drug binding and detailed kinetic and thermodynamic properties.

In consequence, classical and/or QM/MM MD simulations are no more assumed inhibitory for drug design. Alternatively, this is advancing the boundaries of computationally accelerated drug designing in both industry and academia (Borhani and Shaw 2012; Mortier et al. 2015). Some programs MD simulation includes GROMACS, CHAMM, Amber, NAMD, CPMD, CP2K etc. Their documentation can be found on their websites and other web resources (Phillips et al. 2005; Van Der Spoel et al. 2005).

28.6 Advantages of Molecular Dynamics Simulation

MD simulations are normally conducted at a normal temperature, comparatively low energy barricades, for example, 0.6 kcal can be simply negotiated. Therefore if the opening drug-receptor complex configuration following from binding is detached from the very steady configuration by this low barrier, molecular dynamics can reach over the barrier. Molecular simulations might recognize additional stability, hence are very realistic, conformational positions of ligand-receptor complexes (Mortier et al. 2015). Moreover, they may give unique knowledge about conformational alterations of the receptor because of ligand binding; shedding light on the close mechanisms of receptor inhibition or activation which presently cannot be investigated by some other method. Ultimately, molecular simulations usually integrate solvent, therefore, permit the including of solvent properties in the attention. The current investigation is showing the significance of MD simulation to examine the biomolecular adaptability connected with ligand identification (Nair et al. 2011, 2012; Nair and Miners 2014). Investigating the flexibility of the target receptor allow the enhanced strategy of drugs in comparison with basic lock and key concept of the static receptor.

28.7 Combined Docking and MD Simulations

Quick and reasonable binding rules could be coupled along with precise but additionally time-consuming MD procedures to forecast extra dependable receptor-ligand complexes. The power of these arrangements resides in the corresponding strengths and weaknesses. Although, docking procedures are utilized to discover the large conformational space of ligands in a very short time, permitting the inspection of huge libraries of drug-like compounds at a sensible cost. The key disadvantages are the absence, or reduced flexibility of the proteins, that is not permissible to regulate its conformation when joined with a ligand, and the lack of a distinctive and extensively appropriate scoring function, required to create a dependable grading of the final complexes. So, MD simulations could utilize both (ligand and protein) in a flexible manner, permitting an induced fit of the receptor-binding site nearby the newly presented ligand. Also, the consequence of explicit water molecules could be studied through, and very precise binding free energies could be gained. Though, the main problem with MD simulation is that they take time and that the arrangement could get stuck in local minima (Khandelwal et al. 2005). So, the permutation of the two procedures in a module in which docking is applied for the quick screening of huge libraries and MD simulations are then used to explore conformations of the target, optimization of the structures of the ultimate complexes, and estimate exact energies, is a reasonable strategy to refining the drug-design process (Alonso et al. 2006; Khandelwal et al. 2005). Another approach is to perform short MD simulation of the starting target is to obtain diverse conformations of the target, perform molecular docking simulations of the library against representatives of different conformations before subjecting the energetically favorable conformation(s) into MD simulations. Ligand-dependent approaches lacking the structural data of the target, ligand-based technique makes application of the data delivered by identified inhibitors for the target receptor. Structures similar to the identified inhibitors are recognized from chemical databases through a range of approaches; few of the approaches extensively applied are similarity and substructure searching, pharmacophore matching or 3D shape matching.

28.8 Similarity and Substructure Searching

QSAR (Quantitative structure-activity relationship) is a numerical tactic that tries to conclude the physical and chemical characteristics of molecules to their biological features. The goal of QSAR is the estimate of molecular characteristics based on structure deprived of the necessity to accomplish the experiment using in-vitro or in-vivo. It does not take times and resources. Numerous descriptors like number of rotatable bonds, molecular weight, LogP etc. are usually utilized. Numerous QSAR methods are in exercise depending on the data dimensions. It lies between 1D-QSAR to 6D-QSAR. These approaches are dependent on the postulation that the activity of some chemical compound to its structure (Damale et al. 2014; Lill 2007). Very exactly, this method states that the action, or the characteristics, for example, the toxic effect, is connected to the chemical structure via a definite mathematical algorithm, or rule. It is supposed that the existence of a particular feature in the chemical compound, that is residing or not in the structure. For example, it is widely accepted that if in the chemical compound there are some groups, like an aromatic amine, or an epoxide, then the chemical compound is genotoxic.

28.9 Pharmacophore Mapping

This is the method for originating 3D –pharmacophore. A pharmacophore is a feature composed with their comparative spatial alignment that is considered able of interacting with a specific biological target for example positively and negatively charged groups, donors and acceptors, hydrophobic regions and aromatic rings. A pharmacophore plot differentiates the bioactive conformation of every active molecule and entitles in what way to superimpose, relate in 3D, the different active compounds. The plot distinguishes the types of points match in what confirmation of the target. It depends on element types, chemical connectivity (Debnath 2002). A produced by the superposition of active identifies their general characteristics. Founded on the pharmacophore plot either de novo design or 3D database examining can be done. Structure bioactivity relationships may serve as a subsidiary probe of 3D structure and chemical characteristics of the macromolecular identification of a site for ligands. The purpose of pharmacophore mapping is to convert such bioactivity relations data into 3D for binding to the drug target. This to find out 3D databases molecules which match this 3D plot to create novel active molecules (Liu et al. 2010; Marchand-Geneste et al. 2002).

28.10 Free Energy Calculations

Energy calculations from molecular docking have limited accuracy. Hence, more precise energy predictions can be achieved by employing various strategies of molecular simulations. Methods of free energy calculations can be divided into (a) relative free energy methods (b) relative and absolute binding free energy methods.

28.10.1 Approximate Free Energy Methods

28.10.1.1 Linear-Response Approximation Methods

This method is based on a sampling of end states of target-ligand complex, unliganded target, and ligand. The most simple method of binding free energy estimation is the linear-response calculation where the electrostatic free alteration is predicted on the principle of electrostatic interaction energy among the ligand and the environment around it (drug target or solvent) (Tao et al. 1996). The linear-response approximation methods of energy computation can be classified into (a) linear interaction energy (LIE) method, (b) semi-macroscopic protein-dipoles Langevin-dipoles method, (c) molecular mechanics with Poisson-Boltzmann (or Generalized Born) and surface area solvation (MM-PBSA/MM-GBSA). The foremost contemplation of the LIE technique is only convergent average of interaction energies among the ligand and environment around it, essential to be estimated to get an estimate of binding free energies (de Amorim et al. 2008). The master LIE equation is based on the electrostatic and van der Waals energies for the ligand which are calculated from molecular simulations (molecular dynamics (MD) or Monte Carlo (MC) simulations). Implementation of LIE method requires MD/MC simulations the ligand-free in solution and one solvated ligand bound to the drug target. The most comprehensive endpoint methods currently are molecular mechanics along with Poisson-Boltzmann (or generalized Born) and surface area (MM-PBSA/MM-GBSA).MM-PBSA/MM-GBSA method utilizes a range of solvent model to substitute water by handling it as a universal medium (implicit solvent model).In this method, average solvation characteristics of water are collected deprived of averaging over the interactions of thousands of real water molecules that could cause notable instabilities in solute-solvent and solvent-solvent energies (Carlsson and Åqvist 2006). Ligand-solvent interaction energies could be calculated accurately through a solution of the Poisson–Boltzmann equation, or roughly by using the generalized Born theory. Since the generalized Born model is less computationally-intensive, it is more popular for MD simulations (Chen et al. 2008). Fortunately, a lot of improvements have been applied to generalized Born models, and now they are able of reaching the same level of accurateness as Poisson-Boltzmann models. These methods have been utilized in many settings, like protein design, protein-protein interactions, conformer stability and re-scoring (Brandsdal et al. 2003; Chen et al. 2008).

In the MM-PBSA approach, the free energy of a state, which is, P (free protein), L (free ligand) or PL (complex) is calculated on the basis following sum:

$$G = E\ \text{bond} + E\ \text{el} + E\ \text{vdW} + G\ \text{pol} + G\ \text{npol} - TS,$$

Wherever E bond, E el and E vdW are standard molecular mechanics energy expressions from bonded, electrostatic and van der Waals interactions, G pol and G npol are the polar and non-polar assistances to the salvation free energies. Polar terms are calculated via solving of Poisson-Boltzmann (MM-PBSA) or generalized Born equation (MM-GBSA), while nonpolar terms are obtained from a linear relation to the solvent available surface area (Hou et al. 2010). The procedures involve several huge estimates, for instance, a questionable entropy, missing the conformational input and disappeared effects from binding-site water molecules. Moreover, the approaches often overestimate differences between sets of ligands. However, since MM-PBSA and MM-GBSA invest extra effort in sampling and entropies, they are nearer to a true free energy scheming than docking (Mobley and Dill 2009). Average free energy of unbound ligand (G L), unbound protein (G P) and the complex (G PL) are usually estimated from the separate MD or MC simulations for each of them. This approach is called three-average MM-PBSA (3A-MM-PBSA). Though, it is very common to simulate just the complex (PL) and create the collective average of the unbound receptor and ligand by only eliminating the suitable atoms; such approach is called one-average MM-PBSA (1A-MM-PBSA). In a typical scenario, simulations used to estimate the energy terms employ explicit solvent models, but since implicit solvent models (GBSA/PBSA) are used, later all solvent molecules are deleted from each trajectory snapshot. It was also recommended that MM-PBSA calculations could be dependent only on only reduced structures instead of a wide number of MD/MC-trajectory prints. And in practice, minimized structures often give results comparable with those obtained with MD/MC-simulations. At the same time, the results of such calculations are powerfully reliant on the initial structure and ignore the dynamic effects.

Software implementation of the MM-PBSA and MM-GBSA methods. The MM-PBSA method was initially established for the AMBER software and currently is available for free in the Amber Tools. During the past decade, automatic scripts were also created for popular free simulation packages Desmond, NAMD and GROMACS, as well as for APBS software (Genheden and Ryde 2015).

Genheden and Ryde showed that LIE is two to seven times more effective in comparison MM-PBSA, because of the time-taking entropy approximation (Genheden and Ryde 2011). At the same time, MM-PBSA was shown to have better overall performance than MM-GBSA (Homeyer et al. 2014). Poor precision is one of the main problems of the MM-PBSA and MM-GBSA methods, thus sometimes making them useless upon a comparison of ligands with similar affinities. The problem with the precision is usually resolved via computing just interaction energies, investigating various MD-snapshots as possible and utilizing numerous simulations (Genheden and Ryde 2015). Generally, the accuracy of end point methods (correlation coefficients related with experiments of r 2 = 0.0–0.9, based on the protein) is usually better than for molecular docking, but worse than for alchemical perturbation algorithms (Genheden and Ryde 2015). According to an expert opinion of Genhenden and Ryde, end point methods (particularly, MM-PBSA) might be suitable to advance the outcomes of docking and virtual screening or to know detected affinities and trends. Nevertheless, they are not precise enough for later states of drug design.

28.10.2 Relative and Absolute Binding Free Energy Methods

Alchemical free energy approaches could be utilized to calculate either absolute binding affinities (for any particular ligand to a receptor) or binding affinities (an alteration among two or more interrelated ligands). In top optimization efforts, wherever optimization via minor, the sequential chemical alteration is of main attention, precise relative free energies can conclude wherever alterations have enhanced affinity and selectivity (Chodera and Pande 2011). Thermodynamic integration (TDI) and free energy perturbation (FEP) are normally reliable but are also more time consuming than the endpoint or docking approaches (Homeyer et al. 2014). These techniques are usually called alchemical methods, since as a replacement for simulating the binding/unbinding methods that would need a simulation several times the lifetime of the complex; the ligand is alchemically transmuted in either alternative chemical species or a nonreacting molecule via intermediate, probably nonphysical stages (Chodera et al. 2011).

Alchemical methods

(a) Thermodynamic integration (TDI)
(b) Free energy perturbation (FEP)

TDI gives the benefit that the accuracy of binding free energy forecasts can be improved via consequently comprising extra intermediate states (Michel and Essex

2010), so this strategy proposes the probability to begin calculations at the reduced level of accuracy and just conduct sampling wherever required. It is important due to the opposite relationship among accurateness and essential calculating time, which needs one to discover an optimal equilibrium among estimation value and computational demand (Homeyer et al. 2014). TDI transformations of one ligand in the alternative are generally directed via simulations at distinct steps. The free energy alteration ΔG for the transformation is calculated by integrating the average potential functions of the two states at each n step. To conclude the variation in the binding free energy $\Delta\Delta G$ among two ligands, transformations are implemented for both the complex-bound ligands and the solvated ligands. $\Delta\Delta G$ is computed as the variation among the individual free energies: $\Delta\Delta G = \Delta G$ bound – ΔG solvated (Homeyer et al. 2014).

FEP simulations entrenched in statistical mechanics yield a path to take missing effects in the calculations, e.g., conformational sampling, explicit solvent, and the shift of protonation states on the binding, nonetheless they usually need widespread computational sources and skill. FEP methodology has been recognized for more than 20 years currently, but its influence on drug discovery is being recognized (Acevedo et al. 2012). The main obstacle for implementing FEP as a daily procedure in CADD is gaining dependable ΔG approximations for complex bimolecular systems contained by a rational computational time. FEP approach utilizes the classic Zwanzig expression to describe the free energy alteration via building a nonphysical path combining the desired early and last state of a system. On behalf of comparative free energies of binding, single or double topology perturbations could be prepared to change one.

Acknowledgments We would like to thank Central University of Punjab, Bathinda, Punjab, (India) and Director in-charge, National Institute of Pathology, New Delhi (India) for supporting this study with infrastructural requirements. This study was also supported by a Centenary-Post Doctoral Research Fellowship Grant-in-Aid from the Indian Council of Medical Research (ICMR), Government of India awarded to PS.

Conflict of Interest The authors declare that no financial or commercial conflict of interest.

References

Acevedo O, Ambrose Z, T Flaherty P, Aamer H, Jain P, V Sambasivarao S. Identification of HIV inhibitors guided by free energy perturbation calculations. Curr Pharm Des. 2012;18 (9):1199–216.

Alonso H, Bliznyuk AA, Gready JE. Combining docking and molecular dynamic simulations in drug design. Med Res Rev. 2006;26(5):531–68.

Barlow DH, Nock M, Hersen M. Single case experimental designs: strategies for studying behavior for change. Boston: Pearson/Allyn and Bacon; 2009.

Bleicher KH, Böhm H-J, Müller K, Alanine AI. A guide to drug discovery: hit and lead generation: beyond high-throughput screening. Nat Rev Drug Discov. 2003;2(5):369.

Borhani DW, Shaw DE. The future of molecular dynamics simulations in drug discovery. J Comput Aided Mol Des. 2012;26(1):15–26.

Boehr DD, Nussinov R, Wright PE. The role of dynamic conformational ensembles in biomolecular recognition. Nat Chem Biol. 2009;5(11):789–96.

Brandsdal BO, Österberg F, Almlöf M, Feierberg I, Luzhkov VB, Åqvist J. Free energy calculations and ligand binding. Adv Protein Chem. 2003;66:123–58.. Elsevier

Buntrock RE. ChemOffice ultra 7.0. J Chem Inf Comput Sci. 2002;42(6):1505–6.

Carlsson J, Åqvist J. Calculations of solute and solvent entropies from molecular dynamics simulations. Phys Chem Chem Phys. 2006;8(46):5385–95.

Changeux JP, Edelstein S. Conformational selection or induced fit? 50 years of debate resolved. F1000 Biol Rep. 2011;3:1–15.

Chen J, Brooks CL III, Khandogin J. Recent advances in implicit solvent-based methods for biomolecular simulations. Curr Opin Struct Biol. 2008;18(2):140–8.

Chodera JD, Mobley DL, Shirts MR, Dixon RW, Branson K, Pande VS. Alchemical free energy methods for drug discovery: progress and challenges. Curr Opin Struct Biol. 2011;21 (2):150–60.

Chodera JD, Pande VS. The social network (of protein conformations). Proc Nat Acad Sci. 2011;108(32):12969–70.

de Amorim HL, Caceres RA, Netz PA. Linear interaction energy (LIE) method in lead discovery and optimization. Curr Drug Targets. 2008;9(12):1100–5.

Damale MG, Harke SN, Kalam Khan FA, Shinde DB, Sangshetti JN. Recent advances in multidimensional QSAR (4D-6D): a critical review. Mini Rev Med Chem. 2014;14(1):35–55.

Debnath AK. Pharmacophore mapping of a series of 2, 4-diamino-5-deazapteridine inhibitors of Mycobacterium avium complex dihydrofolate reductase. J Med Chem. 2002;45(1):41–53.

Dixon SL, Smondyrev AM, Knoll EH, Rao SN, Shaw DE, Friesner RA. PHASE: a new engine for pharmacophore perception, 3D QSAR model development, and 3D database screening: 1. Methodology and preliminary results. J Comput Aided Mol Des. 2006;20(10–11):647–71.

Durrant JD, McCammon JA. Molecular dynamics simulations and drug discovery. BMC Biol. 2011;9(1):71.

Eswar N, Webb B, Marti-Renom MA, Madhusudhan M, Eramian D, Shen MY, et al. Comparative protein structure modeling using Modeller. Curr Protocol Bioinforma. 2006;15(1):5.6. 1–5.6. 30.

Ferreira L, dos Santos R, Oliva G, Andricopulo AJM. Molecular docking and structure-based drug design strategies. Molecules. 2015;20(7):13384–421.

Genheden S, Ryde U. A comparison of different initialization protocols to obtain statistically independent molecular dynamics simulations. J Comput Chem. 2011;32(2):187–95.

Genheden S, Ryde U. The MM/PBSA and MM/GBSA methods to estimate ligand-binding affinities. Expert Opin Drug Discov. 2015;10(5):449–61.

Guner OF. Pharmacophore perception, development, and use in drug design. La Jolla: International University Line; 2000. p. 29.

Harvey MJ, De Fabritiis G. High-throughput molecular dynamics: the powerful new tool for drug discovery. Drug Discov Today. 2012;17(19–20):1059–62.

Homeyer N, Stoll F, Hillisch A, Gohlke H. Binding free energy calculations for lead optimization: assessment of their accuracy in an industrial drug design context. J Chem Theory Comput. 2014;10(8):3331–44.

Hou T, Wang J, Li Y, Wang W. Assessing the performance of the MM/PBSA and MM/GBSA methods. 1. The accuracy of binding free energy calculations based on molecular dynamics simulations. J Chem Inf Model. 2010;51(1):69–82.

Iman M, Saadabadi A, Davood A. Molecular docking analysis and molecular dynamics simulation study of ameltolide analogous as a sodium channel blocker. Turk J Chem. 2015;39(2):306–16.

Jaworska J, Nikolova-Jeliazkova N, Aldenberg T. QSAR applicability domain estimation by projection of the training set descriptor space: a review. Altern Lab Anim. 2005;33(5):445.

Jorgensen WL. The many roles of computation in drug discovery. Science. 2004;303 (5665):1813–8.

Khandelwal A, Lukacova V, Comez D, Kroll DM, Raha S, Balaz S. A combination of docking, QM/MM methods, and MD simulation for binding affinity estimation of metalloprotein ligands. J Med Chem. 2005;48(17):5437–47.

Kitchen DB, Decornez H, Furr JR, Bajorath J. Docking and scoring in virtual screening for drug discovery: methods and applications. Nat Rev Drug Discov. 2004;3(11):935.

Klabunde T, Hessler G. Drug design strategies for targeting G-protein-coupled receptors. Chembiochem. 2002;3(10):928–44.

Kuntz ID. Structure-based strategies for drug design and discovery. Science. 1992;257 (5073):1078–82.

Li AP. Screening for human ADME/Tox drug properties in drug discovery. Drug Discov Today. 2001;6(7):357–66.

Lill MA. Multi-dimensional QSAR in drug discovery. Drug Discov Today. 2007;12 (23–24):1013–7.

Lipinski CA. Lead-and drug-like compounds: the rule-of-five revolution. Drug Discov Today Technol. 2004;1(4):337–41.

Liu X, Ouyang S, Yu B, Liu Y, Huang K, Gong J, et al. PharmMapper server: a web server for potential drug target identification using pharmacophore mapping approach. Nucleic Acids Res. 2010;38(suppl_2):W609–14.

Marchand-Geneste N, Watson KA, Alsberg BK, King RD. New approach to pharmacophore mapping and QSAR analysis using inductive logic programming. Application to thermolysin inhibitors and glycogen phosphorylase B inhibitors. J Med Chem. 2002;45(2):399–409.

Medina-Franco JL, López-Vallejo F, Kuck D, Lyko F. Natural products as DNA methyltransferase inhibitors: a computer-aided discovery approach. Mol Divers. 2011;15(2):293–304.

Meng X-Y, Zhang H-X, Mezei M, Cui M. Molecular docking: a powerful approach for structure-based drug discovery. Curr Comput Aided Drug Des. 2011;7(2):146–57.

Michel J, Essex JW. Prediction of protein–ligand binding affinity by free energy simulations: assumptions, pitfalls and expectations. J Comput Aided Mol Des. 2010;24(8):639–58.

Mobley DL, Dill KA. Binding of small-molecule ligands to proteins:"what you see" is not always "what you get". Structure. 2009;17(4):489–98.

Mortier A, Faria JP, Correia CM, Santerne A, Santos NC. BGLS: a Bayesian formalism for the generalised Lomb-Scargle periodogram. Astron Astrophys. 2015;573:1–6.

Nair PC, Malde AK, Drinkwater N, Mark AE. Missing fragments: detecting cooperative binding in fragment-based drug design. ACS Med Chem Lett. 2012;3(4):322–6.

Nair PC, Malde AK, Mark AE. Using theory to reconcile experiment: the structural and thermodynamic basis of ligand recognition by phenylethanolamine N-methyltransferase (PNMT). J Chem Theory Comput. 2011;7(5):1458–68.

Nair PC, Miners JO. Molecular dynamics simulations: from structure function relationships to drug discovery. In Silico Pharmacol. 2014;2(1):1–4.

Phillips JC, Braun R, Wang W, Gumbart J, Tajkhorshid E, Villa E, et al. Scalable molecular dynamics with NAMD. J Comput Chem. 2005;26(16):1781–802.

Raut R, Beesetti H, Tyagi P, Khanna I, Jain SK, Jeankumar VU, et al. A small molecule inhibitor of dengue virus type 2 protease inhibits the replication of all four dengue virus serotypes in cell culture. Virol J. 2015;12(1):16.

Senn HM, Thiel W. QM/MM methods for biomolecular systems. Angew Chem Int Ed Engl. 2009;48(7):1198–229.

Singh P, Bast F. Multitargeted molecular docking study of plant-derived natural products on phosphoinositide-3 kinase pathway components. Med Chem Res. 2014;23(4):1690–700.

Singh P, Bast F. High-throughput virtual screening, identification and in vitro biological evaluation of novel inhibitors of signal transducer and activator of transcription 3. Med Chem Res. 2015a;24(6):2694–708.

Singh P, Bast F. Screening and biological evaluation of myricetin as a multiple target inhibitor insulin, epidermal growth factor, and androgen receptor; *in silico* and in vitro. Investig New Drugs. 2015b;33(3):575–93.

Singh P, Bast F. Screening of multi-targeted natural compounds for receptor tyrosine kinases inhibitors and biological evaluation on cancer cell lines, *in silico* and in vitro. Med Oncol. 2015c;32(9):233.

Singh P, Bast F, Singh R. Natural compounds targeting transforming growth factor-β: *in silico* and in vitro study. Elect J Biol. 2016;13:6–13.

Singh P, Kumar S, Bast F. Natural compounds are smart players in context to anticancer potential of receptor tyrosine kinases: an *in silico* and in vitro advancement. In: Wei DQ, Ma Y, Cho W, Xu Q, Zhou F, editors. Translational bioinformatics and its application. Dordrecht: Springer; 2017. p. 177–202.

Tao H, Chu ZT, Warshel A. Quantitative studies of ligand-receptor interactions: a rapid evaluation of binding free energies of endothiapepsin to its inhibitors. Pac Symp Biocomput. 1996:752–5.

Thomsen R, Christensen MH. MolDock: a new technique for high-accuracy molecular docking. J Med Chem. 2006;49(11):3315–21.

Van Der Spoel D, Lindahl E, Hess B, Groenhof G, Mark AE, Berendsen HJ. GROMACS: fast, flexible, and free. J Comput Chem. 2005;26(16):1701–18.

Vogt AD, Di Cera E. Conformational selection or induced fit? A critical appraisal of the kinetic mechanism. Biochemistry. 2012;51(30):5894–902.

Verma J, Khedkar VM, Coutinho EC. 3D-QSAR in drug design-a review. Curr Top Med Chem. 2010;10(1):95–115.

Zhang M-Q, Wilkinson B. Drug discovery beyond the 'rule-of-five'. Curr Opin Biotechnol. 2007;18(6):478–88.

Chapter 29
Study of Plant Exclusive Virus-Derived Small Interfering RNAs

Ajeet Singh and Shailesh Kumar

29.1 Introduction

The emergence of high-throughput, fast, and cost-effective next-generation sequencing (NGS) technology has facilitated the study of small non-coding ribonucleic acids (RNAs) in eukaryotes and their role in RNA silencing mechanisms as a defense response during pathogen infection. Among different causal agents of infection, virus-mediated infections have a tremendous impact on the physiological system, nutritional value, and yield of crop plants (Diener 1963; Bos 1982). Thus, it becomes important to focus on the underlying anti-viral defense mechanisms in plants. Some important natural anti-viral defenses exploit small RNAs in combating the infection in host plants (Hamilton and Baulcombe 1999), preferably termed as virus-induced gene silencing (VIGS). VIGS also assists in the process of chromatin modification, translation process and thus a potent mediator for gene expression regulation bestowing the overall resistance in host plants against the viral defense. This have gained considerable attention in recent years by plant researchers and small interfering RNAs (siRNAs) being an integral component of VIGS have been extensively investigated and studied by plant scientists (Velásquez et al. 2009; Zhu and Guo 2012). Virus induced infection leads to the production of small non-coding RNA molecules in plants and other diverse eukaryotes as well. This may result in either acquirement of anti-viral immunity or pathogenesis in few cases (Ding and Lu 2011). Major portion of these generates small RNA pool comprises of small interfering RNAs with the length ranging from 21–24 nucleotides (nt); bearing unphosphorylated overhangs of 2 nt at 3′-end. They are considered to be the probable gene expression regulators and component of anti-defence machinery in the host plant (Guo et al. 2016).

In plants, virus-derived siRNAs (vsiRNAs) can be generated from either ribonucleic acid (RNA) or deoxyribonucleic acid (DNA) viruses (Szittya et al.

A. Singh · S. Kumar (✉)
National Institute of Plant Genome Research (NIPGR), New Delhi, India
e-mail: shailesh@nipgr.ac.in

S. Kumar, C. Egbuna (eds.), *Phytochemistry: An in-silico and in-vitro Update*,
https://doi.org/10.1007/978-981-13-6920-9_29

2010). These vsiRNAs can be produced by processing of hairpin-shaped single-stranded RNA (folded structure) or double-stranded RNA genome (for RNA viruses). In the case of DNA viruses, they can be generated from replicative intermediates produced from double-stranded DNA genome in the host cells (Moissiard and Voinnet 2006; Donaire et al. 2009). In plants, the biogenesis of small interfering RNAs takes place with the help of the processing enzyme 'Dicer-like enzymes' (DCLs) (Chapman and Carrington 2007; Chen 2010). Specifically the homologs viz. DCL-2, 3 and 4 are participating directly in the vsiRNAs production while DCL-1 are indirectly involved in the biogenesis of plant vsiRNAs (Zhu and Guo 2012). These DCL homologs contribute in multifarious ways for vsiRNAs production while maintaining mutual balance and coordination with each other. There are two categories of vsiRNAs: primary and secondary vsiRNAs. The primary vsiRNAs are generated by the direct action of DCLs. The association of Argonaute proteins with vsiRNAs leads to the formation of the RNA-induced silencing complex (RISC) (Malpica-Opez et al. 2018). The RISC complemented further with plant RNA-dependent RNA polymerases (RDRs) attacks viral genome. During the initial phase, the viral genome is disintegrated into small fragments of dsRNA by the action of DCLs and further, RDRs convert these primary vsiRNAs into the highly active secondary vsiRNAs during the secondary amplification phase (Vazquez and Hohn 2013).

vsiRNAs assembles with RISC in a sequence-specific fashion and pair with its homologous complementary viral RNA or DNA genomic transcript strand and thus aid in silencing expression of the viral genome and in this way, they impart anti-viral resistance in the host plant (Szittya et al. 2010; Zhang et al. 2015). Although this pathway of anti-viral defense is vaguely explored, they are supposed to regulate cellular activities epigenetically by mediating DNA methylation in gene promoters (Rodríguez-Negrete et al. 2009). In addition to their crucial role in transcriptional gene silencing (TGS) and post-transcriptional gene silencing (PTGS), artificially synthesized siRNAs can also be very useful for gene knockout and pathways studies associated with gene silencing during varied stress conditions in plants (Guo et al. 2016). The rapid development of Next Generation Sequencing (NGS) technology has been heavily exploited for studying viral genomics, viral ecological studies, virus-host interactions, and evolution of viruses utilizing RNA interference technology with aid of vsiRNAs (Stobbe and Roossinck 2014; Skums et al. 2015). The hairpin construct approach for designing artificial vsiRNAs can be expressed in plant cells and used for targeting against the specific pathogen (Mansoor et al. 2006; Shimizu et al. 2012). By designing multiple hairpin constructs with different viral sources, transgenic plants have been developed which are resistant to a number of viruses (Prins et al. 1995; Bucher et al. 2006). Thus RNA interference (RNAi) technology has proved to be a boon in horticulture and agriculture for developing plants immune to pathogenic viruses (Duan et al. 2012). The biogenesis and mode of action of vsiRNAs are illustrated in Fig. 29.1.

At present, many databases pertaining to siRNAs and virus-derived siRNAs are available. However, these databases are largely focussed on human diseases caused by viruses. For e.g., HIVsirDB (Tyagi et al. 2011), VIRsiRNAdb (Thakur et al. 2012) and siRNAdb (Chalk et al. 2004). Nevertheless, the wide impact of plant vsiRNAs on the physiology of plants cannot be ignored. There is a need of knowledge base dedicated

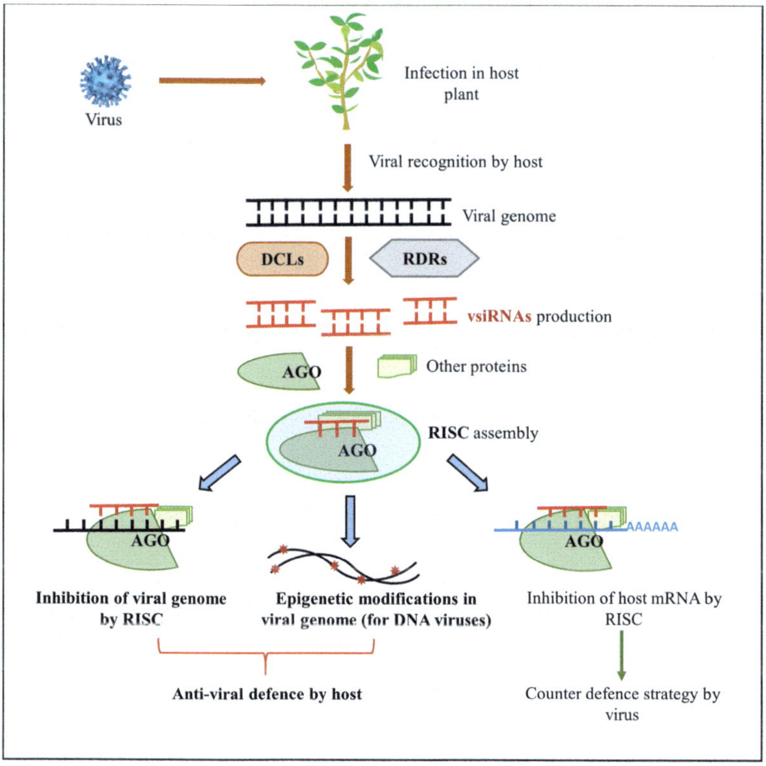

Fig. 29.1 Biogenesis and mode of action of plant vsiRNAs

only for the plant vsiRNAs. This chapter discusses about the database, PVsiRNAdb exclusively for plant vsiRNAs (Gupta et al. 2018). PVsiRNAdb (http://www.nipgr.res.in/PVsiRNAdb) is developed by extensive data mining and harboring information of plant vsiRNAs from literature available till date. The resources available online pertaining to vsiRNAs hold predicted as well as annotated sequences detected in virus-infected plants. This database is developed in such a user-friendly manner for convenience.

29.2 Materials

For this study, data subjected to virus interaction with the plant was collected by data mining of PubMed literature by Gupta et al. and developed a web-based platform named as PVsiRNAdb. It contains information regarding vsiRNA sequences from 20 different viruses infecting 12 different plants which are listed in Table 29.1 with total number of vsiRNAs.

Table 29.1 List of viruses, host plant and total vsiRNAs stored in PVsiRNAdb (Gupta et al. 2018)

Virus	Host plant	Total vsiRNAs
Bamboo mosaic virus	*Dendrocalamus latiflorus* (bamboo)	18
Brassica yellow virus	*Nicotiana benthamiana* (tobacco)	143
Chinese wheat mosaic virus	*Triticum aestivum* (wheat)	19,536
Cotton leaf curl Multan virus	*Gossypium hirsutum* (cotton)	4736
Cucumber green mottle mosaic virus	*Cucumis sativus* (cucumber)	92
Cucumber mosaic virus	*Arabidopsis thaliana* (thale cress)	47
Cymbidium ringspot virus	*Nicotiana benthamiana* (tobacco)	12,305
Grapevine fleck virus and Grapevine rupestris stem pitting associated virus	*Vitis vinifera* (grapevine)	62
Maizechlorotic mottle virus and Sugarcane mosaic virus	*Zea mays* (maize)	260
Pea enation mosaic virus 2	*Nicotiana benthamiana* (tobacco)	137
Potato virus Y	*Solanum tuberosum* (potato)	46,435
Prunus necrotic ring spot virus	*Prunus avium* (cherry)	152
Rice black streaked dwarf virus	*Oryza sativa* (rice)	468
Southern Rice black streaked dwarf virus	*Oryza sativa* (rice)	20,876
Sugarcane mosaic virus	*Zea mays* (maize)	100
Tobacco mosaic virus	*Nicotiana benthamiana* (tobacco)	682
Tobacco rattle virus	*Nicotiana benthamiana* (tobacco)	142
Wheat yellow mosaic virus	*Triticum aestivum* (wheat)	216,018
Zucchini yellow mosaic virus	*Citrullus lanatus* (watermelon)	5

29.3 Methods

29.3.1 Data Collection

An extensive literature search was carried out to excerpt the relevant articles from PubMed (https://www.ncbi.nlm. nih.gov/pubmed). This was carried out by searching queries using a different combination of keywords e.g. viral siRNAs, plant-virus interaction, siRNA, plant-viral siRNAs etc. Relevant experimental information was extracted by a manual screening of articles. Literature lacking relevant information regarding this study were excluded. Full-text search was done for each of the relevant article having the information of plant-specific vsiRNAs. In addition to this, the relevant information associated with the plant, tissue, PubMed ID (PMID) and PVsiRNA-ID was incorporated along with the collected information of vsiRNAs.

29.3.2 PVsiRNAdb Web Platform

PVsiRNAdb web-interface was built on an Apache Hypertext Transfer Protocol server by using Hypertext Markup Language (HTML), Cascading Style Sheets (CSS), Hypertext Preprocessor (PHP) and JavaScript. MySQL, an object-relational database management system (RDBMS), was used to manage all the data in the backend. It provides commands to retrieve and store the data in the database. All common gateway interface and database interfacing scripts were written in the Hypertext Preprocessor (PHP), and Practical Extraction and Reporting Language (PERL).

29.3.3 Organization of Database

The information in PVsiRNAdb is organized at two levels, primary and secondary (Fig. 29.2). At the primary level queries are searched by specific plant name, virus name, PMID or other options as per the users' requirement.

The information will be displayed according to the number of fields selected by the user. The user can also search multiple queries for virus, plant or PMIDs by performing a batch search. The data at the secondary level can be utilized for the retrieval of further information about primary data. At the secondary level, additional information like experimental details, sequence-related information and details of virus-like name, type of genome and classification can also be fetched for each viral strain. The virus name, genome type as well as classification can also be retrieved for each viral strain. The specific details about any experiment can be inquired by clicking on PMID hyperlink, which will direct the user to the original link of that research article. As structure plays an important role in determining the function of any sequence, the secondary structure of vsiRNA sequences was added to the database using in-house generated PERL scripts for running the Mfold (Zuker 2003) and RNA structure (Reuter and Mathews 2010) packages. Mfold was utilized for calculating minimized energy for the folded structure and structure coordinates were predicted by Draw utility of RNA structure software.

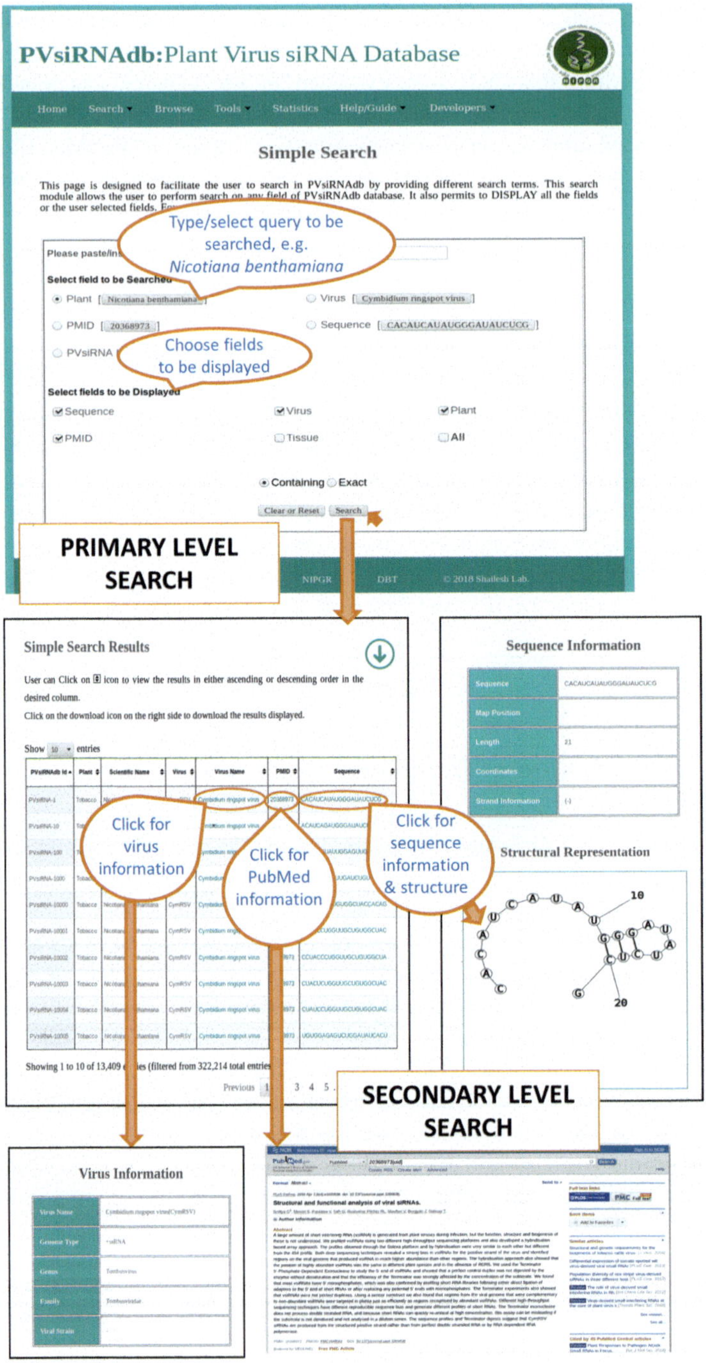

Fig. 29.2 Information at the primary and secondary level of search

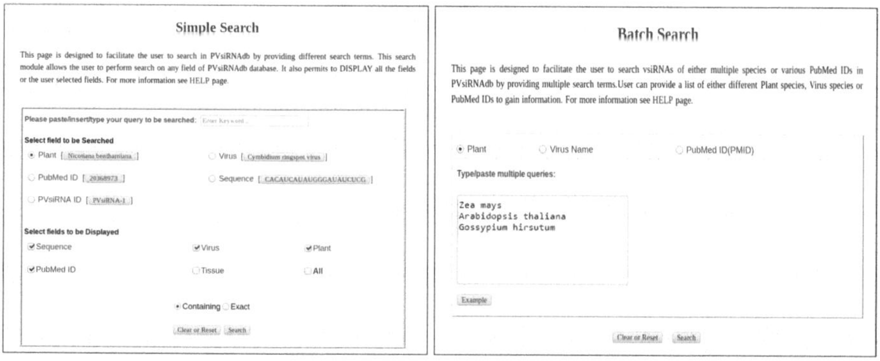

(a) (b)

Fig. 29.3 Illustration of 'Search' option in PVsiRNAdb (**a**) The representation of 'Simple Search' module. (**b**) The window showing 'Batch Search' module with query example

29.3.4 Features and Tools

In PVsiRNAdb, detailed and comprehensive information is incorporated for each siRNA entry. Apart from the core information including the siRNA sequence, siRNA length, virus name, and plant name, additional information like PMID, plant tissue, mapping coordinates of siRNA to the plant genome and the predicted secondary structures of siRNA may be of high utility to the user. PVsiRNAdb provides two user-friendly options to search for siRNA information i.e., 'Simple Search', and 'Batch Search' (Fig. 29.3a, b).

'Simple Search' allows the user to search the query by providing different search terms like the name of the virus, plant name (scientific or common name), siRNA sequence, PMID, and PVsiRNA-ID. For providing the flexibility in the search module, the 'containing' and 'exact' option has been incorporated. This option also facilitates the user to select the fields to be displayed. A total of five display fields are available for a search term. Three display fields namely the 'Virus name', 'PMID' and 'Sequence' are further linked with their corresponding information. Second option to search in PVsiRNAdb is that of the 'Batch Search' providing the facility to search for multiple queries at a time. The user can extract the information of siRNAs by providing a list of plant names, virus names or PMIDs. In this module, an example list of all the three search terms is provided for the users. The PVsiRNAdb information can be browsed by virus name, plant name or PubMed ID by expanding the respective option in 'Browse' section (Fig. 29.4).

'Tools' section contains three module – 'BLAST', 'SW Align', and 'Mapping' (Fig. 29.5). BLAST module is developed by *blastn* utility of ncbi-blast 2.6.0 (https://blast.ncbi.nlm.nih.gov/Blast.cgi) standalone version. The user provided query sequence(s) can be aligned to blast database i.e., PVsiRNAdb. This module is also provided with the option to select the virus genome and change the E-value for alignment. BLAST result, besides the alignment result, also gives each hit

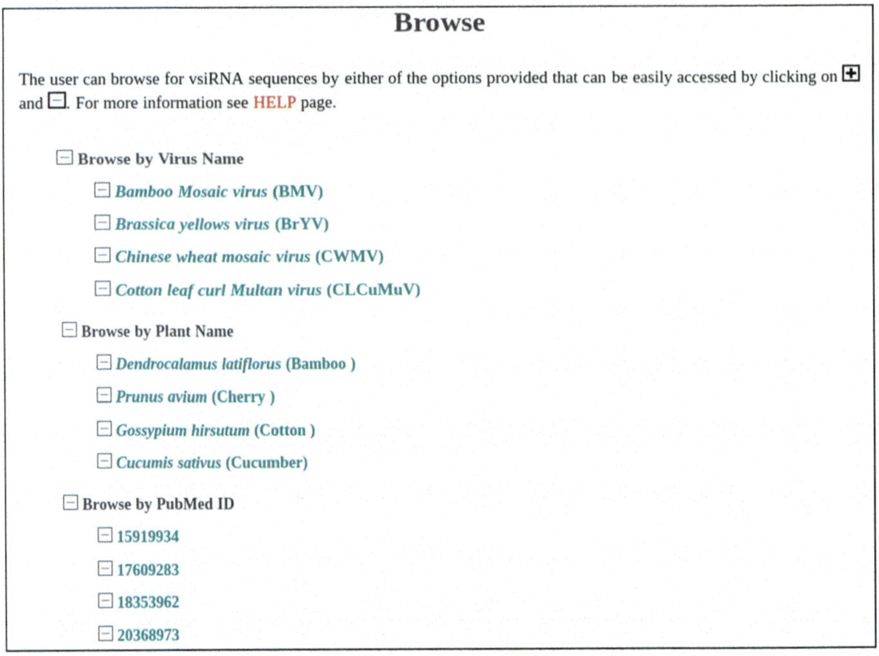

Fig. 29.4 The 'Browse' section of PVsiRNAdb displaying three different options provided

information by a click on it (Fig. 29.5a). 'SW Align' module uses the *water* utility of EMBOSS-6.6.0 (http://emboss.open-bio.org/) to align the query to selected virus siRNA dataset. In-house developed PERL scripts are integrated with the Smith-Waterman algorithm to take the result in the desired pattern (Fig. 29.5b). The 'Mapping' module is designed for the mapping of siRNA sequences, available at PVsiRNAdb to the user-provided sequences e.g. messenger RNA sequences or genomic sequences (Fig. 29.5c). This module uses the *makeblastdb* and *blastn* utility of ncbi-blast 2.6.0 with the PERL script integration. 'Mapping' facility is useful for the designing of specific siRNAs corresponding to the specific viral genome.

Overall statistics of PVsiRNAdb are illustrated in 'Statistics' section in the form of tables and histogram. Any query regarding the use of PVsiRNA web interface is answered by 'Help/Guide' section. This section apart from 'Help' sub-section, also contains two more sub-section i.e., 'Links' and 'References'. From the 'Help' of PVsiRNAdb, the user can understand the working of this database with the help of self-explanatory figures. 'Links' directs the user to important web resources contains information on viral siRNAs. In the 'References' of PVsiRNAdb web interface, all the articles related to vsiRNAs involved plant-virus interaction have been incorporated.

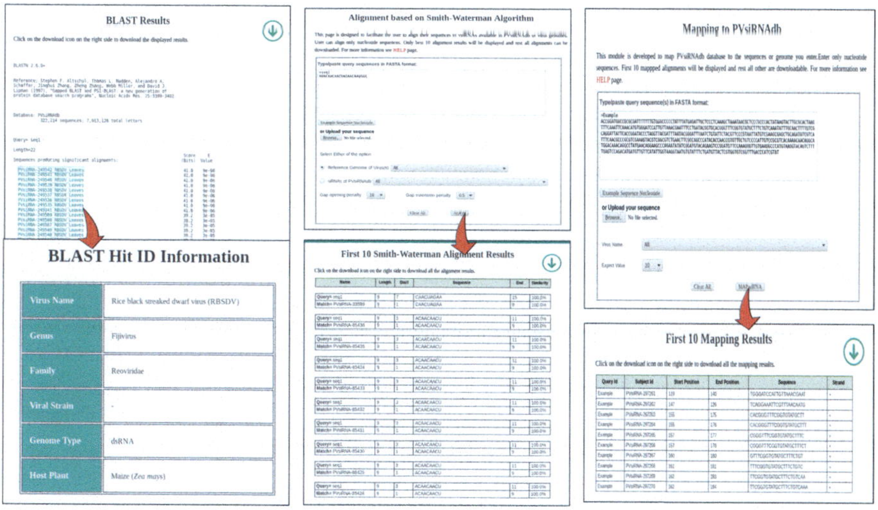

(a) (b) (c)

Fig. 29.5 Three modules in 'Tools' section of PVsiRNdb (**a**) 'BLAST' module window showing alignment result and blast hit information. (**b**) The output of 'SW Align' result after a query search. (**c**) A query and their mapping result by 'Mapping' module

Acknowledgment PVsiRNAdb database is developed by Dr. Shailesh Kumar research group (http://www.nipgr.res.in/research/dr_shailesh.php) at National Institute of Plant Genome Research (NIPGR), New Delhi, India.

References

Bos L. Crop losses caused by viruses. Crop Prot. 1982;1:263–82. https://doi.org/10.1016/0261-2194(82)90002-3.

Bucher E, Lohuis D, van Poppel PMJA, et al. Multiple virus resistance at a high frequency using a single transgene construct. J Gen Virol. 2006;87:3697–701. https://doi.org/10.1099/vir.0.82276-0.

Chalk AM, Warfinge RE, Georgii-Hemming P, Sonnhammer ELL. siRNAdb: a database of siRNA sequences. Nucleic Acids Res. 2004;33:D131–4. https://doi.org/10.1093/nar/gki136.

Chapman EJ, Carrington JC. Specialization and evolution of endogenous small RNA pathways. Nat Rev Genet. 2007;8:884–96. https://doi.org/10.1038/nrg2179.

Chen X. Small RNAs – secrets and surprises of the genome. Plant J. 2010;61:941–58. https://doi.org/10.1111/j.1365-313X.2009.04089.x.

Diener TO. Physiology of virus-infected plants. Annu Rev Phytopathol. 1963;1:197–218. https://doi.org/10.1146/annurev.py.01.090163.001213.

Ding S-W, Lu R. Virus-derived siRNAs and piRNAs in immunity and pathogenesis. Curr Opin Virol. 2011;1:533–44. https://doi.org/10.1016/j.coviro.2011.10.028.

Donaire L, Wang Y, Gonzalez-Ibeas D, et al. Deep-sequencing of plant viral small RNAs reveals effective and widespread targeting of viral genomes. Virology. 2009;392:203–14. https://doi.org/10.1016/j.virol.2009.07.005.

Duan C-G, Wang C-H, Guo H-S. Application of RNA silencing to plant disease resistance. Silence. 2012;3:5. https://doi.org/10.1186/1758-907X-3-5.

Guo Q, Liu Q, Smith NA, et al. RNA silencing in plants: mechanisms, technologies and applications in horticultural crops. Curr Genomics. 2016;17:476–89. https://doi.org/10.2174/1389202917666160520103117.

Gupta N, Singh A, Zahra S, Kumar S. PtRFdb: a database for plant transfer RNA-derived fragments. Database. 2018;2018:63.

Hamilton AJ, Baulcombe DC. A species of small antisense RNA in posttranscriptional gene silencing in plants. Science (New York, NY). 1999;286:950–2.

Malpica-Opez N, Rajeswaran R, Beknazariants D, et al. Revisiting the roles of tobamovirus replicase complex proteins in viral replication and silencing suppression. MPMI. 2018;311094:125–44. https://doi.org/10.1094/MPMI-07-17-0164-R.

Mansoor S, Amin I, Hussain M, et al. Engineering novel traits in plants through RNA interference. Trends Plant Sci. 2006;11:559–65. https://doi.org/10.1016/J.TPLANTS.2006.09.010.

Moissiard G, Voinnet O. RNA silencing of host transcripts by cauliflower mosaic virus requires coordinated action of the four Arabidopsis Dicer-like proteins. Proc Natl Acad Sci. 2006;103:19593–8. https://doi.org/10.1073/pnas.0604627103.

Prins M, de Haan P, Luyten R, et al. Broad resistance to tospoviruses in transgenic tobacco plants expressing three tospoviral nucleoprotein gene sequences. Mol Plant-Microbe Interact. 1995;8:85–91. https://doi.org/10.1094/MPMI-8-0085.

Reuter JS, Mathews DH. RNAstructure: software for RNA secondary structure prediction and analysis. BMC Bioinform. 2010;11:129. https://doi.org/10.1186/1471-2105-11-129.

Rodríguez-Negrete EA, Carrillo-Tripp J, Rivera-Bustamante RF. RNA silencing against geminivirus: complementary action of posttranscriptional gene silencing and transcriptional gene silencing in host recovery. J Virol. 2009;83:1332–40. https://doi.org/10.1128/JVI.01474-08.

Shimizu T, Nakazono-Nagaoka E, Akita F, et al. Hairpin RNA derived from the gene for Pns9, a viroplasm matrix protein of Rice gall dwarf virus, confers strong resistance to virus infection in transgenic rice plants. J Biotechnol. 2012;157:421–7. https://doi.org/10.1016/j.jbiotec.2011.12.015.

Skums P, Artyomenko A, Glebova O, et al. Computational framework for next-generation sequencing of heterogeneous viral populations using combinatorial pooling. Bioinformatics. 2015;31:682–90. https://doi.org/10.1093/bioinformatics/btu726.

Stobbe AH, Roossinck MJ. Plant virus metagenomics: what we know and why we need to know more. Front Plant Sci. 2014;5:150. https://doi.org/10.3389/fpls.2014.00150.

Szittya G, Moxon S, Pantaleo V, et al. Structural and functional analysis of viral siRNAs. PLoS Pathog. 2010;6:e1000838. https://doi.org/10.1371/journal.ppat.1000838.

Thakur N, Qureshi A, Kumar M. VIRsiRNAdb: a curated database of experimentally validated viral siRNA/shRNA. Nucleic Acids Res. 2012;40:D230–6. https://doi.org/10.1093/nar/gkr1147.

Tyagi A, Ahmed F, Thakur N, et al. HIVsirDB: a database of HIV inhibiting siRNAs. PLoS One. 2011;6:e25917. https://doi.org/10.1371/journal.pone.0025917.

Vazquez F, Hohn T. Biogenesis and biological activity of secondary siRNAs in plants. Scientifica. 2013;2013:783253. https://doi.org/10.1155/2013/783253.

Velásquez AC, Chakravarthy S, Martin GB. Virus-induced gene silencing (VIGS) in Nicotiana benthamiana and tomato. JoVE. 2009; https://doi.org/10.3791/1292.

Zhang C, Wu Z, Li Y, Wu J. Biogenesis, function, and applications of virus-derived small RNAs in plants. Front Microbiol. 2015;6:1–12. https://doi.org/10.3389/fmicb.2015.01237.

Zhu H, Guo H. The role of virus-derived small interfering RNAs in RNA silencing in plants. Sci China Life Sci. 2012;55:119–25. https://doi.org/10.1007/s11427-012-4281-3.

Zuker M. Mfold web server for nucleic acid folding and hybridization prediction. Nucleic Acids Res. 2003;31:3406–15.

Chapter 30
Association Between Nuclear Receptor Coactivator 1 and Stem Cell Signaling Pathway and Identification of Natural NCOA1 Inhibitors: An in-silico Study

Pushpendra Singh, Prem P. Kushwaha, Atul K. Singh, and Shashank Kumar

30.1 Introduction

The p160 steroid receptor coactivator (SRC) comprise three protein viz. nuclear receptor coactivator (NCOA) 1 and, NCOA2, and NCOA3 (Xu and Li 2003; Xu et al. 2009). SRC play an imperative role in cell growth and development. Further, it has shown that SRC involved in the pathogenesis of various cancers (Xu et al. 2009). The oncogenic role of NCOA3 has been well well-known in various cancers (Anzick et al. 1997; Tien et al. 2013, 2014). NCOA1 and NCOA3 deficiencies affected the expression of various genes including TGFβ, peroxisome proliferator-activated receptor-β, peroxisome proliferator-activated receptor-γ-regulated genes, and glucose transportation including GLUT1. They are recruited to gene promoters by interacting with (a) nuclear hormone receptors (progesterone receptor, estrogen receptor α, and peroxisome proliferator-activated receptor γ and (b) transcription factors (E2F1, and AP1).

Moreover, NCOA1 maintains IGF1 through enhancing VDR regulated IGFBP3 expression. Many studies support that insulin and IGF1 promoted the cancer cell growth that discloses the link between NCOA1, insulin, and IGF1 in various cancers (Liao et al. 2008; Oh 2014; Singh et al. 2014). NCOA1 Led to acylation of histones, which makes more active DNA transcription. Collectively, these enzymes remodel the chromatin to facilitate assembly of transcription factors and initiation of mRNA transcription. Due to the limited cellular concentrations of these, alteration of NCOAs concentrations by cellular or transcriptional activities may have remarkable

P. Singh
Tumor Biology Laboratory, National Institute of Pathology, New Delhi, India

P. P. Kushwaha · A. K. Singh · S. Kumar (✉)
Department of Biochemistry and Microbial Sciences, Central University of Punjab, Bathinda, Punjab, India

S. Kumar, C. Egbuna (eds.), *Phytochemistry: An in-silico and in-vitro Update*,
https://doi.org/10.1007/978-981-13-6920-9_30

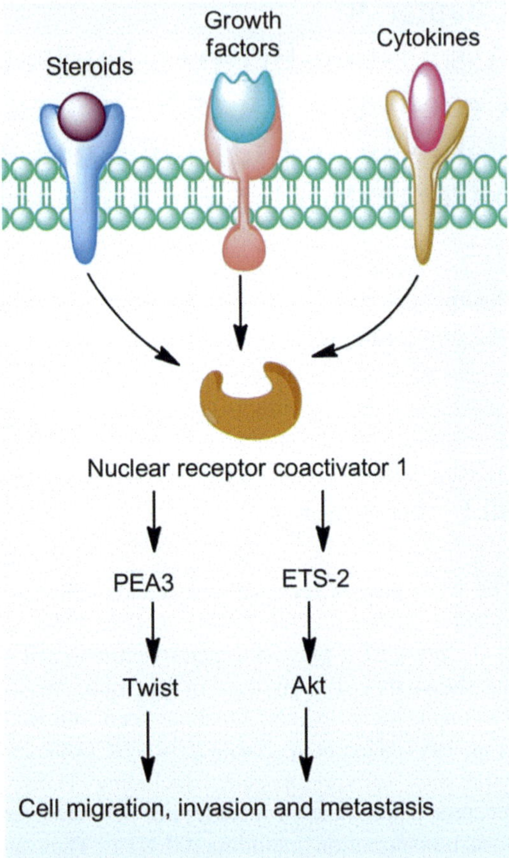

Fig. 30.1 Nuclear receptor coactivator promotes carcinogenesis through multiple pathways. Nuclear receptor coactivator 1 enhances Ets-2-mediated Her2 expression and PEA3 mediated Twist expression and upregulates CSF-1 expression to promote breast tumor cell migration, invasion, and metastasis. Nuclear receptor coactivator 3 upregulates Akt and MAPK activities, resulting in hyperactivation of Akt and MAPK, which contribute to cancer cell proliferation, growth, survival, migration, invasion, and metastasis

effects on the expression of target genes (Chen et al. 2010). NCOA1 over expressed in breast tumors that correlate with HER2 expression, and metastasis (Qin et al. 2015a, b). Moreover, overexpression of NCOA1 in the mouse mammary gland augmented lung metastasis in transgenic mouse mammary tumor virus-polyoma middle T and MMTV-Neu breast cancer mouse models (Qin et al. 2015a, b). It has shown that NCOA1 up-regulate the various gene activity such as Twist1, SDF1, and CXCR4 that promoting metastasis (Fig. 30.1) (Qin et al. 2015a, b; Kishimoto et al. 2005). Thus, due to the precise function of NCOA1 in cancer metastasis may be related to a new target for impeding metastasis. Cancer stem cells are the small population of cells in a tumor have self-renewal, drug evading and cancer growth promoting ability. Disrupt the expression pattern of different proteins have been known to associated with cancer cell stemness property. miRNAs are short stretch RNAs (21-23 nucleotides) and a class of noncoding RNAs. The scientific literature showed that these noncoding RNAs are involved in the maintenance of stemness in

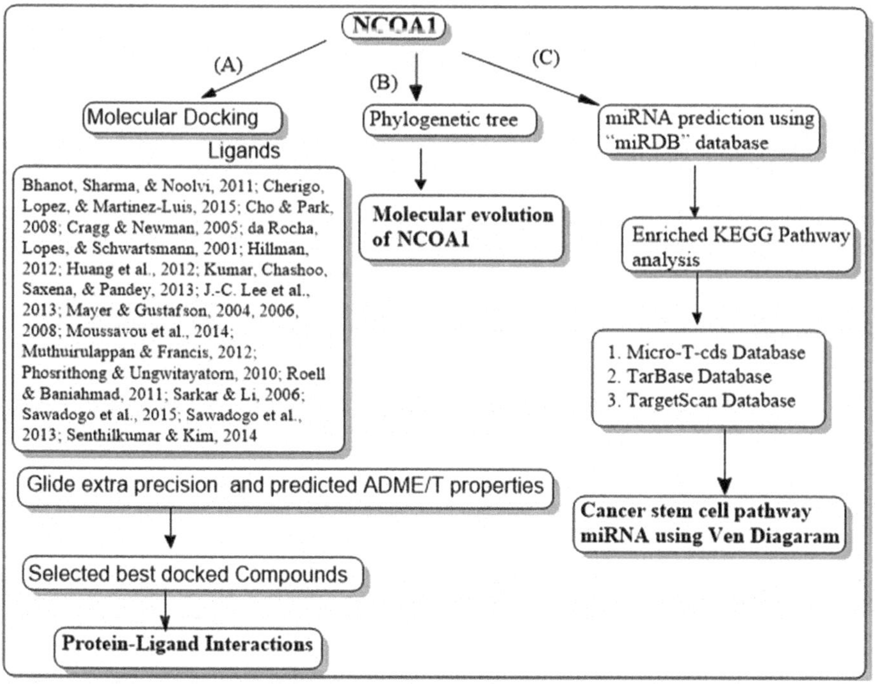

Fig. 30.2 Work flow of study design

cancer stem cells. miRNAs are also involved in drug resistance, disease relapse, and tumor development. Thus in the present study, we identified the common miRNAs involved in NCOA1 gene regulation and cancer stem cell signaling pathways.

A typical docking contains sampling and scoring. Sampling is the creation of ligand orientations in the binding site of a protein, whereas scoring is the prediction of its binding fitness. Docking is now routinely used in hit identification, lead optimization, and virtual screening. Protein-ligands interactions play a vital role in many imperative cellular processes, such as signal transduction, gene expression, enzyme inhibition, etc. and are essential for various biological functions. But, it is quite difficult and expensive to determine the protein-ligand interactions by experimental methods. Thus, docking is a pre-approach for thoughtful the protein-ligands interactions. Advancement in the drug design approaches has discovered numeral drugs like inhibitors of HIV-1 protease, HIV-1 integrase, carbonic anhydrase, neuraminidase, protein kinases, etc. In this context, the present work was conducted to afford molecular evolution and identify possible inhibitors of NCOA1by using in-silico (Fig. 30.2). Over all targeting, the miRNAs involved in NCOA1 gene regulation and cancer stem cell signaling pathway, as well as inhibition of NCOA1 at the protein level by natural inhibitors, might be a fruitful strategy to target cancer drug resistance.

30.2 Methodology

30.2.1 Multiple Alignment and Phylogenetic Analysis

The reviewed amino acid sequence encoded by the NCOA1 was retrieved from the GenBank database of NCBI (http://www.ncbi.nlm.nih.gov/). To search for homolog sequence of NCOA1, BLASTP was carried out against the non-redundant database available at NCBI. Alignment was carried out separately for 14 isolates available with protein database. Selected sequences were first aligned by MUSCLE algorithm which was done using MEGA, a freeware (available at www.megasoftware.net/) and alignments were edited by eye. The Maximum Likelihood (ML) tree was then constructed using MEGA v6. Substitution bias was modeled by Gamma distributed Jones-Taylor-Thornton (JTT) model. A total of 1000 bootstrap replicates were performed under ML criterion to estimate interior branch support (Felsenstein, 1985).

30.2.2 miRNA Identification and Target Pathway Prediction Analysis

miRNAs involved in NCOA1 gene regulation and pathways targeted by these miRNAs were identified by using different online tools. miRNA involved in NCOA1 regulation was predicted from an online database. The DIANA-mirPathv3 online software was used to predict the molecular pathways associated with predicted miRNAs (involved in NCOA1 regulatory) (Vlachos et al. 2015). The software has the capability to analyze the combinatorial effect of different miRNAs on KEGG (Kyoto Encyclopedia of Genes and Genomes) pathways. DIANA-mirPathv3 combines the gene targets of the selected miRs into a superset (union), performs the enrichment analysis and calculates the significance levels (p-values) between each miRNA and every pathway. Statistically significant association ($p > 0.05$) between a selected gene from the list of pathways was measured and not significantly associated pathway were not included in the study. DIANA-microT-CDS and DIANA-Tarbase v7.0 in-silico miRNA target prediction algorithms were used for pathway prediction (miR-gene interactions). The settings utilized were: p-value threshold of 0.05; DIANA-microT-CDS and DIANA-Tarbase v7.0 threshold of 0.85; merging method: pathways union. Fisher's method was used to combine the results of more than one independent test bearing on the same hypothesis. Benjamini and Hochberg's algorithm hypothesis testing was performed to do enrichment analysis (Benjamini and Hochberg 1995; Carotenuto et al. 2016). miR-pathway interaction is presented as "heat maps" (generated by using the DIANA-mirPathv3 online tool) showing a large number of variables.

30.2.3 Preparation of Protein

The X-ray structure of NCOA1 (PDB ID: 1OJ5) have been used as the initial structure in the preparation of protein (Singh and Bast 2014a, b, 2015a, b, c).

30.2.4 Ligand Preparation

GLIDE based molecular docking protocol adapted from our previous published literature with a few modifications (Singh and Bast 2014a, b, 2015a, b, c). Two hundred marine and natural compounds selected from different previous literatures (Bhanot et al. 2011; Cherigo et al. 2015; Jae and Park 2008; Cragg and Newman 2005; DaRocha et al. 2001; Hillman 2012; Huang et al. 2012; Kumar et al. 2013; Lee et al. 2013; Mayer and Gustafson 2004, 2006, 2009; Moussavou et al. 2014; Muthuirulappan and Francis 2013; Phosrithong and Ungwitayatorn 2010; Roell and Baniahmad 2011; Sarkar and Li 2006; Sawadogo et al. 2013, 2015; Kim 2014; Ganatra 2012). LigPrep was used for the preparation of ligands (Singh and Bast 2014a, b, 2015a, b, c; Jorgensen et al. 1996; Jorgensen and Tirado-Rives 1988; Shivakumar et al. 2010).

30.2.5 Receptor Grid Generation and Binding Site Prediction

A sitemap is a vigorous program that analyses the distinctive features of binding sites (Lauria et al. 2009). Therefore, the binding sites of the NCOA1 protein were identified using SiteMap.

30.2.6 Maestro Molecular Docking

Molecular docking was performed by employing GLIDE and outputs represented as the Gscore (Singh and Bast 2014a, b, 2015a, b, c; Friesner et al. 2004, 2006; Halgren et al. 2004).

30.2.7 Absorption, Distribution, Metabolism, Excretion, and Toxicity (ADME/T) Studies

In-silico ADME/T (Absorption, Distribution, Metabolism, Excretion, and Toxicity) analytical tools are a cooperative approach that could eliminate rude compounds.

Thus, various ADME/T parameter of best-docked compounds was predicted (Jorgensen and Duffy 2002; Lu et al. 2004).

30.3 Results and Discussion

30.3.1 Phylogenetic Analysis

The phylogenetic analysis of 14 protein sequences gave a robust relationship between different isolates of nuclear receptor coactivator. The isolates from humans formed a separate clade. Another clade was formed by the isolates from other species like rodents and other higher animals. The tree was rooted with the isolate from Xenopus. The pairwise distance also calculated corroborates with the same. The Table 30.1 shows the distance between the isolates of humans is least ranging from 0 to 0.009 which, of course, is within the intra specific range. The phylogenetic tree in the figure also shows the evolutionary pattern of the protein (Fig. 30.3). The different isoforms of the proteins from humans have evolved simultaneously which resulted in the formation of the conspecific clade. Same is true for isolates from other organisms. The pairwise distance of all the isolates is maximum with Xenopus, which is also confirmed by the phylogenetic tree.

30.3.2 miRNA Identification and Target Pathway Prediction Analysis

List of miRNA involved in NCOA1 (nuclear receptor coactivator 1) gene regulation and their target score was predicted and given in Table 30.2. Tables 30.3, 30.4 and 30.5, depicts the statistically-significant enriched KEGG pathways for miRs involved in different molecular pathways and have an association (predicted) with NCOA1 gene by using Micro-T-cds database, TarBase, and TargetScan databases respectively. The visual representation of the binary heat map showing the miRNAs/pathways interaction is reported in Fig. 30.4. Different types of pathways (affected by miRNAs involved in NCOA1 gene) were predicted by using Micro-T-cds database, TarBase, and TargetScan databases. We found 7 KEGG pathways significantly related to genes targeted by predicted miRs involved in regulation of NCOA1. A consistent number of pathways involved in energy metabolism (such as glycosphingolipid biosynthesis lacto and neolacto series and fatty acid biosynthesis pathways) and cancer-related pathways were identified (Tables 30.3, 30.4 and 30.5). Cancer-related pathways were also present, including proteoglycans in cancer, viral carcinogenesis, and TGF-beta signaling pathways. In addition, we found significant pathways relevant in stem cell biology and tissue homeostasis, such as "signaling pathways regulating pluripotency of stem cells" and the "Hippo signaling

Table 30.1 The pairwise distance between the isolates

gi767915344refXP_011531443.1_PREDICTED: _nuclear_receptor_coactivator_1_isoform_X4_Homo_sapiens													
gi53068248refXP_005264685.1_PREDICTED: _nuclear_receptor_coactivator_1_isoform_X3_Homo_sapiens	0.009												
gi53068244refXP_005264683.1_PREDICTED: _nuclear_receptor_coactivator_1_isoform_X2_Homo_sapiens	0.000	0.009											
gi530368242refXP_005264682.1_PREDICTED: nuclear_receptor_coactivator_1_isoform_X1_Homo_sapiens	0.000	0.009	0.000										
gi22538457refNP_671756.1_nuclear_receptor_coactivator_1_isoform_2_Homo_sapiens	0.009	0.000	0.009	0.009									
gi22538459refNP_671766.1_nuclear_receptor_coactivator_1_isoform_3_Homo_sapiens	0.000	0.009	0.000	0.000	0.009								
gi22538455refNP_003734.3_nuclear_receptor_coactivator_1_isoform_1_Homo_sapiens	0.000	0.009	0.000	0.000	0.009	0.000							
gi34105530gbAAI11534.1_Nuclear_receptor_coactivator_1_Homo_sapiens	0.000	0.009	0.000	0.000	0.009	0.000	0.000						
gi5754800refNP_035011.1_nuclear_receptor_coactivator_1_Mus_musculus	0.083	0.072	0.083	0.083	0.072	0.083	0.083	0.083					
gi157819661refNP_001101482.1_nuclear_receptor_coactivator_1_Rattus_norvegicus	0.067	0.076	0.067	0.O67	0.076	0.067	0.067	0.067	0.042				
gi58534998refNP_001020399.1_nuclear_receptor_coactivator_1_Sus_scrofa	0.034	0.044	0.034	0.034	0.044	0.034	0.034	0.034	0.095	0.078			
gi51097961refNP_001012900.1_nuclear_receptor_coactivator_1_Gallus_gallus	0.302	0.319	0.302	0.302	0.319	0.302	0.302	0.302	0.331	0.312	0.308		
gi163915047refNP_001106333.1_nuclear_receptor_coactivator_1_Xenopus_(Silurana) _tropicalis	0.420	0.441	0.420	0.420	0.441	0.420	0.420	0.420	0.462	0.446	0.445	0.369	
gi329664314refNP_001193144.1_nuclear_receptor_coactivator_1_Bos_tauras	0.041	0.051	0.041	0.041	0.051	0.041	0.041	0.041	0.098	0.076	0.031	0.310	0.446

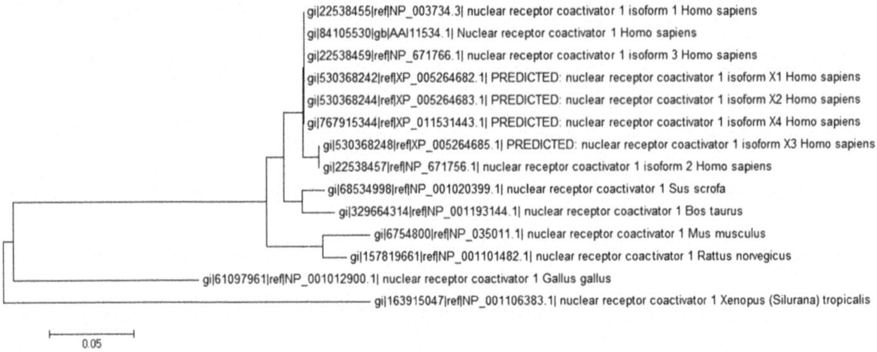

Fig. 30.3 Phylogenetic tree. The Maximum Likelihood (ML) tree was constructed using MEGA v6. Substitution bias was modeled by Gamma distributed Jones-Taylor-Thornton (JTT) model. A total of 1000 bootstrap replicates were performed under ML criterion to estimate interior branch support

pathway". The pathways significantly enriched ($p < 0.05$) for the target of miRs involved in regulation of NCOA1 reported in Tables 30.3, 30.4 and 30.5. Heat map of the pathways related to the common microRNAs involved in NCOA1 gene regulation is depicted in Fig. 30.4. The miRNAs associated with NCOA1 gene regulation and signaling pathways regulating pluripotency of stem cells were identified using Micro-T-cds database (n = 11 miRNAs) and TarBase database (n = 09 miRNAs). Over all five miRNAs involved in NCOA1 gene regulation and signaling pathways regulating pluripotency of stem cells were identified from both the databases (Fig. 30.5).

30.3.3 Molecular Docking and Interaction Analysis

This study explores the potent NCOA1 inhibitor employing molecular docking technology. Molecular docking of NCOA1 against marine and natural compounds has been carried out. Furthermore, the structure of the NCOA1(1OJ5) complex with LXXLL motif of the STAT6 emphasized that lipophilic, electrostatic, hydrogen bond, π-π stacking, and cation–π interactions are a key contributor in interactions (Razeto et al. 2004). In the present investigation, a set of 20 top scoring molecules analyzed and protein-ligands interactions were represented. An analog of kaempferol (CID5465152) and hesperidin (CID3594) have the best docking score against NCOA1. Two compounds with least Gscore for NCOA1, named as CID5465152 and CID3594 were then chosen for further consideration. The binding interaction of these compounds with NCOA1 protein is shown in Fig. 30.6a and 30.7a respectively.

These molecules had a high binding affinity for the NCOA1 binding pocket. CID5465152 and CID3594 had a docking score of −8.88 and −6.83 kcal/mol

Table 30.2 List of miRNA involved in NCOA1 (nuclear receptor coactivator 1) gene regulation and their target score

miRNA name (target score)
hsa-miR-4443 (100), hsa-miR-4775 (99), hsa-miR-3612 (99), hsa-miR-650 (99), hsa-miR-1277-5p (98), hsa-miR-670-3p (98), hsa-miR-2278 (97), hsa-miR-590-3p (97), hsa-miR-4735-3p (96), hsa-miR-548e-5p (96), hsa-miR-18b-5p (96), hsa-miR-18a-5p (96), hsa-miR-4428 (96), hsa-miR-4703-5p (95), hsa-miR-148b-3p (95), hsa-miR-4786-3p (95), hsa-miR-222-5p (95), hsa-miR-148a-3p (95), hsa-miR-3942-5p (95), hsa-miR-152-3p (95), hsa-miR-875-3p (94), hsa-miR-661 (94), hsa-miR-3679-3p (93), hsa-miR-5093 (93), hsa-miR-7111-5p (93), hsa-miR-520d-5p (93), hsa-miR-5692b (93), hsa-miR-5698 (93), hsa-miR-551b-5p (93), hsa-miR-524-5p (93), hsa-miR-6870-5p (93), hsa-miR-5692c (93), hsa-miR-4723-5p (93), hsa-miR-374b-5p (92), hsa-miR-374a-5p (92), hsa-miR-513c-3p (90), hsa-miR-5011-5p (90), hsa-miR-889-5p (90), hsa-miR-6825-5p (90), hsa-miR-513a-3p (90), hsa-miR-6507-5p (89), hsa-miR-647 (89), hsa-miR-4422 (88), hsa-miR-4728-3p (88), hsa-miR-1255b-5p (87), hsa-miR-1255a (87), hsa-miR-4708-5p (87), hsa-miR-130a-3p (87), hsa-miR-6875-3p (85), hsa-miR-7853-5p (85), hsa-miR-4766-3p (85), hsa-miR-2115-3p (85), hsa-miR-6757-5p (85), hsa-miR-548t-5p (84), hsa-miR-6852-5p (84), hsa-miR-548az-5p (84), hsa-miR-377-3p (83), hsa-miR-3158-5p (82), hsa-miR-376c-3p (82), hsa-miR-1915-3p (81), hsa-miR-6764-5p (81), hsa-miR-3936 (81), hsa-miR-548c-3p (81), hsa-miR-493-5p (80), hsa-miR-488-5p (80), hsa-miR-8485 (78), hsa-miR-4778-3p (78), hsa-miR-23a-3p (77), hsa-miR-377-5p (77), hsa-miR-23b-3p (77), hsa-miR-4635 (77), hsa-miR-130a-5p (77), hsa-miR-6086 (77), hsa-miR-23c (77), hsa-miR-26b-3p (77), hsa-miR-6128 (76), hsa-miR-4652-3p (76), hsa-miR-5584-5p (76), hsa-miR-3166 (75), hsa-miR-6882-3p (75), hsa-miR-6124 (74), hsa-miR-186-3p (73), hsa-miR-6888-5p (72), hsa-miR-140-5p (72), hsa-miR-7152-5p (72), hsa-miR-4500 (71), hsa-miR-5688 (71), hsa-miR-495-3p (71), hsa-miR-1275 (70), hsa-miR-27b-3p (70), hsa-miR-27a-3p (70), hsa-miR-3606-3p (69), hsa-miR-320e (69), hsa-miR-6738-3p (68), hsa-miR-454-3p (67), hsa-miR-301b-3p (67), hsa-miR-130b-3p (67), hsa-miR-4295 (67), hsa-miR-3666 (67), hsa-miR-4668-5p (67), hsa-miR-6514-5p (67), hsa-miR-5001-3p (67), hsa-miR-301a-3p (67), hsa-miR-93-3p (67), hsa-miR-3136-5p (66), hsa-miR-4325 (66), hsa-miR-4439 (66), hsa-miR-216a-5p (64), hsa-miR-498 (64), hsa-miR-7151-5p (64), hsa-miR-6782-5p (63), hsa-miR-487a-5p (62), hsa-miR-3653-3p (62), hsa-miR-487b-5p (62), hsa-miR-507 (61), hsa-miR-7-1-3p (61), hsa-miR-7-2-3p (61), hsa-miR-4800-5p (61), hsa-miR-3142 (61), hsa-miR-3622a-3p (61), hsa-miR-1245b-5p (61), hsa-miR-3920 (61), hsa-miR-3622b-3p (61), hsa-miR-557 (61), hsa-miR-3646 (61), hsa-miR-4651 (60), hsa-miR-5580-3p (60), hsa-miR-144-5p (60), hsa-miR-608 (60), hsa-miR-7106-5p (60), hsa-miR-4796-3p (60), hsa-miR-374c-5p (59), hsa-miR-655-3p (59), hsa-miR-103a-2-5p (59), hsa-miR-4670-3p (59), hsa-miR-6833-5p (58), hsa-miR-1252-5p (57), hsa-miR-145-3p (57), hsa-miR-489-3p (56), hsa-miR-1273h-3p (56), hsa-miR-3154 (55), hsa-miR-3179 (55), hsa-miR-7114-5p (55), hsa-miR-103-3p (54), hsa-miR-519d-5p (53), hsa-miR-143-5p (53), hsa-miR-4279 (53), hsa-miR-138-5p (53), hsa-miR-3124-3p (53), hsa-miR-3614-3p (52), hsa-miR-1299 (52), hsa-miR-4482-3p (51), hsa-miR-6886-3p (50), hsa-miR-4483 (50)

Data generated by using an online "miRDB" database. Target score of the respective miRNAs has been given in parentheses

respectively. The ligplus program was then used to study the pattern of molecular interactions between NCOA1 and the ligands (Fig. 30.6b and 30.7b). A comparative observation of G scores and interacting residues between binding partners in all the complexes are given in Table 30.6.

CID5465152 was forming a three side chain hydrogen bond with LeuB795 with a bond length of 2.29 Å, 2.84 Å, and 0.94 Å. Moreover, various other protein residues interacting hydrophobically with CID5465152 were Ile272, AlaA310,

Table 30.3 Enriched KEGG pathways for predicted targets of the microRNAs involved in NCOA1 regulation using Micro-T-cds analysis

KEGG pathway	P value	# Genes	# miRNA
Prion diseases (hsa05020)	<1e-325	11	7
Glycosphingolipid biosynthesis – lacto and neolacto series (hsa00601)	<1e-325	12	9
Mucin type O-glycan biosynthesis (hsa00512)	<1e-325	19	13
Signaling pathways regulating pluripotency of stem cells (hsa04550)	<1e-325	87	16
TGF-beta signaling pathway (hsa04350)	<1e-325	52	22
Glioma (hsa05214)	3.519407e-14	43	17
Proteoglycans in cancer (hsa05205)	1.54543e-13	121	17
Morphine addiction (hsa05032)	1.499789e-11	53	14
Hippo signaling pathway (hsa04390)	6.338501e-09	83	16
FoxO signaling pathway (hsa04068)	9.174981e-07	80	13
Lysine degradation (hsa00310)	3.349079e-06	23	11
Amphetamine addiction (hsa05031)	1.667362e-05	39	11
GABAergic synapse (hsa04727)	5.768701e-05	49	10
ECM-receptor interaction (hsa04512)	7.367286e-05	44	10
Estrogen signaling pathway (hsa04915)	9.528214e-05	37	9
Biosynthesis of unsaturated fatty acids (hsa01040)	0.0002167153	10	7
Wnt signaling pathway (hsa04310)	0.0009746286	71	8
ErbB signaling pathway (hsa04012)	0.002929616	54	11
Renal cell carcinoma (hsa05211)	0.02551611	42	10
Pathways in cancer (hsa05200)	0.0272884	152	6
Thyroid hormone signaling pathway (hsa04919)	0.04500115	59	7

P-value threshold=0.05, MicroT threshold=o.8, Enrichment analysis method=Fischer's Exact Test.

PheA314, ProA796, and ProA797. Furthermore, various other bonds such as electrostatic, hydrogen bond, π-π stacking, and Cation–π interactions were ThrA268, ThrA268, ArgB311, Phe314, Leu795, AspB801 in protein-ligand interactions. Other protein residues interacting CID5465152 was forming the side-chain hydrogen bond with Arg311.

30.3.4 Physico-chemical Properties and Pharmacokinetics of Lead Molecules

QikProp predicts considerable physical descriptors and pharmaceutically pertinent properties of natural compounds and results are summarized in Table 30.7. It also provides noteworthy values for comparing these molecular properties with those of 95% of previously known pharmaceutical drugs. CID5465152 with a molecular

Table 30.4 Enriched KEGG pathways for predicted targets of the microRNAs involved in NCOA1 regulation using TarBase analysis

KEGG pathway	P value	# Genes	# miRNA
Fatty acid biosynthesis (hsa00061)	<1e-325	7	11
ECM-receptor interaction (hsa04512)	<1e-325	33	11
Prion diseases (hsa05020)	<1e-325	14	11
Proteoglycans in cancer (hsa05205)	<1e-325	141	24
Viral carcinogenesis (hsa05203)	2.220446e-16	121	18
TGF-beta signaling pathway (hsa04350)	9.590106e-13	57	16
Hippo signaling pathway (hsa04390)	3.451306e-11	81	10
Adherens junction (mmu04520)	4.235363e-09	55	15
p53 signaling pathway (hsa04115)	5.511812e-08	54	15
Glioma (hsa05214)	6.009146e-08	48	19
Lysine degradation (hsa00310)	1.483986e-07	31	15
Chronic myeloid leukemia (hsa05220)	1.873852e-07	59	20
Hepatitis B (hsa05161)	7.187088e-07	96	15
Transcriptional misregulation in cancer (hsa05202)	3.017093e-06	77	9
Cell cycle (hsa04110)	4.829396e-06	89	14
Fatty acid metabolism (hsa01212)	5.555207e-06	21	12
Pathways in cancer (hsa05200)	0.0001717289	213	14
FoxO signaling pathway (hsa04068)	0.0003814438	94	15
Signaling pathways regulating pluripotency of stem cells (hsa04550)	0.005582075	86	13
Thyroid hormone signaling pathway (hsa04919)	0.006597679	68	10
Colorectal cancer (hsa05210)	0.01158144	44	10
Bacterial invasion of epithelial cells (hsa05100)	0.02378233	44	9
Oocyte meiosis (hsa04114)	0.04334604	68	11
Prostate cancer (hsa05215)	0.04967047	60	10

P-value threshold=0.05, MicroT threshold=o.8, Enrichment analysis method=Fischer's Exact Test.

Table 30.5 Enriched KEGG pathways for predicted targets of the microRNAs involved in NCOA1 regulation using TargetScan

KEGG pathway	P value	# Genes	# miRNA
Prion diseases (hsa05020)	<1e-325	1	2
Mucin type O-glycan biosynthesis (hsa00512)	7.96867e-11	9	7
Fatty acid biosynthesis (hsa00061)	5.276477e-05	1	2

P-value threshold=0.05, MicroT threshold=o.8, TargetScan score type=Context +, TargetScan context score=-0.4, Enrichment analysis method=Fischer's Exact Test

mass of 726.641 g/mol, ten hydrogen bond donors, 25.65 hydrogen bond acceptors and an octanol/water partition coefficient of −2.524 accepted all the conditions of the rule of five as per QikProp application of Maestro9.6.

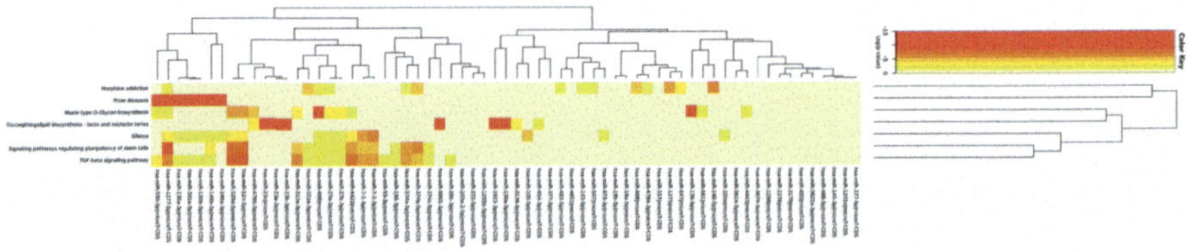

Fig. 30.4 Binary heat map of pathways related to the common microRNAs involved in NCOA1 gene regulation and different signaling pathways. Heat map calculation is based on binary p-values. The significantly targeted pathways are marked with deep red color

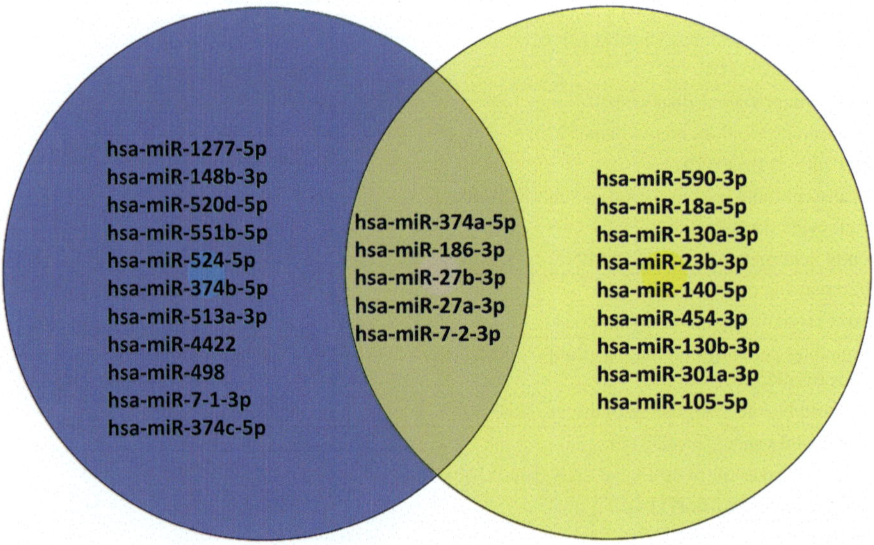

Fig. 30.5 Venn diagram showing miRNAs involved in NCOA1 gene regulation associated with cancer stem cell pathways. The miRNAs were predicted by using Micro-T-cds, and Tarbase database. Micro-T-cds (blue), Tarbase (yellow) and hybrid Micro-T-cds-Tarbase (brown) region

CID3594 had a molecular mass of 610.568 g/mol. The number of hydrogen bond donors and acceptors was 10 and 20.05. Predicted octanol/water partition coefficient was −1.178. CID42607750, 75660876, 89716662, and 70678574 also satisfied the Lipinski's rule. Another parameter for compound 5465152, such as dipole, Solvent accessible surface area (SASA), FOSA, FISA, PISA, and volume are 12.398, 964.31, 331.78, 401.06, 231.46, 1907.50 calculated by QikProp application of Maestro9.6. Moreover, another parameter for compound CID3594, such as dipole, SASA, FOSA, FISA, PISA, and volume are 10.3889, 12.334, 391.692, 352.341, 168.3, and 1694.216 calculated by QikProp application of Maestro9.6.

Kaempferol is a type of flavonoid several studies imply that taking kaempferol may diminish the risk of different cancers. Many preclinical studies have revealed that kaempferol and glycosides of kaempferol have widespread pharmacological activities, including anticancer and antidiabetic (Calderon-Montano et al. 2011).

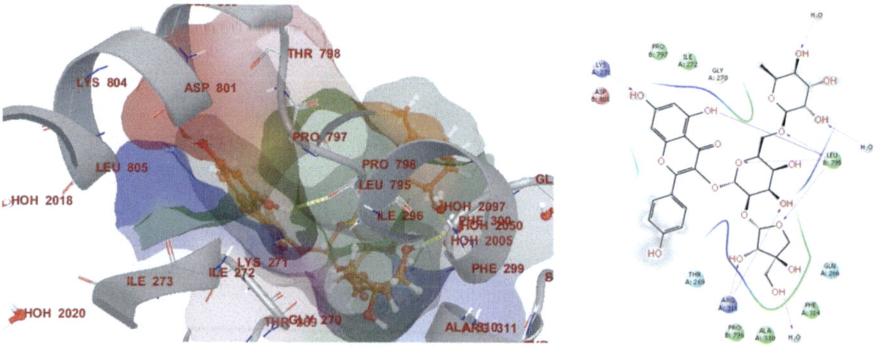

Fig. 30.6 (**a**) Ribbon presentation of NCOA1 (PDB; 1OJ5) protein molecule with CID 5465152 (**b**) Protein-ligand interactions profile of 1OJ5 with CID 5465152

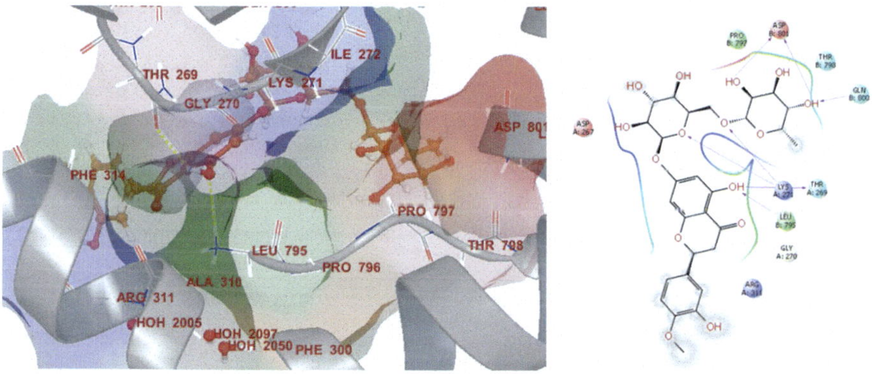

Fig. 30.7 (**a**) Ribbon presentation of NCOA1 (PDB; 1OJ5) protein molecule with CID3594 (**b**) Protein-ligand interactions profile of 1OJ5 with CID3594

Furthermore, the study also demonstrated that kaempferol has low cytotoxicity and impede angiogenesis HIF dependent and HIF independent pathways (Luo et al. 2009). Kaempferol reduced cancer cell growth by inhibiting ER α and γ activities (Wang et al. 2013). Hesperidin inhibited the proliferation of MCF7breast cancer cells through mechanisms other than antimitotic (Lee et al. 2010). Synergism effect is seen when the combined treatment of doxorubicin and hesperidin gave in MCF-7 cells (Febriansah et al. 2014). Moreover, hesperidin in-vivo experiment in mice showed that hesperidin increased lipid peroxides and lactate dehydrogenase. Moreover, it decreases levels of tissue antioxidants like superoxide dismutase, catalase, glutathione peroxidase, reduced glutathione, vitamin E and vitamin C thereby showing the potent anticancer effect in lung cancer (Kamaraj et al. 2009). Binding of drugs to serum albumin influences their pharmacokinetic behavior. Thus, in-silico calculation of HSA binding contribute considerably to the breakthrough of new drug candidates (Vallianatou et al. 2013). Despite various advancements incorrect binding site prediction, use of undesirable small-molecule databases, conflicting docking

Table 30.6 Lowest binding energy for the ligand- NCOA1 (PDB; 1OJ5) protein interaction as detected by Maestro

S. N.	ligand	GScore	Li EvdW	HBond	Elecro	Protein-ligands interactions
1	5465152	−8.88	−2.65	−5.26	−1.97	LysA;271, Arg B;311, Leu795
2	5465152-2	−8.47	−1.6	−5.28	−2	ThrA;268, ThrA;268, LysA;271, Arg B;311, Leu795
3	5465152-3	−7.23	1.96	−5.55	−1.03	ThrA;268, ThrA;268, LysA;271, Arg B;311, Leu795
4	5465152-4	−7.15	1.71	−4.61	−1.42	ThrA;268, ThrA;268, Arg B;311, Leu795
5	5465152-5	−7.11	1.57	−4.39	−1.54	ThrA;268, ThrA;268, Arg B;311, Phe314, Leu795
6	5465152-6	−7.1	1.48	−4.4	−1.59	ThrA;268, ThrA;268, Arg B;311, Phe314, Leu795
7	5465152-7	−7.07	1.88	−4.33	−1.84	ThrA;268, ThrA;268, Arg B;311, Phe314, Leu795, Asp B; 801
8	5465152-8	−7.04	1.68	−4.83	−1.55	ThrA;268, ThrA;268, Arg B;311, Phe314, Leu795, Asp B; 801
9	5465152-9	−6.86	2.65	−3.67	−0.7	ThrA;268, ThrA;268, Arg B;311, Phe314, Leu795
10	3594	−6.8	1.42	−4.19	−1.71	Gly270, Lys271, Leu B;295, and, Asp B; 801
11	5465152-10	−6.75	−2.4	−4.35	−1.24	ThrA;268, ThrA;268, Arg B;311, Phe314, Leu795
12	42607750	−6.49	2.16	−3.52	−1.73	ThrA;268, ThrA;268, Arg B;311, Phe314, Leu795
13	75660876	−6.42	−0.7	−4.68	−1.37	ThrA;268, ThrA;268, LysA;271, Arg B;311, Leu795
14	5465152-11	−6.42	−2.2	−4.15	−1.59	ThrA;268, Arg B;311, Leu795
15	89716662	−6.38	−0.97	−4.27	−1.43	ThrA;268, ThrA;268, LysA;271, Arg B;311, Leu795
16	5465152-12	−6.35	−1.87	−4.01	−1.19	ThrA;268, Arg B;311, Leu795
17	70678574	−6.35	−0.71	−4.39	−1.5	ThrA;268, ThrA;268, LysA;271, Arg B;311, Leu795
18	75660876-2	−6.3	−0.38	−4.29	−2	ThrA;268, Arg B;311, Leu795
19	89716662-2	−6.27	−0.57	−4.63	−1.32	ThrA;268, ThrA;268, LysA;271, Arg B;311, Leu795
20	89716662-3	−6.25	−0.56	−4.24	−1.66	ThrA;268, ThrA;268, Arg B;311, Phe314, Leu795

Ligand CID; Pubchem IDs
GScore; Glide extra precision scores (kcal/mol)
Lipophilic E Vdw; Chemscore lipophilic pair term and fraction of the total protein-ligand vdw energy
HBond; Hydrogen-bonding term
Electro; Electrostatic rewards
Protein-ligands interaction; π-π stacking, π –cat interaction and a hydrogen bond between the ligands and protein

Table 30.7 Evaluation of drug-like properties of the lead molecules by Qikprop Maestro

S. N.	Ligand (CID)	QP log P o/w (−2.0 to 6.5)	QP log Herg (acceptable range: above −5.0)	QPP Caco (nm/sec) (<25-poor; >500-great)	QP log BB (−3-1.2)	QPP MDCK (nm/sec) (<25-poor; >500-great)	QPlogKp	QPlog Khsa (Acceptable range: −1.5 to 1.5)
1	5465152-1	−2.524	−6.335	1.558	−5.075	0.457	−6.271	−1.712
2	5465152-2	−2.329	−6.296	3.092	−4.646	0.958	−5.635	−1.701
3	5465152-3	−2.872	−6.497	0.488	−5.876	0.13	−7.312	−1.747
4	5465152-4	−2.559	−6.593	1.011	−5.485	0.286	−6.625	−1.728
5	5465152-5	−2.691	−6.727	1.069	−5.54	0.304	−6.556	−1.799
6	5465152-6	−3.034	−6.03	0.59	−5.452	0.16	−7.202	−1.731
7	5465152-7	−2.564	−6.889	2.185	−5.157	0.659	−5.838	−1.844
8	5465152-8	−2.645	−6.129	1.87	−4.864	0.556	−6.144	−1.73
9	5465152-9	−2.546	−7.352	1.166	−5.825	0.334	−6.31	−1.856
10	3594	−1.173	−6.326	4.515	−4.419	1.443	−6.076	−1.193
11	546515210	−2.792	−6.573	0.708	−5.656	0.195	−6.922	−1.762
12	42507750	−0.44	−5.306	19.875	−2.905	7.162	−4.84	−0.852
13	755608761	−4.49	0.901	0.012	−4.63	0.007	−9.141	−2.287
14	5455152	−2.521	−6.355	0.881	−5.416	0.247	−6.784	−1.663
15	897166621	−3.494	0.311	0.159	−3.799	0.117	−6.906	−2.505
16	5465152	−2.628	−6.459	1.058	5.432	0.301	−6.658	−1.74
17	70678574	−4.021	0.84	0.024	−4.375	0.016	−8.352	−2.305
18	756608762	−3.956	1.516	0.06	−3.664	0.033	−7.883	−2.13

(continued)

Table 30.7 (continued)

S. N.	Ligand (CID)	QP log $P_{o/w}$ (−2.0 to 6.5)	QP log Herg (acceptable range: above −5.0)	QPP Caco (nm/sec) (<25-poor; >500-great)	QP log BB (−3-1.2)	QPP MDCK (nm/sec) (<25-poor; >500-great)	QPlogKp	QPlog Khsa (Acceptable range: −1.5 to 1.5)
19	897166622	−3.371	0.725	0.324	−3.33	0.214	−6.434	−2.438
20	897166623	−3.53	0.413	0.178	−3.674	0.137	−6.782	−2.506

Ligand (CID); PubChem IIDs of the lead molecules
QPogPo/w (−2.0 to 6.5) Predicted octanol/water partition co-efficient log p; (range: −.20 to 6.5)
Predicted IC50 values for blockage of Herg K+ channels; (acceptable range: above −5.0)
QPP Caco Predicted apparent Caco-2 cell permeability in nm/sec. Caco-2 cells is a model for the gut blood barrier; (nm/sec) <25-poor >500- great
QPlog BB Predicted brain/blood partition coefficient;
QPP MDCK Predicted apparent MDCK cell permeability in nm/sec. MDCK cells are considered to be a good mimic for the blood-brain barrier; (nm/sec) <25-poor >500- great
QPlog Khsa Prediction of binding to human serum albumin; (acceptable range: −1.5 to 1.5)
Percentage of human oral absorption; (<25% is poor and >80% is high)

pose, high dock score, failure in MD simulations, lack of clarity whether the compound is an inhibitor or agonist, etc. are still not appropriately answered by docking methodologies and urge further research.

30.4 Conclusion

These SRC co activators play pivotal roles in growth, development and various cancers. Furthermore, oncogenic role of NCOA has been well established in various cancers. Information obtains from these studies can establish protein-ligand interactions that are required to understand cell growth inhibitory activity. The phylogenetic analyses of different isolates of nuclear receptor coactivator show that humans formed a separate clade. The present study revealed that an analog of kaempferol and hesperidin intended excellent dock score with NCOA1 which may be further studied in-vitro and in-vivo for its inhibition activity. Moreover, in-silico data of the present study also convenience that structural alteration of kaempferol and hesperidin skeleton analog may help to improve their anticancer and ADME/T properties. The study concludes that semi-synthetic analogs of naturally occurring compounds *viz.*, kaempferol and hesperidin may provide a lead to work for the new anticancer drug. Although, further in-vitro and in-vivo experimental studies are needed for the experimental validation of our findings.

Acknowledgment PS and PPK acknowledge Indian Council of Medical Research (ICMR), India and UGC-CSIR, India respectively for providing the financial assistance in the form of Postdoc and Junior Research Fellowship. SK acknowledges the Central University of Punjab for providing infrastructure facilities.

References

Anzick SL, Kononen J, Walker RL, Azorsa D, Tanner MM, Guan X, Sauter G, Kallioniemi OP, Trent JM, Meltzer PS. AIB1, a steroid receptor coactivator amplified in breast and ovarian cancer. Science. 1997;277(5328):965–8.

Benjamini Y, Hochberg Y. Controlling the false discovery rate: a practical and powerful approach to multiple testing. J R Stat Soc Ser B. 1995;1:289–300.

Bhanot A, Sharma R, Noolvi MN. Natural sources as potential anti-cancer agents: A review. Int J Phytomed. 2011;3(1):09–26.

Calderón-Montaño JM, Burgos-Morón E, Pérez-Guerrero C, López-Lázaro M. A review on the dietary flavonoid kaempferol. Mini Rev Med Chem. 2011;11(4):298–344.

Carotenuto F, Albertini MC, Coletti D, Vilmercati A, Campanella L, Darzynkiewicz Z, Teodori L. How diet intervention via modulation of DNA damage response through microRNAs may have an effect on cancer prevention and aging, an in silico study. Int J Mol Sci. 2016;17(5):1–21.

Chen X, Liu Z, Xu J. The cooperative function of nuclear receptor coactivator 1 (NCOA1) and NCOA3 in placental development and embryo survival. Mol Endocrinol. 2010;24(10):1917–34.

Cherigo L, Lopez D, Martinez-Luis S. Marine natural products as breast cancer resistance protein inhibitors. Mar Drugs. 2015;13(4):2010–29.

Cragg GM, Newman DJ. Plants as a source of anti-cancer agents. J Ethnopharmacol. 2005;100 (1–2):72–9.

DaRocha AB, Lopes RM, Schwartsmann G. Natural products in anticancer therapy. Curr Opin Pharmacol. 2001;1(4):364–9.

Febriansah R, Putri DDP, Sarmoko, Nurulita NA, Meiyanto E, Nugroho AE. Hesperidin as a preventive resistance agent in MCF-7 breast cancer cells line resistance to doxorubicin. Asian Pac J Trop Biomed. 2014;4(3):228–33.

Felsenstein J. Confidence limits on phylogenies: an approach using the bootstrap. Evolution. 1985;39(4):783–91.

Friesner RA, Banks JL, Murphy RB, Halgren TA, Klicic JJ, Mainz DT, Repasky MP, Knoll EH, Shelley M, Perry JK, Shaw DE. Glide: a new approach for rapid, accurate docking and scoring. 1. Method and assessment of docking accuracy. J Med Chem. 2004;47(7):1739–49.

Friesner RA, Murphy RB, Repasky MP, Frye LL, Greenwood JR, Halgren TA, Sanschagrin PC, Mainz DT. Extra precision glide: docking and scoring incorporating a model of hydrophobic enclosure for protein-ligand complexes. J Med Chem. 2006;49(21):6177–96.

Ganatra S H. Inhibition studies of naturally occurring terpene based compounds with cyclin-dependent kinase 2 enzyme. J Comput Sci Syst Biol. 2012;5(3):68–73.

Halgren TA, Murphy RB, Friesner RA, Beard HS, Frye LL, Pollard WT, Banks JL. Glide: a new approach for rapid, accurate docking and scoring. 2. Enrichment factors in database screening. J Med Chem. 2004;47(7):1750–9.

Hillman GG. Dietary agents in cancer chemoprevention and treatment. J Oncol. 2012;21:209–19.

Huang W, He T, Chai C, Yang Y, Zheng Y, Zhou P, Qiao X, Zhang B, Liu Z, Wang J, Shi C. Triptolide inhibits the proliferation of prostate cancer cells and down-regulates SUMO-specific protease 1 expression. PLoS One. 2012;7(5):1–17.

Jae YC, Park J. Contribution of natural inhibitors to the understanding of the PI3K/PDK1/PKB pathway in the insulin-mediated intracellular signaling cascade. Int J Mol Sci. 2008;9 (11):2217–30.

Jorgensen WL, Duffy EM. Prediction of drug solubility from structure. Adv Drug Deliv Rev. 2002;54(3):355–66.

Jorgensen WL, Tirado-Rives J. The OPLS potential functions for proteins. Energy minimizations for crystals of cyclic peptides and crambin. J Am Chem Soc. 1988;110(6):1657–66.

Jorgensen WL, Maxwell DS, Tirado-Rives J. Development and testing of the OPLS all-atom force field on conformational energetics and properties of organic liquids. J Am Chem Soc. 1996;118 (45):11225–36.

Kamaraj S, Ramakrishnan G, Anandakumar P, Jagan S, Devaki T. Antioxidant and anticancer efficacy of hesperidin in benzo(a)pyrene induced lung carcinogenesis in mice. Investig New Drugs. 2009;27(3):214–22.

Kim SK. Marine carbohydrates: fundamentals and applications. Oxford: Elsevier; 2014.

Kishimoto H, Wang Z, Bhat-Nakshatri P, Chang D, Clarke R, Nakshatri H. The p160 family coactivators regulate breast cancer cell proliferation and invasion through autocrine/paracrine activity of SDF-1α/CXCL12. Carcinogenesis. 2005;26(10):1706–15.

Kumar S, Chashoo G, Saxena AK, Pandey AK. *Parthenium hysterophorus*: a probable source of anticancer, antioxidant and anti-HIV agents. Biomed Res Int. 2013;2013:1–11.

Lauria A, Ippolito M, Almerico AM. Inside the Hsp90 inhibitors binding mode through induced fit docking. J Mol Graph Model. 2009;27(6):712–22.

Lee CJ, Wilson L, Jordan MA, Nguyen V, Tang J, Smiyun G. Hesperidin suppressed proliferations of both human breast cancer and androgen-dependent prostate cancer cells. Phytother Res. 2010;24(S1):S15–9.

Lee J-C, Hou M-F, Huang H-W, Chang F-R, Yeh C-C, Tang J-Y, Chang HW. Marine algal natural products with anti-oxidative, anti-inflammatory, and anti-cancer properties. Cancer Cell Int. 2013;13(1):1–7.

Liao L, Chen X, Wang S, Parlow AF, Xu J. Steroid receptor coactivator 3 maintains circulating insulin-like growth factor 1 (IGF-1) by controlling IGF-binding protein 3 expression. Mol Cell Biol. 2008;28(7):2460–9.

Lu JJ, Crimin K, Goodwin JT, Crivori P, Orrenius C, Xing L, Tandler PJ, Vidmar TJ, Amore BM, Wilson AG, Stouten PF. Influence of molecular flexibility and polar surface area metrics on oral bioavailability in the rat. J Med Chem. 2004;47(24):6104–7.

Luo H, Rankin GO, Liu L, Daddysman MK, Jiang BH, Chen YC. Kaempferol inhibits angiogenesis and VEGF expression through both HIF dependent and independent pathways in human ovarian cancer cells. Nutr Cancer. 2009;61(4):554–63.

Mayer AMS, Gustafson KR. Marine pharmacology in 2001-2: anti-tumour and cytotoxic compounds. Eur J Cancer. 2004;40(18):2676–704.

Mayer AMS, Gustafson KR. Marine pharmacology in 2003–2004: anti-tumour and cytotoxic compounds. Eur J Cancer. 2006;42(14):2241–70.

Mayer AMS, Gustafson KR. Marine pharmacology in 2005–2006: antitumour and cytotoxic compounds. Eur J Cancer. 2009;44(16):2357–87.

Moussavou G, Kwak DH, Obiang-Obonou BW, Maranguy CAO, Dinzouna-Boutamba SD, Lee DH, Pissibanganga OG, Ko K, Seo JI, Choo YK. Anticancer effects of different seaweeds on human colon and breast cancers. Mar Drugs. 2014;12(9):4898–911.

Muthuirulappan S, Francis SP. Anti-cancer mechanism and possibility of nano-suspension formulation for a marine algae product fucoxanthin. Asian Pac J Cancer Prev. 2013;14(4):2213–6.

Oh A. The nuclear receptor coactivator AIB1 mediates insulin-like growth factor I-induced phenotypic changes in human breast cancer cells. Cancer Res. 2014;64:8299–308.

Phosrithong N, Ungwitayatorn J. Molecular docking study on anticancer activity of plant-derived natural products. Med Chem Res. 2010;19(8):817–35.

Qin L, Wu Y, Toneff MJ, Li D, Liao L, Gao X, Bane FT, Tien JC, Xu Y, Feng Z, Yang Z. NCOA1 directly targets M-CSF1 expression to promote breast cancer metastasis. Cancer Res. 2015a;74 (13):3477–88.

Qin L, Xu Y, Xu Y, Ma G, Liao L, Wu Y, Li Y, Wang X, Wang X, Jiang J, Wang J. NCOA1 promotes angiogenesis in breast tumors by simultaneously enhancing both HIF1alpha- and AP-1-mediated VEGFa transcription. Oncotarget. 2015b;6(27):23890–904.

Razeto A, Ramakrishnan V, Litterst CM, Giller K, Griesinger C, Carlomagno T, Lakomek N, Heimburg T, Lodrini M, Pfitzner E, Becker S. Structure of the NCoA-1/SRC-1 PAS-B domain bound to the LXXLL motif of the STAT6 transactivation domain. J Mol Biol. 2004;336 (2):319–29.

Roell D, Baniahmad A. The natural compounds atraric acid and N-butylbenzene-sulfonamide as antagonists of the human androgen receptor and inhibitors of prostate cancer cell growth. Mol Cell Endocrinol. 2011;332(1–2):1–8.

Sarkar FH, Li Y. Using chemopreventive agents to enhance the efficacy of cancer therapy. Cancer Res. 2006;66(7):3347–50.

Sawadogo W, Schumacher M, Teiten MH, Cerella C, Dicato M, Diederich M. A survey of marine natural compounds and their derivatives with anti-cancer activity reported in 2011. Molecules. 2013;18(4):3641–73.

Sawadogo WR, Boly R, Cerella C, Teiten MH, Dicato M, Diederich M. A survey of marine natural compounds and their derivatives with anti-cancer activity reported in 2012. Molecules. 2015;20 (4):7097–142.

Shivakumar D, Williams J, Wu Y, Damm W, Shelley J, Sherman W. Prediction of absolute solvation free energies using molecular dynamics free energy perturbation and the OPLS force field. J Chem Theory Comput. 2010;6(5):1509–19.

Singh P, Bast F. *In silico* molecular docking study of natural compounds on wild and mutated epidermal growth factor receptor. Med Chem Res. 2014a;23(12):5074–85.

Singh P, Bast F. Multitargeted molecular docking study of plant-derived natural products on phosphoinositide-3 kinase pathway components. Med Chem Res. 2014b;23(4):1690–700.

Singh P, Bast F. High-throughput virtual screening, identification and *in vitro* biological evaluation of novel inhibitors of signal transducer and activator of transcription 3. Med Chem Res. 2015a;24(6):2694–708.

Singh P, Bast F. Screening and biological evaluation of myricetin as a multiple target inhibitor insulin, epidermal growth factor, and androgen receptor; *In silico* and *in vitro*. Investig New Drugs. 2015b;33(3):575–93.

Singh P, Bast F. Screening of multi-targeted natural compounds for receptor tyrosine kinases inhibitors and biological evaluation on cancer cell lines, *in silico* and *in vitro*. Med Oncol. 2015c;32(9):1–18.

Singh P, Alex JM, Bast F. Insulin receptor (IR) and insulin-like growth factor receptor 1 (IGF-1R) signaling systems: novel treatment strategies for cancer. Med Oncol. 2014;31(1):1–14.

Tien JC, Liu Z, Liao L, Wang F, Xu Y, Wu YL, Zhou N, Ittmann MM, Xu J. The steroid receptor coactivator-3 is required for the development of castration-resistant prostate cancer. Cancer Res. 2013;73(13):3997–4008.

Tien JCY, Liao L, Liu Y, Liu Z, Lee DK, Wang F, Xu J. The steroid receptor coactivator-3 is required for developing neuroendocrine tumor in the mouse prostate. Int J Biol Sci. 2014;10 (10):1116–27.

Vallianatou T, Lambrinidis G, Tsantili-Kakoulidou A. *In silico* prediction of human serum albumin binding for drug leads. Expert Opin Drug Discov. 2013;8(5):583–95.

Vlachos IS, Zagganas K, Paraskevopoulou MD, Georgakilas G, Karagkouni D, Vergoulis T, Dalamagas T, Hatzigeorgiou AG. DIANA-miRPath v3. 0: deciphering microRNA function with experimental support. Nucleic Acids Res. 2015;43(W1):W460–6.

Wang H, Gao M, Wang J. Kaempferol inhibits cancer cell growth by antagonizing estrogen-related receptor α and ϒ activities. Cell Biol Int. 2013;37(11):1190–6.

Xu J, Li Q. Review of the *in vivo* functions of the p160 steroid receptor coactivator family. Mol Endocrinol. 2003;17:1681–92.

Xu J, Wu R-C, O'Malley BW. Normal and cancer-related functions of the p160 steroid receptor co-activator (SRC) family. Nat Rev Cancer. 2009;9(9):615–30.

Chapter 31
Parthenin and Its Similar Structure as Potential Lead Inhibitors of *Plasmodium vivax* and *Plasmodium falciparum* Lactate Dehydrogenase

Pushpendra Singh, Prem P. Kushwaha, and Shashank Kumar

31.1 Introduction

Malaria is a life-threatening disease caused by *Plasmodium* parasites conveyed to people through malaria vectors. About three million people die and five million have been reported to be infected with malaria annually worldwide (WHO 2011). The dearth of effectual anti-malarial vaccines, chemotherapy assumes a critical part in control of the illness, but unfortunately, drug-resistant strains of *Plasmodium* (*P. falciparum* and *P. vivax*) have shown up against a large portion of antimalarials present till date. Thus, expanded endeavors are instantly desired for antimalarial drug discovery. The objective must be advanced for safe and inexpensive new medications to slug the spread of malaria parasites that are impervious to existing medications. The malarial parasite lives mainly in the host erythrocytes, where they use cell component as a nourishment for their life-cycle growth. These parasites destroyed the hemoglobin fractions of the infected erythrocyte, which leads to severe ailments such as anemia exclusively in both pregnant women and children (Qidwai et al. 2014; Chen 2014). Malarial parasite present in the erythrocyte fundamentally depends on the glycolysis pathway for their energy production. Cytoplasmic pyruvate fermentation and/or mitochondrial electron transport chain are the mechanism by which the consumed NAD^+ during glycolysis is regenerated. In Plasmodia, lactate is the end-product of the glycolytic pathway because pyruvate does not enter the citric acid cycle. In the presence of NADH, reduction reaction converts pyruvate into lactate in the presence of lactate dehydrogenase (LDH) enzyme. Here, *Plasmodium* pyruvate is not able to work as substrate inhibitor, which allows fast energy production, mandatory in the fast-growing malarial parasite (Ramya et al

P. Singh
Tumor Biology Laboratory, National Institute of Pathology, New Delhi, India

P. P. Kushwaha · S. Kumar (✉)
Department of Biochemistry and Microbial Sciences, Central University of Punjab, Bathinda, Punjab, India

S. Kumar, C. Egbuna (eds.), *Phytochemistry: An in-silico and in-vitro Update*,
https://doi.org/10.1007/978-981-13-6920-9_31

2002). Plasmodial LDH differs from its human counterpart by the presence of a five amino acid insertion at the pyruvate binding site. Thus LDH has been used for testing the novel *Plasmodium* LDH inhibitors versus human LDH. The malarial LDH also has a vast cleft at its active site which can accommodate consilient colossal inhibitors like gossypol derivatives. Based on the above argument LDH may act as a latent target for malaria. *Parthenium hysterophorus* is a proverbial weed found throughout the world. *P. hysterophorus* have numerous medical advantages such as medication for rheumatic aching, urinary tract infections, neuralgia, diarrhea, dysentery and skin inflammation. Parthenin is an important ingredient of *P. hysterophorus* stem, blossom, leaf, and root (Kumar et al. 2013). Because of this, we used different parthenin like compounds to target *P. falciparum* and *P. vivax* LDH enzyme in the present study.

In-silico screening approach is the foremost strategy for introductory identification of new inhibitors for target proteins and evaluating their interactive mode. The larger parts of known anti-malarial drugs are small molecules intended to interact and temper the biological mode of action of the distinctive pathogen proteins. Molecular docking comprises three distinctive objectives sequentially posture prediction, virtual screening, and binding affinity calculation. Computer-aided drug designing are able to discriminate hits, pick leads and rationalize leads to convert naturally active compounds into decent medications by upgrading their physicochemical, pharmaceutical and ADME\T (absorption, distribution, metabolism, excretion, and toxicity) properties. Thus, in-silico approach is utilized impressively to reduce the risk, time and asset necessities for both in-vitro and in-vivo chemical synthesis and biological testing (Ekins et al. 2007). In the present study, we employed Maestro 9.6 software for docking studies of 85 parthenin like compounds. Furthermore, Maestro 9.6 QikProp module was used to evaluate the ADME/T properties of the best-docked parthenin like compounds.

31.2 Materials and Methods

31.2.1 In-silico Methodology

31.2.1.1 Selection of Protein and Ligand Molecules

Protocol for the GLIDE based molecular docking took from our earlier published article with a few modifications (Singh et al. 2018) (Fig. 31.1). The present work is the extension of previously published in-silico antimalarial drug discovery from our laboratory (Singh et al. 2018). Same ligands reported in the published article were used to check their multi-targeted antimalarial potential. List of ligands and their structural and physiochemical parameters are reported in previously published article (Singh et al. 2018).

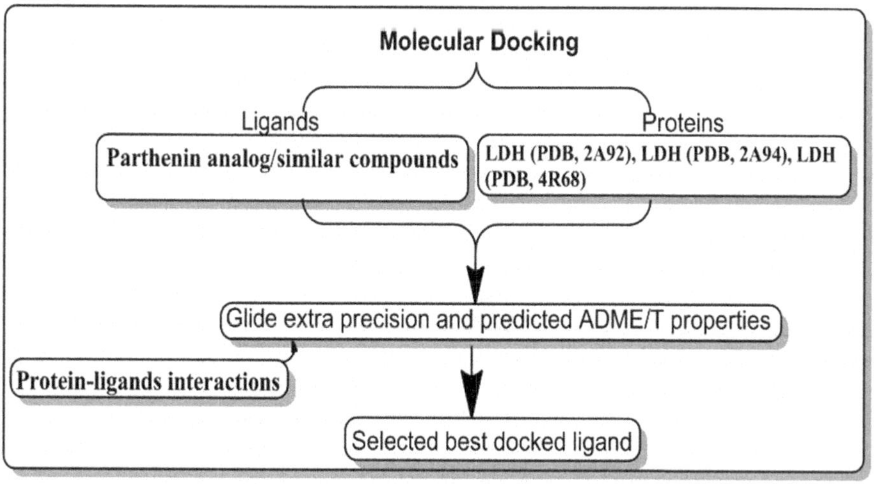

Fig. 31.1 Workflow of study design

Ligprep wizard module of Maestro 9.6 Schrodinger Inc. was used for the ligand preparation. The module accomplishes several modifications in ligands such as hydrogen atom addition, two dimensional to three-dimensional conversion, bond angles and their length corrections, ring conformation, stereochemistry and low energy structure. Ionization was not changed and tautomers were not generated for the ligands. The X-ray crystal structure of test proteins (LDH, PDB: 2A92, LDH PDB: 2A94 and LDH PDB: 4R68) were retrieved from protein data bank (Chaikuad et al. 2005; Labadie et al. 2015). PDB raw structure was modified by using protein preparation wizard module.

31.2.1.2 Molecular Docking

Maestro 9.6 Schrodinger Inc. software package was used for molecular docking studies of selected ligand molecules (Friesner et al. 2004). Ligands were docked against target protein molecules. Ligand dataset and target proteins were prepared by using appropriate modules available in Maestro 9.6 package. The optimized potential for liquid simulations (OPLS_2005) force field was applied for energy minimization and geometry optimization (Shivakumar et al. 2010; Jorgensen and Tirado-Rives 1988; Jorgensen et al. 1996). The one ligand-one conformation was opted for molecular docking protocol. The receptor-grid file was generated to implement the partial atomic charge of 0.25 and 1 Å Van der Waal radii for the receptor atom. After that, ligand-receptor molecular docking was performed. Based on the minimum glide score, ten compounds were selected for the prediction of type of interactions, optimal energy value, bonding potential, and conformations.

31.2.1.3 ADME/T Properties Studies

The ADME/T (Absorption, Distribution, Metabolism, Excretion and Toxicity) properties of lead ligands were evaluated by using the Maestro 9.6 QikProp module. Different pharmacological parameters such as overall CNS activity, log BB, MDCK and Caco-2 cell permeability and logKhsa (human serum albumin binding), etc. of the lead ligand molecules were assessed by QikProp module (Jorgensen and Duffy 2002; Lu et al. 2004).

31.3 Results and Discussion

31.3.1 Analysis of Docking Results of Promising Compounds for **Plasmodium vivax** *LDH*

Crystal structure of *P. vivax* LDH in complex with inhibitor deliver information about position and conformation of enzyme binding site (Kongsaeree et al. 2005). X-ray structure of *P. vivax* LDH (PDB:2A92) was used for docking study. Antimalarial combinatorial possessions of artemisinin derivatives (ACT) with quinoline compounds (amodiaquine and mefloquine) are known for their widespread high resistance. Our selected lactate dehydrogenase of *Plasmodium vivax* structural insights suggest a general approach for developing new generations of antimalarial LDH inhibitors that accommodate the only substrate for their active site, would retain a binding affinity with the mutant enzymes (Kongsaeree et al. 2005). Molecular docking was achieved by using XP mode of GLIDE. Our result highlighted that; CID72786361, 78178433, and 11552273 yielded a pre-eminent dock score for with proteins *Plasmodium vivax* LDH −8.6, −7.73, −7.55, Kcal/mol respectively (Table 31.1). Most of the interactions made by compounds with residues in the active site of LDH (*P. vivax*) seem to be hydrophobic in nature. Protein-ligand interactions of 2A92 with compounds showed that amino acids Met30, Ile31, Leu112, Val54, Phe100, Ala98, Val138, Pro250, Leu163, Leu167 and Pro246 appeared in the hydrophobic interactions. Furthermore, amino acid amino acids Asn140, Ser245, Ile31 and Gly99 involved in back-bone hydrogen bonding of protein-ligand interactions (Fig. 31.2).

31.3.2 Interaction Modes Between the Parthenin Like Compounds and Human LDH

Molecular docking of *Plasmodium vivax* LDH, *Plasmodium falciparum* LDH and human LDH against natural compounds has been carried out. In the present investigation, our result highlighted that; CID 3482907, 70498184, 73199557, 77977597

Table 31.1 Glide binding score for the ligand-LDH (PDB, 2A92) protein interaction

Ligand type	Compounds ID	GScore	Lipophilic Evdw	HBond	Electro	Protein-ligands interactions
LDH inhibitors (control)	89602938	−7.38	−3.7	−2.88	−0.77	Met30, Ile31, Ieu163, and Leu167
	59637586	−7.32	−3.53	−2.81	−0.81	Met30, Val138, Ieu163, and Leu167
	89602952	−7.3	−3.35	−3.15	−0.72	Met30, Ile31, Ala98, and Phe100
	3503	−6.83	−3.61	−2.31	−0.8	Met30, Ile31, Ala98, and Leu167
	86599376	−6.8	−3.56	−2.4	−0.76	Met30, Ile31, Ala98, and Leu167
Anticancer natural compounds	72786361	−8.6	−2.47	−2.77	−1.15	Met30, Ile31, Asn140, Ieu163, and Ser240
	3482907	−8	−2.76	−2.09	−0.77	Met30, Ile31, Val138, and Ieu163
	78178433	−7.73	−2.38	−2.07	−0.83	Met30, Ile31, Val138, Ieu163, and Ser240
	78178433-2	−7.68	−1.39	−2.74	−1.09	Met30, Ile31, Val138, Ieu163, and Gly99
	11552273	−7.55	−2.72	−2.78	−0.99	Met30, Ile31, Val138, and Ieu163
	44583940	−7.55	−2.72	−2.78	−0.99	Met30, Ile31, Val138, and Ieu163
	3482907-2	−7.49	−2.38	−1.98	−0.76	Met30, Ile31, Val138, and Ieu163
	72791246	−7.47	−2.99	−1.87	−0.75	Met30, Ile31, Val138, and Ieu163
	76391862	−7.46	−3.6	−1.04	−0.53	Met30, Ile31, Val138, and Ieu163
	296217	−7.41	−3.03	−1.37	−0.57	Met30, Ile31, Val138, and Ieu163

yielded a pre-eminent dock score for with proteins human LDH −5.72, −5.58, −6.59, −7.64 Kcal/mol respectively (Table 31.2).

Protein-ligand interactions outline emphasized that the hydrogen bonding, lipophilic, and π-π stacking interactions at the active site plays a vital role in protein-ligand interactions. Molecular docking steps recognize the docking free energy value (Gscore) against the target protein biomolecules. Protein-ligands interactions tinted the electrostatic, lipophilic, and hydrogen bond interactions are a key contributor in protein-ligand interactions. The binding modes of parthenin like compounds in the interactive site of human LDH were identified using intermolecular docking simulations by means of Maestro 9.6 program. All the selected compounds were docked

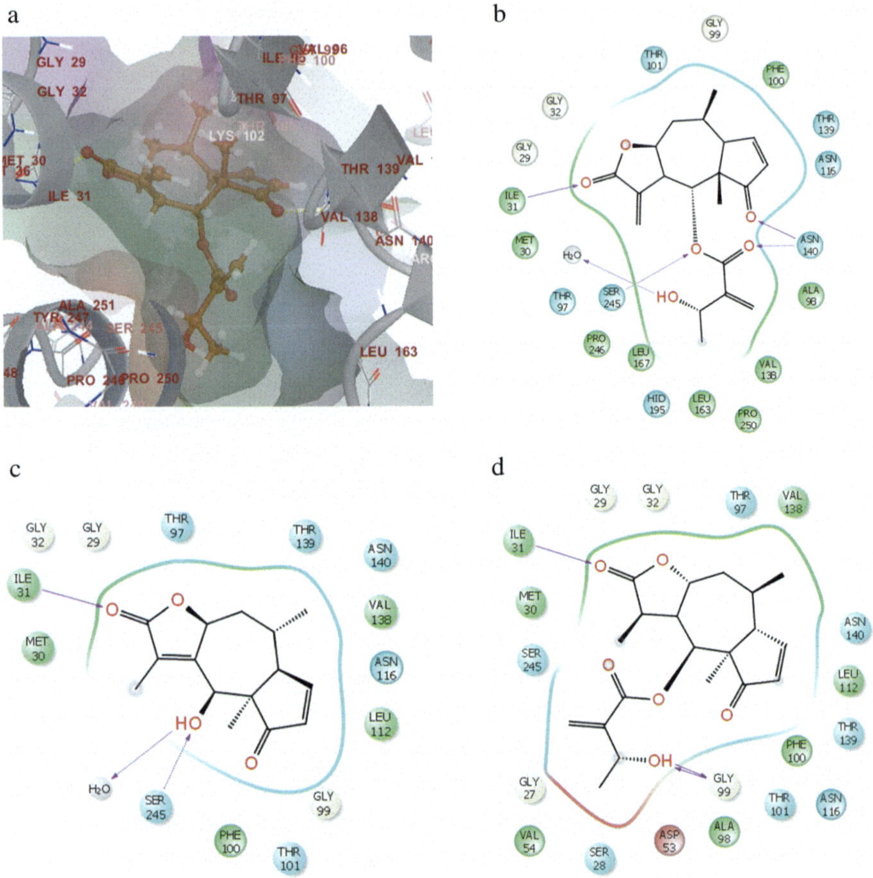

Fig. 31.2 (**a**) Ribbon presentation of LDH (PDB, 2A92) protein molecule with CID 72786361 (**b**) Protein-ligand interactions profile of 2A92 with CID72786361 (**c**) Protein-ligand interactions profile of 2A92 with CID78178433 (**d**) Protein-ligand interactions profile of 2A92 with CID11552273

into the human LDH active site, using the same procedure. Figure 31.3 represents the binding pattern of the parthenin like compounds in the LDH binding pocket. Human LDH active site comprises of mostly hydrophobic amino acids as Pro246, Pro250, Leu167, Ile254, Val138, Leu163, Ile31, Tyr247, Met30, Phe100, and Ala98. These amino acid residues are involved in strong hydrophobic interactions with the parthenin like compounds. As anticipated, inhibitors used in this study interact with the same site like the earlier bounded ligand in the crystallographic complex.

Table 31.2 Glide binding score for the ligand-LDH (PDB, 4R68) protein interaction

Ligand type	Compounds ID	GScore	Lipophilic Evdw	HBond	Electro	Protein-ligands interactions
Compounds	77977597	−7.64	−0.91	−4.24	−2.29	Asn137, Leu164, Hid192, Tyr238, and Ile241
	77977597-2	−5.77	−1.2	−2.69	−1.46	Asn137, Leu164, Hid192, Tyr238, and Ile241
	3482907	−5.72	−1.8	−2.33	−0.98	Leu164, Arg168, Hid192, Tyr238, and Ile241
	73199557	−6.59	−1.5	−2.07	−2.43	Leu164, Arg168, Tyr238, and Ile241
	3482907-2	−5.6	−1.45	−3	−0.64	Leu164, Arg168, Hid192, Tyr238, and Ile241
	70498184	−5.58	−1.13	−2.88	−1.26	Leu164, Tyr238, and Ile241
	73004448	−5.55	−1.85	−2.52	−0.95	Leu164, Tyr238, and Ile241
	77977597-3	−5.54	−1.47	−2.49	−1.21	Asn137, Leu164, Hid192, Tyr238, and Ile241
	56671343	−5.51	−1.73	−2.41	−0.96	Leu164, Tyr238, and Ile241
	3482907-3	−5.49	−1.45	−2.49	−0.87	Leu164, Tyr238, and Ile241

Ligands protein interaction; π-π stacking, π –cat interaction and a hydrogen bond between the ligands and protein

GScore Glide extra precision scores (kcal/mol), *Lipophilic E Vdw* Chemscore lipophilic pair term and fraction of the total protein-ligand vdw energy, *HBond* Hydrogen-bonding, *Electro* Electrostatic bounty

31.3.3 Interaction Modes Between the Parthenin Like Compounds and Plasmodium falciparum *LDH*

We used X-ray structure of *Plasmodium falciparum* LDH in complex inhibitor (PDB code 2A94) for the molecular docking study using GLIDE XP module. Our result highlighted that; CID296217, 3482907, 77977597, and 78178433 yielded a pre eminent dock score for with proteins *Plasmodium falciparum* LDH −6.59, −7.84, −6.73, −6.76 Kcal/mol respectively (Table 31.3). Most of the interactions

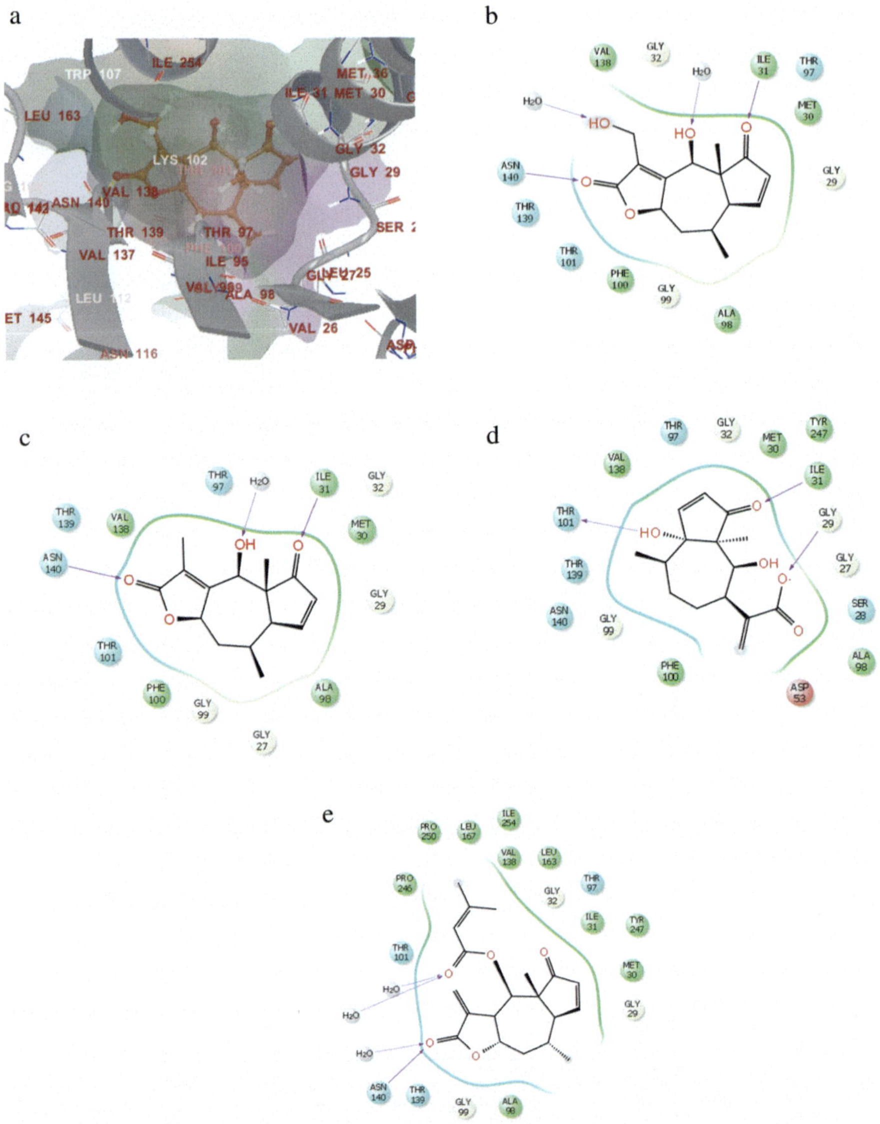

Fig. 31.3 (**a**) Ribbon presentation of LDH (PDB, 2A94) protein molecule with CID 3482907 (**b**) Protein- ligand interactions profile of 2A94 with CID 3482907 (**c**) Protein- ligand interactions profile of 2A94 with CID78178433 (**d**) Protein- ligand interactions profile of 2A94 with CID 77977597 (**e**) Protein- ligand interactions profile of 2A94 with CID296217

made by compounds with residues in the active site of *Plasmodium falciparum* LDH seem to be hydrophobic in nature. Protein-ligand interactions of 2A94 with compounds showed that amino acids Leu164, Ile241, Tyr238, Ala237, Val234, and Pro138 appeared in the hydrophobic interactions. Furthermore, amino acid amino

Table 31.3 Glide binding score for the ligand-LDH (PDB, 2A94) protein interaction

Ligand type	Compounds ID	GScore	Lipophilic Evdw	HBond	Electro	Protein-ligands interactions
Compounds	3482907	−7.84	−1.7	−2.37	−1	Met30, Ile31, Ala98, Phe100, and Asn140
	3482907-2	−7.66	−1.23	−2.68	−0.99	Met30, Ile31, Ala98, Phe100, and Asn140
	78178433	−6.76	−1.47	−1.58	−0.86	Met30, Ile31, Ala98, Phe100, and Thr101
	77977597	−6.73	−1.46	−1.9	−1.07	Met30, Ile31, Ala98, Phe100, and Asn140
	296217	−6.59	−1.91	−2.09	−0.94	Met30, Ile31, Ala98, Phe100, and Val138
	73004448	−6.52	−1.6	−2.3	−1.38	Met30, Ile31, Ala98, Phe100, and Val138
	10265551	−6.32	−2.34	−1.84	−0.6	Met30, Ile31, Ala98, Phe100, and Val138
	13918467	−6.32	−2.34	−1.84	−0.6	Met30, Ile31, Ala98, Phe100, and Val138
	10333765	−6.26	−2.66	−2.02	−0.99	Met30, Ile31, Ala98, Phe00, and Val138
	13918467-2	−6.26	−2.66	−2.02	−0.99	Met30, Ile31, Ala98, Phe100, and Val138

acids Arg168, Asn137 and Hid192 involved in back-bone hydrogen bonding of protein-ligand interactions (Fig. 31.4).

31.3.4 ADME/T Properties of Leads Molecules

ADME/T properties of lead compounds were reviewed by Qikprop module of Maestro 9.6 (Kerns and Di 2010). Compound EGCG was found to be promising based on their docking free energy score and binding mode. A most fascinating aspect of CID73199557, 3482907, 73199557, 72786361, 77977597 are their admirable, Qplogpo/w, QplogHERG, QplogBB, QPP MDCK, Qplogkhsa, and proportion of human oral value which justify the Lipinski's rule of five (Table 31.4). Moreover, high oral bioavailability, polar surface area, H-bond acceptors and donors being imperious criteria for the therapeutic agent's development. It has been suggested that compounds having a polar surface area equal to or less than 140 angstrom and 10 or less rotatable bonds (or 12 or fewer H-bond donors and acceptors) may have a high possibility for best oral bioavailability in-vivo (Veber et al. 2002). Furthermore, it is also reported that the polar surface area is inversely proportional to permeation rate (Becker et al. 1998). These compounds have batter SASA values that are claimed to be suitable for therapeutic agents. These results designate that these compounds will have a better penetration rate.

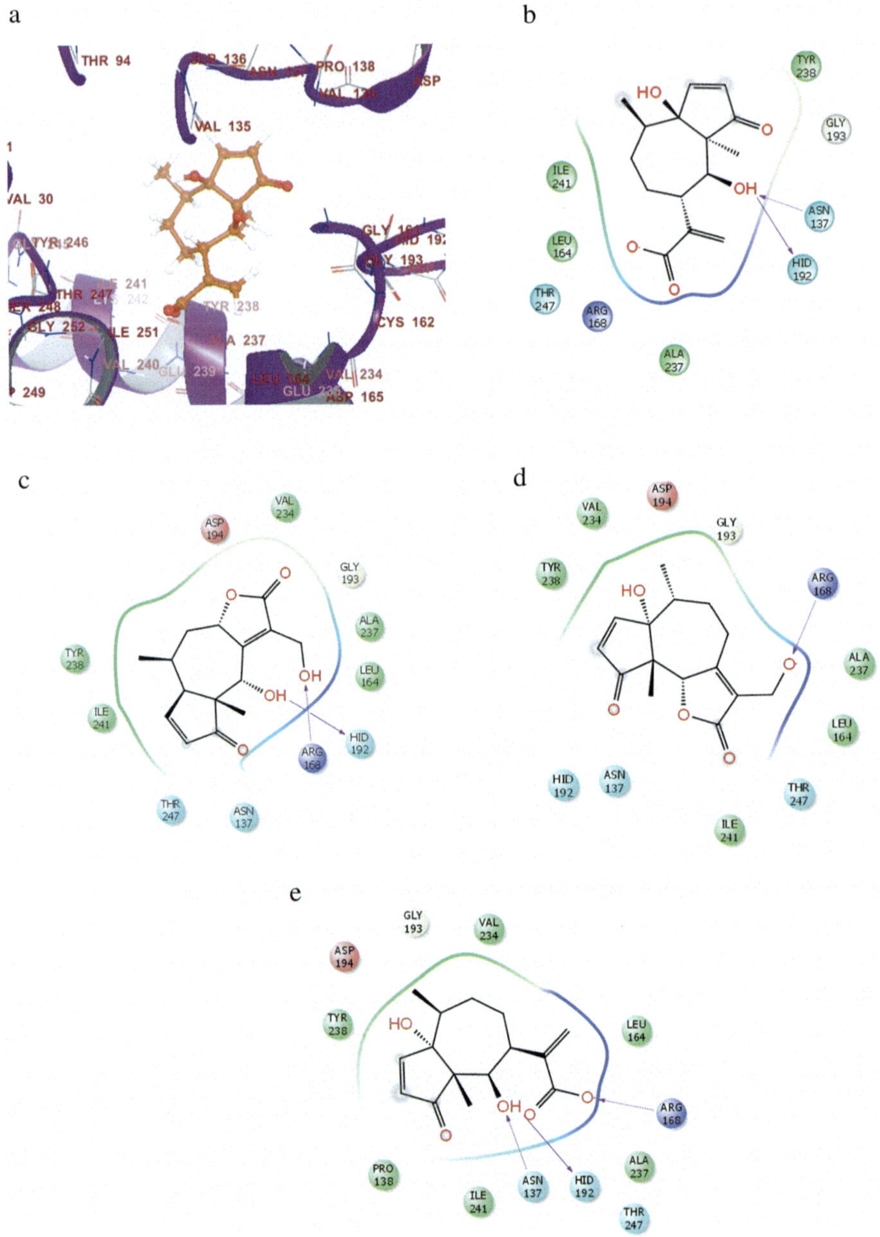

Fig. 31.4 (**a**) Ribbon presentation of LDH (PDB, 4R68) protein molecule with CID77977597 (**b**) Protein-ligand interactions profile of 4R68 with CID77977597 (**c**) Protein-ligand interactions profile of 4R68 with CID3482907 (**d**) Protein-ligand interactions profile of 4R68 with CID73199557 (**e**) Protein-ligand interactions profile of 4R68 with CID70498184

Table 31.4 ADME/T properties. Structural, physicochemical, biochemical, pharmacokinetics and toxicity properties of the compound

S. N	Molecule	QP log $P_{o/w}$ (−2.0 to 6.5)	QPlog HERG (acceptable range: above −5.0)	QPP Caco (nm/s) <25-poor >500- great	QP log BB (−3 to 1.2)	QPP MDCK (nm/s) <25-poor >500- great	QPlog Khsa (acceptable range: −1.5 to 1.5)	Percentage of human oral absorption; (<25% is poor and >80% is high)
1	73199557	0.861	−3.318	259.85	−0.963	115.275	−0.39	75.209
2	155585	0.47	−3.011	92.799	−1.008	83.49	−0.794	70.597
3	3482907	0.47	−3.011	192.799	−1.008	83.49	−0.794	70.597
4	73199557	0.865	−3.253	244.135	−0.972	107.758	−0.38	74.744
5	3482907	0.633	−3.366	256.749	−0.974	113.789	−0.76	73.775
6	72786361	1.522	−3.444	592.017	−0.767	280.718	−0.654	85.479
7	3482907	0.552	−3.063	310.41	−0.845	139.699	−0.826	74.78
8	73199557	0.87	−3.222	271.872	−0.928	121.05	−0.391	75.608
9	3482907	0.52	−2.993	228.077	−0.951	100.118	−0.792	72.196
10	77977597	1.758	−1.654	43.355	−1.185	21.164	−0.325	66.539

QPlogPo/w (−2.0 to 6.5) Predicted octanol/water partition coefficient
QPlogHerg (acceptable range: above −5.0) IC50 for HERG K^+ channels inhibition
QPPCaco (nm/s) <25-poor; >500-great. Apparent Caco-2 cell permeability (nm/s)
QPlogBB (−3 to −1.2) Predicted brain/blood partition coefficient
QPPMDCK (nm/s) <25-poor; >500- great. Apparent MDCK cell permeability (nm/s)
QPlog Khsa- Binding to human serum albumin prediction; (acceptable range: −1.5 to 1.5)
Percentage of human oral absorption; (<25% is poor and >80% is high)

31.4 Conclusion

The present study utilized in-silico approach to search novel *Plasmodium spp.* lactate dehydrogenase inhibitors using parthenin like compounds as a scaffold. Pharmacological and drug-likeness properties of the selected test compounds showed anti-LDH potential. Beside various lead compounds, CID 78178433 showed potentially against both *P. vivax* and *P. falciparum* LDH enzyme. Strong hydrophobic and H-bonding interaction between phytochemicals and pathogen protein was found as revealed by high Dock score. We hereby suggest the in-vitro and in-vivo validation of these lead antimalarial compounds which might provide cost-effective and safer natural antimalarial drug.

Acknowledgments PS and PPK acknowledge Indian Council of Medical Research (ICMR), India and UGC-CSIR, India respectively for providing the financial assistance in the form of Postdoc and Senior Research Fellowship. SK acknowledges the Central University of Punjab for providing infrastructure facilities.

References

Becker S, Groner B, Müller CW. Three-dimensional structure of the Stat3β homodimer bound to DNA. Nature. 1998;394(6689):145–51.

Chaikuad A, Fairweather V, Conners R, Joseph-Horne T, Turgut-Balik D, Leo R. Brady structure of lactate dehydrogenase from *Plasmodium Vivax*: complexes with NADH and APADH†. Biochemistry. 2005;44:16221–8.

Chen C. Development of antimalarial drugs and their application in China: a historical review. Infect Dis Poverty. 2014;3(1):1–10.

Ekins S, Mestres J, Testa B. *In silico* pharmacology for drug discovery: applications to targets and beyond. Br J Pharmacol. 2007;152(1):21–37.

Friesner RA, Banks JL, Murphy RB, Halgren TA, Klicic JJ, Mainz DT, Repasky MP, Knoll EH, Shelley M, Perry JK, Shaw DE. Glide: a new approach for rapid, accurate docking and scoring. 1. Method and assessment of docking accuracy. J Med Chem. 2004;47(7):1739–49.

Jorgensen WL, Duffy EM. Prediction of drug solubility from structure. Adv Drug Deliv Rev. 2002;54(3):355–66.

Jorgensen WL, Tirado-Rives J. The OPLS [optimized potentials for liquid simulations] potential functions for proteins, energy minimizations for crystals of cyclic peptides and crambin. J Am Chem Soc. 1988;110(6):1657–66.

Jorgensen WL, Maxwell DS, Tirado-Rives J. Development and testing of the OPLS all-atom force field on conformational energetics and properties of organic liquids. J Am Chem Soc. 1996;118(45):11225–36.

Kerns E, Di L. Drug-like properties: concepts, structure design and methods: from ADME to toxicity optimization. Amsterdam: Academic Press; 2010.

Kongsaeree P, Khongsuk P, Leartsakulpanich U, Chitnumsub P, Tarnchompoo B, Walkinshaw MD, Yuthavong Y. Crystal structure of dihydrofolate reductase from *Plasmodium vivax*: pyrimethamine displacement linked with mutation-induced resistance. Proc Natl Acad Sci U S A. 2005;102(37):13046–51.

Kumar S, Mishra A, Pandey AK. Antioxidant mediated protective effect of *Parthenium hysterophorus* against oxidative damage using *in vitro* models. BMC Complement Altern Med. 2013;120:1–9.

Labadie S, Dragovich PS, Chen J, Fauber BP, Boggs J, Corson LB. Optimization of 5-(2, 6-dichlorophenyl)-3-hydroxy-2-mercaptocyclohex-2-enones as potent inhibitors of human lactate dehydrogenase. Bioorg Med Chem Lett. 2015;25(1):75–82.

Lu JJ, Crimin K, Goodwin JT, Crivori P, Orrenius C, Xing L, Tandler PJ, Vidmar TJ, Amore BM, Wilson AG, Stouten PF. Influence of molecular flexibility and polar surface area metrics on oral bioavailability in the rat. J Med Chem. 2004;47(24):6104–7.

Qidwai T, Jamal F, Khan MY, Sharma B. Exploring drug targets in isoprenoid biosynthetic pathway for *Plasmodium falciparum*. Biochem Res Int. 2014;2014:1–12.

Ramya TNC, Surolia N, Surolia A. Survival strategies of the malarial parasite *Plasmodium falciparum*. Curr Sci. 2002;83(7):818–25.

Shivakumar D, Williams J, Wu Y, Damm W, Shelley J, Sherman W. Prediction of absolute solvation free energies using molecular dynamics free energy perturbation and the OPLS force field. J Chem Theory Comput. 2010;6(5):1509–19.

Singh P, Kushwaha PP, Kumar S. Novel potent inhibitors of *Plasmodium vivax* dihydrofolate reductase: an *in silico* antimalarial drug discovery. Indian J Pharm Educ Res. 2018;52:1–14.

Veber DF, Johnson SR, Cheng HY, Smith BR, Ward KW, Kopple KD. Molecular properties that influence the oral bioavailability of drug candidates. J Med Chem. 2002;45(12):2615–23.

WHO. World malaria report 2011. Geneva: World Health Organ; 2011.

Correction to: *Bulbine frutescens* Phytochemicals as a Promising Anti-cancer Drug Discovery Source: A Computational Study

Prem P. Kushwaha, Ajay Kumar, Santosh Maurya, Atul K. Singh, Amit Kumar Sharma, and Shashank Kumar

Correction to:
Chapter 26 in: S. Kumar, C. Egbuna (eds.), *Phytochemistry: An in-silico and in-vitro Update*, https://doi.org/10.1007/978-981-13-6920-9_26

The chapter was inadvertently published with an incorrect affiliation of author Ajay Kumar as "A. Kumar, Department of Zoology, Banaras Hindu University, Varanasi, India". It has been updated as "Ajay Kumar, Department of Biochemistry and Microbial Sciences, Central University of Punjab, Punjab, India".

The updated version of this chapter can be found at
https://doi.org/10.1007/978-981-13-6920-9_26

S. Kumar, C. Egbuna (eds.), *Phytochemistry: An in-silico and in-vitro Update*,
https://doi.org/10.1007/978-981-13-6920-9_32

The manufacturer's authorised representative in the EU is Springer Nature Customer Service Centre GmbH, Europaplatz 3, 69115 Heidelberg, Germany. If you have any concerns regarding our products, please contact ProductSafety@springernature.com

Printed and bound by CPI Group (UK) Ltd, Croydon, CR0 4YY
30/09/2026
02225304-0020